Unsteady Aerodynamics and Aeroelasticity of Turbomachines

Unsteady Aerodynamics and Aeroelasticity of Turbomachines

Proceedings of the 8th International Symposium held in Stockholm, Sweden, 14-18 September 1997

edited by

Torsten H. Fransson
Division of Heat and Power Technology,
Royal Institute of Technology,
Stockholm, Sweden

KLUWER ACADEMIC PUBLISHERS
DORDRECHT / BOSTON / LONDON

A C.I.P. Catalogue record for this book is available from the Library of Congress.

ISBN 0-7923-5040-5

Published by Kluwer Academic Publishers,
P.O. Box 17, 3300 AA Dordrecht, The Netherlands.

Sold and distributed in North, Central and South America
by Kluwer Academic Publishers,
101 Philip Drive, Norwell, MA 02061, U.S.A.

In all other countries, sold and distributed
by Kluwer Academic Publishers,
P.O. Box 322, 3300 AH Dordrecht, The Netherlands.

Printed on acid-free paper

Printed in the Netherlands

THE EIGHT INTERNATIONAL SYMPOSIUM
ON
UNSTEADY AERODYNAMICS AND AEROELASTICITY OF TURBOMACHINES (THE 8TH ISUAAT)

Kungliga Tekniska Högskolan
September 14-18, 1997

Organized by
Torsten H. Fransson

Supported by
ABB STAL AB, Finspång (Hans Lennart Olausson)
Elforsk AB, Stockholm (Ulf Arvidsson)
European Commission (Cost Action F1), Belgium (Rainer Gerold)
Forsmarks Kraftgrupp AB, Östhammar (Henning Danielsson)
Martinsson Elektronik AB, Stockholm (Evert Sträng)
Mätforum AB, Solna - Stockholm (Göran Olsson)
NUTEK (The Swedish National Board for Industrial and Technical Development), Stockholm (Sten Åfeldt)
Polytec GmbH, Germany (Dr. Sellbach)
SAS (Scandinavia Airline System), Stockholm (Barbro Roth)
Volvo Aero Corporation, Trollhättan (Ulf Olsson)
Wenner-Grens Center, Stockholm (Torvard Laurent)

The International Scientific Committee

Atassi, H.M. (USA)
Bölcs, A. (Switzerland)
Ffowcs Williams, J.E. (U.K.)
Fransson, T. (Sweden)
Gallus, H.E. (Germany)
Gnesin, V.I. (Ukraine)
Namba, M. (Japan)
Saren, V.E. (Russia)
Széchényi, E. (France)
Tanida, Y. (Japan)
Verdon, J.M. (USA)
Whitehead, D.S. (U.K.)
Zhou, S. (P.R. China)

The 8th ISUAAT in Stockholm, Sweden, September 14-18, 1997

TABLE OF CONTENTS

Preface.. xiii

1:General Unsteady Phenomena..1

Theoretical Analyses of Rotating Cavitation in Inducers..3
S. Watanabe, K. Yokota, Y. Tsujimoto, K. Kamijo

Non-Linear Critical Layer in the Unsteady Boundary Layer induced on a Flat..............19
Plate by Wakes
Z. Wiercinski

A Numerical Analysis of Wakes Propagation in Inviscid Flow of Turbine....................35
P. Ferrand, S. Aubert, G. Oliveira, L. Smati,

2: Experimental Unsteady Flow Around Oscillating Blades-1.................................51

Three Dimensional Unsteady Flow Around a Turbine Blade Oscillating in...................53
Bending Mode: An Experimental and Computational Study
D. L. Bell, L. He

Rotating Blade Row Oscillating Airfoil Aerodynamics...67
K. K. Frey, S. Fleeter

Experimental Investigation of Unsteady Aerodynamic Characteristics of......................83
Transonic Compressor Cascades (Development of Test Facilities)
I. Fujimoto, T. Hirano, H. Tanaka

Experimental Investigation of Unsteady PressureBehaviours......................................103
in a Linear Turbine Cascade
M. Norryd, A. Bölcs

3: Theoretical Studies of Unsteady Flow Around Oscillating Blades-1.................117

Calculation of Unsteady Forces Acting on the Profiles of Double Cascade.................119
Oscillating in a Fluid Flow
O. V. Chernysheva, V. A. Yudin

Unsteady Disturbances in Swirling Turbomachinery Flows...131
H. M. Atassi, V. V. Golubev

Unsteady Forces on Annular Cascade Blades in Subsonic Flow with Swirl................147
V. V. Golubev, H. M. Atassi

Effect of Acoustic Control on the Flutter Boundaries of Supersonic Cascade..............165
K. Nagai, M. Namba

4: Theoretical Studies of Unsteady Flow Around Oscillating Blades-2.................181

Nonlinear Harmonic Analysis of Unsteady Transonic Inviscid and Viscous Flows.....183
L. He, W. Ning

Flutter Analysis of Two-Dimensional Viscous Subsonic and Transonic Flow in.........195
Turbomachines Using the Advection Upstream Splitting Method (AUSM)
W. Höhn, T. H. Fransson

Three Dimensional Linearized Navier-Stokes Calculations for Flutter........................211
and Forced Response
D. G. Holmes, B. E. Mitchell, C. B. Lorence

Some Applications of a Time-Linearized Euler Method to Flutter & Forced...............225
Response in Turbomachinery
J. G. Marshall, M. B. Giles

5: Compressor Flow Instabilities: Experiments..241

An Experimental Flow Investigation of an HP Five- Stage Compressor......................243
Exhibiting Rotating Stall Due to Distorted Inlet Flow Conditions
W. Jahnen, T. Peters, L. Fottner

High Speed Centrifugal Impeller and Diffuser Interaction Near Stall Conditions........259
T. Yamane, T. Nagashima

Unsteady Flow and Turbulence in a Low Speed Axial Compressor.............................273
A. Sentker, W. Riess

Unsteady Inlet/Compressor Interaction Experiment to Support the Modeling of.........287
Compressor-Face Boundary Conditions
M. Sajben, D. D. Freund

6: Aeroelastic Coupling-1...301

Tree-Dimensional Coupled Model for Aeroelastic Analysis of Turbomachine............303
Blade Vibrations. Its Application to a Hydraulic Turbine Rotor
V. B. Kurzin, S. N. Korobeinikov, V. P. Ryabchenko, L. A. Tkacheva

The 2D Flutter of a Bladed Disc in an Incompressible Flow.......................................317
R. Rządkowski, V. Gnesin, A. Kovalyov

Structural Mistuning and Aerodynamic Coupling in Turbomachinery Bladings..........335
G. Kahl

Whole-Assembly Flutter Analysis of a Low Pressure Turbine Blade...........................347
A. I. Sayma, M. Vahdati, J. S. Green, M. Imregun

7: Experimental Blade Row Interactions..361

Passages of Preceding Blade Wakes and their Effects in the Rotor Passages of...........363
a Single Stage Axial-Flow Fan
T. Adachi, Y. Yamashita

Forced Response Vibrations of a Low Pressure Turbine due to Circumferential.........379
Temperature Distortions
S. R. Manwaring, K. L. Kirkeng

Investigation of the Boundary Layer Development on a Highly Loaded Low 393
Pressure Turbine Cascade under the Influence of Unsteady Flow Conditions
P. Acton, L. Fottner

Experimental and Numerical Investigation of Unsteady Rotor-Stator Interaction........407
on Axial Compressor Stage (with IGV) Performance
V. E. Saren, N. M. Savin, D. J. Dorney, R. M. Zacharias

8: Theoretical Studies of Unsteady Flow Around Oscillating Blades-3................425

A 3D Linearized Euler Analysis for Blade Rows,..427
Part 1: Aerodynamic and Numerical Formulations
M. D. Montgomery, J. M. Verdon

A 3D Linearized Euler Analysis for Blade Rows,..445
Part 2: Unsteady Aerodynamic Response Predictions
J. M. Verdon, M. D. Montgomery

Analysis of Nonlinear Oscillating Cascade Aerodynamics Including...........................465
Separated Flow
J. M. Wolff, S. Fleeter

Numerical Solution of the Navier Stokes Equations for Unsteady Unstalled and.........477
Stalled Flow in Turbomachinery Cascades with Oscillating Blades
S. Weber, H. E. Gallus, D. Peitsch

9: Compressor Flow Instabilities: Theory..493

Active Suppression of Compressor Flow Instability..495
A. S. Abo El lail, A. S. Hassan

On Sub-Cell Structure of Deep Rotating Stall in an Axial Compressor........................511
D. Kato, E. Outa, K. Chiba

Stability Analysis of Supersonic Cascade Flow by Actuator Disk Methods.................525
K. Yokota, Y. Iwamoto, Y. Tsujimoto

10: Aeroelastic Coupling-2..539

Semi-Unstructured Mesh Generator for Flow Calculations in Axial............................541
Turbomachinery Blading
L. Sbardella, A. I. Sayma, M. Imregun

Aeroelastic Tailoring for Rotor Blades of Transonic Swept Fans................................555
S. Zhou

Methods for Aeroelasticity in Propulsion...565
H. Mårtensson, P. Groth

11: Theoretical Blade Row Interactions-2...583

Preliminary Results of an Unsteady Throughflow Code to Simulate the......................585
Start-up of a Multistaged Turbopump
M. Debard, P. Ferrand, F. Leboeuf

Investigation of Stator-Rotor Interaction Phenomena in a Centrifugal Pump...............599
Y. Marx

The 3D-Unsteady Aerodynamic Forces - The Forced Vibration of Bladed Discs........613
V. Gnesin, R. Rządkowski

Parallel Computation of Rotor-Stator Interaction...633
P. Cizmas, R. Subramanya

12: Experimental Unsteady Flow Around Oscillating Blades-2..............................647

Correlation of Pressure and Hot Film Data on an Oscillating 2D..................................649
NACA63A006 Single Airfoil at High Incidence
S. Svensdotter, U. Johansson, A. Kemppainen, T. Fransson,

A Basic Forced Response Experiment..665
M. Nowinski, P. Ott

Vibration Characteristics of a Transonic Turbine Cascade...679
T. Watanabe, M. Aotsuka, Y. Machida

Periodic Fluctuation of Shock Waves in Transonic Cascade Flows.............................693
T. Shiratori, M. Matsushita, Y. Noguchi

13: Theoretical Studies of Unsteady Flow Around Oscillating Blades-4...............705

Part-Speed Flutter Analysis of a Wide-Chord Fan Blade...707
J. W. Chew , J. G. Marshall, M. Vahdati, M. Imregun

The Effect of the Blade Vibration Mode on a Flutter in a Transonic Fan.....................725
K. Isomura

Role of Shock Structures in Transonic Fan Rotor Flutter...733
T. Shibata, S. Kaji

Comparison of Numerical Schemes to Investigate Blade Flutter..................................749
L. Smati, S. Aubert, P. Ferrand, F. Massão

14: Theoretical Studies of Unsteady Flow Around Oscillating Blades-5...............765

Eigenmode Analysis of Unsteady Viscous Flows in Turbomachinery Cascades..........767
R. Florea, K. C. Hall, P. G. A. Cizmas

Transonic Cascade Flutter in Combined Bending-Chordwise Translational Mode......783
S. Kaji

On the Prediction of Dynamic Stall Onset on Airfoils in Low Speed Flow..................797
K. D. Jones, M. F. Platzer

15: Theoretical Studies of Unsteady Flow Around Oscillating Blades-6...............813

Flutter of Aircraft Engine Low Pressure Turbine Blades: Oscillating Cascade............815
Experiments and Analysis
J. Panovsky, M. Nowinski, A. Bölcs

Comparison of DLT and CFD Predictions of Unsteady Aerodynamic Force on..........831
Vibrating Supersonic Through- Flow Fan Cascade
M. Namba, N. Yamasaki, T. Otsuka

Chairperson index..847

Author Index..848

List of participants..850

Preface

Twenty-one years have passed since the first symposium in this series was held in Paris (1976). Since then there have been meetings in Lausanne (1980), Cambridge (1984), Aachen (1987), Beijing (1989), Notre Dame (1991) and Fukuoka (1994). During this period a tremendous development in the field of unsteady aerodynamics and aeroelasticity in turbomachines has taken place. As steady-state flow conditions become better known, and as blades in the turbomachine are constantly pushed towards lower weight, and higher load and efficiency, the importance of unsteady phenomena appear more clearly.

The 8th Symposium was, as the previous ones, of high quality. Furthermore, it presented the audience with the latest developments in experimental, numerical and theoretical research. More papers than ever before were submitted to the conference. As the organising committee wanted to preserve the uniqueness of the symposium by having single sessions, and thus mingle speakers and audience with different backgrounds in this interdisciplinary field, only a limited number of papers could be accepted. 54 papers were accepted and presented at the meeting, all of which are included in the present proceedings.

These papers cover a broad range of unsteady flow and aeroelastic phenomena in turbomachines. The state-of-the-art is clearly recognised in this volume. Comparing this with the knowledge in 1976, it is clearly seen that the aeroelastic community has come a long way in these years. Experimental data at high incidence, transonic flow conditions have now been obtained at mid-span on configurations in linear and annular test facilities. Numerical prediction models are capable of predicting fully three-dimensional unsteady viscous flow conditions. Analytical aspects related to the two-dimensional far-field boundary conditions and coupling between blade rows are much better known today.

There are still considerable challenges throughout the whole domain of unsteady effects in the broad sense. Mathematical aspects related to the formulation of three-dimensional wave propagation and boundary conditions, and experiments related to the three-dimensional propagation of unsteady perturbations at high incidence and transonic flow are two of these. Wake and potential influence on the unsteady blade load can still not be calculated with a high degree of accuracy. Various flow-structure coupling schemes will be of higher significance as new materials will be used in the blades. Faster computations with vectorized and parallel prediction models will be necessary. Furthermore, indications of "time-averaged" efficiency gains will certainly mean that a lot of effort will be put into the field of unsteady effects on the "steady-state" flow in the future. Finally, although existing prediction models are applied towards unsteady flow under real operating conditions, it is not yet possible to fully trust the results.

It will be a challenging and interesting period in Unsteady Aerodynamics and Aeroelasticity in Turbomachines over the next few years. The 9th Symposium will be held at the Ecole Centrale in Lyon, France, in the year 2000. The Scientific Committee of the Symposium series wishes you an interesting research and development phase until then, and look forward to meeting you in Lyon.

Last but not least, the Scientific Committee would like to express its gratitude to the organisations and individuals who financially supported the 8th Symposium, as well as the supporters of the previous meetings. These scientific high quality meetings could not have taken place without these contributions.

Torsten H. Fransson
Kungliga Tekniska Högskolan

1: General Unsteady Phenomena

THEORETICAL ANALYSES OF ROTATING CAVITATION IN INDUCERS

S. WATANABE*, K. YOKOTA*, Y. TSUJIMOTO*
and K. KAMIJO**
* *Osaka University, Graduate School of Engineering Science*
1-3 Machikaneyama, Toyonaka 560, Japan
** *Tohoku University, Institute of Fluid Science*
2-1-1 Katahira, Aoba, Sendai 980-77, Japan

1. Introduction

Rotating cavitation (R.C.) is one of the most important problems in the development of modern rocket turbopump inducers (Kamijo et al., 1993, Goirand et al., 1993, Ryan et al., 1994). By 2-D linear analysis (Tsujimoto et al., 1993), rotating cavitation was found to be caused by the positive mass flow gain factor, namely, the increase of cavity volume due to the decrease of flow rate, and to be completely different from rotating stall (R.S.) which is caused by the positive slope of the head characteristic. Moreover, 2-D analysis predicted that there can be two modes of rotating cavitation: the forward rotating mode which propagates faster than the impeller, and the backward rotating mode which propagates in the direction opposite to that of the impeller. Nonlinear (Tsujimoto et al., 1996) and 3-D (Watanabe et al., 1997) analyses resulted in basically the same results as those obtained for the previous 2-D cases and predicted forward and backward propagating modes. These were reported at the last meeting of the conference (Tsujimoto et al., 1995). However, most R.C. found in experiments propagates forward, although the backward mode was also found recently (Hashimoto et al., 1996).

In the above-mentioned studies, flow was assumed to be incompressible. However, real flow has compressibility because vapor and bubbles are contained in the flow. In this paper, a 2-D inviscid compressible linear analysis was carried out to clarify the influence of flow compressibility on R.C. To facilitate consideration of flow compressibility, we herein assume that, (i) the flow is a gas-liquid 2-phase flow, (ii) there are no slips and no phase changes between the gas and liquid phases, and (iii) the

T.H. Fransson (ed.), Unsteady Aerodynamics and Aeroelasticity of Turbomachines, 3-17.

flow field is isothermal because of the large thermal capacity and conductivity of the liquid phase. We also assumed that the upstream flow is irrotational, that the downstream flow is rotational due to the unsteadiness of the impeller, and that the flow in the impeller is perfectly guided by the blades. Thus, the semi-actuator disk method was employed.

2. Outline of Analysis

2.1 ANALYTICAL MODEL

As shown in Fig. 1, we consider a cascade with chord length l, pitch h and stagger β, moving in the direction of the y-axis with velocity U_T. We divide the flow into unsteady flow components and steady components, and assume that the unsteady components are small compared with the steady components.

We assume that the flow is a homogeneous gas-liquid 2-phase flow. We also assume that there are no slips and no phase changes (the quality χ=const.) between the gas and liquid phases, and that the flow field is isothermal because of the large thermal capacity and conductivity of the liquid phase. As a result, the pressure p becomes a function of total density ρ as follows:

$$p=\frac{RT\chi\rho_l\rho}{\rho_l-(1-\chi)\rho} \tag{1}$$

Here, T, R, and ρ_l are the temperature, the gas constant, and the density of the liquid phase, respectively. Equation (1) indicates that the flow is barotropic. Thus, the flow field is determined from the linearized Euler's equation, the continuity equation, and Eq. (1).

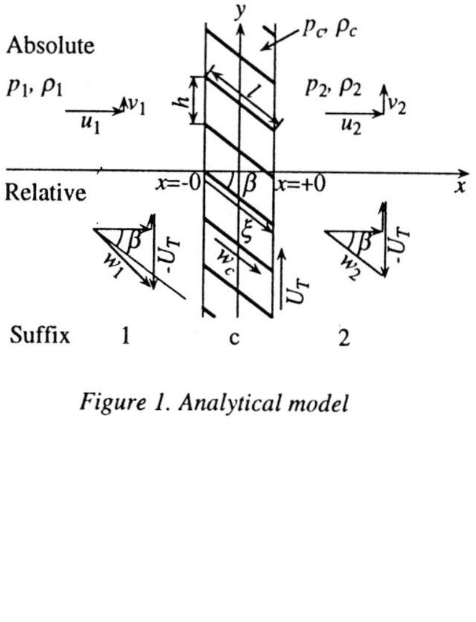

Figure 1. Analytical model

2.2 UNSTEADY FLOW FIELDS

2.2.1 *Unsteady Flow Components of Upstream and Downstream Flow Fields*

The fundamental equations of unsteady flow components can be linearized as follows:

$$\left(\frac{\partial}{\partial t}+U\frac{\partial}{\partial x}+V\frac{\partial}{\partial y}\right)\delta u=-\frac{1}{\overline{\rho}}\frac{\partial \delta p}{\partial x}$$
$$\left(\frac{\partial}{\partial t}+U\frac{\partial}{\partial x}+V\frac{\partial}{\partial y}\right)\delta v=-\frac{1}{\overline{\rho}}\frac{\partial \delta p}{\partial y} \tag{2}$$

$$\left(\frac{\partial}{\partial t}+U\frac{\partial}{\partial x}+V\frac{\partial}{\partial y}\right)\delta \rho+\overline{\rho}\left(\frac{\partial \delta u}{\partial x}+\frac{\partial \delta v}{\partial y}\right)=0 \tag{3}$$

$$\delta p=\frac{1}{\overline{f}}\frac{P}{\overline{\rho}}\delta \rho=\overline{a}^2\delta \rho \tag{4}$$

Here, the quantities with δ mean the unsteady flow components, and both the capital letters and overlines indicate steady components. f and a are the void fraction and the speed of sound, respectively.

For unsteady flow components, we assume the following functional form:

$$\exp 2\pi j\left(nt-y/s\right)\exp\left(2\pi\lambda x/s\right) \tag{5}$$

where, n, s and λ are the complex frequency, the wavelength in the direction of the y-axis, and the wave number in the direction of the x-axis, respectively. Equation (5) allows the disturbance to propagate in the direction of the y-axis.

From equations (2) - (4), we can obtain the following partial differential equation for δp.

$$\frac{1}{\overline{\rho}}\left[\nabla^2-\frac{1}{\overline{a}^2}\left(\frac{\partial}{\partial t}+U\frac{\partial}{\partial x}+V\frac{\partial}{\partial y}\right)^2\right]\delta p=0 \tag{6}$$

The solutions of the above equation are irrotational disturbance components, and their wave numbers in the x-direction are represented as follows:

$$\lambda_{\pm}=\frac{jM_x(k-\tan\gamma)\pm\sqrt{(1-M_x^2)-M_x^2(k-\tan\gamma)^2}}{1-M_x^2} \tag{7}$$

Here, γ is the absolute main flow angle, and $M_x = U/\bar{a}$ is the axial Mach number, which is assumed to be less than unity in this study. Moreover, $k=k_R+jk_I=sn/U$ is the normalized complex frequency, where k_R and k_I mean the normalized frequency and the normalized decay rate, respectively. We should note that there are two wave numbers, $\lambda_{+,-}$ in Eq.(7), indicating two modes of pressure fluctuation. One, the λ_- mode, consists of a pressure wave propagating upstream, which is emitted from the disturbance source moving along the y-axis. The other mode, λ_+, consists of a pressure wave propagating downstream. However, it is not easy to distinguish these two modes only from Eq. (7).

We consider the neutrally stable condition with $n_I=k_I=0$. Figure 2 shows the wave surface emitted from the disturbance source travelling along the y-axis in the case that the axial Mach number is less than unity. There can be three types of flow field which depend on the flow compressibility (Kaji and Okazaki, 1971).

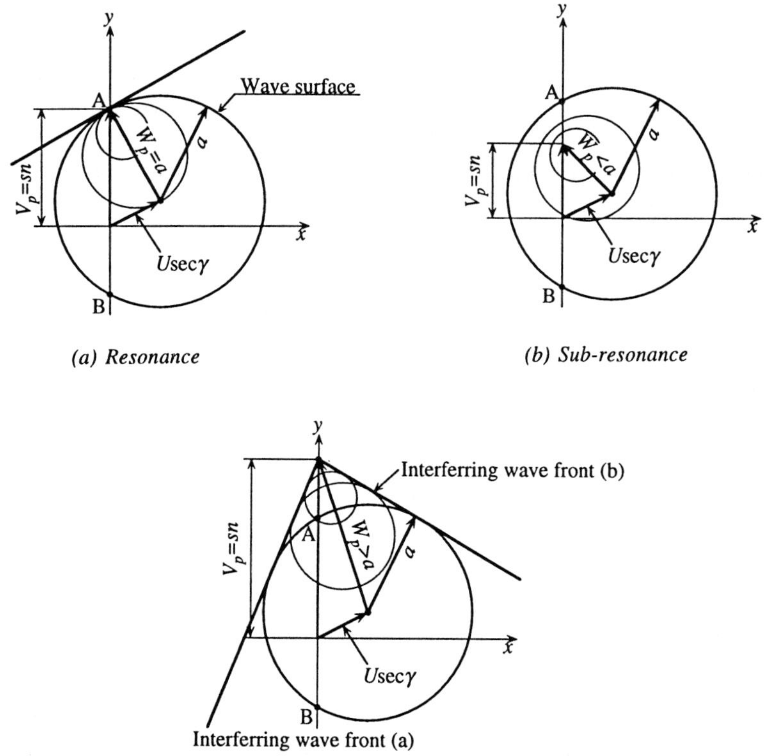

(a) Resonance *(b) Sub-resonance*

(c) Super-resonance

Figure 2. Relationship between wave surface and disturbance source travelling along y-axis

First, we consider the resonance where the velocity of disturbance source relative to the flow media, $|W_p|$, equals the speed of sound, a, as shown in Fig. 2(a). The normalized resonance frequency, k_{res}, is represented as follows:

$$k_{res\pm}=\frac{V_p}{U}=\tan\gamma\pm\frac{\sqrt{1-M_x^2}}{M_x^2} \tag{8}$$

Here, $V_p=sn$ is the velocity of disturbance source in the direction of the y-axis.

The second is the sub-resonance state as shown in Fig. 2(b), where the frequency k is between k_{res-} and k_{res+} with the relative velocity $|W_p|$ smaller than a. One mode of the previous pressure fluctuations becomes infinity at the upstream infinity and the other at the downstream infinity. In the upstream/downstream flow field, we should choose the mode which is finite at the upstream/downstream infinity, because all the pressure waves emitted from the sinusoidal disturbance source should cancel each other out.

The third is the super-resonance state as shown in Fig. 2(c), where the normalized frequency is larger than k_{res+} or smaller than k_{res-} with the relative velocity $|W_p|$ being larger than a. In this case two wave fronts, (a) and (b), are formed as a result of interference of disturbances emitted at different times. Wave front (a) extends upstream of the cascade and wave front (b) extends downstream. Thus, we should choose the upstream extending wave front (a) for the upstream flow field and the downstream extending wave front (b) for the downstream flow field. We can easily choose them by referring to the isophase surface of pressure fluctuation because its slope completely agrees with that of the wave fronts.

Next, we consider the cases when the pressure fluctuation is amplifying/attenuating ($k_I<0/k_I>0$) in time. In these cases, the relationship between the relative velocity of disturbance sources and the speed of sound is same as that for the case of neutral stability. So, for the case of sub-resonance, the wave number can be chosen so that the pressure remains finite in each flow field. For the case of super-resonance, we can choose the wave number which agrees with that for $k_I=0$ (neutral stability case) determined as stated above, within the limit of $k_I=0$.

Table 1 summarizes the above results.

It can be shown that the disturbance field determined above is irrotational. Since the downstream flow of cascade is rotational, we should obtain another type of disturbance. This can be easily done by putting $\delta p=0$ in the linearized Euler's equation assuming disturbances with the functional form of Eq.(5). The result is as follows:

TABLE 1. Selection of axial wave number λ

	Amplifying ($k_I<0$)	Neutrally stable ($k_I=0$)	Attenuating ($k_I>0$)
Superresonance ($k_R<k_{res-}$ or $k_{res+}<k_R$)	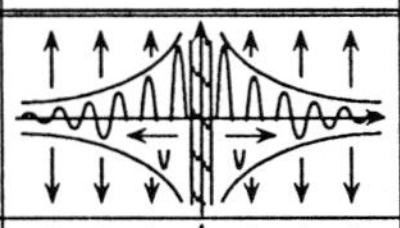	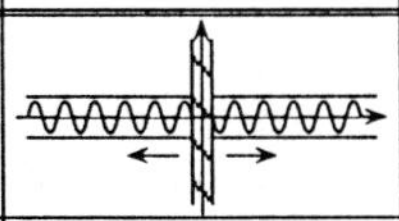	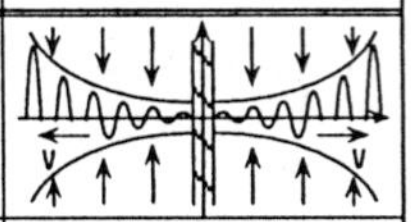
Subresonance ($k_{res-}<k_R<k_{res+}$)	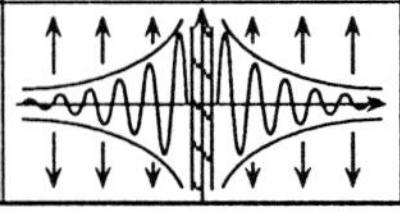	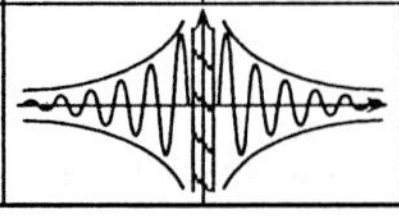	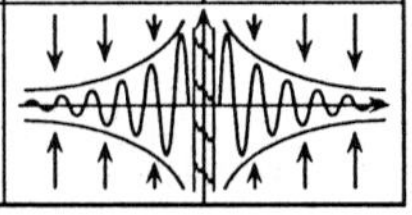

Note: Vertical arrows '↑↑↑' mean time-amplifying or attenuating state.
Horizontal arrows '→' mean the extending direction of the interferring wave fronts (a) or (b).

$$\lambda=-j(k-\tan\gamma) \tag{9}$$

The disturbance flow fields are described by combining the irrotational fluctuations with the pressure fluctuation and the rotational one with the vorticity fluctuation.

Upstream, we assume the flow is irrotational. The unsteady flow components can be expressed as follows, which satisfies Eqs. (2) - (4).

$$\begin{aligned}
\frac{\delta p_1}{\bar{\rho}\, U^2}&=-\frac{jk+\lambda_1}{\lambda_1}A_1\exp 2\pi j\left(nt-\frac{y}{s}\right)e^{2\pi\lambda_1 x/s}\\
\frac{\delta u_1}{U}&=A_1\exp 2\pi j\left(nt-\frac{y}{s}\right)e^{2\pi\lambda_1 x/s}\\
\frac{\delta v_1}{U}&=-\frac{j}{\lambda_1}A_1\exp 2\pi j\left(nt-\frac{y}{s}\right)e^{2\pi\lambda_1 x/s}\\
\frac{\delta \rho_1}{\bar{\rho}}&=-\frac{jk+\lambda_1}{\lambda_1}M_{x1}^2A_1\exp 2\pi j\left(nt-\frac{y}{s}\right)e^{2\pi\lambda_1 x/s}
\end{aligned} \tag{10}$$

Here, A_1 represents the normalized complex amplitude of the axial velocity fluctuation. We should obtain the wave number λ_1 using the upstream axial Mach number M_{x1} in Eq. (7).

Downstream, the flow is rotational due to the vortices being shed from the cascade. The unsteady components can be expressed as follows, which satisfies Eqs. (2) - (4).

$$
\begin{aligned}
\frac{\delta p_2}{\overline{\rho}\,U^2} &= -\frac{j(k-\tan\gamma)+\lambda_2}{\lambda_2} A_2 \exp 2\pi j\left(nt-\frac{y}{s}\right) e^{2\pi\lambda_2 x/s} \\
\frac{\delta u_2}{U} &= \exp 2\pi j\left(nt-\frac{y}{s}\right)\left[A_2 e^{2\pi\lambda_2 x/s} + B_2 e^{2\pi\lambda_3 x/s}\right] \\
\frac{\delta v_2}{U} &= \exp 2\pi j\left(nt-\frac{y}{s}\right)\left[-\frac{j}{\lambda_2} A_2 e^{2\pi\lambda_2 x/s} - (k-\tan\gamma) B_2 e^{2\pi\lambda_3 x/s}\right] \\
\frac{\delta \rho_2}{\overline{\rho}} &= -\frac{j(k-\tan\gamma)+\lambda_2}{\lambda_2} M_{x2}^2 A_2 \exp 2\pi j\left(nt-\frac{y}{s}\right) e^{2\pi\lambda_2 x/s}
\end{aligned}
\tag{11}
$$

Here, A_2 and B_2 represent the normalized complex amplitude of the irrotational and rotational unsteady components of the axial velocity, respectively. We should obtain λ_2 from Eq. (7) using the downstream axial Mach number M_{x2} and λ_3 from Eq. (9).

2.2.2 *Unsteady Flow Components in the Cascade*

In the cascade, the flow is assumed to be perfectly guided by the blades and is a 1-D flow in the direction of the ξ–axis. We represent the unsteady flow components as follows:

$$
\exp 2\pi j n^* t^* \exp\left(2\pi\lambda_c \xi / s\right) \tag{12}
$$

Here, t^* and $n^* = n\text{-}U_T/s$ are the time and the complex frequency in the frame moving with cascade, respectively. We can obtain the following expressions of unsteady flow components from the Euler's equation, the continuity equation, and Eq. (4).

$$
\begin{aligned}
\frac{\delta p_c}{\overline{\rho}\,U^2} &= \exp 2\pi j n^* t^* \left[A_{c1} e^{2\pi\lambda_{c1}\xi/s} + A_{c2} e^{2\pi\lambda_{c2}\xi/s}\right] \\
\frac{\delta w_c}{W_c} &= \exp 2\pi j n^* t^* \left[-M_c A_{c1} e^{2\pi\lambda_{c1}\xi/s} + M_c A_{c2} e^{2\pi\lambda_{c2}\xi/s}\right] \\
\frac{\delta \rho_c}{\overline{\rho}} &= \exp 2\pi j n^* t^* \left[M_c^2 A_{c1} e^{2\pi\lambda_{c1}\xi/s} + M_c^2 A_{c2} e^{2\pi\lambda_{c2}\xi/s}\right]
\end{aligned}
\tag{13}
$$

Here, both A_{c1} and A_{c2} represent the unsteady components of pressure fluctuation. The wave numbers in the ξ–direction, λ_{c1} and λ_{c2}, and the Mach number M_c are represented as follows:

$$\lambda_{c1}=\frac{j\frac{sn^*}{W_c}M_c}{1-M_c}=j\frac{M_c}{1-M_c}\cos\beta\left(k-\tan\overline{\beta}\right)$$

$$\lambda_{c2}=-\frac{j\frac{sn^*}{W_c}M_c}{1+M_c}=-j\frac{M_c}{1+M_c}\cos\beta\left(k-\tan\overline{\beta}\right) \tag{14}$$

$$M_c=\frac{W_c}{\overline{a_c}}=W_c\sqrt{\frac{\overline{f}\,\overline{\rho}}{P_c}}$$

2.3 BOUNDARY CONDITIONS

2.3.1 *Continuity Relation at Cascade Inlet*

We consider the continuity of the flow through the throat at the cascade inlet. By representing the cavity volume which appears at the cascade inlet, V_c, we can obtain the following continuity equation including the cavity volume change:

$$\rho_1 u_1 h-\rho_c u_c h\cos\beta=\frac{\partial^*\rho_c\left(S-V_c\right)}{\partial t^*} \tag{15}$$

where it is assumed that the density of flow through the throat equals that at the inlet, and $S=h^2\cos\beta\sin\beta/2$ is the area of the throat.

We assume that the cavity volume is a function of the incidence angle α and the cavitation number $\sigma=2(p-p_v)/\rho W_1^2$ (p_v:vapor pressure, W_1:upstream relative velocity), and that the cavity volume responds to these parameters quasi-steadily. Then, we can represent the cavity volume change as follows:

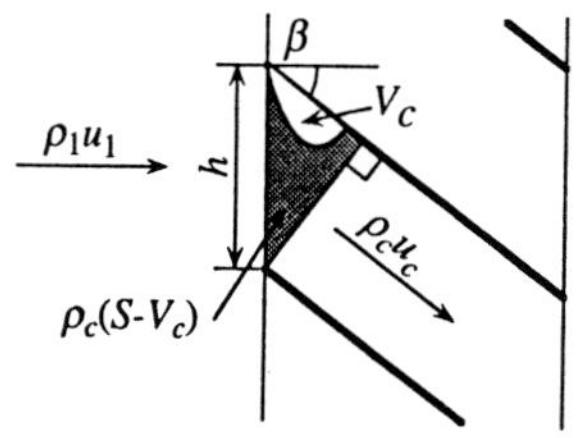

Figure 3. Continuity equation at inlet of cascade

$$\delta V_c = h^2(-K\delta\sigma + M\delta\alpha)$$

$$K = -\frac{1}{h^2}\frac{\partial V_c}{\partial \sigma}, \qquad M = \frac{1}{h^2}\frac{\partial V_c}{\partial \alpha} \tag{16}$$

Here, K and M are called the cavitation compliance and the mass flow gain factor, respectively. With expressions (10), (11), and (16), Eq.(15) can be reduced to

$$\left(\frac{\delta u_1}{U} + \frac{\delta \rho_1}{\overline{\rho}}\right) - \left(\frac{\delta w_c}{W_c} + \frac{\delta \rho_c}{\overline{\rho}}\right) = 2\pi j \frac{h}{s}(k - \tan\overline{\beta}) \times$$

$$\left[\frac{s}{h^2}\frac{\delta \rho_c}{\overline{\rho}} - \left(G_1\frac{\delta u_1}{U} + G_2\frac{\delta v_1}{U} + G_3\frac{\delta p_1}{\overline{\rho}U^2} + G_4\frac{\delta \rho_1}{\overline{\rho}}\right)\right] \tag{17}$$

$$G_1 = 2\overline{\sigma}K\cos^2\overline{\beta} - M\sin\overline{\beta}\cos\overline{\beta}, \qquad G_2 = -2\overline{\sigma}K\sin\overline{\beta}\cos\overline{\beta} - M\cos^2\overline{\beta},$$

$$G_3 = -2K\cos^2\overline{\beta}, \qquad G_4 = \overline{\sigma}K$$

2.3.2 *Pressure Rise at Cascade Inlet*

The flow is assumed to be barotropic and we ignore the frictional loss. We can therefore apply Bernoulli's equation at the cascade inlet. After linearization, the following equation is obtained.

$$U\delta u_1 - U_T\delta v_1 + \frac{\delta p_1}{\overline{\rho}} = W_c\delta w_c + \frac{\delta p_c}{\overline{\rho}} \tag{18}$$

2.3.3 *Conditions at Cascade Exit*

At the cascade outlet, the continuity of pressure, the mass continuity equation, and Kutta's condition must be satisfied:

$$\delta p_c = \delta p_2$$

$$\delta w_c \cos\beta = \delta u_2 \tag{19}$$

$$\delta v_2 = -\delta u_2 \tan\beta$$

2.4 ANALYTICAL METHODS AND CONDITIONS

By substituting the unsteady flow components (10), (11) and (13) into boundary conditions (17) - (19), we can obtain a set of homogeneous linear equations in terms of the normalized complex amplitudes of unsteady components. From the coefficient matrix of these linear equations, we obtain a characteristic equation.

Complex propagation velocity ratio k^*,

$$k^*=k_R^*+jk_I^*=\frac{sn}{U_T}=k\frac{U}{U_T} \tag{20}$$

is determined from the solutions of the characteristic equation. Here, k_R^* and k_I^* represent the propagation velocity ratio and the decay rate, respectively. The negative decay rate, $k_I^*<0$, means that the disturbance is amplifying, i.e., the disturbance is unstable.

Default values of the present analysis are shown in Table 2. Cavitation compliance K and mass flow gain factor M show the effects of cavitation, while the relative inflow Mach number M_{rel} shows the effects of flow compressibility.

$$M_{rel}=\sqrt{\frac{U^2+U_T^2}{\overline{a}_l^2}} \tag{21}$$

We assumed that the flow was isothermal before. However, we can apply other thermal conditions within the assumptions of barotropic fluid, because the assumption of isothermal fluid is only used in the definition of the speed of sound.

TABLE 2. Conditions for present analysis

Design parameters of cascade	
Stagger, β	79.73[deg.]
Chord, l/s	0.815
Pitch, h/s	0.333
Flow condition	
Flow rate, U/U_T	0.110
Cavitation number, σ	0.04

3. Results and Discussion

3.1 SOLUTIONS OF CHARACTERISTIC EQUATION

Figure 4 shows examples of solutions of the characteristic equation with $(K, M)=(0.1, 0.1)$ and $M_{rel}=0.0$ - 0.4. Six solutions are found in this range.

Solutions I, II and III correspond to the conventional forward travelling R.C., backward travelling R.C., and R.S., respectively, because those solutions approach the solutions of incompressible analysis (Tsujimoto et al., 1993) with M_{rel} approaching 0.

Solutions IV is a mode for which the wave number of irrotational disturbance, λ_2, equals that of rotational disturbance, λ_3. The solution is as follows:

$$k^* = (\tan\gamma - j)U / U_T \tag{22}$$

In this mode, we can easily see from Eq.(11) that both downstream pressure and density fluctuations vanish. It can also be shown that the irrotational and rotational components of downstream velocity cancel each other out, i.e., all the downstream fluctuations vanish. From the boundary conditions, we can find that all the fluctuations in each flow field are 0. So, we do not go into the detail about this mode.

Solutions V and VI are the acoustic modes of 1-D flow in the cascade affected by upstream and downstream fluctuations. It was also found that there are many solutions with higher frequency, corresponding to the higher order acoustic modes of 1-D flow in the cascade. These modes are always decaying in the region of (K,M) which we examined.

Generally, six characteristic solutions are found. However, it was found that a 7th solution (VII) exists in special cases, especially in the case when the value of M/K is large .

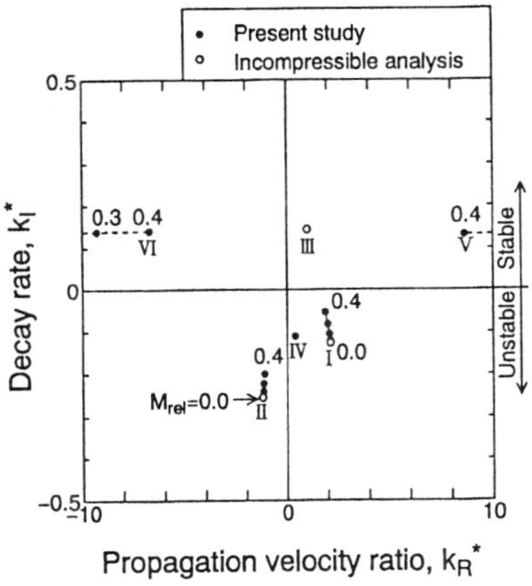

Figure 4. Solutions of characteristic eq. [(K,M)=(0.1,0.1), M_{rel}=0.0 - 0.4]

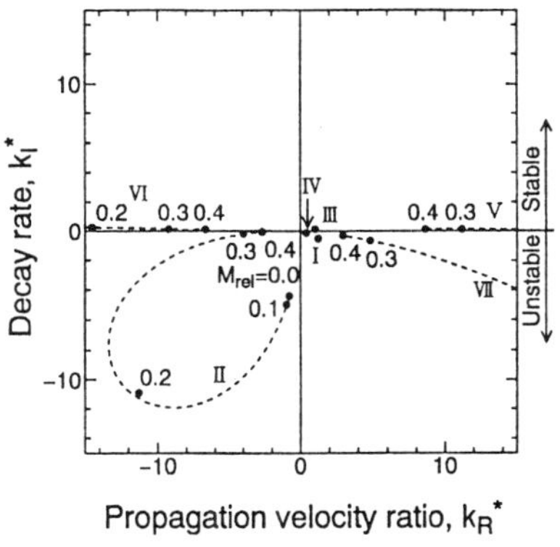

Figure 5. Solutions of characteristic eq. [(K,M)=(0.1,1.0), M_{rel}=0.0 - 0.4]

Figure 5 shows examples of solution VII with (K, M)=(0.1, 1.0). Solution VII is a forward rotating mode, which exists for the case with larger values of M_{rel}. By comparing its propagation velocity ratio with the resonance frequency, we can see that the upstream flow of this mode is in super-resonance condition, while those of conventional R.C. I and II are in sub-resonance. So, solution VII is a forward R.C. different from conventional mode I. However, no evidence of forward R.C. VII has been found in experiments. We can also see that the forward R.C. I is not significantly influenced by the flow compressibility, while the backward R.C. II significantly changes with M_{rel}. The upstream flow of R.C. II changes from sub-resonance to super-resonance around M_{rel}=0.2.

Solutions I - VII are summarized in Table 3.

TABLE 3. Classification of solutions

I	Conventional forward rotating cavitation
II	Conventional backward rotating cavitation
III	Rotating stall
IV	Solution with no disturbances
V,VI	Acoustic modes of 1-D flow in the cascade coupled with upstream and downstream flow
VII	Unknown forward rotating cavitation

3.2 INFLUENCE OF FLOW COMPRESSIBILITY

Figure 6 shows the neutrally stable points of conventional R.C. I and II, with K=0.1 and M_{rel} = 0.0 - 0.4. The regions with values of M smaller than these points are stable region. We can see that the stable region of R.C. slightly increases due to the flow compressibility. The increase of flow compressibility gives an additional compliance to the compliance of cavity. This may be the reason why R.C. is stabilized by the flow compressibility.

Under R.C., it was found that the pressure fluctuation upstream of the impeller is larger than that downstream (Kamijo et al., 1993). Tsujimoto et al. (1993) have shown that R.C. can be simulated even without downstream fluctuations. In the present case we obtain the following equation by substituting δp_c=0 and δw_c=0 into the continuity equation (17).

$$2\pi j\frac{h}{s}\left(k-\tan\overline{\beta}\right)\left(G_1-\frac{j}{\lambda_1}G_2-\frac{jk+\lambda_1}{\lambda_1}G_3-\frac{jk+\lambda_1}{\lambda_1}M_{x1}^2G_4\right)+\left(1-\frac{jk+\lambda_1}{\lambda_1}M_{x1}^2\right)=0 \quad (23)$$

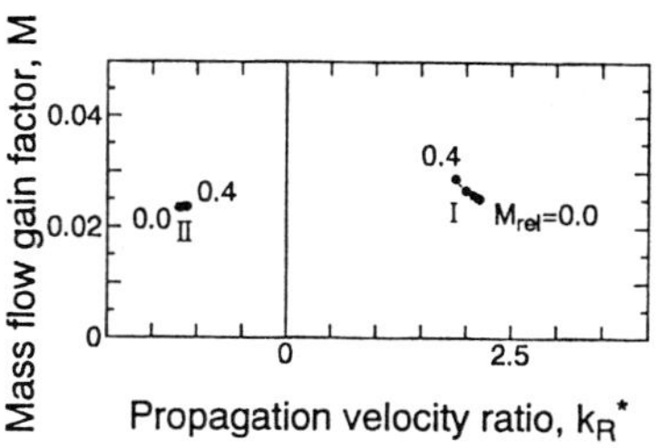

Figure 6. Neutrally stable points of conventional R.C., I and II [K=0.1, M_{rel}=0.0 - 0.4]

Eq. (23) can be reduced to a 4th order polynomial equation by introducing the practical expression of λ.

Figure 7 (a) and (b) show examples of the solutions of Eq. (23) with (K,M)= (0.1,1.0) and M_{rel}=0.0 - 0.4. These figures also show solutions I , II and VII of the characteristic equation. We can see that these solutions agree with the solutions of Eq. (23).

From Fig. 7(a), we can see that solution VII exists as a stable backward travelling mode when M_{rel} is approximately between 0.19 and 0.20. When M_{rel} approaches 0.20, the propagation velocity ratio of solution VII tends to be $-\infty$. Then, the solution of the backward mode disappears, and at the same time solution VII appears as a unstable forward travelling mode. This behavior is closely related to the vanishing of the 4th order coefficient of the final expression of Eq. (23):

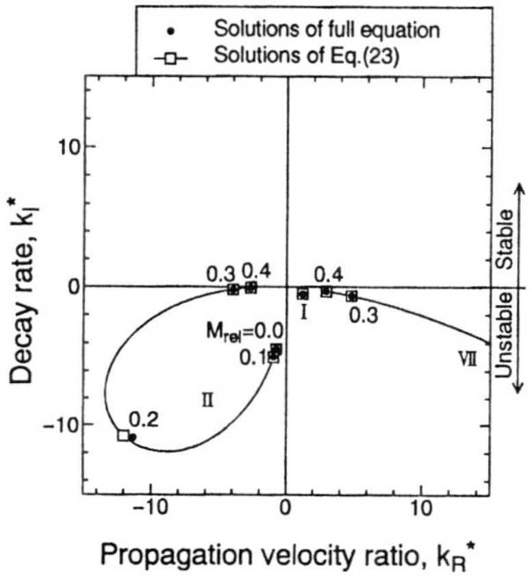

(a) Solutions I, II and VII

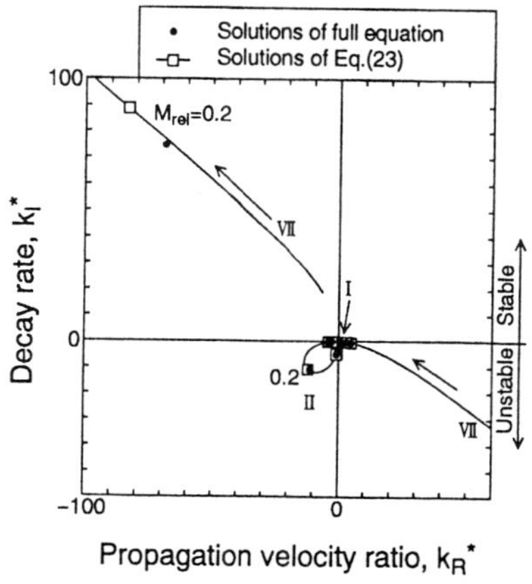

(b) Enlarged figure of (a)

Figure 7. Solutions of Equation (23), compared with those of characteristic eq.

$$\left(\frac{2\pi h}{s}\right)^2\left[(G_1M_{x1}^2-G_3)^2-M_{x1}^2(G_1-G_3)^2\right] \tag{24}$$

Because solution VII exists when M/K is large and the cavitation number $\overline{\sigma}$ is small (usually about 0.01 - 0.1), we substitute $\overline{\sigma}$=0 into Eq. (24)=0 and obtain the following equation.

$$M_{x1}=2\frac{K}{M}\cot\overline{\beta} \tag{25}$$

When the upstream axial Mach number is larger than this value, the unstable forward rotating mode VII exists. So, the condition for the existence of unstable mode VII is as follows.

$$M_{rel}>2\frac{K}{M}\cot\overline{\beta}\sqrt{1+\tan^2\overline{\beta}} \tag{26}$$

Equation (26) gives M_{rel}>0.201 for the conditions used for Fig. 7, which agrees with the results in Fig. 7.

4. Conclusion

We summarize the results obtained in this study as follows.

(1) The characteristic equation obtained in this study generally has 6 types of solutions, corresponding to conventional forward and backward R.C., I and II, R.S., III, solution with no flow disturbances, IV, and the acoustic modes of 1-D flow in the cascade coupled with the upstream and downstream fluctuations, V and VI.

(2) In some special cases, especially in the case with a larger value of M/K, a 7th solution exists as a forward rotating mode, where the upstream flow is in super-resonance, while the upstream flow is in sub-resonance under the conventional R.C. I. This mode is a forward R.C. different from conventional R.C. I. However, no evidence of R.C. VII has been found in experiments.

(3) The flow compressibility can make conventional R.C. modes, I and II, slightly more stable, but does not have a large effect. This is because the flow compressibility adds to the compliance of the cavity, which has stabilizing effects.

The authors are grateful to Mr. Miyamoto, graduate student of Osaka University, for his kind support in the present study. This study was partly supported by the Grant-in-

Aid for Scientific Research of the Ministry of Education, Science, Sports and Culture. One of the authors was supported by the Research Fellowships of the Japan Society for the Promotion of Science for Young Scientists.

5. References

Goirand, B., Mertz, A., Joussellin, F. and Rebattet, C. (1992) Experimental Investigations of Radial Loads Induced by Partial Cavitation with a Liquid Hydrogen Inducer, *IMechE* C453/056, 263-269

Hashimoto, T., Yoshida, M., Watanabe, M., Kamijo, K. and Tsujimoto, Y. (1996) Experimental Study on Rotating Cavitation of Rocket Propellant Pump Inducers, *AIAA Paper* 96-2674.

Kaji, S. and Okazaki, T. (1971) Flutter of Axial Compressor Blade in Compressible Flow, *Trans. JSME Ser. B* **38**-309 (in Japanese), 1023-1033.

Kamijo, K., Yoshida, M. and Tsujimoto, Y. (1993) Hydraulic and Mechanical Performance of LE-7 LOX Pump Inducer, *AIAA J. Propulsion and Power* **9**, 819-826.

Ryan, R. S., Gross, L. A., Mills, D. and Mitchell, P. (1994) The Space Shuttle Main Engine Liquid Oxygen Pump High-Synchronous Vibration Issue, the Problem, the Resolution Approach, the Solution, *AIAA Paper* 94-3153.

Tsujimoto, Y., Kamijo, K. and Yoshida, Y. (1993) A Theoretical Analysis of Rotating Cavitation in Inducers, *Trans. ASME J. Fluids Engineering* **115**, 135-141.

Tsujimoto, Y., Kamijo, K. and Yoshida, Y. (1995) On Theoretical Simulations of Rotating Cavitation, in Y. Tanida and M. Namba (eds.), *Unsteady Aerodynamics and Aeroelasticity of Turbomachines*, Elsevier Science B. V., Amsterdam, pp. 641-657.

Tsujimoto, Y., Watanabe, S., Kamijo, K. and Yoshida, Y. (1996) A Nonlinear Calculation of Rotating Cavitation in Inducers, *Trans. ASME J. Fluids Engineering* **118**, 589-594.

Watanabe, S., Yokota, K., Tsujimoto, Y. and Kamijo, K. (1997) Three-Dimensional Analysis of Rotating Cavitation in Inducers - Linear Cascade Model Spanning between Two Parallel Plates-, *Proc. JSME International Conference on Fluid Engineering* ICFE-97-720, 1377-1382.

NONLINEAR CRITICAL LAYER IN THE UNSTEADY BOUNDARY LAYER INDUCED ON A FLAT PLATE BY WAKES

Z. Wierciński
Institute of Fluid-Flow Machinery PASci.
Fiszera 14
80-952 Gdańsk, Poland

Summary: The paper reports results of experimental investigations of the wake - boundary layer interaction in the region of transition on a flat plate in a wind tunnel of low level of turbulence, Tu≤0.0008. Correlation analysis of velocity fluctuations across the induced boundary layer has revealed a region where a phase shift of external disturbances takes place. Spectral analysis of velocity fluctuations showed also strong attenuation of wakes harmonics in this region. The region exists for values of the Reynolds number Re_x from $8\cdot10^4$ to $2\cdot10^5$, is located at $y_c = 0.32$ and has thickness $d_c = 0.22$, which corresponds to the mean velocity $(U/U_e)_c$ equal to 0.74 in the boundary layer. The location of this region (higher than the highest possible critical layer location for the Tollmien-Schlichting waves) and the level of external disturbances ε which is equal to 0.03 suggest that could be the nonlinear critical layer which has been analysed in the solution of the nonlinear Orr-Sommerfeld equation for the parameter $\lambda=1/(Re^{*}\varepsilon^{3/2}) = O(1)$ (Benney&Bergeron, 1969, and Haberman, 1972).

1.Introduction

Delimitation of the region of the laminar-turbulent transition on a profile is extremely important for determination of the profile losses, and hence the efficiency of a stage. The laminar-turbulent transition in the boundary layer is an especially complex phenomenon due to the pressure gradient along the layer, the high level of turbulence of the external flow, the potential flow interaction, shock waves and so on. Its analysis remains complex even when the flow is not regarded as threedimensional and the system of vortices associated with the rotor and stator as well as the mixing in the blade-to-blade channel are omitted. The threedimensional vortex structures developing in the transitional boundary layer cause an additional difficulty. It is well known from experiments that the transition does not consist in an abrupt jump from the laminar to the turbulent state of flow but occurs gradually, starting with the Tollmien-Schlichting waves, coming through the threedimensional peak and valley structures and ending with high frequency bursts.

Since the works of Reynolds, Rayleigh, Tollmien and Schlichting (see e.g. Schlichting, 1965) it has generally been accepted that the transition is a consequence of the flow instability in the laminar boundary layer. Landau, 1944, gave the general

T.H. Fransson (ed.), Unsteady Aerodynamics and Aeroelasticity of Turbomachines, 19-33.

outline of the consecutive instabilities of the laminar flow with the Tollmien-Schlichting waves at the very beginning. The laminar boundary layer can be regarded as a very complicated nonlinear oscillator and the transition can be treated as its response to various disturbances. The amplitude and frequency of these disturbances can be assumed to vary from very small to quite large values, i.e. from ε=0.0001 to 0.10 and more. The boundary layer response to infinitesimal disturbances is well described by the linearized Orr-Sommerfeld (OS) equation. The initial stage of linear instability, called the primary instability, can be predicted by the solution of the OS equation for modes growing with time in a locally parallel flow. The influence of wakes, resulting in large amplitude disturbances requires taking into consideration nonlinear terms in the OS equation. Benney&Bergeron, 1969, proposed to solve the nonlinear OS equation by means of the method of matched expansions. According to their analysis the balance between the nonlinear and viscous terms in the nonlinear OS equation can be described by a dimensionless parameter

$$\lambda = 1/(\mathrm{Re}^{*}\varepsilon^{3/2}) \qquad (1)$$

where

$$\varepsilon = u'/U_e \qquad (2)$$

is the ratio of the disturbance amplitude to the mean velocity outside the boundary layer. In the literature, e.g Drazin&Reid, 1981, estimations of the three terms of the nonlinear OS equation: convectional, viscous and nonlinear for the region close to the critical layer can be found It can be easily seen that the λ parameter is the cube of the ratio of the thickness of a linear viscous critical layer to that of a nonlinear inviscid one.
Analysing values of the parameters λ, ε and Re^{*} one can conclude that the linearized OS equation fulfils the following conditions:

$$\lambda \ll 1,\ \mathrm{Re}^{*} \gg 1 \text{ and } \varepsilon \ll 1 \qquad (3)$$

Benney & Bergeron, 1969, considered the stability of a twodimensional boundary layer for conditions corresponding to (3), but with a following additional condition:

$$\varepsilon^{1/2} \gg \mathrm{Re}^{*\,-1/3} \qquad (4)$$

indicating that nonlinear effects prevailed over the viscous effects.

On the other hand, Haberman, 1972, analysed the stability of a two-dimensional boundary layer for the case, when nonlinear and viscous effects are of the same order:

$$\lambda = O(1) \text{ and } \varepsilon^{1/2} = O(\mathrm{Re}^{*\,-1/3}) \qquad (5)$$

In the analysis of Benney & Bergeron, 1969, neutral curves c=f(α) for three different velocity profiles in the boundary layer were given. Furthermore, the eigenvalue problem was formulated to determine the location of the nonlinear critical layer for a given wave

number α. The locations y_c of the critical layer, however, were given only for the asymptotic values of the wave number $\alpha \to 0$ and $\alpha \to \infty$ While there exists a critical Reynolds number associated with the growth initiation of a disturbance for the linear problem, no critical Reynolds number exists for the nonlinear problem, and the only conditions are that the Reynolds number be large and the amplitude of disturbance sufficient. In the opinion of those authors, a finite amplitude disturbance producing such a wave might induce the transition without the necessity of passing through the Tollmien-Schlichting wave regime. While in the solution of Benney & Bergeron there is no phase shift of disturbances across the critical layer, such phase shift appears in the solution of Haberman. Thus, the taking into account of both the nonlinearity and viscosity leads to the phase shift across the nonlinear critical layer.
It has been found, when comparing the solutions to the linear and nonlinear OS equations that the „new class of waves" - named so by Benney & Bergeron - of the nonlinear problem may have larger wave numbers than the Tollmien-Schlichting waves. While the Tollmien-Schlichting waves have a length not smaller than about six boundary layer thicknesses, the „new waves" of Benney & Bergeron may have a length of about one boundary layer thickness or even less. Thus, these are waves of much greater wave number and therefore their critical layers have to be located closer to the outer edge of the boundary layer. Such waves can not appear in the natural transition. Benney&Bergeron emphasized in their discussion the possibility of exciting them by disturbances convected with a free stream. They also pointed out the need for experimental confirmation of the occurrence of those waves in the induced boundary layer.

Critical layers play a very important role both in the linearized and nonlinear OS equations. A very good review of theoretical and experimental work concerning the critical layers in different types of flow was given by Maslowe, 1986. In his opinion little had been done with regard to experimental investigations of nonlinear critical layers, although suitable facilities had been available for boundary layer studies.

The critical layer of an induced boundary layer, the main subject of the investigations reported in the present paper, is worth giving more attention to its properties in both the linear and nonlinear form:

1. in a critical layer the local average velocity of flow $\overline{u}$ is equal to the phase velocity c of the wave

$$u - c = 0 \qquad (6)$$

2. a wave phase shift equal to $-\pi$ occurs in a linear critical layer. It is seen only if in the neighbourhood of the critical layer the influence of viscosity is taken into account. The occurence of this linear phase shift has been confirmed experimentally for infinitesimal disturbances.

3. there is no phase shift for a nonlinear critical layer if the condition $\lambda<<1$ is fulfilled as in the analysis of Benney&Bergeron. For a nonlinear critical layer the phase shift may take place only for the case considered by Haberman, 1972, that is for $\lambda=O(1)$.

4. the Kelvin „cat's eyes" pattern of the streamlines is the dominant structure in both the linear and nonlinear critical layer for an observer moving with the wave.

5. according to the theorem of vorticity inside the closed streamlines, the vorticity inside the „cat's eyes" is constant if the flow is assumed to be inviscid.

6. the thickness of a linear critical layer d_{cl} is of the order of $O((\alpha Re^*)^{-1/3})$ and the layer is located at $y_{cl} = O((\alpha Re^*)^{-1/3})$, so it overlaps a wall layer of thickness $d_p = O((\alpha Re^*)^{-1/2})$.

7. the thickness of a nonlinear critical layer d_{cn} is of the order of $O(\varepsilon^{1/2})$ and the layer lies higher above the plate surface than the linear one, so that the wall layer of thickness $d_p = O((\alpha Re^*)^{-1/2})$ does not overlap it. The nonlinear critical layer lies therefore closer to the outer edge of the boundary layer than the linear one. According to Haberman's analysis (Haberman, 1972) the thickness of a nonlinear critical layer is given by the following formula:

$$d_{cn} = \left| \frac{\varepsilon B}{u_c} \right|^{1/2} \tag{7}$$

where B is a constant determined e.g. from the experiment. Haberman assumed that B is equal to unity.

8. a critical layer is also a turning point of the mathematical solution where its form changes.

9. a solution to the Rayleigh equation is singular in a critical layer.

10. in an inviscid critical layer a discontinuity of the Reynolds stresses appears.

Both Benney&Bergeron, 1969, and Haberman, 1972, solved the homogeneous (with zero boundary conditions) twodimensional stability problem with a nonlinear disturbance of the boundary layer, i.e. the eigenvalue problem of a nonlinear OS equation.This problem is only loosely related to the experimental work reported here, which deals with a three-dimensional flow with nonzero bondary conditions, including large nonlinear velocity fluctuations passing as a running.wave the outer edge of the boundary layer. The problem investigated experimentally in this paper is much closer to the boundary layer receptivity problem. The receptivity problem differs both mathematically and physically from the stability problem. In the receptivity problem a physical disturbance is imposed on the edge of the boundary layer, and mathematically, as mentioned before, the problem is no more homogeneous, Reshotko, 1976.

The work of Nishioka and Morkovin, 1986, is devoted to the problem of the boundary layer receptivity to unsteady pressure gradients. The role of the Tollmien-Schlichting waves in the transition of the flow in the boundary layer induced by the near-isotropic free stream turbulence was examined in the work of Boiko et al., 1994.

2. The aim of the experiment

The aim of the experiment reported in this paper was to demonstrate the existence of a nonlinear critical layer with a phase shift and damping of wake harmonics by the action of viscosity in a boundary layer induced by wakes. That indicates a possibility of existence of a „new class of waves", which might excite a transition without going through the Tollmien-Schlichting wave regime.

3. The experimental rig and apparatus

The experimental investigations were carried in a subsonic wind tunnel of low level of turbulence, $Tu<0.0008$, at the Institute of Fluid-Flow Machinery, Gdańsk. A description of this tunnel, its measuring apparatus and a data acquisition system was given by Wierciński, 1995. In the working section of the wind tunnel a flat plate 0.7 m long, 0.6 m wide and 0.014 m thick was mounted. To avoid flow separation from the leading edge the plate was inclined at an incidence angle α of -2.5 °. This resulted in a favourable pressure gradient along the plate.
The wake was generated with a circular cylinder of 4 mm diameter fastened to an arm 200 mm long, which was set in a pendular movement up and down at a frequency of 4 Hz by a crank-and-rocker mechanism. Recording of hot-wire velocity fluctuation measurements was triggered by the movement of the crank-and-rocker mechanism, which ensured a good repeatibility.

4. Programme of investigations

Before starting the investigations of the critical layer in the induced boundary layer measurements of characteristics of both the natural and induced transition on the flat plate were carried out. They were made for three different velocities of flow oncoming the plate, U_o = 10, 15 and 20 m/s, at various distances from the leading edge of the plate, namely for x = 90, 115, 140, 165, 190, 215, 240, 265, 290, 315, 340, 360, 390, 570 and 630 mm. On the grounds of these measurements the thickness δ of the boundary layer, the displacement thickness δ^* and the momentum thickness δ^{**}, the shape factor H, the skin friction parameter C_f and the Reynolds numbers, Re^* and Re^{**} for the beginning and end of the natural and induced transitions were determined. Besides, measurements of the mean velocity, velocity fluctuation' profiles and of the intermittency factor for the transitional boundary layer were performed. The results of these investigations can be found in reports of Wierciński, 1995, Suprun & Wierciński, 1994,1995 and Juszyna & Wierciński, 1995.
Investigations of the critical layer in the induced boundary layer were carried out under the same geometrical and flow conditions as the measurements mentioned above, but

only for two oncoming velocities U_o= 10 and 15 m/s. This part of the investigation programme consisted of measurements of the velocity fluctuation correlation across the boundary layer. Measurements of velocity fluctuation spectra across the boundary layer were also performed. Furthermore, a conditional analysis of the mean velocity and velocity fluctuations was made. The conditional analysis consisted in partitioning of the signal recording of one cycle of the running wake into four sections according to the wake position: two wake passages (up and down) and two time intervals between them, and carrying out the measurements separately for each section.

Measurements for the critical layer were made for values of the parameter $\lambda=1/(Re^{*}\ \varepsilon^{3/2})$ within the range 0.2 ÷ 0.5, large enough to take into consideration simultaneously both the nonlinearity and viscous effects in the induced boundary layer.

5. Results of the measurements

5.1 Natural and induced transition

Because of the non zero incidence angle of the plate as mentioned above the boundary layer on the plate was characterized by a small pressure gradient along it, described by the Pohlhausen parameter Λ=1÷1.5, depending on the flow velocity. The transition in the boundary layer was localized and, in principle, two departures from the well known and

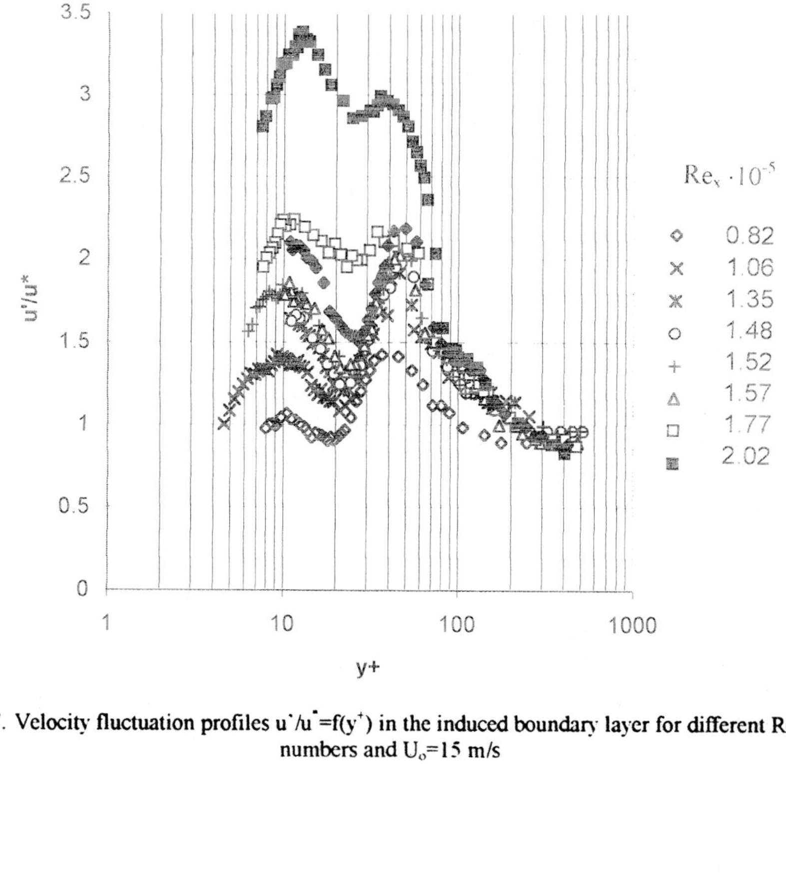

Figure 1. Velocity fluctuation profiles $u'/u^{*}=f(y^{+})$ in the induced boundary layer for different Reynolds numbers and U_o=15 m/s

accepted picture of the natural transition were observed in the investigated boundary layer. The Reynolds number of the natural transition inception , $Re_x=3.1\cdot10^5$ to $4.1\cdot10^5$, attracts attention since it is lower than the Reynolds number, $Re_x= 10^6$ to $2\cdot10^6$, known from the literature for such a small turbulence level, $Tu\leq0.0008$. This departure could have been caused, first of all, by the shape of the plate leading edge. In the theoretical work of Hammerton & Kerschen, 1996, the influence of the leading edge curvature on the appearance of so called precursors of the Tollmien-Schlichting waves, caused by external acoustical disturbances, was indicated. The second departure concerns the turbulent part of the boundary layer: the maximum value of the velocity fluctuations u_{max}'/u^* in the fully turbulized boundary layer was lower than expected. However, measurements of the local skin friction at the surface by means of the Preston tube confirmed its larger value as measured by means of the hot-wire and in that way confirmed the lower value of the u_{max}'/u^*, Wierciński, 1995. Nonetheless, other parameters of the boundary layer, such as the thickness δ, displacement thickness δ^* and momentum thickness δ^{**}, local skin friction coefficient C_f and the shape coefficient H corresponded well to the values known from the literature for the case of a favourable pressure gradient with the following relation between Reynolds numbers $Re^*=1.34\cdot Re_x^{1/2}$. Measurements of the induced boundary layer parameters such as δ, δ^* and δ^{**}, C_f and H, as well as the mean velocity and velocity fluctuation profiles showed how much they depend on the instationarity and pressure gradient. The Reynolds number of the induced transition inception Re_x is equal to $1.8\cdot10^5$.

5.2 Velocity fluctuation profile in the induced boundary layer

The velocity fluctuation profiles in the natural and induced boundary layers differ fundamentally. In the induced boundary layer two peaks in the velocity fluctuation profiles were observed, figure 1, while for the natural transition only one peak was noticed. One of the peaks is located close to the plate surface, in the same place as the peak in the natural transition (one can call it a „natural" peak), while the second of them lies closer to the outer edge of the boundary layer (one can call it an „induced" peak). It is well known that in the natural transition there is only one peak in the velocity fluctuations, namely that near the plate surface. Thus the second, „induced" peak is associated with the action of disturbances running along the outer edge of the boundary layer. At the beginning of the transition the induced peak is higher than the natural one. In the course of transition development the natural peak grows, levels with the induced peak and finally exceeds it, until the induced peak dissappears in the background of the natural one.

5.3 Nonlinear critical layer in the boundary layer

5.3.1 *Phase shift of the velocity across the critical layer*

In Fig. 2 a-c plots of velocity fluctuation traces in the induced boundary layer for a velocity $U_o= 15$ m/s and distance $x = 115$ mm from the leading edge of the plate are displayed for three different points across the layer. The plots show clearly that the phase inversion has taken place across the boundary layer between the points y/δ =0.475 and y/δ =0.161. This phase change may be an evidence for the existence of the critical layer. In Fig.3 the correlation coefficient of velocity fluctuations as a function of the coordinate across the boundary layer is plotted for a velocity U_o = 15 m/s and x = 90 mm. It can be seen that the correlation coefficient equals zero for point y/ δ ≅ 0.32; supposedly, the centre of the critical layer was located at this point. From this plot the thickness of the critical layer can be established as a region of an abrupt change of the correlation coefficient between Corr = 0.5 and Corr = - 0.5. In this way the critical layer thickness can be estimated as d_c = 0.22. The results of the conditional analysis (not shown here) confirmed the results of the correlation analysis with respect to the location of the critical layer.

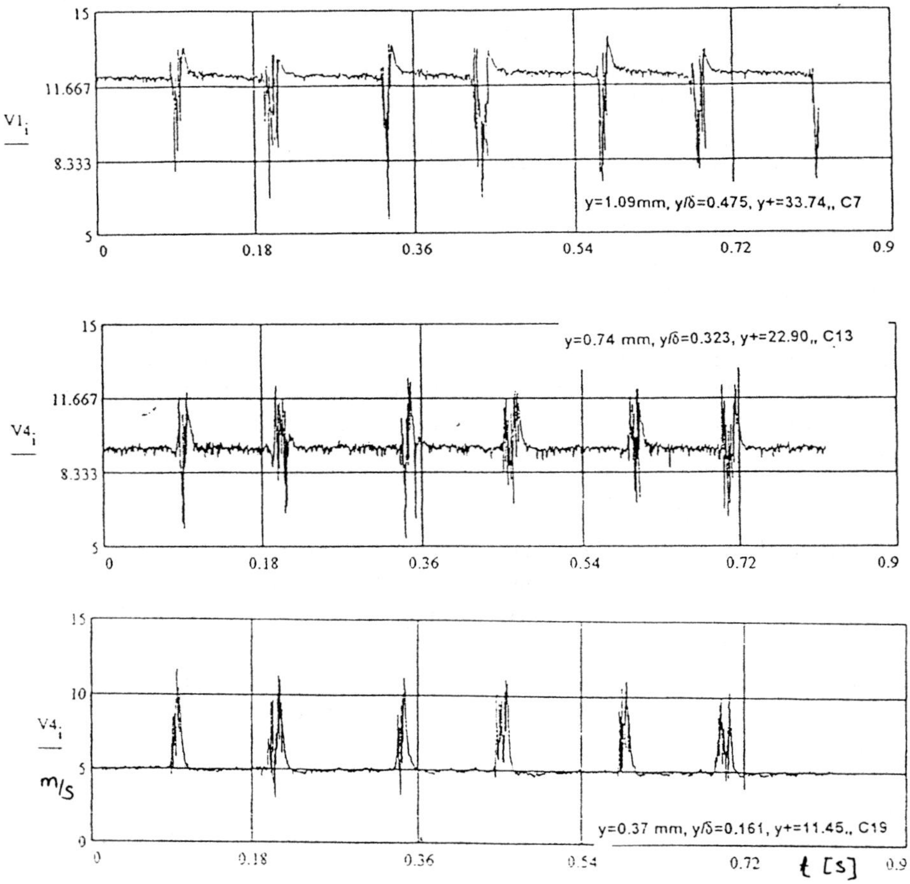

Figure 2. Velocity fluctuations u'=f(t) at three different points y/δ in the induced boundary layer for U_o=15 m/s and x = 115 m/s

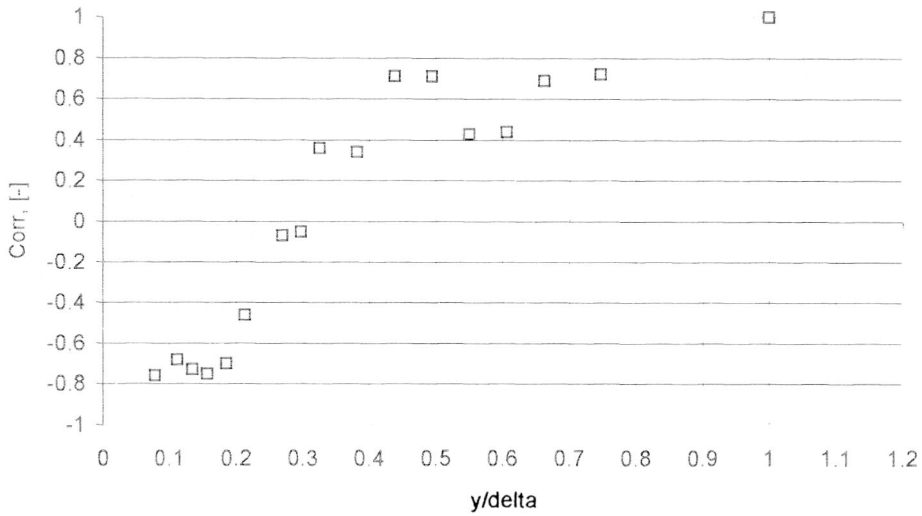

Figure 3. Correlation coefficient of the velocity fluctuations C=f(y/δ) across the induced boundary layer for x = 90 mm and U_o=15 m/s

5.3.2 *Energy spectra of the velocity fluctuations*

To describe the external disturbances running along the boundary layer edge the spectra of velocity fluctuations near it were calculated from the recorded data. The data in figure 4 represent the velocity fluctuation spectra for the velocity U_o=15 m/s, x = 115 mm and y/δ ≈ 3. In the table of this figure numerical values of the relative energy $\varepsilon_f = u_f'/U_e$ contained in the successive harmonics of the wake are given. The total level of

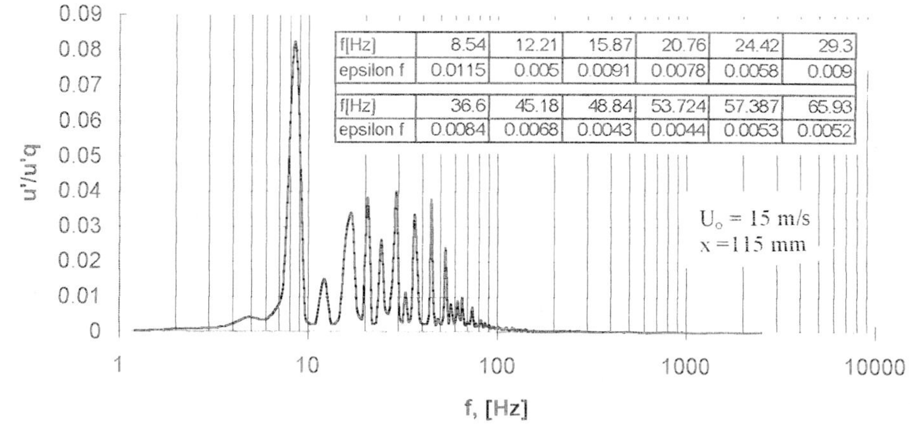

Figure 4. Spectrum of the velocity fluctuations outside the boundary layer at y/δ ≈ 3

disturbance $\varepsilon = u'/U_e$ at this point is equal to 0.031. The total level of disturbances measured at other points at the boundary layer edge was almost the same as at that point. The same concerns the disturbance level ε_f for the respective harmonics of the wake.

Spectra measurements across the boundary layer revealed a very interesting property of the critical layer. In figure 5 a summary of velocity fluctuation spectra is shown for the velocity U_o = 15 m/s and distance x = 90 mm. In this figure the relative mean square values u_f'/u_q' are plotted for the frequency range from 0 to 100 Hz and across the boundary layer for y/δ variable from 0 to 1 and beyond. The values of u_f'/u_q' have dropped markedly in the range y/δ from 0.25 to 0.45, which may be an evidence for a strong attenuation of the wake harmonics in this region. Note that in figure 5 the scale on the boundary layer thickness axis, unlike those of the other two axes, is not uniformly linear.

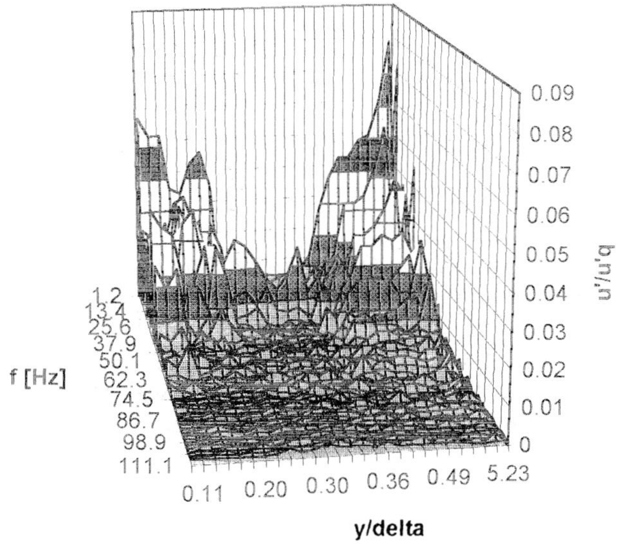

Figure 5. Spectra of the velocity fluctuations across the boundary layer for U_o=15 m/s and x=90mm

In figure 6, the spectral data are used a little bit differently, to show the attenuation u_f'/U_e for the particular wake frequency harmonic 8.54 Hz across the boundary layer, for the velocity U_o = 15 and distance x = 90 mm. Once again a drop of velocity fluctuations in the region of the critical layer is seen from this plot. From these measurements the location and thickness of the critical layer can be estimated.

The lowest point of the velocity fluctuation plot is found to lie in principle in the same place where the velocity fluctuations correlation is equal to zero and where the local minimum of velocity fluctuations between the two peaks is observed, figure 1.

5.3.3 *Location and thickness of the critical boundary layer*

A plot of locations of the centre of critical layer y_c versus Reynolds number Re_x is shown in figure 7, where additionaly the layer thickness d_c is marked. This plot was based on similar measurements as those shown in figure 6. The location of the center of the critical layer y_c is approximately equal to 0.32. Thus it lies higher in the boundary layer than the linear critical layer in the natural transition. The highest possible position of the linear critical layer in the Blasius boundary layer is equal to 0.20, which can be determined from the calculations of Jordinson, 1970, Lewczenko et al., 1975, and Healey, 1995. The highest location of the citical layer for the Pohlhausen boundary layer with $\Lambda=1.5$ differs only a little from the case of the Blasius boundary layer, Lewczenko et al., 1975.

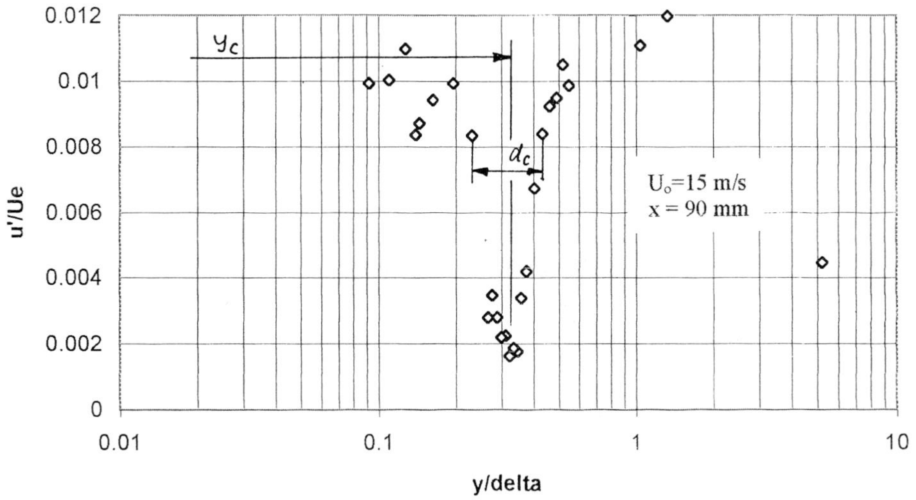

Figure 6. Attenuation coefficient $u_f'/U_e=f(y/\delta)$ for f = 8.54 Hz across the induced boundary layer

Besides, it is obvious from the plot in figures 6 and 7 that the critical layer does not touch the surface of the plate, since it begins at y/δ equal to 0.21. On the grounds of these results the mean velocity in the critical layer $(U/U_e)_c$ can be estimated to be 0.72 - 0.76, figure 8. It is thus higher than expected for the highest possible point of instability of the Blasius boundary layer. Such location and velocity value of the critical layer permit the existence of a new class of waves of lower wave number according to the expectations of Benney&Bergeron, 1969.

The stability of the flow was not directly under consideration here as the growth or decay of external disturbances along the boundary layer were not measured. It is only expected that development of the critical layer may be caused by external disturbances. Despite that restriction use of formula (7) for determination of the constant B is proposed. It is known from the experimental results that the velocity profile in the boundary layer can be approximated by polynomial expansion with Pohlhausen parameter Λ equal to 1.5, hence it is easy to calculate that $|U_c'| \cong 2$. As the total value of external disturbances ε is equal to 0.03, the constant B takes approximately the value of 3.2. One should recollect that both Benney &Bergeron and Haberman assumed that one critical layer only existed

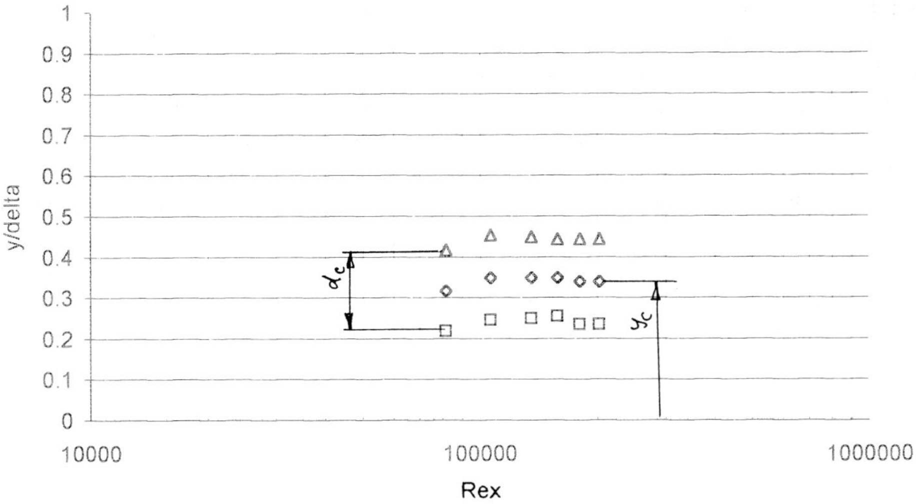

Figure 7. Location y_c and thickness d_c of the nonlinear critical layer in the induced boundary layer

and their analysis concerned only one mode of disturbance. In our experiment one dealt with a wave packet consisting of several harmonics of the wake frequency.

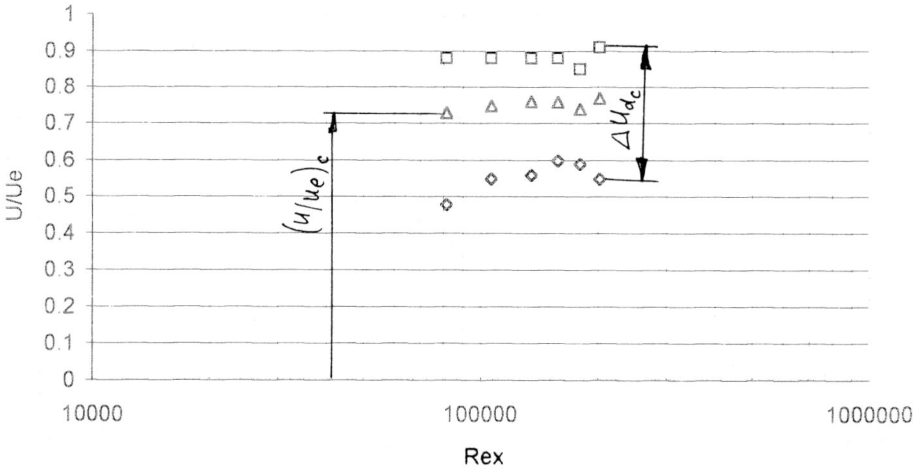

Figure 8. Velocity distribution in the nonlinear critical layer of the induced boundary layer

Making use of the results of the analysis of Benney & Bergeron for three different velocity profiles in the boundary layer (parabolic, error function and integral of the exponential function exp($-z^3$)), one finds easily the wave number of the neutral mode $\alpha\delta^* \cong 1.5 \div 2.1$ for the phase velocity $c = (U/U_e)_c = 0.70$. Further, the minimal value of the neutral wave length λ_{min} may be estimated as being equal to approximately $(0.85 \div 1.15)\delta$ for these three velocity profiles. Hence, this may be waves about six times shorter than the Tollmien-Schlichting waves. It seems that the frequency of these waves may also be larger, equal to approximately $f = (0.82 \div 0.61)\ U_e/\delta$. Thus, experimental confirmation of the evidence for the existence of the „new class of waves" of Benney&Bergeron is stil to be given. The same concerns the neutral curve for the nonlinear disturbances..

6. Conclusions

Investigations of the induced boundary layer on a flat plate were carried out in the working section of a wind tunnel with small level of turbulence, $Tu \le 0.0008$. The small velocity gradient existed along the flat plate with the value of the Polhlhausen parameter $\Lambda=1.5$. The overall level of disturbances ε generated by a swinging cylinder and measured at the outer edge of the boundary layer was equal to approximately 0.03. The spectral analysis of the velocity fluctuations in the vicinity of the boundary layer made it possible to treat the disturbances as a running wave packet with the wake frequency harmonics. Each harmonic ε_f of the packet was not greater than 0.01. The Benney&Bergeron coefficient $\lambda=1/(Re^*\varepsilon^{3/2})$ for the investigated transitional boundary layer was approximately within the range $0.2 \div 0.5$, hence it was of the order of unity, $\lambda=O(1)$.

On the grounds of the presented measurements one can conclude that a nonlinear critical layer due to external disturbances was found in the range of y/δ from 0.2 to 0.4, because across this layer the phase shift was observed. Such critical layers had previously been observed only for the Tollmien-Schlichting waves, that is for a small level of disturbances ε of 0.0001. However, in the case reported here the critical layer was observed, surely for the first time, for a level of disturbances more than one order greater, $\varepsilon \cong 0.03$. Furthermore, this critical layer was located at a distance $y_c=0.32$, for the velocity $U/U_e=0.7$, i.e. in the other region that the critical layer due to linear disturbances. This location of the critical layer confirms the expectations of the nonlinear analysis of Benney&Bergeron, 1969.

Additionaly, in this region attenuation of harmonics of the wake frequency was revealed, they were seen only above and beneath the critical layer. It may be an indication for the damping influence of the molecular viscosity on the disturbance. It is well known from theoretical results for the linear critical layers that viscosity influences the fluid motion. Another explanation would be the viscosity driven absorption of energy in the critical layer from the external disturbances and its transmission to other forms of fluid motion e.g. high frequency turbulent fluctuations or growth of the mean velocity in this region or its neihgbourhood. As far as the nonlinearity of the critical layer in these experiments is concerned, one can state that the level of disturbances was already not linear because it was much higher than the level of disturbances which cause the Tollmien-Schlichting waves in a laminar boundary layer.

The passing wakes bring out large velocity disturbances first in the part of boundary layer which lies close to its outer edge. This results in a peak in the velocity fluctuation profiles there. Next, the disturbances move to the deeper parts of the boundary layer, causing the peak in the velocity fluctuations revealed near the plate surface. This peak grows and later on exceeds the peak near the outer edge of the boundary layer.

If one regards a transition to turbulence without the necessity of passing through the Tollmien-Schlichting waves regime as a „by-pass", this experiment is supposed to show one of its possible sources.

It seems that a good continuation of these investigations would be an attempt to disturb the boundary layer with an external monochromatic wave of great amplitude. The aim of this experiment would be the determination of the neutral curve for nonlinear instability problem of the boundary layer, if these nonlinear waves exist at all.

References

Benney, D.J., Bergeron, R.F., 1969, A New Class of Nonlinear Waves in Parallel Flows, Studies Appl. Math., 48, 181-204

Boiko, A.V., Westin, K.J.A., Klingman, B.G.B., Kozlov, V.V., Alfredson, P.H., 1994, Experiments in a boundary layer subjected to free stream turbulence, Part 2. The role of TS waves in the transition process, J. Fluid Mech., 281, 219-245

Drazin, P.G., Reid, W.H., 1981, Hydrodynamic Stability, Cambridge, University Press

Haberman, R., 1972, Critical Layers in Parallel Flows, Studies in Appl. Math., LI, No.2, 139-161

Hammerton, P.W., Kerschen, E.J., 1996, Boundary - layer receptivity for a parabolic leading edge. J. Fluid Mech., 310, 243-267

Healey, J.J., 1995, On the natural curve of the flat-plate boundary layer: comparison between experiment, Orr -Sommerfeld theory and asymtotic theory, J. Fluid Mech., 288, 59-73

Jordinson, R., 1970, The flat plate boundary layer. Part 1. Numerical integration of the Orr-Sommerfeld equation, J. Fluid Mechanics, 43, 801-811

Juszyna, L., Wierciński, Z., 1995b, Measurements of the mean velocity and velocity fluctuation profiles in a region of natural transition in a boundary layer of a flat plate. IMP PAN, Rep. Nr 246/95, IMP PAN, Gdańsk (in Polish)

Landau, L.D., 1944, On turbulence problem, C.R. Acad. Sci. URSS, Vol. 44, pp..311 (in Rushian)

Lewczenko, W.Ja., Wołodin, A.G., Gaponow, S.A., 1975, Stability charakteristics of boundary layers, Nauka, Nowosibirsk, (in Russian)

Maslowe, S.A., 1986, Critical layers in shear flows, Ann.Rev.Fluid Mech., 18, 405-432

Nishioka, M., Morkovin, M.V., 1986, Boundary layer receptivity to unsteady pressure gradients: experiments and overview, J. Fluid Mech., 171, 219-261

Reshotko, E., 1976, Boundary Layer Stability and Transition, Ann. Rev. Fluid Mechanics, 311-349

Schlichting, H., 1965, Grenzschicht-Theorie, Verlag G. Braun, Karlsruhe,

Suprun, T., Wierciński, Z., 1995, Investigation of the unsteady boundary layer on the flat plate, Part I: Boundary induced by the wake of the moving cylinder, IMP PAN, Rep. Nr 93/95, Gdańsk, (in Polish)

Suprun, T., Wierciński, Z., 1996, Measurements of the velocity profiles in the Pohlhausen boundary layer on the flat plate, MP PAN, Rep. Nr 273/96, Gdańsk, (in Polish)

Wierciński, Z., 1994a, Auswirkungen der Nachlaufdellen hinter einem quer zur Strömung pendelnden Zylinder auf den laminar-turbulenten Umschlag der Grenzschicht auf der flachen Platte, VDI Berichte, Nr. 1109, 1994

Wierciński, Z., 1995a, Measurement of the coefficient C_f in the turbulent boundary layer on a flat plate by means of the Preston tube, IMP PAN, Rep. Nr 310/95, Gdańsk,(in Polish)

Wierciński, Z. 1995b, The Influence of the Periodic Wake Behind a Cylinder Pendulating in the Cross-Stream Direction on the Laminar-Turbulent Transition in the Boundary Layer of a Flat Plate, in Unsteady Aerodynamics and Aeroelasticity of Turbomachines, Elsevier, Amsterdam, Eds. Y. Tanida, M. Namba, pp.461-479

Wierciński, Z., 1996, Measurements of the spectra of velocity fluctuations in the boundary layer induced by the wakes, IMP PAN, Rep. Nr 517/96 , Gdańsk, (in Polish)

A NUMERICAL ANALYSIS OF WAKES PROPAGATION IN INVISCID FLOW OF TURBINE

P. FERRAND, S. AUBERT, G. OLIVEIRA AND L. SMATI
Ecole Centrale de Lyon / Université Claude Bernard,
Lab. Mécanique des Fluides et Acoustique, UMR CNRS 5509,
ECL, BP 163, 69131 Ecully Cedex, FRANCE

1. Introduction

The motion of an airfoil or a cascade blade in a nonuniform or unsteady flow can produce important fluctuating forces along the surfaces. This is, for example in turbomachinery, what happens in the case of rotor - stator interactions or inlet distortion or aeroelasticity phenomena. Unsteady wakes passing through cascade channels have a great influence on the flow field [1, 2], especially because they affect boundary layer transition, shock waves and boundary layer stall. Hodson has provided a great amount of information concerning the mechanisms of wakes propagation [3, 4]. The disturbances introduced by the passing wakes cannot be considered as free stream turbulence since the passing wake frequency is much lower than the characteristic frequency of the turbulence. The understanding of the unsteady effects requires reliable models to study the interaction and reflexion effects. The transition will not be considered here. The objective of this study is to understand the physical mechanisms of perturbation propagation through a mean flow field (in a fixed or relative frame) in an interblade channel of a turbomachine. For this purpose we use an algorithm (PROUST) to solve the unsteady 2D compressible Euler equations. As a preliminary test, we applied it to simulate the unsteady flow in the T106 turbine cascade.

2. Algorithm

PROUST is based on a finite volume formulation with MUSCL technique over structured quadrangular meshes. This code has been especially built to obtain a high accuracy for unsteady flow when wave reflections appear

T.H. Fransson (ed.), Unsteady Aerodynamics and Aeroelasticity of Turbomachines, 35-49.

on the blades possibly through the boundary layer; the complete description is detailed in Smati et al [5]. The time discretization uses an explicit Runge Kutta integration with five steps. The spatial distretization of convective fluxes is a third or fourth order scheme in each direction. The Mixed AUSM+ Liou scheme, a new numerical scheme, has been introduced. It combines the robustness of the flux splitting technique in capturing non linear waves and the precision of the central scheme in representing boundary layers. An hyperbolic tangent limiter is used. It is only sensitive to strong discontinuities. Compatibility relations or non reflective conditions are introduced at all boundaries. For steady compressible flow, the inlet subsonic conditions are the total pressure, the total temperature and two angles in 3D, and the outlet subsonic condition is the static pressure.

3. Wake expression

For wake study, non reflective condition is imposed at the outlet, and a specific treatment to specify the unsteady inlet conditions has been implemented. Two (or three) components of the velocity, the total or static pressure, the total or static temperature can be imposed. Any combination of these variables can be chosen to prescribe vorticity, entropy or acoustic waves. We define the following analytical formulation to simulate a vortical wave :

$$U_x(y,t) = \overline{U_x}\left\{1 - \delta U_x \exp\left[-\alpha\left(\frac{y - y_w(t)}{L}\right)^2\right]\right\} \tag{1}$$

$$\begin{aligned} U_y(y,t) &= \overline{U_y}\left\{1 - \delta U_y \exp\left[-\alpha\left(\frac{y - y_w(t)}{L}\right)^2\right]\right\} - \\ &\quad U_w \exp\left[-\alpha\left(\frac{y - y_w(t)}{L}\right)^2\right] \end{aligned} \tag{2}$$

and for an entropic wave :

$$\rho(y,t) = \overline{\rho}\left\{1 + \delta\rho \exp\left[-\alpha\left(\frac{y - y_w(t)}{L}\right)^2\right]\right\} \tag{3}$$

with :

$$y_w(t) = y_{\text{initial}} - U_w\, t \tag{4}$$

where x is the axial direction, y the tangential direction, L the width of the wake, $\overline{U_x}$ and $\overline{U_y}$ the axial and tangential mean velocities, U_w the wake velocity. This last term appears explicitly in the expression of the tangential velocity fluctuation only if the wake deficit is imposed in the relative frame. The density is not directly introduced as boundary condition, but used to determine the total temperature which is imposed to limit the reflection effect. In fact the computational domain is closed by artificial inlet and outlet surfaces. Non reflective conditions must be used on these surfaces to limit the numerical reflections. This raises no problem at the outlet, but at the inlet the difficulty is to impose the unsteady wake without reflection. To estimate the effect of reflection, a combination of compatibility and non reflective conditions can be imposed, what is called "permeability condition". Through a wake, one part of the information must be specified (a deficit of velocity for example) but another part result from outgoing waves (an acoustic wave can go out through the wake). We define a permeability ratio to simulate the respective weight of each part. Similar results have been achieved with or without permeability condition, so we can assume that inlet reflected waves are negligeable in the studied configuration. A passive scalar is introduced in the wake to trace the fluid particles. It constantly matches the entropy variation for an inviscid configuration. This point will be interesting in viscous flow to dissociate the wake itself from the entropy interactions with the boundary layers.

4. Tested configuration

The conditions applied on AGARD T106 cascade [6] are summarized in table 1. Several time steps and time schemes have been used to verify that the results are independent of the numerical formulation. A H grid is used with 235 points in the axial direction and 99 in the transversal direction. Figure 1 shows the mesh for two channels with only every other node in each direction. The steady Mach number varies between 0.3 at inlet to 0.6 at the outlet with a large acceleration on the suction side as presented in figure 2.

Tests have been performed for different maximum velocity deficits from 5% to 30% of the inlet mean velocity. No significant difference has been observed. The results are presented for velocity deficits δU_x, δU_y and density fluctuation $\delta \rho$ equal to 5%.

5. Results analysis

Basically we begin this study with inviscid calculations to compare later with viscous results. That must separate the convective effects from the viscous interactions. We assume that the main deformation of the wake through the interblade channel is due to the convective terms. If this assumption is verified, it will be easier to focus later on wake/boundary layer interactions. To dissociate the interactions between wakes, we firstly study

TABLE 1. Geometry and Flow conditions

T106 Cascade	Conditions
Blade Chord C	0.198 m
Pitch to Chord ratio P/C	0.8
Wake Pitch / Cascade Pitch	1.0
Inlet Flow Angle	$-38.2°$
Outlet Flow Angle	$63.6°$
Outlet Mach number	0.6
Wake width to Pitch ratio L/P	0.125
Wake shape coefficient α	0.693
Axial Velocity / Wake Velocity U_x/U_w	0.633
Wake Frequency	787 Hz

the effect of an isolated wake, then we analyse the case when the wake pitch is equal to the cascade pitch.

5.1. ISOLATED WAKE

The wake appears at the inlet at non dimensionalized time (by the wake periode) $T = 1/18$, goes out at $T = 95/18$, and only small pressure fluctuations remain after $T = 112/18$. We present the results at two times, $T = 50/18$ (odd figure numbers) and $T = 65/18$ (even figure numbers). All figures present the fluctuation values which are the difference between the instantaneous flow field and the mean flow field. Like the amplitude of the fluctuations is always lower than 30% of the mean value, vortices represent only the unsteady contribution, and there is no recirculation of the particules looking at the instantaneous flow field. Figures 3 and 4 show the wake shape corresponding to these times using the passive scalar. The velocity deficit can be observed in figures 5 and 6 and the velocity fluctuation vector in figures 7 and 8. The points A and B correspond to the center of fluctuating vortices, downstream of the wake (B) and upstream (A). The points C and D indicate the positions where the velocity deficit is equal to zero, corresponding to the inversion of the sign of velocity fluctuations on suction side (C) and on pressure side (D). We will call *sink point* D because velocity fluctuation vectors converge to this point and, by opposition, *source point* C. Points A', B', C', D' are equivalent for time $T = 65/18$. The vortices are produced by the interactions between the wake and the walls. The velocity deficit generates an ejection of the wake particles on the suction side and their aspiration on the pressure side. So the mass flow continuity imposes the formation of the two vortices A and B. The center of the vortices follow the wake which is convected by the mean flow velocity: by definition the velocity fluctuations are zero at the vortex center. In the turbine channel, acceleration being positive, the distance between the two vortex centers increases with time, as well as the wake width between AB and A'B'. Because of the mass flow direction in the wake, the wake particles have a slip velocity from pressure side to suction side. So the more the wake propagates, the more it moves closer to the suction side. The same displacement of the centers of the vortices could be supposed. In fact, this movement is smaller than expected and the centers of the vortices remain around mid pitch. The displacements of the points A, B, C and D explain the deformation of the wake observed between figures 3 and 4.

An other aspect concerns the evolution of points C and D. The displacement of these points seems to be defined by the middle point, called M of segment AB. An imaginery straight line can be drawn between C, M and D. This line is convected at the mean velocity of point M and remains normal to the mean direction. To obtain this evolution, a potential effect must be dominating, the large distance between D and D', compared to C and C' or M and M', can be explained only by waves going at sonic velocity. The results as a whole show that the flow fluctuations do not remain in the wake and large acceleration can be observed in front of the wake in the area of the trailing edge.

We can complete this analysis on isolated wake with some remarks that we can not show on the presented figures. When the wake reaches the trailing edge, it is laid over the suction side and because of the deceleration of the mean flow, its width is shorter and shorter and the velocity deficit intensity increases. Channel contraction limits the amplitude of velocity fluctuations normal to the walls, so at the leading edge, the vortices are like ellipses. Consequently, the exit angle fluctuations remain very much lower than 0.5% and the time averaged angle is close to the mean angle : the direction of the velocity is imposed by the suction side.

5.2. COMPARISON BETWEEN ISOLATED WAKE AND MULTI WAKES

Computations have been performed for periodic wakes configuration with the same wake velocity corresponding to a frequency of 787 Hz and with a wake pitch to cascade pitch ratio equal to one. In this case, always two wakes are in the blade channel (at mid distance between pressure and suction sides) . So in the following, we will call *up wake* the wake coming in the channel and *down wake* the wake going out of the channel.

Figures 9 and 10 show the evolution of the entropy contours, the velocity fluctuation direction (vector modulus is equal to one for all nodes) and the static pressure fluctuation for normalized times $T = 18/51$ and $T = 5/51$

respectively. These times have been chosen because the up wake at $T = 18/51$ can be compared to the isolated wake at $T = 50/18$ and the down wake at $T = 5/51$ can be compared to the isolated wake at $T = 65/18$. The wake shapes are very similar and the most important differences are concentrated near pressure side. As the mean velocity is low on pressure side, the wake is convected slowly, and three or four wakes interact on there. So the shape is modified at the outlet of the channel, the points A' and B' being pushed closer to suction side. As a result, the sink point D appears again on pressure side, the amplitude of the flow fluctuations is reduced on pressure side and the point D has no signifiant effects. So, in this case, more than in the isolated case, the pressure fluctuations on pressure side depend on suction side.

5.3. STATIC PRESSURE FLUCTUATIONS

Figures 11 and 12 show the static pressure fluctuation as a function of time on the upstream and downstream periodic lines and on the blade (pressure side and suction side respectively). The highest fluctuations can be observed on suction side. These fluctuations can be separated in two parts. The first one corresponds to a wave with wake periodicity, starting at leading edge at about $T = 0.5$. This wave follows approximately the wake and is convected by the mean flow. The second one has a negative level and starts at about 25% of chord. One part of this second wave has the same periodicity as the wake and is also convected. The other part has higher time mode - the most important mode is the 7th mode - and propagates upstream and downstream. We can observe some numerical reflection at the inlet, but the amplitudes are small and do not change the main structure of waves.

The static pressure fluctuations on pressure side are mainly produced by a combination of a reflective effect of suction side and a focus point on pressure side at about mid chord. This conjonction produces a stationnary wave on the first three quarters of chord. Just along the last quarter, a convective wave appears that can be explained by the fact that channel walls are almost parallel in this region and that suction side waves and pressure side waves are connected. A part of the pressure fluctuations can be associated to the wakes interactions.

5.4. WAKES INTERACTIONS

The wakes interactions produce a sink point E that appears on suction side at about 25% of chord at $T = 1$. Large oscillations can be observed at this time on figure 13 before occurs a separation in one source point C, existing in isolated wake configuration, and one sink point E, due to the opposite velocity direction of the up and down wakes. We can verify that pressure fluctuations are directly connected to these points by comparison of static pressure contours on figure 12. Figure 14 presents the evolution in time of the vortices centers A and B, the source point C and the sink point E in the blade channel. The same time step separates each letter, but a time step three time lower separates the crosses, the points and the stars. These three symbols correspond to the generation of points C and E. The propagation velocity is much faster than the convective velocity and explains the high modes of the pressure fluctuations presented before. Figure 15 describes the velocity direction for three times corresponding to a cross point, a star point and the C and E points generation. First step, the down wake arrives on cross point with a direction normal to the direction of the up wake. The amplitude of the down wake velocity is small compared to the up wake velocity, so the up wake is not affected. Second step, the amplitude of the vortex A increases, the down wake velocity increases too because the star point is closer to the wall. So the interaction with the vortex A produces a small vortex at the star point in the clockwise direction. Third step, this second vortex arrives on the suction side and generates a source point C and a sink point E. Then the point E follows the down wake and is convected, whereas the point C, after the impact of the vortex, goes upstream before to be convected too. This last effect explains the pressure waves associated with high time modes which go upstream.

6. Conclusion

A numerical study has been presented to try to explain unsteady flow produced by wakes in turbine. A comparison between isolated wake and multi wakes has been performed to dissociate the different phenomena. The results show the main effect of the suction side and the influence of the pressure side on reflective waves structure. The connection between velocity field and pressure field has been established, especially associated with high time modes. More investigations and experimental comparisons are now necessary to confirm these results.

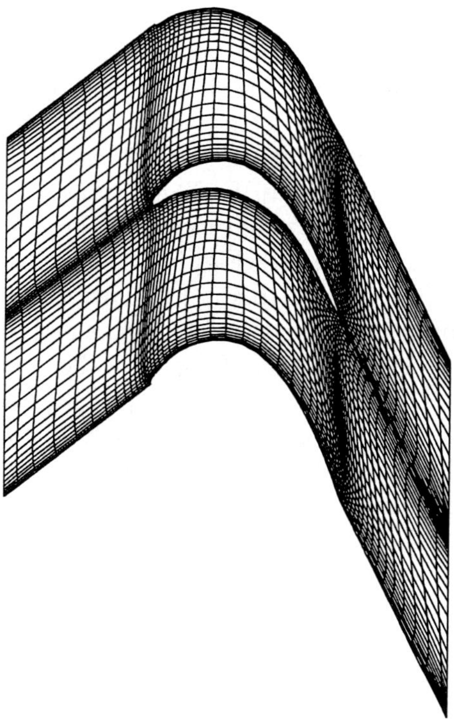

Figure 1. Mesh of the T106

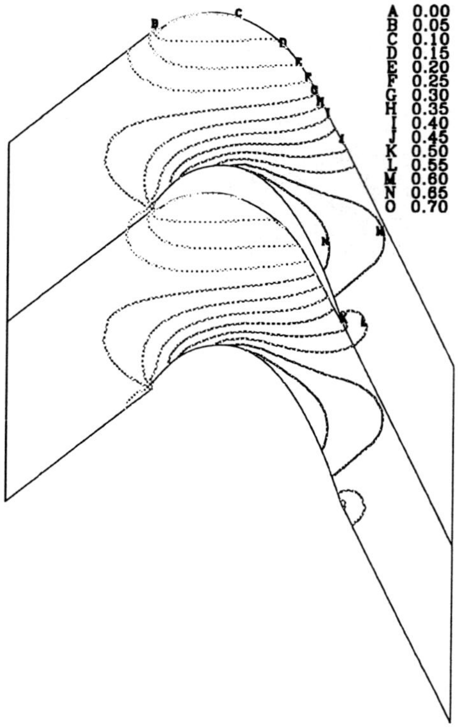

Figure 2. Steady Mach number contours

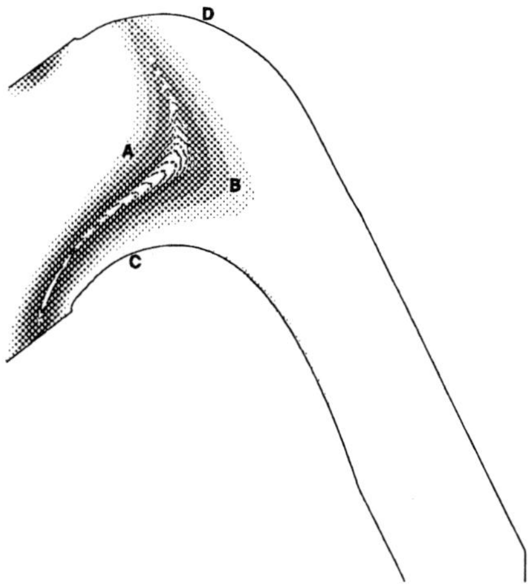

Figure 3. Passive Scalar at T=50

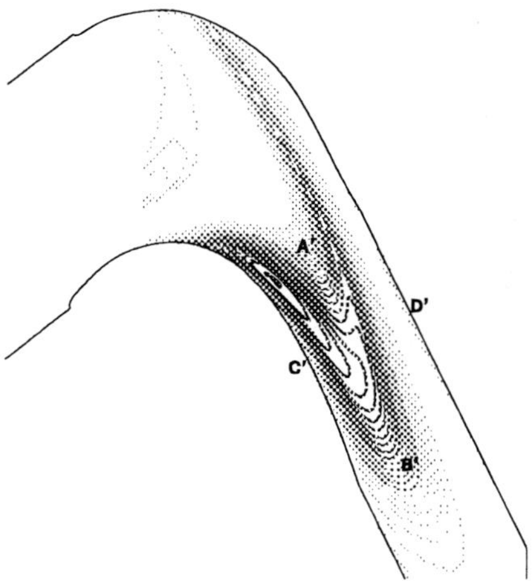

Figure 4. Passive Scalar at T=65

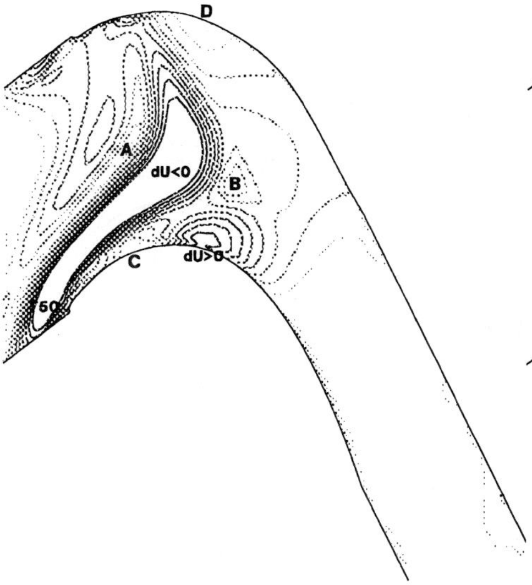

Figure 5. Velocity Deficit contour at T=50

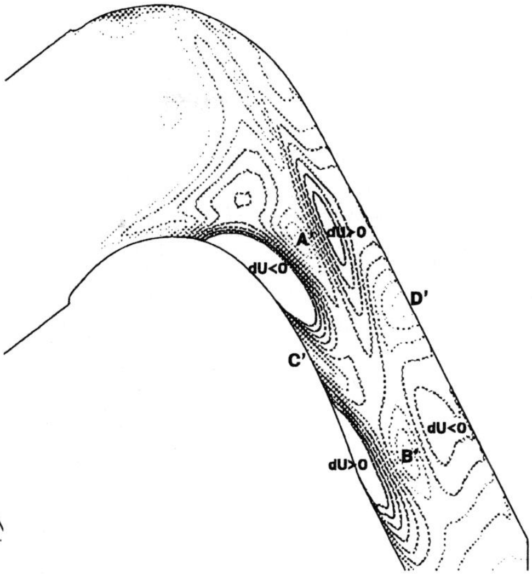

Figure 6. Velocity Deficit contour at T=65

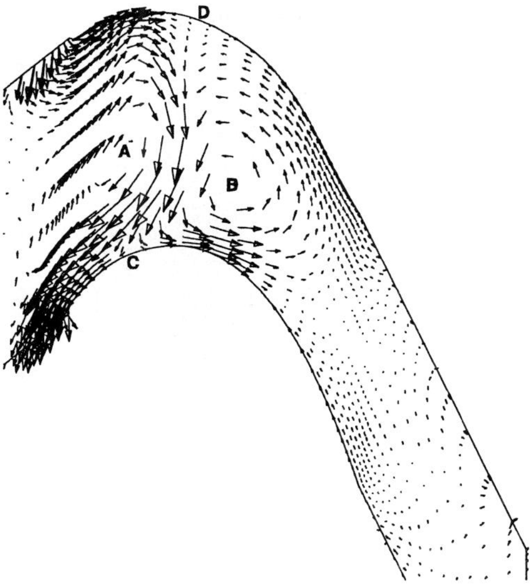

Figure 7. Velocity Fluctuation Vector at T=50

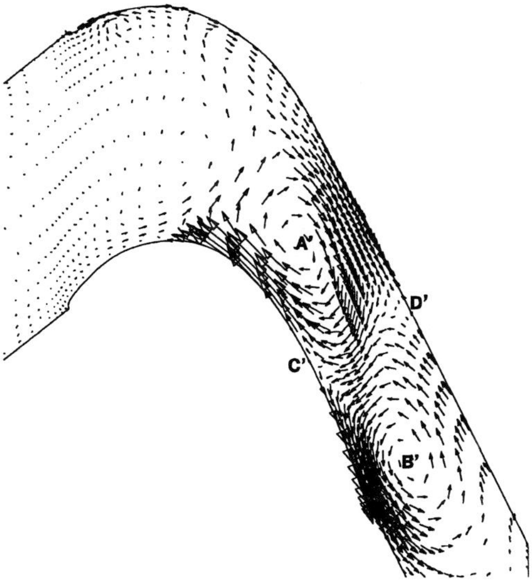

Figure 8. Velocity Fluctuation Vector at T=65

Entropy

Velocity direction

Pressure

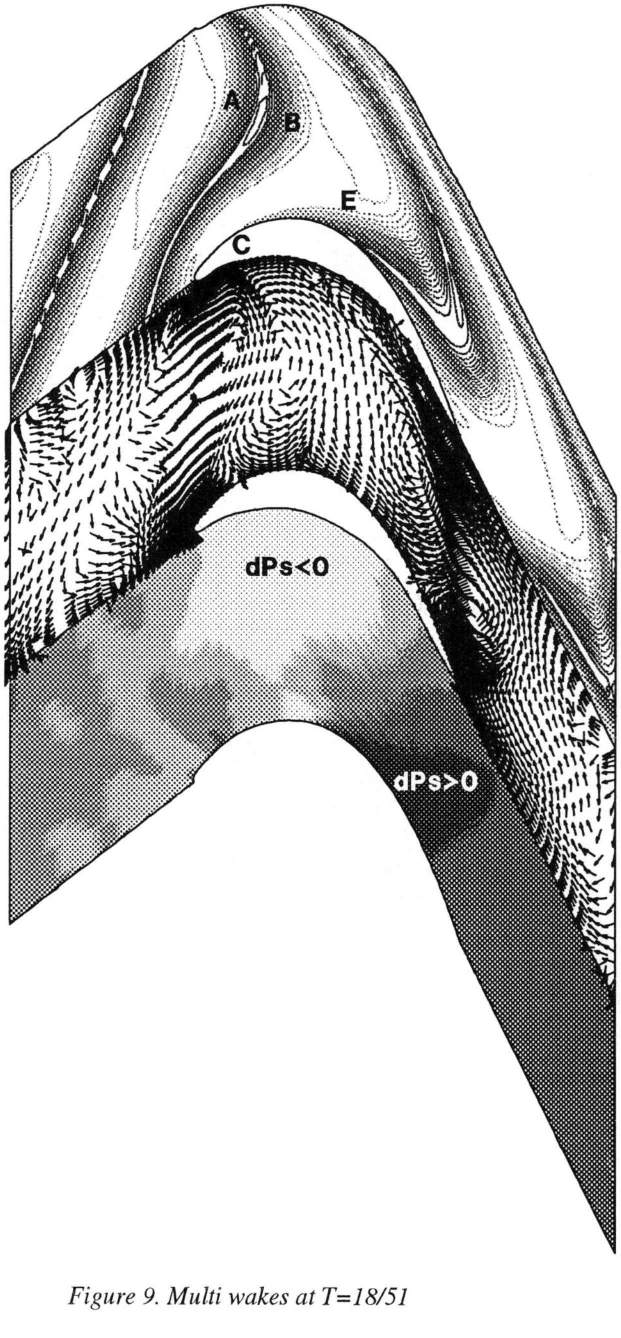

Figure 9. Multi wakes at T=18/51

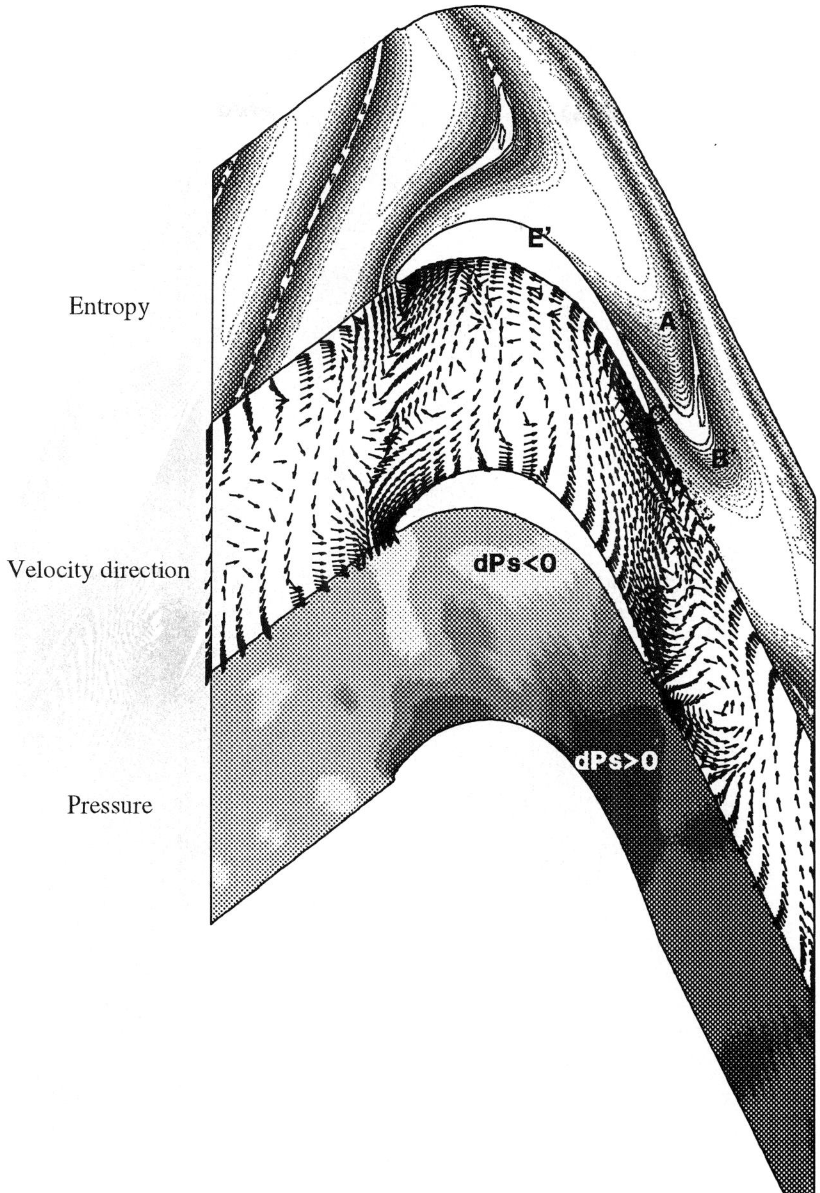

Figure 10. Multi wakes at T=5/51

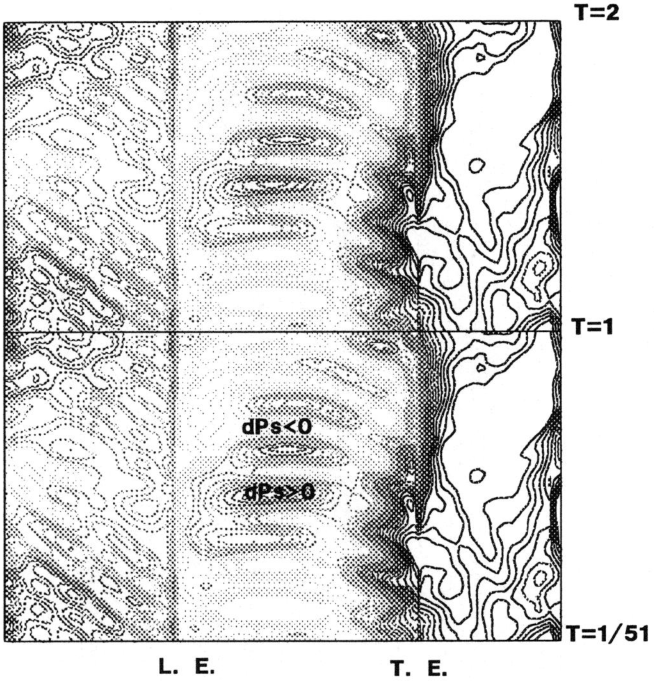

Figure 11. Static Pressure contours on pressure side

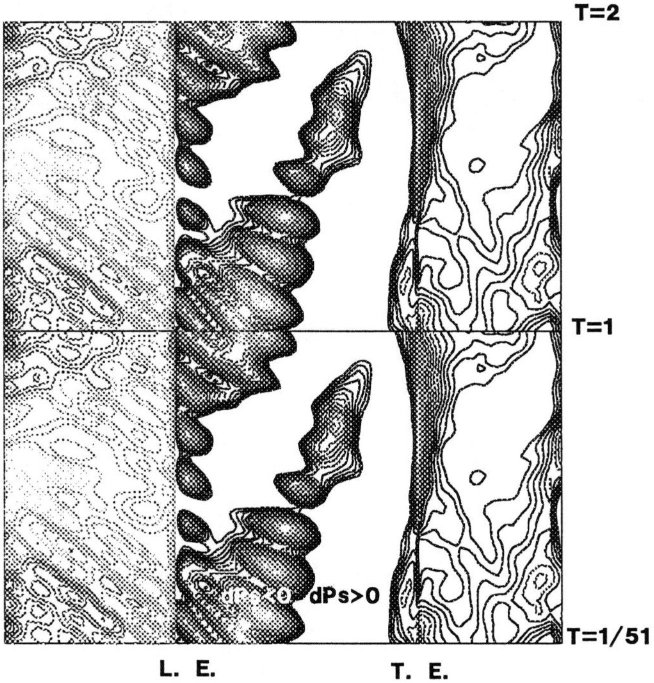

Figure 12. Static Pressure contours on suction side

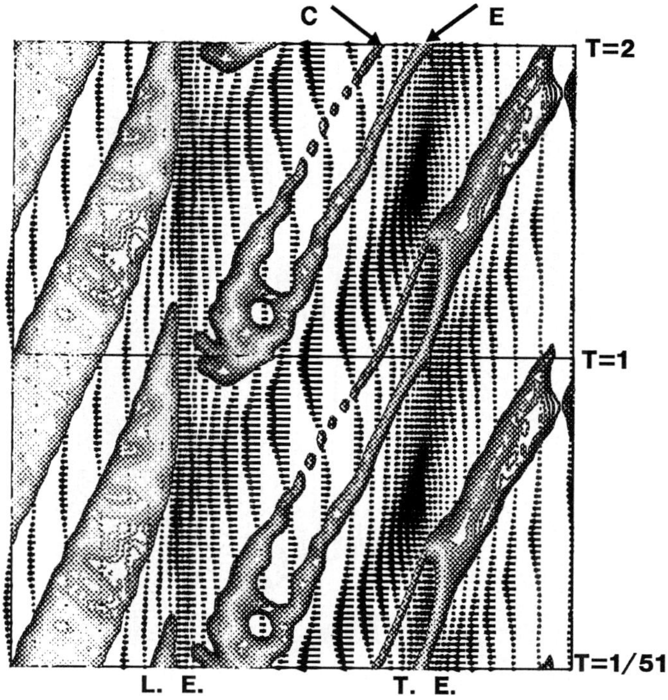

Figure 13. Evolution of points C and E on suction side

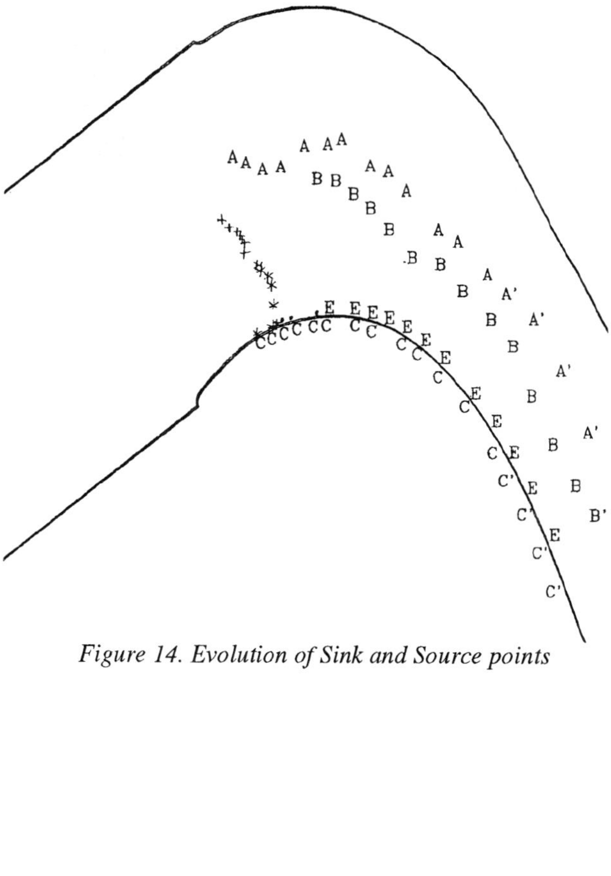

Figure 14. Evolution of Sink and Source points

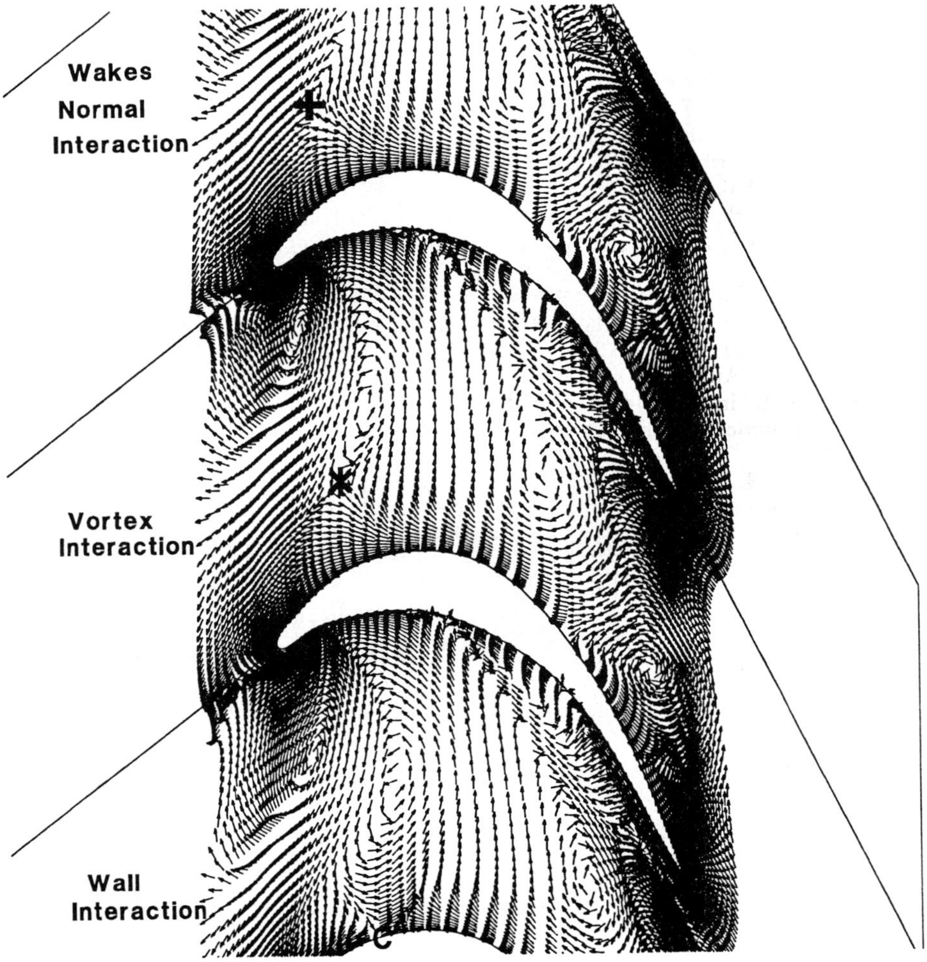

Figure 15. Source and Sink points generation

References

1. Curtis, E. M., Hodson, H. P., Banieghbal, M. R., Denton, J. D., and Howell, R. J., "Development of Blade Profiles for Low Pressure Turbine Applications", ASME Paper No. 96-GT-358, 1996.
2. Cicatelli, G., and Sieverding, C. H., "The Effect of Vortex Shedding on the Unsteady Pressure Distribution around the Trailing Edge of a Turbine", ASME Paper No. 96-GT-359, 1996.
3. Hodson, H.P., "An Inviscid Blade to Blade Prediction of a Wake-Generated Unsteady Flow", *Journal of Eng. for Gas Turbine and Power*, Vol. 107 No. 2, 1985, p. 337.
4. Hodson, H.P., "Measurements of Wake-Generated Unsteadiness in the Rotor Passages of Axial Flow Turbines", *Journal of Eng. for Gas Turbine and Power*, Vol. 107 No. 2, 1985, p. 467.
5. Smati, L., Aubert, S., Ferrand, P., and Massao F., "Comparison of numerical schemes to investigate blade flutter", *8th International Symposium on Unsteady Aerodynamics and Aeroelasticity of Turbomachines*, Stockholm, 14-18 September 1997.
6. Hoheisel, H., Kiock, R., Lichtfuss, H. J., and Fottner, L., "Influence of Free Stream Turbulence and Blade Pressure Gradient on Boundary Layer and Loss Behaviour of Turbine Cascades", ASME Paper No. 86-GT-234, 1986.

2: Experimental Unsteady Flow Around Oscillating Blades - 1

THREE DIMENSIONAL UNSTEADY FLOW AROUND A TURBINE BLADE OSCILLATING IN BENDING MODE - AN EXPERIMENTAL AND COMPUTATIONAL STUDY

D.L. BELL & L. HE
School of Engineering, University of Durham,
Durham, DH1 3LE, United Kingdom

Abstract

An experimental and computational investigation has been conducted on the aerodynamic response of a turbine blade oscillating in a three dimensional bending mode. Detailed blade surface unsteady pressure measurements have been obtained for a range of reduced frequency, from a test facility with clearly defined boundary conditions as described by Bell & He (1997). The measurements presented herein provide reliable 3D test cases which display significant three dimensional effects. These test cases have been utilised towards the validation of a 3D time-marching Euler method, and the ability of this numerical scheme to capture the relevant unsteady aerodynamics is examined. Comparisons between the predicted and measured unsteady pressure response show a satisfactory agreement over the full range of reduced frequency. Furthermore, the linearity of the test configuration is also investigated, and both experimental and numerical results demonstrate a quite linear behaviour of the unsteady aerodynamics.

Nomenclature

B_C	Bending amplitude at tip, non-dimensionalised with chord
B_L	Local bending amplitude, non-dimensionalised with chord
C	Blade chord, m
Cp	Steady pressure coefficient, $Cp = \dfrac{P - \overline{P_2}}{(\overline{P_{01}} - \overline{P_2})}$
$\lvert Cp_n \rvert$	Amplitude of nth harmonic pressure coefficient, $\lvert Cp_n \rvert = \dfrac{\lvert \tilde{P}_n \rvert}{(\overline{P_{01}} - \overline{P_2})B_C}$
dA	Projected blade surface area (unit span), normal to the direction of bending, m
h	Blade span, m
k	Reduced frequency, based on chord and reference velocity
V_{ref}	Reference velocity, ms^{-1}, $V_{ref} = \sqrt{2(\overline{P_{01}} - \overline{P_2}) / \rho}$

T.H. Fransson (ed.), Unsteady Aerodynamics and Aeroelasticity of Turbomachines, 53-65.

x	Chordwise position, m
z	Spanwise position, m
ϕ_n	Phase angle of the nth harmonic pressure response, degrees
ξ_C	Local aerodynamic damping coefficient, $\xi_C = \int_C \frac{-\pi B_L \lvert Cp_1 \rvert Sin\phi_1}{CB_C}.dA$
ξ	Aerodynamic damping coefficient, $\xi = \frac{1}{h}\int_h \xi_C.dz$

Subscripts

0	Stagnation parameter
1	Inlet parameter; 1st harmonic
2	Outlet parameter; 2nd harmonic

1. Background

A significant effort has been made in recent years to develop numerical methods for the prediction of unsteady flows around oscillating turbomachinery blades, with the intent to accurately determine flutter stability margins under practical working conditions. Due to increases in blade loading and aspect, accompanied by a reduction in mass ratio, research has progressed to include problems encountered in turbines as well as fans and compressors.

The present state of art numerical methods for oscillating turbomachinery flows largely consists of two approaches; non-linear time marching methods such as those presented by He & Denton (1994) and Gerolymos (1993); and time-linearized methods, e.g. Hall & Lorence (1993). Both approaches have been developed into fully three dimensional methods which can be applied to the complex 3D blade geometries which characterise the modern turbomachine. It is unfortunate then, although not unexpected, that published experimental programmes have not similarly progressed to provide challenging three dimensional test cases. The lack of 3D test data not only poses the obvious validation difficulty, but more crucially it limits our understanding. This has severe implications, because the understanding gained from experimental investigations plays an important role in the development of such prediction tools.

It is the purpose of the present work to provide detailed and reliable 3D test data for a turbine blade oscillating in bending mode, and examine the ability of a 3D time-marching Euler solver to capture the relevant unsteady aerodynamics. The measurements and predictions, which cover a range of reduced frequency, are also studied in order to determine the linearity of the test configuration.

2. Experimental Test Facility & Procedures

In this section the experimental test facility and procedures employed are described in brief, however, a more thorough account is provided by Bell & He (1997).

2.1. DESCRIPTION

A single, purpose built, turbine blade with constant section was used for the experimental investigation. The blade profile exhibits some quite typical characteristics, however, the deflection at the operational point was restricted to 45° in order to reduce the three dimensional effects from blade passage secondary flows. The blade profile is shown in fig. 1., and the specification along with the operational conditions is provided in table 1.

The test facility operates at low speed with a typical exit Mach number of 0.1. The working section, shown in fig. 2., comprised of the single blade mounted in a profiled duct. The test section side walls were specified by the adjacent blades in cascade, with plane sections extending one axial chord upstream of the leading edge and downstream of the trailing edge. The test section was positioned in the exhaust of a low speed wind tunnel and the tunnel boundary layers were bled.

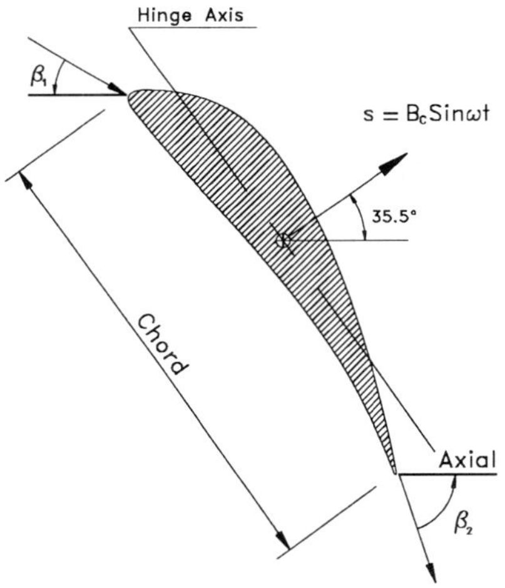

Figure 1: Blade Specification

Blade	
Blade chord, C	0.200 m
Aspect ratio	1.0
Solidity	1.35
Maximum thickness	0.175 C
Blade inlet angle, β_1'	25.0°
Blade outlet angle, β_2'	74.3°

Experiment Conditions	
Inlet flow angle, β_1	31.0°
Nominal exit flow angle, β_2	71.3°
Typical reference velocity, Vref	33.4 m/s
Bending mode direction	35.5°
Bending amplitude at tip, B_C	0.055 C
Bending amplitude at hub	0.006 C
Expt. Reynolds number, Re	4.5×10^5
Expt. Reduced frequency, k	0.15-0.75

Table 1: Blade specification and operational conditions

A three dimensional bending mode was realised by hinging the blade at the root/hub section and driving the tip at a sinusoidal rate. This provided a linear variation in bending amplitude along the blade span, which increased from 0.6% chord at the hub to 5.5% chord at the tip. To accommodate the motion of the blade tip, a tip clearance of 0.5% chord was included in the design. The sinusoidal drive was produced by a single bar crank mechanism attached to a con-rod extending from the blade tip, and the power was delivered by a thyristor controlled d.c. shunt wound motor. The direction of bending was nominally set at 35.5° as specified in fig. 1.

The testing environment described was designed to provide detailed and reliable test data with clearly defined boundary conditions. This was intended to ease the process of interpretation, and provide test cases which can serve in the validation of three dimensional numerical schemes.

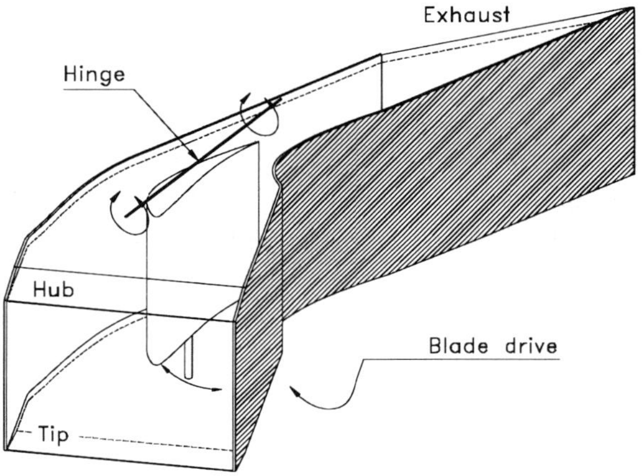

Figure 2: The test facility working section

2.2. INSTRUMENTATION AND DATA ACQUISITION

The blade was instrumented with 110 static tappings, evenly distributed over five spanwise sections, 10%, 30%, 50%, 70% and 90% of blade span. At each section ten tappings were positioned on both the pressure and suction surface, with an extra tapping at the leading and trailing edge. These tappings were used for both steady and unsteady pressure measurements. The working section was also instrumented for fivehole probe traverses at two axial measurement planes, the first at inlet to the test section and the second 75% chord downstream of the blade trailing edge.

The low speed and scale of the test facility, meant that realistic reduced frequencies were reproduced using low oscillation rates, 4 to 21 Hz. This allowed the unsteady pressure signals from the blade surface to be recorded with externally mounted pressure transducers. Five calibrated Sensym 142C01D pressure transducers were used to monitor the unsteady pressure response, and the unsteady voltage output from the transducers was discretized and recorded by a PC mounted Amplicon PC-74 data-logging card. The data acquisition was triggered by an optical trigger which referenced a fixed phase in the blade motion.

The unsteady pressure signals recorded from each tapping were ensemble averaged over 40 periods, and reduced into their harmonic components by a Fourier transform. The final harmonic components were then corrected for phase shift and attenuation along the tube lengths which separated the blade surface tapping points and the transducers, Bell & He (1997).

3. Numerical Method

3.1. DESCRIPTION OF NUMERICAL SCHEME

The experimental test cases were numerically reproduced by an extended three dimensional time-marching Euler method. In this method the flow is governed by the unsteady three dimensional inviscid Euler equations. The equations were discretized in space by a cell-vertex finite volume scheme, Denton (1983), and temporally integrated with the explicit 5-step Runge Kutta method. The 2nd and 4th order adaptive smoothing proposed by Jameson et al (1981) was adopted in the streamwise direction, with simple 2nd order smoothing in the circumferential and radial directions. The scheme was applied to a 'h-type' mesh with a zonal moving grid to accommodate the motion of the blade.

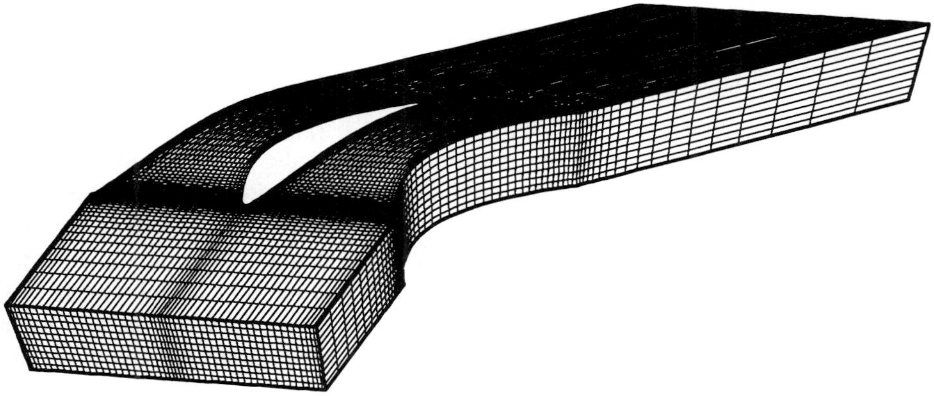

Figure 3: Computational domain (81x54x11 nodes)

3.2. COMPUTATION OF EXPERIMENTAL TEST CASES AND BOUNDARY CONDITIONS

In order to model the experimental working section a single passage computational domain was extended to encompass two blade passages. This domain, as shown in fig. 3., was a geometric reproduction of the experimental test section - the sidewalls were incorporated. The mesh used for both steady and unsteady computations consisted of 81 nodes in the axial direction, 54 nodes circumferentially and 11 nodes in the radial direction. The mesh was refined in the axial and circumferential directions to resolve the leading and trailing edge regions, and a single node cusp was attached to the trailing edge. Downstream of the trailing edge, the side and endwalls of the computational domain were linearly contracted to achieve the correct mass flow in the inviscid calculations. The extent of the contraction was determined by displacement thickness measurements performed at the downstream measurement plane in the test section. At the downstream boundary each endwall had been contracted by 0.75% chord in the radial direction, and each sidewall by 3.0% chord in the circumferential direction. A tip clearance was not included in the computational domain.

The inlet boundary condition was stipulated by uniform inlet flow conditions, previously obtained from the test facility, whilst the downstream boundary was set by a fixed back pressure condition. The reflecting downstream boundary condition was adopted, because the experimental facility exhausted to atmosphere.

To obtain a converged solution, the computations were performed at a higher Mach number than found in the test facility. The exit Mach number for the computations was in the region of 0.3, compared to 0.1 in the test facility. This should not, however, introduce any significant compressibility effects into the computations. The reduced frequency was matched for each unsteady test case.

4. Experimental and Numerical Results

4.1. STEADY FLOW RESULTS

Figure 4 demonstrates quite reasonable agreement between the predicted and measured steady flow blade pressure distribution. Although the results in fig. 4. are just taken from the midspan, they are representative of the full blade surface. Previous steady flow results from the test facility, Bell & He (1997), demonstrated negligible three dimensional effects in terms of the blade pressure distribution, at least between 10% and 90% span. And without the inclusion of tip clearance, there was no source of three dimensional effects in the inviscid steady flow computations given the uniform inlet conditions.

The steady flow solution obtained from the 3D Euler solver was found to deteriorate quite markedly when the contraction of the computational domain downstream of the trailing edge was omitted. Although this is largely due to the blockage from the contraction setting a more realistic mass flow, the way in which the

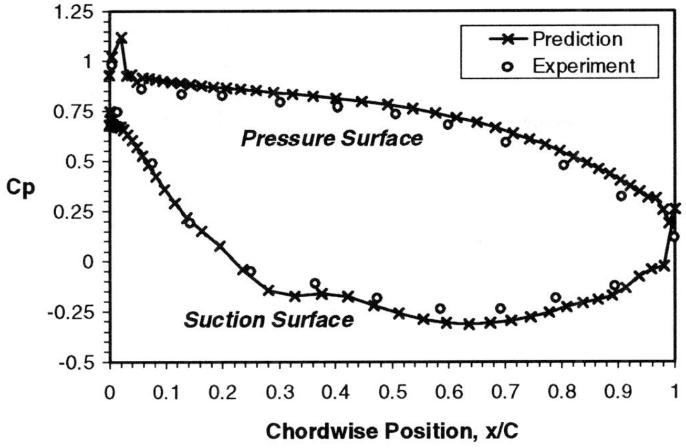

Figure 4: Steady flow blade surface pressure distribution (midspan)

contraction was applied, (from the trailing edge downstream), helped to relieve the diffusion towards the trailing edge of the blade suction surface. It is thought that this contributed to producing a more realistic pressure at the trailing edge, through the simulation of blockage provided by the growth in boundary layer towards the trailing edge of the blade suction surface in the test facility.

4.2. UNSTEADY FLOW RESULTS

4.2.1. *Linearity*

Before the experimental and numerical results are discussed, the linearity of the test configurations will be examined. This is of general interest in terms of the unsteady flow physics and to gauge the suitability of the test cases for time-linearized numerical schemes. Moreover, it was important to ascertain the extent of the linearity for the purposes of this investigation, because the unsteady calculations were performed at half the experimental bending amplitude. This situation was imposed due to difficulties in obtaining a converged solution for the full amplitude. This is presumed to be associated with the direction of bending and large amplitude at the tip, causing excessive skewing of the 'h-type' zonal mesh. Furthermore, it was noticeable that the stability limit of the numerical scheme, in terms of bending amplitude, could be extended by reducing the CFL number.

It was felt that the subsonic attached flow, which characterised the operation of the test facility, would demonstrate quite linear behaviour. However, both numerical and experimental tests were performed in order to verify this. Numerical test cases were calculated for three different bending amplitudes at a reduced frequency of 0.75, and experimental measurements were taken at both the normal bending amplitude and a reduced value, used for the computations, at a reduced frequency of 0.50. For the experimental test, fig. 5., lines denote the normal amplitude and symbols the reduced.

Bending Amplitude, B_C	0.0275 Chord	0.0165 Chord	0.0055 Chord
Overall damping:	5.13	5.13	5.21

Table 2: Numerical test for linearity - damping versus amplitude, (k: 0.75)

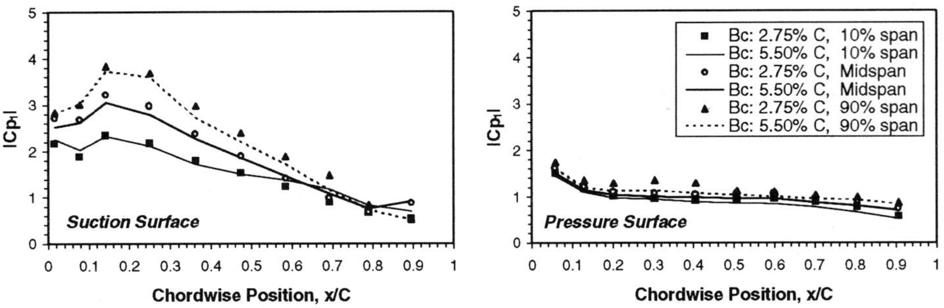

(i) Amplitude, (Expt., Bc: 0.055 & 0.0275C)

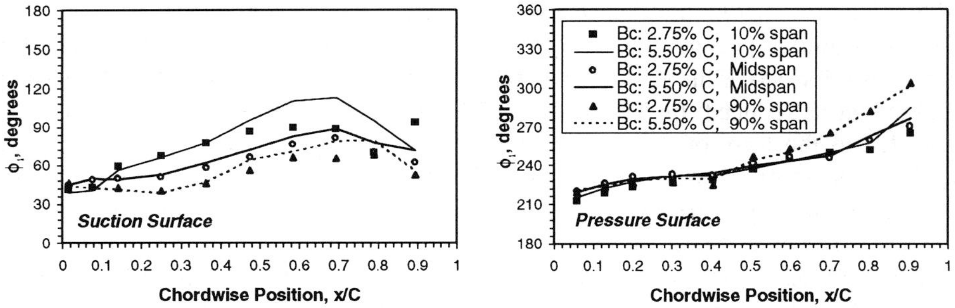

(ii) Phase angle, (Expt., Bc: 0.055 & 0.0275C)

Figure 5: Measured first harmonic pressure for different bending amplitudes, k: 0.50

Bending amplitude, B_C:	0.0275 Chord *(low)*	0.055 Chord *(high)*
Overall damping, ξ:	4.32	4.11

Table 3: Experimental test for linearity - damping versus amplitude, (k: 0.50)

The numerical test for linearity, table 2., shows the inviscid aerodynamic response to the blade motion to be linear. The overall aerodynamic damping was identical for the two highest bending amplitudes tested, and only deviated very slightly for the low amplitude (0.0055 Chord) where the effect of machine error was more appreciable. However, these tests alone were not sufficient to prove linearity, because it was considered that the most likely source of non-linear behaviour would arise from unsteady viscous effects in the test facility.

The experimental tests also demonstrates quite linear behaviour. As can be seen in figure 5(i), the amplitude of the first harmonic pressure response is almost identical over the full blade surface for both bending amplitudes. The phase also shows excellent agreement for the whole pressure surface and from the leading edge to midchord on the suction surface. Aft of 60% chord on the suction surface the phase measurements tend to deviate beyond the envelope of experimental error (+/- 10°), most notably towards the hub section (10% span). Two possible sources are suggested for this apparent non-linear behaviour; either the unsteady behaviour of an external leakage flow that is thought to have entered the test section through a gland at the hub of the suction surface; or the response of the suction surface boundary layer which is subject to diffusion in this region. Interaction between the two may equally have caused this. It should be noted, however, that in the area of non-linear activity the amplitude of the unsteady pressure response, and the contribution from the deviation in phase towards the aerodynamic damping, is relatively small. This is reflected by the similar values of aerodynamic damping obtained for the two bending amplitudes, table 3.

The largely excellent agreement between the two experimental tests described here, is also a clear indication of the reliability and repeatability of the experimental test cases presented.

4.2.2. *Evaluation of Unsteady Experimental & Numerical Results*

Figures 6 and 7 present the experimental measurements, and relevant numerical predictions, for two of the four test cases conducted. These two figures show the results obtained at a reduced frequency of 0.15 and 0.50, respectively. The results are presented in the form of the first harmonic pressure response - the amplitude, phase and local damping distributions are provided for both cases. The measurements and predictions of amplitude and phase of the first harmonic pressure response are provided for just three spanwise sections in order to preserve clarity. These are the hub section (10% span), midspan and the tip section (90% span). Contributions towards damping from measurements at other spanwise sections is, however, provided in the plots of local damping coefficient.

Over the full range of reduced frequency tested, the agreement between the predicted and measured first harmonic pressure response is generally very good. This suggests that the unsteady flow in the test facility behaves in a dominantly inviscid manner. The amplitude of the unsteady pressure response is predicted exceptionally well by the 3D Euler solver for the suction surface, although slightly under predicted, but with the correct trends, for the pressure surface.

The predicted phase angle of the first harmonic response also demonstrates excellent agreement with the experimental measurements for all spanwise positions on the pressure surface. However, the prediction of phase for the suction surface does not display quite the same level of agreement. From the leading edge to around 40% chord there is reasonable agreement between the inviscid predictions of phase and those measured for the suction surface, however, past this point, and towards the trailing edge, the agreement does tend to deteriorate. In the predictions the unsteady pressure response of the suction surface is led by that at the trailing edge, whereas for the experiment the phase at the trailing edge lags the response at midchord. In the inviscid case, the unsteady lift and therefore blade pressure distribution is largely determined

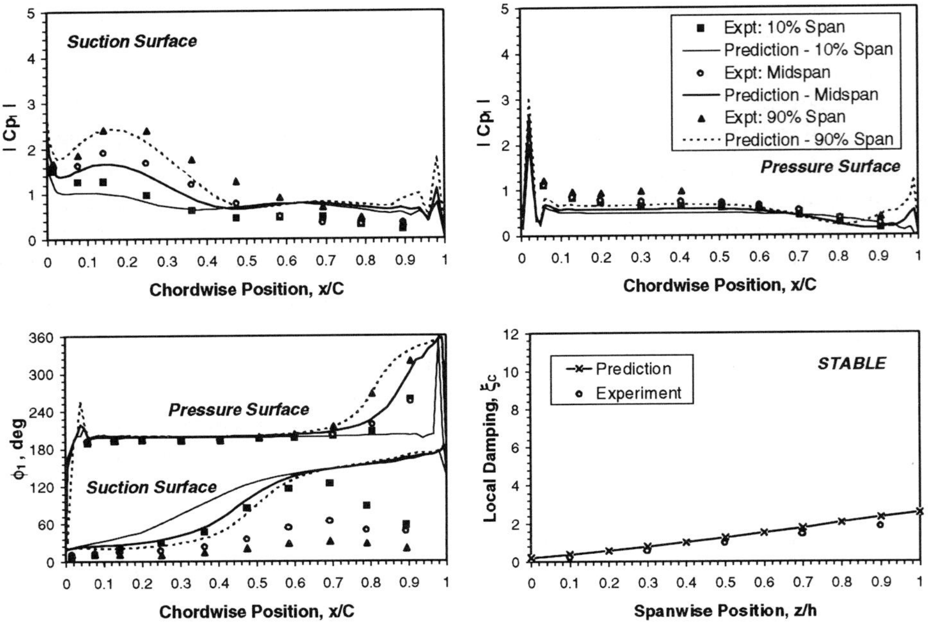

Figure 6: Predicted and measured first harmonic pressure response ***(k: 0.15)***

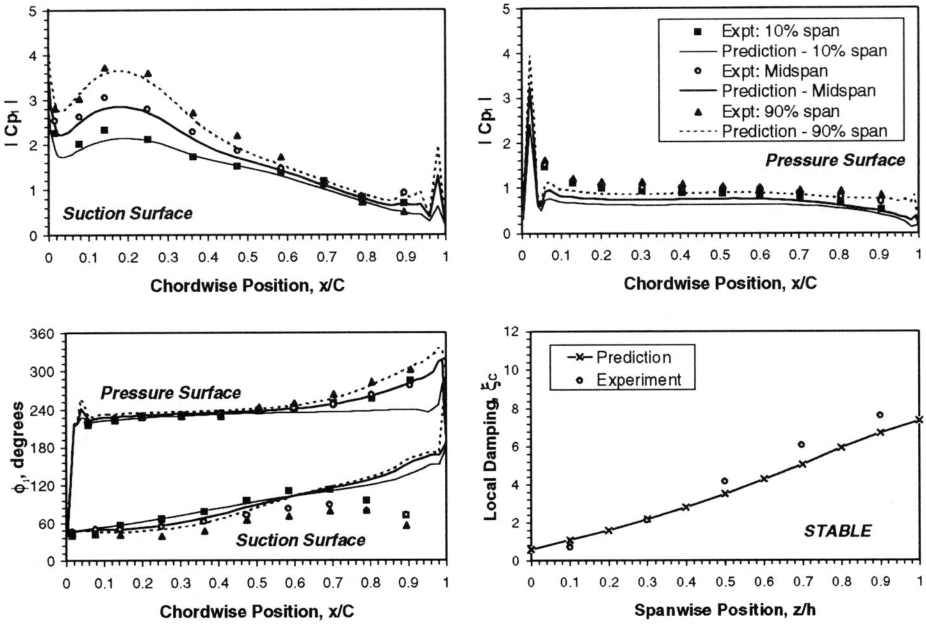

Figure 7: Predicted and measured first harmonic pressure response ***(k: 0.50)***

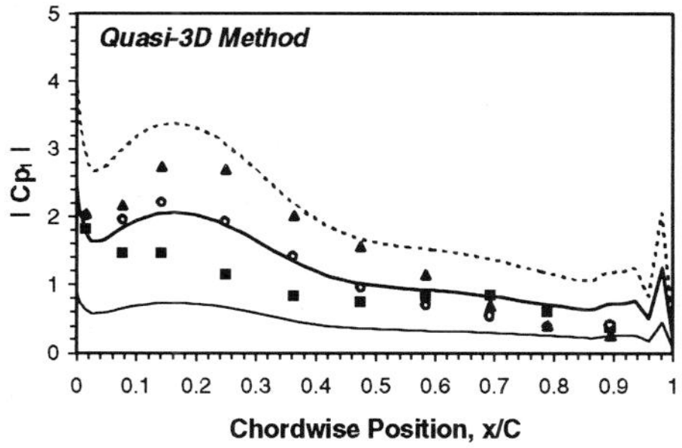

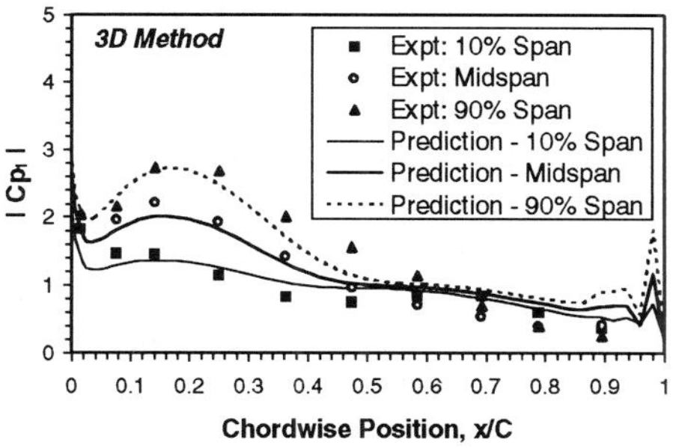

Figure 8: A comparison of predictions from quasi-3D and full 3D Euler solver (First harmonic amplitude on suction surface, k: 0.25)

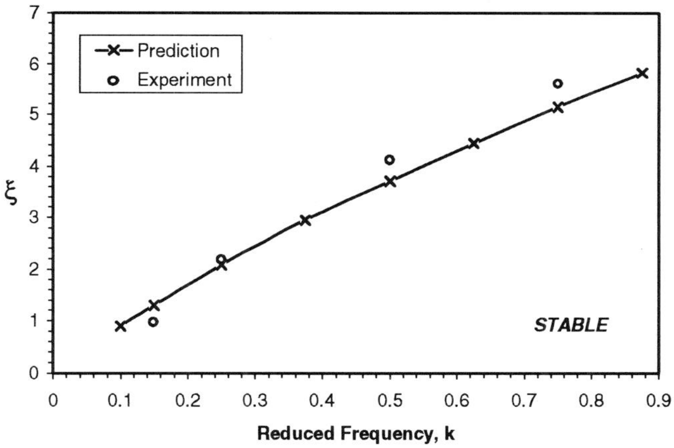

Figure 9: Overall aerodynamic damping against reduced frequency

by the behaviour of flow at the trailing edge. Information propagates upstream from the trailing edge and manifests in the phase lead of the unsteady response from this position, as demonstrated by the inviscid solutions and the response measured for the pressure surface. However, in the test facility, due to the relatively strong diffusion of the suction surface, the pressure towards the trailing edge might be sensitive to information convected from upstream, since this will determine the behaviour of the relatively thick boundary layer found here. This may explain the drop in phase observed in the experimental results towards the trailing edge of the suction surface.

The discrepancies described occur in regions where the amplitude of the unsteady response is relatively small, and consequently the predicted local damping also demonstrates a high level of agreement with the experimental test cases, as shown in figures 6 and 7.

Figure 8 helps to demonstrate the ability of the 3D method to capture three dimensional effects produced by the test facility. In this figure, a comparison is made between a prediction from the 3D Euler solver already described and a quasi-3D version of this method. Figures 6 through 8 all show that the amplitude of the unsteady pressure response, measured at different spanwise locations, displays a degree of insensitivity to the local bending amplitude. This three dimensional behaviour could not be adequately captured by the quasi-3D method, as shown in figure 8.

Possibly the most important feature of the numerical predictions is the ability to capture the trends in the unsteady pressure response over the range of reduced frequency. The numerical solution consistently picks up the increases in both phase and amplitude as the reduced frequency is increased, as seen in figures 6 and 7. And this observation is reflected by the good comparison between the predicted and measured variation in the overall aerodynamic damping with reduced frequency, shown in figure 9.

5. Concluding Remarks

Detailed and reliable experimental test cases have been presented for a turbine blade oscillating in a three dimensional bending mode at subsonic flow conditions. The test cases cover a range of reduced frequency and demonstrate significant three dimensional effects in terms of the measured unsteady pressure response on the blade surface. The amplitude of the unsteady pressure response at different spanwise positions is largely insensitive to the local bending amplitude.

In addition, the test cases have been computed with an extended time-marching three dimensional Euler solver. Predictions from this method exhibit very good agreement with the experimental test data for the full range of reduced frequency tested. The inviscid numerical scheme proved quite capable of capturing the relevant 3D unsteady aerodynamic phenomenon observed in these test cases. Some discrepancies in the predictions were, however, observed towards the trailing edge of the suction surface. This has been attributed to unsteady viscous behaviour in the test facility, induced by the relatively strong diffusion towards the trailing edge of the blade suction surface.

The issue of linearity was also addressed, and both numerical and experimental tests demonstrated a predominantly linear behaviour of the unsteady aerodynamics.

Acknowledgements

The research is sponsored by GEC ALSTHOM Turbine Generators Limited, and the authors would like to acknowledge Drs. B. Haller, P. Walker and M. Hart of GEC ALSTHOM for their technical assistance.

References

Bell, D.L. and He, L. (1997) Three Dimensional Unsteady Pressure Measurements for an Oscillating Turbine Blade, *ASME Paper* 97-GT-105.

Denton, J.D. (1983) An Improved Time-Marching Method for Turbomachinery Flow Calculations, *ASME Journal of Engineering Power*, Vol. 105, pp 514-524.

Gerolymos, G.A. (1994) Advances in the Numerical Integration of the Three-Dimensional Euler Equations in Vibrating Cascades, *ASME Journal of Turbomachinery*, Vol. 115, pp 781-790.

Hall, K.C. and Lorence, C.B. (1993) Calculation of Three-Dimensional Unsteady Flows in Turbomachinery using the Linearized Harmonic Euler Equations, *ASME Journal of Turbomachinery*, Vol. 115, pp 800-809.

He, L. and Denton, J.D. (1994) Three Dimensional Time-Marching Inviscid and Viscous Solutions for Unsteady Flows Around Vibrating Blades, *ASME Journal of Turbomachinery*, Vol. 116, pp 469-476.

Jameson, A., Schmidt, W. & Turkel, E. (1981) Numerical Solutions of the Euler Equations by Finite Volume Method using the Runge-Kutta Timestepping Scheme, *AIAA Paper* 81-1259

ROTATING BLADE ROW OSCILLATING AIRFOIL AERODYNAMICS

KUK KIM FREY
Allison Engine Company
Dynamics & Acoustics
Indianapolis, IN 46206

SANFORD FLEETER
McAllister Distinguished Professor
Purdue University
School of Mechanical Engineering
West Lafayette, Indiana 47907

Abstract

A series of experiments is performed in a research compressor to investigate the aerodynamic damping unsteady aerodynamics of a compressor rotor blade row. Specifically, the torsion mode unsteady aerodynamics of the first stage rotor are investigated and quantified, including the effects of oscillation amplitude and steady aerodynamic loading. An experimental influence coefficient technique is implemented in conjunction with a unique torsion mode drive system that provides rotor blade torsion mode oscillations without other motor drive systems. The oscillating blade unsteady pressure influence coefficient data are acquired with dynamic pressure transducers embedded in rotor blades. The resulting rotor blade aerodynamic damping data are then calculated and correlated with linear theory predictions. Oscillation magnitude and steady loading have a significant effect on the equivalent all airfoils oscillating unsteady pressure data, with nonlinear effects evident at high oscillation amplitude.

Nomenclature

c	rotor blade chord	Δp	unsteady pressure difference (psi)
$\bar{C}_p$	steady pressure coefficient	Δp^n	reference blade unsteady pressure difference for blade n oscillating (psi)
$\Delta\bar{C}_p$	steady pressure difference coefficient	U	rotor blade midspan wheel speed
ΔC_p	unsteady pressure difference coefficient	$\bar{V}_x$	mean axial velocity, determined by venturi orifice meter
$C_p{}^n$	unsteady pressure influence coefficient for blade n oscillating	$\overline{W}$	mean rotor inlet relative velocity
$\Delta C_p{}^n$	unsteady pressure difference influence coefficient for blade n oscillating	$\bar{y}$	half height of the cam follower rise
l	length of cylindrical joint lever arm	α'	time variant rotor blade angular position
n	rotor blade number or harmonic number	$\bar{\alpha}$	torsion oscillation magnitude or mean absolute flow angle
p	time variant rotor blade surface pressure	α^+	complex torsion oscillation amplitude
$\bar{P}$	rotor blade surface steady pressure	ϕ	steady flow coefficient, $\bar{V}_x / U$
$\bar{P}_{e,c}$	corrected rotor blade exit midspan static pressure	σ	interblade phase angle
p	unsteady pressure (psi)	ω	rotor blade oscillation frequency or forcing function frequency (r/s)

T.H. Fransson (ed.), Unsteady Aerodynamics and Aeroelasticity of Turbomachines, 67-82.

1. Introduction

Blade high cycle fatigue is caused by flow induced vibrations, with rotor blades particularly susceptible because of the large mean stresses from centrifugal forces. For flutter, a self-excited vibration, the unsteady aerodynamic forces acting on the blading are dependent only on the motion, with the flow field supplying the energy to sustain that motion. Forced vibrations are externally excited wherein the unsteady aerodynamic forces that cause the blade motion are independent of that motion.

The unsteady aerodynamics associated with flutter are analyzed by considering an harmonically oscillating cascade to determine the motion-induced unsteady aerodynamics or aerodynamic damping. With regard to forced response, the unsteady aerodynamic blade row response to a forcing function is comprised of two components: the gust response and the aerodynamic damping. With mechanical damping considerably reduced in advanced rotor designs, the aerodynamic damping determines the response level. Thus, it is important to investigate and quantify the aerodynamic damping for both flutter and flow induced vibrations.

Unsteady aerodynamic modeling is continually improving, largely due to the rapid advancements in computing power and efficiency. However, the validity of the physical and mathematical assumptions inherent in these models, their limitations and range of applicability have yet to be established. For the most part, new models are verified by correlating predictions with those of simpler models or with lightly loaded thin airfoil cascade data. This is due to the lack of appropriate data for model verification, determining necessary modeling enhancements, or to direct advanced model development. Also, the increase in modeling level from linearized analyses to Navier-Stokes methods is accompanied by the need for significantly increased computational resources. Thus the establishment of the applicability and the limitations of models is also necessary to determine the most efficient analysis for a particular application.

Limited oscillating cascade data exists, with these experiments directed at flutter, even though these aerodynamic damping data are also of significance to forced response. This is due to the fundamental differences in aerodynamic damping for flutter and forced response. The issue for flutter is stability, with aerodynamic damping data needed over the complete range of interblade phase angles. This requires extensive experimental studies utilizing complex and expensive airfoil drive systems. For forced response, vibration amplitude is key, with the aerodynamic damping required at a particular interblade phase angle value at a frequency matching that of the forcing function.

The difficulty in oscillating a cascade of airfoils at a constant interblade angle has been overcome by an experimental influence coefficient technique [1]. The unsteady aerodynamics of the airfoil row oscillating at a constant interblade phase angle are obtained by means of influence coefficients in which only a single airfoil is oscillated, with the airfoil response measured on the oscillating airfoil and its stationary neighbors.

However, linear cascades are subject to endwall effects which can contaminate the oscillating cascade data [2]. Although able to be overcome, often with a great deal of effort, this difficulty is not present in a compressor blade row. Oscillating blade row unsteady aerodynamics were investigated in an isolated low speed rotor, with the blades oscillated with a constant amplitude at different interblade phase angles [3]. However, even though the amplitude of oscillation was small, the correlation of these data with linear theory [4] was poor as a result of leading edge separation which was not modeled.

Thus, it is important to investigate and quantify the aerodynamic damping unsteady aerodynamics of realistic airfoil configurations without endwall effects. For these unique data to be appropriate for both flutter and forced response model verification and direction, the effects of oscillation amplitude and steady loading over the complete interblade phase angle range must be considered.

This paper addresses these needs, describing a series of experiments performed in a research compressor directed at investigating the aerodynamic damping unsteady aerodynamics of a compressor rotor blade row. Specifically, the torsion mode unsteady aerodynamics of the first stage rotor are investigated and quantified, including the effects of oscillation amplitude and steady aerodynamic loading. An experimental influence coefficient technique is implemented in conjunction with a unique torsion mode drive system that provides rotor blade torsion mode oscillations without other motor drive systems, i.e. as long as the compressor is operating, the mechanism oscillates a rotor blade. The oscillating airfoil surface unsteady pressure influence coefficient data are acquired with dynamic pressure transducers embedded in the rotor blade. The resulting oscillating airfoil data are then correlated with linear theory.

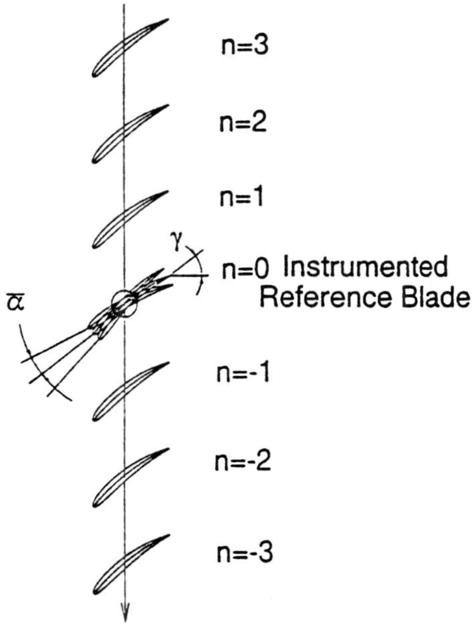

Figure 1. Finite compressor blade row with instrumented reference airfoil oscillating

2. Experimental Influence Coefficient Technique

The unsteady aerodynamics of the blade row oscillating at a constant interblade phase angle are obtained by an influence coefficient technique in which a single instrumented airfoil (n = 0) and its adjacent airfoils, Figure 1, are individually oscillated. The influence of the nth blade oscillating is measured on the instrumented reference blade as an unsteady aerodynamic influence coefficient $\Delta C_p{}^n(x)$. The unsteady aerodynamics of an equivalent blade row oscillating at a constant interblade phase angle σ are then determined.

$$\Delta C_p(x,\sigma) = \sum_{n=-N}^{N} \Delta C_p{}^n(x)\, e^{in\sigma} \quad (1)$$

As this technique involves oscillating a finite number of blades, the drive system cost and complexity are minimized. However, there is a 2N-1 increase in required experiments to individually oscillate each blade.

3. Research Compressor

The Purdue Axial Flow Research Compressor models the fundamental turbomachinery unsteady aerodynamic phenomena, including incidence angle, velocity and pressure variations, reduced frequency, and rotor and stator row geometries and flow conditions. The compressor is driven by a 15 HP DC motor at 2,000 RPM. Each identical stage contains 43 rotor blades and 31 stator vanes having a British C4 airfoil profile.

4. Rotor Blade Oscillations

The torsion mode drive system consists of the blade assembly, the cam follower assembly, and the cam, Figure 2. As the compressor rotates, the cam follower rides along the sinusoidal surface of a stationary cam, with its linear motion translated into rotary motion via a lever arm fixed to the blade. The elegance of this system is that the torsion motion of the blades is achieved without any other motor drive systems - the mechanism oscillates the rotor blade as long as the compressor is operating. Also, the sinusoidal cam surface produces a single frequency harmonic motion.

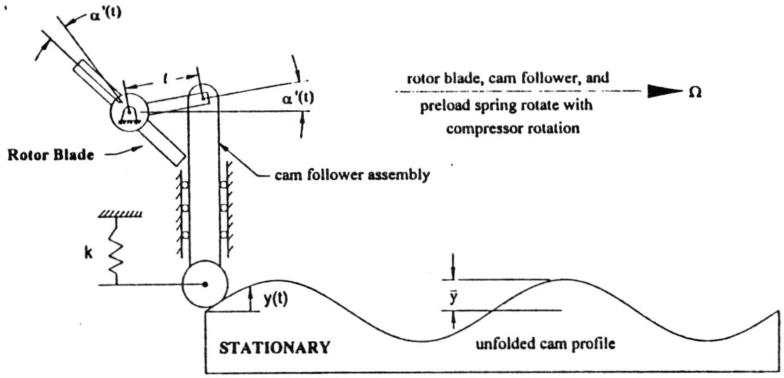

Figure 2. Schematic of rotor blade oscillation mechanism

The angular deflection amplitude for these experiments is 10° or less. Thus, the rotor blade torsion motion $\alpha'(t)$ is described by Equation 2.

$$\alpha'(t) \cong \frac{\bar{y}}{l} \sin \omega t = \bar{\alpha} \sin \omega t \tag{2}$$

Four blades are instrumented with dynamic pressure transducers. Hence, blades n = -2, -1, 0, and 1 are individually oscillated, accomplished with four identical mechanisms operated simultaneously. Individual blade oscillations are achieved by indexing the endplate with the cam follower assemblies to the appropriate blades. First, the inner parts of the blade assembly are indexed to the particular blades to be oscillated. The plate is then engaged by rotating the endplate until the cylindrical joint slips onto the cam follower shaft slot. Different vertical height cams change the blade oscillation amplitude. Three oscillation amplitudes are considered in these experiments: 3.5°, 5°, and 10°.

5. Instrumentation, Data Acquisition and Analysis

The instantaneous oscillating rotor blade motion $\alpha'(t)$ is determined by a precision rotary potentiometer with voltage output linearly proportional to angular position. These data are accurate to 0.15°.

The first stage rotor inlet flow is measured with a rotating cross hot-wire mounted at blade midspan, 18.8% chord upstream of the leading edge and 65.5% span circumferentially from an adjacent blade. The hot-wire is calibrated for velocities from 100 ft/s (30.5 m/s) to 200 ft/s (61.0 m/s) and flow angles ranging from -40° to +40°, with the velocity and flow angle uncertainty estimated to be 5% and ± 0.5°.

The rotor blade steady loading is obtained with a rotor based Scanivalve system. Two blades are instrumented with 10 midspan static pressure taps, one pressure surface and one suction surface. The blade unsteady aerodynamics are obtained with 20 ultra miniature high response pressure transducers installed at midspan on four blades. To minimize the possibility of flow disturbances, a reverse transducer mounting technique is utilized. These transducers can operate to 10,000 g's steady acceleration and 100 g's of 2 kHz dynamic acceleration. Thus the transducers are suitable for the steady rotating (1018 g's) and dynamic (50 g's at 66.67 Hz) environments of these experiments. The static transducer calibration uncertainty is estimated to be ±4%. Transducer dynamic calibration for frequencies ranging from 40 Hz to 2,700 Hz revealed a maximum gain and phase error of ± 0.60 dB and ± 1.5°.

A computer controlled digital data acquisition and analysis system is based on the detailed instrumentation shown in Figure 3. The rotor based steady and unsteady data are transmitted to the stationary frame by a mercury wetted slip ring, with low noise-to-signal ratio achieved by amplifying the dynamic pressure voltages 200 times. The steady data are determined by averaging 100 samples of the Scanivalve pressure signals and 200 samples of the hot-wire signals. Uncertainties in the mean values are calculated from the Student-t 95% confidence distribution. The time-variant cross hot-wire and dynamic pressure transducer signals are remotely multiplexed and then amplified before being digitized. The acquisition of the time-variant signals is controlled by an optical trigger mounted on the compressor shaft.

The rotor blade loading is characterized by a steady pressure coefficient $\overline{C}_p(x)$, with the mean inlet relative velocity adjusted for the blade potential field.

$$\overline{C}_p(x) = \frac{\overline{P}(x) - \overline{P}_{e,c}}{\varrho \overline{W}_{adj}^2} \tag{3}$$

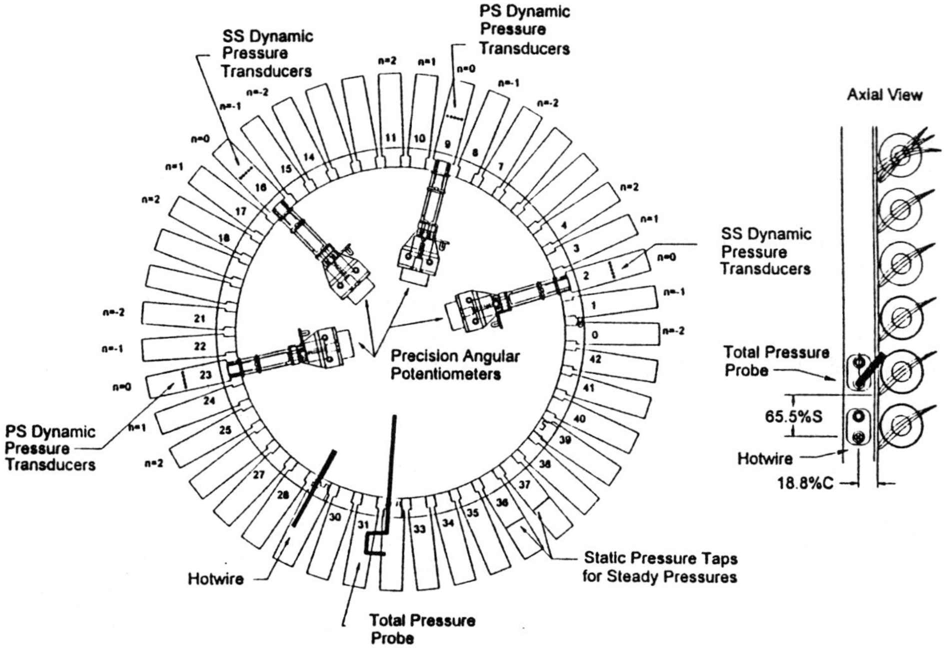

Figure 3. Rotor instrumentation

The unsteady pressure influence coefficient and the equivalent all blades oscillating unsteady pressure coefficient, Equations 4 and 5, characterize the blade torsion unsteady aerodynamics. The unsteady pressure difference coefficients are determined by subtracting the pressure surface coefficients from the suction surface coefficients.

$$C_p^{\,n}(x) = \frac{p^n(x)}{\varrho \overline{W}^2 \alpha} \tag{4}$$

$$C_p(x,\sigma) = \sum_{n=-N}^{N} C_p^{\,n}(x)\, e^{in\sigma} \tag{5}$$

The unsteady lift and moment coefficients are determined from the rotor blade surface unsteady pressure differences.

$$C_l * = \frac{1}{\Delta c} \int_{5.5\%c}^{81.0\%c} \Delta C_p * dx \tag{6}$$

$$C_m * = \frac{1}{(\Delta c)^2} \int_{5.5\%c}^{81.0\%c} \Delta C_p * (x_{ea} - x) dx \tag{7}$$

where $\Delta C_p * = \dfrac{p_{ss} - p_{ps}}{\varrho \overline{W}^2 \frac{\alpha}{\bar{\alpha}}}$ and $\Delta c = 81.0\%c - 5.5\%c$.

6. Results

A series of experiments is performed in a research compressor to investigate and quantify the aerodynamic damping unsteady aerodynamics of the first stage rotor, including the effects of oscillation amplitude and steady loading. The resulting steady and unsteady aerodynamic data are correlated with linear theory predictions.

6.1 STEADY PRESSURE DATA

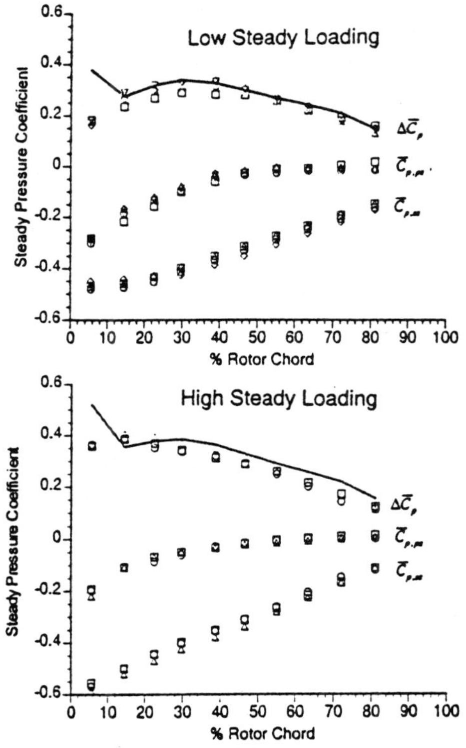

Figure 4. Rotor blade surface steady pressure data

The effect of steady loading on the rotor blade surface steady pressure distribution is shown in Figure 4. Also presented are predictions from the incompressible inviscid small camber airfoil cascade analysis INCMCSCD [5].

The steady pressure distribution is a function of the steady loading. The steady pressure on both blade surfaces are affected predominantly in the front half chord by the loading level, with the steady pressure increased on the pressure surface and decreased on the suction surface in this region. Increased loading results in increased steady pressure difference over the entire blade, particularly in the front half chord. Overall, the steady data agree well with the INCMCSCD predictions, except in the leading edge region where the steady pressure gradient correlation is poor. The correlation is better at low loading and INCMCSCD correctly predicts the location of the maximum pressure difference at low steady loading, but not at high steady loading.

6.2 ROTOR BLADE TORSION MOTION

Harmonic rotor blade torsion oscillations are achieved mechanically. However, transmission losses, manufacturing tolerances, and wear, may affect the blade motion. Thus, a Fourier transform of the blade motion $\alpha'(t)$ is performed. Figure 5 shows the 10° amplitude, cam driven motion of the four rotor blades and their corresponding Fourier transforms. The motion is predominantly sinusoidal, with the second and third harmonics being 0.4% - 3.6% and 7% - 12.6% of the first. Thus, linear theory, single harmonic torsion mode blade oscillations are experimentally well modeled.

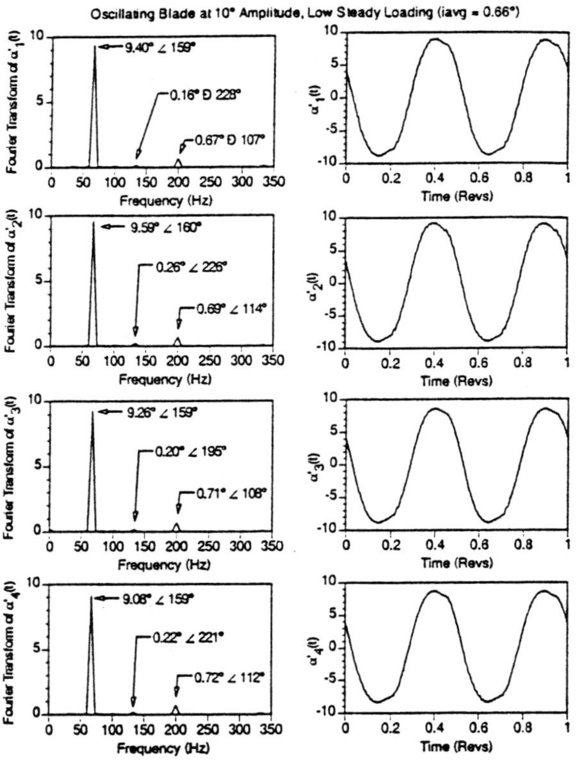

Figure 5. Rotor blade 10° torsional oscillation amplitude

6.3 EQUIVALENT ALL AIRFOILS OSCILLATING 1st HARMONIC DATA

Figure 6 shows oscillation amplitude and steady loading effects on the suction surface, pressure surface and pressure difference data for the equivalent rotor with all blades oscillating for $\sigma = -16.74°$ at low and high steady loading. Increased oscillation amplitude is denoted by larger symbols. High steady loading is denoted by solid symbols.

There is a pronounced effect of oscillating amplitude on the pressure surface data. Namely, a region of large unsteady pressure is evident over the front half of the blade, with the size largely dependent on oscillation amplitude. With a 3.5° oscillation amplitude, the large unsteady pressure occurs over the front 20% chord followed by a region of reduced unsteady pressure over the next 20% chord. A 5° oscillation amplitude extends this region to 40% chord, and a 10° oscillation amplitude extends this region further to 55% chord. At all three oscillation amplitudes, the pressure surface unsteady pressure is small and relatively constant beyond this region. Steady loading reduces the region of large unsteady pressure. The unsteady pressure phase is affected by the oscillation amplitude when the value is changed from 3.5° to 5°, with no effect when the oscillating amplitude is increased to 10°. Increased steady loading causes a large phase response over the aft chord region.

The entire suction surface is affected by both the oscillation amplitude and the steady loading, but these effects are smaller than on the pressure surface. In the leading edge region, the unsteady pressure decreases with increased oscillation amplitude and increased steady loading. Over the front 20% chord, steady loading has a larger effect on the unsteady pressure magnitude than oscillation amplitude. However, in the midchord region, oscillation amplitude effects dominate. In this region, the unsteady pressure decreases, with the decrease being more rapid with increased oscillation amplitude. The minimum unsteady pressure occurs at 72% chord, 65% chord, and 40%

chord for oscillation amplitudes of 3.5°, 5°, and 10°, respectively, at low steady loading. The effect of steady loading on this minimum is mixed. Large effects of oscillation amplitude on the suction surface phase data are evident, with a large phase shift in occurring earlier across the chord with increased oscillation magnitude. In contrast, steady loading delays the onset of the sudden phase shift.

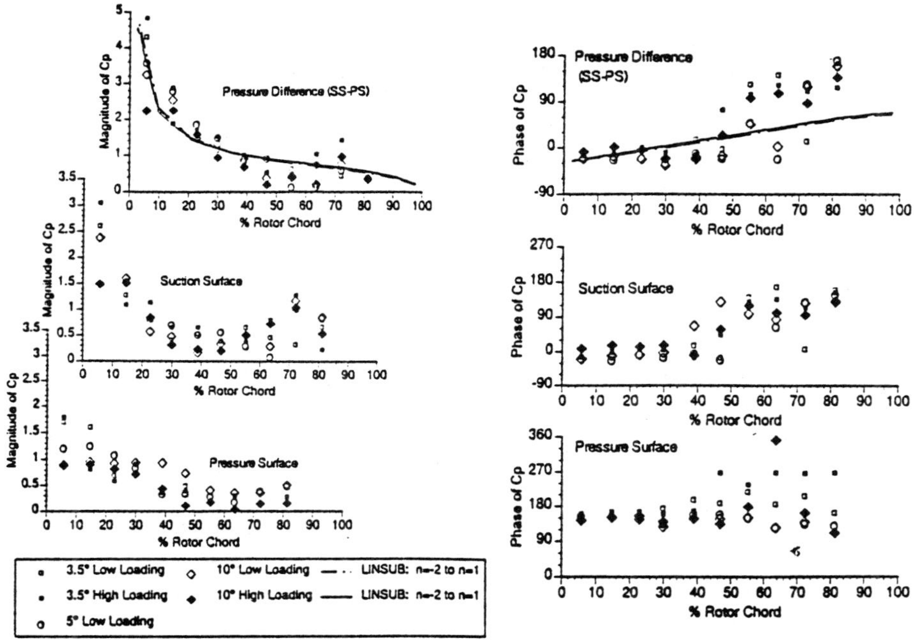

Figure 6. Equivalent all airfoils oscillating unsteady pressure data

For the unsteady pressure difference, oscillation amplitude and steady loading effects on the individual airfoil surfaces are manifest. The primary effect of increased oscillation amplitude is to decrease the unsteady pressure difference in the leading edge region. Increased oscillation amplitude also suppresses the local unsteady pressure difference increase at 45% chord for 3.5° and 5° of oscillation. Increased steady loading results in decreased unsteady pressure difference over the front half chord. The effects of oscillation amplitude and steady loading are also evident on the unsteady pressure difference phase. Over the front 40% chord, the phase is relatively constant for the 5° and 10° oscillation amplitudes, but linearly increases for the 3.5° amplitude. However, aft of 40% chord, the phase is affected by both steady loading and oscillation amplitude.

Figure 6 also shows the correlation of the 3.5°, 5°, and 10° oscillation amplitude all blades oscillating data with LINSUB [6] predictions for low and high steady loading. Note that in perturbation analysis such as this, the magnitude of the unsteady aerodynamics is linearly proportional to the oscillation amplitude. Hence, only one prediction is presented at a particular steady loading, with the pressures nondimensionalized by the oscillation amplitude.

The unsteady pressure difference data generally correlate in trend with LINSUB. However, a larger magnitude decrease is predicted over the front half chord and a smaller magnitude decrease over the aft chord. Also, the large phase shift over the aft chord is not predicted. Overall, the 3.5° oscillation amplitude data at low steady loading exhibit the best correlation. However, the chordwise magnitude and phase data do not match the predictions because even at low steady loading, the blade camber results in steady lift which is not considered in the flat plate model.

6.4 1st HARMONIC UNSTEADY LIFT AND MOMENT COEFFICIENTS

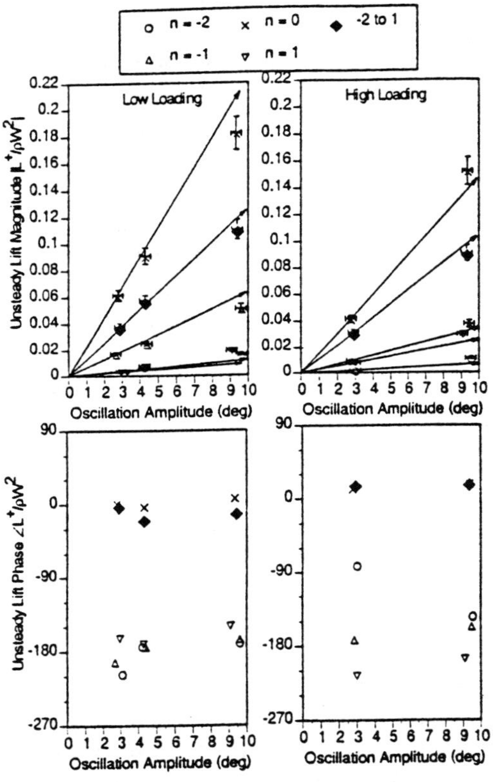

Figure 7. Oscillating amplitude effect on 1st harmonic unsteady lift

The linearity of the oscillating blade unsteady aerodynamics is addressed by considering oscillation amplitude effects on the unsteady lift and moment. The phase of the lift and moment influence coefficients should be constant with oscillation amplitude because they are nondimensionalized by the complex blade motion. However, the magnitudes of the lift and moment coefficients are nondimensionalized by the rotor relative dynamic pressure only. Thus, they should increase linearly with increasing blade oscillation amplitude $\bar{\alpha}$, indicated by the linear reference line extended from zero oscillation amplitude to the low oscillation amplitude data.

Figures 7 and 8 show the 1st harmonic unsteady lift and moment influence coefficient magnitude and phase for low and high steady loading. For blade n = 0 oscillating, the phase data are relatively constant as the oscillation amplitude is increased for both loadings, in agreement with linear theory. With blades in other positions oscillating, the phase data shift 30° or less between the 3.5° and 5° oscillation amplitudes, but are relatively constant for the 5° and 10° oscillation amplitudes at low steady loading. The phase data trends with oscillation amplitude are similar for high steady loading except for n= -2 blade oscillating which experiences an unsteady lift phase shift of approximately 60° between the low and high oscillation amplitude.

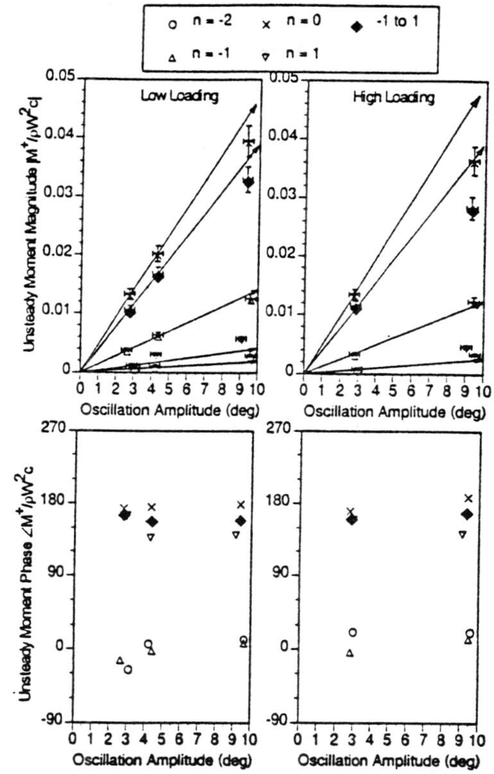

Figure 8. Oscillation amplitude effect on 1st harmonic unsteady moment

At low loading, the 5° and 10° oscillation lift and moment influence coefficients and the equivalent all airfoils oscillating magnitude data are consistently below the 3.5° linear reference line, except for small unsteady lift and moment values with blades n = -2 and 1 oscillating.

At high loading, the unsteady lift magnitudes at the higher oscillation amplitude for blades oscillating at all positions are larger than 3.5° linear reference line. However, for equivalent all blades oscillating, the summed unsteady lift magnitude is smaller than the 3.5° linear reference line. The moment magnitudes are reduced as compared to the linearly extended line except at low moment magnitudes, but the deviations are much larger than at low loading. Comparing the unsteady lift and moment coefficient data at the two loading conditions, it is clear that the unsteady lift and moment magnitudes are significantly reduced at the high steady loading, as previously noted on the blade surface pressure plots. Unsteady lift and moment data clearly show that nonlinear effects are present for blades oscillating with amplitudes larger than 5°.

The equivalent all blades oscillating 1st harmonic instantaneous lift and moment coefficients are shown in Figure 9. The periodic unsteady lift and moment are nondimensionalized by the mean rotor relative dynamic pressure only and then plotted versus the corresponding incidence angle. Thus, the unsteady lift and moment magnitudes are proportional to the oscillation amplitude. The effect of oscillation amplitude on the unsteady lift and moment magnitude and phase are presented together, with the area of the instantaneous lift-incidence curves corresponding to the magnitude and the ellipse skewness the phase. Note that with linear theory, the lower oscillation amplitude data should be enclosed within the larger oscillation amplitude results, with the loop size proportionally larger as the oscillation amplitude is increased. The oscillation amplitude affects both the magnitude and phase of the equivalent all blades oscillating data, with the low loading unsteady lift ellipse variations larger than the moment ones. At high loading, both the lift and moment loops differ, further verifying the significant nonlinear effects for oscillation amplitudes greater than 5°.

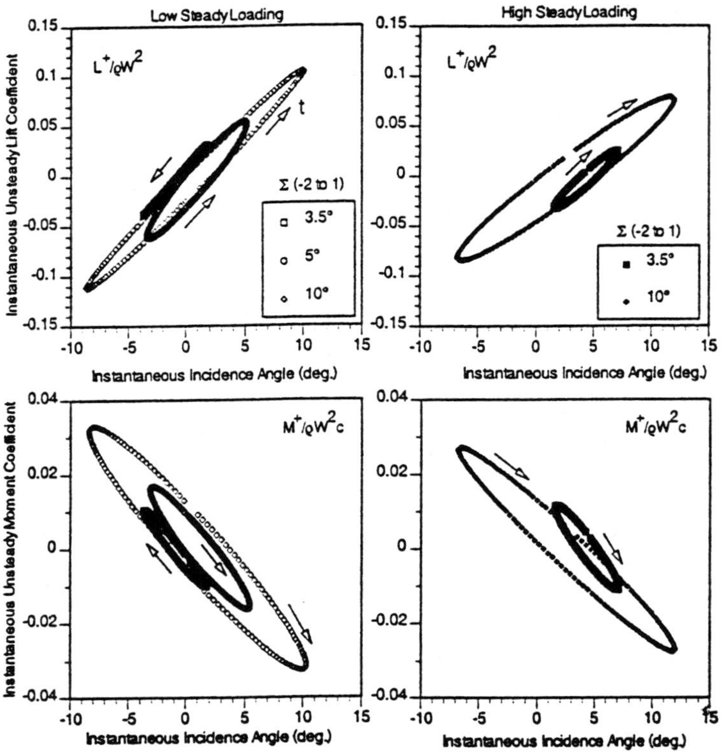

Figure 9. Equivalent all airfoils oscillating 1st harmonic instantaneous lift and moment

6.5 HIGHER HARMONIC UNSTEADY AERODYNAMICS

Figures 10 and 11 present the periodic variation of the instantaneous lift and moment coefficients derived from the sum of the first two and the first five harmonics. As the blade motion is essentially a 1st harmonic motion, the sum of the first two and first five harmonics lift and moment data being significantly different from the 1st harmonic data implies that the unsteady aerodynamic are nonlinear. With only two harmonics summed, there is a significant difference between the instantaneous $i - C_l$ and $i - C_m$ loops and the fundamental harmonic loops of Figure 9. The two harmonics summed unsteady lift loops are skewed but the direction of loop traversal is maintained except at the 10° oscillation amplitude with low loading. For this case, the counterclockwise loop is such that a crossover occurs from a rate of change in the unsteady lift variation with incidence. Unsteady lift changes faster with incidence during the downstroke (nose up) than the upstroke (nose down) of the blade motion. The unsteady moment loops are also skewed with the clockwise loop direction maintained except at the 10° oscillation amplitude with high loading. In this case, the different rates of change of moment with incidence during the upstroke and downstroke portions of the blade motion cause a crossover.

With five harmonics summed, the unsteady lift and moment loop edges are no longer smooth because of significant higher harmonics. Two crossovers occur in the 10° oscillation low loading lift. The most significant unsteady lift change is the slope change at high incidence for the 10° oscillation amplitude at high steady loading. This slope change at 9° incidence indicates unsteady lift stall, a phenomena not able to be discerned with only the 1st harmonic lift. These five harmonics summed results are also very different from the 1st harmonic responses, thereby indicating the importance of higher harmonics.

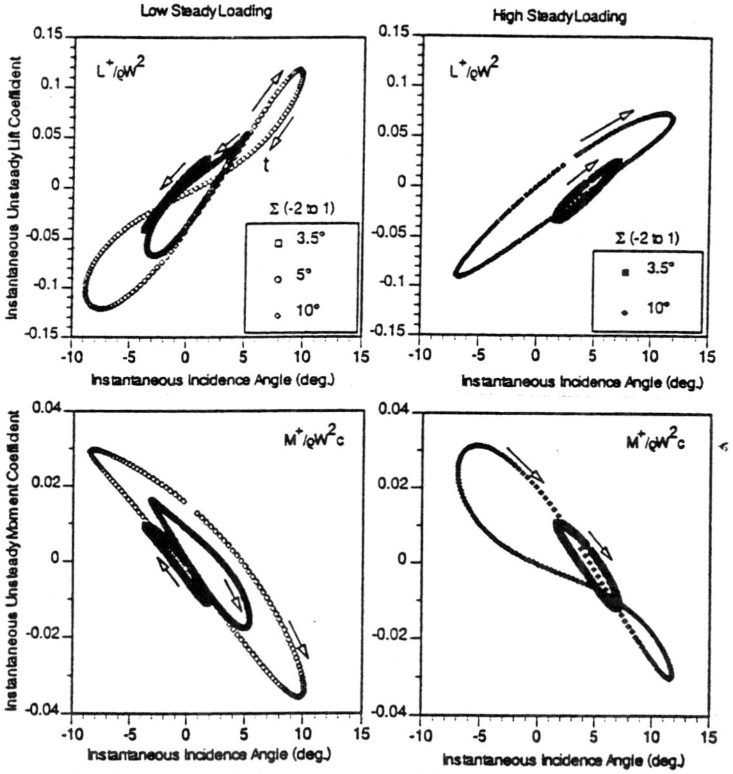

Figure 10. Equivalent all airfoils oscillating instantaneous lift and moment variation with incidence, 2 harmonics summed

To further examine these nonlinearities, the unsteady aerodynamic work is calculated from the areas of the instantaneous lift and moment incidence loops, Figure 12. The best indicator of nonlinearities is a significant increase in the unsteady lift and moment work summed from the first two harmonics since the second harmonic of the oscillation magnitude is virtually nonexistent.

At low loading with a 3.5° oscillation amplitude, the unsteady lift and moment work change by 6% and 0.43% of their 1st harmonic values when the first two harmonics are used. However, because the absolute change is so small, 0.0005, the percent change may not be an accurate reflection of real effects.

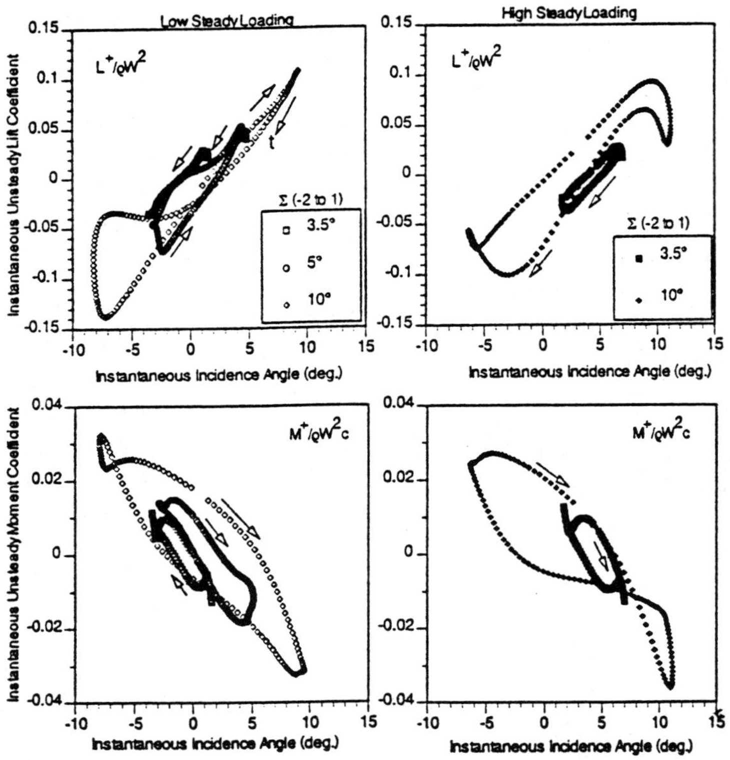

Figure 11. Equivalent all airfoils oscillating instantaneous lift and moment variation with incidence, 5 harmonics summed

At high loading with a 3.5° oscillation, the lift and moment work change by 0.5% and 0.12%. At both loading levels with a 3.5° oscillation, there is a noticeable unsteady lift and moment work change with two or three harmonics summed. This is attributed to the blade motion 3rd harmonic and an upstream strut. Hence, this does not indicate a nonlinearity. However, this is not the case at 10° oscillation amplitude. There is a 9% increase in the unsteady lift work at 10° oscillation amplitude even though the 2nd harmonic of the blade motion is only 2% of the 1st harmonic. The unsteady moment work changes by 0.7% with the 2nd harmonic added. At high steady loading, the unsteady lift work changes by 0.3% but the unsteady moment work increases by 7% with the first two harmonics summed. A 2.2% addition of the 1st harmonic oscillation amplitude cannot account for this increase. Also, these differences cannot be attributed to other harmonic excitations. Thus, strong nonlinearities in the unsteady aerodynamics are present at an oscillation amplitude of 10°.

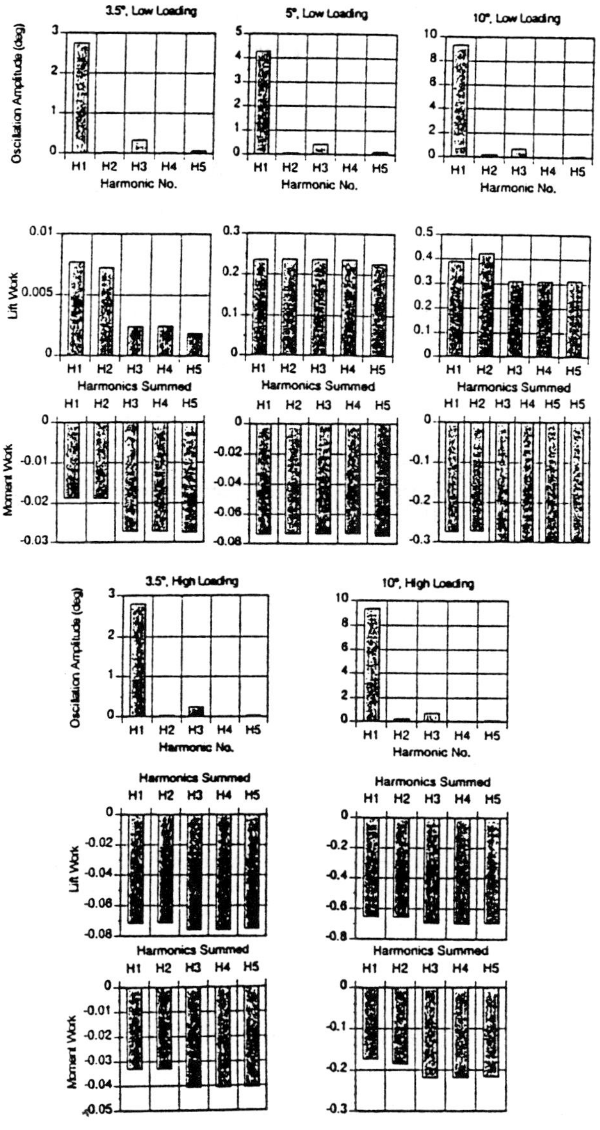

Figure 12. Oscillating blade row unsteady lift and moment work

7. SUMMARY & CONCLUSIONS

A series of experiments was performed to investigate the aerodynamic damping unsteady aerodynamics of the first stage rotor of a research compressor, including the effects of oscillation amplitude and steady loading. The resulting aerodynamic damping data were then correlated with linear theory.

Equivalent all airfoils oscillating unsteady pressure data were obtained by summing the influence coefficient data. Steady loading had a significant effect on all the unsteady pressure data, Although oscillation amplitude effects were evident on both blade surfaces, it had a larger effect on the pressure surface magnitude. In general, increased oscillation amplitude and increased steady loading resulted in reduced unsteady pressure magnitude and larger chordwise phase shifts. Correlation with LINSUB predictions was relatively good. However, the chordwise magnitude and phase unsteady pressure difference data did not match the flat plate cascade predictions.

The unsteady lift and moment harmonics indicated nonlinearity in the unsteady aerodynamics for oscillation amplitudes greater than 5°. The nonlinear response resulted in a decreased 1st harmonic unsteady lift and moment at the higher oscillation amplitudes. Thus, a 1st harmonic analysis may not be sufficient to determine oscillating airfoil aerodynamics at high oscillation amplitudes, as significant, higher harmonic nonlinear effects might be neglected.

Acknowledgments

This research sponsored, in part, by the Army Research Office (ARO). This support is most gratefully acknowledged. The truly exceptional design and manufacturing support from Ronnie D. McGuire is also well appreciated and acknowledged.

References

1. Buffum, D. H. and S. Fleeter (1990) Oscillating Cascade Aerodynamics by an Experimental Influence Coefficient Technique *Journal of Propulsion* **6**, 612-620.
2. Buffum, D.H. and S. Fleeter (1993) Wind Tunnel Wall Effects in a Linear Oscillating Cascade *Journal of Turbomachinery* **115**, 147-156.
3. Hardin, L.W., R.O. Carta, J.M. Verdon (1987) Unsteady Aerodynamic Measurements on a Rotating Compressor Blade Row at Low Mach Number *ASME Paper No. 87-GT-221*
4. Verdon, J. M. and J. R. Caspar (1980) Subsonic Flow Past an Oscillating Cascade with Finite Mean Flow Deflection *AIAA Journal* **18**, 540-548.
5. Henderson, G.H. and Fleeter, S. (1993) Oscillating Aerodynamics and Flutter of an Aerodynamically Detuned Cascade in an Incompressible Flow *International Journal of Turbo & Jet Engines* **10**.
6. Whitehead, D. S. (1987) Classical Two-Dimensional Methods *AGARDograph No. 298, AGARD Manual on Aeroelasticity in Axial Flow Turbomachines, Volume 1: Unsteady Turbomachinery Aerodynamics*, pp. 3.1-3.30.

EXPERIMENTAL INVESTIGATION OF UNSTEADY AERODYNAMIC CHARACTERISTICS OF TRANSONIC COMPRESSOR CASCADES:

DEVELOPMENT OF TEST FACILITIES

I.FUJIMOTO
Takushoku University
815-1 Tatemachi, Hachioji-shi, Tokyo 193, Japan
T.HIRANO
Takushoku University
815-1 Tatemachi, Hachioji-shi, Tokyo 193, Japan
H.TANAKA
Professor Emeritus of University of Tokyo
6-15-11 Minami Oizumi, Nerima-ku, Tokyo 178, Japan

ABSTRACT

To elucidate the unsteady aerodynamic characteristics of a transonic compressor cascade, test facilities are developed. A high-speed mechanical drive system is developed which provides controlled pitching oscillation of a blade in the linear cascade. A traverse unit with dual turntables is developed in order to survey the entire flow field around the cascade. Indirect pressure measurement techniques are applied to obtain the unsteady blade surface pressures when only one blade in the cascade is forced to vibrate. The effect of the change in attack angle of one blade in the cascade looks limited within the flow passages adjacent to the relevant blade. For capturing the high-speed periodic shock wave motion, an accurate photographing and image processing system is developed using a video camera with a moderate frame rate instead of using a high-speed video camera. The feasibility for generating successive pictures in one period whose frequency is up to 500 Hz is confirmed. Although the amplitude of shock wave displacement does not change much within the range of the present experiment, the phase lag relative to the blade oscillation reaches around $\pi/2$ as the blade oscillation frequency increases to 200 Hz.

T.H. Fransson (ed.), Unsteady Aerodynamics and Aeroelasticity of Turbomachines, 83-102.

1. Introduction

High performance of advanced turbomachinery requires operation at high speed, which causes shock waves to occur in part of the stages. The shock waves cause aspects of aerodynamic complication having a mixture of subsonic and supersonic flow properties around a blade, such as a shock wave reflection from/to blade surfaces and its interference with the boundary layer in the flow passage, even from the viewpoint of steady aerodynamics. In such a transonic flow region from an unsteady aerodynamic point of view, it is important that the shock wave movement along the blade surface induces a large fluctuation of aerodynamic loads due to a significant chordwise change in surface pressure across the shock wave. The shock wave thus plays an important role in the unsteady aerodynamic characteristics of transonic cascades.

Experimental investigations have been carried out to examine the unsteady transonic cascades concerning flutter characteristics [1]-[13]. In an aeroelastic sense, the behavior of shock waves, the amplitude and/or the phase shift of shock wave motion relative to blade motion, is a key factor. Although it is thus necessary to clarify the behavior of shock waves in transonic cascades, little information has been obtained to date.

The goal of the present study is to elucidate experimentally the effects of shock waves on the unsteady aerodynamic characteristics of the transonic compressor cascades. It is therefore necessary to examine the shock wave motion in the flow passage of cascades of blades oscillating at high frequency. One possible method to capture rapid motion of complete shock waves is through flow visualization with a high-speed video camera. The normal frame rate of around 10000 fps (frames per second) of conventional high-speed video cameras is not however fast enough to take successive pictures within a single period at a blade oscillation frequency of several hundred hertz. An extremely high-speed video camera with a frame rate of up to 40000 fps, which is adequate for this experiment, has recently become available on the market. However, installation of this type of photography system is extremely costly.

In this paper, an accurate photographing and image processing system for capturing the high-speed motion of shock waves in unsteady transonic flow cascades is developed using a video camera which, while still being a high-speed camera, has a comparatively low frame rate of 240 fps. As a first step, the shock wave motion is analyzed when only one blade in the cascade is subjected to high-frequency pitching oscillation. Also indirect pressure measurement techniques are applied for measuring the unsteady blade surface pressure. A traverse unit with dual turntables is developed in order to survey the entire flow field around the cascade as well.

2. Experimental Apparatus and Technique

2.1. WIND TUNNEL AND TEST CASCADE

Figure 1 shows a closed-type transonic air flow wind tunnel of about 10 meters in length and about 4 meters in height. The wind tunnel is set in operation after its internal pressure is reduced to less than 80 kPa. A centrifugal turbo blower ② shown in Figure 1 is driven by a 580 kW motor. It is possible to adjust the mean flow Mach number from M=0.3 to 1.2 at the test section ⑦. A turntable structure is installed in the side walls. The test section has a rectangular section with a span of 100 mm and with a height variable in the range of 200 to 400 mm by flexible upper and lower wall plates. In the downstream of the test section, upper and lower end plates are installed to adjust the exit flow direction. These plates are able to move vertically. Air temperature and wind velocity can be controlled during operation of the wind tunnel.

Figure 2 shows a schematic of the transonic linear cascade consisted of nine blades. As shown in Figure 3, two-dimensional, cambered double-circular-arc blades of 67.25 mm in chord and of 99 mm in span are installed with a blade spacing of 40 mm corresponding to a solidity of 1.68 and with a stagger angle of 30°. The maximum blade thickness is 5.6 mm which corresponds to a thickness-to-chord ratio of 0.083. The center blade(No.0) is made of titanium containing 6% aluminum and 4% vanadium and the other blades are made of duralumin. Only the center blade is forced to vibrate and the others are fixed. The center blade is supported at midchord by a single trunnion connected to the high-speed vibrator at one end and by a pivot at the other. Five blades from No.-2 to No.+2 are used to measure blade surface pressures. Figure 3 also shows pressure holes and pressure leads of a test blade for the measurement of the blade surface pressures at thirteen points on the suction surface and eleven on the pressure surface. The location of the pressure holes is shown in Table 1.

2.2. HIGH-SPEED PITCHING VIBRATOR

Figure 4 shows a high-speed pitching vibrator with a groove of eight periods of sinusoidal waves per turn in a cam (a rotary drum). A roller slides along the groove, thus giving a pitching vibration with an angular amplitude of 2° to a vibration axis. Maximum frequency of this vibrator is 500 Hz. The center (No.0) blade is mounted on one side of the vibration axis and a potentiometer to detect angular displacement of the blade is placed on the other.

2.3. TRAVERSE UNIT WITH DUAL TURNTABLES

A general view of a traverse unit with dual turntables is shown in Figure 5. This unit comprises a flange (540mm in diameter), a large turntable (442mm in diameter), a small turntable (252mm in diameter) and a probe holder (80mm in diameter). This unit is mounted in the side wall of the test section. Rotation of both the large and small turntables enables the probe position to be set at any location in the dotted region, within a circle with a diameter of 342mm, as shown in the figure. The probe holder also rotates to adjust the pitching angle of the probe. Plexiglas observation windows are installed in both the large and small turntables.

A cross section of the traverse unit is shown in Figure 6. The large turntable is built in the flange and the small turntable is built in the large turntable with ball bearings and miniature cam followers. The probe holder is built in the small turntable with a cross roller ring.

A schematic view of the measuring points (x, y) and the turning angles α and β is shown in Figure 7, where α is a turning angle of the large turntable and β of the small one. The relation between (x, y) and (α, β) is given as follows.

$$x = r\cos\alpha + r\cos(\alpha+\beta), \quad y = r\sin\alpha + r\sin(\alpha+\beta) \tag{1}$$

or

$$\alpha = \cos^{-1}\left(\frac{x}{Ar}\right) - \theta_0, \quad \beta = \cos^{-1}\frac{\left(\frac{x}{r}\right)^2 + \left(\frac{y}{r}\right)^2 - 2}{2} \tag{2}$$

where

r : length between centers of large and small turntables

$$A = \sqrt{(1+\cos\beta)^2 + \sin^2\beta}$$

$$\theta_0 = \tan^{-1}\frac{\sin\beta}{1+\cos\beta}$$

A five-hole pressure probe as shown in Figure 8 is installed in the probe holder. The measurement position of the probe coincides with the center of the holder. The nose of the probe is made of brass and shaped into a conical trapezoid with a slope angle of 45°. The probe has five pressure holes with a diameter of 0.4 mm. One pressure hole is on the front end and the others are on the conical surface as shown in the figure. The support of the probe is designed to have a diamond-shaped cross section to minimize its effect on the supersonic flow.

2.4. HIGH-SPEED PHOTOGRAPHING AND IMAGE PROCESSING SYSTEM

Schlieren optical devices, a high-speed video camera and image processing equipment, are used to analyze the shock wave motion in the transonic cascade. Figure 9 shows a block diagram of this system. The light of a 2 kW xenon lamp is led through a half mirror and a concave mirror of 250 mm aperture to the test section. The reflected light from a flat mirror in the test section returns in a reverse direction to a high-speed digital video camera via a knife edge. The specifications for this video camera are listed in Table 2. The maximum frame rate is 240 frames per second, shutter speed is 1/70000 sec and the recording pixel array is $256 \times 256 \times 8$ bit per frame. Up to 512 frames can be recorded in an IC memory. The video camera is controlled by a micro computer as shown in Fig.9. Image data is stored in an optical magnetic disk through a data transfer board and also recorded on a video tape as analog data.

Three kinds of data are essential in this system. First, data on blade displacement is given through a potentiometer installed on the vibrator pitching axis. Second, a reference pulse, which indicates the moment when the blade attack angle exactly coincides with its mean value during a nose-up period of blade oscillation, is generated by a photo micro sensor and a pulse disk of the vibrator. Finally, a pulse of 240 Hz is generated from the video camera and used as a trigger pulse for photographing. These three data, i.e., the oscillating blade displacement, the reference pulse representing the instant of a mean attack angle and the photographing trigger pulse, are A/D converted and stored in a high-speed multi-channel volatile memory. The minimum sample period of the volatile memory is 3 μs so a sampling error will be less than 1° at the maximum blade oscillation frequency of 500 Hz. The accumulated images are processed by the microcomputer through the image processing board.

2.5. INSTRUMENTATION

2.5.1 *Capturing Techniques for High-Speed Periodic Shock Wave Motion*

This subsection describes how to obtain successive pictures in a single period of blade oscillation by using a video camera with a moderate frame rate, instead of successively taking whole pictures within a single period by using a high-speed video camera with an extremely high frame rate.

The shock wave motion around the oscillating blade is quite periodic. It is therefore conceivable that successive pictures in a single period can be obtained by taking pictures for multiple periods at a moderate frame rate and then rearranging them by considering the phase angle at which each picture is taken. Figure 10 shows a method for generating successive pictures at a blade frequency of 50 Hz, for example.

If whole pictures are taken from the beginning with a constant interval of 1/240 s, five periods of blade oscillation are equally divided into 24 parts. When these pictures are taken out in due order and rearranged, successive pictures for a single period of blade oscillation consequently be formed with every 15° of phase difference, as shown at the lower left in Figure 10. Obviously in the case of higher frequencies where the number of pictures that can be taken in a period of blade oscillation is reduced, formation of successive pictures for one period is possible in the same manner if the number of periods is increased.

2.5.2 *Unsteady Pressure Measurements*

Figure 11 shows a schematic of a measurement system for the blade surface pressure. Both the steady and the unsteady data for five blades are measured through three 48 port scanning valves controlled with a microcomputer. The unsteady pressure data are reduced by a method of enhancing and corrected in accordance with a transfer function of each pressure lead. Three commercial high-response semiconductor pressure transducers are mounted flush with the suction surface of blade No.-1 in the cascade. The data from these surface pressure transducers are also utilized to reconfirm the pressure correction process. The locations of these transducers are shown in Table 1.

3. Results

3.1. TOTAL PRESSURE MEASUREMENTS

Total pressures in the entire flow field can be obtained by using the traverse unit with dual turntables and the five-hole pressure probe. As an example, the total pressure loss at a half chord downstream of the cascade is measured for the span of two pitches as shown in Figure 2 at a Mach number of 0.43. The results with the variation in attack angle of only the center blade are shown in Figure 12 where the attack angle of the cascade is 5°. The definition of the coefficient of total pressure loss ζ is as below.

$$\varsigma = \frac{P_{t0} - P_{t1}}{P_{t0} - P_{s0}} \tag{3}$$

P_{t0} : upstream total pressure
P_{t1} : downstream total pressure
P_{s0} : upstream static pressure

In the figure, the left and right ordinates are the total pressure loss and the outflow angle, respectively. The abscissa is the pitchwise position along the cascade. The uniformity of the cascade is observed when the deviation in attack angle of blade No.0, $\Delta\alpha_0$, equals to 0°, although the total pressure loss of blade No.-1 is slightly large. In the case of $\Delta\alpha_0$=-2°, the effect of a decrease in attack angle of blade No.0 is not obvious. In the case of $\Delta\alpha_0$=+2°, the total pressure loss of blade No.0 decreases and that of blade No.+1 increases.

3.2. SHOCK WAVE MOTION

Because the purpose of this report is to develop a photographing and image processing system, a simple and basic cascade composed of flat-plate blades with the stagger angle of 0 and the attack angle of 0 is initially applied as a test cascade.

Shock wave motion during one blade oscillation at frequencies of f=23, 100, 150 and 200 Hz is analyzed through the present system. These frequencies correspond to 0.033, 0.142, 0.213 and 0.284 in terms of the reduced frequency $k=2\pi fc/U$, where c denotes the blade chord length and U the upstream velocity.

Figure 13 represents the results showing successive pictures with a phase difference of $\pi/3$ radian, which are rearranged and selected from accumulated data for the case of f=150 Hz. The center blade is forced to oscillate, and the upper and lower blades are kept stationary. It is found that the shock wave motion, though smeared, seems periodic following the blade motion.

Figure 14, which shows the results of image processing, illustrates the streamwise displacement of shock waves during three periods of blade oscillation at f=23, 100 and 200 Hz. The notations of U1, U2, U3, L1, L2 and L3 in Figure 14 represent the pitch-wise measurement sections, as specified in the schematic at the top of this figure. The shock wave displacement at these sections is normalized by the static data. From a geometrical point of view, the phase difference in shock wave displacement should be exactly π radian between the upper side (U1, U2, U3) and the lower side (L1, L2, L3).

In the case of f=23 Hz, it is found from Figure 14(a) that the shock wave displacements at U3 and at L3 near stationary blades have a phase difference of almost π, although both shock waves do not move with an identical amplitude due to slight inaccuracy concerning the installation of the cascade.

The shock wave displacements at U1 and at L1 near the oscillating blade seem to have almost twice the frequency of blade oscillation. Furthermore, both waves look inphase although both should have a phase difference of π as described above. The both shock waves near the trailing edge thus vary in the identical direction as described in the case of the steady state, and they also vary in anti-phase, i.e., the phase shifts by π. Consistent interpretation of these phenomena leads to the result that the shock waves in the vicinity of the trailing edge of the oscillating blade vary at twice the frequency of the blade motion. In other words, twice the frequency simultaneously meets two conditions of inphase and anti-phase variations.

Similarly, the shock wave displacements at U2 and at L2 also vary with a phase difference of π. An arithmetic mean value of displacement at U1 and U3, i.e., (U1+U3)/2, is indicated by a broken line in the figure of U2 for the purpose of rough comparison, and the mean at L1 and L3, i.e., (L1+L3)/2, is shown in the figure of L2. There is hardly any difference between this mean value and the actual displacement, which indicates that the shock wave varies in a simple manner as though it were a rigid body.

In the cases of f=100 Hz and 200 Hz in Figure 14(b) and (c), the shock wave motion at each section displays the same features as the case of f=23 Hz. However, it contains a phase lag of about $\pi/6$ and $\pi/2$ relative to the blade motion, respectively.

Figure 15 shows the phase lag of shock wave motion relative to the blade motion. It is evident that the phase lag ϕ increases with an increase in blade oscillation frequency f. The value of ϕ reaches around $\pi/2$ as f increases to the level of 200 Hz. Concerning the amplitude of shock wave motion, there is hardly any variation as is obvious from Figure 10, within the limits of this experiment.

3.3. PRESSURE MEASUREMENT

The pressure data are expressed in the form of the pressure coefficient Cp given as below.

$$Cp = \frac{2}{\gamma M^2}\left(1 - \frac{P_\infty}{P}\right) \tag{4}$$

γ : Specific heat ratio
M : Inlet Mach number
P_∞: Upstream static pressure
P : Blade surface pressure

3.3.1. *Steady State*

As an example of pressure measurements, the steady pressure distributions on each blade surface at the inlet Mach number M=0.445 are shown in Figure 16 with variation in the attack angle of blade No.0. From Figure 16(b) of $\Delta\alpha_0$=0°, the uniformity of the cascade is observed. The effects of $\Delta\alpha_0$, the deviation in attack angle of blade No.0, are seen from Figures 16(a) and 16(c). A significant change in pressure distribution is observed on the center blade on both the suction and the pressure surfaces around the leading edge. The deviation in attack angle of the center blade causes a significant change in surface pressure, especially on the suction side of blade No.-1 located upstream of the center blade and on the pressure side of blade No.+1 downstream. The effect of the change in attack angle of one blade in the cascade looks limited within the flow passages adjacent to the relevant blade. As a next example, the case of M=0.76 is shown in Figure 17. As in the case of a lower Mach number M=0.445, the effect of the change in attack angle of one blade is not significant on the blades located far from the relevant blade.

3.3.2. *Unsteady State*

Chordwise expressions of unsteadiness are informative to analyze the effects of shock wave motion. As examples, time histories in surface pressure are shown in Figure 18 at the Mach number of 0.43 and the blade vibration frequency of 25.8 Hz which corresponds to the reduced frequency of 0.075. Sinusoidal variation in surface pressure is seen over the whole range of chordwise location on both suction and pressure sides of blade No.0. The variation is also seen on the pressure surface of the leading edge side of blade No.+1 located downstream of blade No.0 and on the suction surface of the trailing edge side of blade No.-1 upstream. Similar results are obtained as shown in Figure 19 for the case of M=0.75 and the frequency is 26.5 Hz (0.043 in reduced frequency). As a next step, an example of chordwise distributions of unsteady pressure amplitudes and phase angles relative to the motion of the center (No.0) blade are shown in Figure 20. Furthermore, Figure 21 shows an example of the unsteady moment in the form of the amplitude and the phase angle relative to the displacement of the center blade. These figures indicate even for the unsteady case that the effect of the change in attack angle of one blade in the cascade looks limited within the flow passages adjacent to the relevant blade.

4. Conclusions

To elucidate the unsteady aerodynamic characteristics of a transonic compressor cascade, test facilities were constructed. A high-speed mechanical drive system was developed which provides controlled pitching oscillation of one blade in the linear cascade. A traverse unit with dual turntables was developed to survey the entire flow field around the cascade. For measuring the unsteady blade surface pressure, indirect pressure measuring techniques were applied. For capturing the periodic shock wave motion, a high-speed photographing and image processing system was developed using a video camera with a moderate frame rate. The conclusions so far are:

1. The feasibility of formation of successive Schlieren pictures in a single period of high-frequency blade oscillation of up to 500 Hz was confirmed by rearranging the pictures taken during multiple periods at a moderate frame rate.

2. Behavior of shock waves was captured through the photographing and image processing system in the range up to 200 Hz of the blade oscillation frequency.

3. In the vicinity of the trailing edge of the oscillating blade, the shock waves moved at twice the frequency of the blade oscillation, which consistently meets two conditions. One is the existence of a phase difference of π between upper and lower shock wave displacements, and the other is the existence of the same streamwise direction of shock wave movements.

4. Although the amplitude of shock wave displacement did not change much within the range of the present experiment, the phase lag relative to the blade oscillation increases up to almost $\pi/2$ radian as the blade oscillation frequency increases to 200 Hz.

5. The deviation in attack angle of the center blade caused a significant change in surface pressure, especially on the suction side of blade No.-1 located upstream of the center (No.0) blade and on the pressure side of blade No.+1 downstream. The effect of the change in attack angle of one blade in the cascade looks limited within the flow passages adjacent to the relevant blade.

References

1. Hanamura, Y., Yamaguchi, K., Hirano, T. and Suzuki, K.: Effects of Mach Number on the Aerodynamic Interblade Interactions and Aerodynamic Damping of the High Turning Turbine Cascade in Transonic Flow, Trans. JSME, Series B, 55-512(1989), 951-959. (in Japanese)

2. Hanamura, Y., Yamaguchi, K., Hirano, T. and Suzuki, K.: An Experimental Investigation on Flutter Characteristics and Interblade Interactions of Turbine Cascade with High Turning in the Incompressible to Supersonic Flow, J. GTSJ, 17-67(1989), 66-75. (in Japanese)

3. Kobayashi, H.: Unsteady Aerodynamic Characteristics of Annular Cascade Oscillating in Transonic Flow (3rd Report, Low Back-Pressure Supersonic Compressor Blade Flutter), Trans. JSME, Series B, 52-480(1986), 2920-2929. (in Japanese)

4. Kobayashi, H.: Effects of Shock Waves on Aerodynamic Instability of Annular Cascade Oscillation in a Transonic Flow, Trans. ASME, J. Turbomachinery., 111(1989), 222-230.

5. Kobayashi, H.: Annular Cascade Study of Low Back-Pressure Supersonic Fan Blade Flutter, ASME Paper, No.89-GT-297(1989).

6. Kobayashi, H.: Unsteady Aerodynamic Damping Measurement of Annular Turbine Cascade with High Deflection in Transonic Flow, Trans. ASME, J. Turbomachinery, 112(1990), 732-740.

7. Adamczyk, J. J., Stevans, W. and Jutras, R.: Supersonic Stall Flutter of High-Speed Fans, Trans. ASME, J. Eng. for Power., 104(1982), 675-682.

8. Shaw, L. M., Boldman, D. R., Buggele, A. E. and Buffum, D. H.: Unsteady Pressure Measurements

on a Biconvex Airfoil in a Transonic Oscillating Cascade, Trans. ASME, J. Eng. for Gas Turbines and Power, 108(1986), 53-59.

9. Davies, M.R.D. and Bryanston-Cross, P. J.: Holographic Measurements and Theoretical Predictions of the Unsteady Flow in a Transonic Annular Cascade, ASME Paper, No.84-GT-174(1989).

10. Buffman, D.H. and Fleeter, S.: Experimental Investigation of Oscillating Cascade Aerodynamics, Unsteady Aerodynamics and Aeroelasticity of Turbomachines and Propellers, Beijing University, Beijing, China, September, 1989, 223-238.

11. Buffman, D.H. and Fleeter, S.: The Aerodynamics of an Oscillating Cascade in a Compressible Flow Field, ASME Paper, No.89-GT-271(1989).

12. Boldman, D.R., Buggele, A. E. and Michalson, G. M.: Stall Flutter Experiment in a Transonic Oscillating Linear Cascade, NASA-TM-82655(1981).

13. Fleeter, S., Novick, A. S., Riffel, R. E. and Caruthers, J. E.: An Experimental Determination of the Unsteady Aerodynamics in a Controlled Oscillating Cascade, ASME Paper, No.76-GT-17(1976).

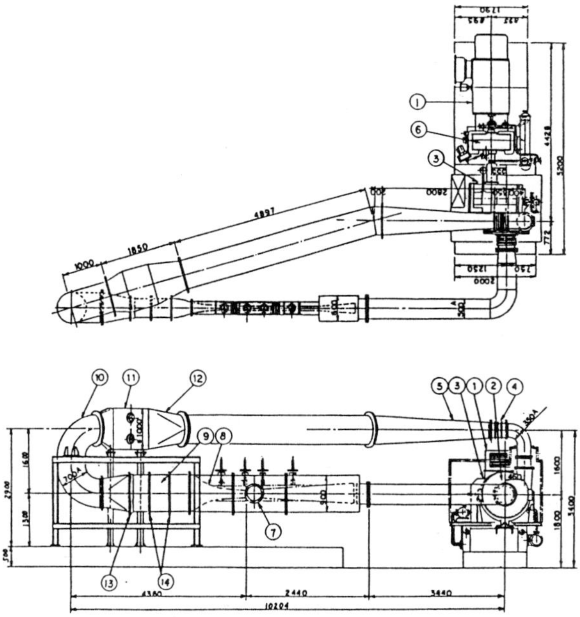

Figure 1. Transonic wind tunnel.

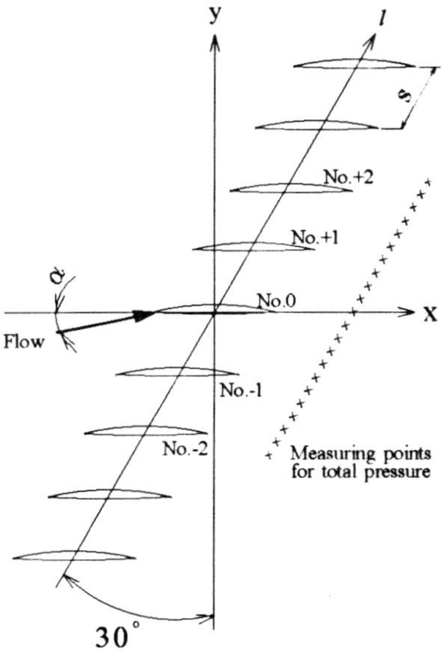

Figure 2. Linear cascade geometry.

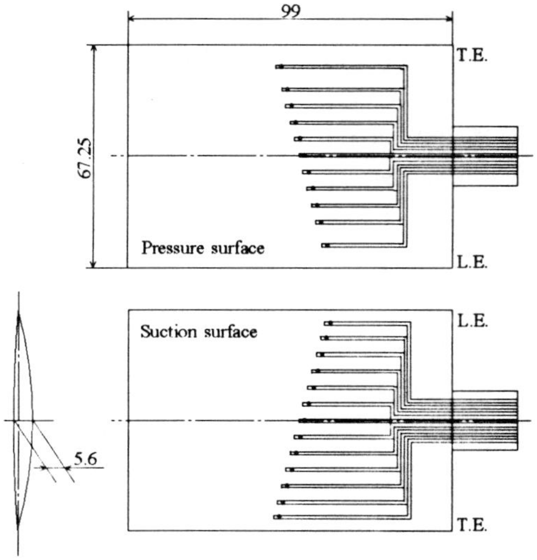

Figure 3. Test blade.

TABLE 1. Location of pressure holes (% chord)

Suction surface						Pressure surface					
1	6.9	6	42.6	⑪	79.7	1	9.9	6	50.0	11	90.1
2	13.6	⑦	50.0	12	86.4	2	20.3	7	57.4		
③	20.3	8	57.4	13	93.1	3	27.7	8	64.9		
4	27.7	9	64.9			4	35.1	9	72.3		
5	35.1	10	72.3			5	42.6	10	79.7		

③⑦⑪: high-response pressure transducer is installed at the same location.

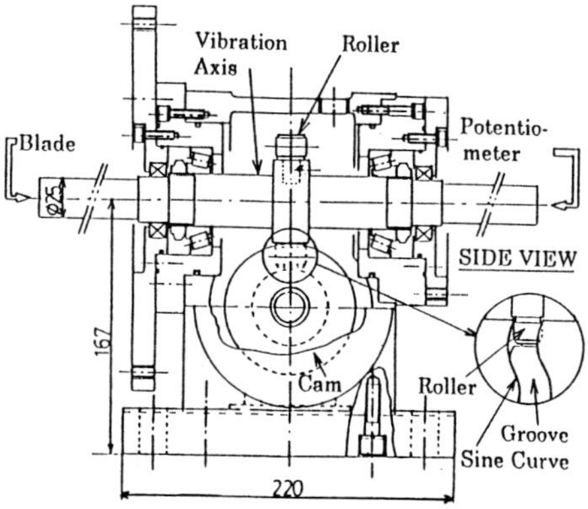

Figure 4. High-speed pitching vibrator.

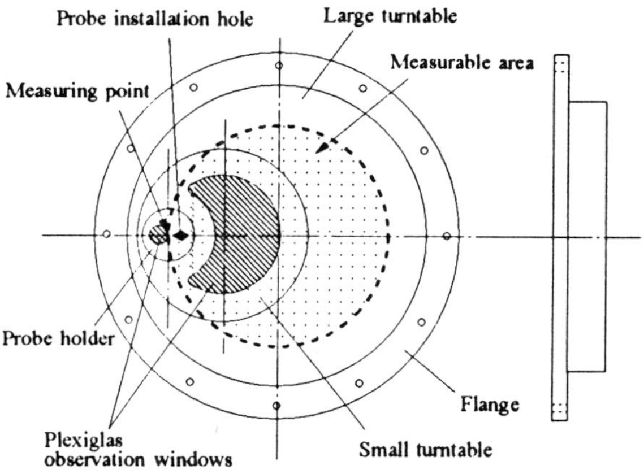

Figure 5. Traverse unit with dual turntables.

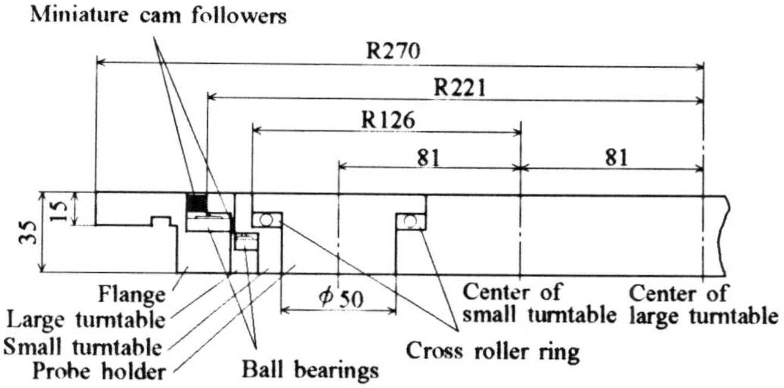

Figure 6. Cross section of traverse unit.

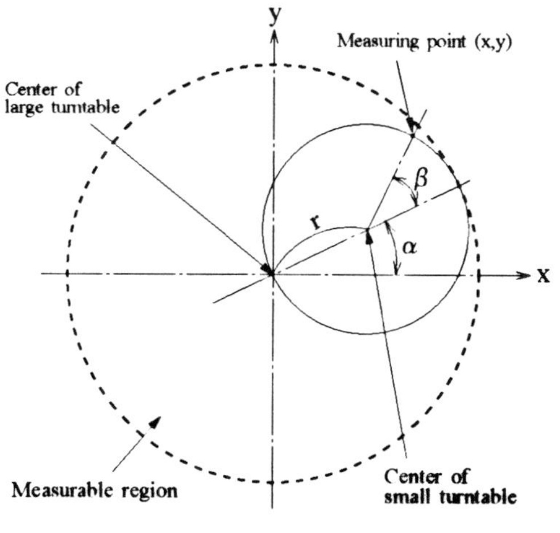

Figure 7. Schematic of measuring point and turning angle.

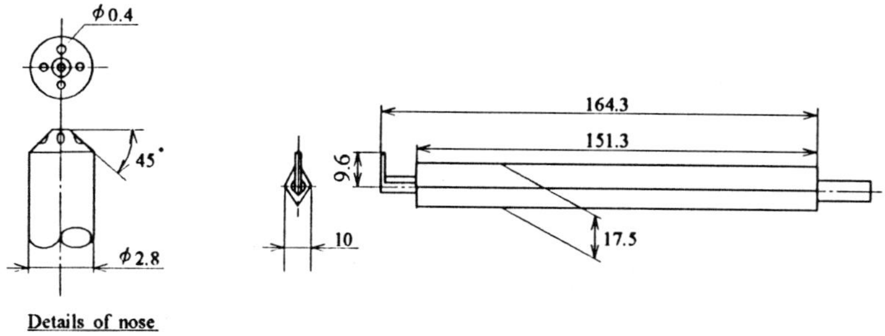

Figure 8. Scheme of five-hole pressure probe.

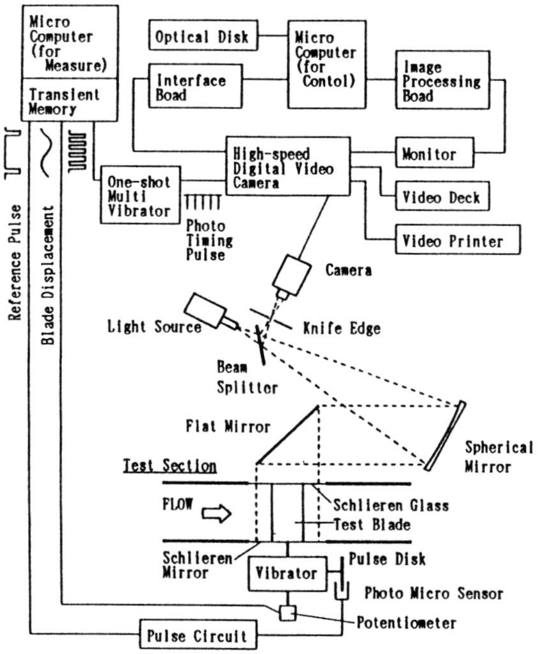

Figure 9. Block diagram of data acquisition and image processing system.

TABLE 2. Specification of Video Camera.

Lens mount	C-mount
Image sensor	CCD array
Video memory	IC memory(Dynamic RAM)
Memory capacity	32 MB
Frame rate	30, 60, 120, 240 pps
Shutter speed	1/70000 s
Resolution	256 × 256 pixels
Number of gray	256 (8 bit)
Play speed	2, 5, 10, 15, 30 pps
Maximum number of frame	512
Recording time	2.13 s (for 240 pps)

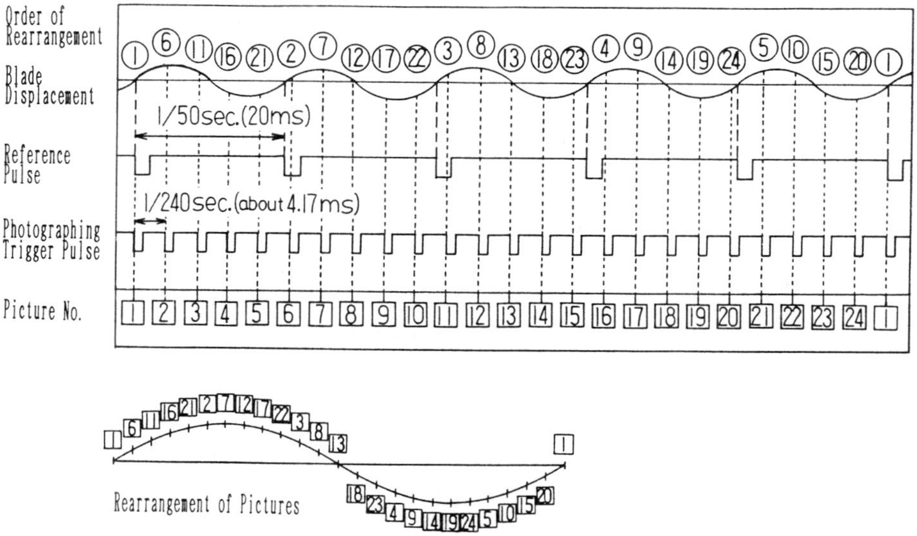

Figure 10. Formation principle of high-speed successive pictures.

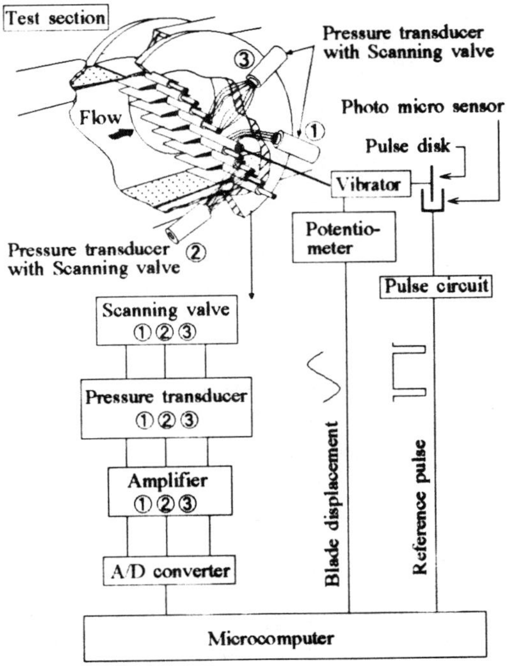

Figure 11. Schematic of blade surface pressure measuring system.

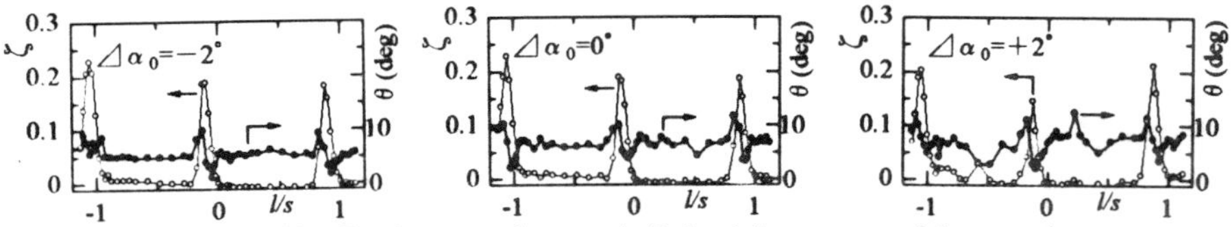

Figure 12. Total pressure loss at a half chord downstream of the cascade.

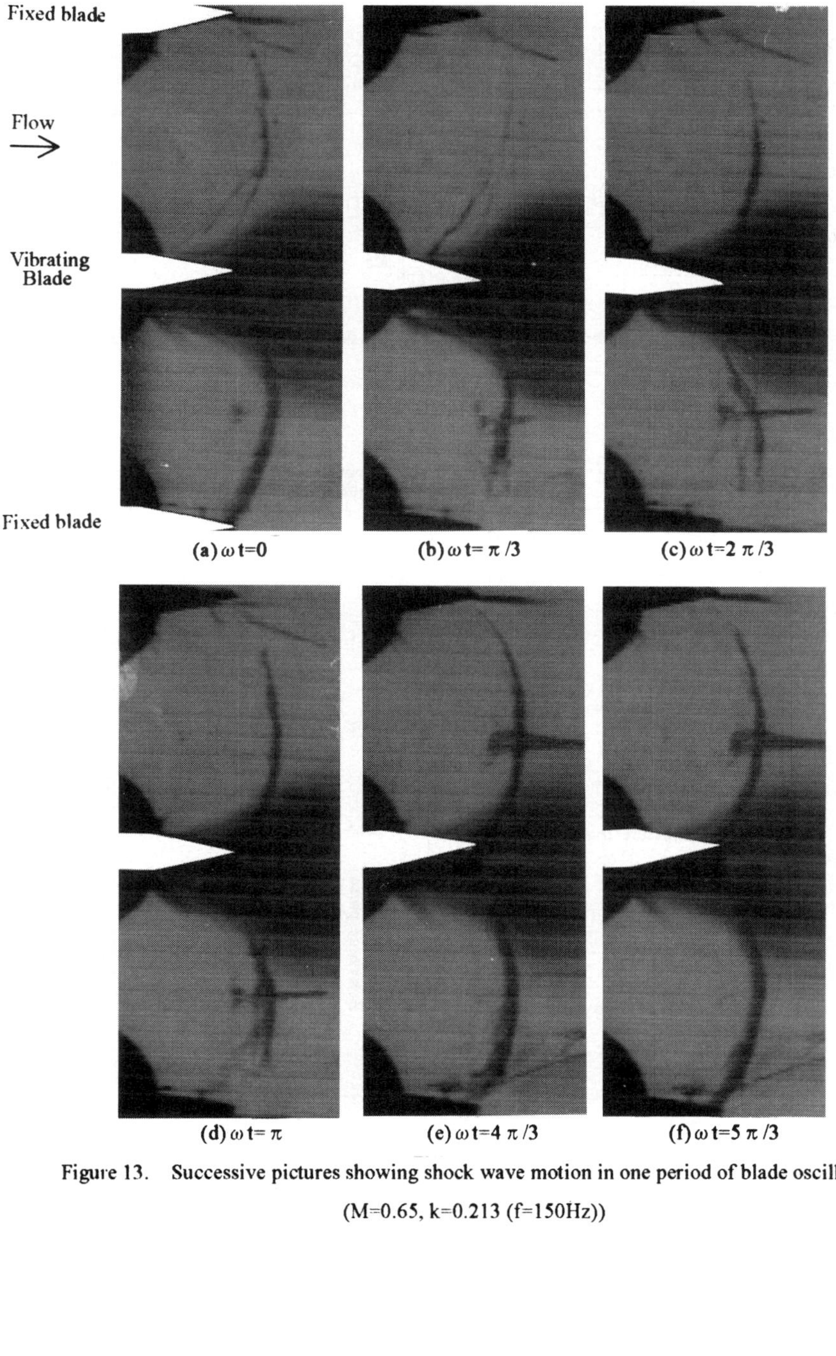

Figure 13. Successive pictures showing shock wave motion in one period of blade oscillation. (M=0.65, k=0.213 (f=150Hz))

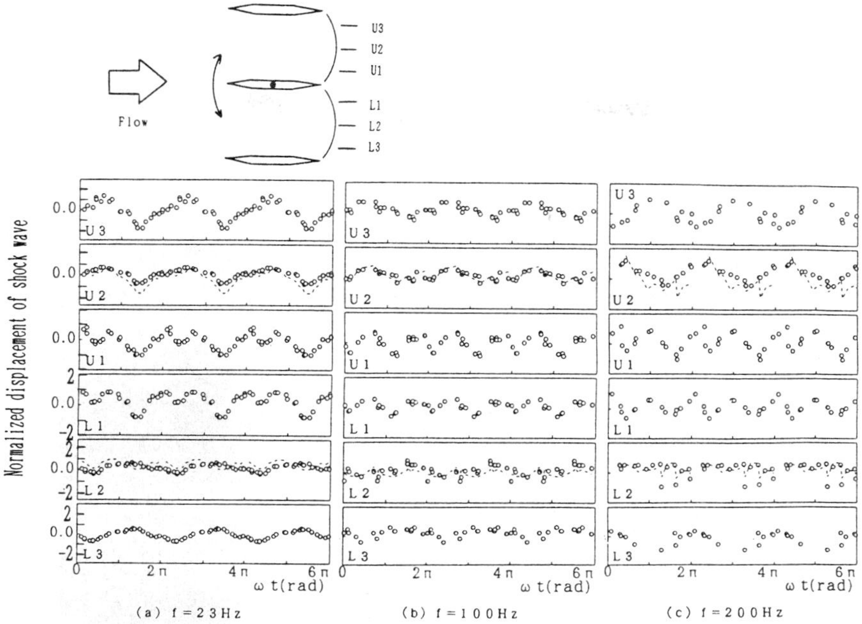

Figure 14. **Streamwise displacement of shock wave.**

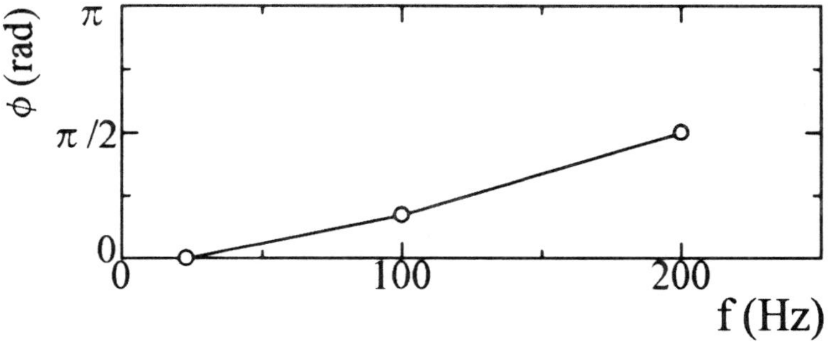

Figure 15. **Phase lag of shock wave motion relative to blade oscillation.**

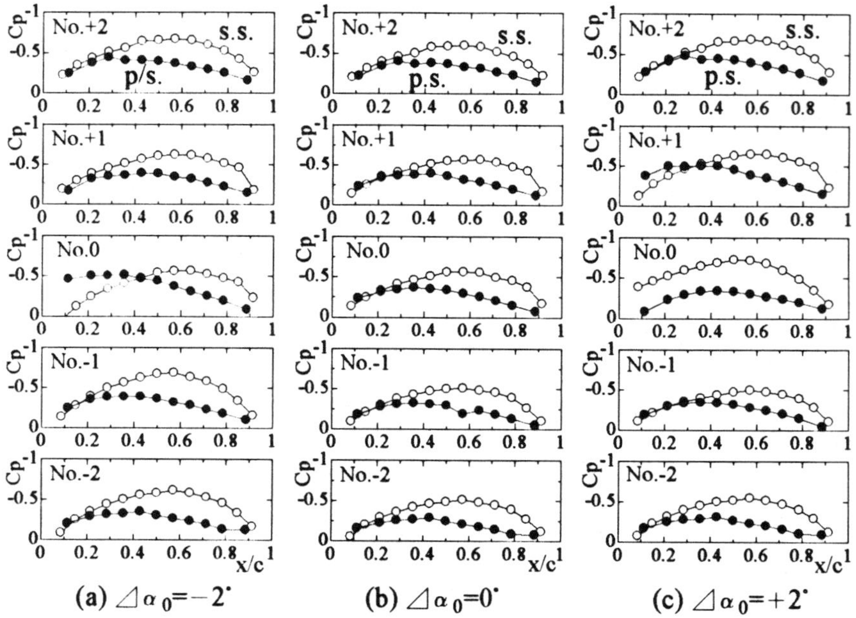

Figure 16. Steady pressure distribution of blades at M=0.43

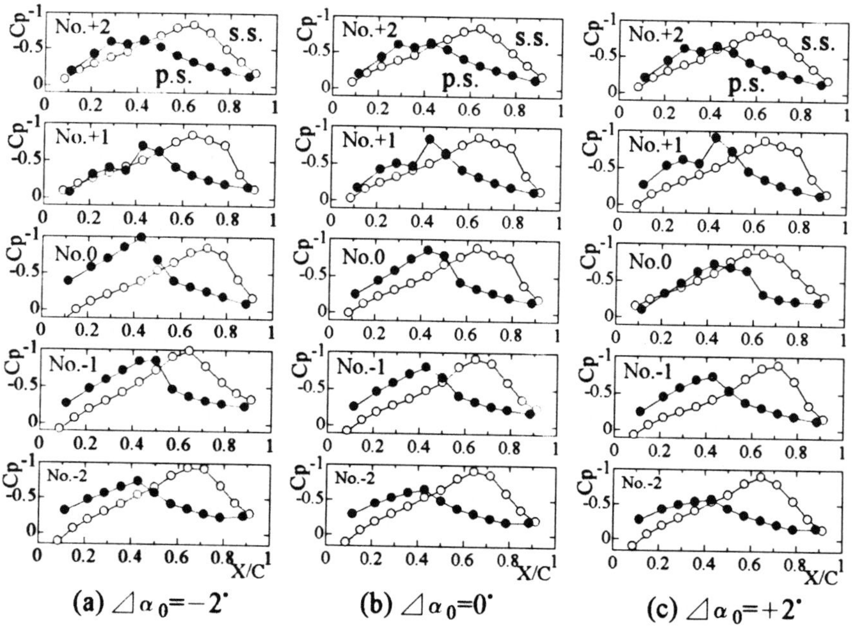

Figure 17. Steady pressure distribution of blades at M=0.75

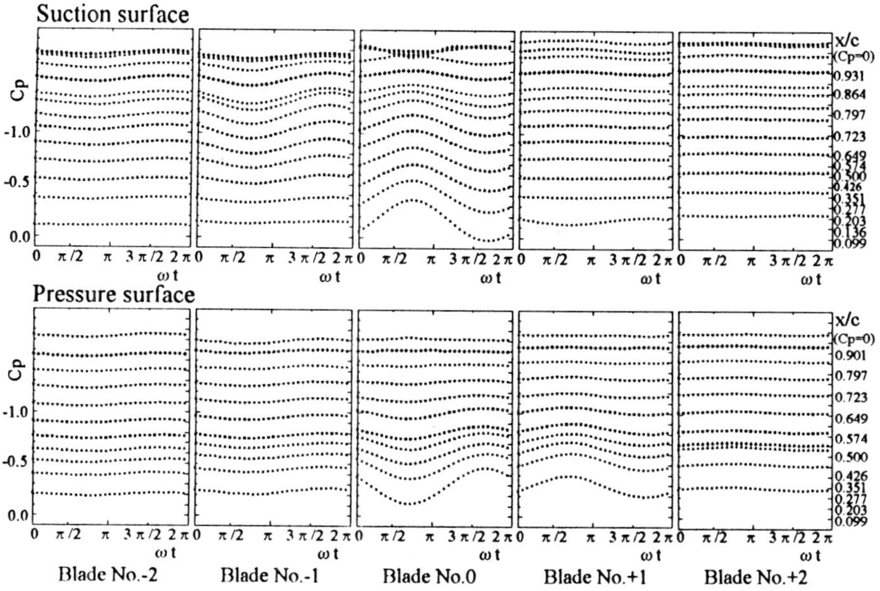

Figure 18. Surface pressure time histories at M=0.43 and f=25.8 Hz

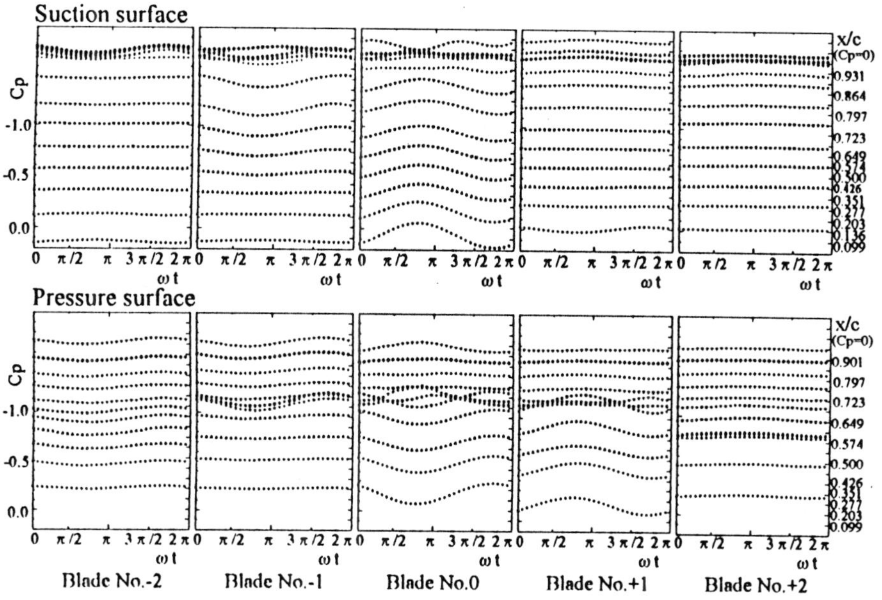

Figure 19. Surface pressure time histories at M=0.75 and f=26.5 Hz

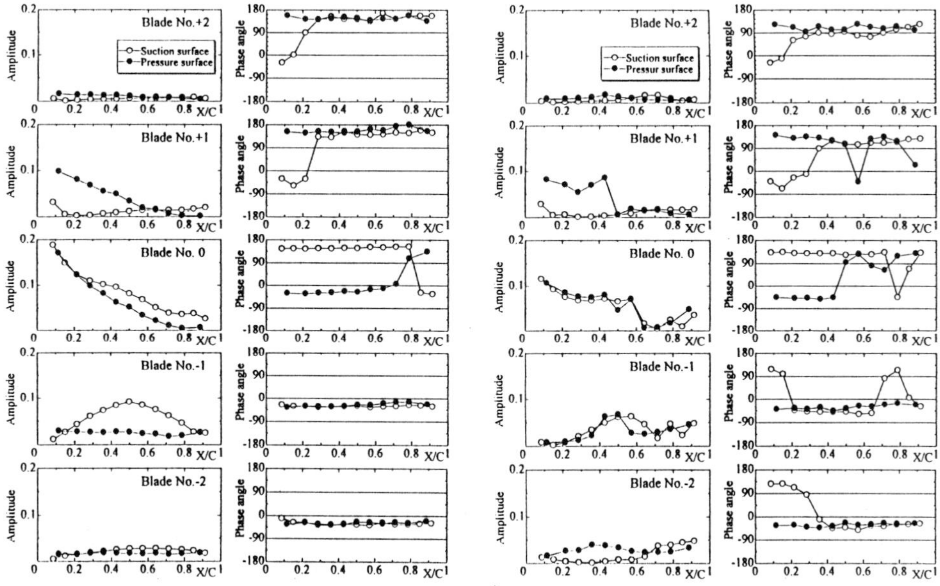

(a) M=0.43, f=25.8 Hz (k=0.075)

(b) M=0.75, f=26.5 Hz (k=0.043)

Figure 20. Chordwise distributions of unsteady pressure amplitudes and phase angles relative to blade oscillation.

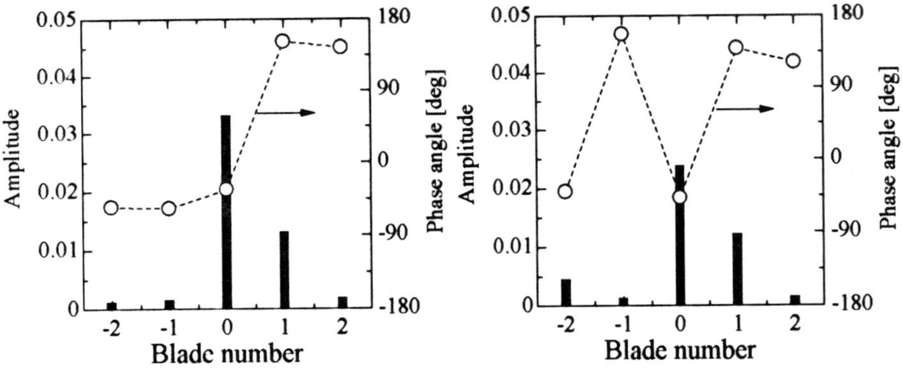

(a) M=0.43, f=25.8 Hz (k=0.075)

(b) M=0.75, f=26.5 Hz (k=0.043)

Figure 21. Amplitudes and phase angles of unsteady moment.

EXPERIMENTAL INVESTIGATION OF UNSTEADY PRESSURE BEHAVIOURS IN A LINEAR TURBINE CASCADE

M. NORRYD, A. BÖLCS
Laboratory of Applied Thermodynamics and Turbomachinery (LTT)
Swiss Federal Institute of Technology
1015 Lausanne, Switzerland

Abstract

Unsteady flow effects have been investigated for sub-, trans- and supersonic exit flow conditions in a turbine cascade. Specifically, experimental forced bending blade vibrations—with one oscillating blade—were performed in a linear cascade. The blade frequency was 160 Hz. The resulting unsteady pressures were measured with piezo-electric pressure transducers along blade surfaces at mid-span and on the facility side-wall. The cascade is composed of five blades. The aims were: 1) to achieve a better physical understanding of unsteady pressure propagations within cascades; 2) to achieve a better understanding of disturbing influences on measurements in linear cascades; 3) to evaluate possible influences of tip clearance on unsteady measurements at blade mid-span; 4) comparison with existing prediction models for design and analysis and creation of a database of experimental data. Systematic investigations have been conducted in a linear facility for sub-, trans- and supersonic exit flow conditions to achieve the aims above. It was found that side-wall unsteady pressure measurements can provide a good indication of the unsteady behaviour throughout the cascade. For transonic exit flow conditions a tailboard influence can occur in the outlet cascade region. Single blade vibrations in turbine cascades produce a different pressure propagation behaviour in each of the two blade passages around the oscillating blade. The local damping at mid-span is not significantly influenced by a change in tip clearance or a slightly non-periodic flow.

Nomenclature

c	blade chord length [m]
$\tilde{c}_p$	unsteady pressure coefficient
f	blade vibration frequency [Hz]
h	bending vibration amplitude dimensionless (with chord)
H	blade span; $H / c = 1.14$

Greek letters:

β	flow angle, against the axis [deg]
γ	stagger angle [deg]
δ	blade vibration direction [deg]
Φ_p	phase angle; positive when disturbance leads blade "m" [deg]

T.H. Fransson (ed.), Unsteady Aerodynamics and Aeroelasticity of Turbomachines, 103-116.

i	incidence angle, positive against pressure side [deg]	σ	interblade phase angle; σ is positive when blade "+1" leads blade "0" [deg]
k	reduced frequency, $k = \omega c/U_2$		
M	Mach number	τ	tip clearance, % of total width H = 100 mm [% or mm]
p	pressure [N/m^2]		
$\tilde{p}(x,t)$	perturbation pressure [N/m^2]	Ξ	aerodynamic damping coefficient
t	cascade pitch; t / c = 0.73		**subscript:**
	superscript:	1	upstream conditions
$-$	mean value	2	downstream conditions
$\sim$	perturbation value	ic	influence coefficient
	abbreviations:	p	pressure related
TB	a calculation with two simulated tailboards downstream of the cascade	s	isentropic
		loc	local
TEO	distance between the blade trailing edge and the grid boundary outlet in the Euler calculations, non dimensionalised with the axial chord length		

Introduction

In structure/fluid interactions, relatively small structural oscillations can develop into resonance, a phenomenon capable of generating powerful and destructive forces, which is able to damage almost any structure. Interaction between such structural oscillations and the resulting induced unsteady pressure forces constitutes the phenomenon called aeroelasticity which can potentially lead to flutter. Flutter can completely destroy constructions such as large bridges, aircraft-wings and turbomachinery blades. These aeroelastic phenomena should of course be avoided, and hence it is extremely important that one is capable of predicting such unsteady aerodynamic interactions. To achieve a better physical understanding of unsteady pressure perturbations, measurements of unsteady pressure flow fields were conducted using piezo-electric pressure transducers on the test facility side-wall.

In an annular facility the flow periodicity is automatically fulfilled, however in a linear facility a limited number of blades are used and the periodicity is influenced. Linear turbine cascade testing require tailboards downstream of the cascade to obtain periodic flow though, up- and downstream of the cascade. Satisfyingly good periodicity is generally attained in the centre blade passages. Measured results are repeatedly used to validate flow solvers. Therefore, it is extremely important that test results are as accurate as possible. Numerical calculations with and without simulated downstream tailboards were conducted to study the influence of downstream tailboards acting on unsteady measurements within the cascade.

To conduct forced bending blade oscillation testing a tip clearance generally exists between the vibrating blade and the facility side-wall, which produces a tip clearance vortex in the cascade. Such vortices induce 3D flow effects and the unsteady forces acting on the blades are also influenced. To obtain an unsteady pressure distribution along the blades equivalent to the real distribution, the unsteady pressures are generally measured at mid-span. However, the tip clearance can drastically vary (about τ = 0 - 2.5%) depending on facilities and setups. The influence of such tip clearance on the unsteady behaviour along the blade surfaces at mid-span can be important with regards to the unsteady pressure behaviour and stability conditions. The tip clearance influence on the unsteady behaviour along the blade surfaces at mid-span was investigated for varying tip clearance using piezo-electric pressure transducers embedded into a blade.

Still today a large requirement for extensive unsteady data for design and analysis exists. Steady and unsteady flow fields and blade surfaces data for sub-, trans- and supersonic exit flow conditions cover some of these domains where extensive data is needed for comparison with flow solvers. The experimental results along the blade surfaces at mid-span and on the side-wall were compared with three existing 2D flow solvers. A database of the steady and unsteady data was assembled.

The present paper represents a concise summary of a dissertation work (Norryd, 1997) conducted at LTT in Lausanne.

Test facility and cascade geometry

Essentially unsteady flow effects have been investigated in the linear facility shown in Figure 1a. The dimensions of the inlet flow section are 100 x 340 mm. The cascade (53% overlapping) is composed of five turbine blades numbered "-2", "-1", "0", "+1" and "+2". There are two adjustable tailboards and valves used to ensure a satisfyingly uniform flow through, up- and downstream of the two centre blade passages. The inlet incidence angle is varied by turning the round plate (No. 5) on which the cascade is mounted. The static pressure distributions in pitch-wise direction up- and downstream of the cascade together with the upstream stagnation pressure is used to determine the in- and outlet isentropic Mach number. The isentropic exit Mach number in the case of supersonic outlet flow is determined by the mass flow and the known flow quantities (M_r, β_r) in a reference position on a blade surface (Bölcs et al., 1986 and Norryd, 1997). Figure 1b shows a developed double-sided labyrinth (Norryd et al., 1994) around the oscillating blade. This design reduces the leakage secondary flow through the gap between the side-wall and the vibrating blade. For the unsteady investigations the centre blade "0" was forced to vibrate in a pure bending mode. The "unsteady" measuring blade, Figure 1b, contains 12 piezoelectric transducers—10 at the suction side and two at the pressure side—in addition to four static pressure taps. The piezo-electric transducers are embedded at mid-span. The measured signals were assumed to be unaffected by the transducer mounting methods. The blade surface is smoothed with a two-component epoxy after the mounting. A perfectly smooth blade surface finishing was obtained.

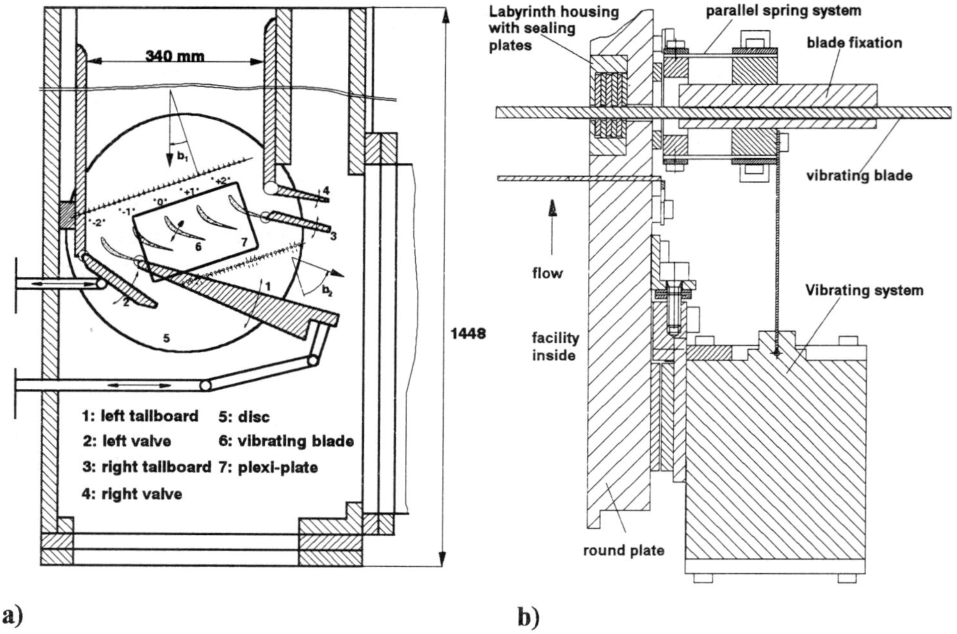

Figure 1. Linear Test Facility.

Figure 2 shows the side-wall measuring grid and the dimensions of a moveable tap construction. Eight pressure transducers—embedded in moveable tap constructions with screw-mounts—were utilised to conduct the side-wall time-dependent measurements.

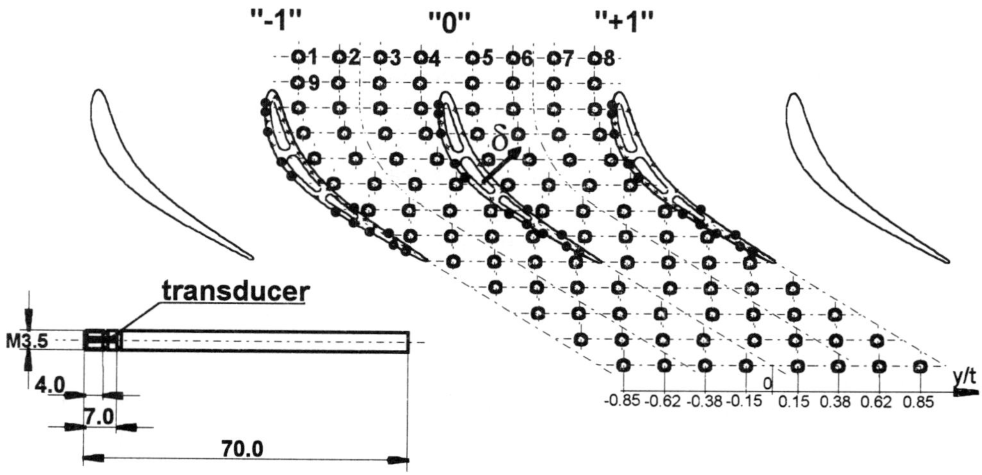

Figure 2. Side-wall measuring grid and a moveable tap construction.

Evaluation and analysis of the time-dependent data

The procedure to evaluate the measured signals is the ensemble averaging technique. This method requires a well-defined trigger signal which in this case is the blade oscillation. With a known blade frequency each sampled data point is corrected for the time-shift effects due to the components in the measuring chains, according to the calibrations, and placed in the corresponding place in the first period of each signal. The first period is subsequently divided into 128 intervals and the local averages are calculated in each interval. For these local averages much of the data spread is cancelled out, which is the great advantage with this evaluating technique. Approximately 25000 data points are taken per channel, which results in about 200 sampled data points in each interval. This was found to be a sufficient number to conduct statistical analysis on the measured data. An FFT-function is used to determine the exact amplitudes and phase shift of the averaged signals. The first harmonic of each signal is considered. It can be shown that the damping behaviour of a vibrating blade depends only on the unsteady pressure components with the same frequency as the blade oscillation (Schläfli, 1989 and Széchényi et al. 1983), as the total energy input into the vibrating blade from the fluctuating pressure components with frequencies separated from the blade frequency result in zero over time.

Definitions and data presentations

The oscillating blade is considered as a rigid body undergoing a sinusoidal motion. This defines the blade oscillation non dimensionlised by the blade chord as,

$$h(t) = \mathrm{Re}[\, h\, e^{i(\omega t)}] = h \cos(\omega t) \tag{1}$$

where the motion is defined as positive in δ-direction against the neighbouring blade "+1", Figure 2. In Eq. 1, h is the blade vibration amplitude of the 1st harmonic of the reference frequency. The reduced frequency represents a comparison between down-stream flow velocity (in a turbine), blade chord and blade oscillation frequency as:

$$k = \frac{\omega c}{U_2} \tag{2}$$

The reduced frequency in the present work was between $0.22 < k < 0.61$. The unsteady pressure coefficient is defined as,

$$\tilde{c}_p(x,t) = \frac{\tilde{p}(x,t)}{(p_{w1} - p_1)} \frac{1}{h} \tag{3}$$

where $\tilde{p}(x,t)$ is the fluctuation amplitude of the time-dependent pressure at the reference frequency, p_{w1} is the upstream stagnation pressure, p_1 is the upstream static pressure and h is the dimensionless blade amplitude. The unsteady pressure amplitude $\tilde{p}$ together with a phase angle, Φ_p, are derived from an FFT transform for each time-dependent signal. The phase angle indicates the angle between the blade motion and the induced unsteady pressure and is defined positive when the unsteady pressure leads blade motion. This phase angle relation is essential for the evaluation of unsteady data. The instantaneous unsteady pressure coefficient can be expressed by:

$$\tilde{c}_p \sin(\Phi_p + \omega t) \tag{4}$$

where, $\omega t = 0, 45, \ldots, 315°$. The blade motion is represented by $h(t) = h \sin(\omega t)$ and the oscillating blade moves against the pressure side for a positive blade motion.

Error estimation of the measured signals

The transducers signals: The piezoelectric transducers were calibrated and re-calibrated (before and after measurement tests) regarding the sensitivity and phase shift. Systematic errors of less than 1% in the sensitivity and 0.2° in the phase angle were determined. The error bars of 95% confidence interval (random error) of the unsteady pressure coefficients measured along the blade surfaces and on the side-wall depend on the measuring position. They are evaluated as follows; (1) in the inlet region: ±3% along the blade surfaces, ±8% on the side-wall; (2) in the outlet region: ±20% along the blade surfaces, ±30 - 40% on the side-wall. The larger errors in the outlet region are a result of the relatively small unsteady pressure amplitudes measured there.

Results from tests regarding unsteady pressure propagation behaviour

Side-wall measurements have been interpreted for a subsonic flow condition at nominal incidence angle and minimum tip clearance. The aim was to investigate the unsteady pressure propagation behaviour within the cascade.

The oscillating blade produces propagating pressure waves. A wave, produced from the pressure side of the blade "0" propagates in the inlet of the pressure side passage against the flow and in the outlet with the flow. For this reason the wave is accumulated near the leading edge of blade "+1" and high unsteady pressure magnitudes are formed. However, the pressure wave near the trailing edge is taken with the flow and small unsteady pressure magnitudes are developed. Furthermore, however, Figure 3 and Figure 4 show that the unsteady pressure propagates in an almost quasi-steady manner in this pressure side passage, as the phase angles in the flow field are relatively constant. This indicates an almost instantaneous propagation behaviour, related to the blade frequency.

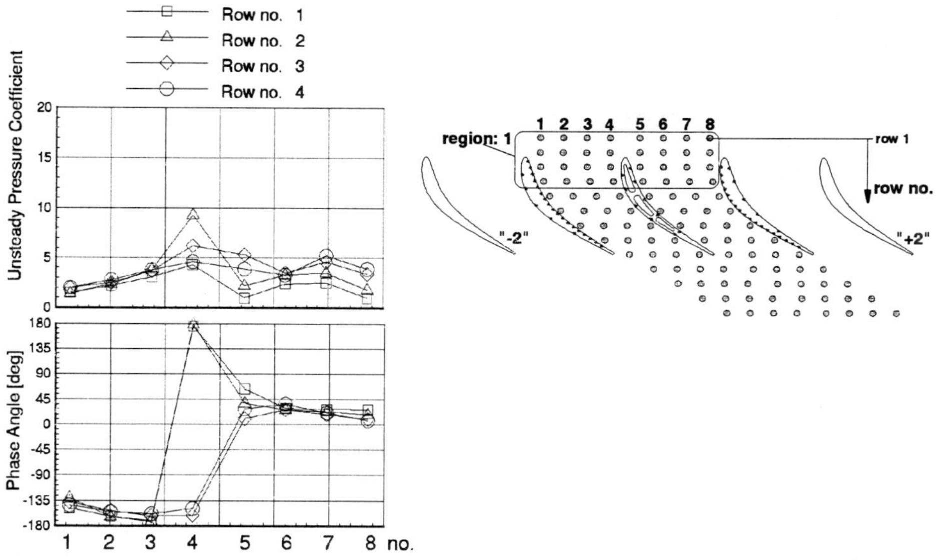

Figure 3. Unsteady pressure coefficient and phase angle distributions measured at the facility side-wall in the leading edge region: $M_{2S} = 0.75$ and $i_1 = 12.0°$.

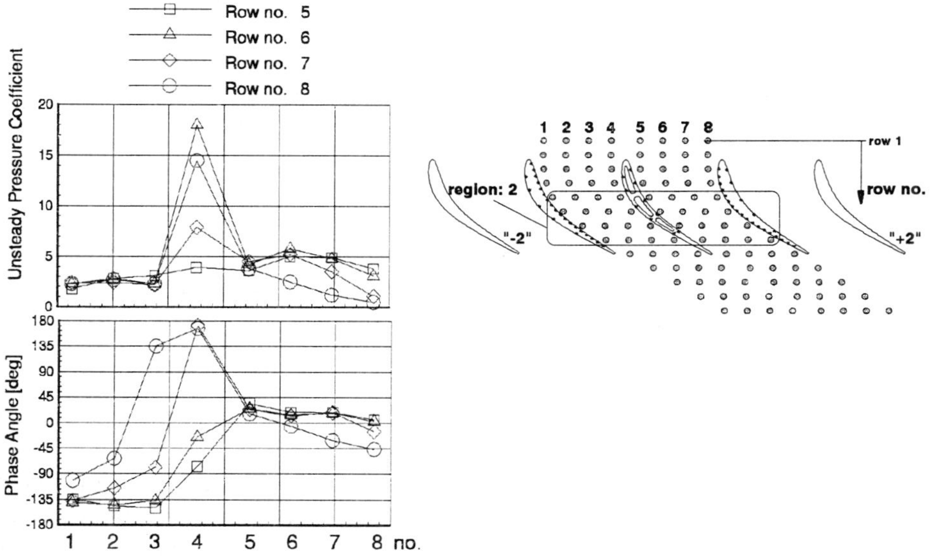

Figure 4. Unsteady pressure coefficient and phase angle distributions measured at the facility side-wall in the overlapped cascade part region: $M_{2S} = 0.75$ and $i_1 = 12.0°$.

A pressure wave produced along the suction side of the oscillating blade is always propagating downstream with the flow. Hence, there are only small unsteady pressure magnitudes produced along the pressure side of blade "-1". Figure 4 indicates that the

time-dependent effects are important in this blade passage, especially in the outlet region of the cascade (see Figure 4). The high unsteady pressure coefficients near the oscillating blade in the suction side passage arise due to the tip clearance leakage flow, although the tip clearance is the minimum one (0.5%).
From these observations it was concluded that the unsteady pressure propagation behaviour in the cascade is not symmetric in the two passages around the centre blade and the total propagation behaviour cannot be described using quasi-steady models.

Results from numerical tests regarding tailboard influences on cascade measurements

The blade surface steady and unsteady mid-span measurements have been compared using three existing 2d flow solvers; FINSUP (a potential code), EULERLTT (an Euler based code), UNSFLO (a coupled Navier-Stokes/Euler code). The aims were: 1) to predict the measurements with the existing flow solvers; 2) to investigate the influence of downstream tailboards on the steady and unsteady pressure distributions along the blade surfaces; 3) to analyse stability (local damping) tendencies between the measurements and calculations. The UNSFLO and EULERLTT calculations were compared with and without simulated tailboards downstream of the cascade for two flow conditions (Norryd, 1997). The FINSUP calculations were conducted only with periodic boundaries. The study treated sub-, trans- and supersonic exit flow conditions with two inlet flow angles.
Figure 5 and Figure 8 show the isentropic Mach number distributions comparisons between the measurements and the calculations. It is observed that the agreement is generally good, except near shocks.
Figure 6, Figure 7 and Figure 10 show unsteady features along blade “0”. The overall unsteady pressure magnitude is well predicted for a nominal inlet flow angle from the leading edge to about 60% of the chord along the blade suction side. The corresponding phase angle distributions differ by about 30 - 40° between the measurements and the calculations. This can be important for the local damping, especially for Φ_p around 180°, since the sign of the local damping is determined by the phase angle, and therefore determines if a flow condition is aerodynamically damped or excited. For subsonic flow (see Figure 6) downstream of 60% chord along the blade suction side the unsteady pressure magnitudes are low, which results in a small influence on blade stability.
In Figure 5 and Figure 8 comparisons between simulated tailboards and periodic boundaries downstream of the cascade can be studied. Concerning the isentropic Mach number distributions no significant influence occurs for the subsonic flow, as well as for transonic exit flow in the inlet and overlapped cascade regions. The normal shock position on the centre blade seems to be slightly influenced by the tailboard for both the calculation as well as the measurement (see Figure 8). On blade “+1” no significant influence of shock position is observed. For the periodic downstream boundaries (without tailboards), the normal shocks appear at about 80% of the chord. Also, for transonic exit flow a tailboard influence on the measurements can occur in the outlet cascade region.

Figure 6 and Figure 10 exhibit the blade surface unsteady features along blade "0". It is observed that no significant influence occur between the calculations with and without simulated tailboards downstream of the cascade for the overall subsonic flow and for the transonic exit flow condition in the inlet and overlapped cascade regions. In the outlet region for the transonic exit flow conditions differences are observed when, comparing the unsteady UNSFLO blade surface calculations with and without the downstream tailboards (Figure 10). A tailboard influence is also observed from the EULERLTT comparisons in this outlet region. The UNSFLO calculation without tailboards predict relatively strong peaks in the unsteady pressure coefficient distributions on both blade 0 and +1 at about 80% of the chord (steady-state shock positions). The calculation with tailboard shows a small visible peak in unsteady pressure amplitude at about 95% of the chord, corresponding to the shock position. This suggests that the downstream tailboards stabilise the motion of the normal shock. It was found that a tailboard influence on the measurements can occur in the outlet cascade region.
Figure 7 shows the local damping of the measurement and calculations along the oscillating blade for the transonic as well as the subsonic flow condition for nominal inlet flow angle (i_1 = 12.0°). The calculations result in damped flow conditions in the inlet along both the suction and pressure side, but the measurements indicate an excited flow condition near the leading edge along the blade suction side. The measured transonic case is unstable for almost the whole blade suction side. Differences between measurements and calculations can occur when the calculation method cannot predict separations in a proper way.

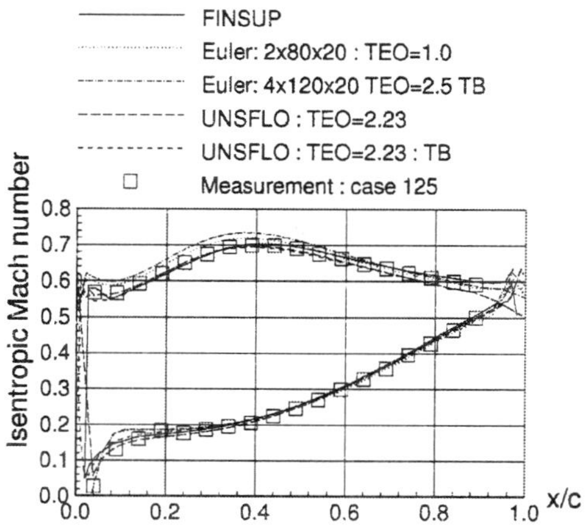

Figure 5. Comparison of FINSUP, EULERLTT and UNSFLO calculations together with measured data showing isentropic blade surface Mach number distributions along blade "+1", i_1 = 12.0°, M_{2s} =0.58.

It is found that the steady and unsteady blade surface measurements are generally well predicted by the existing flow solvers, apart from within the regions around shocks and flow separation. Hence, the accuracy of the measurements is supported by good comparison with the calculations. A tailboard influence on the measurements can occur in the outlet flow region for transonic exit flow condition. Even though the unsteady pressure coefficient and phase angle distributions are relatively well predicted, differences in the local aerodynamic damping between the measurements and the calculations occur.

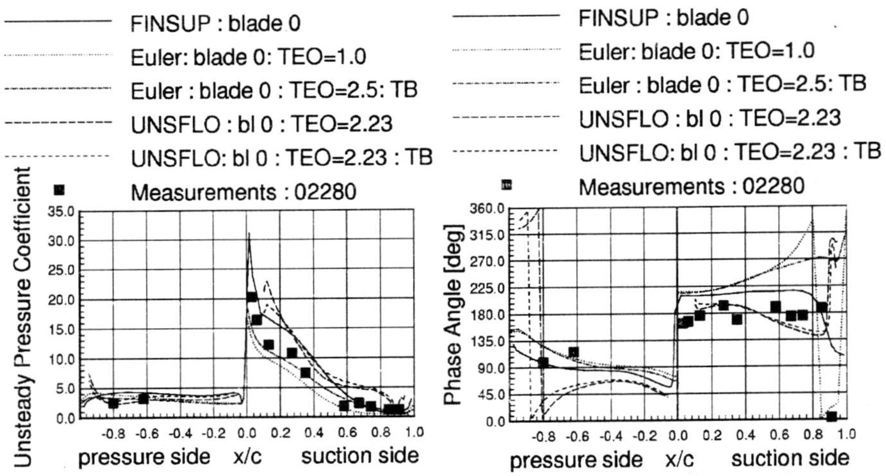

Figure 6. Comparisons of blade surface unsteady pressure magnitude and phase angle distributions between FINSUP, EULERLTT and UNSFLO calculations and measured data along the blade "0", $i_1 = 12.0°$, $M_{2s} = 0.58$.

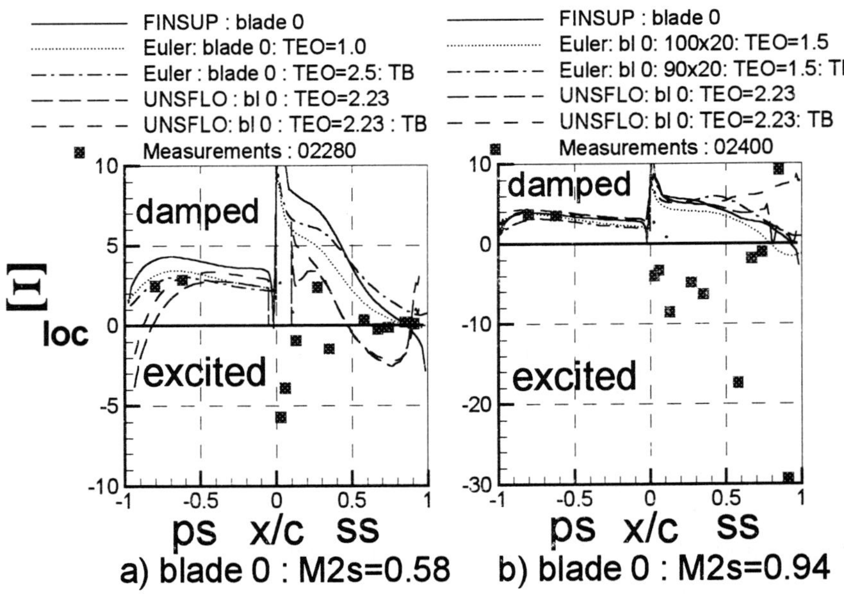

Figure 7. Comparison of the local damping along blade "0" between measurements and calculations for a) a subsonic, $M_{2s} = 0.58$, and a transonic case, $M_{2s} = 0.94$.

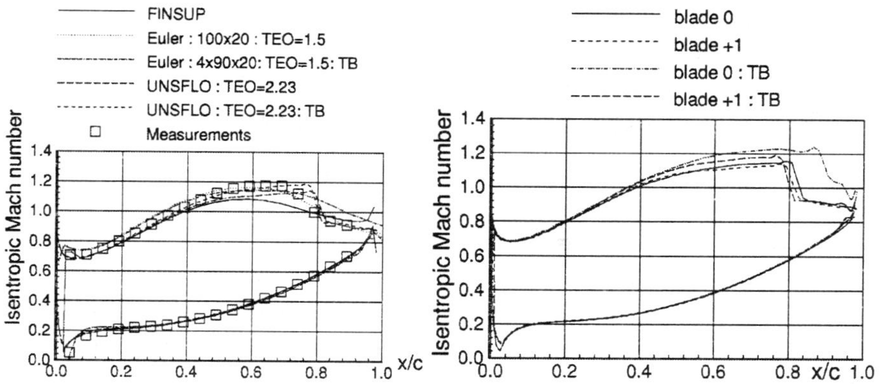

Figure 8. Calculations and measured data comparisons of isentropic blade surface Mach number distributions, $i_1 = 12.0°$, $M_{2s} = 0.94$.

Figure 9. Schlieren image of the flow condition, $i_1 = 12.0°$, $M_{2s} = 0.94$.

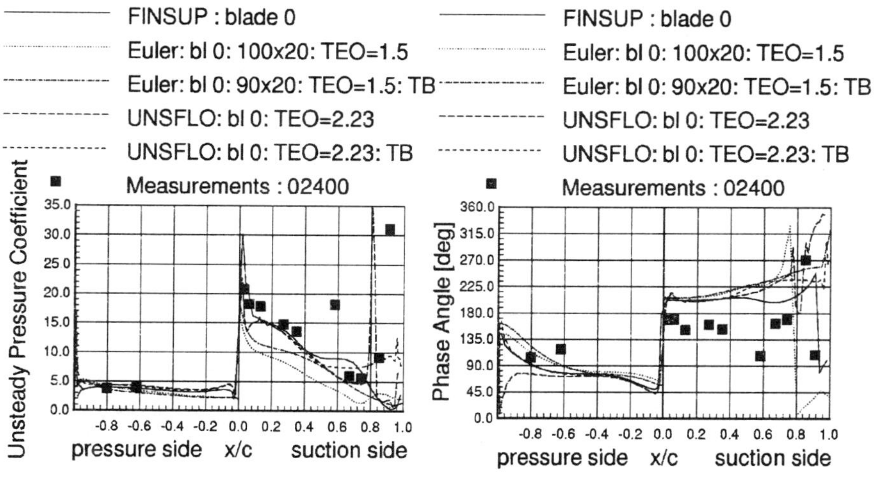

Figure 10. Comparisons of blade surface unsteady distributions between the calculations and the measured data along blade "0", $i_1 = 12.0°$, $M_{2s} = 0.94$.

Results from tests regarding the influence of tip clearance on blade surface mid-span measurements

For the measurements with varying tip clearance the tip clearance is only changed at blade "0". The neighbouring blades "1/-1" have no tip clearance between the wall and the blade profile. The aim was to investigate the influence of tip clearance on the isentropic Mach number and the unsteady features along the blade surfaces at mid-span within the cascade passages on both sides of blade "0".
Figure 11 exhibits the isentropic Mach number distributions along blade "0" at mid-span for a transonic exit flow condition at normal incidence angle. It is observed that the distribution along blade "0" is influenced by a change in tip clearance, while the neighbouring blade suction side distributions are almost unaffected. Along the suction side of blade "0" the isentropic Mach number distribution is decreased for an increased tip clearance. The change in Mach number distribution is strongest for transonic exit flow condition. The reason for that is the reduced flow surface because of the arising tip clearance vortex. The decrease in isentropic Mach number because of the reduced flow surface is strongest for transonic flow velocities, as a small change in flow surface changes the flow velocity drastically. The tip clearance flow changes the cascade exit pressure, which also influences the position of the shock at transonic exit flow conditions.

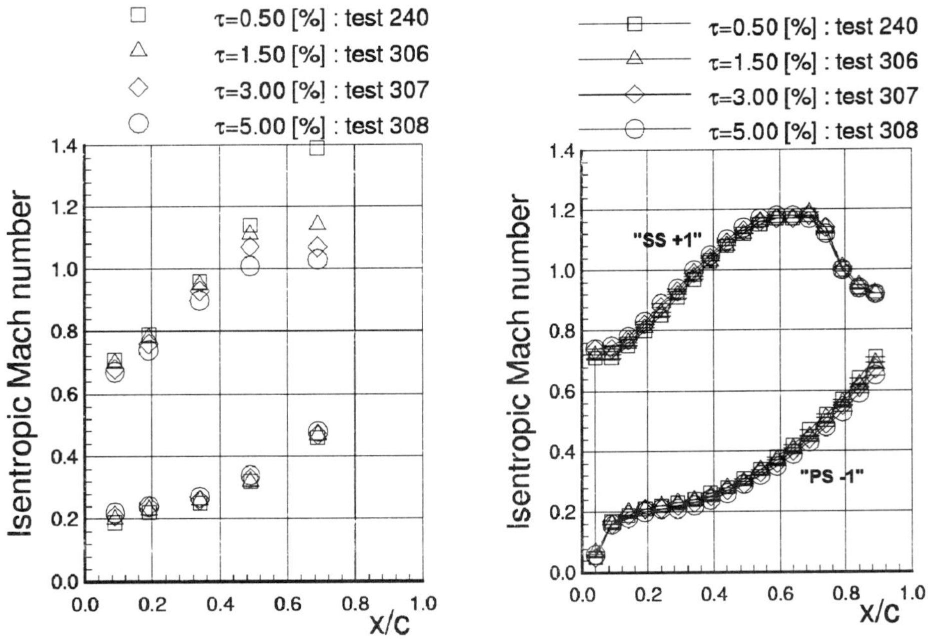

a.) Blade "ss 0" and "ps 0"   b.) Blade "ss +1" and "ps -1"

Figure 11. Blade surface isentropic Mach number distributions along the three centre blades for four tip clearances for transonic exit flow, $M_{2s} = 0.94$: $i_1 = 12.0°$.

Figure 12 shows the unsteady mid-span blade surface distributions for the corresponding transonic exit flow case. The unsteady measurements generally show a reduced unsteady pressure magnitude at mid-span for an increased tip clearance, as the pressure waves from the oscillating blade are dispersed over the tip clearance at the blade tip. The dispersion of the unsteady pressure magnitude is observed to be most pronounced near the leading edge in the suction side region of the oscillating blade. The mid-span phase angle distribution along the vibrating blade suction side is not significantly influenced by a change in tip clearance. Since the phase angle is very dominating for the calculation of the local damping, the local damping is as well not significantly influenced by a change in tip clearance. The unsteady pressure features along the neighbouring blades at mid-span are not importantly influenced by a change in tip clearance on blade "0", as they have no tip clearance. For transonic flow condition, the shock position and the shock influenced unsteady pressure magnitudes are modified by a change in tip clearance, since it changes the steady exit flow conditions.

It was found that a change in tip clearance can effect both steady- and unsteady blade surface pressure distributions at mid-span, however, the local damping is not significantly affected, as the phase angle remains relatively constant.

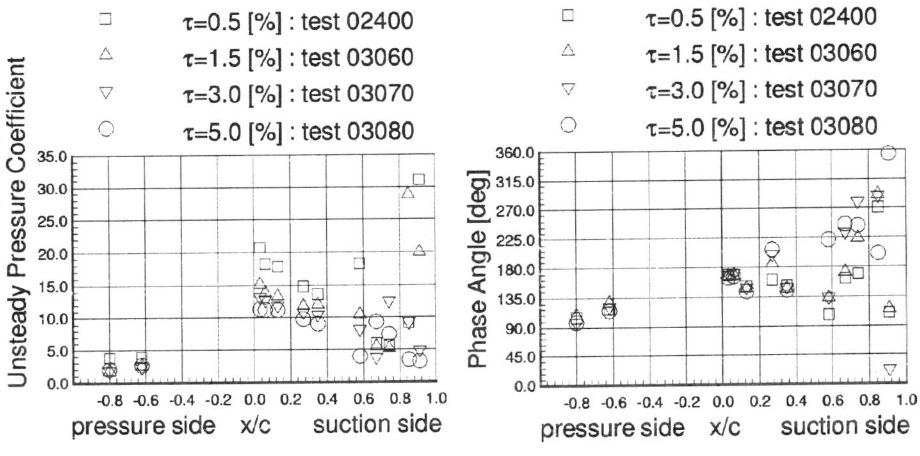

Figure 12. Unsteady pressure coefficient and phase angle distributions along blade "0" for transonic downstream flow conditions ($M_{2S} = 0.94$) for four different tip clearances.

Final conclusions

- Unsteady flow field investigations can be conducted applying piezo-electric transducers at the facility side-wall to measure unsteady pressure. Specifically, it was found that side-wall measurements can provide a good indication for the unsteady behaviour throughout the cascade.

- Calculations show a negligible tailboard influence on the unsteady features for subsonic flow. However, for transonic exit flow condition an tailboard influence can occur in the outlet cascade region, without disturbing the inlet and overlapped cascade regions.

- Single blade vibrations in turbine cascades produce a different pressure propagation behaviour in each of the two blade passages around the oscillating blade. The total propagation behaviour cannot be described using quasi-steady models.

- A change in tip clearance influences both steady- and unsteady blade surface pressure distributions at mid-span. The local damping at mid-span is not significantly influenced by a change in tip clearance and a slightly non-periodic flow.

Acknowledgements

The experimental data in the present paper has been obtained at the Laboratory of applied Thermodynamics and Turbomachinery (LTT) at the Ecole Polytechnique Fédérale de Lausanne in Switzerland during a joint scientific research project "Unsteady Flow Around Vibrating Cascades At or Near Stall" financed by ABB (ASEA-Brown Boveri, Baden, CH), RR (Rolls Royce plc., Derby, UK), VFA (Volvo Aero Engines, Trollhättan, S) and CERS (Swiss Research Foundation) in collaboration with KTH, (Royal Institute of Technology, Stockholm, S).

References

Bölcs A., Suter P. (1986) Transsonische Turbomaschinen, Wissenschaft und Technik : Taschenausgabe, Karlsruhe, G. Braun.

Norryd M., Bölcs A. (1994) Development of a Sealing construction to Prevent Leakage Flow During Flutter Tests in a Linear Cascade, 12th Symposium on Measuring Techniques for Transonic and Supersonic Flow in Cascades and Turbomachines, Prague, Czech Republic, 12-13 September.

Norryd M. (1997) Experimental Investigation of Unsteady Flow Effects including Tip Clearance in a Turbine Cascade, PH. D. These No. 1661, Presentée au Département de génie Mécanique, EPF-Lausanne.

Schläfli D. (1989) Experimentelle Untersuchung der instationären Strömung in oszillierenden Ringgittern, PH. D. These No. 800, Presentée au Département de Mécanique, EPF-Lausanne.

Széchényi E., Girault J. P. (1983) An Aeroelastic Testing Technique in a Straight Cascade Wind Tunnel, Aeroelasticity In Turbomachines, Proceedings of the Symposium "Measuring Techniques in Transonic and Supersonic Cascades and Turbomachines", Aachen, Germany.

3: Theoretical Studies of Unsteady Flow Around Oscillating Blades - 1

CALCULATION OF UNSTEADY FORCES ACTING ON THE PROFILES OF DOUBLE CASCADE OSCILLATING IN A FLUID FLOW

O.V.CHERNYSHEVA*, V.A.YUDIN
Lavrentyev Institute of Hydrodynamics
Lavrentyev pr. 15,
Novosibirsk, 630090,
Russia

Introduction. During turbine operation the unsteady hydrodynamic forces on each blade depend not only on an shape, a stagger angle and oscillation law of considered blade, but also on the geometrical parameters and oscillation laws of the rest of turbine blades. For the turbine vane, which is constructed from the double blades (main and auxiliary) (Fig.1), this dependence is essential, because of the distance between the main and the auxiliary blades is small enough. To describe the hydrodynamic influence of other blades on the considered blade the aerodynamic coefficients of influence (ACI) are comfortable to use.

In the present work on the basis of the flat model an algorithm for the calculation of the steady and unsteady forces and moment on profiles of the double cascade oscillating in a flow of ideal incompressible liquid is constructed. And also the algorithm for the calculation of the ACI for bending (normal to profile chord) and torsional oscillations (around leading edge) has been done.

For a single cascade these algorithms were constructed [1] and it reduces to the solution of one integral equation at the profile of cascade. For double cascade these algorithms may be generalised, but the problem now reduces to the solution of two integral equations at main and auxiliary profiles. Above algorithms were realised as the computer program at AT-486.

The investigation of the dependencies of hydrodynamic characteristics (steady and unsteady forces and moment) on main and auxiliary profiles of double cascade on different blade shapes, mutual position as well as different kind of blade oscillation laws was done.

It was shown, in particular, that one of basic parameters essentially influenced on the level of unsteady forces is inter-blade phase angle between oscillations of main and auxiliary profiles of cascade.

Statement of the problem. In the plane of complex variable $z = x + iy$

* Presently: Chair of Heart and Power Technology, the Royal Institute of Technology, Stockholm, Sweden

T.H. Fransson (ed.), Unsteady Aerodynamics and Aeroelasticity of Turbomachines, 119-129.

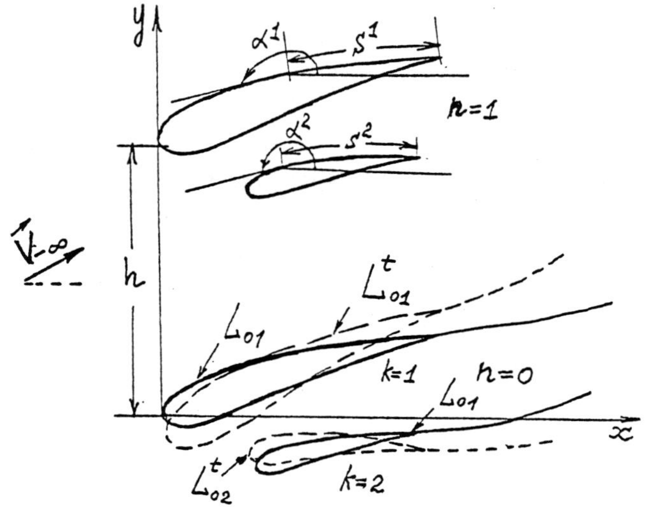

Fig.1. The double cascade of vibration profiles

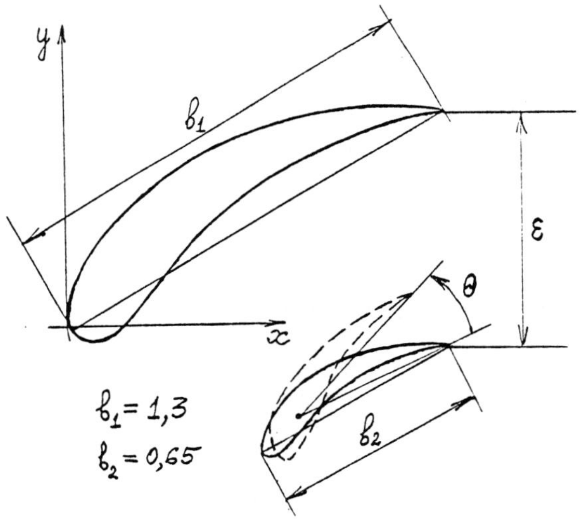

Fig.2

consider the continuous flow around the double cascade of oscillating profiles by a potential flow of ideal incompressible liquid. For the cascade an upstream flow velocity $V_{-\infty}$ is assumed constant. The profiles of main and auxiliary cascades are suggested arbitrary with sharp trailing edges. Assume that the profile oscillations occur with small amplitudes A according to the harmonic law of the same frequency ω with constant inter-blade phase angle μ between the neighbouring profiles :

$$z_n^k(s,t) = z_0^k(s) + inh + Ag^k(s)\exp(j[\omega t + n\mu]) \qquad (1)$$

Here $k = 1$ and 2 corresponds to main and auxiliary profiles, respectively; $z_0^k(s)$ is the complex co-ordinate of the contour point s of initial cascade profile in average (undeformed) state;

$$g^1 = g_x^1 + ig_y^1$$

$$g^2 = \left(g_x^2 + ig_y^2\right) e^{j\mu_{12}}$$

are the forms of the main and auxiliary profile oscillations, respectively; μ_{12} is the inter blade phase angle between them; h is the cascade pitch; n is the number of the double profile;

$$\mu = 2\pi m/N, \ \left(m = \overline{1, N-1}\right)$$

Hereinafter the imaginary value j introduced for description of time processes does not interact with imaginary value i $(i \cdot j \neq -1)$.

Account should be taken of the vortex traces behind the profiles which are simulated by the lines of velocity contact discontinuity.

By virtue of potential character of the flow the complex flow velocity $\overline{V}_\Sigma(z,t)$ is analytical function with respect to z everywhere beyond the oscillating profiles of double cascade and vortex traces behind them. It may be presented as follows:

$$\overline{V}_\Sigma(z,t) = \overline{V}_0(z) + \overline{V}(z)e^{j\omega t},$$

where $\overline{V}_0(z)$ is the velocity of flowing past immovable double cascade; $\overline{V}(z)$ is unsteady addition to velocity due to small oscillations.

The function $\overline{V}_\Sigma(z,t)$ should satisfy the following conditions:

1. At the infinity ahead of the cascade:

$$\overline{V}_0(-\infty) = \overline{V}_{-\infty}; \ \ \overline{V}(-\infty) = 0$$

2. Periodicity condition:

$$\overline{V}_0(z + inh) = \overline{V}_0(z); \ \overline{V}(z + inh) = \overline{V}(z)e^{jn\mu}$$

3. Impermeability condition on the profiles:

$$\overline{V}_{\Sigma rn}^{k}(\varsigma,t) = V^k(s,t)\, e^{-i\alpha_n^k(s,t)}$$

where

$$V^k(s,t) = V_0^k(s) + V^k(s)\ e^{j(\omega t + n\mu)}$$

is the modulus of relative flow velocity;

$$\overline{V}^k_{\Sigma rn} = \overline{V}^k_{\Sigma n} - j\omega A \overline{g}^k(s)\ e^{j(\omega t + n\mu)}$$

is the relative flow velocity at n-profile of the cascade;

$$\alpha_n^k(s,t) = \theta^k(s) + \alpha^k(s) e^{j(\omega t + n\mu)}$$

is the angle between the tangent to the contour L'_{nk} (oscillating contour) and OX-axis, where $\theta^k(s)$ is the angle between the tangent to contour L_{nk} (immovable contour) and OX-axis; s is the length of the contour arc L_{nk} calculated from the trailing edge in the direction of positive bypass of the contour.

4. At vortex traces: continuous pressure (dynamic condition) and normal constituent of flow velocity (kinematic condition) should retain when passing through the vortex traces. Owing to the smallness of the oscillation amplitude we consider [1] that the vortex traces are placed along the streamline of steady flow passing from the profile trailing edges.

5. For sharp trailing edges the Kutt-Joukovski condition, which is written with regard for dynamic condition as follows:

$$V_0^k(+0) + V_0^k(-0) = 0$$

$$V^k(+0) + V^k(-0) = -\frac{j\omega}{V_0^k(0)} \int_{L'_{0k}} V^k(\sigma) d\sigma$$

(velocity jump condition at sharp trailing edge)

The solution method. According to the generalised Cauchy formula with regard for the upstream condition for the cascade the following presentation is obtained:

$$\overline{V}_\Sigma(z,t) = \frac{1}{2iNh} \sum_{k=1}^{2} \sum_{n=0}^{N-1} \int_{L'_{0k}} \overline{V}_\Sigma(\varsigma,t) \left(cth\left[\frac{\pi}{Nh}\left(z - \varsigma^k - inh\right) \right] + 1 \right) d\varsigma^k +$$

$$\frac{1}{2iNh} \sum_{k=1}^{2} \sum_{n=0}^{N-1} \int_0^\infty \eta_n^k(\tau,t) \left(cth\left[\frac{\pi}{Nh}\left(z - \varsigma^{*k}(\tau) - inh\right) \right] + 1 \right) d\tau^k + \overline{V}_{-\infty}$$

where the value of velocity contact discontinuity at the vortex trace is:

$$\eta_n^k = -\frac{1}{V_0^k(\tau)} \frac{\partial}{\partial t} \left(\mathrm{Re}_i \int_{L'_{nk}} V_{\Sigma n}(\varsigma^k, t) d\varsigma^k \right) \Bigg|_{t=t^k}, \quad t^k = t - \int_0^\tau \frac{d\tau}{V_0^k(\tau^k)}$$

Using the Sokhotsky formulae pass to the limit from the analyticity domain of $\overline{V}_\Sigma$ to the double cascade profiles. By virtue of the smallness of the oscillations linearize around the stationary solution V_0. In doing so, take the boundary conditions on steady profile position, and replace the integration by moving contours L'_{0k} with that by immovable ones L_{0k}, neglecting the values of the second order of smallness as compared to the oscillation amplitude A. Then collect terms with $e^{j\omega t}$ multiplier. Taking into account the Kutt-Joukovski condition to order to obtain the unique solution of our problem and taken the real part with respect to i, obtain the system of integral equations for the values of stationary velocities $V_0^k(s)$ and amplitudes of unsteady parts of relative velocities $V^k(s)$ at the cascade profiles, which has the unique solution:

$$\begin{cases} V_0^p(s) - \sum_{k=1}^{2} \int_{L_0^k} V_0^k \Phi_{10}^{pk}(s,\sigma) d\sigma^k = \Pi_{10}^p(s), \qquad p = 1,2 \\ V^p(s) - \sum_{k=1}^{2} \int_{L_0^k} V^k(\sigma) \Phi_1^{pk}(\mu,s,\sigma) d\sigma^k = A \cdot \Pi_1^p(s) \end{cases} \tag{2}$$

$$\Phi_{10}^{pk}(s,\sigma) = \Phi_0^{pk}(s,\sigma) - 0.5\left(\Phi_0^{pk}(+0,\sigma) + \Phi_0^{pk}(-0,\sigma)\right)$$

$$\Phi_0^{pk}(s,\sigma) = \mathrm{Re}_i \left\{ \frac{e^{i\theta^p(s)}}{ih} \left(cth\left[\frac{\pi}{h}\left(z_0^p - \varsigma_0^k\right)\right] + 1 \right) \right\}$$

$$\Phi_{10}^p(s) = \Phi_0^p(s) - 0.5\left(\Phi_0^p(+0) + \Phi_0^p(-0)\right), \qquad \Phi_0^p(s) = \mathrm{Re}_i \left\{ 2 e^{i\theta^p(s)} \overline{V}_{-\infty} \right\}$$

$$\Phi_1^{pk}(\mu,s,\sigma) = \Phi^{pk}(\mu,s,\sigma) - 0.5\left(\Phi^{pk}(\mu,+0,\sigma) + \Phi^{pk}(\mu,-0,\sigma)\right)$$

$$\Phi^{pk}(\mu,s,\sigma) = \mathrm{Re}_i \left\{ -\frac{e^{i\theta^p(s)}}{iNh} \sum_{n=0}^{N-1} e^{jn\mu} \left\{ cth\left[\frac{\pi}{Nh}\left(z_0^p - \varsigma_0^k - inh\right)\right] + 1 \right\} - \right.$$

$$\left. - j\omega \int_0^\infty \frac{\exp\left(-j\omega T^k(\tau)\right)}{V_0^k(\tau)} \left\{ cth\left[\frac{\pi}{Nh}\left(z_0^p - \varsigma_0^{*k} - inh\right)\right] + 1 \right\} d\tau \right\}$$

$$\Pi_1^p(\mu,s,\sigma) = \Pi^p(\mu,s,\sigma) - 0.5\left(\Pi^p(\mu,+0,\sigma) + \Pi^p(\mu,-0,\sigma)\right)$$

$$\Pi^p(\mu,s,\sigma) = \sum_{k=1}^{2} \int_{L_0^k} V_0^k(\sigma) D^{pk}(\mu,s,\sigma) d\sigma^k + j\omega G^p(\mu,s)$$

$$+\, j\omega \int_{L_0^p} \left\{ V^{p}(s) + j\omega\, \mathrm{Re}\left(g^{p}(s) e^{-\alpha^{p}(s)} \right) \right\} \left(\left|z^{p}\right|^{2} - \left|z_*^{p}\right|^{2} \right) ds^{p}$$

Here z_*^p is the complex co-ordinate of the sharp trailing edge of L_0^p profile. These formulates are comfortable to use because of they include only $V_0^p(s)$, $\mathrm{V}^p(s)$ (solutions of integral equations (2)).

Further consider

$$\Delta C_X^p = \sum_{q=1}^{2} \sum_{n=0}^{N-1} \left(k_{nX}^{pq} \frac{x_n^q}{h} + k_{nY}^{pq} \frac{y_n^q}{h} + k_{n\alpha}^{pq} \alpha_n^q \right)$$

$$\Delta C_Y^p = \sum_{q=1}^{2} \sum_{n=0}^{N-1} \left(l_{nX}^{pq} \frac{x_n^q}{h} + l_{nY}^{pq} \frac{y_n^q}{h} + l_{n\alpha}^{pq} \alpha_n^q \right) \qquad (3)$$

$$\Delta C_M^p = \sum_{q=1}^{2} \sum_{n=0}^{N-1} \left(m_{nX}^{pq} \frac{x_n^q}{h} + m_{nY}^{pq} \frac{y_n^q}{h} + m_{n\alpha}^{pq} \alpha_n^q \right)$$

where $x_n^q(t)$, $y_n^q(t)$, $m_n^q(t)$, respectively, bending displacement of n^q - profile along X,Y-axis's and its turning around some fixed point. Coefficients k_{nX}^{pq}, k_{nY}^{pq}, ..., $m_{n\alpha}^{pq}$ define the part of hydrodynamic reactions on the initial q-profile, which causes by "unit" displacement of n^q - profile along X,Y-axis's or "unit" turning around the same point. These coefficients are called ACI.

To order to obtain formulae for the calculation of these ACI it is enough to express $x_n^q(t)$, $y_n^q(t)$, $m_n^q(t)$ from (1) and calculate $\Delta C_X^p(x_n^q, y_n^q, \alpha_n^q; \mu, \mu_{12})$, $\Delta C_Y^p(x_n^q, y_n^q, \alpha_n^q; \mu, \mu_{12})$, $\Delta C_M^p(x_n^q, y_n^q, \alpha_n^q; \mu, \mu_{12})$ for values set $\mu = \mu_{12} = \dfrac{2\pi k}{N}$, $k = 1, \ldots, N-1$ and for two values of μ_{12} (for example, for $\mu_{12} = 0$ and $\mu_{12} = \pi$). Substitution of obtained values in (3) takes the system and after inversion of this system the following formulae for calculation of ACI are

$$k_{nX}^{pq} = \frac{1}{N} \sum_{k=0}^{N-1} C_{XX}^{pq}(\mu_k) e^{-jn\mu_k}$$

$$k_{nY}^{pq} = \frac{1}{N} \sum_{k=0}^{N-1} C_{XY}^{pq}(\mu_k) e^{-jn\mu_k}$$

$$\vdots$$

$$k_{n\alpha}^{pq} = \frac{1}{N} \sum_{k=0}^{N-1} C_{M\alpha}^{pq}(\mu_k) e^{-jn\mu_k}$$

$$D^{pk}(\mu, s, \sigma) =$$

$$= \mathrm{Re}_i\left\{-\frac{\pi e^{i\theta^p(s)}}{iN^2h^2}\sum_{n=0}^{N-1}\left(g(s) - g(\sigma)e^{jn\mu}\right)sh^{-2}\left[\frac{\pi}{Nh}\left(z_0^p - \varsigma_0^k - inh\right)\right]\right\}$$

$$G^p(\mu, s) = \mathrm{Re}_i\left\{2e^{i\theta^p(s)}\left(-\overline{g}^p + \sum_{k=1}^{2}\int_{L_0^k}\overline{g}^k(\sigma)\Phi^{pk}(\mu, s, \sigma)d\varsigma_0^k\right)\right\}$$

The system of integral equations (2) was solved numerically. The profile was substituted by the polygon inscribed in it [3], while the integral over the trace is calculated using the quadrature Filon formulae for the integrals of oscillating series [4]. Thus obtained system of linear algebraic equations was solved by the Gauss method with selecting of the maximum element.

Calculation of the forces, moment and ACI. At the initial $(n = 0)$ profile the formulae to calculate forces

$$X^p = \frac{1}{2}\rho\overline{V}_{-\infty}^2 h\left(C_{0X}^p + C_X^p A e^{j\omega t}\right)$$

$$Y^p = \frac{1}{2}\rho\overline{V}_{-\infty}^2 h\left(C_{0Y}^p + C_Y^p A e^{j\omega t}\right)$$

and moment

$$M^p = \frac{1}{2}\rho\overline{V}_{-\infty}^2 h\left(C_{0M}^p + C_M^p A e^{j\omega t}\right)$$

(where ρ - flow density, C_{0X}^p, C_{0Y}^p, C_{0M}^p, C_X^p, C_Y^p, C_M^p - respective nondimensional coefficients) are obtained by the Cauchy-Langrange integral and are [1,3]:

$$C_{0X}^p - iC_{0Y}^p = i\int_{L_0^p}\left(V_0^p(s)\right)^2 e^{-i\alpha^p(s)}ds^p \qquad C_{0M}^p = \mathrm{Re}\int_{L_0^p}\left(V_0^p(s)\right)^2 \overline{z}^2 d\overline{z}^p$$

$$C_X^p - iC_Y^p = 2i\int_{L_0^p}\left\{V_0^p(s)V^p(s) + j\omega V_0^p(s)\,\mathrm{Re}\left(g^p(s)e^{-i\alpha^p(s)}\right)\right\}ds^p -$$

$$- i\int_{L_0^p}\left(V_0^p(s)\right)^2\frac{dg^p}{dz^p}e^{-i\alpha^p(s)}ds^p -$$

$$- 2ij\omega\int_{L_0^p}\left\{V^p(s) + j\omega\,\mathrm{Re}\left(g^p(s)e^{-\alpha^p(s)}\right)\right\}\left(\overline{z^p - z_*^p}\right)ds^p$$

$$C_M = 2\int_{L_0^p}\left\{V_0^p(s)V^p(s) + j\omega V_0^p(s)\,\mathrm{Re}\left(g^p(s)e^{-i\alpha^p(s)}\right)\right\}\mathrm{Re}\left(z^p d\overline{z}^p\right)ds^p +$$

$$+ \int_{L_0^p}\left(V_0^p(s)\right)^2\mathrm{Re}\left(\frac{dg^p}{dz^p}z^p dz^p\right) - \int_{L_0^p}\left(V_0^p(s)\right)^2\mathrm{Re}\left(g^p d\overline{z}\right) +$$

where

$$C_{XX}^{pq}(\mu_k) = \frac{1}{2}\left(C_X^p(x_n^p,0,0;\mu_k,0) - (-1)^q C_X^p(x_n^p,0,0;\mu_k,\pi)\right)$$

$$C_{XY}^{pq}(\mu_k) = \frac{1}{2}\left(C_X^p(0,y_n^p,0;\mu_k,0) - (-1)^q C_X^p(0,y_n^p,0;\mu_k,\pi)\right)$$

$$\vdots$$

$$C_{M\alpha}^{pq}(\mu_k) = \frac{1}{2}\left(C_X^p(0,0,\alpha_n^p;\mu_k,0) - (-1)^q C_X^p(0,0,\alpha_n^p;\mu_k,\pi)\right)$$

Results. The above algorithm was realised as the computer program for steady and unsteady hydrodynamic characteristics and ACI of the double cascade profiles. Usually one should specify in calculations no more than 30 points at main and auxiliary profiles and no more than 100 points in the traces behind them to obtain the accuracy to the third decimal place. The duration of data processing at AT-486 does not exceed 3 minutes. To verify the validity of the program performance it was compared with [2] for the case when the auxiliary profile coincides with the main one and was located under it at the distance of half-pitch of double cascade. The oscillation laws were also set identical $g^1(s) = g^2(s)$. In this case the double cascade became equivalent to single cascade with the pitch reduced twice. For all compared calculations the values of velocity and forces (both steady and unsteady) and ACI coincide almost completely (the difference is no more than 0.1%).

To investigate the behaviour of steady and unsteady hydrodynamic characteristics the double cascade with upstream flow $\overline{V}_{-\infty} = 1 \cdot e^{-i\pi/4}$, pitch h=1 and initial profiles represented on the Fig.2 was used.

As an example of the calculations Fig.3 presents the dependencies of stationary circulation Γ_0^k characterising the flow turn by main ($k = 1$) and auxiliary ($k = 2$) cascades. The values $\Gamma_0 = \Gamma_0^1 + \Gamma_0^2$ and $\gamma_0 = \Gamma_0^1 / \Gamma_0^2$ will presumably characterise the work of the double cascade and relative contribution to this work of each of the cascades. It is evident from the plot that both Γ_0 and γ_0 depend significantly on the position and stagger angle of auxiliary cascade with respect to the main cascade, especially when they are close to each other.

Fig.4 presents the dependence of unsteady forces amplitude on size of auxiliary profile (b_2 chord) for different ω frequencies of bending oscillations ($g^1 \equiv i,\ g^2 \equiv i,\ \mu_{12} = 0$). As it was expected with growth of ω the level of unsteady interaction of profiles is increased. And the dependence of unsteady forces amplitudes on profile size has complicated character and, in particular, may be non-monotonous.

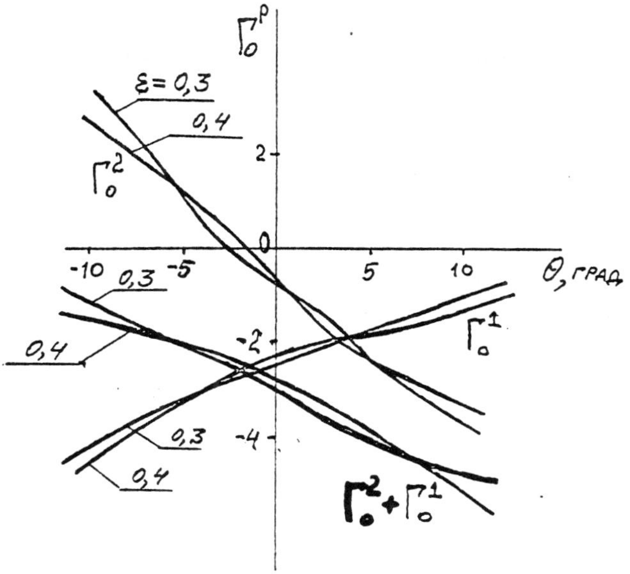

Fig.3. The dependence of circulation on the position of auxiliary cascade

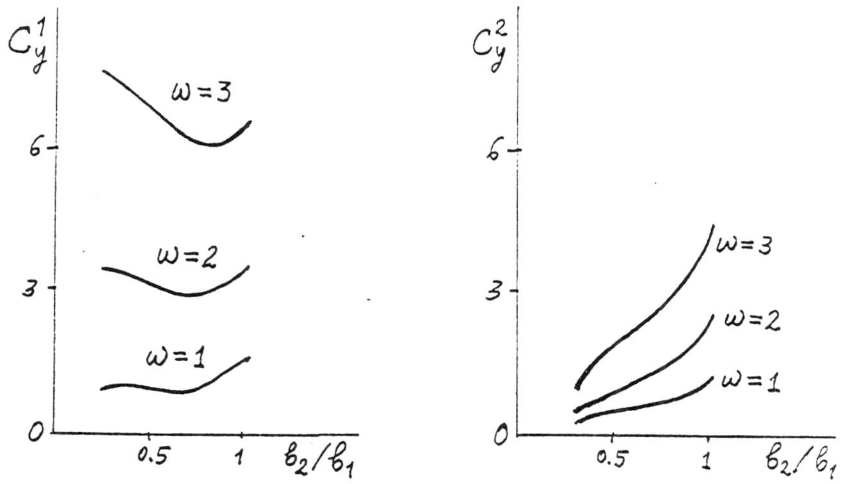

Fig.4. Dependencies of unsteady force amplitudes on size of auxiliary profile for different frequencies. Bending oscillations.

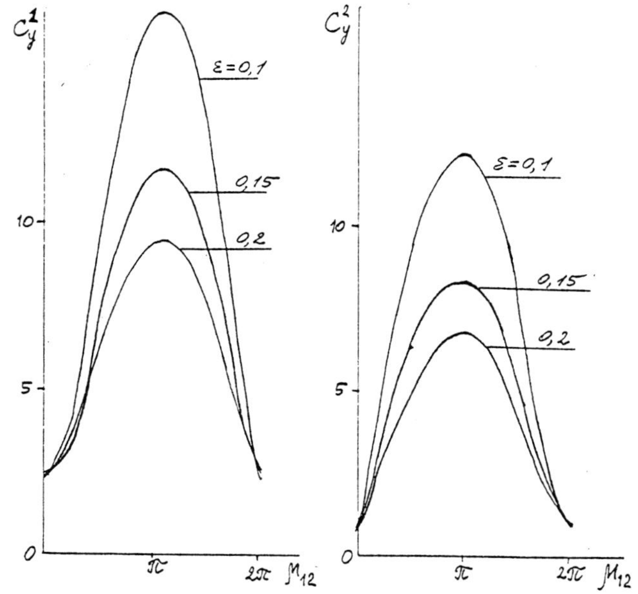

Fig.5. Dependencies of unsteady force amplitudes on inter-blade phase angle for different position of auxiliary profile. Bending oscillations.

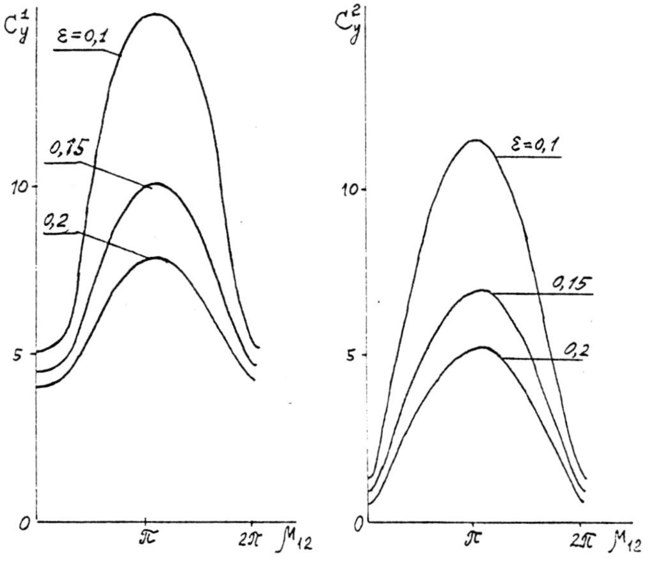

Fig.6. Dependencies of unsteady force amplitudes on inter-blade phase angle for different position of auxiliary profile. Torsional oscillations.

Fig.5 and Fig.6 present, respectively, for bending ($g^1 \equiv i,\ g^2 \equiv i \cdot e^{j\mu_{12}}$) and torsional oscillations ($g^1 = i(z_0^1(s) - z_{**}^1)$, $g^2 = i(z_0^2(s) - z_{**}^2)e^{j\mu_{12}}$, z_{**}^1, z_{**}^2 - co-ordinates of trailing edges for initial profiles) the dependencies of unsteady forces amplitudes on inter-blade phase angle μ_{12} for different positions of auxiliary profile with respect to main one ($\varepsilon = 0.1,\ 0.15,\ 0.2$). It is evident from the plots that amplitude maximum has been observed when oscillations of main and auxiliary profiles are out-of-phase and this maximum is increased significantly with decreasing of ε distance between them. It is happened because of change of inter-blade channel width between main and auxiliary profiles during the blade oscillations. Obviously that for the fixed ε it will be maximum for out-of-phase oscillations ($\mu_{12} \cong \pi$). And this relative change of inter-blade channel width is increased with decreasing of ε.

REFERENCES

1.Gorelov, D.N., Kurzin, V.B., Saren, V.E. (1971) *Aerodynamics of Cascades in Unsteady Flow.* Nauka, Novosibirsk.

2. Ryabchenko, V.P. (1974) Unsteady aerodynamic characteristics of arbitrary profiles oscillating in potential flow of incompressible liquid // *Izv. AN SSSR*, 1.

3. Ryabchenko, V.P., Saren, V.E. (1972) On calculation of aerodynamic characteristics of arbitrary form cascade profiles // *Izv. AN SSSR*, 2.

4. Hamming, R.V. (1971) *Numerical Methods.* Moscow (Russian translation).

UNSTEADY DISTURBANCES IN SWIRLING TURBOMACHINERY FLOWS

H.M. ATASSI AND V.V. GOLUBEV
Department of Aerospace and Mechanical Engineering
University of Notre Dame, Notre Dame, IN 46556, USA

Abstract

The propagation of small disturbances in an annular duct with a mean swirling flow is analyzed. It is shown that because of the centrifugal and Coriolis forces created by the mean swirl, the upstream disturbances can no longer be decomposed into distinct potential, rotational and entropic modes. A normal mode analysis shows that, for a rotational mean swirl, there are two distinct sets of discrete acoustic-vorticity eigenmodes. A pressure-dominated set whose eigenmodes propagate at near sonic speed and carry only small amount of vorticity, and a vorticity-dominated set of modes with only a small pressure associated with them and which accumulates with increasing density at the edges of a critical layer. The normal modes do not form a *complete* set and must be complemented with initial-value type solutions. A generalized definition for incident rotational waves (gusts) is proposed which accounts for both the eigenmodes and the initial-value solutions.

1. Introduction

A major source of noise and forced vibration in a fan engine is due to the interaction of flow disturbances behind the rotor with the downstream guide vanes. Analytical and numerical modeling of impinging unsteady flow disturbances invariably considers such disturbances to be imposed on an upstream uniform mean flow. It is then possible to split the unsteady upstream disturbances into distinct potential, entropic and vortical modes obeying distinct equations, and to study their propagation and interaction with downstream blading [1]. For periodic disturbances, a uniform mean flow upstream uncouples the unsteady disturbances from mean flow effects

T.H. Fransson (ed.), Unsteady Aerodynamics and Aeroelasticity of Turbomachines, 131-146.

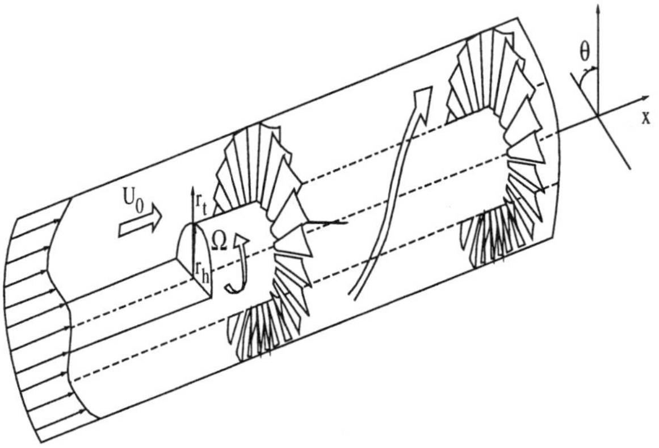

Figure 1. Unsteady swirling flow in a turbomachine.

and leads to steady-state oscillatory motions which can then be expanded in terms of fundamental eigensolutions. As an example, an impinging vortical disturbance can be represented in terms of harmonic disturbances and one needs to determine only the interaction of these harmonic disturbances with the downstream blading. A review of these methods and application to turbomachine flows is given in [2].

In turbomachines, the mean flow leaving the rotor stage has a significant swirl (the swirl angle could be as small as 12° and may reach 45°) which strongly affects the aerodynamic and acoustic phenomena associated with rotor/stator systems. Thus wake nonuniformities and secondary flows downstream of a rotor stage evolve in a mean swirling motion (Figure 1). The swirl produces centrifugal and Coriolis forces which deflect the fluid motion and couple the various modes [3]. The magnitude and phase of the unsteady disturbances is therefore constantly changing as they evolve downstream.

The present paper examines the effects of mean flow swirl on the evolution and propagation of unsteady disturbances. If the mean swirl is potential, i.e., it can be represented by a free vortex of the form Γ/r, where Γ and r represents the circulation strength and the radius, respectively. The effect of the potential swirl is to stretch the vorticity as it propagates downstream and to modify the acoustic impedence of the flow. It can be readily shown that the vorticity will increase linearly with $(M_S/M_0)x$, where M_0 and M_S represent the mean axial and swirl Mach numbers, respectively and x is the axial distance. The swirl also changes the system impedence and modifies the cut-on conditions for the acoustic modes [4].

For the general case of a vortical swirl, the Coriolis force per unit mass is proportional to ζu^*, where ζ is the local mean vorticity and u^*, is a representative of the unsteady disturbance velocity. The Coriolis force deflects the unsteady velocity and creates a force imbalance resulting in a pressure gradient in the flow. As a result, it is not possible to have a purely convected disturbance as in the case of uniform or potential mean flows. Oscillatory disturbances nearly convected by the the mean flow will be most affected by the Coriolis force and thus may be amplified leading to instability. The time scale associated with this amplification is of the order of $1/\zeta$. Note that the time scale associated with the blade passing frequency is of the order of $1/(\Omega B_r)$, where Ω is the angular frequency of the rotor and B_r is the number of the rotor blades. Since $\zeta \approx 2\Omega$, the time scale for the instability to occur in a typical fan or compressor is large compared with the oscillatory period of the disturbances. The vortical swirl also changes the system acoustic characteristics. Thus, there are no potential acoustic waves but rather near sonic propagating waves with a small vorticity [5].

These results suggest that it is important to determine the fundamental eigensolutions of waves evolving and propagating in a mean swirling flow. However, the eigensolutions do not form a *complete* set which suggests that an initial-value analysis must be carried out to represent and determine the evolution of all possible disturbances.

In section 1, the basic mathematical formulation is presented for both potential and vortical swirls. A velocity splitting tailored after Goldstein's splitting [6] is used to describe the basic modes of the disturbances. In section 2, results for the eigensolutions are presented from [4] and [5]. In section 3, the initial-value analysis is presented paving the way to the annular cascade analysis given in [7]. The conclusions show the importance of the vortical swirl to the aerodynamic and aeroacoustics of annular cascades.

2. Mathematical Formulation

We consider the propagation of small-amplitude unsteady disturbances in an annular inviscid swirling flow (Figure 2). We assume that the total velocity and pressure fields can be expanded as

$$\vec{V}(\vec{x},t) = \vec{U}(\vec{x}) + \vec{u}(\vec{x},t), \tag{1}$$

$$p(\vec{x},t) = p_0(\vec{x}) + p'(\vec{x},t), \tag{2}$$

and a similar expansion for the other flow quantities. $\vec{U}(\vec{x})$ and p_0 are the mean flow velocity and pressure and $\vec{u}(\vec{x},t)$ and p' represent the disturbance velocity and pressure such that $|\vec{u}(\vec{x},t)| \ll |\vec{U}(\vec{x})$ and $|p'| \ll |p_0|$. Since we intend to determine the essential effects of a swirling mean flow on the unsteady velocity $\vec{u}(\vec{x},t)$, we consider, for simplicity, the mean flow to be composed of an axial velocity U_0 with rigid body rotation and a free axial

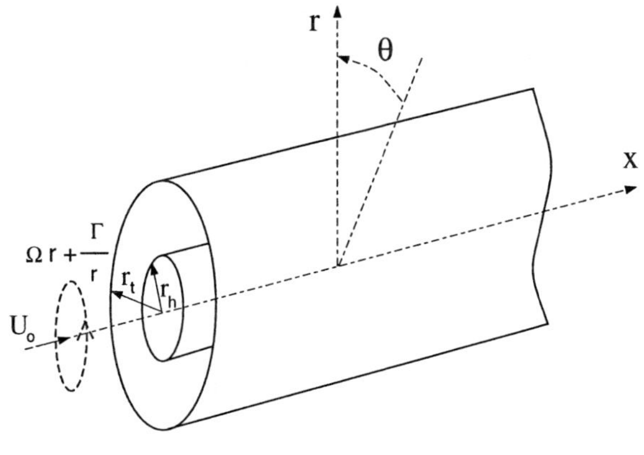

Figure 2. Swirling mean flow in an annular duct.

vortex. Thus, we can write

$$\vec{U}(\vec{x}) = U_0\hat{e}_x + (\Omega r + \frac{\Gamma}{r})\hat{e}_\theta, \tag{3}$$

where Ω and Γ are two constants representing the angular velocity of the solid body rotation and the axial vortex circulation, respectively; r is the radial distance, $\hat{e}_x$ and $\hat{e}_\theta$ represent unit vectors in the axial and circumferential directions, respectively. This mean flow has been shown to represent turbomachinery flows [3] and has the advantage of depending solely on two independent parameters, Ω and Γ. Note that if the stagnation enthalpy of the mean flow is constant, the assumption of a rigid body swirl implies radial variation of the mean flow temperature.

In order to compare with the classical results and to gain a physical insight in the coupling process due to the mean swirl, we now decompose the disturbance velocity into potential and rotational parts,

$$\vec{u} = \nabla\phi + \vec{u}^{(R)}. \tag{4}$$

Without loss of generality, we assume that

$$p' = -\rho_0\frac{D_0\phi}{Dt}, \tag{5}$$

where ρ_0 is the mean flow density. As a result, the linearized Euler equations are reduced to the following two *coupled* equations [5]:

$$\frac{D_0\vec{u}^{(R)}}{Dt} + (\vec{u}^{(R)}\cdot\nabla)\vec{U} = -\vec{\zeta}\times\nabla\phi, \tag{6}$$

$$\mathcal{L}\phi = \frac{1}{\rho_0}\nabla\cdot(\rho_0\vec{u}^{(R)}), \tag{7}$$

where the mean flow vorticity is $\vec{\zeta} = 2\Omega\hat{e}_x$, D_0/Dt is the convected derivative

$$\frac{D_0}{Dt} = \frac{\partial}{\partial t} + U_0\frac{\partial}{\partial x} + (\frac{\Gamma}{r^2} + \Omega)\frac{\partial}{\partial \theta}, \tag{8}$$

and $\mathcal{L}$ is the convected wave operator defined as

$$\mathcal{L} \equiv \frac{D_0}{Dt}\frac{1}{c_0^2}\frac{D_0}{Dt} - \frac{1}{\rho_0}\nabla \cdot (\rho_0 \nabla), \tag{9}$$

where c_0 is the mean flow speed of sound. Assuming rigid duct walls, the boundary conditions at the hub and tip radii r_h and r_t are

$$\frac{\partial \phi}{\partial r} + u_r^{(R)} = 0. \tag{10}$$

When the upstream mean flow and density are uniform, the incident unsteady vortical and potential velocities are uncoupled and independently propagate downstream. The vortical velocity is solenoidal and purely convected by the uniform mean flow. When the upstream mean flow is *not* uniform, then equations (6) and (7) show that an incident vortical disturbance may *not* be solenoidal and will then induce an unsteady "hydrodynamic" pressure field. If the mean flow is potential, i.e., $\vec{\zeta} = 0$, an incident acoustic wave will not create a vortical disturbance. However, if the mean flow is rotational, i.e., $\vec{\zeta} \neq 0$, then there is coupling between both types of disturbances. This coupling is caused by the local Coriolis forces represented by the right-hand-side of equation (6). Incident acoustic waves will then produce vorticity and vorticity will induce pressure. Thus, there are no purely vortical or purely acoustic modes. The advantage of the velocity splitting (4) will be justified later as we show that initially the coupling between the two modes is rather weak.

3. Normal Mode Analysis

A first and important step toward understanding the physics of disturbances propagating in swirling flows is to carry out a normal-mode analysis to obtain the spectrum of propagating and evanescent acoustic-vorticity modes of equations (6-7). To this end, we assume the following Fourier expansion:

$$\{u_x^{(R)}, u_r^{(R)}, u_\theta^{(R)}, \phi\}(x, r, \theta; t) =$$

$$\int_{-\infty}^{\infty} \sum_{m=-\infty}^{\infty} \sum_{n=1}^{\infty} \{X_{mn}(r), R_{mn}(r), T_{mn}(r), \varphi_{mn}(r)\} e^{i(k_{mn}x + m\theta - \omega t)} d\omega, \tag{11}$$

where m and n are integer modal numbers characterizing the circumferential and radial eigenmodes, respectively. Due to the linear character of the problem, each Fourier component of the solution vector can be considered separately. Using (5), the pressure modes are given by

$$p'_{mn}(r) = -i\Lambda_{mn}\rho_0(r)\varphi_{mn}(r). \tag{12}$$

where we have introduced the non-constant eigenvalue of the convected operator D_0/Dt

$$\Lambda_{mn} = k_{mn}U_0 + m(\Omega + \frac{\Gamma}{r^2}) - \omega \tag{13}$$

The coupled system of equations (6-7) can now be represented in terms of normal modes as

$$\begin{aligned}
\Lambda_{mn}X_{mn}(r) &= 0,\\
\Lambda_{mn}R_{mn}(r) + 2iT_{mn}(r)(\Omega + \frac{\Gamma}{r^2}) - 2\Omega\frac{m}{r}\varphi_{mn}(r) &= 0,\\
i\Lambda_{mn}T_{mn}(r) + 2\Omega R_{mn}(r) + 2\Omega\frac{\partial\varphi_{mn}(r)}{\partial r} &= 0, \qquad (14)\\
\frac{\partial^2\varphi_{mn}(r)}{\partial r^2} + (\frac{1}{r} + \frac{d\ln\rho_0}{dr})\frac{\partial\varphi_{mn}(r)}{\partial r} + (\frac{\Lambda_{mn}^2}{c_0^2} - k_{mn}^2 - \frac{m^2}{r^2})\varphi_{mn}(r) &\\
+(\frac{1}{r} + \frac{d\ln\rho_0}{dr})R_{mn}(r) + \frac{\partial R_{mn}(r)}{\partial r} + \frac{im}{r}T_{mn}(r) + ik_{mn}X_{mn}(r) &= 0,
\end{aligned}$$

It is also convenient to introduce the modified frequency, $\tilde{\omega}_m = \omega - m(\Omega + \Gamma/r^2)$. We assume a duct with rigid walls, and therefore the boundary conditions for (14) are

$$R_{mn}(r) + \frac{\partial\varphi_{mn}(r)}{\partial r} = 0 \tag{15}$$

at hub and tip radii $r = r_h$ and $r = r_t$. We investigate the properties of numerical solutions of the eigenvalue problem (14-15) for different kinds of mean swirl. The eigenvalues k_{mn} represent allowed axial wave numbers of the acoustic-vorticity modes.

3.1. NON-CONVECTED MODES

If the normal mode is *not* convected, i.e., $\Lambda_{mn} \neq 0$, then the first equation of system (14) shows that the axial convected mode $X_{mn} = 0$. Thus, *non-convected normal modes represent only the asymptotic coupling between the pressure and vorticity modes.* We now introduce the transformation

$$\psi_{mn}(r) = \varphi_{mn}(r)[r^2\Lambda_{mn}]^{\frac{m\Omega}{-\omega+M_0k+m\Omega}}, \tag{16}$$

and then eliminate $R_{mn}(r)$ and $T_{mn}(r)$ from equations (14). The boundary-value problem reduces to a single equation:

$$\psi''_{mn}\{\frac{\Lambda^2_{mn}}{D(\Lambda_{mn})}\} + \psi'_{mn}\{(\frac{1}{r} + \frac{d(\ln \rho_0)}{dr})\frac{\Lambda^2_{mn}}{D(\Lambda_{mn})} + \frac{4\Lambda_{mn}\Omega[m\Lambda^2_{mn}r^2 + 2\Lambda_{mn}\Gamma - 4mr^2(\Omega + \Gamma/r^2)^2]}{r^3 D^2(\Lambda_{mn})}\} + \psi_{mn}\{\frac{m^2}{r^2} + k^2_{mn} - \frac{\Lambda^2_{mn}}{c_0^2}\} = 0 \tag{17}$$

and boundary conditions

$$\frac{\partial \psi_{mn}(r)}{\partial r} = 0 \tag{18}$$

at $r = r_h$ and $r = r_t$, where $D(\Lambda_{mn}) = 4\Omega(\Omega + \Gamma/r^2) - \Lambda^2_{mn}$.

Two important results can be inferred from (17). First, the eigenvalue k_{mn} appears in a nonlinear way in all the coefficients of equation (17), and as a result, the boundary-value problem (17-18) is not a Sturm-Liouville problem (not self-adjoint). Second, for a rotational mean flow, i.e., $\Omega \neq 0$, the solution of (17) and (18) may have a singular critical layer corresponding to $\Lambda_{mn} = 0$. Asymptotic and numerical solutions to this eigenvalue problem for potential and rotational swirls were given by Golubev and Atassi [4] and [5]. Their numerical scheme used a pseudo-spectral method based on Chebyshev expansions with exponential convergence in order to resolve the large gradients near the critical layer. This numerical scheme is complemented by a multiple scale analysis for the eigenmodes near the critical layer.

For a potential swirl, the mean flow is represented as a combination of a uniform flow and a free vortex swirl, $\vec{U} = U_0\hat{e}_x + (\Gamma/r)\hat{e}_\theta$. Equations (14) show that all rotational terms, X_{mn}, R_{mn} and T_{mn} must vanish and the non-convective modes are potential. The eigenvalue problem reduces to that of discrete sound pressure modes:

$$\mathcal{L}\phi = 0. \tag{19}$$

and the eigenvalue problem (17) and(18), reduces to

$$\{\frac{d^2}{dr^2} + (\frac{1}{r} + \frac{d(\ln \rho_0)}{dr})\frac{d}{dr} + (\frac{\Lambda^2_{mn}}{c_0^2} - k^2_{mn} - \frac{m^2}{r^2})\}\varphi_{mn}(r) = 0, \tag{20}$$

with $\partial\varphi_{mn}(r)/\partial r = 0$ at $r = r_h$ and $r = r_t$. Note that this boundary-value problem does not have a critical layer as Λ_{mn} is no longer a singular parameter of the problem. However, k_{mn} still appears nonlinearly in (20) as a result of refraction of sound waves by the mean flow.

3.2. CONVECTED MODES

For a potential mean swirl, there is a continuous spectrum of convected eigenvalues, with the acting centrifugal forces giving rise to an algebraic growth of the corresponding vorticity modes as shown later in section 4. On the other hand, for a vortical mean swirl Coriolis forces deflect a perturbed fluid particle from its convected path, thus prohibiting a purely convected solution. This produces a mutual coupling of potential and vortical disturbances, which leads to a *critical layer* in the region of nearly-convected eigenvalues. However, this coupling does not occur for the axial component of the disturbance which may be purely convected.

3.3. NUMERICAL RESULTS

We now present numerical results of the normal mode analysis for different types of swirl and compare these results with those of a uniform flow in a duct. The annular duct has hub and tip radii $r_h = 4$ and $r_t = 6$, respectively. A non-dimensional frequency $\overline{\omega} = \omega r_m/\overline{c}_0$ is introduced, where r_m is the mean radius of the annulus and $\overline{c}_0$ is the stagnation speed of sound. The results are presented for $\overline{\omega} = 8$, and an axial flow Mach number $M_0 = 0.3$. In what follows we also introduce the swirl Mach numbers, $M_\Gamma = \Gamma/r_m\overline{c}_0$ and $M_\Omega = \Omega r_m/\overline{c}_0$. Figure 3 shows four plots of the real and imaginary values of the eigenvalues k_{mn} for four different swirl velocities. On every plot, results are shown for three circumferential modal numbers $m = -5, 0, 5$, and 15 radial modal numbers $n = 1, ..., 15$. Real eigenvalues correspond to propagating modes, while complex eigenvalues represent evanescent and amplifying modes. The amplifying modes in the acoustic region are discarded as non-physical. The evanescent modes are not physically significant but play an important role in the formulation of numerical inlet/outlet conditions [8].

Figure 3(a) shows the case of a uniform axial flow, $M_0 = 0.3$, with no swirl. There are four axisymmetric acoustic modes ($m = 0$), two propagating upstream and two downstream, and two non-axisymmetric acoustic modes for $m = \pm 5$. Note that the axial component of the mean flow produces a Doppler shift in $Re(k_{mn})$, so that all evanescent modes will propagate only upstream. In the frame of reference moving with the local mean velocity, such modes do not propagate. We also note that there is a single mode convected at the mean flow velocity.

Figure 3(b) shows the case of a potential swirling flow with $M_0 = 0.3$ and $M_\Gamma = 0.3$. For $m = 0$, again there are four acoustic modes, two propagating upstream and two downstream. For non-axisymmetric modes ($m \neq 0$), because of the Doppler effect due to the swirling component of the mean flow, there are two acoustic modes propagating in the upstream and down-

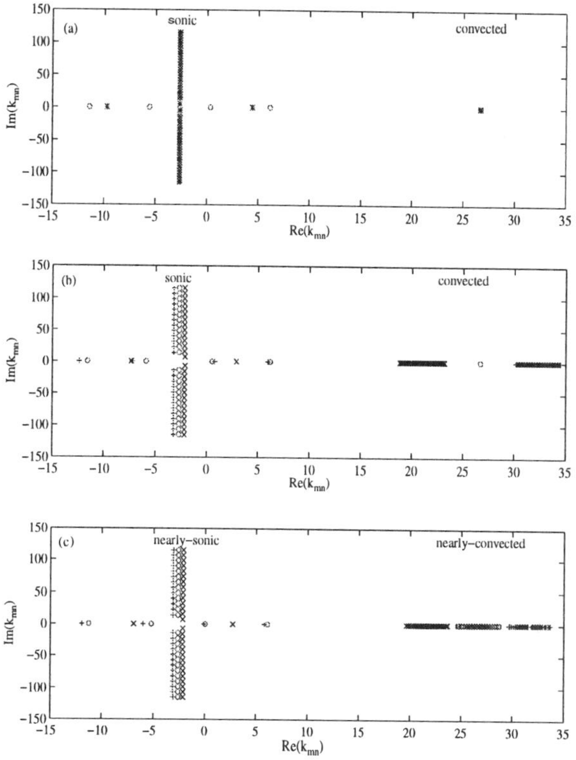

Figure 3. Eigenvalues $k_{mn}^{(u,d)}$ for $\tilde{\omega} = 8, m = -5('+'), m = 0('\text{o}'), m = 5('\text{x}'), n = 1, .., 15$. (a) : $M_0 = 0.3$, (b) : $M_0 = 0.3, M_\Gamma = 0.3$, (c): $M_0 = 0.3, M_\Omega = 0.3$.

stream directions for $m = 5$, but four such modes appear for $m = -5$. This results from the fact that for such modes the modified frequency $\tilde{\omega}_m$ is larger than ω, when m is negative. In addition to the *discrete* set of acoustic eigenvalues, there is one single convected mode for axisymmetric waves ($m = 0$), and a *continuous* spectrum of convected eigenvalues for the nonaxisymmetric modes ($m = \pm 5$).

Figure 3(c) shows the eigenvalues of (14) for a rigid body rotation swirl with $M_\Omega = 0.3$. Two *distinct* sets of eigenvalues are seen. A set of acoustic-vorticity eigenmodes, which propagate with phase speeds close to the speed of sound, is *pressure-dominated.* Modes corresponding to the second set are nearly-convected and *vorticity-dominated.* Note that as k_{mn} gets closer to the critical value k_c corresponding to purely convected modes, the eigenvalues form an accumulation pattern with increasing density near k_c. Figure 4(a) shows details of this accumulation pattern for the rigid body swirl of Figure 3(c). For the case of a general swirl with $M_0 = 0.3$, $M_\Omega = 0.2$, and $M_\Gamma = 0.1$, there is a range of critical values for k_{mn} corresponding to pure

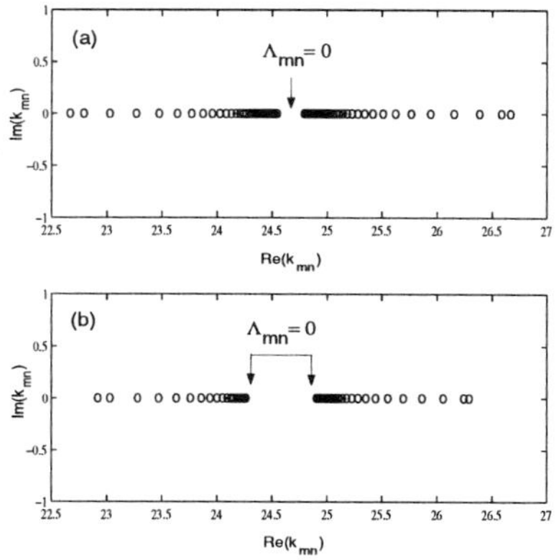

Figure 4. Nearly-convected eigenvalues for $\bar{\omega} = 8, m = 2$. (a) $M_0 = 0.3, M_\Omega = 0.3$; (b) $M_0 = 0.3, M_\Omega = 0.2, M_\Gamma = 0.1$.

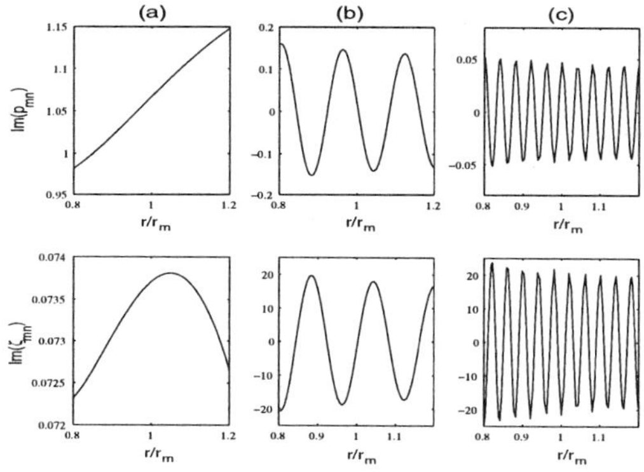

Figure 5. Pressure and axial vorticity eigenmodes for $M_0 = 0.3, M_\Omega = 0.3, \bar{\omega} = 8, m = 2$. Pressure-dominated eigenmodes: (a) $k_{mn} = 5.37$. Vorticity-dominated eigenmodes: (b) $k_{mn} = 23.64$, (c) $k_{mn} = 24.36$.

convection ($\Lambda_{mn} = 0$). Figure 4(b) shows how the eigenvalues accumulate near the right and left borders of this critical zone.

Figure 5 shows plots of the imaginary part of the pressure and the axial vorticity modes for $M_0 = 0.3$, $M_\Omega = 0.3$, $\overline{\omega} = 8$, and $m = 2$. For a pressure-dominated mode corresponding to $k_{mn} = 5.37$, Figure 5(a) shows that the pressure is of order one while the vorticity carried by this mode is of the order of 0.07. For a vorticity-dominated mode, Figures 5(b) shows that the value of the vorticity is large while that of the pressure is relatively small. As k_{mn} gets closer to the critical value k_c corresponding to purely convected modes, the modes exhibit oscillatory behavior with larger gradients and increasing frequency of oscillation (Figure 5(c)).

We now consider the case of a general swirl where $M_0 = 0.3$, $M_\Omega = 0.2$,

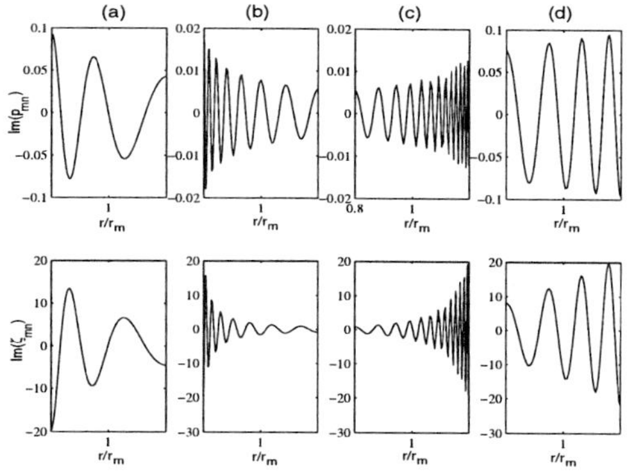

Figure 6. Nearly-convected pressure and axial vorticity eigenmodes for $M_0 = 0.3, M_\Omega = 0.2, M_\Gamma = 0.1, \bar{\omega} = 8, m = 2$. (a) $k_{mn} = 23.63$, (b) $k_{mn} = 24.19$, (c) $k_{mn} = 24.93$, (d) $k_{mn} = 25.35$.

$M_\Gamma = 0.1$, $\overline{\omega} = 8$, and $m = 2$. Figure 6 shows plots of the imaginary part of the pressure and the axial vorticity modes as k_{mn} approaches the critical layer from the left and then from the right. As we move along the left branch of eigenvalues toward the left edge of the critical layer, the corresponding eigenfunctions become highly oscillatory near the hub radius as shown in Figures 6(a) and (b). A strong radial variation of amplitudes of the modes can also be noted. The same pattern, but near the tip radius, is observed in Figures 6(c) and (d) for eigenmodes corresponding to the right branch of eigenvalues. Thus, as k_{mn} moves from the left to the right edge of the critical layer, the singularity travels from hub to tip.

4. Initial-Value Analysis

The normal mode solutions do not represent all possible solutions to the system of equations (6) and (7). For example, we have seen in 3.1 that for a non-convected mode the axial component $u_x^{(R)}$ or X_{mn} must vanish. In order to represent the most general type of a disturbance, we need to consider the response of equations (6) and (7) to imposed convected disturbances. This leads to an initial-value problem for the evolution of the disturbances. Without loss of generality we can consider a single harmonic component of the upstream disturbance

$$\vec{u}^{(R)}(x, r, \theta; t) = \vec{A}(\tilde{x}, r) e^{i[\alpha_r \tilde{x} + m\theta - \omega t]}, \tag{21}$$

where $\vec{A}(\tilde{x}, r)$ is an amplitude vector, $\tilde{x} = x - x_{in}$ and

$$\alpha_r = k_x - \frac{m}{r} \frac{M_s}{M_0} \tag{22}$$

where k_x is a constant, and M_s is the mean swirl Mach number. x_{in} is the location where the initial disturbance is measured. It is advantageous to consider first the case of a potential mean swirl.

4.1. POTENTIAL MEAN SWIRL

In this case the initial-value problem is significantly simplified because the vortical velocity governed by equation (6) is uncoupled from the pressure. Solutions to (6) of the form (21) can be readily obtained. For $t > 0$ and $\tilde{x} \geq 0$

$$\vec{A}(x,r) = \vec{a}(r) + \tilde{x}\,\vec{b}(r), \tag{23}$$

$$\alpha_r = k_x - m\frac{\Gamma}{U_0 r^2}, \tag{24}$$

where $\vec{a}(r)$ and $\vec{b}(r)$ are expressed in terms of the initial conditions:

$$\begin{aligned} \vec{a}(r) &= \{A_x(r), A_r(r), A_\theta(r)\}, \\ \vec{b}(r) &= \{0, \frac{2\Gamma}{U_0 r^2}A_\theta(r), 0\}, \end{aligned} \tag{25}$$

The wavenumber k_x is related to the frequency of ω by $k_x = \omega/U_0$. Note that A_r must vanish at the hub and tip. It is convenient to expand A_r in series of Hankel functions [7]. We then solve the initial-value problem defined by

$$\mathcal{L}\phi = \frac{1}{\rho_0}\nabla \cdot (\rho_0 \vec{u}^{(R)}), \tag{26}$$

with $p' = 0$ at $\tilde{x} = 0$ and a vanishing pressure far downstream.

The results of the numerical solutions are illustrated in Figure 7 for $M_0 = 0.3$, $M_\Gamma = 0.3$, $k_x = 26.7$ ($\omega = 8$), and $m = 5$. The initial velocity is assumed solenoidal ($\nabla \cdot \vec{u}^{(R)} = 0$) for comparison with a uniform axial flow. Figure 7 shows plots versus r at two axial locations, $\tilde{x} = 0$ and $\tilde{x} = 2r_m$, of the real parts of the vortical and induced potential velocities in dashed and dashed-dotted lines, respectively, and the magnitude of the total unsteady velocity in solid line. It also shows plots of the real part (dashed line) and the magnitude (solid line) of the unsteady pressure. It is interesting to note, in particular, the significant change of the total axial and circumferential velocities between the initial axial location $\tilde{x} = 0$ and the location $\tilde{x} = 2r_m$, since these two velocities determine the blade boundary condition for the gust response problem. Finally, we note that the magnitude of the pressure induced by the incident rotational velocity remains relatively small. Figure

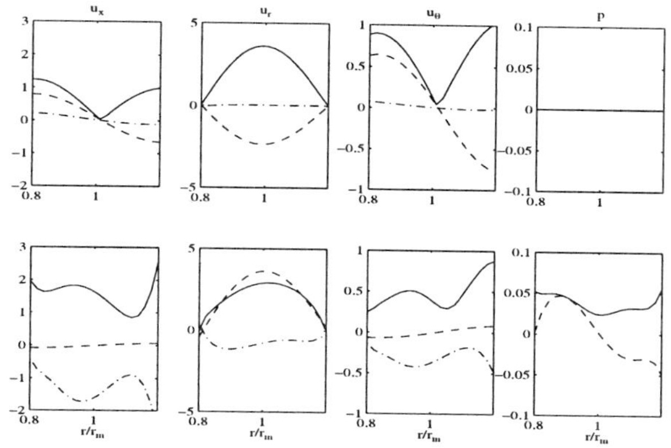

Figure 7. Radial profile of convected velocity and pressure perturbations at $\tilde{x} = 0$ (top plots) $\tilde{x} = 2$ (bottom plots). $M_0 = 0.3, M_\Gamma = 0.3, \bar{\omega} = 8, m = 5$.

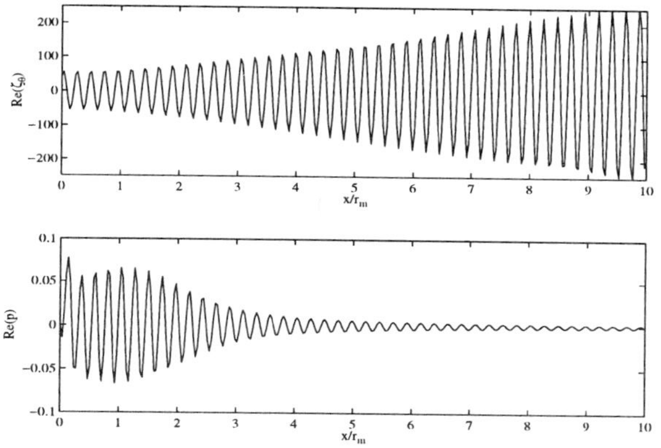

Figure 8. Evolution of convected tangential vorticity and pressure perturbations. $M_0 = 0.3, M_\Gamma = 0.3, \bar{\omega} = 8, m = 5, r/r_m = 0.9$.

8 shows the long term evolution of the vorticity and pressure. The vorticity increases linearly with the distance $\tilde{x}$, while the pressure vanishes. Note that this evolution is relatively slow and scales with r_m which would be large compared to a blade chord length.

4.2. GENERAL SWIRL

The homogeneous solution of equation (6) for $t > 0$ and $x \geq x_{in}$, gives the amplitude vector $\vec{A}_h(x, r)$ in terms of the initial gust conditions at $x = x_{in}$

$$\begin{aligned} A_{xh} &= A_x(r), \\ A_{rh} &= A_r(r) \cos \frac{\lambda}{U_0}\tilde{x} + A_\theta(r) \frac{\lambda}{2\Omega} \sin \frac{\lambda}{U_0}\tilde{x}, \end{aligned} \tag{27}$$

$$A_{\theta h} \quad = \quad A_\theta(r)\cos\frac{\lambda}{U_0}\tilde{x} - A_r(r)\frac{2\Omega}{\lambda}\sin\frac{\lambda}{U_0}\tilde{x}.$$

The gust axial phase is given by

$$\alpha_r = k_x - m\frac{\Omega + \Gamma/r^2}{U_0}, \tag{28}$$

and we also introduce the *Coriolis frequency* $\lambda(r)$,

$$\lambda(r) = 2\sqrt{\Omega(\Omega + \Gamma/r^2)}. \tag{29}$$

We recall from the results of the normal mode analysis that all the *nearly-convected* eigenvalues cluster in the region between the convected *critical layer*, $\Lambda_{mn} = 0$, and the boundary of $|\Lambda_{mn}| = \lambda(r)$ (the latter corresponds to $D(\Lambda_{mn}) = 0$). Thus, the radial and circumferential components of homogeneous solution (27) of the initial-value problem may tend asymptotically to the spectrum of nearly-convected modes whose eigenvalues accumulate at the critical layer. On the other hand, the axial component A_{xh} of (27) is certainly independent of the eigenmodes.

Note that the Coriolis frequency $\lambda(r)$ in (29) becomes purely imaginary if $\Omega(\Omega + \Gamma/r^2) < 0$, which corresponds to Rayleigh criterion for centrifugal instability in the flow. As expected, the homogeneous solution (27) amplifies exponentially in this case.

To obtain a complete solution to the system (6) and (7), we need to solve the initial-value problem

$$\frac{D_0\vec{u}_p^{(R)}}{Dt} + (\vec{u}_p^{(R)} \cdot \nabla)\vec{U} + \vec{\zeta} \times \nabla\phi \quad = \quad 0, \tag{30}$$

$$\mathcal{L}\phi - \frac{1}{\rho_0}\nabla \cdot (\rho_0\vec{u}_p^{(R)}) \quad = \quad \frac{1}{\rho_0}\nabla \cdot (\rho_0\vec{u}_h^{(R)}), \tag{31}$$

with $p' = 0$ at $x = x_{in}$ and a weak condition far downstream.

The results of the numerical solutions are illustrated in Figure 9 for $M_0 = 0.3$, $M_\Omega = 0.2$, $M_\Gamma = 0.1$, $k_x = 26.7$ ($\omega = 8$), and $m = 5$. The initial velocity is assumed solenoidal ($\nabla \cdot \vec{u}^{(R)} = 0$) for comparison with a uniform flow case. Figure 7 shows plots versus r, at two axial locations, $\tilde{x} = 0$ and $\tilde{x} = 2r_m$, of the real parts of the vortical and induced potential velocities in dashed and dashed-dotted lines, respectively, and the magnitude of the total unsteady velocity in solid line. It also shows plots of the real part (dashed line) and the magnitude (solid line) of the unsteady pressure. It is interesting to note, in particular, the evolution of the axial and circumferential velocities since these two velocities determine the boundary condition

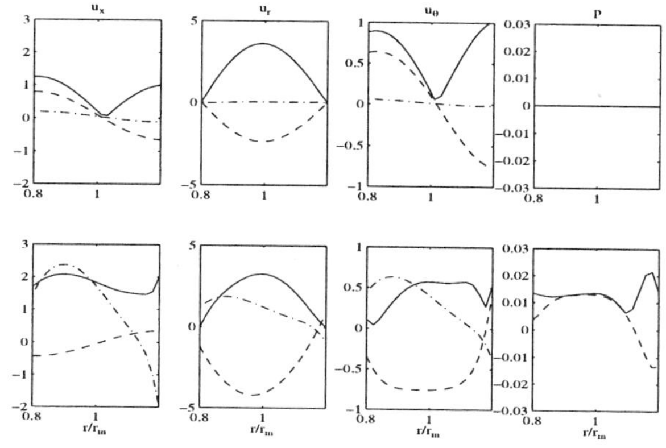

Figure 9. Radial profile of nearly-convected velocity and pressure perturbations at $\tilde{x} = 0$ (top plots) $\tilde{x} = 2$ (bottom plots). $M_0 = 0.3, M_\Omega = 0.2, M_\Gamma = 0.1, \bar{\omega} = 8, m = 5$.

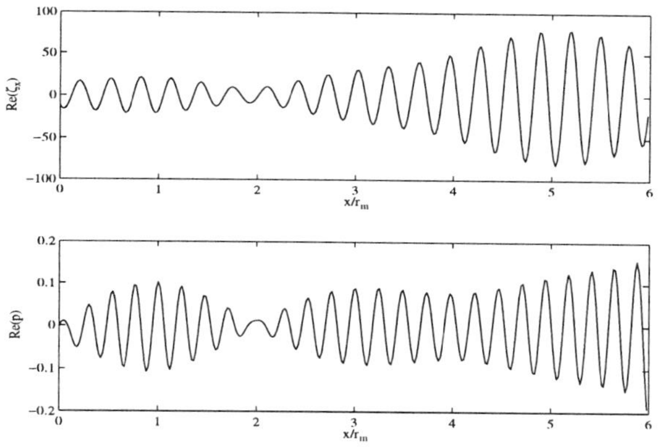

Figure 10. Evolution of nearly-convected tangential vorticity and pressure perturbations. $M_0 = 0.3, M_\Omega = 0.2, M_\Gamma = 0.1, \bar{\omega} = 8, m = 5, r/r_m = 0.9$.

along a blade for the gust response problem. Finally, we note that the magnitude of the pressure induced by the incident rotational velocity remains relatively small. Figure 10 shows the long term evolution of the vorticity and pressure. Both exhibit cyclical variations with the Coriolis frequency. Again we note that this dependence on the Coriolis frequency scales with the duct radius r_m.

5. Conclusions

The propagation of small disturbances in an annular duct with a mean swirling flow has been analyzed. It is shown that the centrifugal and Coriolis forces due to the mean swirl couple the pressure and vortical modes. The upstream disturbances can no longer be decomposed into distinct potential, rotational and entropic modes obeying distinct equations. The swirl also affects the duct acoustic impedence and modifies the acoustic cut-on conditions. The normal mode analysis shows that, for a rotational mean swirl, there are two distinct sets of discrete acoustic-vorticity eigenmodes. A pressure-dominated set whose eigenmodes propagate at near sonic speed and which carries only small amount of vorticity, and a vorticity-dominated set of modes with only a small pressure associated with them which accumulates with increasing density at the edges of a critical layer. The normal modes do not form a *complete* set and must be complemented with initial-value type solutions. A generalized definition for incident rotational waves (gusts) is proposed which accounts for both the eigenmodes and the initial-value solutions. Solutions to the gust interaction problem is given in [7].

Acknowledgements

The research was supported by the Office of Naval Research Grant No. N00014-92-J-1165, and monitored by Mr. L.Patrick Purtell, and NASA Lewis Research Center Grant No. NAG3-732, and monitored by Mr. Dennis L. Huff.

References

1. Atassi, H.M. (1994) Unsteady Aerodynamics of Vortical Flows: Early and Recent Developments, in K.-Y.Fung (ed.), *Aerodynamics and Aeroacoustics*, World Scientific, pp.121-172.
2. Verdon, J.M. (1993) Unsteady Aerodynamic Methods, in H.M. Atassi (ed.), *Unsteady Aerodynamics, Aeroacoustics and Aeroelasticity of Turbomachines and Propellers*, Springer-Verlag, pp.3-42.
3. Kerrebrock, J.L. (1977) Small Disturbances in Turbomachine Annuli with Swirl, *AIAA J.* **15**, pp.794-803.
4. Golubev, V.V and Atassi, H.M. (1996) Sound Propagation in an Annular Duct with Mean Potential Swirling Flow, *J. Sound and Vibration* **198**, pp.601-616. .
5. Golubev, V.V. and Atassi, H.M. (1997) Acoustic-Vorticity Waves in Swirling Flows, *J. Sound and Vibration* (to appear).
6. Goldstein, M.E. (1978) Unsteady Vortical and Entropic Distortions of Potential Flows Round Arbitrary Obstacles, *J. Fluid Mechanics* **89**, pp.433-468.
7. Golubev, V.V. and Atassi, H.M. (1997) Unsteady Forces on Annular Cascade Blades in Subsonic Flow with Swirl, *8th International Symposium on Unsteady Aerodynamics and Aeroelasticity of Turbomachines*, to appear in Proceedings.
8. Fang, J. and Atassi, H.M. (1993) Compressible Flows with Vortical Disturbances Around a Cascade of Loaded Airfoils, in H.M. Atassi (ed.), *Unsteady Aerodynamics, Aeroacoustics and Aeroelasticity of Turbomachines and Propellers*, Springer-Verlag, pp.149-176.

UNSTEADY FORCES ON ANNULAR CASCADE BLADES IN SUBSONIC FLOW WITH SWIRL

V.V. GOLUBEV AND H.M. ATASSI
Department of Aerospace and Mechanical Engineering
University of Notre Dame, Notre Dame, IN 46556, USA

Abstract

The effect of mean flow swirl on unsteady response of unloaded annular cascade subject to an incident convected disturbance is examined as a model for a compressor stage flow. The concept of a generalized gust is introduced which describes the evolution of an upstream nearly-convected perturbation to the point of its interaction with downstream cascade. The correspondence to classical gust definition is established followed by comparison of predicted unsteady forces with two-dimensional cascade results. In Region I corresponding to IGV/rotor interaction with a 26.6^0 swirling wheel flow, the effect of swirl on the amplitude of cascade unsteady response is limited to the change of the far field radiation conditions. In contrast, in Region II corresponding to rotor/stator interaction with a 45° mixed swirling flow, a remarkable amplification of the upwash component of the gust is noted. The predicted increment of cascade unsteady lift is obtained as a function of the generalized gust evolution distance.

1. Introduction

Unsteady fluctuating forces induced on annular cascades are often a result of blade interaction with flow non-uniformities which are nearly convected by the mean motion. In the classical treatment established for uniform compressible flows, such non-uniformities are represented by "frozen" convected vorticity modes which are completely uncoupled from pressure and entropy modes in the governing equations of motion.

The effect of non-uniform mean flows on the evolution of convected flow disturbances has been investigated mainly for *weakly* rotational flows around lifting airfoils and cascades [1]. The studies indicate a close link

T.H. Fransson (ed.), Unsteady Aerodynamics and Aeroelasticity of Turbomachines, 147-164.

of the evolution characteristics to local flow non-uniformities, and propose an effective approach [2], [3] which accounts for rapid distortion of the incident vortical disturbances by the locally non-uniform flow. Based on this approach, the problem of unsteady gust-cascade interaction has been successfully treated by the *strip* theory wherein a two-dimensional unsteady flow around an unrolled loaded cascade is considered [4], [5].

Such approach, however, is not justified for three-dimensional flows in turbomachines which usually exhibit a strong swirling mean motion. The amount of swirl velocity in a fan engine stage varies but can be of the order 15% – 45% to the mean axial velocity. In the particular example of compressor stage flow considered by Kerrebrock [6], two regions were identified (Figure 1). In Region I corresponding to IGV/rotor interaction, the steady flow passes through the guide vanes acting as an actuator disk and subsequently acquires a swirling free wheel motion, with tangential velocity $U_s = \Omega r$. In Region II corresponding to rotor/stator interaction, the wheel flow persists from Region I with the addition of a free vortex flow generated by rotor to provide a radially constant stagnation temperature rise, which gives $U_s = \Omega r + \Gamma/r$. The general form of the swirling mean flow velocity thus can be described as follows,

$$\mathbf{U}(\mathbf{x}) = U_0 \hat{e}_x + (\Omega r + \Gamma/r)\hat{e}_\theta \tag{1}$$

where Ω is the angular velocity and Γ is the strength of the free vortex.

The eigenmode description of unsteady swirling motion given by Kerrebrock [6] indicated coupling between all modes of unsteady motion as a result of unbalanced centrifugal and Coriolis forces. A more complete analysis of this coupling is given in this volume by Atassi and Golubev [7]. The present paper is mainly concerned with analyzing the interaction of unsteady disturbances with a blade row in a mean swirling flow as shown in Figure 1.

The analysis is based on the description given in [7] and [8] for a generalized gust. The incident gust is decomposed into vortical and potential components, which preserves the major advantage of the classical splitting theory related to the formulation of unsteady interaction problems. The concept of a generalized gust follows the eigenmode analysis of the coupled equations of unsteady swirling motion [9]. In this analysis, two distinct regions of eigenvalues are identified corresponding to pressure-dominated nearly-sonic and vorticity-dominated nearly-convected eigenmodes. A compact discrete spectrum of nearly-convected eigenvalues clusters with increasing density approaching a convected *critical layer*. However, as shown in [7], the normal mode analysis may not represent the most general solution to unsteady Euler equations. In addition, an initial-value analysis shows that such solutions can significantly modify the magnitude and phase

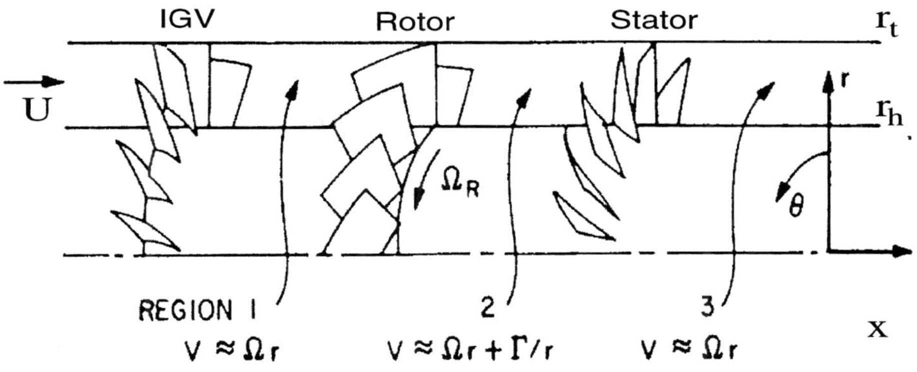

Figure 1. Schematic of an axial compressor flow (adapted from [6]).

of the gust. Moreover, those initial-value solutions evolve at a scale of the order of mean radius of an annulus, r_m, which is much larger than the distance separating two blade rows, while the normal modes can only represent the far downstream asymptotic behavior.

In Section 2, we present the mathematical model and introduce the basic formulation for the generalized gust. The boundary-value problem for the interaction with an anuular cascade is formulated in Section 3, with application to the rotor/stator interaction presented in Section 4.

2. Mathematical Model

We assume that the unsteady velocity component represents a small-amplitude unsteady perturbation of the steady swirling mean flowfield, so that the total flow velocity can be represented as follows:

$$\mathbf{V}(\mathbf{x}, t) = \mathbf{U}(\mathbf{x}) + \mathbf{u}(\mathbf{x}, t), \tag{2}$$

with the mean velocity described by (1). The following equations are derived [8] which govern the propagation and scattering of small-amplitude unsteady disturbances in a swirling mean flow:

$$\frac{D_0 \mathbf{u}^{(R)}}{Dt} + (\mathbf{u}^{(R)} \cdot \nabla)\mathbf{U} = -\zeta_0 \times \nabla\phi, \tag{3}$$

$$\frac{D_0}{Dt}\frac{1}{c_0^2}\frac{D_0\phi}{Dt} - \frac{1}{\rho_0}\nabla \cdot (\rho_0 \nabla)\phi = \frac{1}{\rho_0}\nabla \cdot (\rho_0 \mathbf{u}^{(R)}), \tag{4}$$

where the perturbation velocity field is decomposed in terms of the vortical and potential parts,

$$\mathbf{u} = \mathbf{u}^{(R)} + \nabla\phi, \tag{5}$$

and $\zeta_0 = 2\Omega\hat{e}_x$ is the mean flow vorticity. The unsteady pressure field is determined solely in terms of the potential function,

$$p' = -\rho_0 \frac{D_0\phi}{Dt}. \tag{6}$$

In equations (3-4), $D_0/Dt = \partial/\partial t + \mathbf{U}\cdot\nabla$ and ρ_0 and c_0 are the density and speed of sound of the mean flow, respectively.

Solution of the wave-structure interaction problem first requires the initial-value modal analysis of couped equations (3-4). Such analysis provides the evolution characteristics for an imposed upstream modal perturbation, and thus generalizes the classical concept of a convected *gust* established for uniform compressible flows.

2.1. GENERALIZED GUST SOLUTION

The generalized gust solution defines the evolution of imposed at a given upstream axial location $x = x_{in}$ velocity and pressure disturbances. For $t > 0$ and $x \geq x_{in}$, a modal Fourier component of the disturbance velocity field is sought in the form:

$$\mathbf{u}_g(x, r, \theta; t) = \mathbf{A}(x, r)e^{i[\alpha_r x + m\theta - \omega t]} \tag{7}$$

where $\mathbf{A}(x, r)$ is the generalized gust amplitude vector and $\alpha_r = [\omega - m(\Omega + \Gamma/r^2)]/U_0$ is the axial wavenumber of the convected phase of the gust which may depend on the radius of the duct.

In the description of the generalized gust solution, the cases of potential and vortical swirling flows are considered separately since they show different types of evolution for the travelling modal perturbation.

2.1.1. *Potential Swirl*

In the case of a potential swirl, $\Omega = 0$, the vortical gust velocity satisfies the homogeneous equation,

$$\frac{D_0\mathbf{u}_g^{(R)}}{Dt} + (\mathbf{u}_g^{(R)}\cdot\nabla)\mathbf{U} = 0, \tag{8}$$

which suggests that for $t > 0$ and $x \geq x_{in}$, the amplitude vector $\mathbf{A}(x, r)$ is of the form,

$$\mathbf{A}(x, r) = \mathbf{a}(r) + \tilde{x}\,\mathbf{b}(r), \tag{9}$$

$\tilde{x} = x - x_{in}$, where the amplitudes $\mathbf{a}(r)$ and $\mathbf{b}(r)$ can be expressed in terms of the initial conditions,

$$\begin{aligned} \mathbf{a}(r) &= \{A_x(r), A_r(r), A_\theta(r)\}, \\ \mathbf{b}(r) &= \{0, \frac{2\Gamma}{U_0 r^2} A_\theta(r), 0\}. \end{aligned} \tag{10}$$

This implies that the centrifugal forces will induce an additional radial perturbation of the vortical gust velocity which will linearly increase as the disturbance propagates downstream. This further suggests that a balance should be created by the corresponding radial component of the potential gust velocity governed by the non-homogeneous convective wave equation (4). The far field asymptotic expansion of (4) also confirms this assumption. The convective wave equation is then solved in the frequency domain with wall conditions,

$$\partial\phi_g/\partial r + u_{gr}^{(R)} = 0, \tag{11}$$

imposed at the hub and tip annulus radii, $r = r_h$ and $r = r_t$, respectively. The total generalized gust solution includes both vortical and potential parts, $\mathbf{u}_g = \mathbf{u}_g^{(R)} + \nabla\phi_g$. Note that in view of (6), the potential part determines the pressure carried by the gust.

2.1.2. *Vortical Swirl*

In the general case of (1), the coupling in equations (3-4) induced by local Coriolis forces suggests the following decomposition of the vortical component of the generalized gust velocity,

$$\mathbf{u}_g^{(R)} = \mathbf{u}_{gh}^{(R)} + \mathbf{u}_{gp}^{(R)}. \tag{12}$$

The nearly-convected part of the vortical gust velocity $\mathbf{u}_{gh}^{(R)}$ satisfies the equation,

$$\frac{D_0 \mathbf{u}_{gh}^{(R)}}{Dt} + (\mathbf{u}_{gh}^{(R)} \cdot \nabla)\mathbf{U} = 0, \tag{13}$$

which solution for $t > 0$ and $x \geq x_{in}$ gives the amplitude vector $\mathbf{A}_h(x, r)$ in terms of the initial conditions at $x = x_{in}$:

$$\begin{aligned} A_{xh} &= A_x(r), \\ A_{rh} &= A_r(r) \cos\frac{\lambda}{U_0}\tilde{x} + A_\theta(r)\frac{\lambda}{2\Omega} \sin\frac{\lambda}{U_0}\tilde{x}, \\ A_{\theta h} &= A_\theta(r) \cos\frac{\lambda}{U_0}\tilde{x} - A_r(r)\frac{2\Omega}{\lambda} \sin\frac{\lambda}{U_0}\tilde{x}, \end{aligned} \tag{14}$$

where the frequency $\lambda(r) = 2\sqrt{\Omega(\Omega + \Gamma/r^2)}$. Note that $\lambda(r)$ becomes purely imaginary if $\Omega(\Omega + \Gamma/r^2) < 0$, which corresponds to Rayleigh criterion for centrifugal instability in the flow. As a result, solution (14) would amplify exponentially, and in the further analysis this case will be excluded from consideration.

Note that in the eigenmode analysis of unsteady swirling motion [9], the radial and circumferential components of solution (14) correspond to the nearly-convected spectrum which clusters in the region surrounding the convected critical layer. The behavior of the eigensolution is singular at the critical layer, hence indicating an instability development for the corresponding initial-value problem. By construction, the initial-value solution is initiated by $\mathbf{u}_{gh}^{(R)}$ and governed by coupled equations (3-4) which are solved simultaneously by iterations, with the wall conditions (11).

2.1.3. *Initial Gust Conditions*

The initial conditions for the generalized gust solution of equations (3-4) at the point of origin (or measurement) should include the hydrodynamic pressure distribution and the amplitude vector of nearly-convected gust velocity. The gust pressure evolution appears to be rather insensitive to its initial distribution, and for simplicity a pressure-less imposed gust is assumed in numerical examples below. However, the imposed pressure amplitude does affect the potential component of the initial gust velocity distribution and thus should be accounted for in comparison with experimental data. In the present setup of the numerical code, the input gust velocity is vortical and divergence-free (which is the case in the limit of a uniform mean flow), and imposed in terms of the amplitude vector $\mathbf{A}(x_{in}, r)$ in (10) and (14). The potential component develops in the calculations, and the initial input can be iteratively adjusted to achieve the desired gust velocity distribution.

To represent the amplitude vector $\mathbf{A}(x_{in}, r)$, one may choose any complete set of radial basis functions. To provide a continuous limit to the classical vortical gust imposed on a uniform flow, and thus facilitate the comparison with linear cascade results, the following expansion in terms of Hankel functions is considered:

$$\begin{aligned}
A_x(x_{in}, r) &= A_{xn}[H_0^{(1)}(\mu_n r) + \kappa H_0^{(2)}(\mu_n r)]. \\
A_r(x_{in}, r) &= A_{rn}[H_1^{(1)}(\mu_n r) + \kappa H_1^{(2)}(\mu_n r)]. \\
A_\theta(x_{in}, r) &= A_{\theta n}[H_0^{(1)}(\mu_n r) + \kappa H_0^{(2)}(\mu_n r)] r / r_m,
\end{aligned}$$

where A_{xn}, A_{rn} and $A_{\theta n}$ are the imposed constant modal amplitudes which are assumed coupled by the divergence-free condition,

$$\frac{i\omega}{U_0} A_{xn} + \mu_n A_{rn} + \frac{im}{r_m} A_{\theta n} = 0, \tag{15}$$

r_m is the mean radius of the annulus, $\kappa = -H_1^{(1)}(\mu_n r_h)/H_1^{(2)}(\mu_n r_h)$, and the values of μ_n for radial modal numbers n are determined from the wall condition, $u_{gr}^{(R)} = 0$. Note that for $r \gg 1$ taken for numerical comparison with two-dimensional linear cascade model below, the radial amplitude, for instance, can be approximated using expansions of Hankel functions, as follows:

$$A_r(x_{in}, r) \simeq A_{rn}(\frac{8}{\pi\mu_n r})^{1/2} \sin\mu_n(r - r_h)[\sin(\mu_n r_h + \frac{\pi}{4}) + i\sin(\mu_n r_h - \frac{\pi}{4})], \quad (16)$$

which reduces to the classical sinusoidal gust representation given below. For the present numerical calculations, $\mu_1 \simeq \pi/(r_t - r_h)$ will only be considered.

3. Generalized Gust Problem

The generalized gust problem is formulated for an incident nearly-convected wave evolving from an upstream location $x = x_{in}$ and interacting with an annular cascade of blades. This interaction produces a near-field unsteady aerodynamic response coupled with far-field hydrodynamic and acoustic pressure fields.

The boundary-value problem is governed by equations (3-4). Without loss of generality, a single harmonic component of upstream unsteady flow excitation is considered, with a given frequency ω. Due to the linearity of the problem, the sound radiated from blades will be at the same frequency.

By construction, the total solution of the boundary-value problem is equal to the sum of the generalized gust solution $\mathbf{Z}_g = \{\mathbf{u}^{(R)}, \phi\}_g$, and a scattered solution $\mathbf{Z}_s = \{\mathbf{u}^{(R)}, \phi\}_s$ which should have an outgoing-wave behavior far upstream and downstream from the blade row. Both parts are coupled along the blade surface by the impermeability condition.

The gust-cascade interaction will produce coupled scattered vorticity and pressure modes described by $\mathbf{u}_s^{(R)}$ and ϕ_s satisfying equations (3-4). However, this coupling is weak [9], and the scattered pressure field induced by source term in (4) which represents the effect of scattered vorticity distortion by non-uniform mean flow, is not significant. Hence, for unloaded blades with a good approximation, the cascade unsteady pressure response is governed by a boundary-value problem for the *homogeneous* convective wave equation,

$$\frac{D_0}{Dt}\frac{1}{c_0^2}\frac{D_0}{Dt}\phi_s - \frac{1}{\rho_0}\nabla \cdot (\rho_0\nabla)\phi_s = 0. \quad (17)$$

This equation is solved numerically with boundary conditions of flow impermeability across the blades and continuity of perturbation pressure and normal velocity across the wakes. The unsteady response to a single harmonic perturbation of the form (7) is obtained in frequency domain by taking

$$\phi_s(x,r,\theta;t) = \hat{\phi}_s(x,r,\theta)e^{-i\omega t}. \tag{18}$$

Since the approach is linear, it is advantageous to employ the cascade quasiperiodicity condition,

$$\hat{\phi}_s(x,r,\theta) = \hat{\phi}_s(x,r,\theta+2\pi/N_b)e^{-i\sigma}, \tag{19}$$

where $\sigma = 2\pi m^*/N_b$ is the interblade phase angle, and m^* is a multiple integer of the gust circumferential modal number, m (N_b is the number of blades in a row). This condition allows to consider a single interblade passage domain in numerical computations.

Far away upstream and downstream from the blade row, an asymptotic behavior of scattered unsteady disturbances is governed by the normal mode analysis of governing equations [9]. The *exact* inflow/outflow conditions assume a normal mode representation of outgoing unsteady pressure waves, and set to zero all the modes entering the computational domain, to provide a non-reflective solution. As a result of surface coupling between incident and scattered acoustic-vorticity waves, the far field expansion of the scattered solution contains, in general, nearly-sonic and nearly-convected pressure modes propagating away from the blade row. The nearly-sonic modes are analogous to acoustic modes in a uniform flow and dominate the cascade pressure response in the far field. The nearly-convected modes generally possess a small pressure component and may be neglected in formulation of outflow conditions. Numerical implementation of these conditions employs a pseudospectral technique for the normal mode analysis of governing equations.

Note that the physical domain aligned with swirling streamlines of the mean flow will turn as it extends downstream. Numerical solution of the boundary-value problem is thus obtained in the transformed coordinate system $\{\zeta = x, \xi = r, \vartheta = \theta - (\Omega + \Gamma/r^2)x/U_0\}$. The computational domain comprises a single-passage annular sector with two adjacent blades centered between the specified x-locations upstream and downstream.

4. Application to Compressor Stage Flow

In this section, the generalized gust problem formulation is applied to model the unsteady flow interactions in Regions I and II of a fan or a compressor stage described in Figure 1. A schematic view from above of the IGV, rotor and stator rows is shown in Figure 2 at a radius r of the turbomachinery annulus. The rotor is rotating with an angular velocity Ω_R so that the linear

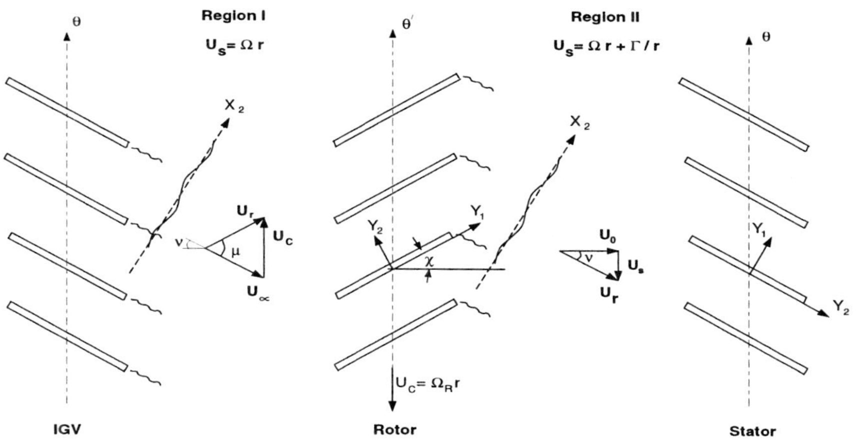

Figure 2. Unrolled compressor stage cascades.

cascade moves with velocity $U_c = \Omega_R r$ between two infinite lines connecting leading and trailing edges of the blades.

In both Regions, a wake-induced gust velocity field $\mathbf{u}_{g\infty}$ is viewed as stationary in the upstream cascade frame of reference, and periodic in the angular direction. Hence, $\mathbf{u}_{g\infty} \sim \exp(im\theta)$, where m is equal to the number of blades in the upstream cascade.

In the classical two-dimensional cascade theory (e.g., [10]), the stationary velocity perturbation is represented in the coordinates $\mathbf{X} = (X_1, X_2, X_3)$ aligned with the mean flow, as follows:

$$\mathbf{u}_{g\infty}^{(2D)} = \mathbf{A}^{(2D)}(X_3) e^{imX_2/r\cos\nu}, \tag{20}$$

where X_3 is the spanwise coordinate, $\mathbf{A}^{(2D)}(X_3)$ is the "frozen" (non-evolving) gust velocity amplitude vector which can be written in the form,

$$\mathbf{A}^{(2D)}(X_3) = (A_{m,n}\widehat{\mathbf{I}}_1 + B_{m,n}\widehat{\mathbf{I}}_2)\cos(\frac{\pi n X_3}{b}) + (A'_{m,n}\widehat{\mathbf{I}}_1 + C_{m,n}\widehat{\mathbf{I}}_3)\sin(\frac{\pi n X_3}{b}), \tag{21}$$

where $\hat{\mathbf{I}}_1, \hat{\mathbf{I}}_2$ and $\hat{\mathbf{I}}_3$ are unit vectors in the X_1, X_2, and X_3 directions; $A_{m,n}$, $A'_{m,n}$, $B_{m,n}$ and $C_{m,n}$ are constant complex amplitudes of the gust harmonic coupled by the divergence-free condition,

$$\frac{im}{\pi r \cos\nu} B_{m,n} + \frac{n}{b} C_{m,n} = 0, \tag{22}$$

n is the radial modal number; $b = r_t - r_h$ is the radial spacing of the annulus. The term with $A'_{m,n}$ is commonly omitted in two-dimensional theories since it does not contribute to cascade unsteady response, because of the radial symmetry (note, however, that it does contribute in three-dimensional theory). Usually the first term of the imposed disturbance velocity (21) models velocity defects in wakes of an upstream cascade, while the other two terms may represent an unsteady secondary flow produced in a turbomachinery annulus.

For the two-dimensional model, the boundary-value gust-cascade interaction problem reduces to an integral equation for the unsteady pressure jump across the blade surface [10]. This integral equation is solved numerically in [11] in the frame of reference fixed with the downstream cascade blades, $\mathbf{Y} = (Y_1, Y_2, Y_3)$, with a unit upwash velocity component of the sinusoidal gust,

$$\mathbf{u}_g^{(2D)} = \mathbf{a}e^{i(\mathbf{k}\cdot\mathbf{Y}-\omega t)}, \tag{23}$$

where $\mathbf{a} = \{a_1, a_2, a_3\}$ and $\mathbf{k} = \{k_1, k_2, k_3\}$ represent, respectively, the gust amplitude and wave vectors in the $\mathbf{Y}$-coordinates. Note that only the *upwash* gust component a_2 will contribute to the unsteady response of the blades. The results of the integral solver [11] are used below for comparison.

Since formulations of the generalized gust interaction problems are somewhat different for Regions I and II, there is need to first examine the interaction models separately for each Region.

4.1. REGION I

Upstream of the rotor stage in Region I, a specified *stationary* perturbation $\mathbf{u}_{g\infty}$ is imposed on a swirling mean motion, with flow swirl agle $\nu = tan^{-1}U_s/U_0$ to the duct axis (Figure 2). In the particular example considered, the mean flow velocity is $\mathbf{U}_\infty = U_0\hat{e}_x - \Omega r\hat{e}_\theta$, and the unsteady perturbation models the wake velocity defects induced by the IGV blades.

In the rotor frame of reference (x, r, θ'), the total mean velocity $\mathbf{U}_r = U_0\hat{e}_x + (\Omega_R - \Omega)r\hat{e}_{\theta'}$ comprises an angle μ with the oncoming flow, and thus a cascade stagger angle χ is introduced to provide zero angles of attack of unloaded blades ($\chi = \mu + \nu$). In these coordinates, the stationary gust perturbation appears unsteady in view of the relation, $\theta = \theta' - \Omega_R t$.

For the two-dimensional model, the following transformation to the local system of coordinates fixed with the rotor blades is carried out,

$$X_2 = Y_1 \sin\mu + Y_2 \cos\mu - \Omega_R r t \cos\nu, \tag{24}$$

so that at the rotor stage the gust velocity (20) transforms to

$$\mathbf{u}_g^{(2D)}(Y_1, Y_2, Y_3) = \mathbf{a}e^{im\Omega_R(\frac{Y_1}{U_r}+\frac{Y_2 cot\mu}{U_r}-t)}. \tag{25}$$

The contributing upwash component of the local gust velocity is easily deduced,

$$a_2 = -A_{m,n} \sin \mu + B_{m,n} \cos \mu, \tag{26}$$

with the wave vector components $k_1 = m\Omega_R/U_r, k_2 = k_1 \cot \mu, k_3 = \pi n/b$ (U_r is the magnitude of $\mathbf{U}_r$).

The stationary generalized gust evolution is obtained in Region I in the duct coordinates (x, r, θ) in which (7) reduces to

$$\mathbf{u}_g(x, r, \theta) = \mathbf{A}(x, r)e^{im(\theta + x\Omega/U_0)}. \tag{27}$$

Subsequently, the boundary-value interaction problem is solved numerically in the moving rotor frame of reference by transforming the gust solution to the computational domain with $\vartheta = \theta - (\Omega_R - \Omega)x/U_0 - \Omega_R t$.

4.2. REGION II

Downstream of the rotor, the mean flow velocity is given by $\mathbf{U}_\infty = U_0\hat{e}_x - U_s\hat{e}_\theta$, where $U_s = \Omega r + \Gamma/r$ (Figure 2). As before, we consider an imposed perturbation velocity $\mathbf{u}_{g\infty}$ which models the wake velocity defects induced by the upstream rotor cascade. However, the disturbance will now appear stationary in the rotor frame of reference where we assume it of the form, $\mathbf{u}_{g\infty} \sim \exp(-im\theta')$ (m is now equal to the number of rotor blades, and minus sign is taken for convenience of the further analysis).

In the stationary duct coordinates, the wake-induced perturbation is viewed as unsteady since $\theta' = \theta + \Omega_R t$. The unsteady disturbance evolves in the swirling mean motion and interacts with the stator row of blades aligned with the swirling streamlines of the mean flow ($\tilde{\chi} = \tilde{\nu}$).

In the two-dimensional cascade model, the following corresponding form of the imposed gust velocity should be assumed ($\mathbf{X} = (X_1, X_2, X_3)$ are now rotor coordinates aligned with the downstream flow streamlines):

$$\mathbf{u}_{g\infty}^{(2D)} = \mathbf{A}^{(2D)}(X_3)e^{-imX_2/r \cos \tilde{\nu}}, \tag{28}$$

Transformation to the stationary coordinates $\mathbf{Y}$ and (x, r, θ), fixed, respectively, with the stator blades and the duct, gives:

$$X_2 = Y_2 + \Omega_R r \tau \cos \tilde{\nu} = x \sin \tilde{\nu} + r\theta \cos \tilde{\nu} + \Omega_R r \tau \cos \tilde{\nu}, \tag{29}$$

where the *retarded* time is introduced, $\tau = t - x/U_0$. Since $\mu = 0$ in this case, the upwash amplitude of the sinusoidal gust (23) is $a_2 = B_{m,n}$, and the wave vector components are $k_1 = m\Omega_R/U_r, k_2 = k_1 \tan \tilde{\nu} - m/r \cos \tilde{\nu}, k_3 = \pi n/b$ (these parameters are used in the integral solver calculations [11]).

The initial-value problem for the generalized gust evolution is solved in the duct coordinates in which the disturbance is of the form,

$$\mathbf{u}_g(x,r,\theta) = \mathbf{A}(x,r)e^{im[-\theta+(\Omega_R/U_0-U_s/rU_0)x-\Omega_R t]}. \tag{30}$$

The generalized gust boundary-value problem is formulated in the transformed coordinates aligned with the swirling streamlines of the mean flow, with $\vartheta = \theta + xU_s/rU_0$.

5. Results and Discussion

Numerical code GENGUS solving the generalized gust problem in a swirling mean flow has been previously validated in [8]. For the present numerical example, response predictions are obtained for the compressor stage cascades installed in an infinite, uniform annular duct with hub and tip radii $r_h = 4.25$ and $r_t = 5.25$, respectively (normalized by the half-chord $c/2$ of unturned cascade blades). Each cascade in the stage has $N_b = 30$ blades, which provides a mid-span spacing $\tilde{s} \approx 2\pi r_m/cN_b = 0.5$ taken for comparison in the two-dimensional cascade calculations [11].

The following parameters determining mean flow quantities are considered, $M_0 = U_0/\bar{c}_0 = 0.6, M_\Omega = \Omega r_m/\bar{c}_0 = 0.3, M_\Gamma = \Gamma/r_m\bar{c}_0 = 0.3$, $\Omega_R/\Omega = 2$ ($\bar{c}_0$ is the reference stagnation speed of sound). The frequency of unsteady response is determined by $\omega = m\Omega_R$, which gives the reduced frequency of the problem, $k_g = m\Omega_R c/U_0 = 12.6$.

Unsteady blade forces are characterized in terms of the normalized real parts of the unsteady pressure jump across the blades,

$$C_p^s(x) = \frac{Re[\Delta p'(x,r)]}{\bar{\rho}_0 cU_0|A_n^{le}|}, \tag{31}$$

non-dimensionalized by U_0, the blade chord, c, the reference flow density, $\bar{\rho}_0$, and the generalized gust upwash magnitude at the blade leading edge, $|A_n^{le}|$. The blade strip s is defined by $s = (r - r_h)/(r_t - r_h)$. The strip unsteady lift coefficient is also introduced,

$$|C_L^s| = \frac{|L'|}{\pi\bar{\rho}_0 cU_0|A_n^{le}|}. \tag{32}$$

To simplify the numerical formulation of cascade radiation conditions, the boundary-value problems for each Region assume the same mean flow upstream and downstream of blade rows.

5.1. REGION I

Numerical results for Region I (Figures 3-6) first show the evolution of the upwash component of the generalized gust velocity which excites the unsteady response of the rotor. Notably, the stationary gust convecting in a 26.6^0 swirling wheel flow practically does not change the radial profile and amplitude of its upwash component as shown in Figure 3. The initial profile at $x = x_{in}$ is fixed for the amplitude ratio of the total gust velocity components, $|u_{g\theta}|/|u_{gx}| = 0.5$, and accounts for the potential part which develops in calculations. However, the contribution of the latter to the total upwash amplitude is not very significant. The downstream evolution of the vortical, potential and total components of the generalized gust upwash over the distance $\Delta x = c$ from the point of origin to mid-chords of the rotor blades is further illustrated in Figure 4. An appreciable radial gust component does not contribute to unsteady response in this particular case.

It follows that the effect of swirl on the rotor unsteady response should be mainly limited to the change of the far field radiation conditions. Figure 5 shows the branches of complex acoustic eigenvalues $k_{\nu_l^\pm,n}$ representing the axial wavenumbers of propagating pressure modes in the moving rotor frame of reference. Circumferential modal numbers $\nu_l^\pm = m \pm lN_b$ for all integer $l = 0, 1, 2, ...$ are selected by the cascade quasiperiodicity condition, and integer $n = 1, 2, ...$ correspond to radial modal numbers. In numerical computations, the far field series expansion of the scattered solution over all l and n is truncated by $l = 5$ and $n = 10$, with the higher order modes rapidly decaying and thus practically not contributing. The branches of eigenvalues from the first quadrant of the complex $k_{\nu_l^\pm,n}$-plane correspond to pressure modes which propagate or decay downstream; those from the third quadrant, appear upstream. For all *acoustic* modes, $Im(k_{\nu_l^\pm,n}) = 0$, otherwise the mode is *evanescent* and decays exponentially as it propagates away from the cascade. Nevertheless, since in the present numerical computations the far field boundaries of the computational domain are only half-chord away from the edges of the blades, the evanescent modes play an important role in the formulation of outflow conditions.

As shown in Figure 5, the far field cascade response is dominated by four upstream-propagating and two downstream-propagating acoustic modes. The asymmetry is caused by Doppler effect which changes the effective modal frequency in presence of the mean flow. The same effect caused by the mean swirl splits the branches of evanescent modes which appear both upstream and downstream (in contrast to the uniform flow where all evanescent modes are shifted upstream). Additional swirl effects may be related to modal refraction by the non-uniform medium which seems to produce a small congestion of evanescent branches, and acoustic-vorticity coupling in presence of the mean flow vorticity. These effects are not very pronounced in the present example.

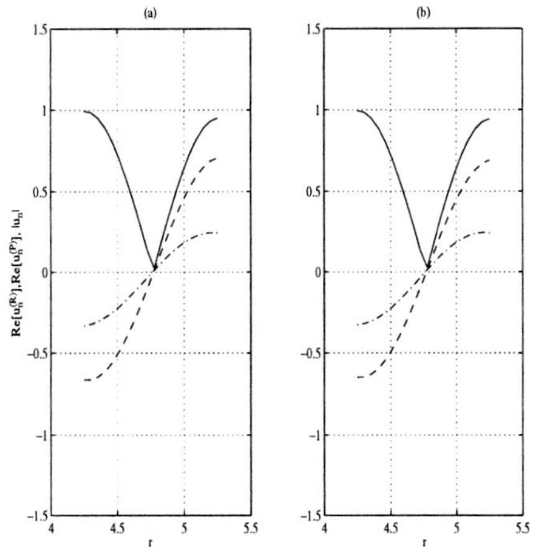

Figure 3. Radial profile of the vortical, potential, and total components of the generalized gust upwash at the initial station (a) and the blade leading edge (b): --, $Re[u_n^{(R)}]$, -.-, $Re[u_n^{(P)}]$, -, $|u_n|$. $M_0 = 0.6, M_\Omega = 0.3, k_g = 0, m = 30$.

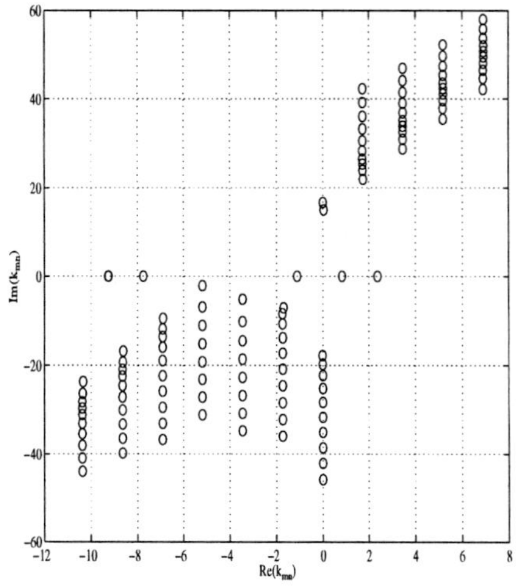

Figure 5. Eigenvalues for upstream/downstream unsteady response. $M_0 = 0.6, M_\Omega = 0.3, k_g = 12.6, m = 30$.

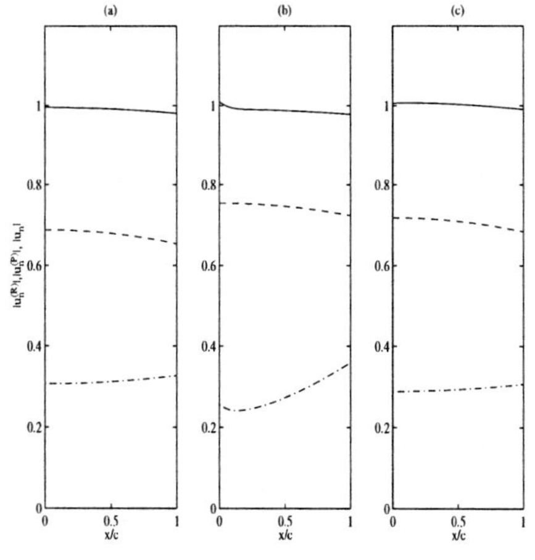

Figure 4. Downstream evolution of the vortical, potential, and total components of the generalized gust upwash at strips $s = 0.25(a)$, $s = 0.5(b)$, $s = 0.75(c)$: --, $Re[u_n^{(R)}]$, -.-, $Re[u_n^{(P)}]$, -, $|u_n|$. $M_0 = 0.6, M_\Omega = 0.3,\ k_g = 0, m = 30$.

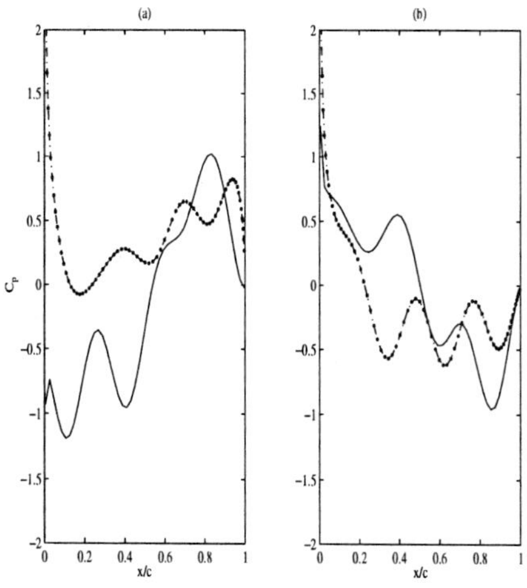

Figure 6. Aerodynamic response C_p at blade strips $s = 0.25(a)$; $s = 0.75(b)$: numerical analysis (solid line), two-dimensional cascade analysis (circles). $M_0 = 0.6, M_\Omega = 0.3, k_g = 12.6, m = 30$.

The far field unsteady response is closely coupled to cascade aerodynamic response which is illustrated in Figure 6 for two strips along the blade span, $s = 0.25$ and $s = 0.75$. The circles mark the results obtained from the linear cascade model which match the oscillatory behavior of the aerodynamic loading along the chord obtained in the numerical computations. The deviations are more notable at the hub sections of the blades, near the leading edge, where the phase shift in the unsteady responses is more pronounced. Apart from differences in cascade radiation conditions, these deviations are attributed to variations of the generalized gust amplitudes over the surface of the blades, and to significant spanwise redistribution of unsteady aerodynamic loading.

5.2. REGION II

For Region II downstream from the rotor, the results are presented in Figures 7-12. These results indicate a remarkable amplification of the upwash velocity component of the generalized gust. The disturbance propagates in the 45° swirling flow and carries an unsteady excitation which induces the unsteady response of the downstream stator blades.

The initial velocity perturbation is provided for the amplitude ratio of the total gust velocity components, $|u_{g\theta}|/|u_{gx}| = 1$, and thus have *no* contribution to the upwash velocity component as expected for the two-dimensional cascade model. In this case, the induced imaginary component of the radial gust velocity, not shown in Figure 7, becomes the most important due to its contribution to the upwash magnitude in presence of the free vortex flow. This explains the developed hat-shaped radial profile in Figure 7. Note also that the vortical and potential gust velocity components have comparable amplitudes.

The further rapid growth of the upwash magnitude in Figure 8 is attributed to the amplification of the vortical tangential velocity component (compare with Figure 9c), which in turn is related to the growth of hydrodynamic pressure carried by the gust (Figure 9d). This amplification results from the large-scale interchange between the radial and tangential velocity perturbations (compare plots (b) and (c) in Figure 9) which is produced by Coriolis forces as indicated in equation (14). Note that the potential gust component becomes less significant.

As shown in Figure 10, there are plenty of acoustic modes which are cut-on upstream from the radiating stator but only few appear downstream as a result of higher shifted frequencies for modes spinning opposite to the swirl direction ($\nu_l^- < 0$). The effect of mean flow refraction is also significant and results in distorted branches of evanescent eigenvalues.

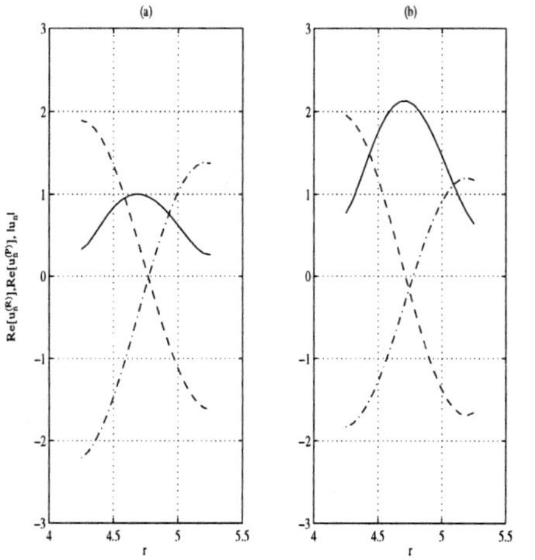

Figure 7. Radial profile of the generalized gust upwash at the initial station (a) and the blade leading edge (b): --, $Re[u_n^{(R)}]$, -.-, $Re[u_n^{(P)}]$, -, $|u_n|$. $M_0 = 0.6, M_s = 0.6, k_g = 12.6, m = 30$.

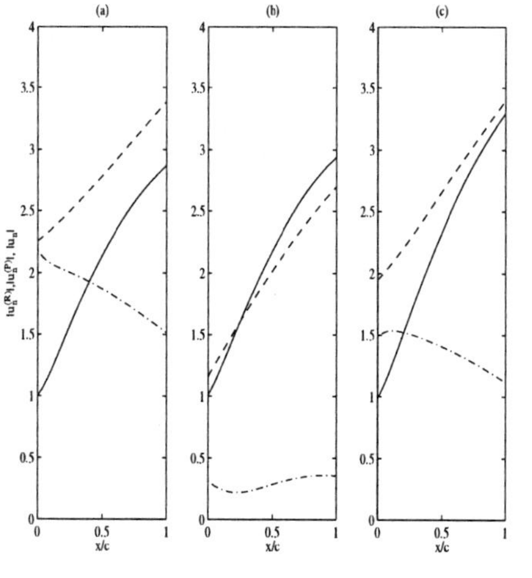

Figure 8. Downstream evolution of the vortical, potential, and total components of the generalized gust upwash at strips $s = 0.25(a)$, $s = 0.5(b)$, $s = 0.75(c)$: --, $Re[u_n^{(R)}]$, -.-, $Re[u_n^{(P)}]$, -, $|u_n|$. $M_0 = 0.6, M_s = 0.6, k_g = 12.6, m = 30$.

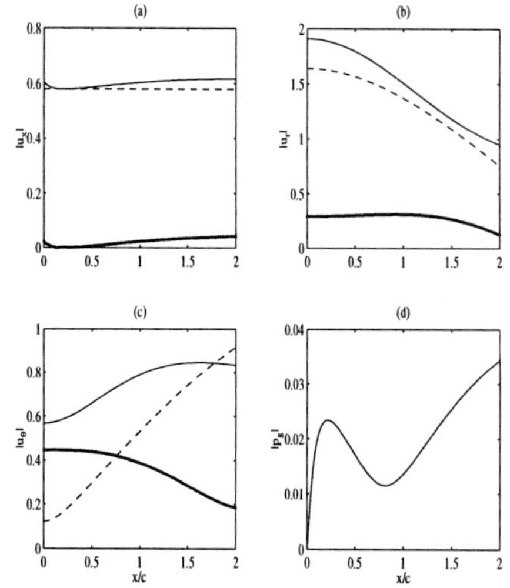

Figure 9. Downstream evolution of the generalized gust velocity and pressure components at strip $s = 0.75$: --, $Re[u^{(R)}]$, '.', $Re[u^{(P)}]$, -, $|u_n|$. $M_0 = 0.6, M_s = 0.6, k_g = 12.6, m = 30$.

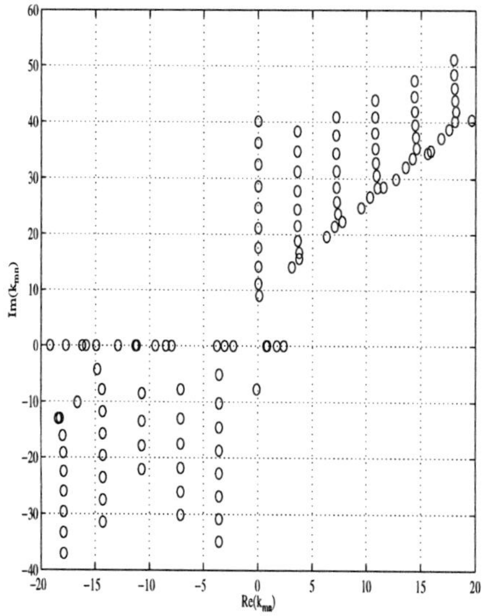

Figure 10. Eigenvalues for upstream/downstream unsteady response. $M_0 = 0.6, M_s = 0.6, k_g = 12.6, m = 30$.

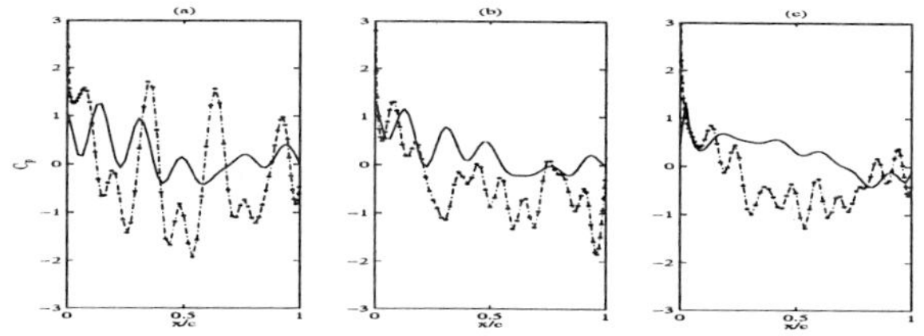

Figure 11. Aerodynamic response C_p at blade strips $s = 0.25(a)$, $s = 0.5(b)$, $s = 0.75(c)$: numerical analysis (solid line), two-dimensional cascade analysis (circles). $M_0 = 0.6, M_s = 0.6, k_g = 12.6, m = 30$.

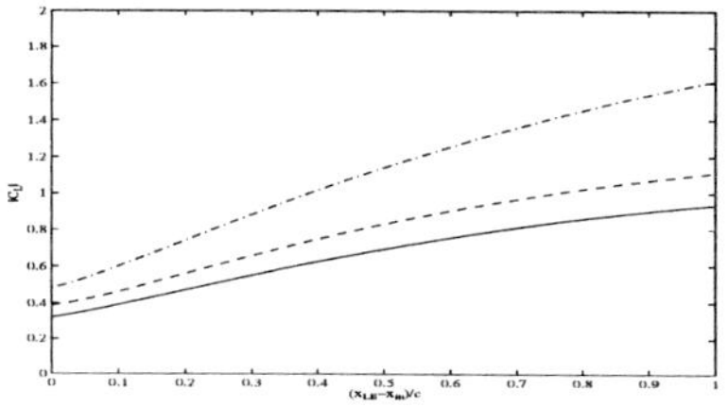

Figure 12. Unsteady lift coefficient C_L vs. generalized gust evolution distance, at blade strips: --, $s = 0.25$; -, $s = 0.5$, -.-, $s = 0.75$. $M_0 = 0.6, M_s = 0.6, k_g = 12.6, m = 30$.

The aerodynamic blade response is shown in Figure 11 for three blade strips, $s = 0.25, 0.5$ and $s = 0.75$. The results of the two-dimensional cascade calculations are very sensitive to the varying parameters, the spacing $\tilde{s}$ and the wavenumber k_2. The highly oscillatory behavior of the two-dimensional and three-dimensional solutions is in good qualitative agreement. Complex variations of the generalized gust amplitude and the change in far field radiation conditions contribute to the observed deviations and stress the effect of swirl on cascade unsteady response. This effect is further illustrated in Figure 12 which shows the predicted increment of the unsteady lift coefficient at different blade strips, as a function of the generalized gust evolution distance (compare with Figure 8).

Conclusions

The paper examined the model of unsteady interaction of a compressor stage with a strong swirling mean motion using the framework of the generalized gust problem. In contrast to the classical gust definition, the evolving generalized gust solution of the initial-value problem included both vortical and potential components, with a small hydrodynamic pressure carried downstream.

In Region I corresponding to IGV/rotor interaction with a 26.6^0 swirling wheel flow, the stationary wake-induced perturbation produced by the IGV blades revealed a practically "frozen" type of evolution. The effect of swirl on the amplitude of rotor unsteady response was mainly limited to the change of the far field radiation conditions.

In contrast, in Region II corresponding to rotor/stator interaction with a 45° mixed swirling flow, a remarkable amplification of the upwash component of the generalized gust was noted as a result of developing tangential perturbation. Cascade far field response showed plenty of acoustic modes cut-on upstream, but only few appeared downstream as a result of higher shifted frequencies for modes spinning opposite to the swirl direction. The highly oscillatory behavior of unsteady blade loading appeared in good qualitative agreement with two-dimensional cascade predictions. The effect of swirl was illustrated for predicted increment of cascade unsteady lift as a function of the generalized gust evolution distance.

Acknowledgements

The research was supported by the Office of Naval Research Grant No. N00014-92-J-1165, and monitored by Mr. L.Patrick Purtell, and NASA Lewis Research Center Grant No. NAG3-732, and monitored by Mr. Dennis L. Huff.

References

1. Atassi, H.M. (1994) Unsteady Aerodynamics of Vortical Flows: Early and Recent Developments, in K.-Y.Fung (ed.), *Aerodynamics and Aeroacoustics*, World Scientific, pp.121-172.
2. Goldstein, M.E. (1978) Unsteady Vortical and Entropic Distortions of Potential Flows Round Arbitrary Obstacles, *J. Fluid Mechanics* **89**, pp.433-468.
3. Atassi, H.M. and Grzedzinski, J. (1989) Unsteady Disturbances of Streaming Motions Around Bodies, *J. Fluid Mechanics* **209**, pp.385-403.
4. Fang, J. and Atassi, H.M. (1993) Compressible Flows with Vortical Disturbances Around a Cascade of Loaded Airfoils, in H.M. Atassi (ed.), *Unsteady Aerodynamics, Aeroacoustics and Aeroelasticity of Turbomachines and Propellers*, Springer-Verlag, pp.149-176.
5. Hall, K.C. and Verdon, J.M. (1991) Gust Response Analysis for Cascade Operating in Nonuniform Mean Flows, *AIAA J.* **29**, pp.1463-1471.
6. Kerrebrock, J.L. (1977) Small Disturbances in Turbomachine Annuli with Swirl, *AIAA J.*, **15**, pp.794-803.
7. Atassi, H.M. and Golubev, V.V. (1997) Unsteady Disturbances in Swirling Turbomachinery Flows, *8th International Symposium on Unsteady Aerodynamics and Aeroelasticity of Turbomachines*, to appear in Proceedings.
8. Golubev, V.V. and Atassi, H.M. (1997) Interaction of Acoustic-Vorticity Waves with an Annular Cascade in a Swirling Flow, *AIAA Paper* No. 97-1635, *Third AIAA/CEAS Aeroacoustics Conference.*
9. Golubev, V.V. and Atassi, H.M. (1997) Acoustic-Vorticity Waves in Swirling Flows, *J. Sound and Vibration* (to appear).
10. Goldstein, M.E. (1976) *Aeroacoustics*, McGraw-Hill.
11. Ventres, C.S. (1980), A Computer Program to Calculate Cascade 2D Kernel, *NASA Technical Memorandum.*

EFFECT OF ACOUSTIC CONTROL ON THE FLUTTER BOUNDARIES OF SUPERSONIC CASCADE

K. NAGAI AND M. NAMBA
Department of Aeronautics and Astronautics
Kyushu University
6-10-1, Hakozaki, Higashi-ku, Fukuoka 812, Japan

Abstract

The active acoustic control of the supersonic cascade flutter under steady loading is studied. A three-dimensional linear cascade of blades oscillating in a supersonic flow between parallel walls is considered. The effect of steady loading, i.e., angle of attack, camber and thickness on the unsteady aerodynamic force is dealt with on the basis of the double linearization theory. One of the walls is partially made of actuator surface like a loudspeaker which generates sound waves at the same frequency as that of the blade oscillation, so that the unsteady aerodynamic force due to acoustic disturbances generated from the actuator surface cancels the existing unsteady force on the blades due to blade motions themselves. The optimum condition of the actuator surface geometry and motion is investigated and the capability to enlarge the flutter free domain is demonstrated.

1. Introduction

The acoustic control technology based on the anti-sound concept proposed by Ffowcs Williams[1][2] deserves widespread attention, because many undesirable unsteady fluidmechanical phenomena originate in interactions with acoustic disturbances. Suppression of aeroelastic instabilities is one of the most attractive applications.

Of particular interest will be the acoustic control of cascade flutter, because the blade-to-blade aerodynamic interaction is nothing else than an acoustic phenomenon and the instability of blade vibration mainly depends upon whether the acoustic waves generated from vibrationg blades them-

T.H. Fransson (ed.), Unsteady Aerodynamics and Aeroelasticity of Turbomachines, 165-179.

selves act beneficially or detrimentally to the blade vibration. Then comes out the concept of suppressing the cascade flutter by artifitially generating acoustic waves which can give rise to damping forces on blades.

One of the advantages of using acoustic devices for flutter suppression is that the systems can be simple and of light weight. Furthermore, the adaptable frequency range of flutter is very wide and its response is much quicker than the conventional aero-mechanical control devices.

The authors have made a series of studies of active acoustic control of cascade flutter based on the model consisting of a linear cascade between parallel walls equiped with an actuator surface which oscillates like a loudspeaker. The theory have been applied to the subsonic cascade flutter without and with steady loading effects[3][4][5] and the supersonic cascade flutter without steady loading[6]. Results indicate that the phase angle between oscillations of blades and the actuator surface, the location, the width and the vibration modes of the actuator surface are key parameters for effective achievement of control. In the case of subsonic cascade flutter problems, the dominant acoustic wave modes are usually of cut-off state, and the effectiveness of the actuator surface exponentially decreases with increasing distance between the actuator surface and the cascade. On the other hand in the case of supersonic cascade flutter problems, the acoustic waves are usually dominated by cut-on modes which propagate without decaying, and the aerodynamic force exerted on the cascade by action of a remote actuator surface is not so weak as in the case of subsonic cascades. The optimum parameters of the actuator surface geometry and motion are well correlated to the wavelength and the phase of the effective and dominant acoustic wave mode.

The present paper deals with the problem of the supersonic cascade flutter under steady loading. The steady loading effect on the supersonic cascade flutter is much larger than that on the subsonic cascade flutter. The paper demonstrates how the flutter boundary expressed in terms of the blade geometry parameters can be influenced by the acoustic control. Explicit relations between the optimum condition of the actuator surface and the acoustic wave mode parameters are presented.

2. Outline of Theory

2.1. ANALYTICAL MODEL

The model considered is a linear cascade of blades with the stagger angle γ, the pitch-chord ratio s, the aspect ratio h and the angle of attack α_0 as shown in Figure 1. In this paper, lengths, velocities, pressures and times are scaled with respect to c (the blade chord), U_0 (the time and space averaged fluid velocity), $\rho_0 U_0^2$ and c/U_0, respectively, where ρ_0 is the time and space averaged fluid density. The profiles of the camber and the thickness of the cascade blades are assumed to be parabolic and spanwisely uniform, i.e. expressed by $\mu_0(1-4\xi^2)$ and $\tau_0(1-4\xi^2)$. Here μ_0 and τ_0 denote the camber ratio and the thickness ratio respectively. The end walls are assumed to be rigid except the actuator surface location domain, i.e., $x_c - l \leq X \leq x_c + l, \quad -\infty \leq Y \leq \infty$. Here x_c and l denote the center line position and the semi-width of the actuator surface respectively, and an oblique coordinate system (X, Y) is defined by $X = x - y\tan\gamma, \;\; Y = y/\cos\gamma$.

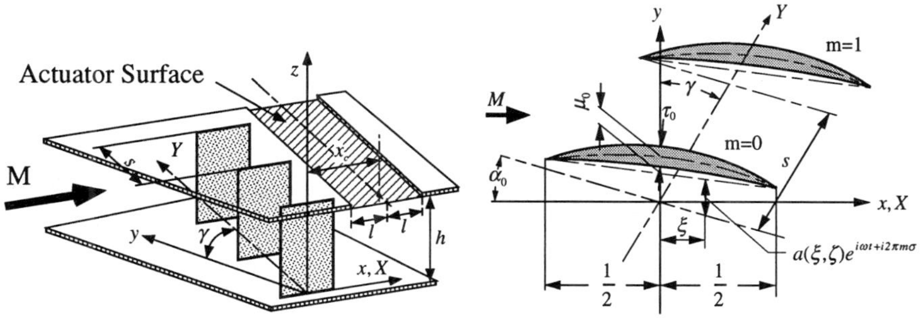

Figure 1. The analytical model and the cascade configuration.

The steady disturbance quantities are denoted by symbols with an over-bar $\bar{(\;)}$, while the unsteady disturbance quantities are denoted by symbols with a tilde $\tilde{(\;)}$ times $e^{i\omega t}$.

All blades are assumed to be undergoing a simple harmonic bending oscillation with normal displacement $\tilde{a}(\xi, \zeta)e^{i\omega t + i2\pi m\sigma}$, where ω is the reduced frequency, $2\pi\sigma$ is the interblade phase angle, and m is the blade number. In the present analysis the spanwise distribution of the blade vibration amplitude is assumed to be coincident with a natural mode shape of a uniform cantilever beam fixed at $z = 0$. From the standpoint of the modal analysis it is beneficial to express the amplitude of the vibrating blade $\tilde{a}(x, z)$ in terms of the spanwise acoustic mode functions (cosine functions):

$$\tilde{a}(x, z) = H \sum_{n=0}^{\infty} Z_n^b \cos \frac{n\pi z}{h}, \tag{1}$$

where Z_n^b is n-th Fourier cosine component of the normalized spanwise mode shape and H denotes the amplitude at the tip, which is real and positive.

The actuator surface is also assumed to be making oscillation with normal displacement $\tilde{g}(X, Y)e^{i\omega t}$. Here the displacement amplitude $\tilde{g}(X, Y)$

is assumed to be circumferentially periodical so that the phase angle between motions of two arbitrary points separated by the distance equal to the cascade pitch s in the circumferential direction Y is $2\pi\sigma$. Therefore it is expressed as

$$\tilde{g}(X,Y) = \sum_{\nu=-\infty}^{\infty} \sum_{\mu=1}^{\infty} |B_{\mu\nu}| e^{i\Psi_{\mu\nu}} \sin \mu\Theta e^{iq_\nu Y}; \quad \Theta = \cos^{-1}\left(\frac{X - X_c}{l}\right), \quad (2)$$

where $q_\nu = 2\pi(\sigma + \nu)/s$, $\nu + \sigma$ denotes the circumferential wave number, $|B_{\mu\nu}|$ is the modal amplitude of the actuator surface oscillation and $\Psi_{\mu\nu}$ is the phase angle between the oscillation of the $m = 0$ blade and the motion of the modal component of the actuator surface at $Y = 0$.

2.2. METHOD OF ANALYSIS

To deal with the effect of steady loading on the cascade flutter, it is assumed on the basis of the double linearization concept[7] that $\alpha_0, \mu_0, \tau_0 = O(\bar{\epsilon})$, and $\tilde{a}(\xi, \zeta) = O(\epsilon)$. Furthermore, $1 >> \bar{\epsilon} >> \epsilon > 0$: i.e., the steady disturbance is small enough to be governed by linearized equations, and the unsteady disturbance is far smaller than the steady disturbance. Therefore we can express the disturbance velocity $\boldsymbol{q}(x, y, z, t)$ in the form

$$\boldsymbol{q}(x, y, z, t) = \bar{\boldsymbol{q}}(x, y, z) + \tilde{\boldsymbol{q}}(x, y, z)e^{i\omega t}, \quad (3)$$

where the velocity vector components are denoted by $\bar{\boldsymbol{q}} = (\bar{u}, \bar{v}, \bar{w})$, and $\tilde{\boldsymbol{q}} = (\tilde{u}, \tilde{v}, \tilde{w})$.

According to the mathematical formulation of the double linearization[7], the blade surfaces are represented as surfaces of singularities composed of monopoles, dipoles and quadrupoles. Furthermore the strengths of singularities are expressed in terms of the pressure difference across a blade surface $\Delta p(x, z, t)$, which is composed of steady and unsteady components as follows:

$$\Delta p(x, z, t) = \Delta\bar{p}(x, z) + \Delta\tilde{p}(x, z)e^{i\omega t}. \quad (4)$$

Owing to the double linearization the steady pressure difference $\Delta\bar{p}(x, z)$ can be expressed in the form of linear combination of the effects of angle of attack $\alpha_0\Delta\bar{p}_\alpha(x, z)$, camber $\mu_0\Delta\bar{p}_\mu(x, z)$ and thickness $\tau_0\Delta\bar{p}_\tau(x, z)$, while the unsteady pressure difference $\Delta\tilde{p}(x, z)$ is composed of the component due to blade oscillation without steady loading $\Delta\tilde{p}_a(x, z)$, the component due to actuator surface motion $\Delta\tilde{p}_m(x, z)$ and the component dependent on the steady loading, which can further be decomposed into contributions of the angle of attack $\alpha_0\Delta\tilde{p}_\alpha(x, z)$, camber $\mu_0\Delta\tilde{p}_\mu(x, z)$ and thickness $\tau_0\Delta\tilde{p}_\tau(x, z)$ as follows:

$$\Delta\bar{p}(x, z) = \alpha_0\Delta\bar{p}_\alpha(x, z) + \mu_0\Delta\bar{p}_\mu(x, z) + \tau_0\Delta\bar{p}_\tau(x, z), \quad (5)$$

$$\Delta\tilde{p}(x,z) = \Delta\tilde{p}_a(x,z) + \alpha_0\Delta\tilde{p}_\alpha(x,z) + \mu_0\Delta\tilde{p}_\mu(x,z) + \tau_0\Delta\tilde{p}_\tau(x,z) + \Delta\tilde{p}_m(x,z). \tag{6}$$

On the other hand the actuator surface on the wall can be represented as a surface of unsteady mass sources[8] with a surface density of $\tilde{m}(X,Y)e^{i\omega t}$. In the present problem all disturbances originate in the singularities on the blades and the actuator surface. Then by making use of the kernel function method the steady disturbance velocity $\bar{\boldsymbol{q}}(x,y,z)$ can be expressed by an integral form containing the function $\Delta\bar{p}(x,z)$, while the unsteady disturbance velocity $\tilde{\boldsymbol{q}}(x,y,z)$ can be expressed as combinations of integrals containing the functions $\Delta\bar{p}(x,z)$, $\Delta\tilde{p}(x,z)$ and $\tilde{m}(X,Y)$[7][8].

Then the boundary conditions on the actuator surface at the endwall $z = h$ is given by

$$\tilde{w}(x,y,h)e^{i\omega t} = \frac{D}{Dt}[\tilde{g}(X,Y)e^{i\omega t}]; \quad x_c - l \le X \le x_c + l. \tag{7}$$

Substitution of the integral representation of $\tilde{w}(x,y,z)$ into equation (7) yields an explicit expression of the mass source function $\tilde{m}(X,Y)$ in terms of the parameters contained in equation (2) representing $\tilde{g}(X,Y)$[3][4][5][6].

On the other hand the instantaneous position of the zeroth blade surface is expressed by

$$y = -\alpha_0 x + \mu_0(1-4x^2) \pm (\tau_0/2)(1-4x^2) + \tilde{a}(x,z)e^{i\omega t}, \tag{8}$$

as shown in Fig.1. Then the flow tangency condition on the blade surface is split into steady and unsteady parts, which are given by

$$\bar{v}(x,\pm 0,z) = -\alpha_0 - 8\mu_0\xi \mp 4\tau_0\xi, \tag{9}$$

and

$$\begin{aligned}\tilde{v}(x,\pm 0,z) &= \frac{\partial\tilde{a}(x,z)}{\partial x} + i\omega\tilde{a}(x,z) - \tilde{a}(x,z)\frac{\partial\bar{v}(x,\pm 0,z)}{\partial y} \\ &+ \frac{\partial\tilde{a}(x,z)}{\partial x}\bar{u}(x,\pm 0,z) + \frac{\partial\tilde{a}(x,z)}{\partial z}\bar{w}(x,\pm 0,z),\end{aligned} \tag{10}$$

respectively. Substitution of the integral representations of the disturbance velocities into the boundary conditions (9) and (10) gives integral equations for the pressure difference functions $\Delta\bar{p}(x,z)$ and $\Delta\tilde{p}(x,z)$ respectively, which can be solved numerically.

The unsteady aerodynamic work exerted by the fluid on a vibrating blade per cycle can be obtained from

$$W = \frac{\pi}{h}\int_0^h dz \oint dt \int_{-1/2}^{1/2} \Re\left[\Delta\tilde{p}(x,z)e^{i\omega t}\right] \Re\left[\frac{\partial}{\partial t}\left\{\tilde{a}(x,z)e^{i\omega t}\right\}\right] dx. \tag{11}$$

Furthermore W can be divided into two components; unsteady aerodynamic work exerted by the aerodynamic force due to the blade vibration; W_A, and unsteady aerodynamic work exerted by the aerodynamic force due to the excitation of the actuator surface; W_m. According to the double linearization theory [7], W_A can further be decomposed into four components:

$$W_A = W_a + \alpha_0 W_\alpha + \mu_0 W_\mu + \tau_0 W_\tau. \tag{12}$$

Here W_a denotes the component devoid of the steady loading effect, while $\alpha_0 W_\alpha$, $\mu_0 W_\mu$ and $\tau_0 W_\tau$ denote contributions of the steady angle of attack, camber and thickness to the unsteady aerodynamic work respectively. Note that W_A is dependent on the blade motion $\tilde{a}(\xi, \zeta)$ but independent of the actuator motion $\tilde{g}(X, Y)$. On the other hand W_m is dependent not only on $\tilde{a}(\xi, \zeta)$ but also on $\tilde{g}(X, Y)$.

If $W_A > 0$ and the control is switched off, the cascade will undergo self-excited vibration. If the control is switched on in such a way that the condition $W_A + W_m < 0$ is satisfied, the flutter will be suppressed. One should note, however, that once the flutter is suppressed, neither the blades are vibrating nor the actuator surface is in oscillating motion. From the standpoint of the feedback control system, therefore, the ratio $\tilde{g}(X, Y)/\tilde{a}(\xi, \zeta)$ or $|B_{\mu\nu}|/H$ gives a measure of the feedback gain factor.

In order to obtain a distinct expression of the flutter boundary in terms of the design parameters, define the work coefficients of the components of W_A by

$$\begin{Bmatrix} C_{Wa} \\ C_{W\alpha} \\ C_{W\mu} \\ C_{W\tau} \end{Bmatrix} = \begin{Bmatrix} W_a/H^2 \\ W_\alpha/H^2 \\ W_\mu/H^2 \\ W_\tau/H^2 \end{Bmatrix}. \tag{13}$$

On the other hand one can decompose W_m into modal components corresponding to the modes of the actuator motion (equation (2)) as follows:

$$W_m = \sum_{\mu=1}^{\infty} \sum_{\nu=-\infty}^{\infty} W_{m(\mu\nu)}. \tag{14}$$

Then defining

$$C_{Wm(\mu\nu)} = W_{m(\mu\nu)}/(H\,|B_{\mu\nu}|), \tag{15}$$

one can express the the critical flutter condition $W_A + W_m = 0$ in the form,

$$C_{Wa} + \alpha_0 C_{W\alpha} + \mu_0 C_{W\mu} + \tau_0 C_{W\tau} + \sum_{\mu=1}^{\infty} \sum_{\nu=-\infty}^{\infty} \frac{|B_{\mu\nu}|}{H} C_{Wm(\mu\nu)} = 0. \tag{16}$$

Note that the coefficients C_{Wa}, $C_{W\alpha}$, $C_{W\mu}$, and $C_{W\tau}$ are dependent on M, γ, s, h, ω, σ and the mode of the blade motion, while each of $C_{Wm(\mu\nu)}$ is dependent on the same parameters and further on μ, ν, $\Psi_{\mu\nu}$, l and x_c. It is worth emphasizing that equation (16) gives an explicit expression of the flutter boundary in terms of α_0, μ_0, τ_0 and $|B_{\mu\nu}|/H$.

3. Numerical Examples and Discussion

3.1. SPECIFIED CONDITIONS

The important parameters with respect to the actuator surface oscillation are the amplitude ratio $|B_{\mu\nu}|/H$, the phase angle $\Psi_{\mu\nu}$, the center line location x_c and the semi-width l. In the following numerical examples, the Mach number $M = 1.3$, the angle of stagger $\gamma = 60°$, the pitch/chord ratio $s = 1.0$, the span/chord ratio $h = 3.0$ and the reduced frequency $\omega = 0.05$ are fixed. For simplicity, the geometry of blades are assumed to be spanwisely uniform. Furhtermore calculations in this paper are confined to the blade vibration of the first bending mode and the actuator surface motion of the fundamental streamwise and circumferential modes ($\mu = 1, \nu = 0$).

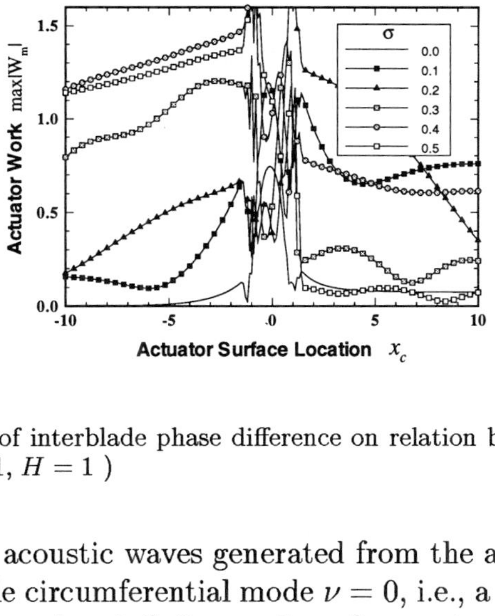

Figure 2. The effect of interblade phase difference on relation between max $|W_m|$ and x_c. ($2l = 2.0$, $|B_{\mu\nu}| = 1$, $H = 1$)

In this case the acoustic waves generated from the actuator surface are composed of a single circumferential mode $\nu = 0$, i.e., a single circumferential wave number σ, and an infinite number of spanwise modes $n = 0, 1, \ldots$. Accordingly the work component W_m can be decomposed into spanwise mode components:

$$W_m = \sum_{n=0}^{\infty} W_m^n. \tag{17}$$

Note that in general cases each of $W_{m(\mu\nu)}$ in equation (14) can further be decomposed into spanwise modes $W^n_{m(\mu\nu)}$: ($n = 0, 1, 2, \ldots$).

3.2. DEPENDENCE OF WORK ON ACTUATOR SURFACE LOCATION

In general the aerodynamic work component W_m is heavily dependent on the phase angle $\Psi_{\mu\nu}$ of the actuator surface motion and the location x_c in such a way that W_m varies periodically with change in $\Psi_{\mu\nu}$ for constant x_c or with change in x_c for constant $\Psi_{\mu\nu}$. There is an optimum phase angle $\Psi_{\mu\nu}$ which makes $|W_m|$ maximum at each x_c. In the following discussions the maximum $|W_m|$ implies that the optimum value of $\Psi_{\mu\nu}$ is selected at each x_c.

Figure 2 shows the maximum $|W_m|$ as a function of actuator location x_c for various interblade phase differences. Note that $x_c = -0.5$ or $x_c = 0.5$ implies that the center line of the actuator surface is coincident with the leading edge locus or the trailing edge locus of the cascade respectively.

For all interblade phase differences, $|W_m|$ sharply fluctuates with changing x_c in the domain where the actuator surface and the cascade overlap with one another $(-0.5 - l \leq x_c \leq 0.5 + l)$. On the other hand $|W_m|$ by and large decreases as the actuator becomes more distant from the cascade. $(x_c < -0.5 - l, x_c > 0.5 + l)$. But the decreasing rate is not so high as in subsonic cases[5].

Furthermore, it is worth noting that the maxmum value of $|W_m|$ is highly dependent on the interblade phase angle σ.

Figures 3 and 4 show the modal components of $|W_m|$ written in equation (17) as functions of x_c for the cases $\sigma = 0.2$ and $\sigma = 0.5$, respectively. Left-hand sides of those figures show the maximum value of $|W_m^n|$ for each spanwise mode n, i.e., $|W_m^n|$ with the phase angle $\Psi_{\mu\nu}$ optimized for each spanwise mode n at each location x_c. On the other hand right-hand sides of those figures show W_m^n with a constant phase angle $\Psi_{\mu\nu} = 0°$.

Firstly, it is observed that in the case of the first order bending oscillation, substantial contributions to the aerodynamic work on the blades are made only by the acoustic modes of lowest two orders $n = 0$ and $n = 1$. Consequently only W_m^0 and W_m^1 are predominant.

Secondly, in most cases the acoustic modes of lower orders are of cut-on state in which acoustic disturbances propagate up to far-field without decaying. Therefore each W_m^n periodically varies with change in x_c under fixed $\Psi_{\mu\nu}$ or with change in $\Psi_{\mu\nu}$ under fixed x_c, but the maximum value of each W_m^n with appropriately selected $\Psi_{\mu\nu}$ at each x_c is invariant with x_c as long as the actuator surface is located away from the cascade, i.e. $x_c > l + 0.5$ or $x_c < -0.5 - l$.

Thirdly, the total maximum work $|W_m|$ shown in left side figures, which is obtained by appropriately selecting $\Psi_{\mu\nu}$ at each x_c, is, however, not constant but decreases as the actuator surface goes away from the cascade. This is because the optimum phase angle for W_m^0 is generally different from

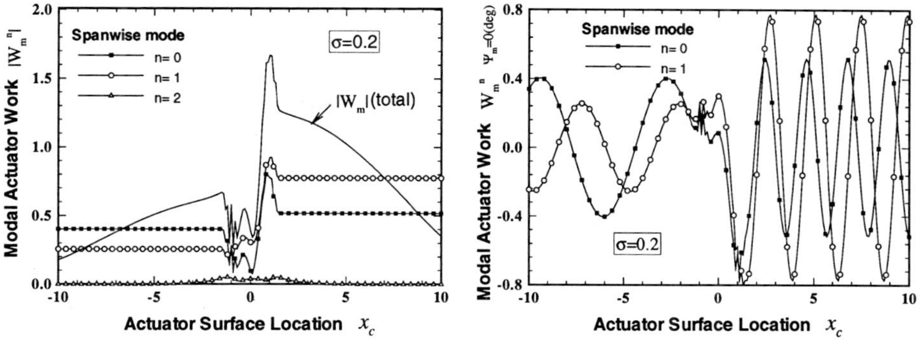

Figure 3. Modal components of the actuator surface work; $\max|W_m^n|$ and $W_m^n(\Psi_{\mu\nu} = 0°)$. $(2l = 2.0, |B_{\mu\nu}| = 1, H = 1\ , \sigma = 0.2)$

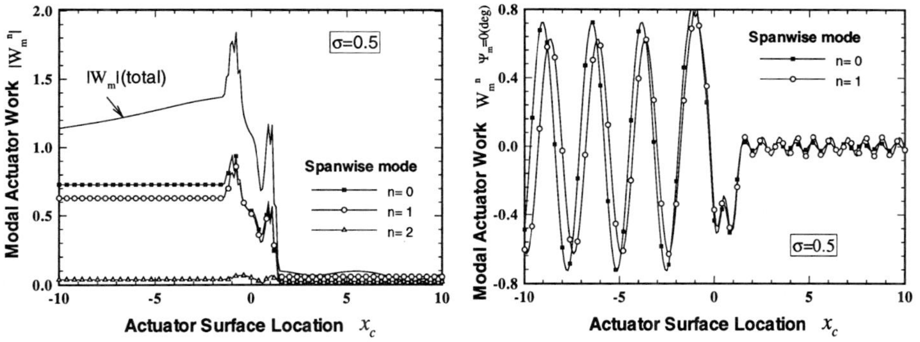

Figure 4. Modal components of the actuator surface work; $\max|W_m^n|$ and $W_m^n(\Psi_{\mu\nu} = 0°)$. $(2l = 2.0, |B_{\mu\nu}| = 1, H = 1\ , \sigma = 0.5)$

that for W_m^1, and the difference in phase between those two modes increases with increasing $|x_c|$ as is observed in right side figures. The difference in the periodicity between those two modes for the case $\sigma = 0.2$ is much larger than that for the case $\sigma = 0.5$. This is the reason why $|W_m|$ for the case $\sigma = 0.2$ more rapidly decreases with increasing $|x_c|$ than that for the case $\sigma = 0.5$. The reason of distinct sensitivity of the aerodynamic work to the interblade phase angle σ is discussed later.

From the practical point of view it is inconvenient to locate the actuator surface by overlapping the cascade not only because the structure becomes complicated, but also because the aerodynamic work varies too sharply with a small change in x_c. It will be most desirable to locate the actuator surface upstream or downstream of the cascade with small gap ($x_c \simeq -0.5-l$ or $0.5+l$).

3.3. DEPENDENCE OF WORK ON ACTUATOR SURFACE WIDTH

Figure 5 shows dependence of the maximum $|W_m|$ on the actuator surface width $2l$. The most arresting feature is a periodical variation of $|W_m|$ with changing width $2l$. The periodicity is longer for the actuator surface located upstream of the cascade than that located downstream. The peak value of $|W_m|$ decreases as the actuator width becomes larger. In this case ($\sigma =$ 0.2), the actuator located downstream of the cascade gives larger work than that located upstream of the cascade. But the situation depends on the interblade phase angle σ, and for example in the case of $\sigma = 0.5$ the upstream actuator provides higher work as seen in Figures 2 and 4. A distinct interpretation of the dependences on σ and $2l$ is given in the next subsection.

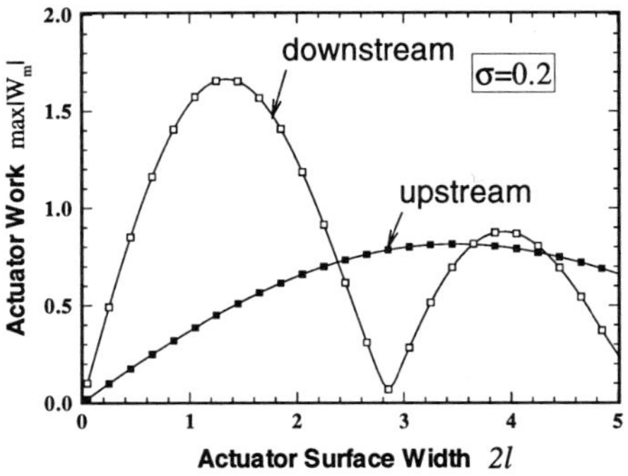

Figure 5. The dependence of max $|W_m|$ on the actuator surface width $2l$. ($|B_{\mu\nu}| = 1$, $H = 1$, X_s (gap between the actuator surface and the cascade)$= 0.5$)

3.4. CORRELATION WITH ACOUSTIC WAVELENGTH

The streamwise wavelength of a cut-on acoustic mode is given by

$$\lambda_n^{\pm} = 2\pi / \left[\frac{q_\nu \sin\gamma + \omega M_a^2}{\beta_a^2} \pm \Omega_{\nu,n} \right], \tag{18}$$

where

$$\Omega_{\nu,n} = -\text{sgn}(q_\nu \sin\gamma + \omega)\frac{1}{\beta_a}\sqrt{\frac{M_a^2}{\beta_a^2}(q_\nu \sin\gamma + \omega)^2 - (q_\nu^2 + n^2\pi^2/h^2)\cos^2\gamma}, \tag{19}$$

$$M_a = M\cos\gamma, \quad \beta_a^2 = 1 - M_a^2, \tag{20}$$

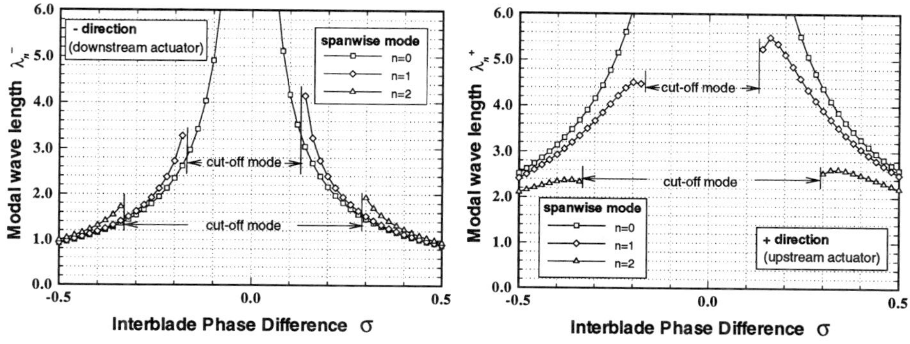

Figure 6. The modal wavelength of streamwise direction $\lambda_n^{\pm}$. ($\nu = 0$).

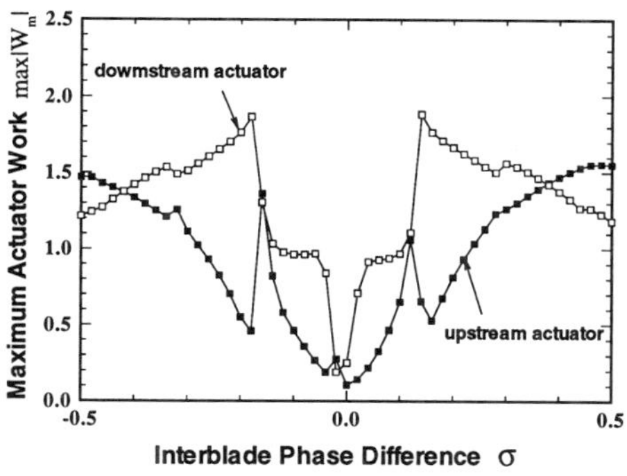

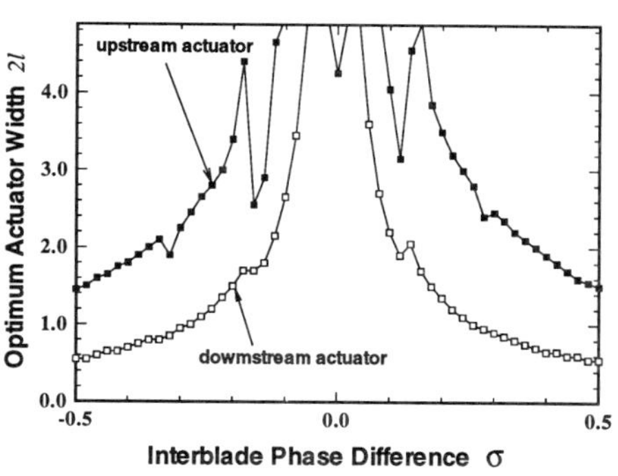

Figure 7. [Left] The maximum $|W_m|$ dependent on the interblade phase difference σ. ($\mu = 1$, $\nu = 0$, $|B_{\mu\nu}| = 1$, $H = 1$)

Figure 8. [Right] The optimized width $2l$ of the actuator surface to maximize $|W_m|$ at each interblade phase difference σ. ($\mu = 1$, $\nu = 0$)

and superscripts + and − denote waves propagating in the downstream and upstream directions respectively.

Figure 6 shows the modal streamwise wavelength $\lambda_n^{\pm}$ ($\nu = 0$) of acoustic disturbances as functions of the interblade phase angle σ. The right side graph indicates the wavelength of the acoustic waves propagating downstream λ_n^+ and this corresponds to the case where the cascade is given a work from the actuator surface located upstream. On the other hand the left side graph indicating the wavelength of the modes propagating upstream λ_n^- corresponds to the opposite case. It should be noted that only wavelengths of cut-on modes are plotted in these figures. The lowest order mode $n = 0$ is unconditionally of cut-on state for the whole interblade phase angle, while $n = 1$ mode is of cut-on state only when $\sigma < -0.18$ and $\sigma > 0.24$. Furthermore the extent of the cut-on region diminishes as the

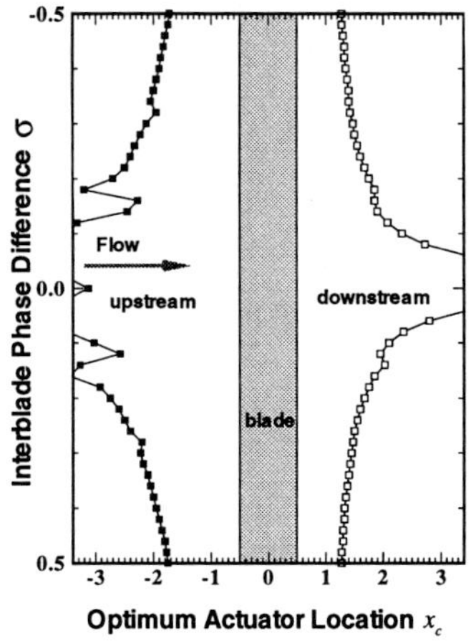

Figure 9. The optimized actuator location x_c at each interblade phase difference σ. ($\mu = 1$, $\nu = 0$)

spanwise mode number becomes higher.

The periodicity of W_m^n with respect to the location x_c seen in Figures 3 and 4 exactly corresponds to the wavelength λ_n. Figure 7 indicates that the difference in the wavelength between $n = 0$ and $n = 1$ modes increases as the interblade phase angle approaches from $\sigma = \pm 0.5$ to $\sigma = 0$. Therefore the difference of wavelengths in the case $\sigma = 0.2$ is much larger than that in the case $\sigma = 0.5$ as can be seen in Figures 3 and 4.

The optimum value of the actuator width $2l$ can also be related to the streamwise wavelength of the dominant acoustic mode. Comparison between Figures 5 and 6 indicates that the optimum width $2l$ that makes $|W_m|$ maximum is approximately equal to a half of the streamwise wavelength λ_n of the most dominant mode. On the other hand the actuator is very inefficient when the width $2l$ is approximately equal to an integral number times as large as the wavelength of the most dominant mode.

3.5. OPTIMUM CONDITION

Figure 7 shows the maximum works $|W_m|$ for the upstream and downstream actuators as functions of the interblade phase angle σ. It should be noted that optimum values of $2l$ and x_c are selected at each point in Figure 7. Those optimum values of $2l$ and x_c are shown in Figures 8 and 9, respectively.

The sharp variations of $|W_m|$ around $\sigma = -0.16$ and $\sigma = 0.12$ originate in the fact that the $n = 1$ mode turns from cut-on to cut-off across these points as seen in Figure 6. One can again confirm from comparison between Figures 8 and 6 that the optimum actuator width which gives largest $|W_m|$ is nearly equal to a half of the streamwise wavelength of the predominant acoustic mode ($2l = \lambda_n/2$).

The reason why $|W_m|$ for $\sigma = 0.5$ is quite small when the actuator is located downstream of the blade ($x_c > 1.5$) as seen in Figure 4 is now clear and it is that the actuator width of $2l = 2.0$ is nearly equal to twice the wavelength of $\lambda_n^- \simeq 1.0$, when $\sigma = 0.5$ as seen in Figure 6.

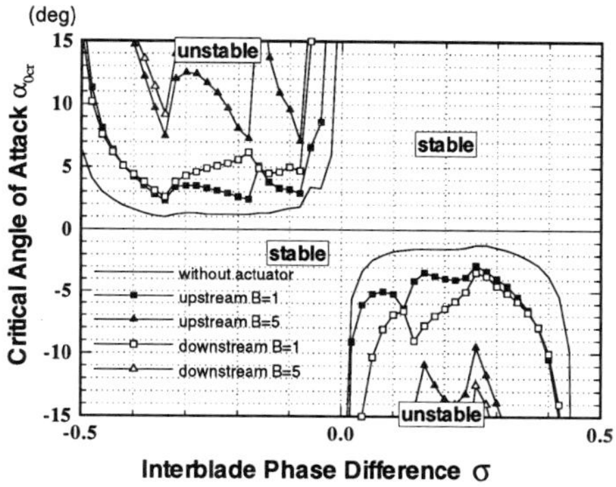

Figure 10. The critical angle of attack α_{0cr} for a supersonic cascade. ($M = 1.3$, $\mu_0 = \tau_0 = 0$, $\mu = 1$, $\nu = 0$). The parameter B denotes $|B_{\mu\nu}|/H$.

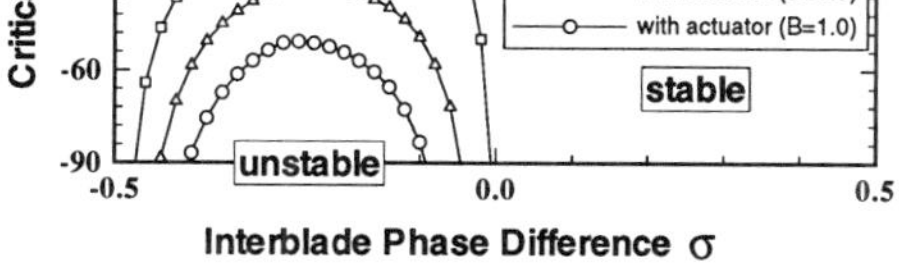

Figure 11. The critical angle of attack α_{0cr} for a subsonic cascade. ($M = 0.6$)

3.6. EFFECT ON THE FLUTTER BOUNDARIES

Figure 10 shows the critical angle of attack α_{0cr}, i.e., α_0 which satisfies equation (16), for a flat plate cascade ($\mu_0 = 0, \tau_0 = 0$) dependent on the gain factor $|B_{1,0}|/H$ and the interblade phase parameter σ. The parameters l, x_c and $\Psi_{\mu\nu}$ are optimized at each σ.

The downstream actuator can enlarge the flutter free domain more effectively than the upstream actuator. This result corresponds to Figure 7, which shows that the downstream actuator provides larger work on the blades than the upstream actuator for a main part of the interblade phase angle.

The domain of the flutter free angle of attack without control is $-1.2° < \alpha_0 < 1.0°$. But when the control is on, the domain is distinctly enlarged, i.e., $-3.4° < \alpha_0 < 2.6°$ for $|B_{1,0}|/H = 1$ and $-12.5° < \alpha < 8.5°$ for $|B_{1,0}|/H = 5$. The flutter free domain for $|B_{1,0}|/H = 1$ is about three times as large as that for no control, and the flutter free domain for $|B_{1,0}|/H = 5$ is enlarged by about ten times. For comparison a case of subsonic cascade for Mach number $M = 0.6$ is shown in Figure 11. In general the effect of steady loading on supersonic cascade flutter is much more significant than that on subsonic cascade flutter, and the flutter free condition for supersonic cascades is severely restricted. Therefore the active control will be very useful for supersonic cascades.

4. Conclusions

The results obtained from the present study are summarized as follows:

1. Essentially important parameters to attain efficient control of supersonic cascade flutter are the width and the location of the actuator surface.
2. The optimum width of the actuator surface is nearly equal to a half of the streamwise wavelength of the predominant acoustic mode, which is highly dependent on the interblade phase angle.
3. To locate the actuator surface downstream of the cascade is more effective than to locate that upstream for suppression of the bending flutter at low frequencies.
4. The domain of the angle of attack free from the bending flutter can be substantially broadened by the active acoustic control with a reasonable value of the feedback parameter (the amplitude ratio of the actuator vibration and the blade vibration).

References

1. Ffowcs Williams, J.E. (1984) Review Lecture: Anti-Sound, *Proceedings of the Royal Society of London*, **A395**, pp.63–88.
2. Ffowcs Williams, J.E. (1989) Active Control of Engine Instabilities, *Proceedings of 9th International Symposium on Air Breathing Engines*, **Vol.1**, pp.45–48.

3. Nagai, K., Toshimitsu, K. and Namba, M. (1995) Active Supression of Cascade Flutter by Means of Sound Waves from Duct Wall Actuator Surface, *Unsteady Aerodynamics and Aeroelasticity of Turbomachines*, Elsevier Science B.V., pp.565–581.
4. Nagai, K. and Namba, M. (1995) Acoustic Control of Subsonic Cascade Flutter in Bending Motion, *Proceedings of the 12th International Symposium on Air Breathing Engines*, **Vol.2**, pp.992–1000.
5. Nagai, K. and Namba, M. (1996) Active Control of Cascade Flutter by Means of Sound Waves from Duct Wall Sources, *JSME International Journal*, **Vol.39, No.3**, pp.608–614.
6. Nagai, K. and Namba, M. (1996) Active Control of Supersonic Cascade Flutter by Sound Waves from Actuator Surface, *Transactions of JSME*, **Series B, Vol.62, No.604**, pp.4119–4126.
7. Namba, M. and Toshimitsu, K. (1991) Double Linearization Theory of Three-Dimensional Cascades with Vibrating Blades under Spanwise-Nonuniform Mean Loading, II: Supersonic Flow, *Journal of Sound and Vibration*, **Vol.148, No.1**, pp.69-102.
8. Namba, M. and Yamasaki, N. (1987) Effect of Sound Absorbing Wall Liners on Unsteady Aerodynamic Forces of a Cascade in a Supersonic Flow, *Transactions of JSME*, **Series B, Vol.53, No.492**, pp.2281–2288.

4: Theoretical Studies of Unsteady Flow Around Oscillating Blades - 2

NONLINEAR HARMONIC ANALYSIS OF UNSTEADY TRANSONIC INVISCID AND VISCOUS FLOWS

L. HE and W. NING
School of Engineering, University of Durham
Durham DH1 3LE
UK

Abstract

The nonlinear harmonic approach has been recently developed, in order to effectively include nonlinear effects in turbomachinery blade flutter analysis. In this paper, the effectiveness in dealing with nonlinear effects associated with shock oscillations are assessed by examining unsteady flows around an oscillating transonic cascade. The computed results using a nonlinear harmonic Euler method are compared with those by a nonlinear time-marching Euler solver. The capability of the new approach is clearly demonstrated, although its limitation in a very nonlinear situation is also indicated. Using the nonlinear harmonic approach, the inviscid Euler method has been extended to a viscous solver with the Navier-Stokes equations and a calculation of unsteady turbulent flow in a transonic diffuser is presented.

1. Introduction

In recent years, unsteady computational fluid dynamics methods have been gradually introduced in the turbomachine design and development systems to improve aerodynamic performance and structural integrity. It is recognised that typical nonlinear time-domain unsteady flow solutions require much more computing time than their steady flow counterparts. For design purposes, efficient and accurate unsteady flow solvers are much needed.

The linear harmonic frequency-domain methods are computationally very efficient and widely used in the turbomachine design systems, mainly for blade flutter analysis. Notable examples are the time-linearized potential methods (Verdon and Caspar[1], Whitehead[2]) and Euler methods (Hall and Crawley[3], Lindquist and Giles[4]). However, the linear assumption adopted in this kind of harmonic methods might well limit their applications when nonlinearity is strong. For example, unsteady transonic flows with significant shock motions may cause considerable nonlinear effects on aerodynamic damping (Ayer and Verdon[5]). To relax the limitation due to the linear assumption, the

T.H. Fransson (ed.), Unsteady Aerodynamics and Aeroelasticity of Turbomachines, 183-193.

present authors have recently developed a new frequency domain based approach, the nonlinear harmonic method (Ning and He[6]). In this method, unsteady flow is decomposed into a time-averaged flow plus a small harmonically varying unsteady flow. The nonlinear effects are effectively captured by the coupling between the time-averaged flow and the unsteady disturbances.

This paper is intended to report some of the latest work in developing and validating the nonlinear harmonic method. In the following sections, the basic methodology will briefly be described. Then an unsteady flow around an oscillating transonic cascade will be examined by using the inviscid Euler solver. The nonlinear harmonic inviscid flow method has been extended to calculate unsteady viscous flows by using Navier-Stokes Equations with a mixing length turbulence model. Some computational results for an unsteady viscous flow in a diffuser will also be presented.

2. Nonlinear Harmonic Methodology

In the conventional time-linear model, the steady flow field will not be affected by unsteadiness, and therefore the time-averaged flow is the same as the steady one. The nonlinearity manifests itself in the difference of time-averaged flow from the steady one. And the difference in time-averaged flow will in turn affect the unsteady perturbations. A simple example of this nonlinear effect can be found by looking at a sinusoidal oscillating normal shock wave. The time-averaged pressure distribution will show a much smeared pressure jump, compared to the steady one. And the unsteady pressure due to the shock motion will have smaller peak value and cover wider region, compared to what would be predicted by a linear method. The difference between the results from a time-linear method and those from a nonlinear method strongly depends on the amplitude of the shock oscillation.

The basic idea of the nonlinear harmonic method is to include the nonlinear effect on the time-averaged flows, and to calculate the unsteady perturbations based on the time-averaged flow field. To illustrate the methodology in a simple way, we consider a 1-D convection equation:

$$\frac{\partial u}{\partial t} + \frac{1}{2}\frac{\partial uu}{\partial x} = 0 \qquad (1)$$

An instantaneous flow variable can be expressed as:

$$u = \overline{u} + u'(x,t) \qquad (2)$$

where $\overline{u}$ is the time-averaged quantity, u' is a periodic unsteady perturbation. Substitute equation (2) into the equation (1), we have:

$$\frac{\partial u'}{\partial t}+\frac{1}{2}\frac{\partial}{\partial x}(\bar{u}\bar{u}+2\bar{u}u'+u'u')=0 \tag{3}$$

The time-averaged equation can be obtained by time-averaging equation (3):

$$\frac{\partial \bar{u}\bar{u}}{\partial x}+\frac{\partial}{\partial x}(\overline{u'u'})=0 \tag{4}$$

Compared to the equation (1), time-averaging generates an extra term in the time-averaged equation. This extra term $\frac{\partial}{\partial x}(\overline{u'u'})$ is a nonlinear term which is similar to the turbulence (Reynolds) stress term. For the situation of present interest, we call it 'unsteady stress' term because it is generated by a periodic unsteadiness.

The equation for unsteady perturbations can be obtained by the difference between the basic unsteady flow equation (1) and the time-averaged equation (4), e.g.

$$\frac{\partial u'}{\partial t}+\frac{1}{2}\frac{\partial}{\partial x}(2\bar{u}u'+u'u'-\overline{u'u'})=0 \tag{5}$$

But this complete form of the unsteady perturbation equation is not readily solvable if a frequency-domain harmonic approach is to be used. It is assumed that the unsteady perturbation is dominated by the 1st order term. Consequently, the second order terms in the unsteady perturbation equation (5) are neglected. The resultant first order equation is :

$$\frac{\partial u'}{\partial t}+\frac{\partial}{\partial x}(\bar{u}u')=0 \tag{6}$$

It is further assumed that the unsteady perturbation changes in a harmonic mode,

$$u'=\tilde{u}e^{i\omega t} \tag{7}$$

where $\tilde{u}$ is complex amplitude of the unsteady perturbation u', ω is the angular frequency of the periodic unsteadiness. Substituting the above into equation(6), we can obtain the 1st order harmonic perturbation equation, which gives:

$$i\omega\tilde{u}+\bar{u}\frac{\partial\tilde{u}}{\partial x}+\tilde{u}\frac{\partial\bar{u}}{\partial x}=0 \tag{8}$$

The unsteady perturbation equation (8) is in the same form as the perturbation equation in the conventional time-linearized method. However the equation (8) is no longer linear, because the time-averaged variable $\bar{u}$ is unknown, which in turn depends on the unsteady perturbation. Because of the interaction between the time-averaged and the unsteady perturbation equations, a coupled approach needs to be used to solve the system consisting of these two equations, such as the simultaneous time-marching applied to the aerodynamic and structural coupling (He[7]).

It should be mentioned that the nonlinear effect on the time-averaged flow is only included as the product term of the 1st order unsteady perturbations. In most circumstances, the contributions from the 1st order terms are likely to be dominant.

A nonlinear harmonic Euler method has been developed by authors (Ning and He[6]) based on this methodology. Recently the nonlinear harmonic Euler method has been extended to a Navier-Stokes solver by using the Reynolds averaged Navier-Stokes equations, and the Baldwin-Lomax mixing length model is incorporated to deal with the turbulent flow.

3. Numerical Results and Discussions

In the present work, two numerical cases are considered. The first one is a computation of transonic inviscid unsteady flows around an oscillating biconvex cascade by using the nonlinear harmonic Euler method. The second one is a computation of transonic turbulent unsteady flows in a diffuser by using the nonlinear harmonic Navier-Stokes method.

3.1. OSCILLATING BICONVEX CASCADE (EULER SOLUTION)

Calculations of unsteady transonic flows using the Euler equations are particularly useful because they can be used to address the issue of nonlinearity associated with shock oscillation without being confused with that due to turbulence. In this case, the biconvex cascade has a 0 degree stagger angle, with chord of 0.1524m, relative thickness of 0.076, solidity of 1.3. The cascade is oscillating in a torsion mode around the leading edge with Inter-Blade-Phase-Angle (IBPA) of 180 degree, reduced frequency of 1.3. It has been previously illustrated by a nonlinear time-marching calculation(He[9]) that a strong nonlinear behaviour of shock movement could be resulted in when the shock is oscillating near the throat.

In the present calculations, two back pressure conditions are used to set up distinctly different shock positions. Firstly, a lower pressure ratio of $p_b / P_0 = 0.7$ is specified, where p_b is the back static pressure, P_0 is the inlet stagnation pressure. At this condition, a fairly strong shock with an upstream Mach number of 1.3 is situated in the blade passage around 85 % chord, as shown in the steady Mach number contours in Fig.1. In the unsteady calculations, the torsion amplitude is specified as 2 degrees. The unsteady calculations are carried out by three different methods, the nonlinear time-marching method (He[9]), the time-linearized Euler method and the nonlinear harmonic

Euler method (Ning and He[6]). Even though the amplitude of the blade vibration is very big (2°), the shock oscillation is confined in a relatively small region, as indicated by the unsteady pressure 'impulse' shown in Fig.3. The time-averaged Mach number distributions by the two nonlinear methods are very close to the steady solution (Fig.2). The calculated amplitudes and phases of unsteady pressures on blade surfaces by using the three methods (Fig.3) agree well. Overall, the computational results for this strong shock case suggest a very weak nonlinear effect. Thus, the conventional time linear approach should be sufficiently adequate for this case.

Then the pressure ratio is increased to 0.725 to push the shock forward. The steady Mach contours are shown in Fig.4. It can be seen that the steady shock is at around 65 % chord, quite close to the cascade throat. In the unsteady calculations, firstly the torsion amplitude of 0.75° is used. The calculated steady and time-averaged Mach number distributions are presented in Fig.5. At this back pressure condition, the time-averaged pressure distribution is very different to the steady one. The shock moves in a wider range around its mean position when the cascade is oscillating. This smeared time-averaged pressure jump due to the shock oscillation is very well predicted by the nonlinear harmonic method as shown in Fig.5. The unsteady shock impulse predicted by the time-linearized method is much higher and narrower than that predicted by the nonlinear time-marching method (Fig.6). The comparison between the nonlinear time-marching method and the nonlinear harmonic method is excellent. These comparisons demonstrate the effectiveness of the new nonlinear harmonic approach, since the nonlinear time-marching solutions are much more time-consuming than the frequency based harmonic solutions.

At the same pressure ratio of 0.725, further calculations are carried out at a torsion amplitude of 2°. From the nonlinear time-marching calculation we found that the unsteady shock behaves very differently from that in the case with the amplitude of 0.75°. In this case, when the shock moves to the throat, it does not decay into a compression wave, as one would expect in a quasi-steady sense. The unsteady inertia makes the shock move into the subsonic region. Once in the subsonic region, the shock has to propagate upstream and eventually disappears around the leading edge. Then a new shock appears around the throat. This dramatic shock movement generates very strong nonlinear effects. As a result, the time-averaged Mach number distribution is markedly different from the steady one as shown in Fig.7. Due to the strong nonlinearity the calculated unsteady pressure by the linear method and the nonlinear time-marching method are very different as shown in Fig.8. Although improvement can be seen from th result by the nonlinear harmonic method, there is an apparent discrepancy from the nonlinear time-marching calculation. This indicates the limit of the applicabity of the present nonlinear harmonic method.

3.2. UNSTEADY TURBULENT FLOW IN TRANSONIC DIFFUSER (NAVIER-STOKES SOLUTION)

This case was experimentally studied at McDonell-Douglas by Sajben et al[10]. In the present numerical study is carried out with a pressure ratio of 0.826, producing a shock with pre-shock Mach number of 1.24 at $x/h^* = 1.4$, here x is the streamwise distance measured from the throat and h^* is the throat height. The transonic flow is assumed to be fully turbulent from diffuser inlet and the Reynolds number is 1.1×10^6. The steady flow in this transonic diffuser is calculated and the steady Mach number contours are shown in Fig.9. The calculated steady pressure distribution along the diffuser top wall is shown in Fig.10. The experimental shock position and strength are well predicted by the present steady viscous calculation. The marked difference between the viscous and inviscid solutions indicates a significant blockage effect of boundary layers.

For this case, an unsteadiness was introduced by back pressure oscillation. Calculations are carried out at a frequency of 300 Hz. In the nonlinear harmonic method, the steady turbulence viscosity generated by the Boldwin-Lomax model is freezed for unsteady perturbation calculations. The nonlinear harmonic Navier-Stokes computations are compared to the results from the nonlinear time-marching method and available experiment data. The amplitude and phase distributions of the unsteady pressure along the top wall are shown in Fig.11. The comparison between the nonlinear harmonic Navier-Stokes method and the nonlinear time-marching method is very good, and both of them match the experimental results well. Further validations of the nonlinear harmonic Navier-Stokes method are on going.

4. Concluding Remarks

In this paper, some of the recent work in developing and validating efficient unsteady flow computational methods based on a nonlinear harmonic methodology has been reported. The basic intention is to adopt the efficient frequency domain solution method while include the nonlinear coupling between the time-averaged flow and the unsteady disturbances. So far quasi 3 dimensional nonlinear harmonic Euler and Navier-Stokes methods have been developed. Computational results covering a range of cases have clearly demonstrated the effectiveness of this kind of methods.

Acknowledgement

This work is sponsored by European Gas Turbines. The authors wish to acknowledge Roger Wells and Yansheng Li for many helpful discussions.

References

1. Verdon, J.M. and Caspar, J.R., " A Linearized Unsteady Aerodynamic Analysis for Transonic Cascades", *Journal for Fluid Mechanics*, Vol.149, pp.403-429

2. Whitehead, D.S. (1990), " A Finite Element Solution of Unsteady Flow in Cascades", *International Journal for Fluid Mechanics*, Vol.149, pp.403-429

3. Hall, K.C. and Crawley, E.F. (1989), " Calculation of Unsteady Flows in Turbomachinery Using the Linearized Euler Equations", *AIAA Journal*, Vol.27, No.6, pp.777-787

4. Linquist, D.R. and Giles, M.B. (1994), " Validaity of Linearized Unsteady Euler Equations with Shock Capturing", *AIAA Journal*, Vol.32. No.1, pp.46-53

5. Ayer, T.C. and Verdon, J.M., (1996), " Validation of Nonlinear Unsteady Aerodynamic Simulator for Vibrating Blade Rows,", ASME Paper 96-GT-340, to appear in the ASME *Journal of Turbomachinery*

6. Ning, W. and He, L., (1997), " Computation of Unsteady Flows Around Oscillating Blades Using Linear and Nonlinear Harmonic Euler Methods", ASME Paper 97-GT-229, to appear in the *ASME Journal of Turbomachinery.*

7. He, L. (1994), " Integration of 2-D Fluid/Structure Coupled System for Calculations of Turbomachinery Aerodynamics/Aeroelastic Instabilities", *International Journal of Computational Fluid Dynamics*, Vol.3, pp217-231

8. He, L. (1996), " I. Modelling Issues for Computation of Unsteady Turbomachinery Flows", VKI Lectures Series "Unsteady Flows in Turbomachines", 1996-05.

9. He, L. (1990), " An Euler Solution for Unsteady Flows Around Oscillating Blades", *ASME Journal of Turbomachinery*, Vol.112, pp714-722, 1990

10. Sajben, M., Bogar, T.J. and Kroutil, J. C. (1984), " Forced Oscillation Experiments in Supercritical Diffuser Flows", *AIAA Journal*, Vol. 22, No.4, 1984

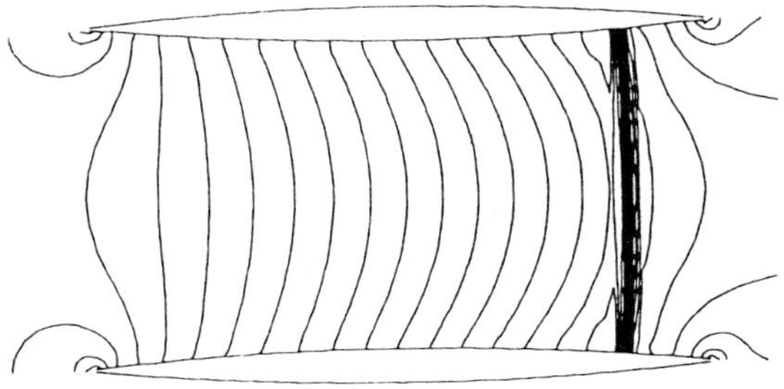

Fig.1 Steady Mach number contours at pressure ratio of 0.7

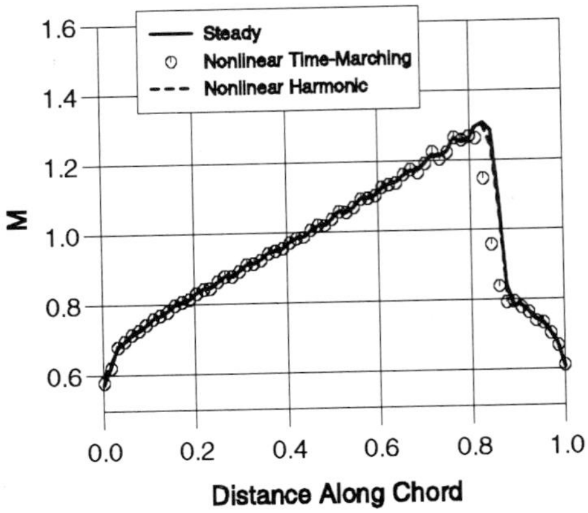

Fig.2 Steady and time-averaged Mach number distributions

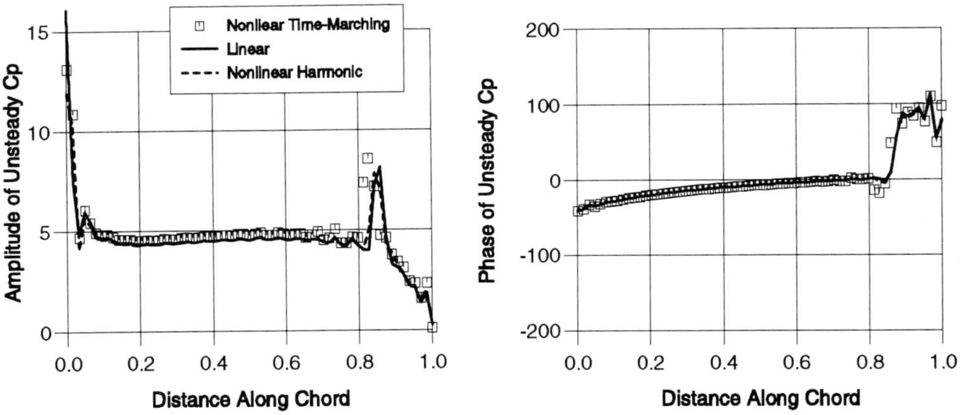

Fig.3 Unsteady pressure distributions at pressure ratio of 0.7

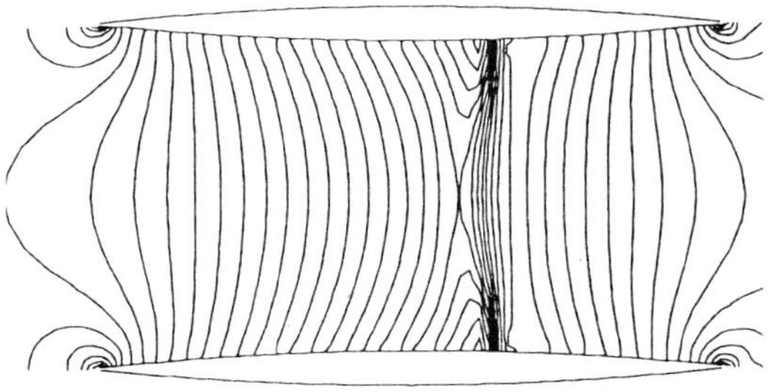

Fig.4 Steady Mach number contours at pressure ratio of 0.725

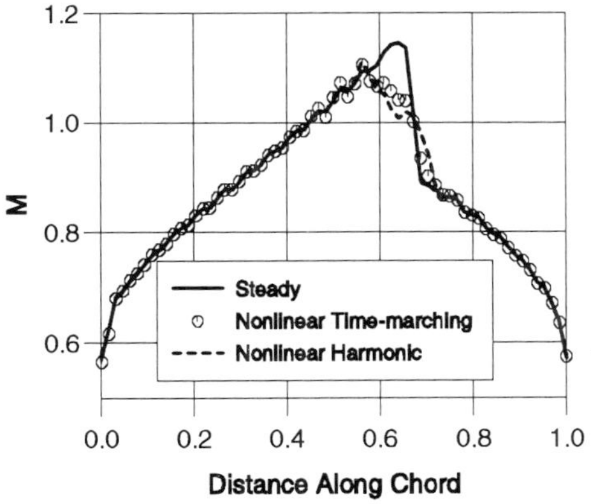

Fig.5 Steady and time-averaged Mach number distributions (torsion amplitude 0.75 deg.)

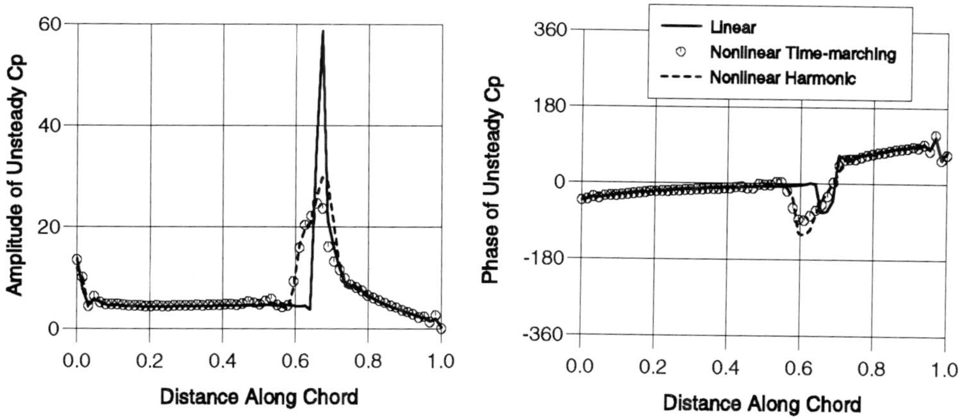

Fig.6 Unsteady pressure distributions at pressure ratio of 0.725 (torsion amplitude 0.75 deg.)

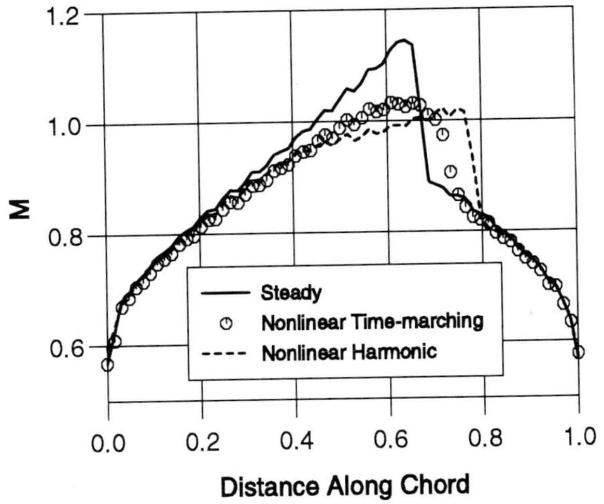

Fig.7 Steady and time-averaged Mach number distributions at pressure ratio 0.725 (torsion amplitude 2 deg.)

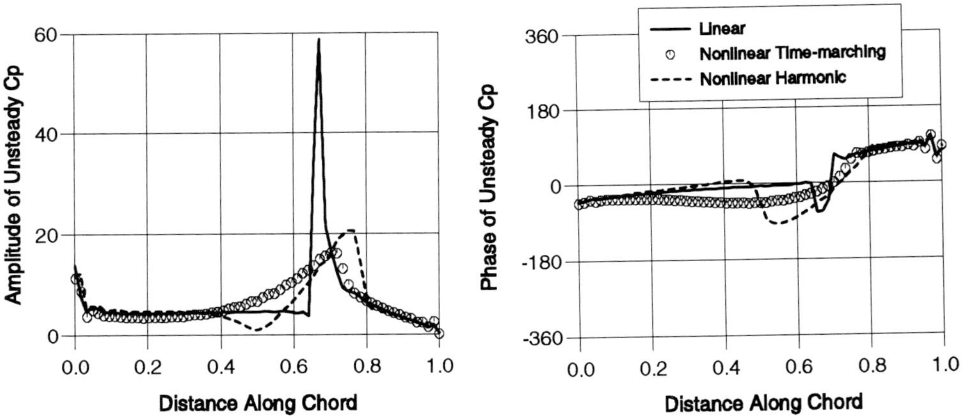

Fig.8 Unsteady pressure distributions at pressure ratio of o.725 (torsion amplitude 2 deg.)

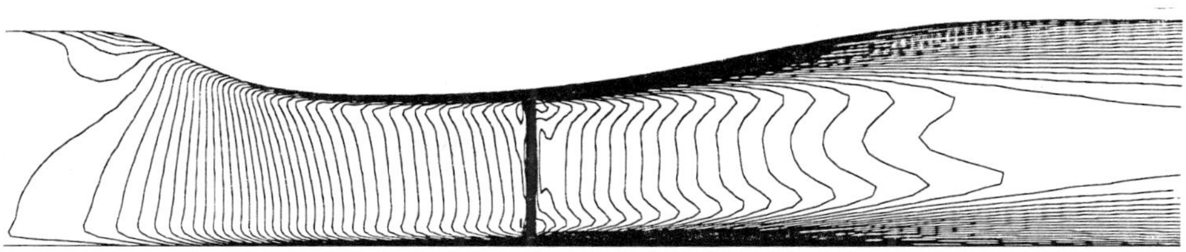

Fig.9 Steady Mach number contours in a transonic diffuser (vertical scale enlarged by a factor of 2)

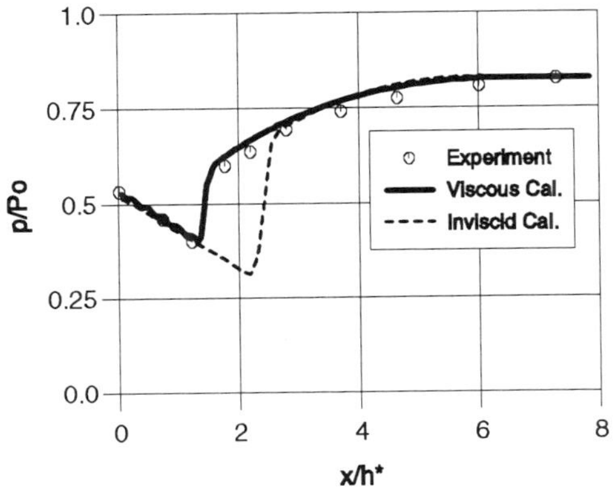

Fig.10 Steady static pressure on top wall of a transonic diffuser

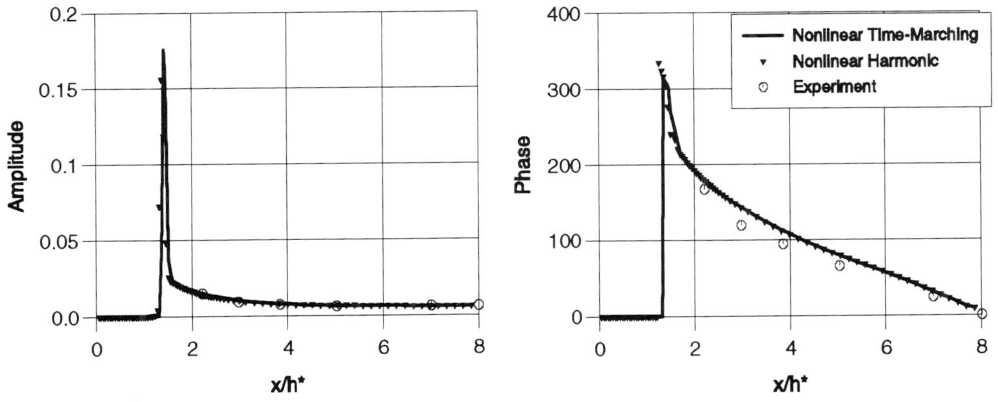

Fig.11 Unsteady pressure distributions on top wall of a transonic diffuser

FLUTTER ANALYSIS OF TWO - DIMENSIONAL VISCOUS SUBSONIC AND TRANSONIC FLOW IN TURBOMACHINES USING THE ADVECTION UPSTREAM SPLITTING METHOD

WOLFGANG HÖHN AND TORSTEN H. FRANSSON
KTH, The Royal Institute of Technology
S-100 44 Stockholm, Sweden

Abstract. An unsteady, compressible, two-dimensional, parallelized Navier Stokes solver is developed to predict the flow around cascaded vibrating blades. The presented method applies the Advection Upstream Splitting Method to discretize the convective terms and central differences for the diffusive terms of the Navier-Stokes equations on a structured H-type mesh in boundary-fitted coordinates. Turbulence is modeled by using the Baldwin-Lomax turbulence model. The time accurate integration of the governing equations is performed by applying an explicit four-stage Runge-Kutta scheme. Some of the calculations are conducted on a IBM/SP2 parallel computer using the message passing system MPI. At the farfield boundary, viscous effects are assumed to be negligible and for steady state and the unsteady flow capacitive boundary conditions based on the theory of characteristics for the locally one-dimensional problem are used. Unsteady inviscid calculations are performed with the same code using non-reflecting boundary condtitions.

Emphasis is laid on non-linear effects within the blade flutter phenomenon, i.e. cases where separation and shock boundary-layer interaction occurs in compressor cascades. In detail the pressure response of the flow on imposed blade motions is presented for these flow situations. The steady state results agree well for viscous flow for the two shown test cases with other viscous and inviscid flowsolvers. Unsteady viscous flow calculations on these flow configurations show the applicability of the presented method to the blade flutter phenomenon. For separated subsonic unsteady viscous flow good agreement with other flow solvers was found, whereas the test case for transonic viscous unsteady flow showed a big influence of the boundary conditions on the results.

T.H. Fransson (ed.), Unsteady Aerodynamics and Aeroelasticity of Turbomachines, 195-210.

1. Introduction

The design of more efficient turbomachines leads to higher loaded blades with increased flow temperatures and velocities. On the other hand, improved lighter blade materials for turbo-engines are introduced as well as more slender blades in order to increase the performance and reduce the weight. This is of particular interest, especially for aircraft engines. Long blades in compressors or in the last stage of turbines therefore become more sensitive to flow-induced vibrations. Consequently, aeroelasticity is an important discipline as the combination of the aerodynamic and the structural dynamics of the blade in turbomachines.

Different kinds of aeroelastic phenomena are defined in the literature [Försching, 1974]. They can be divided into static and dynamic appearances. The latter can be divided in forced vibrations and self-exited vibrations, usually called flutter. In order to avoid flutter occurring in turbomachines, it is useful to have a powerful prediction model to calculate the unsteady flow through cascaded blades. Then, some of the cost- and time consuming experiments can be avoided and the different parameters which influence the flutter phenomena can be studied.

Since the beginning of the 80's a number of methods using linearized models [Whitehead, 1990] have been developed to study the classical blade flutter. In such approaches the unsteady equations of motions are linearized with respect to small harmonic perturbations in time. Consequently, the unsteady surface-pressure distribution is independent of the magnitude of the small harmonic motion. For realistic modern flow configurations, nonlinear and viscous effects are far from negligible, especially in the vicinity of shocks as well as for shock boundary-layer interaction. Due to that the aerodynamic work of the fluid on the structure can be strongly influenced by separated flow situations in compressors and by shocks occurring in turbine cascades. For this purpose several Euler-, for example Fransson and Pandolfi [1986] or Huff et al. [1991], and Navier-Stokes procedures [Siden, 1991] have been developed.

The presented numerical method accounts for unsteady, viscous, fully turbulent flow around cascaded blades towards a better understanding of the above mentioned flow situations and uses the "Advection Upstream Splitting Method" (AUSM) [Wada and Liou, 1994] for the discretization of the convective terms and central differences for the diffusive terms of the Navier-Stokes equations. Advantages of the method are the simple implementation and the fact that there is no need to specify numerical dissipation parameters. The blade flutter phenomenon is simulated by imposing a motion on the blade, which consists of harmonic body translation in two directions and rotation, allowing an interblade phase angle between two neighbouring blades. The validity of the code is illustrated by several test cases, where non-linear effects within the blade flutter phenomena are investigated.

2. Governing Equations

2.1. NAVIER-STOKES EQUATIONS

The non-dimensionalized Reynolds-averaged Navier-Stokes equations for the conservation of mass, momentum and energy read [Höhn, 1996]:

$$\frac{\partial U}{\partial t} + \frac{\partial F}{\partial x} + \frac{\partial G}{\partial y} = \frac{\rho^*_{-\infty} q^*_{-\infty}}{Re_{-\infty}} \left(\frac{\partial R}{\partial x} + \frac{\partial S}{\partial y} \right) \quad (1)$$

with

$$U = \begin{pmatrix} \rho \\ \rho u \\ \rho v \\ \rho e \end{pmatrix}, F = \begin{pmatrix} \rho u \\ \rho u^2 + p \\ \rho u v \\ \rho u H \end{pmatrix}, G = \begin{pmatrix} \rho v \\ \rho u v \\ \rho v^2 + p \\ \rho v H \end{pmatrix},$$

$$R = \begin{pmatrix} 0 \\ \tau_{xx} \\ \tau_{xy} \\ u\tau_{xx} + v\tau_{xy} + \frac{\mu' \partial h}{Pr \partial x} \end{pmatrix}, S = \begin{pmatrix} 0 \\ \tau_{xy} \\ \tau_{yy} \\ u\tau_{xy} + v\tau_{yy} + \frac{\mu' \partial h}{Pr \partial y} \end{pmatrix}. \quad (2)$$

2.2. TRANSFORMATION OF THE GOVERNING EQUATIONS

If new independent variables are introduced, i.e. a boundary-fitted moving co-ordinate system, a strong conservation law form can be maintained. Subject to the general transformation $\xi = \xi(x, y, t), \eta = \eta(x, y, t), \tau = t$ the equation can be written as

$$\partial_\tau \bar{U} + \partial_\xi \bar{F} + \partial_\eta \bar{G} = \frac{\rho^*_{-\infty} q^*_{-\infty}}{Re_{-\infty}} \left(\partial_\xi \left(y_\eta R - x_\eta S \right) \right) + \left(\partial_\eta \left(-y_\xi R + x_\xi S \right) \right) \quad (3)$$

with

$$\bar{U} = \frac{U}{J}, \bar{F} = \left(\left(-x_\tau y_\eta + y_\tau x_\eta \right) U + y_\eta F - x_\eta G \right),$$
$$\bar{G} = \left(\left(-x_\tau y_\xi - y_\tau x_\xi \right) U - y_\xi F - x_\xi G \right). \quad (4)$$

2.3. ALGEBRAIC TURBULENCE MODEL

The extension to the turbulent flow is achieved by implementing the algebraic turbulence model of Baldwin and Lomax [1978]. In stress terms of the laminar Navier-Stokes equations, the molecular coefficient of viscosity μ is replaced by $\mu + \mu_t$. In heat flux the term $k/cp = \mu/Pr$ is changed to $\mu/Pr + \mu_t/Pr_t$.

2.4. BOUNDARY CONDITIONS

The boundary conditions for the calculations are no slip at the adiabatic walls. The pressure is linearly extrapolated. The energy is then determined using the equation of state for an ideal gas and the adiabatic wall conditions. The computations are performed on a single blade passage using periodic boundary conditions [Höhn, 1996]. For steady state- and unsteady flow conditions, boundary conditions based on the theory of characteristics for the locally one-dimensional problem are used [Ott, 1991]. Comparative studies are made with inviscid calculations using non-reflecting boundary conditions by Giles [1991].

3. Spatial Discretization of the Governing Equations

3.1. DISCRETIZATION OF THE CONVECTIVE TERMS

Flux vector splitting schemes such as the Van Leer splitting [van Leer, 1982] have proved to be a simple and useful technique to arrive at upwind differencing. Unfortunately, the simplicity comes at a price of reduced accuracy due to the numerical diffusion. As a result, significant errors appear in the viscous region. One possibility to improve the modelling of boundary-layers is the Advection Upstream Splitting Method (AUSM), see Wada and Liou [1994]. The main step in the splitting method is to recognise that the inviscid flux vector, for instance F, consists of two physically distinct parts, namely the convective and the pressure term:

$$F = u \begin{pmatrix} \rho \\ \rho u \\ \rho v \\ \rho H \end{pmatrix} + \begin{pmatrix} 0 \\ p \\ 0 \\ 0 \end{pmatrix} \tag{5}$$

In a first step the vector U is evaluated at the cell face. Depending on the propagation, the independent flow variables are discretizised in the following way, for instance at the cell-face i+1/2,j [Höhn, 1996]:

$$\begin{aligned} U^-_{i+1/2,j} &= U_{i,j} + \Phi_{i,j}\frac{U_{i,j} - U_{i-1,j}}{2} \\ U^+_{i+1/2,j} &= U_{i+1,j} + \Phi_{i+1,j}\frac{U_{i+2,j} - U_{i+1,j}}{2} \end{aligned} \tag{6}$$

Φ denotes a "Flux-Limiter" which is switched from 1 to 0 where a shock occurs, which leads to 2nd or 1st order accuracy, respectively. Thereafter, the fluxes F^+ and F^- have to be calculated at the cell faces. One can write the F-flux, for example, in the following way in boundary-fitted coordinates:

$$F = \rho \begin{pmatrix} (y_\eta u - x_\eta v + \xi_t) \\ (y_\eta u - x_\eta v + \xi_t)\, u \\ (y_\eta u - x_\eta v + \xi_t)\, v \\ (y_\eta u - x_\eta v + \xi_t)\, H \end{pmatrix} + p \begin{pmatrix} 0 \\ y_\eta \\ -x_\eta \\ -\xi_t \end{pmatrix} \tag{7}$$

The term $(y_\eta u - x_\eta v + \xi_t)$ can be denoted as the contravariant velocity $\bar{u}^\xi$ for a moving grid. After that, the AUSM scheme is applied on this quasi one - dimensional formulation. The flux is splitted according to the direction of propagation into

$$F^{1/2} = \frac{1}{2}\left(\left(\rho\bar{u}^\xi\right)_{1/2}(\Psi_L + \Psi_R)_{1/2} + \left|\left(\rho\bar{u}^\xi\right)_{1/2}\right|(\Psi_L - \Psi_R)_{1/2}\right) + p_{1/2} \tag{8}$$

with $\Psi^T = (1, u, v, H)$. In the presented method the so-called AUSMD/V [Wada and Liou, 1994] splitting is used, it reads

$$\left(\rho\bar{u}^\xi\right)_{1/2} = \bar{u}_L^{\xi+}\rho_L + \bar{u}_R^{\xi-}\rho_R \tag{9}$$

for the mass flux. The splittings for $\bar{u}_L^{\xi+}$, $\bar{u}_R^{\xi-}$ are given as

$$\bar{u}_L^{\xi+} = \begin{cases} \alpha_L\left(\frac{\left(\bar{u}_L^{\xi+}+c_m\right)^2}{4c_m} - \frac{\bar{u}_L^{\xi+}+\left|\bar{u}_L^{\xi+}\right|}{2}\right) + \\ \frac{\bar{u}_L^{\xi+}+\left|\bar{u}_L^{\xi+}\right|}{2}, if \frac{\bar{u}_L^{\xi+}}{c_m} \leq 1 \\ \frac{\bar{u}_L^{\xi+}+\left|\bar{u}_L^{\xi+}\right|}{2}, otherwise \end{cases} \tag{10}$$

$$\bar{u}_R^{\xi-} = \begin{cases} \alpha_R\left(-\frac{\left(\bar{u}_R^{\xi-}-c_m\right)^2}{4c_m} - \frac{\bar{u}_R^{\xi-}-\left|\bar{u}_R^{\xi-}\right|}{2}\right) + \\ \frac{\bar{u}_R^{\xi-}-\left|\bar{u}_R^{\xi-}\right|}{2}, if \frac{\bar{u}_R^{\xi-}}{c_m} \leq 1 \\ \frac{\bar{u}_R^{\xi-}+\left|\bar{u}_R^{\xi-}\right|}{2}, otherwise \end{cases} \tag{11}$$

with

$$\alpha_L = \frac{2\,(p/\rho)_L}{(p/\rho)_L + (p/\rho)_R};\, \alpha_R = \frac{2\,(p/\rho)_R}{(p/\rho)_R + (p/\rho)_L} \tag{12}$$

and $c_m = max\,(a_L; a_R)$. Secondly, the pressure flux is

$$p_{1/2} = p_L^+ + p_R^-, \tag{13}$$

with

$$p^{+/-}_{1/2} = \begin{cases} \frac{1}{4} p_{L/R} \left(\frac{\bar{u}^{\xi +/-}_{L/R}}{c_m} \pm 1 \right)^2 \left(2 \mp \frac{\bar{u}^{\xi +/-}_{L/R}}{c_m} \right) \\ \quad if \frac{\bar{u}^{\xi +/-}_{L/R}}{c_m} \leq 1 \\ p_{L/R} \frac{\bar{u}^{\xi +/-}_{L/R} \pm \left| \bar{u}^{\xi +/-}_{L/R} \right|}{2\bar{u}^{\xi +/-}_{L/R}}, otherwise. \end{cases} \tag{14}$$

3.2. DISCRETIZATION OF THE DIFFUSIVE TERMS

The differencing for the diffusion terms representing shear stress and heat-transfer effects corresponds to a second order accurate central difference in which second derivatives are treated as differences across cell interfaces of first derivative terms, as

$$\partial_\eta S \approx \delta_\eta S_{i,j} = \left(S_{i,j+1/2} - S_{i,j-1/2} \right). \tag{15}$$

For example, the term

$$x_\xi \mu \left(\frac{\partial u}{\partial y} + \frac{\partial v}{\partial x} \right) \tag{16}$$

applying the chain rule under the thin-layer approximation (for the sake of simplicity) becomes

$$x_\xi \mu \left(\eta_y \frac{\partial u}{\partial \eta} + \eta_x \frac{\partial v}{\partial \eta} \right) \tag{17}$$

and is differenced in $S_{i,j+1/2}$ as

$$x_{\xi_{i,j+1/2}} \mu_{i,j+1/2} \left(\begin{array}{c} \eta_{y_{i,j+1/2}} \delta_\eta u_{i,j+1/2} + \\ \eta_{x_{i,j+1/2}} \delta_\eta v_{i,j+1/2} \end{array} \right) \tag{18}$$

where $\delta_\eta u_{i,j+1/2} = u_{i,j+1} - u_{i,j}$.

3.3. TIME DISCRETIZATION

A Runge-Kutta four-step integration scheme is used, which is second order accurate in time [Höhn, 1996].

4. Results

The applicability of the the presented method INSTHPT for simple flow situations, i.e. the laminar and turbulent flow over a flat plate and the laminar flow around a cylinder in 2D was demonstrated by Höhn and Fransson [1996].

In the following results are shown for viscous steady state- and unsteady cascade flow in axial turbomachines. Subsonic and transonic flow cases are studied where non-linear phenomena occur, i.e. separation and shock waves. Therefore, the standard configurations 5 and 10 were chosen as test cases [Fransson and Verdon, 1993].

4.1. STEADY STATE RESULTS

For steady state viscous flow the results of two test cases are presented. First, the results for standard configuration 5 for an angle of incidence of 4 degrees is shown, where separation takes place. Thereafter, viscous transonic flow, which is partly separated, is studied for standard configuration 10.

4.1.1. *Standard Configuration 5, 4 degrees incidence*

Figure 1 shows the geometry and the mesh for a cascade compressor case, i.e. the fifth standard configuration, which is used for calculations of the presented method INSTHPT with 80 points on the blade on the upper and the lower side, respectively and 120x100 meshpoints in total. The Reynolds number based on the inlet velocity and the chord length is 1.400.000 for this flow case and the Mach number for this computation is 0.51. Y^+ is chosen to be 1.5 close to the leading edge for the calculations conducted with INSTHPT. The results for the presented method INSTHPT shown in figure

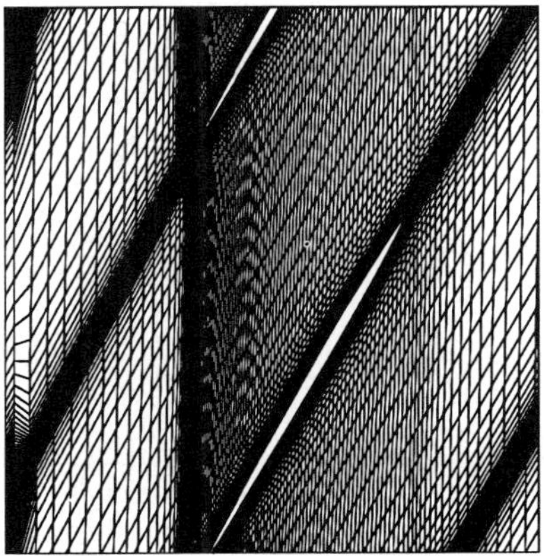

Figure 1. Mesh used for standard configuration 5, 120 x 100 meshpoints

2, the steady state pressure distribution, are compared with experiments and the Navier-Stokes solver VOLFAP [Sidén, 1991] as well as VOLSOL [Groth et al. 1996]. The computations for the presented method INSTHPT as well as VOLFAP are computed with an angle of incidence of four degrees and good agreement was found. The agreement with the experimental data as well as the results obtained with VOLSOL is not good at all. Earlier investigations [Hedberg, 1994] have shown that it is possible to match the experimental data by increasing the angle of incidence to 5 degrees. The computations of VOLSOL were conducted at an angle of incidence of 4.7 degrees and are quite close to the experiments. It can be concluded that the experiments can be matched by increasing the angle of incidence.

However, the results for the presented method were computed with an angle of incidence of 4 degrees and no efforts were made to match the experiments. Figure 3 shows the length of the separation bubble, which occurs

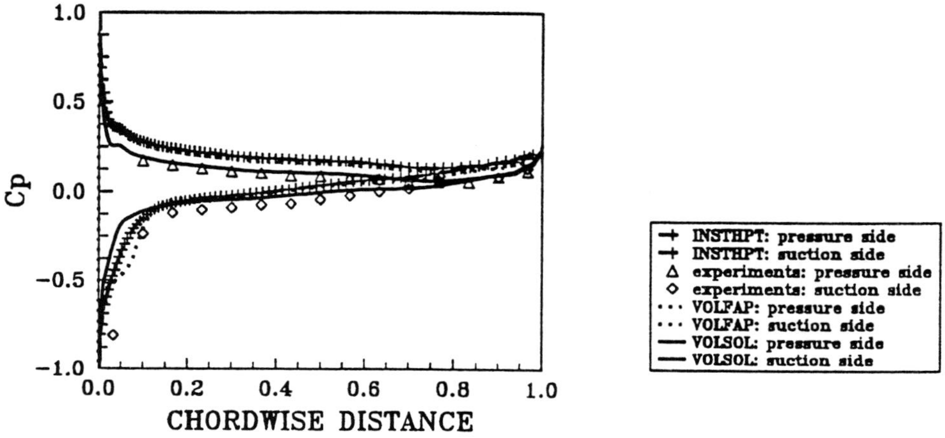

Figure 2. Steady state surface pressure distribution of standard configuration 5, case 2

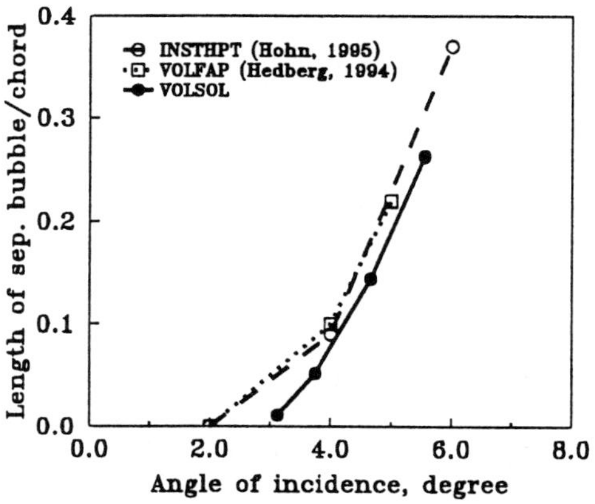

Figure 3. Length of the separation bubble against the angle of incidence, standard configuration 5

on the suction side of the blade close to the leading edge, against the angle of incidence. Comparison is made again with VOLSOL and VOLFAP and good agreement was found. VOLSOL seems to predict smaller separation bubbles than the other two codes, especially for small angles of incidence. The computations for all three codes are performed with the turbulence model by Baldwin and Lomax [1978].

4.1.2. *Standard Configuration 10,* $M_1 = 0.8$

The second test case is a transonic flow situation around the tenth standard configuration [Fransson and Verdon, 1993]. For this calculation a mesh was used, as shown in figure 4, which has 140 x 100 meshpoints for the viscous and 140x40 for the inviscid calculations. The mesh lines were normal to the local blade surface in order to capture the shock reasonably well.

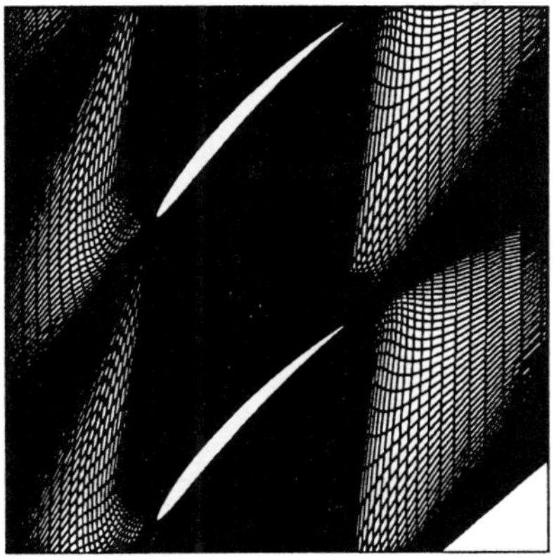

Figure 4. Mesh used for standard configuration 10, $M_1 = 0.8$

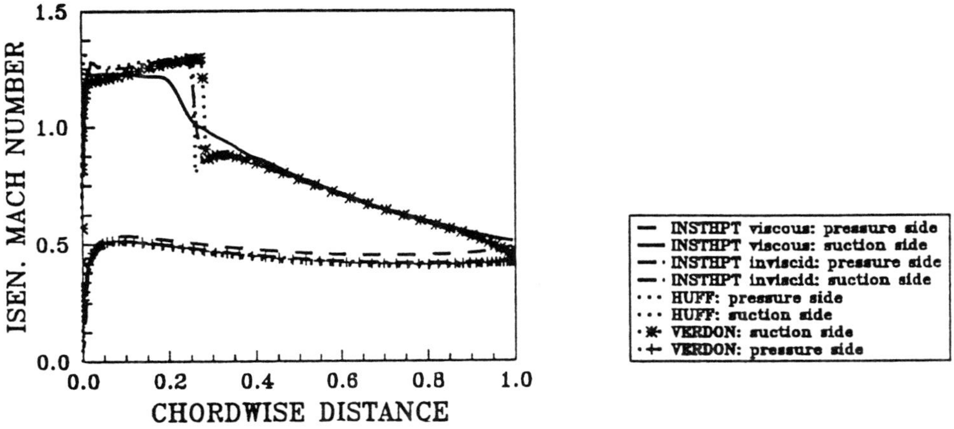

Figure 5. Steady state Mach number distribution of standard configuration 10, $M_1 = 0.8$

Figure 5 shows the steady state isentropic Mach number distribution for an inlet Mach number of 0.8 and an inlet flow angle of 58°. The viscous and inviscid results for the presented method INSTHPT are compared with a 2D linearized inviscid result of LINFLO by Verdon [Fransson and Verdon, 1993] and a 2D cascade solver based on the Euler equations by Huff et al. [1991]. The strength of the shock and the position are reasonably well predicted for the inviscid computations. For the viscous computations the computed Mach number distribution smears out a bit, which indicates shock boundary-layer interaction.

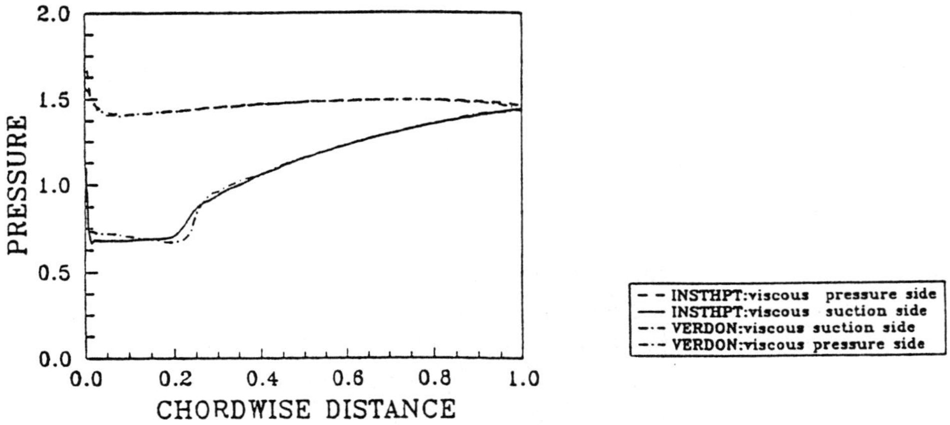

Figure 6. Steady state surface pressure distribution of standard configuration 10, $M_1 = 0.8$

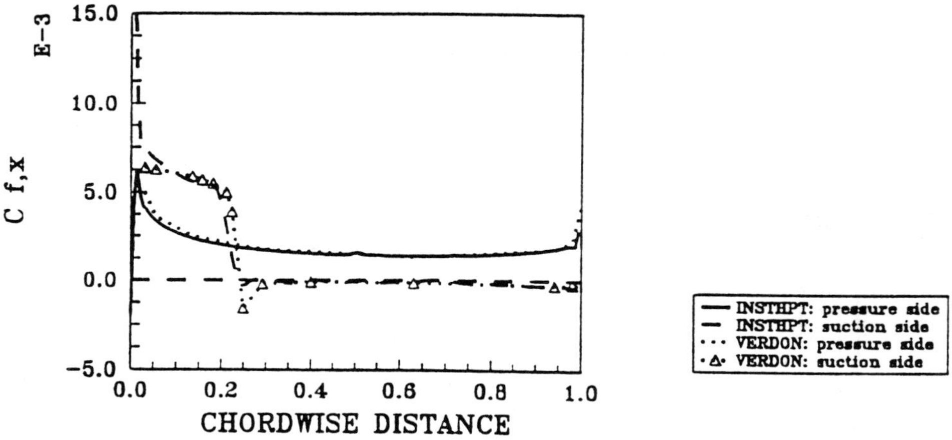

Figure 7. Skin friction of standard configuration 10, $M_1 = 0.8$

The behaviour in the shock region with respect to pressure is shown in figure 6 in greater detail. The computations performed by Ayer and Verdon [1996] with the Navier-Stokes solver NPHASE are slightly more accurate than the viscous predictions with the presented method INSTHPT. However, also the computations conducted by Ayer and Verdon [1996] indicate a shock boundary-layer interaction which results in a Mach number and pressure distribution which is not as steep as for the inviscid predictions in figure 5. Figure 7 shows the computed skin friction against results obtained by Ayer and Verdon [1996] using a Baldwin-Lomax turbulence model as well. The shock-induced separation agrees resonably well.

4.2. UNSTEADY RESULTS

Based on the steady state results of the previous section, unsteady computations were performed.

4.2.1. *Standard Configuration 5, incidence angle of four degrees*

The shown unsteady viscous flow situation was case 7 of standard configuration 5 according to Fransson and Verdon [1993]. The cascade is forced into a pure torsional vibration mode of a frequency of 550 Hz and an amplitude of 0.1^o with an interblade phase angle of 180° and a reduced frequency of 2.04 based on full chord. The steady state initial solution used is characterized by an incidence angle of 4° and an inlet Mach number of 0.5. As the figures 8 and 9 show the results for the unsteady amplitude and phase of the surface pressure distribution are in reasonable agreement with experiments presented by Fransson and Verdon [1993]. The results are shown

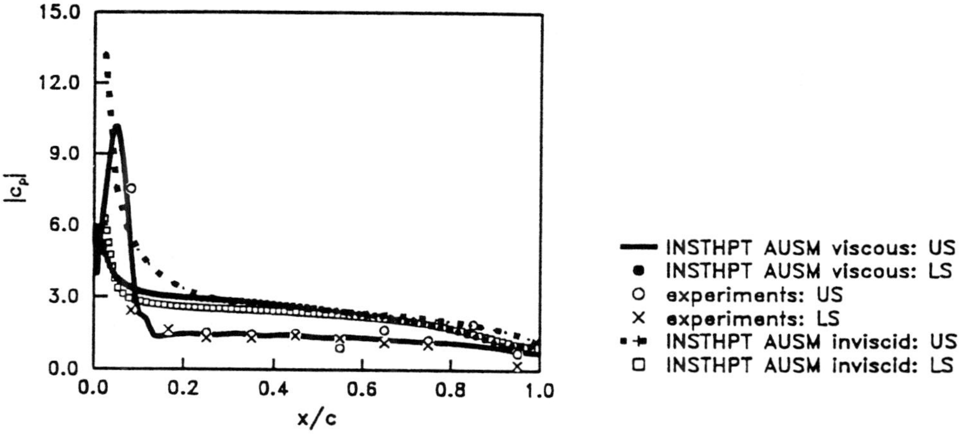

Figure 8. Amplitude of the unsteady pressure over chord for standard configuration 5 case 7, interblade phase angle of 180 degrees: comparison AUSM viscous and inviscid - experiments

for the pressure side (LS) and the suction side (US) of the blade for the viscous and inviscid version of the presented method INSTHPT. The inviscid code, using the same capacitive boundary conditions and running for 12 cycles as well (which was found to be sufficient for an unsteady periodic solution), shows on the pressure side (LS) as expected no big difference to the viscous code for the unsteady amplitude. The difference in the leading edge region is due to a coarser mesh for the inviscid solution. The used mesh had 80 x 40 mesh points for the inviscid computations. On the other hand, the suction side shows a big difference, which is expected due to the fact that separation occurs in the viscous solution. Since, the results

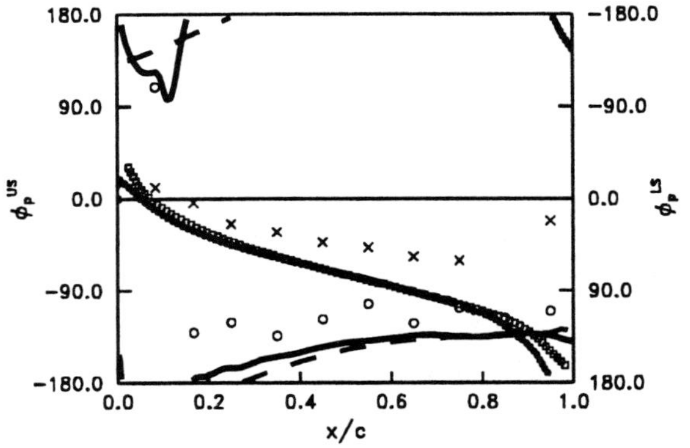

Figure 9. Phase of the unsteady pressure over chord for standard configuration 5 case 7, interblade phase angle of 180 degrees: comparison AUSM inviscid and viscous - experiments

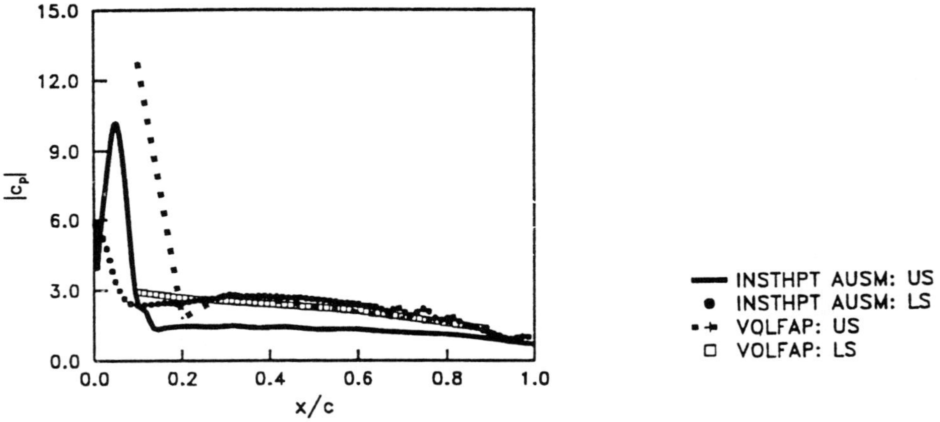

Figure 10. Magnitude for standard configuration 5 case 7, interblade phase angle of 180 degrees for VOLFAP and INSTHPT

cannot really be compared with the experiments where only one blade was excited, a comparison with the Navier-Stokes code VOLFAP [Sidén, 1991] was made under the same conditions as for the results obtained with the presented method. The results obtained with VOLFAP [Sidén, 1991] by Hedberg [1994], see figure 10, agree well with the presented viscous method on the pressure side but generally have higher amplitudes for the unsteady pressure amplitude on the suction side, especially close to the leading edge where separation occurs. The results agree fairly well in phase with the presented method for the suction side as well as for the pressure side, as can be seen in figure 11.

4.2.2. *Standard Configuration 10, $M_1 = 0.8$, (Case 24)*

The cascade is forced to pure pitching vibration of a reduced frequency of 2 based on full chord and an amplitude of 0.1 degrees with an interblade

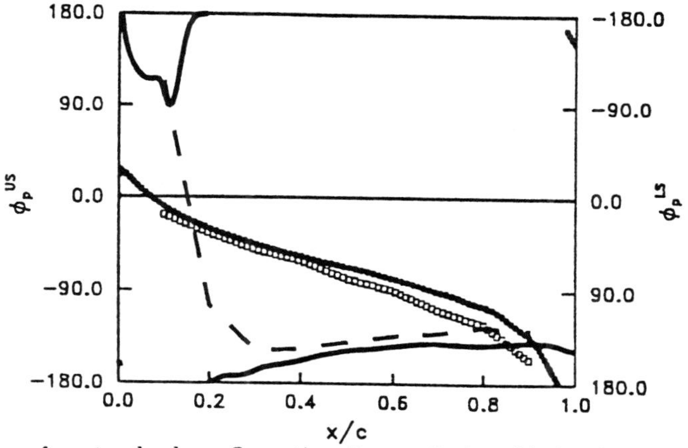

Figure 11. Phase for standard configuration 5 case 7, interblade phase angle of 180 degrees for VOLFAP and INSTHPT

phase angle of 90 degrees. The steady state solution used is characterized by an inlet Mach number of 0.8 and an inlet flow angle of 58 degrees.

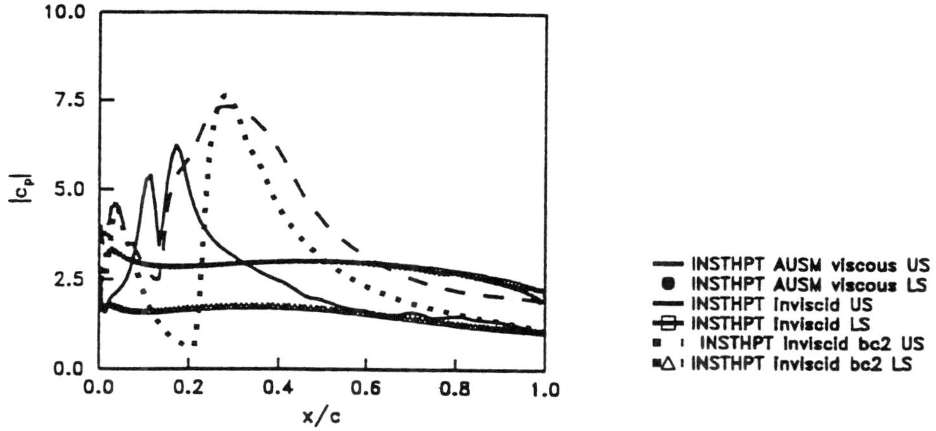

Figure 12. Magnitude for standard configuration 10, interblade phase angle of 90 degrees for INSTHPT, $M_1 = 0.8$

Figure 12 and 13 show the amplitude and phase of the unsteady surface pressure distribution based on the first Fourier component. The pictures show the results obtained with the presented method for viscous and inviscid calculations using capacitive boundary conditions [Ott, 1991] as well as inviscid calculations using approximately 2D non-reflecting boundary conditions developed by Giles [1991]. For the unsteady amplitude the results show a strong influence of the boundary conditions. The predicted results on the pressure side (LS) show an agreement between the capacitive boundary conditions for the viscous and inviscid computations but differ from the

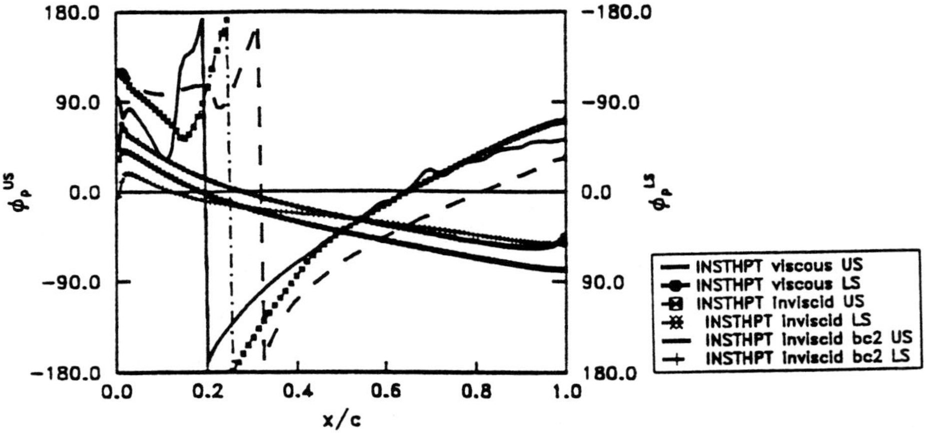

Figure 13. Phase for standard configuration 10, interblade phase angle of 90 degrees for INSTHPT, $M_1 = 0.8$

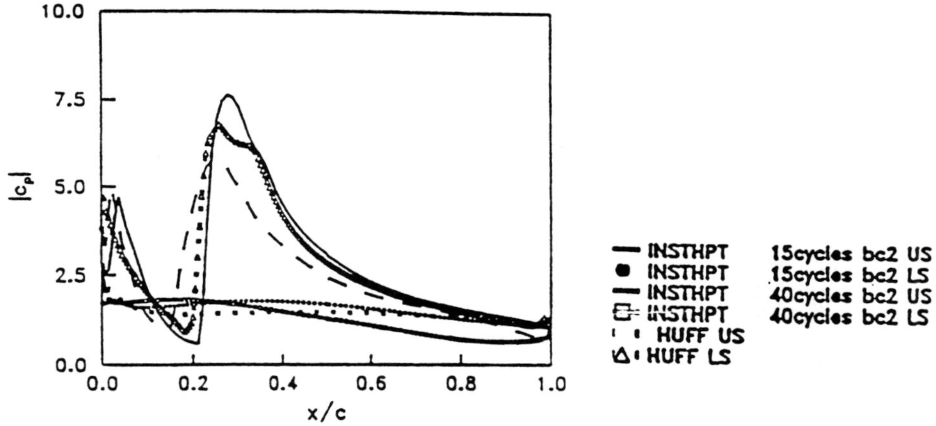

Figure 14. Magnitude for standard configuration 10, interblade phase angle of 90 degrees for INSTHPT, $M_1 = 0.8, inviscid calculation$

results for the inviscid code using 2D non-reflecting boundary conditions. On the suction side (US) the results for the inviscid methods agree in the location of the peak in amplitude due to the shock, but differ towards the trailing edge. The location of the shock for the viscous computations is moved upstream. The agreement in phase between the three different computations is much better. The phase shift due to the shock is predicted at different postitions, which corresponds to the different locations of the pressue peaks in the amplitude diagram. Several researchers have investigated the tenth standard configuration. The following two figures 14 and 15 show the computed results for the presented method for inviscid flow computed for different numbers of cycles against the inviscid results by Huff edited by Fransson and Verdon [1993]. Good agreement was found for the presented

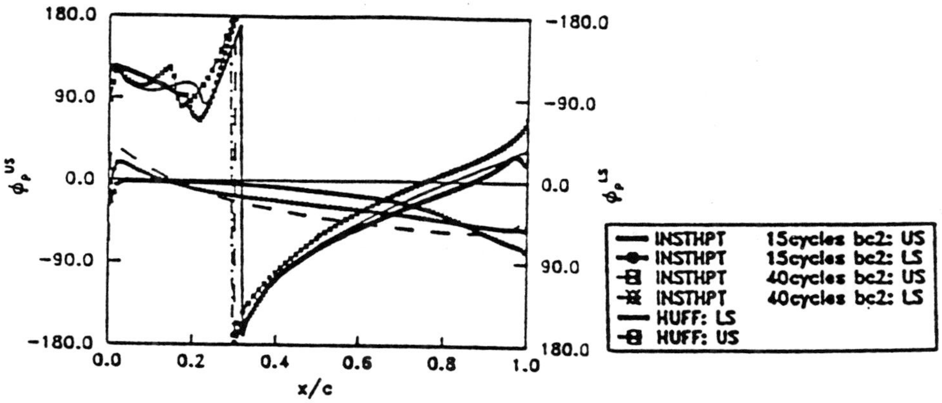

Figure 15. Phase for standard configuration 10, interblade phase angle of 90 degrees for INSTHPT, $M_1 = 0.8$*, inviscidcalculation*

method using the 2D approximately non-relfecting boundary conditions. It may be concluded that the capacitive boundary conditions do not give good results for this unsteady flow case, whereas the results shown for the non-reflecting boundary conditions are much closer to the results published in literature.

5. Conclusions

The presented method shows a good general agreement for the investigated test cases in comparison with experiments and results obtained by other flow solvers for steady state viscous flow in subsonic and transonic compressor cascades.

The unsteady results are in good agreement with other flow solvers for moderately separated flow in a subsonic compressor. The unsteady results for transonic compressor flow show a strong influence of the boundary condition on the unsteady pressure amplitude.

The presented method is easier to implement, computationally less expensive than the flux difference splitting methods and is just as accurate in the boundary layer [Höhn and Fransson, 1996]. Moreover, there is no need to specify a numerical dissipation in the code due to the upwind formulation.

6. Acknowledgements

The financial support of ASEA Brown Boveri (CH), Volvo Aero Corporation AB (S), Rolls Royce plc. (UK), NUTEK (S) and FMV (S) is gratefully acknowledged.

References

Ayer, T. C. and Verdon, J. M.; 1996
"Numerical Unsteady Aerodynamic Simulator for Vibrating Blade Rows." ASME Paper 96-GT-340, IGTI Birminingham, 1996.

Baldwin, B. and Lomax, H.; 1978
"Thin-Layer Approximation and Algebraic Model for Separated Turbulent Flow." AIAA Paper 78-257, 1978.

Förschіng, H.; 1974
"Aeroelastizität.", Springer-Verlag, Berlin-Heidelberg, 1974.

Fransson, T. H. and Pandolfi, M.; 1986
"Numerical Investigation of Unsteady Compressible Flow Through an Oscillating Cascade.", ASME Paper 86-GT-304, 1986.

Fransson, T. H. and Verdon, J.M.; 1993
"Standard Configurations for Unsteady Flow through Vibrating Axial-Flow Turbomachine Cascades.", in: Unsteady Aerodynamics, Aeroacustics and Aeroelasticity of Turbomachines and Propellers, H.M. Atassi (ed.), Springer-Verlag, New York, pp. 859 - 889, 1993.

Giles, M.; 1991
"UNSFLO User Manual.", GTL Report No. 205, MIT GTL, 1991.

Groth, J. P., Martensson, H., and Eriksson, L.-E.; 1996
"Validation of a 4D Finite Volume Method for Blade Flutter.", ASME Paper 96-GT-429 presented at ASME Turbo Expo, Birmingham, 1996.

Hedberg, K.; 1994
"Validation of Flutter Prediction Models.", Master of Science Thesis, Royal Institute of Technology, Chair of Heat and Power Technology, Stockholm, 1994.

Höhn, W. and Fransson, T.H.; 1996
"Flutter Analysis of Two Dimensional Viscous Flow in Turbomachines Using the Advection Upstream Splitting Method (AUSM).", in: Conf. Proceed., ECCOMAS Conference on Comp. Fluid Dynamics, Paris, pp . 917 - 923, Sept. 1996, Wiley & Sons.

Höhn, W.; 1996
"Unsteady Viscous Flow Around Cascaded Vibrating Blades at or Near Stall.", "Teknologi Licenciat", Royal Institute of Technology, Chair of Heat and Power Technology, Stockholm, 1996.

Huff, D.L., Swafford, T.W. and Reddy, T.S.R.; 1991
"Euler Flow Predictions for an Oscillating Cascade Using High Resolution Wave-Split Scheme.", ASME-Paper 91-GT-198, 1991.

Ott, P.; 1991
"Oszillierender senkrechter Verdichtungsstoss in einer ebenen Duese.", Thèse No. 985, Ecole Polytechnique Fderale de Lausanne, 1991.

Sidén, L. D. G.; 1991
"Numerical Simulation of Viscous Compressible Flow Applied to Turbomachinery Blade Flutter.", Ph. D. Thesis, Chalmers University of Technology, Gothenburg, Sweden, 1991.

Van Leer, B; 1982
"Flux Vector Splitting for the Euler Equations." Lecture Notes in Physics, Vol. 170, p. 507, Springer-Verlag, New York-Berlin, 1982.

Wada, Y.; Liou, M. S.; 1994
"A Flux Splitting Scheme with High-Resolution and Robustness for Discontinuities.", AIAA Paper 94-0083, 1994.

Whitehead, D.S.; 1990
"A Finite Element Solution of Unsteady Two-Dimensional Flow in Cascades.", Int. Journal of Numerical Methods in Fluids, Vol. 10, 1990.

THREE DIMENSIONAL LINEARIZED NAVIER-STOKES CALCULATIONS FOR FLUTTER AND FORCED RESPONSE

D. Graham Holmes
Brian E. Mitchell
General Electric Corporate R. & D.
P.O. Box 8
Schenectady, NY 12301 USA

Christopher B. Lorence
General Electric Aircraft Engines
1 Neumann Way
Cincinnati, OH 45215 USA

This paper describes a new computational tool developed for the analysis of two or three dimensional, viscous or inviscid turbomachinery flutter and forced response problems. Time harmonic results are obtained by linearizing the underlying steady flow numerical algorithm, including the k-ω turbulence model. Solutions are obtained via explicit time marching, with multigrid, on multiblock grids. We present results that focus on the role of the linearization of the viscous terms and the turbulence model in the accurate computation of linearized unsteady flows.

1 Introduction

It is largely accepted that the onset of flutter in turbomachinery blade rows can be modeled by computational methods based on a linearization of the fluid flow equations, i.e. by solving for a linear harmonic perturbation about a known time-steady (or time-averaged) mean flow (see for instance Hall & Crawley 1989; Verdon 1993). There is also experience that suggests that forced response can be modeled using linearized methods. Previous work on linearized methods has focused on the development and application of either two and three dimensional inviscid solvers or two dimensional viscous / turbulent solvers. We describe here the development of a linearized computational method for flutter and forced response that is both three dimensional and viscous / turbulent.

Three dimensional effects can be important for a number of reasons:

- Modern blades, which are designed routinely using inviscid and viscous 3D Computational Fluid Dynamics (CFD) solvers, can have highly three dimensional shapes.
- Three dimensional flow features, e.g. hub and tip boundary layers, secondary flows, tip vortices, etc., can occupy a significant portion of the flow field, limiting the region in which the flow can be regarded as two dimensional.
- Transonic flows are inherently three dimensional, with a strong coupling of the flows in adjacent streamsheets due to even small variations in the streamsheet thicknesses.
- For transonic flows, especially in transonic fans, shock patterns can be highly three dimensional.
- Even when the mean flow is two dimensional, the vibrational mode shape of the blade may be three dimensional yielding a three dimensional linear perturbation.

T.H. Fransson (ed.), Unsteady Aerodynamics and Aeroelasticity of Turbomachines, 211-224.

- The inlet and exit disturbances introduced to examine forced response can be highly three dimensional as, for instance, in the case when the forced response is generated by a tip vortex from the upstream blade row.

Although many problems of flutter and forced response can be well understood by considering only the Euler equations, there are instances when the Navier-Stokes equations (plus an appropriate turbulence model) may need to be considered. For example:

- Subsonic stall flutter may involve oscillations of a region of separated flow.
- Even in attached flows, viscous effects play an important role in establishing the mean flow. An example is the flow in a transonic fan where the location of the passage shock depends strongly on the boundary layer blockage.
- Since it is known that the unsteady motion of the shock is a major contributor to the unsteady work (Hall *et al* 1994; Lindquist & Giles 1994), it is possible that the unsteady perturbations of the boundary layer will be important in determining the range of motion of the shock and thus the unsteady work.

For these reasons we are developing a linearized, three dimensional, Navier-Stokes solver to analyze turbomachinery flutter and forced response. Since we are interested in analyzing flows with complex geometries, e.g. fan blades with part-span shrouds, we have written a structured, multi–block code. Furthermore, since it is recognized that not all problems require such generality, the code has efficient inviscid and two dimensional modes. The same code is used to compute the time-steady mean flow about which the unsteady linearized perturbations are sought. This paper describes the essential features of this code, and presents some illustrative examples of its application. The code has been written in Fortran 90, taking full advantage of the modern language features provided by the Fortran 90 standard.

2 Governing Equations

The equations we solve are either the 3D Euler equations (inviscid flows), the 3D Navier-Stokes equations (laminar flows), or the 3D Reynolds–averaged Navier-Stokes with a turbulence model. As seems to be a widespread practice in turbomachinery CFD, we use a two equation turbulence model rather than an algebraic or one equation model. We use the k-ω turbulence model advocated by Wilcox (1988, 1993). Our only significant departure from the standard k-ω model is that we implement the production modification of Launder & Kato (1993). The turbulence production terms in the standard k-ω model are driven by the strain rate tensor

$$P = S_{ij}S_{ij} \tag{1}$$

Use of this production term results in an unphysically high rate of generation of turbulence kinetic energy in regions of high irrotational strain. This is especially noticeable near blade leading edges. The production modification replaces the production term by:

$$P = \sqrt{S_{ij}S_{ij}O_{kl}O_{kl}} \tag{2}$$

where O_{kl} is the rotation rate tensor, which is zero in irrotational parts of the flow field. In a pure shear flow (as in a boundary layer or wake), the strain rate and rotation rate tensors are identical, and the production modification has no effect.

3 Numerical Method

3.1 STEADY FLOW ALGORITHM

The basic steady flow solution algorithm is unremarkable and is essentially that due to Jameson and others (see for instances Jameson *et al* 1981, Jameson 1983, Martinelli & Jameson 1988, and Arnone 1995) which has been in use within the General Electric Company for many years. Solutions are computed in a reference frame rotating with the blade row. We solve for absolute flow quantities in this relative frame.

We use a cell centered finite volume scheme with central differences. For the inviscid terms, the central difference scheme for computing first derivatives requires storage of three families of vector face areas and the cell volumes. Viscous solutions necessitate the computation of derivatives at cell faces using three families of offset control volumes. No metric data is stored for these offset control volumes. Metric data for the offset control volumes can be constructed quickly from the metric data for the primary control volumes.

Viscous fluxes are evaluated only on the first partial step and only on the fine grids. The forcing functions used by the multigrid algorithm take care of the resulting inconsistencies between the fine grid and coarse grid solutions.

Numerical smoothing is provided by a standard adaptive 2nd / 4th order smoother driven by a 2nd order pressure switch in each grid direction. An optional constant coefficient (and thus linear) 4th order smoother is available. This linear smoother has proved to be useful in developing the linearized version of the code – see Section 3.4.1.

The solution is obtained via a multi-step Runge-Kutta explicit time marching scheme with convergence acceleration via local time steps, residual averaging, and V-cycle or W-cycle multigrid.

3.2 TURBULENCE MODEL

The two conservation equations for the turbulence kinetic energy k and the dissipation rate ω are solved by time marching in almost exactly the same way as the five Euler or Reynolds averaged Navier-Stokes equations. The conserved quantities are ϱk and $\varrho\omega$. The conservation equations include terms representing advection, viscous diffusion, and source terms accounting for the production and decay of ϱk and $\varrho\omega$. The same local time stepping, residual averaging, and multigrid algorithms are used. It has been found that the production and decay source terms should be evaluated at every partial step on every grid level to ensure stability. The production term involves the net strain rate $S_{ij}S_{ij}$ at each cell – see Section 2. The strain rate tensor is evaluated only on the first partial step of each full time step, and only on the fine grid. On the coarse grids, the production term is gathered from the fine grid via a volume weighted summation.

We do *not* use the adaptive 2nd / 4th order smoother for the two turbulence equations (Connell, Holmes & Braaten 1993). The turbulent viscosity is determined by the quotient of the turbulence kinetic energy k and the dissipation rate ω. Both of these quantities fall to low values at the edge of a boundary layer. The under- and over-shoots that can be seen when using a 4th order smoother for k and ω can thus give rise to physically incorrect large values for the turbulence viscosity. We use instead a pure second order smoother. At each face we define a first difference smoothing flux with a weight equal to the product of the face area, the local relative velocity normal to the face, and a dimensionless coefficient. When the coefficient is set to unity, this scheme is easily shown to be equivalent to full upwind differencing of ϱk and $\varrho\omega$. We typically use values significantly less than one. When the flow is parallel to a face, the smoothing weight becomes zero. To avoid possible problems, a small additional weight proportional to the local sound speed can be added to prevent the smoothing weight falling all the way to zero. In most applications, viscous diffusion of ϱk and $\varrho\omega$ is sufficient to render this floor under the numerical smoothing unnecessary.

At wall boundaries, wall functions may be used or the equations may be integrated all the way to the wall. One advertised advantage of the k-ω turbulence model is that the equations may be integrated to the wall without the use of damping functions. At the wall, the turbulence kinetic energy is simply set to zero and the turbulent dissipation rate is set to a large value based on an assumed dimensionless wall roughness. When wall functions are used, terms may be added to account for streamwise pressure gradients (Wilcox 1989, 1993). In all of the applications reported here, wall functions were used. It is currently too expensive in computer time to integrate all the way to the wall. This is partly because of the additional mesh required to resolve boundary layers down to the required small values of y^+ – as opposed to the typical $30 < y^+ < 150$ range when using wall functions. More importantly perhaps, the very thin, high aspect ratio cells near the wall render an explicit solution algorithm too stiff to be practical using current computer resources.

The two quantities ϱk and $\varrho\omega$ are inherently positive, yet there is nothing about the time marching solution algorithm to ensure that the computed ϱk and $\varrho\omega$ are in fact positive. Lower bounds are placed under the two quantities to prevent them from going negative. It is important that these clips be based on a combination of small *dimensionless* numbers and the local flow field, rather than on arbitrary dimensional limits. The turbulence kinetic energy is bounded from below by requiring that the Mach number of the turbulence – which is proportional to $\sqrt{k}$ – be above some very small value. The turbulence dissipation rate ω is bounded below by setting a limit proportional to the local strain rate $\sqrt{2S_{ij}S_{ij}}$. If *both* clips are invoked, there is a danger that the eddy viscosity could assume absurd values. We prevent this by modifying the k clip so as to prevent the eddy viscosity from being more than a small multiple to the laminar viscosity when both clips are operative.

In addition, CFD folklore suggests that the ratio of the eddy viscosity to the laminar viscosity be clipped to stay below some fairly large ratio – say 500 or 1000. This is usually done to make the code more robust. It is easy to demonstrate that such a ratio is essentially the Reynolds number of the energy containing turbulent eddies. There is surely no physical justification for believing that the turbulent eddies should be limited to such a Reynolds number, especially for high Reynolds number flows, but clipping the eddy viscosity is sometimes necessary to assure numerical stability.

Numerical evidence obtained to date indicates that the k-ω model, when discretized as described above, lives up to its reputation for robustness. For most blade row applications, it is possible to go directly from a simple, axisymmetric starting guess derived from a throughflow design system to a full multigrid, 3D turbulent solution. Occasionally it is necessary to turn on the multigrid gradually, by running, say, ten time steps on one grid, then ten time steps on two grids, and so on.

3.3 LINEARIZATION

3.3.1 *Approach*

We take the now widely used approach originally due to Ni & Sisto (1976). We solve for a steady mean flow in a blade row, and then solve for linear harmonic unsteady perturbations about the mean flow, driven either by blade motion (flutter) or by inlet or exit boundary disturbances (forced response). Following Ni and Sisto, and many subsequent papers, we solve the resulting system of linear equations by introducing a false time and marching the equations to convergence in false time using all of the convergence acceleration techniques used by the steady flow solver – residual averaging, multigrid, etc.

The non-linear Euler or Navier-Stokes equations must be both linearized and discretized. This can be done in two ways: either (a) one can write down a linearization of the partial differential equations, and then discretize, or (b) one can discretize the non–linear equations and then linearize the discretization. Ideally, both approaches should yield the same result. Holmes and Chuang (1993) took approach (a) (linearize / discretize) in writing a quasi-3D linearized Euler solver. That experience suggested that approach (b) (discretize / linearize) is better. Two advantages of the discretize / linearize approach are that the non–linear discretization needs to be written anyway, in order to provide consistent mean flow solutions, and that the linearization becomes essentially a mechanical coding exercise.

The essential linearization operation can be illustrated as follows. A non–linear flux through a face can be computed as:

$$\mathcal{F} = A(Q_1Q_2 + Q_3Q_4....) \tag{3}$$

where A is a face area and the Q_i are flow variables. The linearization becomes:

$$\mathcal{F}' = A'(Q_1Q_2 + Q_3Q_4....) + A(Q_1'Q_2 + Q_1Q_2' + Q_3'Q_4 + Q_3Q_4'....) \tag{4}$$

where the prime superscript denotes a linear perturbation. The first right-hand-side term involves metric perturbations due to the grid motion (associated with the blade motion) and the mean flow. This term needs to be computed only once and can then be stored as a *source* term. The remaining terms involve steady metrics and combinations of mean flow and perturbed flow quantities. These terms must be computed repeatedly during the false time marching solution.

3.3.2 *Code Testing*

The linearized code that results is complicated, and coding errors are likely. Rigorous testing of the code is difficult. One very powerful, necessary (but not sufficient) check on the accuracy of the coding is obtained if we proceed as follows:

(1) Provide a *steady* linear option in which all perturbations are real, and the angular frequency of the perturbations is zero. This requires a linearization of the steady non-linear boundary conditions.

(2) Compute three non-linear solutions to some flow problem: (i) a baseline non-linear solution S_0, (ii) a non-linear solution S_+ in which either the geometry, e.g. the blade shape, or one of the boundary conditions, is given a small but finite perturbation, characterized by some scalar parameter δ, and (iii) a non-linear solution S_- in which the finite perturbation has the same magnitude but opposite sign.

(3) Compute two divided difference perturbation solutions:

$$\Delta_{\pm} = \frac{\pm S_{\pm} \mp S_0}{\delta} \tag{5}$$

(4) Compare these two divided difference solutions Δ_+ and Δ_- to a steady, linear solution obtained using a unit *linear* perturbation identical to the small, finite perturbations.

The linear perturbation solution should match closely the two divided difference solutions. Care is required in choosing δ. If δ is too small, the two divided difference solutions will be corrupted by roundoff. If δ is too large, non-linear effects will be large and the two divided difference solutions will not agree with each other. When the right perturbation amplitude is selected (and the code is correct!) the linear solution will very neatly split the small difference between the two divided difference solutions. One other requirement for this approach to work is that the three non-linear solutions be converged down to machine zero.

3.3.3 *Grid Motion*

Flutter problems are handled by making the grid move with the blade. Finite element mode shapes are mapped from finite element meshes to the computational grid. This gives grid displacements on the blade surfaces. For cases with non-zero interblade phase lag, the blade surface displacements are phase shifted appropriately. For cyclic symmetry cases, two finite element mode shapes are mapped, corresponding to the cosine and sine components of the mode. The cosine mode is treated as a real displacement, and the sine mode as an imaginary displacement.

Once the (possibly complex) grid displacements are known at the grid points lying on the blade surfaces, these displacements must be interpolated to all grid points – the remaining boundary grid points and all interior grid points. The exact displacements computed are not important, although better quality solutions are probably obtained by smooth interpolations. A sequence of 2D and 3D trans-finite interpolation operations proves to be fast and gives acceptable results.

3.4 LINEARIZATION ISSUES

A complete linearization of the discretized non–linear flow solver may be impossible. Certainly, some portions of the non-linear flow solver can only be linearized at either significant computational cost, or via subtle modifications to the non-linear discretized model to facilitate linearization. One important goal of this work is to determine how complete the linearization has to be.

Some specific areas in which the linearization causes problems are discussed briefly below. Many of these linearization issues can be studied using the divided difference calculations described above. If the divided difference tests show a linearization to be necessary, the linearization must be performed, no matter how unpalatable the linearization may be, in terms of code complexity or computer run time.

3.4.1 *Numerical Smoothing*

The adaptive 2nd / 4th order smoother used on the mean flow equations contains absolute value and maximum operators that inhibit linearization. One way of avoiding the resulting linearization problems is to freeze the adaptive smoothing coefficients to their non–linear values, resulting in a smoothing algorithm that is linear. This produces only a small corruption of the linearization for subsonic problems. For transonic flows with shocks, the corruption may be significant. Lindquist and Giles (1994) and other authors have discussed this issue.

The current second order, partial–upwind smoothing algorithm used on the turbulence model equations is linearized using frozen smoothing coefficients, exactly as for the other equations. Numerical evidence to date suggests that the effects of the partial linearization are minor. Note that fully linearizable smoothers, with coefficients independent of the mean flow field, are available as an option for both the two turbulence and the five mean flow equations, so that divided difference tests of the other linearized terms can be performed. These smoothers, however, tend to be too diffusive for Navier-Stokes solutions.

3.4.2 *Clips on Turbulence Quantities*

The clips on the non-linear turbulence quantities described above raise questions about how the clips interact with the linearization. In an ideal world, the clips need to be invoked rarely, if at all in the converged solution and the issue is moot. Turbulence clips are linearizable expect at the point discontinuity – the junction between the clipped and non-clipped regions. It is possible to conceive of modified, "soft" clips that are fully linearizable.

3.4.3 *Production Modification*

The production modification described in Section 2, Eqn. (2), has a square root singularity at zero strain rate and / or zero rotation rate. A small modification to the production term eliminates the singularity and permits linearization with a negligible effect on the production term.

3.4.4 *Linearization of the Laminar Viscosity*

For compressible flows the laminar viscosity depends on the flow field, e.g. on the static temperature, via Sutherland's law. For most subsonic cases perturbations in the laminar viscosity can probably be ignored. For transonic and supersonic cases, these perturbations *may* be significant. Linearization of the laminar viscosity is optional in the code.

3.4.5 *Linearization of the Turbulence Model*

A related, and far more significant, linearization issue is whether or not the turbulence model itself needs to be linearized. Under the Boussinesq approximation, the primary role of the turbulence is to provide enhanced diffusivities intended to mimic the turbulent mixing. There are other ways in which the turbulence interacts with the mean flow, but these are minor (for moderate Mach numbers and moderate turbulence kinetic energies) compared to the added diffusivity introduced via the eddy viscosity. Avoiding linearization of the turbulence model and simply using frozen mean flow values for the eddy viscosity would save significant arithmetic and a lot of hard to write and to debug linearized code. Linearization of the turbulence model is optional in the code. Hall (1997) and Florea, Hall, & Cizmas (1996) have linearized both algebraic and one equation turbulence models and shown that in some cases a full linearization may be necessary. We have similarly found cases in which a full linearization of the k-ω turbulence model is necessary.

4 Results

4.1 STEADY DIVIDED DIFFERENCES

The issue of the necessary extent of the linearization is illustrated via the Standard Configuration 10 compressor cascade (Fransson & Verdon 1993). We have performed a steady 2D subsonic, turbulent analysis of the flow through this cascade. We have used a 64 by 160 cell grid. The assumed Reynolds number is 10^6, and the inlet turbulent intensity is low. In addition to this baseline solution we have perturbed the exit pressure by 0.1% to provide us with two divided difference calculations of the effects of a steady perturbation in the exit pressure. We have then performed a steady linear calculation based on the baseline mean flow, driven by a unit perturbation in the exit pressure. This linear perturbation has been performed in three ways:

(1) As an inviscid perturbation about the turbulent mean flow.

(2) As a viscous perturbation with the eddy viscosity at each cell frozen at the computed mean flow values. The two turbulence equations are not linearized.

(3) With full linearization of the turbulence model, giving rise to perturbations in the eddy viscosity.

Figure 1 shows how these three linear solutions compare to the divided difference solutions. We plot pressure perturbations on the blade suction and pressure surfaces vs. axial chord. Note that:

(1) The two divided difference solutions are both smooth and close to each other. This indicates that we have picked finite perturbation amplitudes that are small enough to be linear, yet large enough to avoid roundoff errors.

(2) The fully linearized solution lies very close to the divided difference solutions. This indicates that the linearization of the viscous and turbulence terms has been done correctly.

(3) More importantly, the frozen eddy viscosity solution differs significantly from the fully linearized solution. The errors involved in the frozen viscosity solution are comparable to errors involved in computing a purely inviscid perturbation solution. This is a significant result. It implies that even for this rather simple subsonic compressor cascade, with the flow nowhere near separation, a full linearization of the turbulence model is essential.

4.2 UNSTEADY 2D FORCED RESPONSE

One might expect that this conclusion would carry over into a true unsteady solution. We examine whether or not this is so using the same Standard Configuration 10 mean flow solution. We can apply one of the four possible "gust" perturbations: downstream travelling entropy and vorticity waves, a downstream travelling acoustic wave, and an acoustic wave travelling upstream from the exit. We select the upstream travelling acoustic wave as the closest analog of the steady exit perturbation discussed above. An arbitrary interblade phase angle of 90 degrees and a reduced frequency (defined as angular frequency × semi-chord / inlet velocity) of 1.4 have been used in the computation. We use the unsteady 2D non-reflecting boundary conditions of Giles (1990) at both inlet and exit, keeping ± 4 modes. Figure 2 shows plots of the real part of the pressure perturbations on the blade suction and pressure surfaces vs. axial chord, for the following cases:

(1) Inviscid mean flow run to the nominal Configuration 10 exit pressure, plus inviscid perturbations.

(2) Inviscid mean flow run to the same mass flow as the turbulent solution (by adjusting the exit pressure), plus inviscid perturbations.

(3) Turbulent mean flow, plus inviscid perturbations.

(4) Turbulent mean flow, plus turbulent perturbations with frozen eddy viscosity.

(5) Turbulent mean flow, plus turbulent perturbations with perturbed eddy viscosity, i.e. with a full linearization of the turbulence model.

Solution (5) – the solid line in Figure 2 – with the full linearization of the turbulence model, is presumed to be the standard against which the others are to be judged, since for unsteady perturbations we cannot use a divided difference solution as our standard. We observe that the inviscid mean flow / inviscid perturbation solution run to the nominal exit pressure is quite incorrect. The inviscid / inviscid solution run to mass flow is, on the other hand, remarkably good for this case. This suggests that, if one has access to only an inviscid non-linear steady / linear unsteady code, running the mean flow to the desired mass flow is probably the best approach. The usefulness of computing unsteady inviscid perturbations based on an inviscid solution run to the viscous mass flow probably derives from the fact that the *mean* inviscid solution run to the viscous mass flow is quite similar to the viscous solution. Figure 3 shows the mean pressure distributions for the inviscid solution run to exit pressure, the inviscid solution run to the viscous mass flow, and the viscous solution. The pressure distributions obtained from viscous solution and from the inviscid solution run to the viscous mass flow are similar over most of the blade.

The viscous mean flow / inviscid perturbation solution is fairly accurate for this case, but has often proved to be seriously incorrect in other problems. The two viscous mean flow / viscous perturbation solutions (with and without linearization of the turbulence model) are virtually indistinguishable for this problem, suggesting that *for this case* it is not important to perform a full linearization of the turbulence model. This contrasts sharply with the conclusion drawn from the steady perturbation example above.

To see whether the necessity of linearizing the turbulence model depends on reduced frequency, we performed further calculations at much lower reduced frequencies. It was necessary to switch to an interblade phase angle of zero degrees to ensure that at all frequencies considered, the incoming wave was cut on. Solutions are computed only for the two viscous mean flow / viscous perturbation cases, i.e. for a complete linearization of the viscous perturbations, and for a linearization with frozen eddy viscosity. Reduced frequencies vary over a wide range: 3.4, 0.34, 0.034. Fig. 4 shows blade surface distributions of the real part of the pressure perturbation.

We see that, for the lower reduced frequencies, solutions obtained with and without linearization of the turbulence model do differ. Given the linearized *steady* result above, this is perhaps no surprise. What is more surprising is that as the reduced frequency is increased the solutions obtained with and without linearization of the turbulence model merge. This suggests that the eddy viscosity responds to the unsteady disturbance on a much longer time scale than do the inviscid aspects of the flow field.

Note that the unsteady solution does not tend, as the frequency goes to zero, to the steady solution shown above in Figure 1. This is because of the fundamentally different nature of the linearized steady and unsteady boundary conditions: a small change in *steady* exit pressure is *not* equivalent to an incoming acoustic wave of zero frequency.

It is interesting to note that Hall (1997) observed a similar need for a complete linearization of the turbulence model for the same compressor cascade, but *not* at low reduced frequencies, and at a higher incidence, much closer to stall.

5 Summary

This paper has described a new 2D / 3D inviscid / viscous / non–linear / linear code for flutter and forced response calculations. The primary focus of the paper has been on the role of the linearization of the viscous terms and of the turbulence model in the accurate computation of linearized unsteady flows. We have examined a series of test case solutions for a compressor cascade: Standard Configuration 10. For this cascade, viscous effects are significant in establishing the overall flow field. We conclude that:

(1) Solutions made with inviscid mean flows and inviscid perturbations are most useful when the mean flow solution is run to the intended mass flow rather than the intended exit pressure.

(2) Inviscid perturbation solutions computed on viscous mean flows are not reliable and should not be used.

(3) For viscous solutions linearization of the turbulence model has been shown to be necessary at sufficiently low frequencies. This appears to be due to the existence of much slower times scales for the establishment of the boundary layer flow than for the inviscid portion of the flow field.

(4) The issue of the necessity for a full linearization of the turbulence model is by no means resolved.

Further investigations are required. It would be interesting to perform similar calculations for a turbine test case, for which viscous effects may prove to be less significant.

Acknowledgments

This work was jointly sponsored by GEAE and GE CR&D. The authors are grateful to the General Electric Company for permission to publish this paper. In particular, the authors wish to thank Dr. Robert Kielb of GEAE for his guidance and support.

References

ARNONE, A. (1995) Multigrid methods for turbomachinery Navier-Stokes calculations. In *Solution Techniques for Large-Scale CFD problems*, ed. Wagdi G. Habashi, John Wiley & Sons, Chichester.

BÖLCS, A. & FRANSSON, T. H. (1986) Aeroelasticity in turbomachines: comparison of theoretical and experimental cascade results. AFOSR-TR-87-0605.

CONNELL, S. D., HOLMES, D. G. & BRAATEN, M. E. (1993) Adaptive unstructured 2d Navier-Stokes solutions on mixed quadrilateral/triangular meshes ASME paper 93-GT-99.

FLOREA, R., HALL, K. C. & CIZMAS, P. G. A. (1996) Reduced order modelling of unsteady viscous flow in a compressor cascade. AIAA Paper 96-2572.

FRANSSON, T. H, & VERDON, J. M. (1993) Panel discussion of standard configurations for unsteady flow through vibrating axial–flow turbomachines cascades. In *Unsteady Aerodynamics, Aeroacoustics, and Aeroelasticity of Turbomachines and Propellers*, ed. H. M. Atassi, Springer-Verlag, New York.

GILES, M. B. (1990) Nonreflecting boundary conditions for Euler equation calculations. *AIAA Journal,* **28**, 2050-2058.

HALL, K. C. (1997) Personal communiation.

HALL, K. C., CLARK, W. S. & LORENCE, C. B. (1994) A linearized Euler analysis of unsteady transonic flows in turbomachinery. *ASME Journal of Turbomachinery,* **116**, 477—488.

HALL, K. C. & CRAWLEY, E. F. (1989) Calculation of unsteady flows in turbomachinery using the linearized Euler equations. *AIAA Journal,* **27**, 777-787.

HOLMES, D. G. & CHUANG, H. A. (1993) 2D linearized harmonic Euler flow analysis for flutter and forced response. In *Unsteady Aerodynamics, Aeroacoustics, and Aeroelasticity of Turbomachines and Propellers*, ed. H. M. Atassi, Springer-Verlag, New York.

JAMESON, A., *et al* (1981) Numerical solution of the Euler equations by finite volume methods using Runge-Kutta time-stepping schemes. AIAA Paper 81-1259.

JAMESON, A. (1983) Solution of the Euler equations for two dimensional transonic flow by a multigrid method. *App. Math. and Comp.,* **13**, 327-355.

LAUNDER, B. E. & KATO, M. (1993) Modelling flow-induced oscillations in turbulent flow around a square cylinder. *ASME FED*, **57**, 189-199.

LINDQUIST, D. R. & GILES M. B. (1994) Validity of linearized unsteady Euler equations with shock capturing. *AIAA Journal,* **32**, 46-53.

MARTINELLI, L. & JAMESON, A. (1988) Validation of a multigrid method for the Reynolds averaged equations. AIAA Paper 88-0414.

NI, R. H. & SISTO, F. (1976) Numerical computation of nonstationary aerodynamics of flat plate cascades in compressible flows. *Trans. ASME J. Engineering for Power* **98**, 165-170.

VERDON, J. M. (1993) Unsteady aerodynamic methods for turbomachinery aeroelastic and aeroacoustic applications. *AIAA Journal,* **31**, 235-250.

WILCOX, D. C. (1988) Reassessment of the scale-determining equation for advanced turbulence models. *AIAA Journal* **28**, 1299-1310.

WILCOX, D. C. (1989) Wall matching, a rational alternative to wall functions. AIAA Paper 89-0611.

WILCOX, D. C. (1993) *Turbulence Modelling for CFD*. DCW Industries, La Cañada, California.

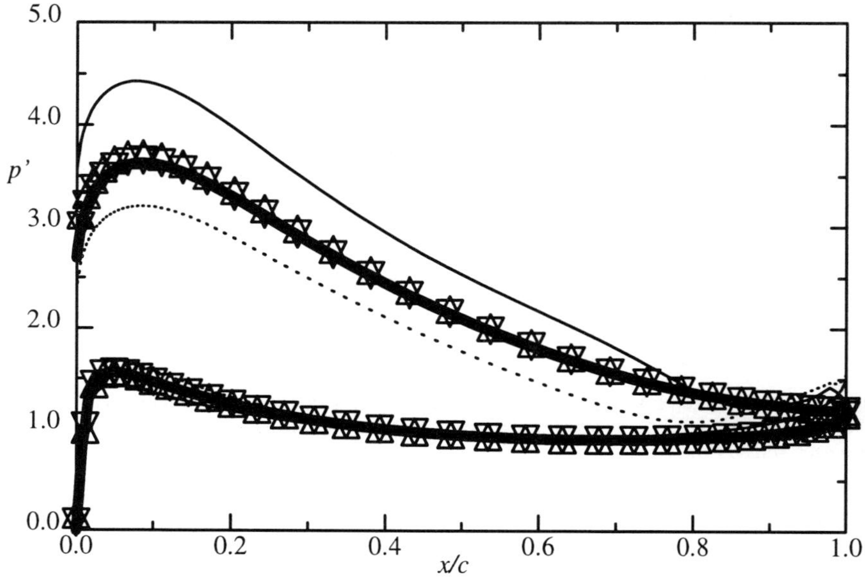

Figure 1. Response to a *steady* change in exit static pressure. Symbols: Δ,∇: divided differences; solid line: inviscid perturbation about a viscous mean flow; short dashed line: frozen turbulence; heavy solid line: full linearization.

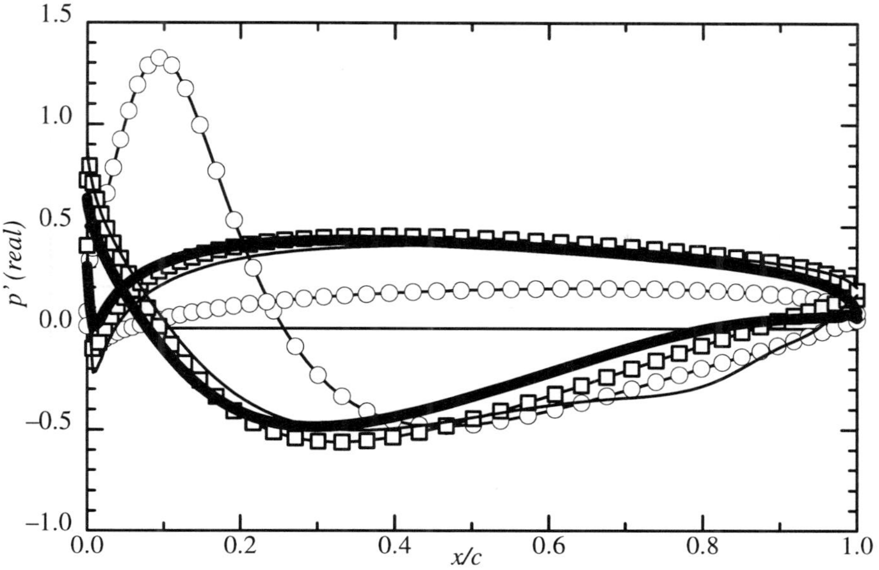

Figure 2. Response to an *unsteady* change in exit static pressure. Symbols: ○: inviscid mean flow & perturbations, run to exit pressure; □: inviscid / inviscid, run to mass flow; solid line: inviscid perturbation about a viscous mean flow; short dashed line: frozen turbulence; heavy solid line: full linearization.

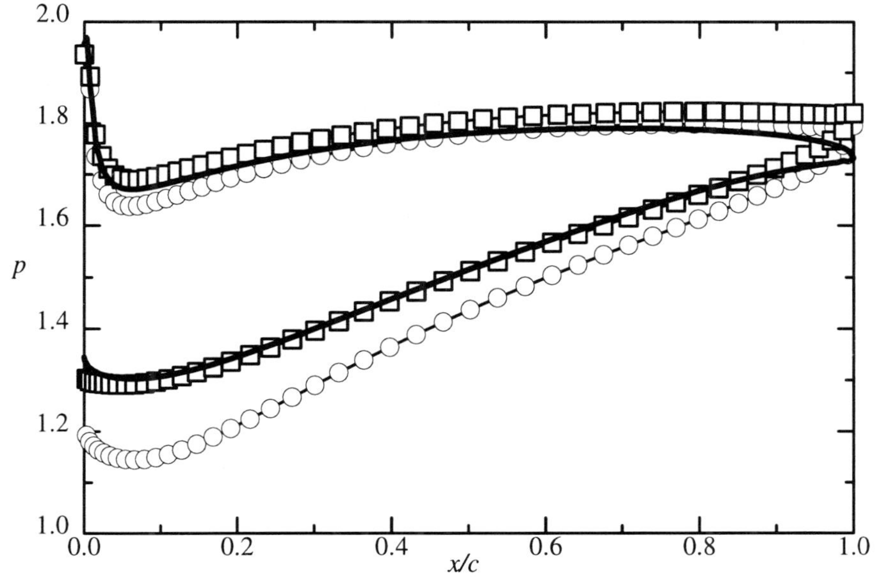

Figure 3. Mean static pressure. Symbols: ○: inviscid, run to exit pressure; □: inviscid, run to mass flow; heavy solid line: viscous.

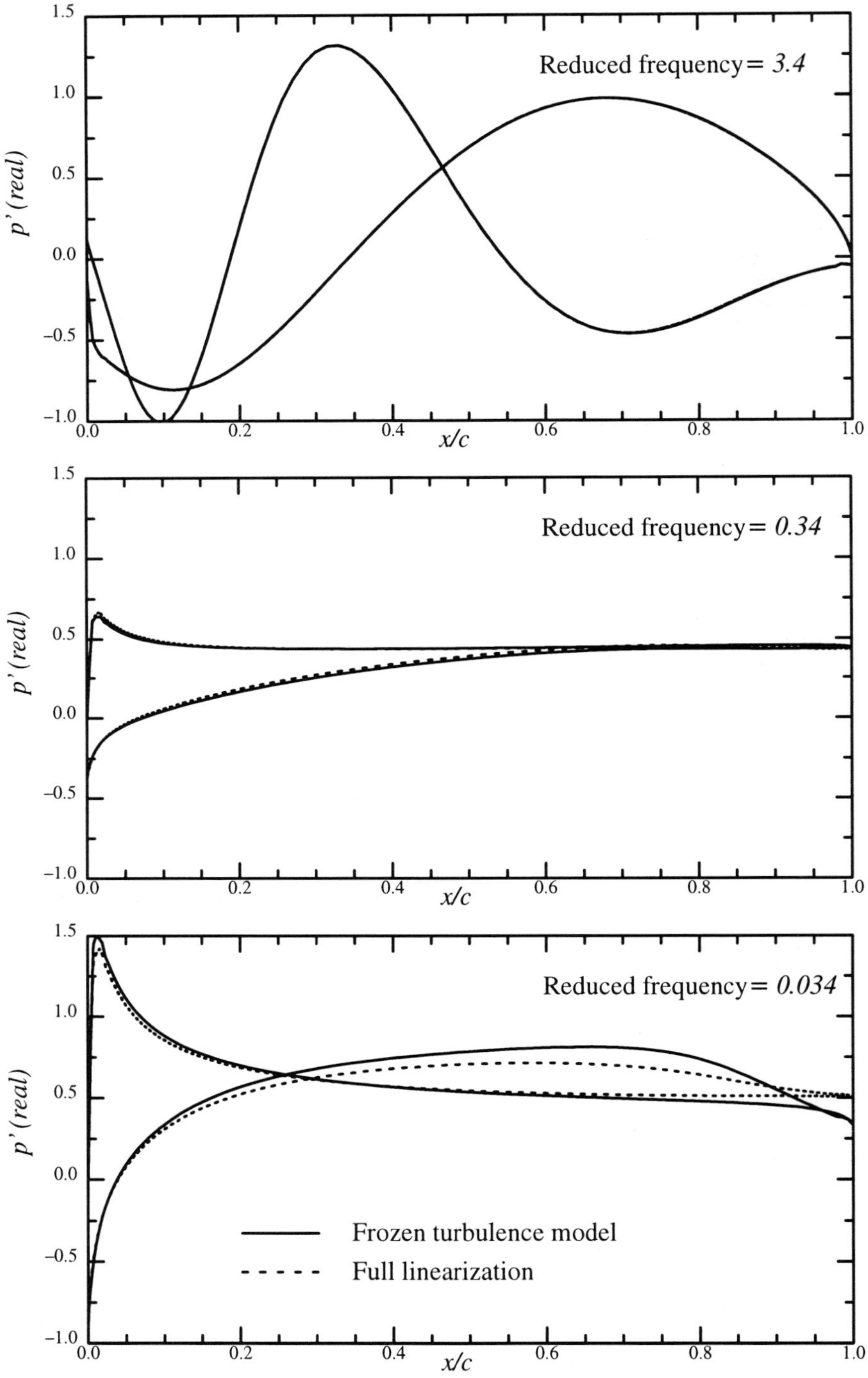

Figure 4. Comparison of the effect of exit pressure perturbations at different reduced frequencies.

SOME APPLICATIONS OF A TIME-LINEARIZED EULER METHOD TO FLUTTER & FORCED RESPONSE IN TURBOMACHINERY

J.G. MARSHALL
Aerothermal Technology
Rolls-Royce plc.
P.O.Box 31, Moor Lane
Derby DE24 8BJ, UK

M.B. GILES
Oxford University Computing Lab.
Wolfson Building
Parks Road
Oxford OX1 3QD, UK

Abstract

This paper presents the theory and a numerical algorithm for the prediction of flutter and forced response using the three-dimensional linearized Euler equations. The non-linear steady-state flow field is first calculated by solving the steady Euler equations using a standard numerical method. The linearized unsteady Euler equations are derived by assuming that the unsteady flow solution is a small perturbation to the steady flow. Due to linear superposition, the unsteady part of the flow can then be split into a sum of components of differing frequencies, each of which can be calculated independently. The harmonic equations for a single frequency are solved by a pseudo-time-marching approach, which allows the use of standard numerical algorithms. Computational results are presented to validate the implementation of the numerical algorithm. Comparisons are made to two-dimensional, linear unloaded flat plate cascade theory for test cases involving forced response due to wakes and pressure waves, as well as bending and torsion. The method is also applied to two 3D test cases – a high pressure turbine stage, and a fan flutter problem.

1. Introduction

Flutter is a linear aeroelastic phenomenon, concerned with the possible exponential growth in the amplitude of structural vibrations which are initially of very small magnitude. Accordingly, the unsteady aerodynamic analysis methods which are used in flutter analysis and prediction are based upon the linear assumption that the unsteadiness is a small perturbation to a steady flow field. Verdon gives a very good review of the history of the development of such methods {Verdon, 1989}. One of the earliest noteworthy achievements was the development by Smith of a method of calculating the linear unsteady flow around a two-dimensional flat plate cascade {Smith, 1971}. In this work the steady flow is taken to be a uniform subsonic flow, and the blade and wake are modelled with an unsteady vortex sheet. Verdon {1977} and others developed corresponding methods for perturbations to uniform supersonic flow.

T.H. Fransson (ed.), Unsteady Aerodynamics and Aeroelasticity of Turbomachines, 225-240.

In addition to being computationally inexpensive, an advantage of these methods is the ability to analyse the unsteadiness due to wake/blade interaction or incoming potential waves, in addition to structural bending or torsion. The weakness is the assumption of the uniform steady flow, which does not take account of the effect of mean loading on the unsteady pressure field. The next major section of work was on linear potential methods, in which the steady flow is taken to be a solution of the non-linear two-dimensional full potential equation. Whitehead {1982} and Verdon & Caspar {1984} developed quite similar numerical methods for solving the linear equations that result from assuming that the unsteadiness is sinusoidal in time. These linear potential methods have been the main flutter prediction technique used in the turbomachinery industry for the last ten years.

Forced response, the generation of unsteady blade forces due to some incoming unsteady flow field, is not a fundamentally linear phenomenon. In every type of forced response, for example stator/rotor interaction, there is a specified finite level of unsteadiness. It is not clear *a priori* that linear perturbation methods should give accurate predictions for such interactions. It is possible that the level of unsteadiness is sufficiently large that the unsteady flow field is truly non-linear, and can only be modelled by unsteady non-linear methods. However, linear methods are being used for forced response prediction, and there is growing evidence that this is appropriate. Suddhoo has recently used a linear potential analysis to calculate the first harmonic of the unsteady response to a transonic turbine stator/rotor interaction {Suddhoo et al. 1991}. There was very good agreement with a time-marching method solving the non-linear unsteady Euler equations. The agreement was particularly surprising because of the presence of complex shock propagation and reflections in the problem. Also, Hall and Verdon have extended the potential analysis method to include the effects of incoming vortical and entropic disturbances, making it possible to analyse wake/rotor interaction in addition to potential/rotor interaction {Hall & Verdon, 1991}.

The linear potential methods are all for two-dimensional flows, with the addition of streamtube thickness and radius change to introduce some three-dimensional effects. Considering the movement to three-dimensional flows, the complexities of extending the potential analysis plus the computational cost of the standard matrix solution methods, suggests that other approaches should be considered. Therefore, there is now a move to consider the solution of the linearized Euler equations, which is the subject of this paper. In 1974, Ni did preliminary research on a numerical method for solving a modified form of the linear unsteady Euler equations {Ni & Sisto 1975}. Later, in 1987, Hall developed a finite element method for solving the linear harmonic Euler equations {Hall & Crawley, 1989}. This work was for subsonic cascade flows, and transonic ducts with shock fitting. It also demonstrated the validity of the linear approximation up to quite substantial levels of unsteadiness. In 1988, Giles proposed that a modified form of Ni's method would be the most efficient way to solve the two-dimensional and three-dimensional linear Euler equations {Giles, 1988}. At least four separate research groups are now working on such an approach {Holmes & Chuang 1991, Hall & Clark 1991, Kahl & Klose 1991}. A key contribution by Lindquist is the proof that if correctly formulated and implemented, shock capturing can produce the same correct answers as shock-fitting {Lindquist & Giles 1991}. This is important because shock capturing is very much simpler to implement, particularly in three dimensions.

2. Body-fitted Deforming Grid

There are two options in constructing linear equations which will allow one to handle the moving boundaries that occur in flutter calculations.

The approach which is standard for linear potential calculations is to consider the grid to be completely stationary {Verdon & Caspar 1984, Whitehead 1982}. A vibrating blade surface therefore moves (in a linear sense) through the grid. Although a stationary grid seems to be the simplest approach, and appeared most natural as an extension of the classical linear perturbation methods, it does cause significant numerical complications at the blade surface. This is best illustrated by examining the post-processing step of computing the unsteady pressure at a particular point on the blade surface, as a prelude to performing a lift or moment integration. If the co-ordinates of the surface point vary according to:

$$\mathbf{x}(t) = \overline{\mathbf{x}} + \widetilde{\mathbf{x}}(t) \qquad (1)$$

then the pressure at that moving point can be linearly approximated as:

$$p(\mathbf{x},t) \approx p(\overline{\mathbf{x}},t) + \widetilde{\mathbf{x}}(t) \bullet \nabla p \approx \overline{p}(\overline{\mathbf{x}}) + \widetilde{p}(\overline{\mathbf{x}},t) + \widetilde{\mathbf{x}}(t) \bullet \nabla \overline{p} \qquad (2)$$

Thus, the unsteady component of the surface pressure has two components, one due to the unsteady pressure field as seen by the grid and the other due to the unsteady motion of the surface through the non-uniform steady pressure field. The latter term causes problems in its numerical approximation because it involves numerical differentiation of the steady flow field. This is particularly difficult at the surface where the numerical scheme for calculating the steady flow is most inaccurate, and it is even more of a problem near leading edges where the mean flow gradients can be extremely large.

Whitehead relieved this problem by using a rigid grid that translated and rotated with the vibrating blade {Whitehead, 1989}. There were still some complications with the periodic boundary and the non-inertial co-ordinates, but the accuracy was greatly increased.

The alternative approach is to use a computational grid which is fitted to the blade surface, and deforms smoothly as it moves. This is the natural approach for those performing time-marching solutions of the non-linear, unsteady Euler and Navier-Stokes equations {Giles & Haimes 1991, He & Denton 1991, Huff et al. 1991, Siden 1991}. It is also being used for the linearized Euler equations by Lindquist and Giles {1991} and Hall & Clark {1991} and for the linear potential equation by Hall {1992}.

With the deforming grid approach, the unsteady pressure computed at a particular grid point is the pressure which would be seen by an observer moving with the grid. If the grid point is attached to a blade surface then it corresponds to the unsteady pressure that the blade surface experiences. Therefore there is much less of a problem in constructing the surface boundary conditions, and no numerical differentiation of the steady state flow filed is required. The linearized moving grid simply corresponds to the non-linear moving grid in the limit of very small changes in grid position as well as in all flow variables.

3. Linearized Equations of Motion

The construction of the linear equations on a deforming grid begins with the integral form of the non-linear, unsteady Euler equations for a deforming control volume.

$$\frac{d}{dt}\iiint_V \mathbf{U}dV = -\iint_{\partial V}(\mathbf{F}-\mathbf{U}\dot{\mathbf{x}})\cdot d\mathbf{A} + \iiint_V \mathbf{S}dV \tag{3}$$

where the state vector **U** is a five-component vector, and the source term **S** containing both centrifugal and Coriolis effects is

$$\mathbf{U} = \begin{pmatrix} \rho \\ \rho u_x \\ \rho u_y \\ \rho u_z \\ \rho E \end{pmatrix}, \quad \mathbf{S} = \begin{pmatrix} 0 \\ 0 \\ \rho(\Omega^2 y + 2\Omega u_z \\ \rho(\Omega^2 z - 2\Omega u_y \\ 0 \end{pmatrix} \tag{4,5}$$

the three flux vectors, $\mathbf{F}_x$, $\mathbf{F}_y$ & $\mathbf{F}_z$ are given by:

$$\mathbf{F}_x = \begin{pmatrix} \rho u_x \\ \rho u_x^2 + p \\ \rho u_x u_y \\ \rho u_x u_z \\ u_x(\rho E + p) \end{pmatrix}, \quad \mathbf{F}_y = \begin{pmatrix} \rho u_y \\ \rho u_x u_y \\ \rho u_y^2 + p \\ \rho u_x u_y \\ u_y(\rho E + p) \end{pmatrix}, \quad \mathbf{F}_z = \begin{pmatrix} \rho u_z \\ \rho u_z u_x \\ \rho u_z u_y \\ \rho u_z^2 + p \\ u_z(\rho E + p) \end{pmatrix} \tag{6}$$

$\dot{\mathbf{x}}$ is the velocity of the moving control volume boundary, which will become the grid velocity in the numerical discretization. Consequently, $\dot{\mathbf{x}} \cdot d\mathbf{A}$ is the rate at which new volume is added to the control volume due to the motion of the bounding surface. To complete the set of equations one needs an equation of state, and here it is assumed that the gas is ideal and calorifically perfect.

$$p = (\gamma - 1)\left(\rho E - \tfrac{1}{2}\rho(u_x^2 + u_y^2 + u_z^2)\right) \tag{7}$$

If the control volume is stationary, then the integral equation reduces to

$$\frac{d}{dt}\iiint_V \mathbf{U}dV = -\iint_{\partial V}\mathbf{F}\cdot d\mathbf{A} + \iiint_V \mathbf{S}dV \tag{8}$$

and this is the equation that is used as the basis of most time-marching methods which use an evolutionary approach to solve the steady-state equations of motion. Because time accuracy is unimportant, a variety of acceleration techniques are commonly employed, including local timesteps and multigrid.

The next step in the linear unsteady analysis is the assumption that both the grid co-ordinates and the flow variables can be expressed as a sum of a steady-state part plus a small amplitude unsteady perturbation.

$$\mathbf{x}(t) = \overline{\mathbf{x}} + \widetilde{\mathbf{x}}(\overline{\mathbf{x}}, t)$$
$$\mathbf{U}(\mathbf{x}(t), t) = \overline{\mathbf{U}}(\overline{\mathbf{x}}) + \widetilde{\mathbf{U}}(\overline{\mathbf{x}}, t) \tag{9}$$

For convenience, we will refer to the three co-ordinates of **x** and the five components of **U** as a single eight-component entity labelled **W**. Thus the last two equations can be expressed jointly as:

$$\mathbf{W}(\mathbf{x}(t), t) = \overline{\mathbf{W}}(\overline{\mathbf{x}}) + \widetilde{\mathbf{W}}(\overline{\mathbf{x}}, t) \tag{10}$$

Following from the perturbations to **W** there are corresponding perturbations to all other variables. For any arbitrary scalar function $A(\mathbf{W})$, the perturbation $\widetilde{A}$ is given by

$$\widetilde{A} = \frac{\partial A}{\partial \mathbf{W}} \widetilde{\mathbf{W}} \tag{11}$$

The perturbations to the velocity components are of the form:

$$\widetilde{u}_x = \frac{\widetilde{\rho u_x} - u_x \widetilde{\rho}}{\rho}, \text{ etc.} \tag{12}$$

with the overbar on the steady terms being omitted for clarity. After some algebra, the pressure perturbation is

$$\widetilde{p} = (\gamma - 1)\left(\widetilde{\rho E} - \widetilde{\rho}\left(E - \frac{p}{(\gamma - 1)\rho}\right) - \rho(u_x \widetilde{u}_x + u_y \widetilde{u}_y + u_z \widetilde{u}_z) + \rho\Omega^2 (y\widetilde{y} + z\widetilde{z}) \right) \tag{13}$$

Finally, the perturbations to $\mathbf{F}_x$, $\mathbf{F}_y$, $\mathbf{F}_z$ and **S** are:

$$\mathbf{F}_x = \begin{pmatrix} u_x \widetilde{\rho} + \widetilde{u}_x \rho \\ u_x \widetilde{\rho u_x} + \widetilde{u}_x \rho u_x + \widetilde{p} \\ u_x \widetilde{\rho u_y} + \widetilde{u}_x \rho u_y \\ u_x \widetilde{\rho u_z} + \widetilde{u}_x \rho u_z \\ u_x(\widetilde{\rho E} + \widetilde{p}) + \widetilde{u}_x(\rho E + p) \end{pmatrix}, \quad \mathbf{F}_y = \begin{pmatrix} u_y \widetilde{\rho} + \widetilde{u}_y \rho \\ u_y \widetilde{\rho u_x} + \widetilde{u}_y \rho u_x \\ u_y \widetilde{\rho u_y} + \widetilde{u}_y \rho u_y + \widetilde{p} \\ u_y \widetilde{\rho u_z} + \widetilde{u}_y \rho u_z \\ u_y(\widetilde{\rho E} + \widetilde{p}) + \widetilde{u}_y(\rho E + p) \end{pmatrix} \tag{14, 15}$$

$$\mathbf{F}_z = \begin{pmatrix} u_z\tilde{\rho} + \tilde{u}_z\rho \\ u_z \tilde{\rho u_x} + \tilde{u}_z\rho u_x \\ u_z \tilde{\rho u_y} + \tilde{u}_z\rho u_y \\ u_z \tilde{\rho u_z} + \tilde{u}_z\rho u_z + \tilde{p} \\ u_z(\tilde{\rho E} + \tilde{p}) + \tilde{u}_z(\rho E + p) \end{pmatrix}, \quad \mathbf{S} = \begin{pmatrix} 0 \\ 0 \\ \Omega^2(\rho\tilde{y} + \tilde{\rho}y) - 2\Omega\,\tilde{\rho u_z} \\ \Omega^2(\rho\tilde{z} + \tilde{\rho}z) + 2\Omega\,\tilde{\rho u_y} \\ 0 \end{pmatrix} \tag{16, 17}$$

The linear perturbation of the unsteady integral equation of motion is:

$$\frac{d}{dt}\iiint_V \tilde{\mathbf{U}}dV + \mathbf{U}\,\tilde{dV} = -\iint_{\partial V}(\tilde{\mathbf{F}} - \mathbf{U}\dot{\tilde{\mathbf{x}}})\cdot d\mathbf{A} + \mathbf{F}\cdot\tilde{d\mathbf{A}} + \iiint_V \tilde{\mathbf{S}}dV + \mathbf{S}\,\tilde{dV} \tag{18}$$

The terms $\tilde{d\mathbf{A}}$ and $\tilde{dV}$ represent the unsteady perturbations to the infinitesimal surface area vector and volume element. The physical interpretation of these is best understood by thinking of the deforming control volume as a gas-filled balloon, in which case $\tilde{d\mathbf{A}}$ corresponds to the unsteady stretching of an infinitesimal piece of the balloon's skin, and $\tilde{dV}$ represents the unsteady compression or expansion of an infinitesimal element of gas within the balloon. The evaluation of these terms will prove to be extremely simple when it comes to the numerical implementation.

The fact that this equation is linear and all of the coefficients depend on **U** and geometric terms, which do not vary in time, means that there are harmonic solutions of the form $\hat{\mathbf{U}}(\mathbf{x})e^{-i\omega t}$. The most general periodic solution can be expressed as the real part of a linear superposition of such solutions for a number of different real frequencies ω, all satisfying the necessary periodicity constraints.

$$\tilde{\mathbf{U}} = \Re\left(\sum_{\omega}\hat{\mathbf{U}}_{\omega}e^{-i\omega t}\right) \tag{19}$$

Since:

$$\Re(\hat{\mathbf{U}}e^{-i\omega t}) = \Re(\hat{\mathbf{U}}^{*}e^{i\omega t}) \tag{20}$$

where $\hat{\mathbf{U}}^{*}$ is the complex conjugate of $\hat{\mathbf{U}}$, the summation need only be over positive values of ω. A key contribution of Ni {1974} is to formulate the harmonic equation by retaining a time dependence in $\hat{\mathbf{U}}$. Substituting

$$\begin{aligned}\tilde{\mathbf{x}} &= \hat{\mathbf{x}}(\mathbf{x})e^{-i\omega t} \\ \tilde{\mathbf{U}} &= \hat{\mathbf{U}}(\mathbf{x},t)e^{-i\omega t}\end{aligned} \tag{21}$$

gives the unsteady, harmonic equation:

$$\frac{d}{dt}\iiint_V \hat{\mathbf{U}}dV = -\iint_{\partial V}(\hat{\mathbf{F}} - i\omega\mathbf{U}\hat{\mathbf{x}})\cdot d\mathbf{A} + \mathbf{F}\cdot\widehat{d\mathbf{A}} + \iiint_V(\hat{\mathbf{S}} + i\omega\mathbf{U})dV + \mathbf{S}\,\widehat{dV} \tag{22}$$

The advantage of retaining the time-dependent term, is that numerical integration of this equation until a `steady-state' solution is achieved will result in the desired harmonic solution. This allows one to employ any of a set of standard numerical techniques and tricks to accelerate the convergence.

4. Boundary Conditions

There are four different sets of boundary conditions: solid wall, periodic boundary, inflow and outflow. The solid wall boundary condition is that there is no flow through the surface of the moving wall. Therefore, the local fluid velocity relative to the moving wall has no component normal to the wall. Applying this condition to the evaluation of the flux term for a solid wall surface yields:

$$\iint_{\partial V} (\hat{\mathbf{F}} - i\omega \mathbf{U}\hat{\mathbf{x}}) \cdot d\mathbf{A} = \iint_S \begin{pmatrix} 0 \\ \hat{p} dA_x + p\, d\hat{A}_x \\ \hat{p} dA_y + p\, d\hat{A}_y \\ \hat{p} dA_z + p\, d\hat{A}_z \\ i\omega p \hat{\mathbf{x}} d\mathbf{A} \end{pmatrix} \tag{23}$$

This treatment of the solid wall is very much simpler than would be necessary if a stationary grid were used.

The periodic boundary condition is simply that the complex harmonic flow solution is periodic when expressed in cylindrical co-ordinates, with the exception of a phase shift due to the inter-blade phase angle.

$$\hat{\mathbf{U}}_c(x, r, \theta + \Delta\theta) = e^{-i\phi} \hat{\mathbf{U}}_c(x, r, \theta) \tag{24}$$

where $\hat{\mathbf{U}}_c$ is the harmonic perturbation to **U** in cylindrical polar co-ordinates.

A complete discussion of the formulation of inflow and outflow boundaries is beyond the scope of this paper. The basic idea follows an approach used for non-linear time-marching computations, namely to split the solution into one part which is the incoming waves plus the remainder which is the outgoing waves. The incoming waves can either be specified by the user, as taken from a steady-state stator/rotor calculation {Giles, 1992}. The outgoing waves are treated with non-reflecting boundary conditions to prevent spurious numerical reflections at the boundary {Giles, 1990}.

5. Numerical Implementation

The steady-state equations are solved on a structured hexahedral grid using a four-step

Runge-Kutta time-marching method of the type developed by Jameson {1986}. The integral form of the steady state equations is approximated on a hexahedral cell labelled *k* as:

$$R_k(\mathbf{U}) = \sum_{faces} \overline{\mathbf{F}} \cdot \mathbf{A} - V_k \overline{\mathbf{S}} = 0 \tag{25}$$

where **A** is the area vector of the face, $\overline{\mathbf{F}}$ is the average value of **F** for the face based on a simple arithmetic average of the values of **F** at the nodes at the corners of the cell, V is the cell volume, and $\overline{\mathbf{S}}$ is the value of **S** averaged over all eight corner nodes.

For a particular node, adding together the surrounding hexahedral cells and adding in an unsteady term yields:

$$\left(\sum_k V_k\right)\frac{\partial \mathbf{U}}{\partial t} = -\sum_k R_k(\mathbf{U}) \tag{26}$$

To suppress high-frequency spatial oscillations, numerical smoothing is added to this equation. The system of non-linear o.d.e.'s at each grid node is then integrated in time using the four-step Runge-Kutta method. Convergence to steady-state is accelerated using local time-steps, and in the future multigrid will be added.

Using exactly the same computational grid and the steady state flow field, the integral form of the linear, harmonic flow equations is discretized on a hexahedral cell as

$$R_k(\hat{\mathbf{U}}) = \sum_{faces}\left((\overline{\hat{\mathbf{F}}} + i\omega\hat{\mathbf{x}})\cdot \mathbf{A} + \overline{\mathbf{F}}\cdot\hat{\mathbf{A}}\right) - \hat{V}_k\overline{\mathbf{S}} - V_k(\overline{\hat{\mathbf{S}}} + i\omega\hat{\mathbf{U}}) = 0 \tag{27}$$

The harmonics variation in the face area vectors $\hat{\mathbf{A}}$ and the cell volume $\hat{V}_k$ are evaluated given knowledge of the harmonic grid motion at each node. Combining all of the cells around one node, and adding back in the pseudo-unsteady term due to Ni, gives:

$$\left(\sum_k V_k\right)\frac{\partial \hat{\mathbf{U}}}{\partial t} = -\sum_k R_k(\hat{\mathbf{U}}) \tag{28}$$

Again numerical smoothing is added to this equation, and for the maximum accuracy in subsonic flow calculations, and for the correct treatment of shocks in transonic flow calculations, it is important for the smoothing in the unsteady calculation to correspond as exactly as possible to the smoothing in the steady flow calculation. An alternative way of expressing this is to say that the linear harmonic discrete equations should be not only a discretization of the linear harmonic Euler equations, but also a linearization of the discretization of the unsteady, non-linear Euler equations. Only in this way can one be certain that the discrete linear equations will yield the same solution as the discrete non-linear equations in the limit of small amplitude unsteadiness.

Lindquist has also shown that in treating shocks it is important that the shock be slightly smeared by the numerical smoothing {Lindquist & Giles, 1991}. This is necessary so that the Euler terms plus numerical smoothing resemble the Navier-Stokes equations with a different viscosity. It can be proved rigorously that for a quasi-1D duct with an oscillating shock, the solution of the linearized Navier-Stokes equations produces the same integrated lift oscillation as the solution of the linear Euler equations with shock fitting. The unsteady impulse term in the Euler analysis due to the oscillation in the shock appears in the Navier-Stokes solution as a smooth unsteady pressure term which has magnitude of order $O(Re)$ over the shock of width which is $O(Re^{-1})$.

6. 2D Test Cases

One difficulty in the development of a three-dimensional, unsteady flow code is finding test cases with which to check that the code is a correct implementation of the discrete equations, and to check the accuracy of the discretization. Experience with a two-dimensional non-linear time-marching code has shown that it is very helpful to compare to the linear flat plate theory of Smith {Giles & Haimes 1991}. Therefore, the same flat plate theory is used here by executing the three-dimensional program with a linear flat plate cascade and just three grid nodes in the third dimension. All of the cases which will be presented have the same computational grid and steady flow field. For these calculations, the stagger angle is 30° and the gap/chord ratio is 0.5. The steady flow is uniform with a Mach number of 0.7.

The first test case is for wake/blade interaction, in which an incoming sinusoidal wake is specified at the inflow boundary. The frequency (non-dimensionalized by the chord and steady flow velocity) is 3.14 and the inter-blade phase angle is 90°. **Figure 1** shows the excellent comparison between the 3D code and the linear theory of Smith as implemented in the program LINSUB by Whitehead {1989}. As in all subsequent figures, the solid line represents results from LINSUB, and the symbols are results from the new 3D code.

The second test case is for potential/blade interaction, in which an incoming pressure wave is specified at the outflow boundary. The frequency is 2.0 and the inter-blade phase angle is 0°. **Figure 2** again shows excellent agreement with LINSUB.

The final test case is for a bending deflection of the blades. The frequency is 2.0 and the inter-blade phase angle is 90°. At this inter-blade phase angle, one of the acoustic modes of the system is almost exactly at cut-off, the transition between being a propagating wave and being an evanescent wave with an amplitude that decays exponentially. A poor prediction was observed when using one-dimensional non-reflecting boundary conditions in this case, while **Figure 3** shows the excellent agreement achieved using two-dimensional non-reflecting boundary conditions. **Figure 4** plots the complex amplitude of the lift for this case as a function of the inter-blade phase angle, keeping the frequency fixed. The data points from the 3D analysis agree well with the results from LINSUB.

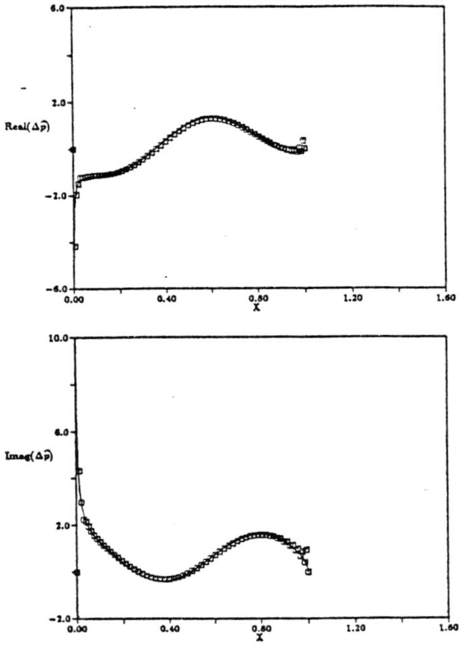

Figure 1: Wake Interaction

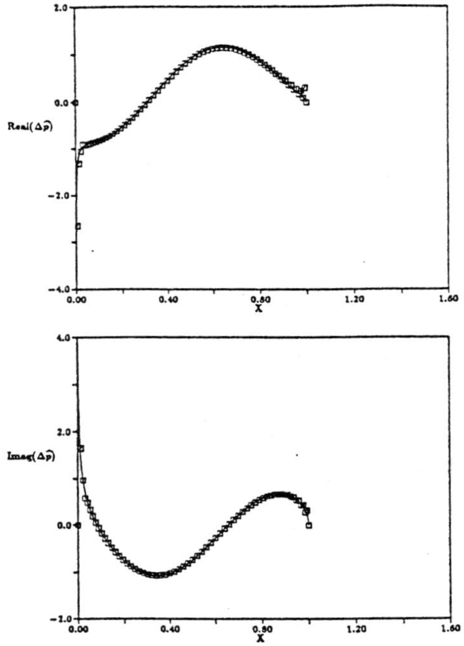

Figure 2: Upstream Running Acoustic Wave Interaction

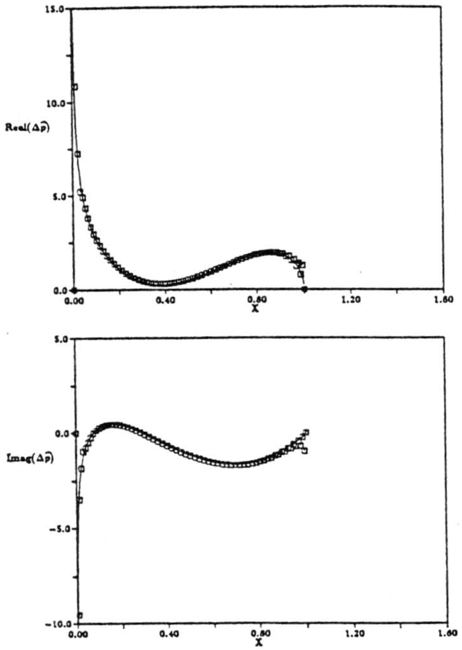

Figure 3: Bending at Cut-off (2D b.c.'s)

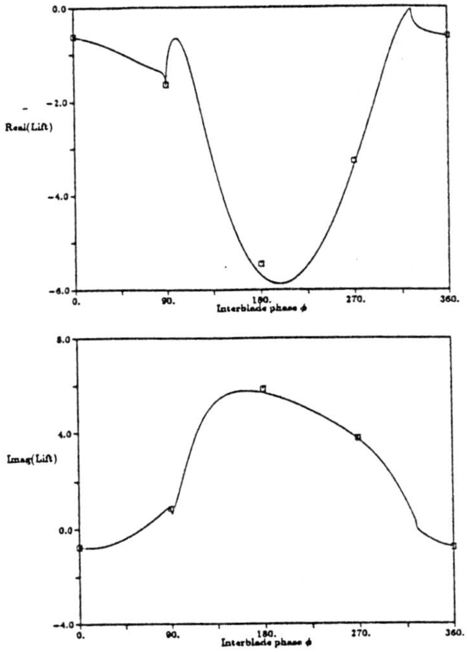

Figure 4: Unsteady Lift due to Bending against IBPA

7. 3D Applications

As a demonstration of the application of this method to some real examples, two test cases have been selected, typical of the aeroelasticity phenomena of a modern turbomachine. The first is the unsteady flow in a high pressure turbine stage, for which the unsteady pressures may be further used in a forced response prediction. The second is the flutter stability of a high aspect ratio fan rotor.

7.1 OXFORD TURBINE STAGE

The high pressure turbine chosen is typical of the design used in contemporary turbomachines, and the flow characteristics have been previously tested in the Oxford rotor facility and reported by Moss et al. {1995}. The test rig consists of a piston tube driving a shroudless turbine stage. The design speed is 8434 rpm, tip diameter: 0.554 m axial chord: 24.35 mm, with 36 vanes and 60 rotor blades. The Reynolds No is 1.55×10^6, with an exit Mach No of 0.959. The instrumentation consists of miniature Kulites on blade as well as a miniature hot wire probe.

The inviscid steady flow part of the method (known as 'SLiQ') was used to determine the rotor steady flow. Various meshes were tried initially, but the final one used has 141 axial by 32 circumferential by 15 radial points, giving about 75,000 points in total. **Figure 5** shows the steady flow mach numbers on this mesh. This mesh required about 20 Mbytes of internal memory and about 2 hours of CPU time to converge on a typical workstation. The steady flow in the vane has been obtained from a 3D viscous calculation. The circumferential non-uniformity in the nozzle passage leads to unsteady flow in the rotor (relative) frame. For this case there are likely to be 2 components of the 'wake' of interest: the pressure non-uniformity (also known as the acoustic or potential field), and the non-uniform velocity field (commonly referred to as the wake, or the unsteady gust or vortical component). The other non-uniformity is the temperature or entropic field, but that is unlikely to be of great importance here.

These different components have different impacts on the SLiQ calculations, as the code models the (linearized) Euler equations. Thus, potential field decay should be correctly handled by SLiQ, equally as in the steady viscous code. However, the vortical field will not decay correctly in SLiQ (as there is no viscosity to mix it out). Thus it should not matter at what axial location the potential field is passed from one domain to the other (as long as it travels the correct physical distance between the bladerows), but the vortical field is more open to interpretation. One answer might be to allow the wake to travel one axial spacing in the viscous steady code, where it should decay, followed by up to another axial spacing in SLiQ - where it will not decay (other than due to numerical viscosity). However, another approach would be to pass the wake approximately half way between the bladerows. This problem is compounded by the fact that real spacings are quite small (often less than 1 chord), and it is not possible to run SLiQ with zero or near zero axial length for the inlet or exit sections; at least half a chord is required to obtain a good mesh.

The potential and vortical fields from the vane are spatially Fourier decomposed, and in this case the first, second and third harmonic components have been retained in the unsteady analysis. Once the unsteady part of SLiQ was converged (which typically takes about twice the time of the equivalent steady calculation), the rotor surface pressure variation can be seen in **Figure 6**, and the entropy animation due to the first harmonic is shown in **Figure 7**. This can be qualitatively compared to the animation from the Q-3D non-linear UNSFLO reported by Moss et al. {1995}. Also the pressure temporal variation at two points on the suction surface of the combination of the 1st and 2nd harmonics from SLiQ are compared to UNSFLO and test data in **Figure 8**. The agreement is quite good – especially in the amplitude, although the mean level from the linear code is slightly out – this may be due to the lack of boundary layer blockage as the code is inviscid.

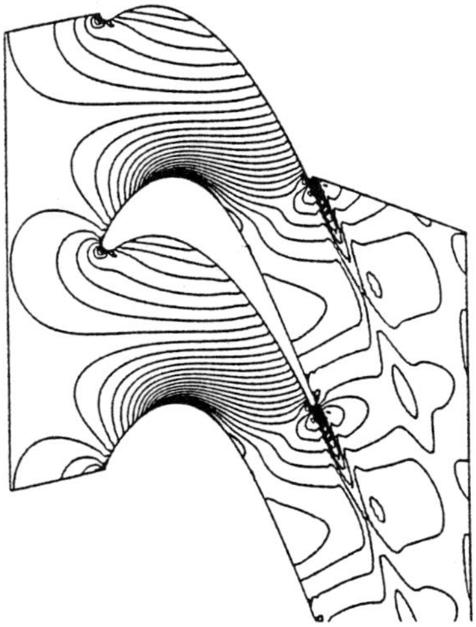

Figure 5: Steady Flow Mach # Contours

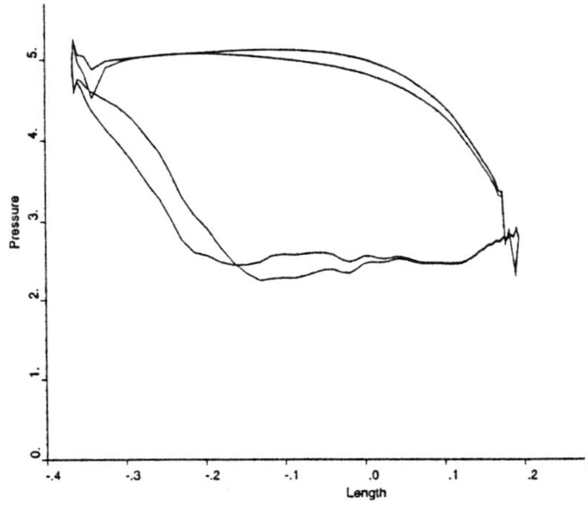

Figure 6: Rotor Unsteady Pressure at Mid-height

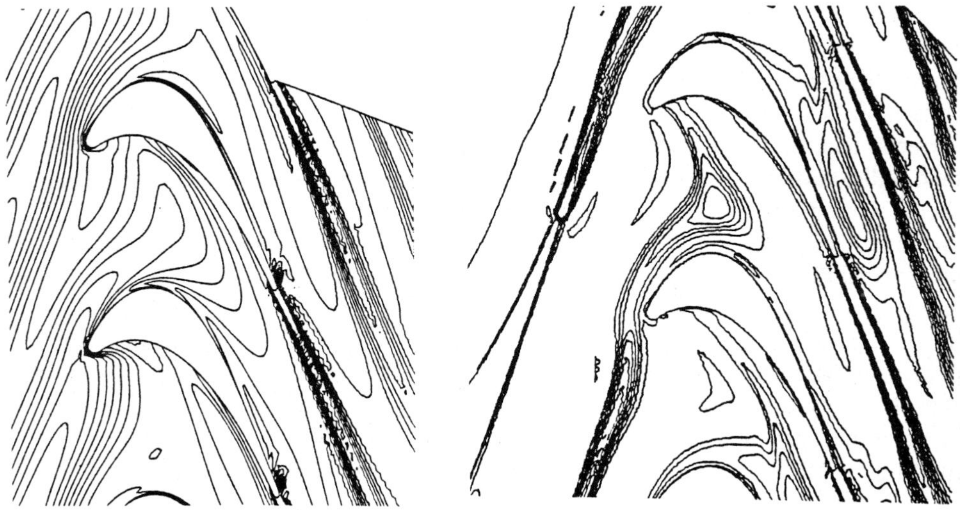

Figure 7: Entropy Annimation - a) 1st Harmonic from SLiQ (linear) b) UNSFLO (non-linear)

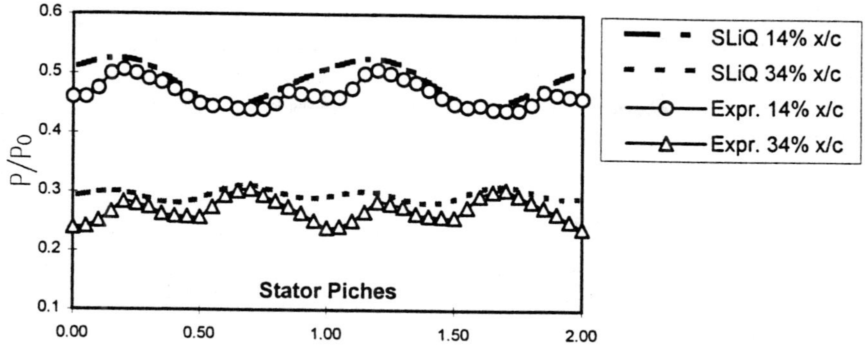

Figure 8: Time History of Pressure on Blade Surface against Experimental Data

7.2 NASA ROTOR 67 FLUTTER

To investigate the flutter performance of the SLiQ linearized code, the case of the NASA Rotor 67 transonic fan has been used. Although this is not a flutter case *per se*, this particular fan was considered to be the best public-domain vehicle to illustrate the capability of the flutter capability. The laser anemometer measurements of the first stage are well documented for two steady-state conditions: near peak-efficiency and towards stall boundary (Pierzga and Wood, 1985). The rotor has 22 low aspect ratio blades, a design speed of 16,043 rpm, a total pressure ratio of 1.63 and a mass flow rate of 33.25 kg/s. This case has been investigated for flutter stability previously by Marshall & Imregun {1996} using a non-linear model.

Using finite element code ANSYS, the first 20 vibration modes of the rotor blade were calculated at the design speed, thus including the centrifugal stiffening effect, assuming the material was Titanium. The results are listed in **Table 1** for the first 5 modes.

TABLE 1: The first 5 modes at 16,043 rpm

Mode	Frequency (Hz)	Description
1	561.5	1st Flap
2	1686.0	2nd Flap
3	1911.7	1st Torsion
4	3516.3	1st Edgewise
5	4314.8	1st Plate

The first flap mode has been investigated for flutter stability, and it is shown in **Figure 9**. The second step in the analysis is to obtain steady flow solutions for given operating conditions. In this case one point near peak efficiency (i.e. on the working line - PE), and a second point near stall (NS) are selected, both at the design speed. These steady flow operating conditions have been exhaustively investigated both experimentally (Stratizar et al. 1981) and numerically (Pierzga & Wood 1985).

In this study, a structured mesh of 73 (axial) by 25 (circum) by 25 (radial) points was used, giving about 2000 points on the two blade surfaces. The inviscid steady state solution was found not to change significantly when a finer mesh was used. The

Reynolds number based on chord at mid-height and relative exit velocity is about 10^6. The blade-to-blade view of the mesh and the pressure contours for the 2 conditions at 30% height from the tip are shown in **Figure 10**. The crucial flow feature is the shock position – at PE the shown is within the passage and thus hits both suction surface (SS) and pressure surface (PS); whereas at NS the shock is expelled and thus only hits the SS.

The flutter analysis is conducted by allowing the blade surfaces to move in the fashion determined by the 1F mode; all possible inter-blade phase angles (IBPAs) representing different nodal diameter modes of the fan assembly can be tried one at a time. The interior points in the fluid mesh are moved in a linear interpolation between SS and PS. The flutter stability is inferred by a 'worksum' analysis: the aerodynamic force due to the unit (mass-normalised) motion of the given modeshape is integrated over the blade surface dotted with the transpose of the modeshape vector. The logdec (positive for stable, decaying modes) is directly related to this aerodynamic work.

The stability (logdec) for the calculated IBPAs at the two flow conditions is shown in **Figure 11**. It is clear that the change of stability with IBPA is more powerful than the change with operating condition. Both conditions show a minimum stability at about 2-3 ND, forward travelling modes (+40° - +50° IBPA). However, although the change with flow condition is small, it is important as the NS condition shows only neutral stability at the +3 ND mode, indicating the possibility of flutter. It should be remembered, however, that there is no mechanical damping included in this analysis, which may explain why this fan was not reported to flutter before it stalled at this 100% speed condition.

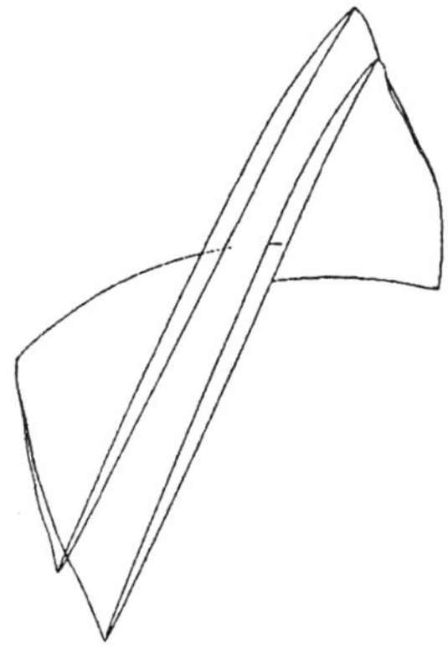

Figure 9: 1st Flap Modeshape

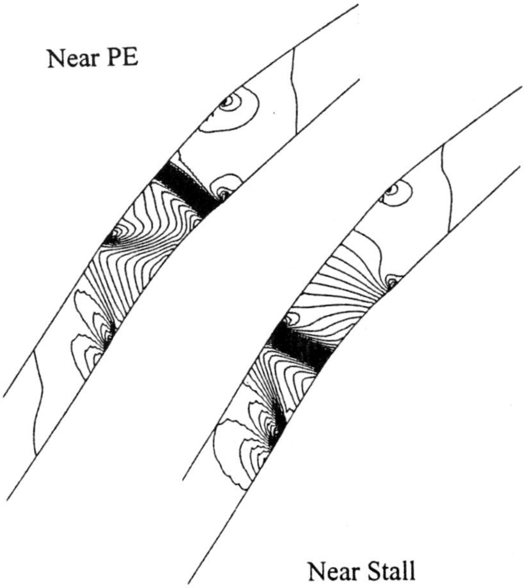

Figure 10: Pressure Contours (10% span from tip)

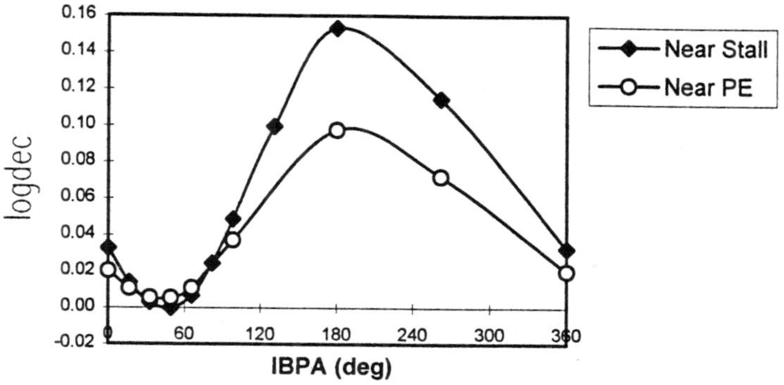

Figure 11: Aero Damping (logdec) against IBPA at 2 flow conditions

8. Conclusions

This paper has presented the analytic formulation of the linearized unsteady equations of motion for an inviscid compressible gas. The linearization is performed by assuming that the unsteadiness is a small perturbation to the non-linear steady flow field. This is valid for flutter analysis which is by definition a linear phenomenon, but the validity for finite amplitude forced response prediction needs to be examined and justified.

An important numerical aspect of the formulation is the inclusion of an oscillating grid whose boundaries move with the vibrating blades. This greatly simplifies the solid wall boundary conditions and improves numerical accuracy by avoiding the need to use spatial derivatives of the steady flow field at the surface. The moving grid is easily incorporated since the numerical scheme simply corresponds to the linearization of existing algorithms for the solution of the non-linear unsteady equations of motion on deforming grids.

The results presented in this paper illustrate the capabilities to do forced response due to wake/rotor and potential/rotor interactions, as well as flutter analysis. Comparison to two-dimensional analytic theory shows that the 3D code is working correctly in a limited two-dimensional mode. Also the method has been demonstrated on realistic 3D cases, typical of contemporary turbomachinery design. These cases show that the new method captures much of the physics of unsteady flow in turbomachines, at a fraction of the cost of equivalent non-linear methods. Future work will concentrate on adding viscous terms, and applying the method to a wide range of applications, to test the limits of applicability of the linearizing assumptions.

9. References

Giles, M.B., & Haimes, R. 1991, 'Validation of a Numerical Method for Unsteady Flow Calculations', *ASME Paper* 91-GT-271

Giles, M.B., 1989, Numerical Methods for Unsteady Turbomachinery Flow', *VKI Lecture Series* 1989-06

Giles, M.B. 1990, 'Non-Reflecting Boundary Conditions for Euler Equation Calculations', *AIAA Journal*

28(12) 2050-2058

Giles, M.B. 1991, 'UNSFLO: A Numerical Method for Unsteady Flow in Turbomachinery', *MIT Gas Turbine Laboratory* Technical Report No. GTL-205

Giles, M.B. 1992, 'An Approach for Multi-Stage Calculations Incorporating Unsteadiness', *ASME Paper* 92-GT-282

Hall, K.C., 1992, 'A Deforming Grid Variational Principle and Finite Element Method for Computing Unsteady Small Disturbance Flows in Cascades', *AIAA* Paper 92-0665

Hall, K.C., & Crawley, E.F. 1989, 'Calculation of Unsteady Flows in Turbomachinery Using the Linearised Euler Equations', *AIAA Journal* **27**(6) 777

Hall, K.C., & Clark, W.S. 1991, 'Prediction of Unsteady Aerodynamic Loads in Cascades using the Linearized Euler Equations on Deforming Grids', *AIAA* Paper 91-3378

Holmes, D.G., & Chuang, H.A. 1991, 'Two Dimensional Linearized Harmonic Euler Flow Analysis for Flutter and Forced Reponse', *Proceedings of 6th UAATP, Notre Dame*

He, L., & Denton, J.D. 1991a, 'Inviscid-Viscous Coupled Solution for Unsteady Flows Through Vibrating Blades: Parts 1 & 2', *ASME Papers* 91-GT-125 & 126

Huff, D.L., Swafford, T.W., & Reddy, T.S.R. 1991, 'Euler Flow Predictions for an Oscillating Cascade Using a High Resolution Wave-Split Scheme', *ASME Paper* 91-GT-198

Hall, K.C. & Verdon, J.M., 1991, 'Gust Response Analysis for Cascades in Non-uniform Mean Flow', *AIAA Journal* 29(9) pp 1463-1471

Jameson, A., 1986, 'Current Status and Future Directions of Computational Transonics', *Computational Mechanics – Advances and Trends*, pp 329-367, *ASME* Ed. A.K. Noor

Marshall, J.G. & Imregun, M., 1996, 'An Analysis of the Aeroelastic Behaviour of a Typical Fanblade with Emphasis on the Flutter Mechanism', *ASME* Paper 96-GT-78 Turbo Expo, Birmingham

Moss, R., Sheldrake, C., Ainsworth, R. Smith, A. & Dancer, S., 1995, 'Unsteady Pressure and Heat Transfer Measurements on a Rotating Blade Surface in a Transient Flow Facility', *AGARD Symposium on Loss Mechanism and Unsteady Flow in Turbomachines, AGARD* CP-571 Paper 22, Derby

Kahl, G. & Klose A., 1991, 'Time Linearized Euler Calculations for Unsteady Quasi-3D Cascade Flows', *6th International Symposium on Unsteady Aerodynamics and Aeroelasticity of Turbomachines and Propellers*

Lindquist, D. & Giles, M.B., 1991, 'On the Validity of Linearized Unsteady Euler Equations with Shock Capturing', *AIAA* Paper 91-1958

Ni, R.H., 1974, 'Non-Stationary Aerodynamics of Arbitrary Cascades in Compressible Flow', *PhD Thesis, Stevens Institute of Technology*

Ni, R.H., & Sisto, F. 1976, 'Numerical Computations of Nonstationary Aerodynamics of Flat Plate Cascades in Compressible Flow', *ASME Journal of Engineering for Gas Turbines and Power* **98** 165

Siden, L.D.G. 1991, 'Numerical Simulation of Unsteady Viscous Compressible Flows Applied to Blade Flutter Analysis', *ASME Paper* 91-GT-203

Pierzga, M.J. & Wood, J.R. (1985), 'Investigation of the Three-Dimensional Flow Field Within a Transonic Fan Rotor: Experiment and Analysis', *ASME Journal of Engineering for Gas Turbines and Power* **107**, p. 436, April 1985

Smith, S.N., 1971, 'Discrete frequency Sound Generation in Axial Flow Turbomachines', *University of Cambridge Department of Engineering* Report CUED/A-Turbo/TR 29

Srivastava, R., Sankar, L.N., Reddy, T.S.R., & Huff, D.L. (1991), 'Application of an Efficient Hybrid Scheme for Aeroelastic Analysis of Advanced Propellors', *AIAA Journal of Propulsion*, **7** (5) p.767, Sept-Oct 1991

Suddhoo, A., Giles, M.B. & Stow, P., 1991, 'Simulation of Inviscid Blade Row Interation Using a Linear and a Non-Linear Method', *ISABE Conference*

Verdon, J.M. & Caspar, J.R. 1984, 'A Linearized Unsteady Aerodynamic Analysis for Transonic Cascades', *Journal of Fluid Mechanics* **149** 403

Verdon, J.M., 1989, 'Unsteady Aerodynamics for Turbomachinery Aeroelastic Applications', *AIAA Progress in Aeronautics and Astronautics Series: Unsteady Transonic Aerodynamics,* Vol. 120

Whitehead, D.S., 1982, 'The Calculation of Steady and Unsteady Transonic Flow in Cascades', *University of Cambridge Department of Engineering* Report CUED/A-Turbo/TR 118

Whitehead, D.S., 1989, 'Classic Two-dimensional Methods, In Aeroelasticity in Axial-Flow Turbomachines', *AGARD Manual on Aeroelasticity in Turbomachines*, AGARD-AG-298, Vol.1

5: Compressor Flow Instabilities: Experiments

AN EXPERIMENTAL FLOW INVESTIGATION OF AN HP FIVE-STAGE COMPRESSOR EXHIBITING ROTATING STALL DUE TO DISTORTED INLET FLOW CONDITIONS

WERNER JAHNEN, THOMAS PETERS, LEONHARD FOTTNER
Institut für Strahlantriebe
Universität der Bundeswehr München
D-85577 Neubiberg

The operational behavior of an HP five stage axial compressor has been investigated at total pressure and swirl distorted inlet flow conditions. The whole flowfield in the compressor inlet and exit and the circumferential static pressure distributions between each blade row have been recorded at 80% and 100% design speed. Evaluating the measured data in multiple segments, the compressor characteristics, all flow variables between each blade row, and the aerodynamic loading parameters have been obtained as a function of the circumferential angle. The inception process of rotating stall due to distorted inlet flow conditions has been investigated with hot wire probes in the inlet and in the outlet of the compressor. The areas of amplification and attenuation of an emerging stall cell have been determined referring to the circumferential position of the distorted area. It is shown that the stall development depends on the local aerodynamic loading and the type of distortion. Non-linear and more-dimensional effects play an important role.

Introduction

The demand for higher pressure ratios and lower weights of modern engine compressors tends to move the aerodynamic load of each compressor stage to its limits. Unfortunately, this leads to an unfavorable response to inlet distortions. Frequently observed inlet distortions of significant performance deterioration for the engine are swirl and inhomogenous distributions of total pressure or total temperature. The distorted inlet flow seen by the compressor generally leads to a decrease of efficiency, pressure ratio, and mass flow rate in connection with a loss of efficiency and thrust of the engine. In addition, it tends to raise premature rotating stall or surge. Since these effects become more violent with the increased compressor stage loading, a design more permissive to inlet distortions has to be considered.

Investigations on the complex flow phenomena caused by the most common inlet distortions, namely swirl and total pressure fluctuations, will provide a better understanding of the causal connections between local blade row loadings and the overall compressor performance. Furthermore the mechanisms resulting in premature compressor instability should be investigated.

T.H. Fransson (ed.), Unsteady Aerodynamics and Aeroelasticity of Turbomachines, 243-257.

Experimental Set Up

COMPRESSOR

The experimental investigations have been performed at the HP 5-stage axial-flow compressor Rig 212 (fig. 1) originating from the RB 199 jet engine development. Each compressor stage is designed for axial inlet flow. All rotor blades are DCA profiles, all guide vanes NACA 65 profiles. At the design point condition the relative inlet Mach number of the first rotor is 0.86 at the mean radius. The compressor design speed is 13860 rpm and the inlet mass flow rate amounts to 4.68 kg/s at INA 0 km inlet conditions. The outlet mass flow rate is reduced to 4.54 kg/s by a bleed behind the third stage. The total pressure ratio is 2.87. The inlet hub to tip ratios are rather high: 0.84 and 0.91, respectively (Jahnen, 1996).

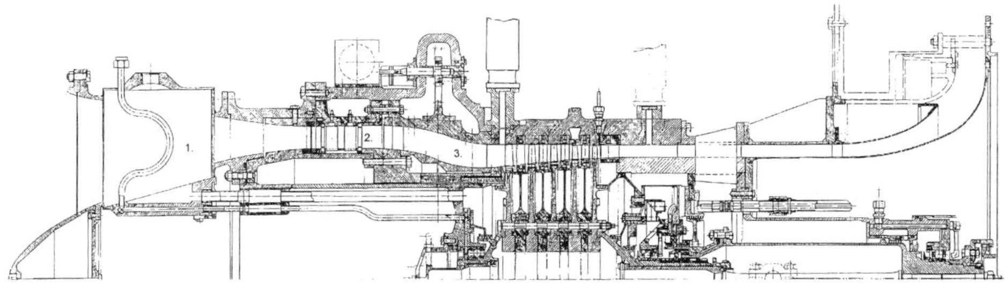

Figure 1. High pressure compressor Rig 212

TEST FACILITY AND INLET DISTORTION GENERATION

The compressor is run on a test facility for multi-stage axial compressors at the Universität der Bundeswehr München. A first throttling device is positioned at the axial-radial diffusor on the right of fig. 1 close to the compressor outlet. Closing this throttle the compressor is forced to exhibit rotating stall because of the small outlet plenum (Jahnen et Fottner, 1996).

Distorted inlet flow conditions have been generated by total pressure screens and guide vanes extending over sectors of the annulus of the compressor inlet duct (position 2. in fig. 1). The measured Rolls-Royce distortion coefficients

$$DC_\varphi = \left[\frac{1}{\varphi} \int_\varphi \left[\frac{1}{r_T - r_H} \int_{r_H}^{r_T} \frac{p_{t2}(r,\varphi) - \bar{p}_{t2}}{\bar{q}_2} dr \right] d\varphi \right]_{min} \tag{1}$$

of the 60°-, 90°- and 120°-total pressure distortions discussed in this report amount to $DC_{60^\circ} = -0.65$, $DC_{90^\circ} = -0.66$ and $DC_{120^\circ} = -0.66$, respectively. The swirl coefficients (Pazur, 1991)

$$SC_\varphi = \left[\frac{1}{\varphi} \int_\varphi \left[\frac{1}{r_T - r_H} \int_{r_H}^{r_T} \frac{c_{u2}(\varphi)}{\bar{c}_2} dr \right] d\varphi \right]_{max} \tag{2}$$

defined for a non-axisymmetric swirl pattern are $SC_{60°} = 0.31$ and $SC_{90°} = 0.37$ at the compressor inlet for 60° and 90° circumferential extent of the installed guide vanes. Thus inhomogenous, counterrotating swirl has been generated.

Pressure distortion generators or guide vanes are mounted to the outer casing and rotated by a small actuator. A rotation of 128° is enabled which is more than the maximum circumferential distance of the probe and pressure tap positions of any measurement plane farther downstream. This way the distorted flow field can be traversed.

MEASUREMENT INSTRUMENTATION

A schematic view of the test facility with an enlarged compressor and the measurement planes is given in fig. 2. Apart from the compressor the test facility has few measurement positions used to provide the values of mass flow rates, of the inlet total temperature, and of the rotational speed. Inside the compressor rig there are measurement planes in front of and behind each blade row. Each plane embodies at least 6 pressure taps evenly distributed over the annulus in the outer casing. At the inlet and outlet there are additional taps at the hub (Jahnen et Fottner, 1996).

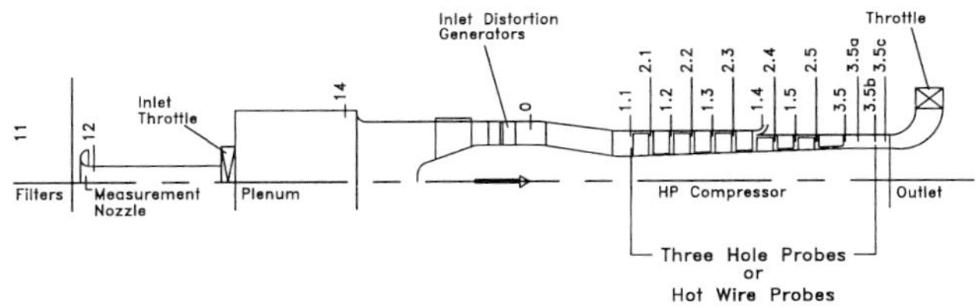

Figure 2. Measurement positions

Apart from the static pressure instrumentation four total temperature rakes and four total pressure rakes apiece are installed permanently in plane 3.5c. During the steady state measurements three hole traverse probes are inserted in the planes 1.1 and 3.5b.

In order to record the development of rotating stall three hot wire probes are inserted in plane 3.5b and four hot wire probes are inserted in plane 1.1. The single-wire probes installed in the inlet are aligned in axial flow direction. All wires at the outlet are oriented angular between axial and circumferential direction. Due to the high dynamic head hot wire measurements at the outlet could only be performed at reduced speeds (80% design speed).

Measurement Results at Steady State Flow Conditions

DISTORTED FLOW PATTERN

Total Pressure Distortion at the Inlet

The generated 60°-total pressure distorted flow pattern at the compressor inlet is shown in fig. 3 in the flow properties total pressure, static pressure, Mach number and swirl angle. It is characterized by a sharp-edged change from the low pressure level of the distorted sector to the high pressure level of the undistorted sector. Since the first blade row is working at high incidence the reduced axial velocity downstream of the screen is partly transformed into a static pressure deficit in front of the first rotor. The

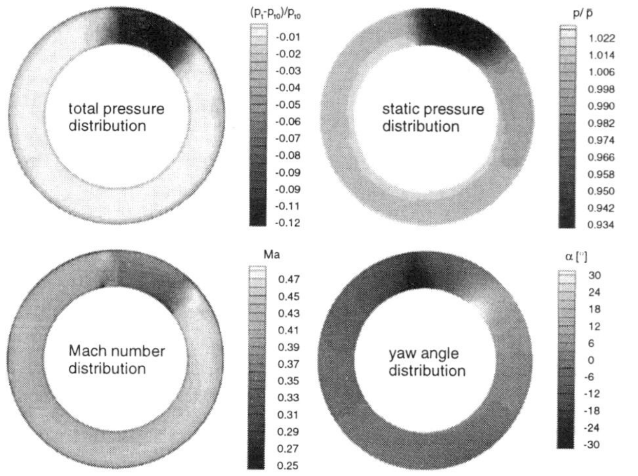

Figure 3. Flow pattern at compressor inlet, 60°-total pressure distortion

resulting flow acceleration from the undistorted to the distorted sector leads to high swirl angles on the border of the distorted sector. The marginal Mach numbers of the undistorted sector have increased. Thus the static pressure field of the undistorted sector is not homogenous either.

Since the fluid passes the curved inlet duct from point 2 to point 3 in fig. 1, secondary flow components are created raising radial gradients in the inlet flow. For the same reason the contraction of the the distorted sector is slightly further advanced at the tip.

Parallel Compressor Model is not suitable because the swirl components at the inlet change the undistorted compressor performance. Though there is approximately axial inlet flow in the center of the distorted sector the flow characteristic does not fit to the Parallel Compressor theory either, demonstrated later on in this paper.

Swirl Distortion at the Inlet

As swirl distortion forms part of the total pressure distortion, total pressure distortion forms part of the swirl distortion. The mounting position of the swirl generator is shown top left of fig. 4. The counterrotating swirl transports the fluid, which passes the guide vane, to the left against the clockwise rotor propagation. On the right edge of the swirl generator, diverging flow components are produced leading to recirculation and high losses in this area. The resulting total pressure deficit can be observed inside the marked guide vane position.

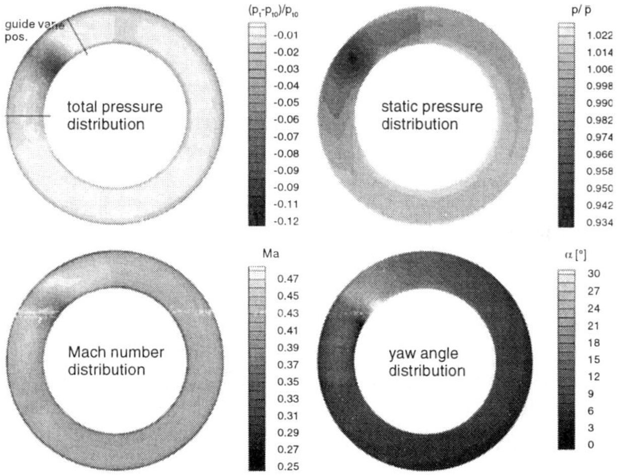

Figure 4. Flow pattern at compressor inlet, 60°-swirl distortion

The segment of maximum swirl components extends also to the right of the total pressure minimum. Once again the flow, that has not passed the distortion generator, is accelerated in circumferential direction. Nevertheless the generated swirl pattern is not at all homogenous. Most of the annulus remains slightly distorted. In the most distorted segments low axial velocities and high swirl angles both raise high incidence angles for the first rotor blade row. The static inlet pressure decreases in these regions.

Static Pressure Distribution inside the Compressor

The increase of static pressure inside the compressor with a 120° inlet total pressure distortion is shown in fig. 5 for a working point near instability and at design speed. On the left side of the distorted sector the strong static pressure gradient at the inlet is transformed to a static pressure minimum due to the corotating swirl components in this area, whereas on the right side a pressure maximum emerges. This flow pattern is maintained to the fifth stage except of a distinct attenuation of the pressure disturbance.

From stage to stage the pressure maximum moves obviously to the right, whereas the pressure minimum remains almost in position. How far flow displacement and flow redistribution are superposed is calculated by the steady state data evaluation.

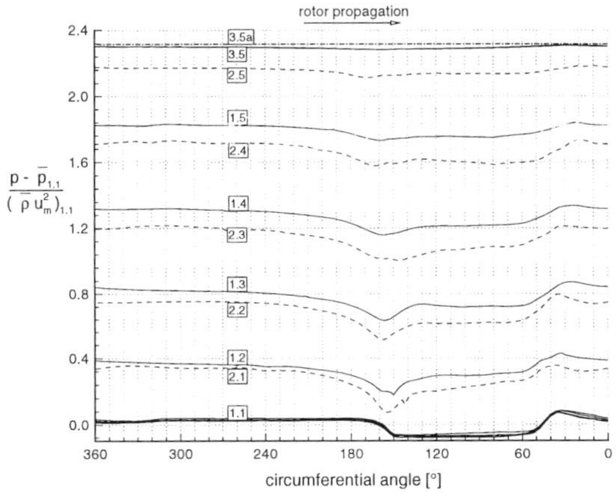

Figure 5. Static pressure distribution, 120°-total pressure distortion, design speed

STEADY STATE DATA EVALUATION

The steady state data evaluation uses mass continuity condition, Euler equation, the definition of stagnation enthalpy, isentropic change of state and blade row angle correlations to calculate temperatures, total pressure and velocities from the measured inlet conditions and the measured static pressure distributions in each measurement plane. The calculation is performed on 5 radial and 180 circumferential positions.

The angular displacement of the flow segments is approximated along the stagger direction. In the rotor blade rows it is also dependent on the fluid velocity. According to Mazzawy (1977) the displacement can be calculated from the relative velocity and the speed of sound as follows

$$\Delta\theta_{\text{Segment}} = \frac{\ell_S}{(w + a)} \cdot \omega \tag{3}$$

which is valid for small disturbances. Since the disturbances are bigger than small, the propagation is slower than the speed of sound. Comparing the measured compressor outlet flow pattern with the calculated one an iterative algorithm corrects for deviations of the disturbance propagation from the speed of sound.

Starting from the mass flow distribution at the inlet, the mass flow redistribution in each measurement plane is determined (fig. 9). A weak digital low pass filter is applied accounting for cross flows inside the blade passages and the axial clearances.

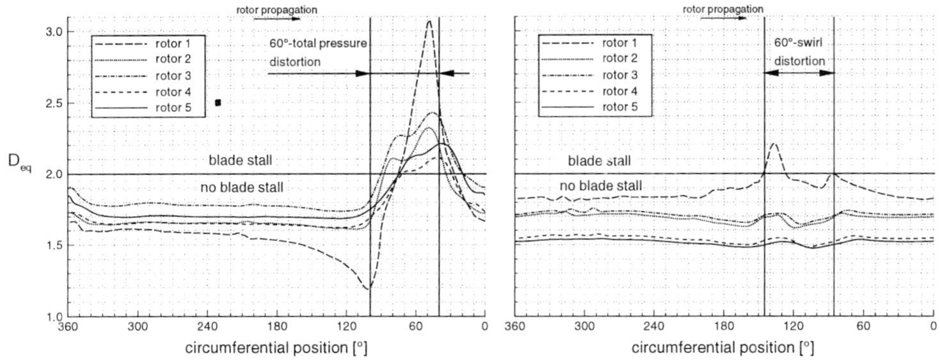

Figure 6. Equivalent diffusion ratios, 60°-total pressure and 60°-swirl distortion

The guide vane outlet angles are assumed to be always axial. Thus the load of each blade row can be calculated. Lieblein's (1957) equivalent diffusion ratios are presented in fig. 6 for a 60°-total pressure and a 60°-swirl distortion. With Rolls-Royce empirical correlations (Fottner, 1967) eventual flow separation can be examined. For the 60°-total pressure distortion at design speed near the instability line all rotor blade rows inside the distorted sector exhibit stall on the suction side. In this case the data evaluation performs an outlet angle correction (fig. 7). Furthermore a time delay for the initiation and a hysteresis for the break down of the blade stall is considered.

Due to the rotor propagation through the different sectors an exchange of momentum takes place between the distorted and the undistorted sectors (Mazzawy, 1977). The rotor blade passages experience an instationary inlet velocity resulting in an instationary total temperature rise and pressure rise across the rotor blade row. According to fig. 8 this effect is strongest in the first

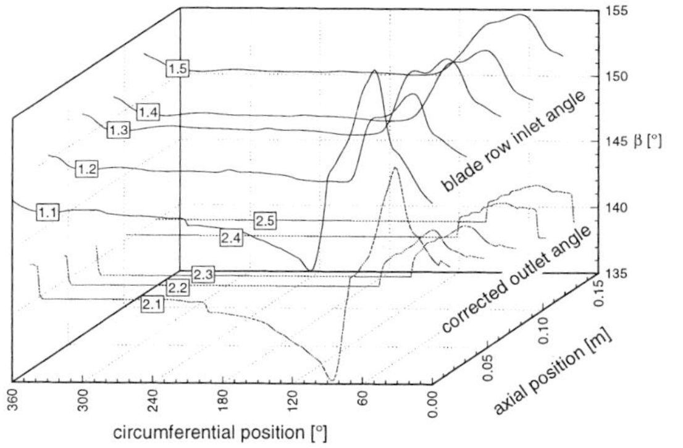

Figure 7. Blade row inlet and corrected outlet angles, 60°-p_t-distortion

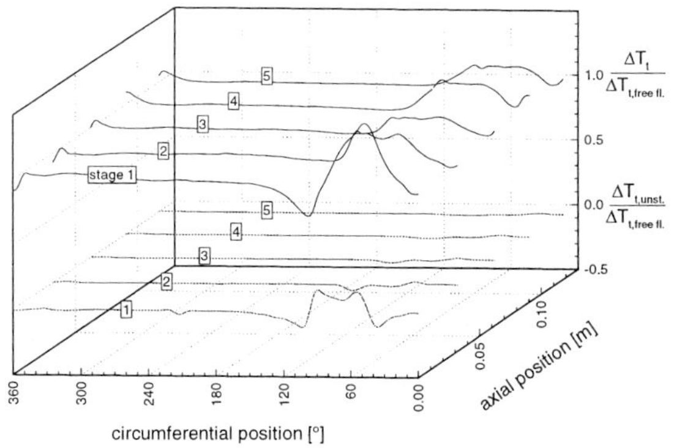

Figure 8. Unsteady total temperature rise

blade row because of the steep gradient of the inlet flow angle.

The conclusive development of the total temperature from inlet to outlet is plotted over the development of the mass flow density in fig. 9. The validation of the presented procedure is given by the comparison of the measured and the calculated outlet temperatures. The total temperature distortion at the outlet is characterized by a steep decrease of temperature when the rotor is leaving the distorted sector. On the one hand the sudden temperature decrease coincides with a similar increase of the local mass flow density, on the other hand the undistorted temperature level is reached before the mass flow density is equal to the undistorted value. The delayed flow reattachment on the blade suction side is responsible for the discrepancy between local mass flow rate and local temperature rise. If flow separation was not taken into account the calculated outlet temperature would by far exceed the measured outlet temperature.

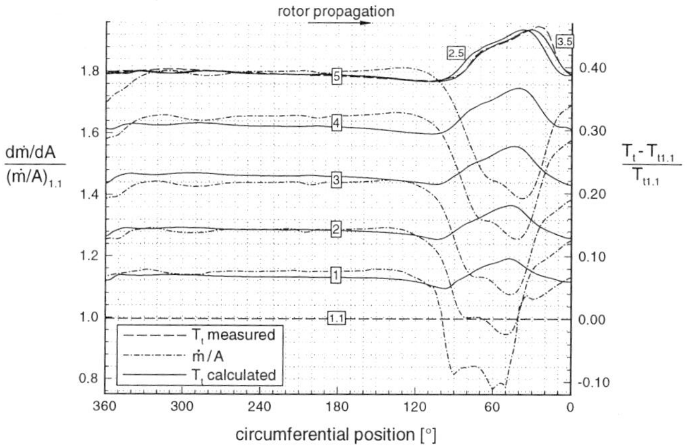

Figure 9. Development of total temperature and mass flow density from inlet to outlet, 60°-p_t-distortion, design speed

Since a digital low pass filter had to be implemented to provide a numerical damping the calculated distribution is smoothed compared to the measured one.

The total pressure distortion has almost completely vanished at the outlet in both the measured and the calculated distributions.

COMPRESSOR CHARACTERISTICS DUE TO DISTORTED INLET FLOW

60°-Total Pressure Distortion

Since all flow data are available over the whole annulus in each blade row plane, any sector can be evaluated for the compressor characteristics. In order to prove the applicability of the Parallel Compressor Model (Pearson et McKenzie, 1959) and in order to distinguish between the most distorted and the almost undistorted sectors the compressor performance data of the undistorted compressor, the 60°-total pressure distorted compressor, the distorted and the undistorted sector are shown in fig. 10. Total pressure ratio and isentropic efficiency characteristics of 80% and 100% design speed are plotted over the reduced mass flow rate. The latter is increased from the sector's value to a representative value for the total circumference. Though the inlet total pressure of the distorted sector is very low the reduced mass flow rate is explicitly smaller than that of the undistorted sector or undistorted compressor, respectively. Because of the flow separation present inside the distorted sector the flow cross section is blocked by a large extent.

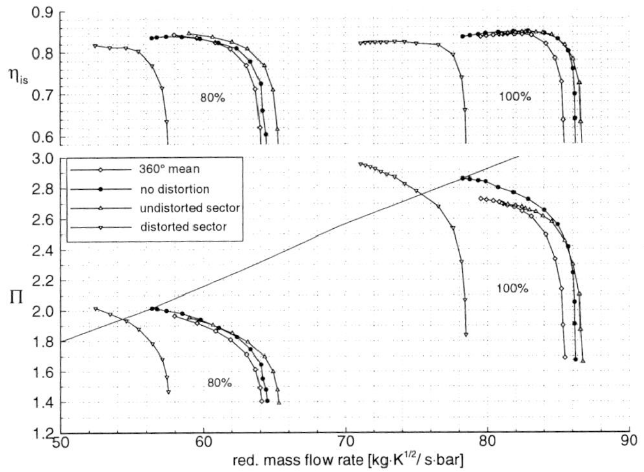

Figure 10. Undistorted and distorted characteristics, 60°-p_t-distortion

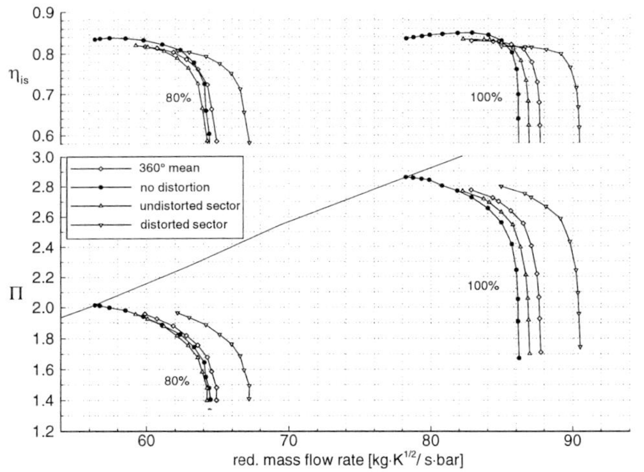

Figure 11. Undistorted and distorted characteristics, 60°-swirl-distortion

Near the choking line the undistorted sector seems to be less throttled than the undistorted compressor due to a contraction of the distorted sector in the back stages. Both sectors are not working on the characteristic of the undistorted compressor.

Concerning the instability onset the distorted compressor exhibits its rotating stall at higher mass flow rates and reduced total pressure ratios. Nevertheless the flow inside the distorted sector is stable far beyond the instabilty line of the undistorted compressor. Neither the distorted nor the undistorted sector is itself responsible for the instability onset.

Except of the undistorted sector the achieved efficiencies are lowered compared to the undistorted compressor. Although the mass flow rate and pressure ratio of the undistorted compressor are strongly reduced the efficiencies show a rather small decrease.

60°-Swirl Distortion

In contrast to the total pressure distortion, efficiencies of the swirl distorted compressor are much more deteriorated (fig. 11). In case of the swirl distortion high losses are produced in the highly loaded and broadly stalled first rotor blade row, whereas losses in the distorted sector of the total pressure distortion are partially avoided by the immediate unloading of the blade passages entering the undistorted sector.

The average high incidence of the first blade row leads to an increased mass flow rate of the distorted compressor. This effect is more evident at design speed than at 80% speed. Obviously the opponent influence of the increased flow separation at lower speeds for the first rotor inhibits higher mass flow rates. The stability properties of the swirl distorted flow are deteriorated. Total pressure ratios at the instability line and the whole stable flow range are reduced.

Development of Rotating stall

UNDISTORTED TRANSIENTS

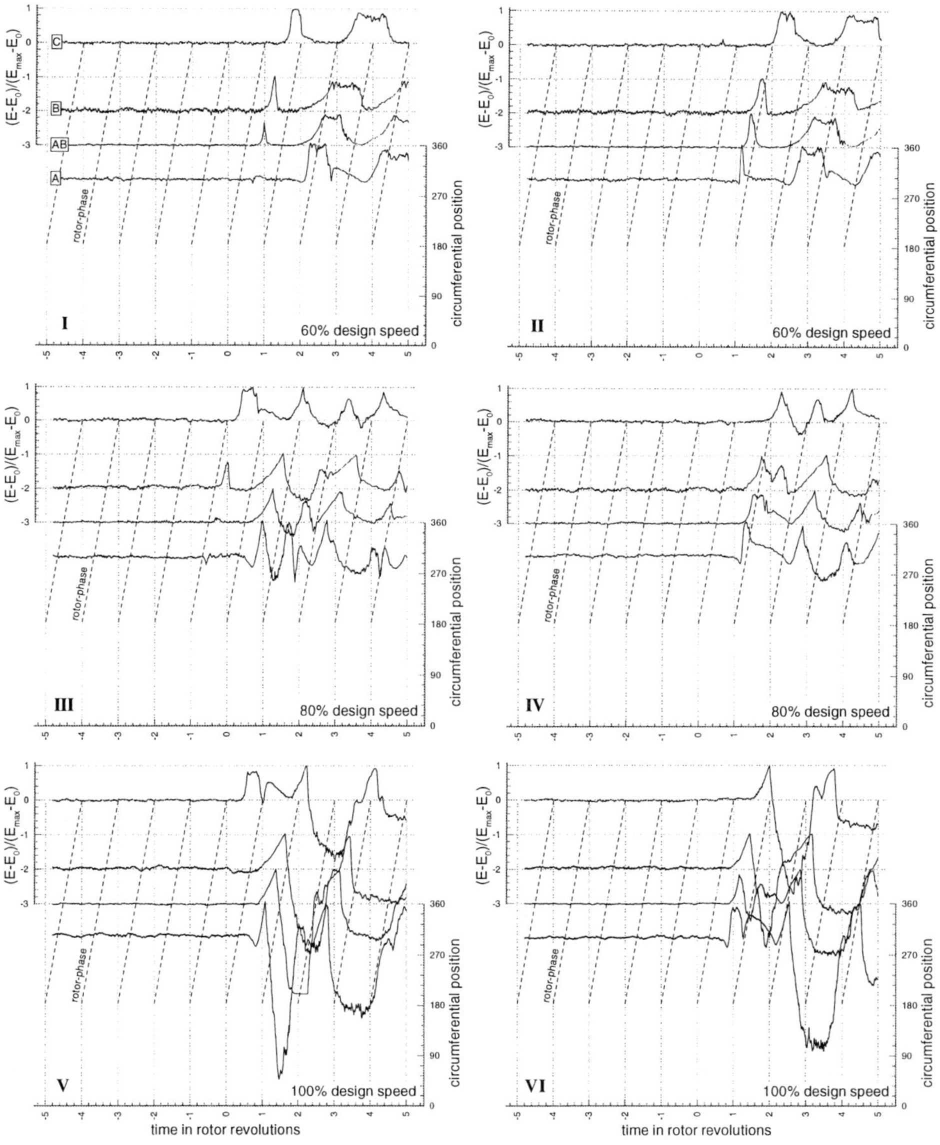

Figure 12. Rotating stall transients, undistorted inlet flow

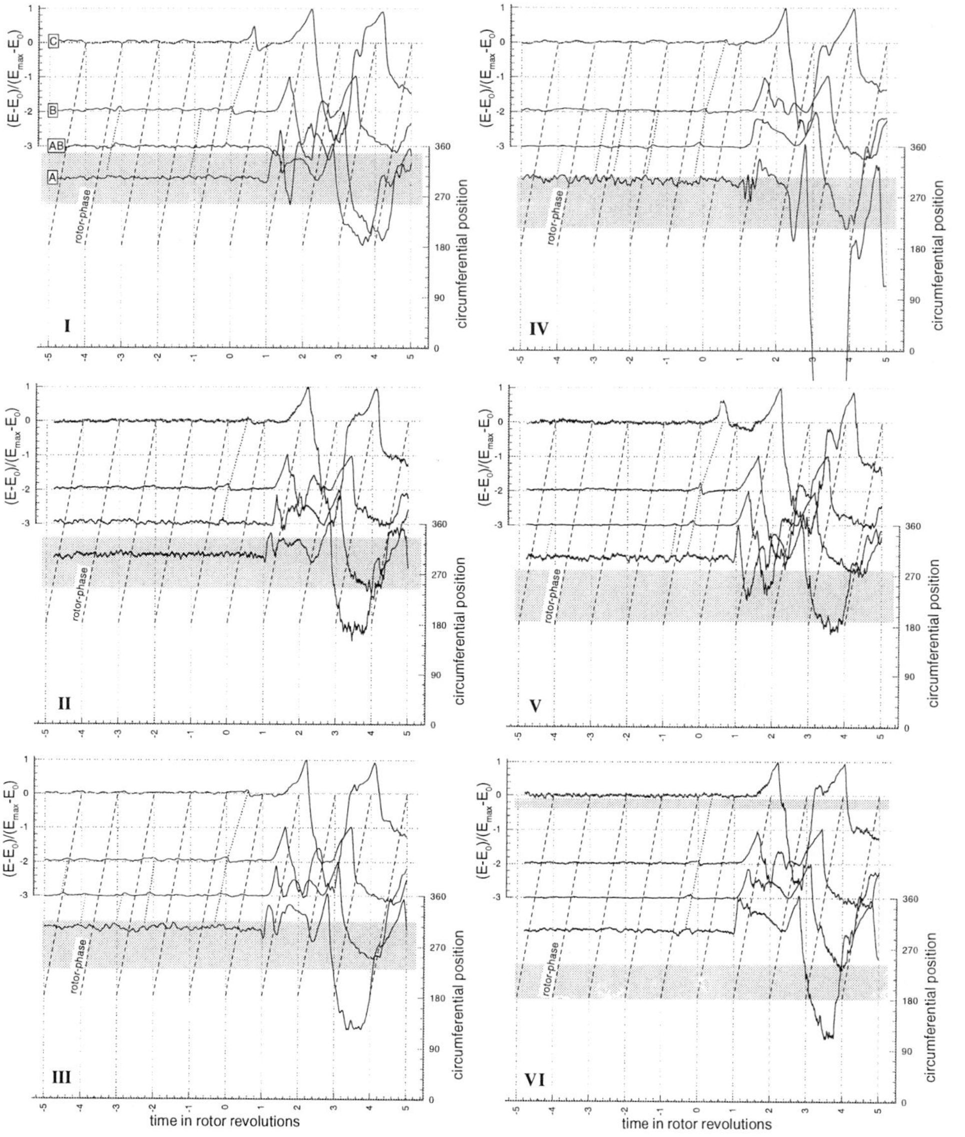

Figure 13. Rotating stall transients, 90°-total pressure distorted inlet flow, design speed

All presented hot wire data have been separated from their DC-offset present at steady flow. Instead of a calibration the data of each probe are related to their maximum positive rotating stall peak.

The stall transients recorded at the compressor inlet at homogenous inlet flow conditions are plotted in fig. 12 for 60%, 80%, and 100% design speed over 10 rotor revolutions. The visible development of the rotating stall cells to full size takes only one rotor revolution, at design speed even less. Small disturbances propagate with rotor speed over the annulus and decelerate to 40% rotor speed growing into fully developped rotating stall. The initial cells mostly occur as a negative signal peak (I, III, IV, V in fig. 12) always changing to positive peaks when passing the next probe. The first appearance of a passage blockage causes a flow stagnation. Because of the fluid inertia the cross acceleration of the fluid is delayed. Therefore at first the axially aligned sensors feel a reduced velocity and afterwards they experience the developped cross flow components. Keeping the hot wire sensitivity in mind, different circumferential positions, where the inital disturbances originate from, are observed.

TRANSIENTS AT 90°-TOTAL PRESSURE DISTORTION

The 90°-total pressure distortion has been rotated from measurement to measurement presented in fig. 13 in order to observe the spatial development of rotating stall relative to the distortion position.

In contrast to the undistorted flow the onset of rotating stall is triggered by the distortion. The disturbance propagation over the circumference is marked by dotted lines from the very beginning. All stall cells originate from the upper border or from inside the distorted sector. Passing the undistorted sector they either are amplified (I, V), damped (II, VI) or almost keep their magnitude (III, IV). Reentering the distorted sector they always grow to fully developped rotating stall within this sector. Since the amplification inside the undistorted sector is rather small the development advances more slowly than in the undistorted case.

Beside of the emerging stall cells small disturbances are additionally denoted in diagram I, III, IV, and V. Starting from the upper border of the distorted sector they are all eliminated inside the undistorted sector. These initial disturbances occur in a sporadic manner and therefore seem to be produced stochastically.

Wether an instability emerges is dependent on wether the initial disturbance survives the pass through the undistorted sector. The cell's size determines the rate of amplification in this area.

Regarding fig. 14 the time domain data of the inlet and of the outlet can be compared. The emerging stall cell is first visible at the inlet seen by probe A (arrow three), whereas at the outlet it appears 240° later the first time (probe C, arrow 1). Besides, the disturbance is seen one third rotor revolution later by probe C at the outlet than by probe C at the inlet. Thus the stall cell axis is

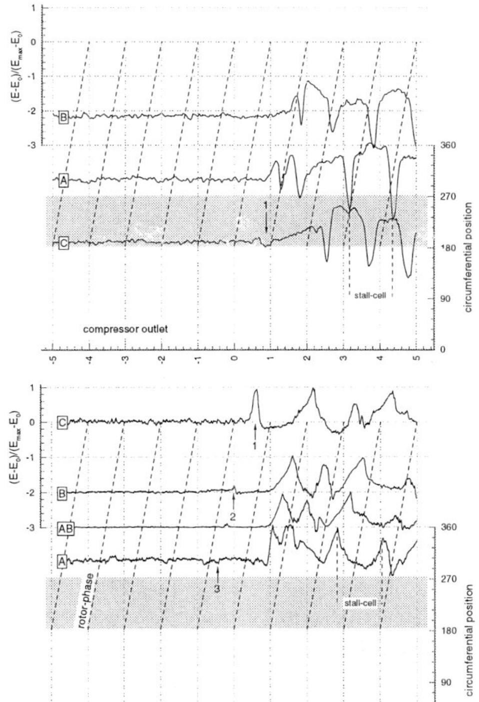

Figure 14. Stall transients at the inlet and at the outlet, 90°-p_t-distortion, 80% speed

aligned angular between inlet and outlet. Taking the maximum load of the first rotor blade row into account (fig. 6) the origin of the emerging stall cell is supposed to be in the first stage.

TRANSIENTS AT 90°-SWIRL DISTORTION

The high load of the first rotor blade row compared to the back stages (on the right of fig. 6) results in a permanent rotating disturbance limited to the first stage before compressor instability emerges. The weak rotating stall is most evident at probe A in fig. 15. The disturbance is coupled to the rotor, i.e. it reoccurs in distances of one rotor revolution. The development leading to rotating stall progresses more slowly than with total pressure distorted flow. Again it is characterized by an alternating amplification and damping of the rotating disturbance in the sectors of different loading.

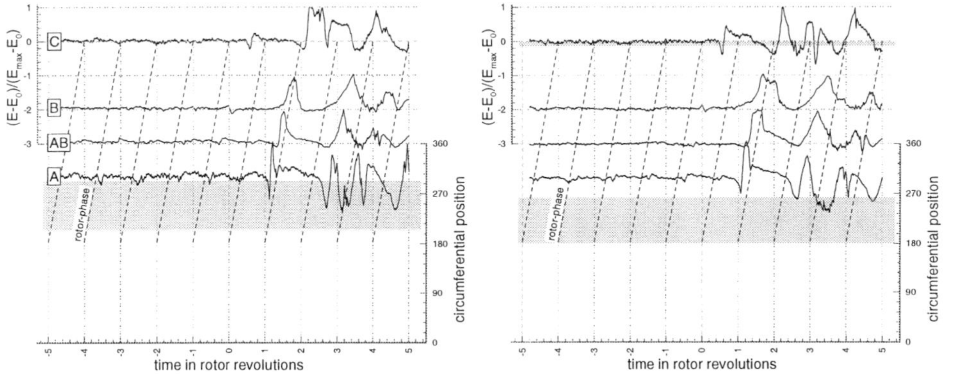

Figure 15. Rotating stall transients, 90°-swirl distorted inlet flow, 80% design speed

The presence of a permanent rotating disturbance in the first rotor blade row, before instability has fully developped, is proven by the frequency spectra of fig. 16. In case of the total pressure distorted inlet flow no eminent frequency exists before instability, whereas in case of the 90°-swirl distortion the rotor frequency is clearly distinguished. The FFT-spectra at compressor outlet do not show any rotating disturbance preceding instability neither for the total pressure distorted nor for the swirl distorted inlet flow.

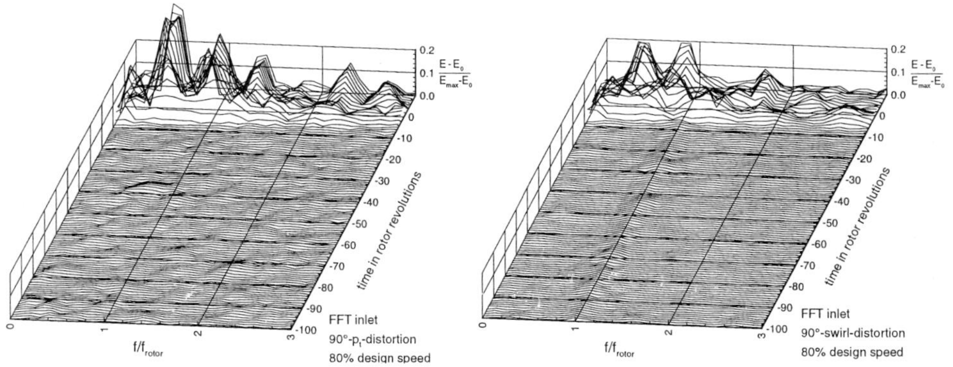

Figure 16. Frequency spectra at the onset of rotating stall, 90°-p_t- and 90°-swirl distortion

While a rotating disturbance is only present in the first stage, the compressor flow is obviously stabilized by the back stages.

Modelling Compressor Instability due to Distorted Inlet Flow Conditions

INFLUENCE OF TOTAL PRESSURE DISTORTED FLOW

The inhomogenous blade row loading, immediately before rotating stall emerges, is shown schematically in fig. 17 on the left. It corresponds to the eq. diffusion ratios of fig. 6 or e.g. the measured loss production of the cascades. At time t = 0 the whole distorted sector is highly loaded. All rotor blades exhibit suction side flow separation in this area. Because of additional counterrotating swirl components leaving the distorted sector, the loading of the first blade row passages is maximal. Furthermore they are exposed to high turbulence levels at the same time resulting in a stochastically produced blockage of a single passage.

Dependent on the cell's size an elimination or an amplification of the small stall cell, at first being attached to the original blade row passage, may take place in the undistorted sector. If the cell is amplified the blockage will extend with speed of sound axially through the compressor (t = 0.8 on the right of fig. 17). Meanwhile the blockage of the first stage moves with almost rotor speed over the annulus. The vector difference of the axial propagation and the circumferential transport results in the angular orientation of the cell inside the compressor (t = 1). The second pass of the distorted sector is always connected with a strong growth of the stall cell.

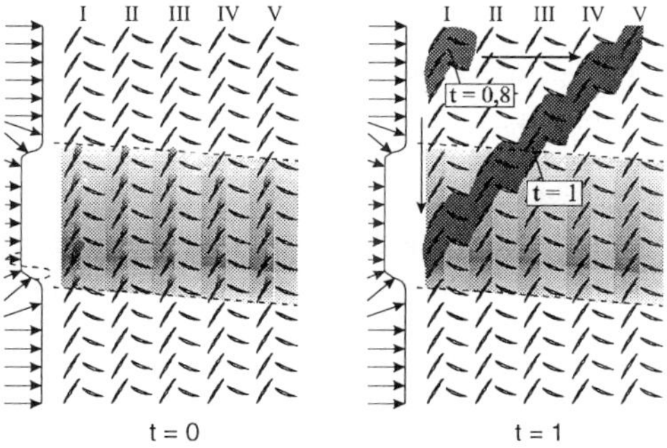

Figure 17. Development of a rotating stall cell due to total pressure distortion

INFLUENCE OF SWIRL DISTORTED FLOW

The local blade row loading of the swirl distorted flow is characterized by a large discrepancy between the loading of the first rotor blade row and of the back stages (fig. 18). The first rotor blade row therefore exhibits suction side flow separation over almost the whole circumference. A permanent rotating disturbance is present in the first stage alternately growing and decaying due to the change of swirl

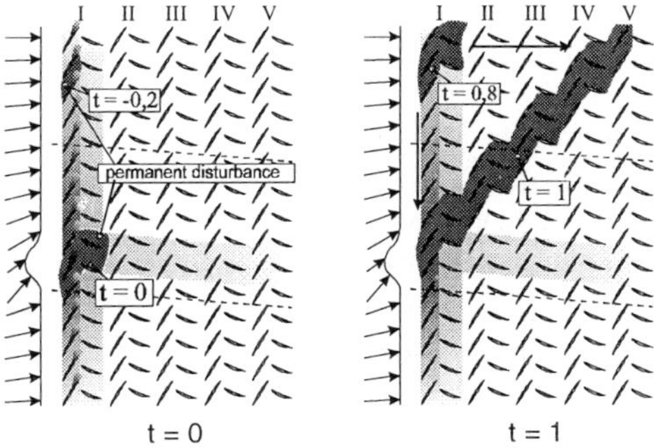

Figure 18. Development of a rotating stall cell due to swirl distortion

angle in front of the first rotor (on the left of fig. 18).

The superposed total pressure distortion generated by the applied distortion configuration is accompanied by intense fluctuations. Similar to the total pressure distorted flow, the size of the disturbance may exceed a certain limit in this circumferential position so that the back stages are not longer able to maintain stable compressor flow. The blockage spreads over all stages (on the right of fig. 18). The cell orientation is the same as for the total pressure distortion.

PREDICTIBILITY OF COMPRESSOR INSTABILITY

Since the onset of rotating stall is determined by the interaction of flow areas of different loading the analysis of local loading parameters will not provide a correct prediction of the overall flow stability. On the other hand global approaches like the Parallel Compressor Model may fail because the segments of different flow conditions inside the compressor work far away from the undistorted characteristics. Though the observed change of amplification and decay of initial disturbances due to the different loadings accords with the model, the quantitative determination of the damping fails for the investigated compressor. The transfer of damping factors of the undistorted compressor is not valid because damping factors show a strong dependence on the cell's initial size, which is very small for the undistorted flow and quite large for the distorted flow.

At least a one-dimensional treatment of the compressor flow has proven to be not correct. In case of the swirl distorted flow, compressor instability is preceded by a rotating stall cell limited to the first rotor. More-dimensional effects are also supposed to be responsible for the non-linear damping behavior.

Summary

Extensive experimental investigations on an HP 5-stage axial-flow compressor at undistorted, total pressure distorted, and swirl distorted flow conditions have been performed. The steady state measurement results presented have improved the understanding of the distorted flow pattern concerning the different performances of distorted and undistorted sectors and the local blade row loadings. Neither distorted nor undistorted sectors coincide with the undistorted characteristic. Especially the distorted sector flow differs significantly from the undistorted compressor flow mainly due to flow separation in the rotor blade rows.

The observed rotating stall transients have shown that the development of compressor instability is characterized by amplification and damping of initial disturbances. The amplification is dependent on the local loading and the size of the stall cell. Due to the differently distributed blade row loading in case of the inlet swirl distortion instability is preceded by a permanent rotating disturbance limited to the first stage. In case of the inlet total pressure distortion stochastically produced, non-permanent rotating disturbances grow very rapidly into fully developped rotating stall.

The development of rotating stall under distorted inlet flow conditions has shown to be a very complex flow phenomenon. As well a global as also a local approach to the problem need further investigations.

References

Fottner, L. (1967) Kennfeldrechnung für Axialverdichter, *MAN Turbo* Technical Report 67/17, Projekt 244

Jahnen, W. (1996) Versuchsmeßtechnik und Meßwerterfassung am Axialverdichterprüfstand für Untersuchungen am 5-stufigen Axialverdichter Rig 212, *Institut für Strahlantriebe* Report LRT-WE12-96/17

Jahnen, W. and Fottner, L. (1996) A New Multi-Stage Compressor Test Facility for Inlet Distortion Investigations *Proceedings of the 13th Symposium on Measuring Techniques for Transsonic and Supersonic Flow in Cascades and Turbomachines*, pp 6-0 ff

Lieblein, S. (1957) Analysis of Eperimental Low-Speed Loss and Stall Characteristics of Two-Dimensional Compressr Blade Cascades, *NACA* Research Memorandum RM E53D01

Pazur, W. (1991) Experimentelle und numerische Untersuchungen zum Einfluß von Eintrittsdrallstörungen auf das Betriebsverhalten von Flugtriebwerksverdichtern, *Universität der Bundeswehr München* Doctor Thesis

Nomenclature

SYMBOLS

a	m/s	speed of sound
c	m/s	velocity
D_{eq}	—	equivalent diffusion ratio
E	V	signal voltage
f	1/s	frequency
l_S	m	chord length
$\dot{m}$	(kg $K^{1/2}$)/(s bar)	reduced mass flow rate
Ma	—	Mach number
p	Pa	pressure
q	Pa	dynamic head
r	m	radius
t	s	time
w	m/s	relative velocity
α	—	yaw angle
β	—	blade row inlet or exit flow angle
η_{is}	—	isentropic efficiency
φ	—	sector angle
Π	—	total pressure ratio
ρ	kg/m^3	density
θ	—	circumferential angle
ω	—	angular velocity

SUBSCRIPTS

0	undistorted
1.1	plane 1.1
2	compressor inlet
3	compressor outlet
free fl.	free flow
H	hub
m	at mean radius
t	total
T	tip
u	circumferential direction
unst.	unsteady

ABBREVIATIONS

DC	Distortion Coefficient
DCA	Double Circle Airfoil
FFT	Fast Fourier Transformation
HP	High Pressure
INA	International Norm Atmosphere
SC	Swirl Coefficient

HIGH SPEED CENTRIFUGAL IMPELLER AND DIFFUSER INTERACTION NEAR STALL CONDITIONS

T. YAMANE
Thermofluid Dynamics Division, National Aerospace Laboratory
7-44-1 Jindaiji-higashi, Chofu, Tokyo 182, JAPAN

AND

T. NAGASHIMA
Department of Aeronautics and Astronautics, University of Tokyo
7-3-1 Hongo, Bunkyo-ku, Tokyo 113, JAPAN

1. Introduction

Amongst the flow instabilities observed in compressors, stall and surge are taken to be the most disastrous effects upon the performance of turbomachinery. It is necessary to secure always the surge margin in compressor map when the operation shifts towards the lower mass flow rate.

In centrifugal compressors, the surge is often found to follow the occurrence of stall in either impeller or diffuser, whilst the complexity comes in due to the interaction between impeller and diffuser of practical high speed turbomachines.

In our previous papers(Yamane *et al.*, 1995)(Yamane and Nagashima, 1995), an implicit TVD Finite Difference code for three dimensional viscous flows has been employed to analyze unsteady flows associated with the impeller/diffuser interaction. The calculation results have been compared with available experimental observations, whereupon a good agreement has been obtained with respect to the changes in the shock wave pattern as well as the instantaneous static pressure on the diffuser shroud wall along a constant speed line. The main interest was, however, unsteady impeller/diffuser interaction around the conditions where the flow is choked in the diffuser, so that less attention has been payed for the lower mass flow rate operations.

T.H. Fransson (ed.), Unsteady Aerodynamics and Aeroelasticity of Turbomachines, 259-271.

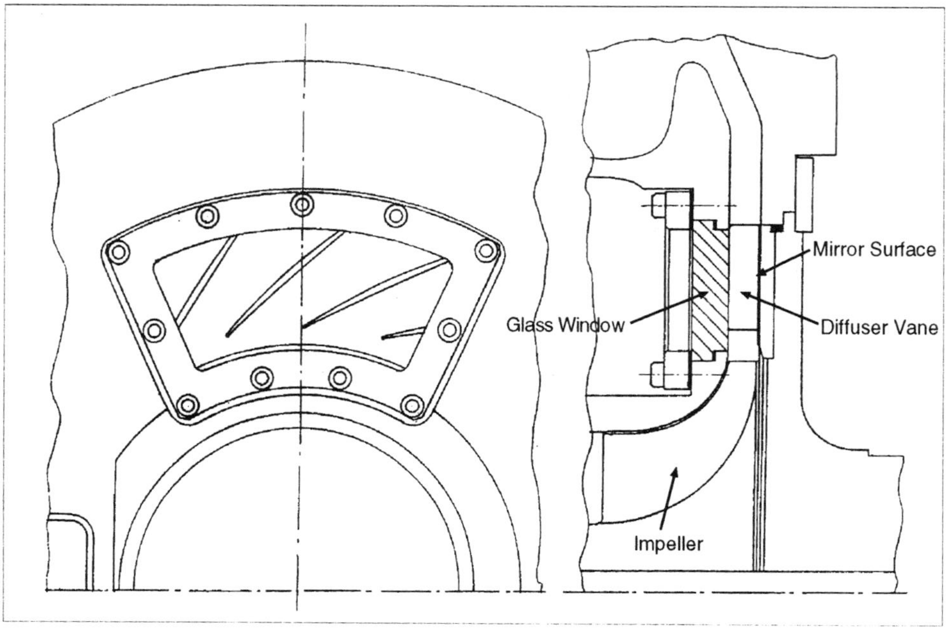

Figure 1. Model Centrifugal Compressor

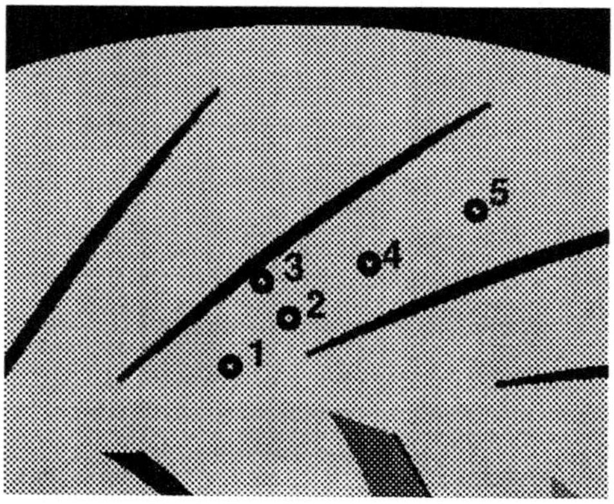

Figure 2. Unsteady Pressure Mesuring Points

In the present study, additional calculations have been performed for better understandings of the flow characteristics. The calculation results are checked against the experimental measurement of pressure distributions. Numerical flow simulation over a wider range of operation will help to provide a clue of clarifying the mechanism of unsteady loss generation.

Impeller		
Outlet diameter	136	mm
Inlet diameter at hub	35.3	mm
Inlet diameter at shroud	94.4	mm
Number of blades		
(in calculations)	14	
(in experiments)	10+10	
Inlet total temperature	288	K
Inlet total pressure	1.013×10^5	Pa
Max. number of revolution	76000	rpm
Max. total pressure ratio	3.6	
Diffuser		
Leading edge diameter	152.3	mm
Trailing edge diameter	214.8	mm
Number of vanes	21	
Chord length	55.16	mm

TABLE 1. Specification of Model Compressor

2. Model Compressor

A small turbocharger with its specifications listed in Table 1 has been used as the model centrifugal compressor. The actual impeller has 10 full blades and 10 splitter blades, which is, however, changed into 14 full blade impeller for the convenience of numerical simulation. Since the number of diffuser vanes is 21, the circumferential periodicity is attainable over 2 blade pitches for the impeller and 3 vane pitches for the diffuser. Experiments are carried out in the arrangement shown in Figure 1. Mass flow rate is checked at the inlet circular duct attached in front of the impeller and the exit total pressure is measured just after the diffuser vanes and inside the exhaust duct. A glass window is equipped on the shroud wall which covers several pitches of diffuser inlet region, while the hub wall is a mirror surface so that schlieren observations can be done. Replacing the glass window by a solid metal wall with static pressure taps enables to obtain static pressure distributions on the shroud wall. These are of time averaged because of tubes which connect the taps and the sensor. In order to investigate instantaneous pressure fluctuations, fast response sensors are directly attached to the shroud wall at 5 locations shown in Figure 2.

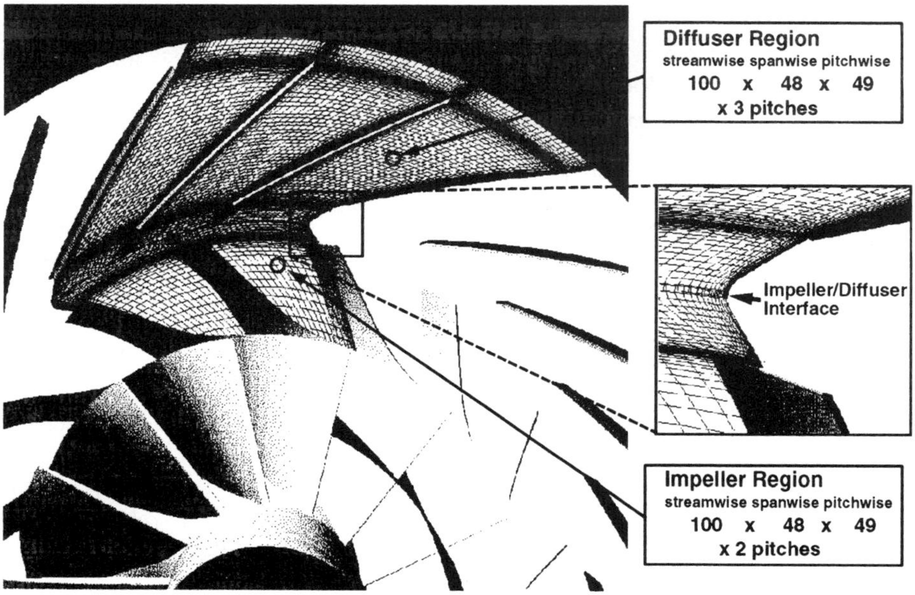

Figure 3. Computational Grid

3. Numerical Method

The numerical method for three dimensional compressible viscous calculations in this study is based upon Chakravarthy-Osher's third order upwind TVD scheme(Chakravarthy and Osher, 1985). The original scheme employs the 'minmod' function for flux limiter but the differentiable limiter(Anderson *et al.*, 1985) has been instead used for better numerical stability. The time integration is based upon an implicit scheme of LU-ADI approximate factorization method(Obayashi *et al.*, 1986). Turbulence is represented by Baldwin-Lomax algebraic model.

Computational grid is shown in Figure 3. It consists of 100 streamwise, 48 spanwise, and 49 pitchwise grid lines for both impeller and diffuser that are algebraically constructed. For impeller/diffuser interaction calculations, density, flux and total energy should be exchanged at the impeller/diffuser intersection. Grid lines are overlapped at this intersection boundary so that the exit boundary values of impeller region and the inlet boundary values of diffuser region are gained from the diffuser internal region and the impeller internal region, respectively. For fully unsteady simulations, 2 pitches for impeller and 3 for diffuser are necessary, whilst only one pitch for each is enough when steady flows are assumed by averaging values along the circumferencial interface of impeller and diffuser.

A vector-parallel computer system made such fully unsteady three di-

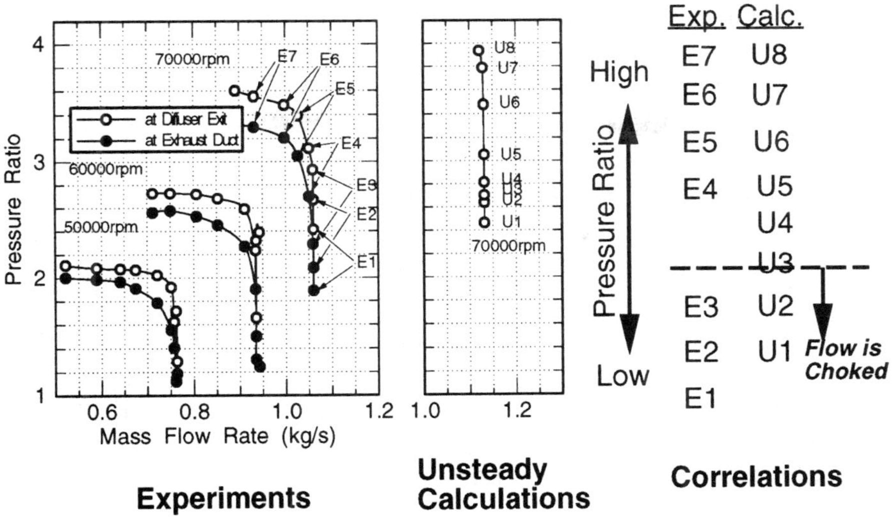

Figure 4. Performance Map of Model Compressor Stage

mensional simulations possible. Up to 24 processor elements of "Numerical Wind Tunnel" at NAL yields the flows in one blade pitch. Presently, 5 pitches are solved at the same time, thus 120 processors are utilised at the maximum speed.

4. Results and Discussion

4.1. COMPARISONS BETWEEN COMPUTATION AND EXPERIMENTS

Figure 4 shows the performance map of the model centrifugal compressor which is obtained from experiments and unsteady calculations. In the numerical simulations the exit total pressure has been calculated by averaging at the diffuser exit boundary. Whilst in the experiments it has been measured both at the diffuser exit and at the exhaust duct, so the difference in pressure ratio is due to the losses at the scroll and duct. At large volume flow rate conditions, the flow at the diffuser exit becomes transonic and quite unsteady, thus the curves show slight oscillations at the lower ends.

Comparisons between the experiments and numerical simulations were made for the operating points along the performance line of 70000 rpm. In the previous ISUAAT paper(Yamane *et al.*, 1995), unsteady calculations at 5 operating points (U1,U3,U5,U6,U7 in Fig.4) have been carried out along the performance line of 70000 rpm. Three more points (U2,U4,U8) are presently added in order to capture the flow pattern changes precisely in which the choking in the diffuser disappears as well as the flow at the diffuser exit becomes unstable near the highest pressure ratio. In the experiments the mass flow rate above the operating condition E5 rapidly decreases while it is not in the calculations.

Correlations between unsteady calculations and experimental operating conditions have been obtained by comparisons of schlieren pictures and time averaged static pressure distributions on the shroud wall, which showed a good agreement(Yamane *et al.*, 1995). Instantanous pressure fluctuations also showed good comparisons. Figure 5 shows a comparison of static pressure time traces between calculations and experiments on the shroud wall at the hole positions indicated in Figure 2. Because the number of impeller blades is different, the passing frequency is 16.3 kHz in the calculations and 23.3 kHz in the experiments. As for the calculation results of Figure 5, fluctuations during only one impeller blade passing are plotted for 3 diffuser blade pitches, thus 3 phase-shifted lines are correspondingly overlapped.

Experiment E2 and Calculation U1 both show similar pressure change and fluctuation amplitude. A shock wave is located between the measuring hole 4 and 5 at this operating condition, which moves towards upstream as the exit pressure goes higher. A large and unstable pressure fluctuation which is observed at hole 4 both in E3 and U2 can be explained by the shock wave motion of back and forth just around this point.

At higher exit pressure conditions which correspond to U7 and U8 where the flow between diffuser vanes is not choked, levels show smooth increase towards the diffuser exit, a similar trend with the experimental conditions E6 and E7. However, the larger fluctuation amplitudes at holes 3, 4, and 5 are observable only in the experiments. As the operating condition approaches the higher exit pressure, that is, slower flow speed, the flow in the diffuser tends to become unstable. There should be many origins of flow instability in actual flows, thus these pressure fluctuations are not necessarily any evidences of diffuser stall. On the other hand, in the calculations, it is likely that the flows near the diffuser exit be more stable due to the boundary condition that fixes the static pressure thereupon. Further examination will be needed to apply for the proper boundary condition at the diffuser exit.

4.2. PRESSURE LOSS DISTRIBUTION

Mach number coutours at an instance on three spanwise planes are drawn in Figure 6 in order to review the flow pattern changes for 4 operating conditions. At condition U1 the flow is choked at the diffuser vane throat because of a large volume flow rate and a shock wave structure appears downstream of the throat. As the pressure ratio goes higher this shock structure disappears and the flow between vanes becomes totally subsonic. Because the wake discharged from the impeller spreads wider near the shroud, the inlet angle varies from the hub to shroud, thus the flow in the diffuser tends to be highly three dimensional as shown in condition U5 where the flow is separated near hub but not near shroud. At condition U7 the flow looks quite smooth where there are no large separation at every spanwise location, but a new low velocity region appears on the opposite side of the vanes at condition U8.

Figure 7 shows total pressure contours in various crosschannel planes. The bottom line, top line and side lines of each plane correspond to the hub, shroud, and blade/vane surfaces respectively. Impeller blade and diffuser vane shapes are also drawn in the figures. In the diffuser section total pressure values are normalized by an averaged value at diffuser inlet, while in the impeller relative total pressures are normalized by an averaged value at the impeller exit. The wake is clearly observed near the impeller exit, which expands wider at low pressure ratio conditions. Two high pressure regions along the shroud and hub walls are observed on the plane which crosses the leading edge of the diffuser vane at condition U1 and U5. These high pressures are dissipated before they reache the mid chord position of the vane. At condition U1 a loss region starts to expand from the position where a shock wave is observed in the Mach number contours (Figure 6(a)). At condition U5 the loss is similarly shown, but it is due to the flow separation from the vane leading edge near the hub wall. The flow angle at the diffuser inlet varies from hub to shroud by the existence of a wake, so the flow is not separated near the shroud. As the exit pressure goes higher, this loss region disappears at condition U7. Near the shroud, a higher distribution at the vane leading edge plane is not dissipated and brought further to the exit. The pressure loss is also small at condition U8, but a new loss region is observed along the diffuser vane lower side. This loss will expand if the pressure ratio becomes still higher and may cause a flow instability at the diffuser exit which determines an operation limit of the present centrifugal compressor, but such extreme calculations have not been accomplished yet.

4.3. OVERALL LOSS IN DIFFUSER

Overall total pressure losses by unsteady calculations are plotted against the averaged inlet flow angle of the diffuser in Figure 8. Results by steady stage calculations with the same boundary conditions are also plotted. An abrupt loss increase in the graph usually indicates an unstable operation point. Both lines show similar curves along the operation line. An irregular change at U3 and S3 is attributed to the disappearance of the flow choking. Concerning the inlet flow angle, the averaged unsteady results show the smaller angles by approximately 2 degrees than the steady stage results.

The inlet flow angle is actually fluctuating so that the averaged values are not the same as the simple numerical averaging taken by the steady simulations.

Figure 9 is another graph which shows the loss changes on three spanwise position. These curves are sometimes clumsy because they are calculated from the information upon the plane thinner than the passage height. Both unsteady calculations and steady calculations generally show similar changes on the planes near the hub and midspan, but completely different from each other on the plane near the shroud. As shown in Figure 8 the inlet flow angle gets smaller as the exit pressure becomes higher, but it is opposite way near the shroud by unsteady calculations. The flows near the shroud are strongly affected by the wake from the impeller thus the averaged flow angle at the inlet is usually smaller than those near the hub and midspan planes. As the operation goes a still higher pressure ratio, the wake becomes weak and the flow angle near the shroud turns to larger values. This effect of the wake is mixed away by steady calculations on the other hand, thus the changes in both the loss and flow angle near the shroud are quite small compared to the unsteady results.

5. Conclusions

Fully unsteady calculations of high speed small centrifugal impeller and diffuser have been carried out to examine the change in performance along the 70000 rpm operation line. Pressure fluctuations on shroud wall showed good agreements with measurements except some larger amplitudes observed only in the experiments at high pressure conditions. Numerical results clearly revealed the total pressure loss distributions and propagations in the diffuser at various flow conditions. A difference in averaged inlet flow angles of the diffuser has been found between the fully unsteady calculations and steady stage calculations.

References

Yamane,T. Fujita,H. Nagashima,T., "An Investigation of Impeller and Diffuser Interaction in a Transonic Centrifugal Compressor", *Unsteady Aerodynamics and Aeroelasticity of Turbomachines*, 1995, pp. 499-513

Yamane,T. Nagashima,T., "Flow Choking and Shock Wave Structure at Diffuser Vanes in a High Speed Centrifugal Impeller", *Proceedings of 12th International Symposium on Air Breathing Engines*, Vol.1, 1995, pp. 135-145

Chakravarthy,S.R. Osher,S., "A New Class of High Accuracy TVD Schemes for Hyperbolic Conservation Laws", AIAA Paper 85-0363, 1985

Anderson,W.K. Thomas,J.L. Van Leer,B., "A Comparison of Finite Volume Flux Vector Splitting for the Euler Equations", AIAA Paper 85-0122, 1985

Obayashi,S. Matsushima,K. Fujii,K. Kuwahara,K., "Improvements in Efficiency and Reliability for Navier-Stokes Computations Using the LU-ADI Factorization Algorithm", AIAA Paper 86-0338, 1986

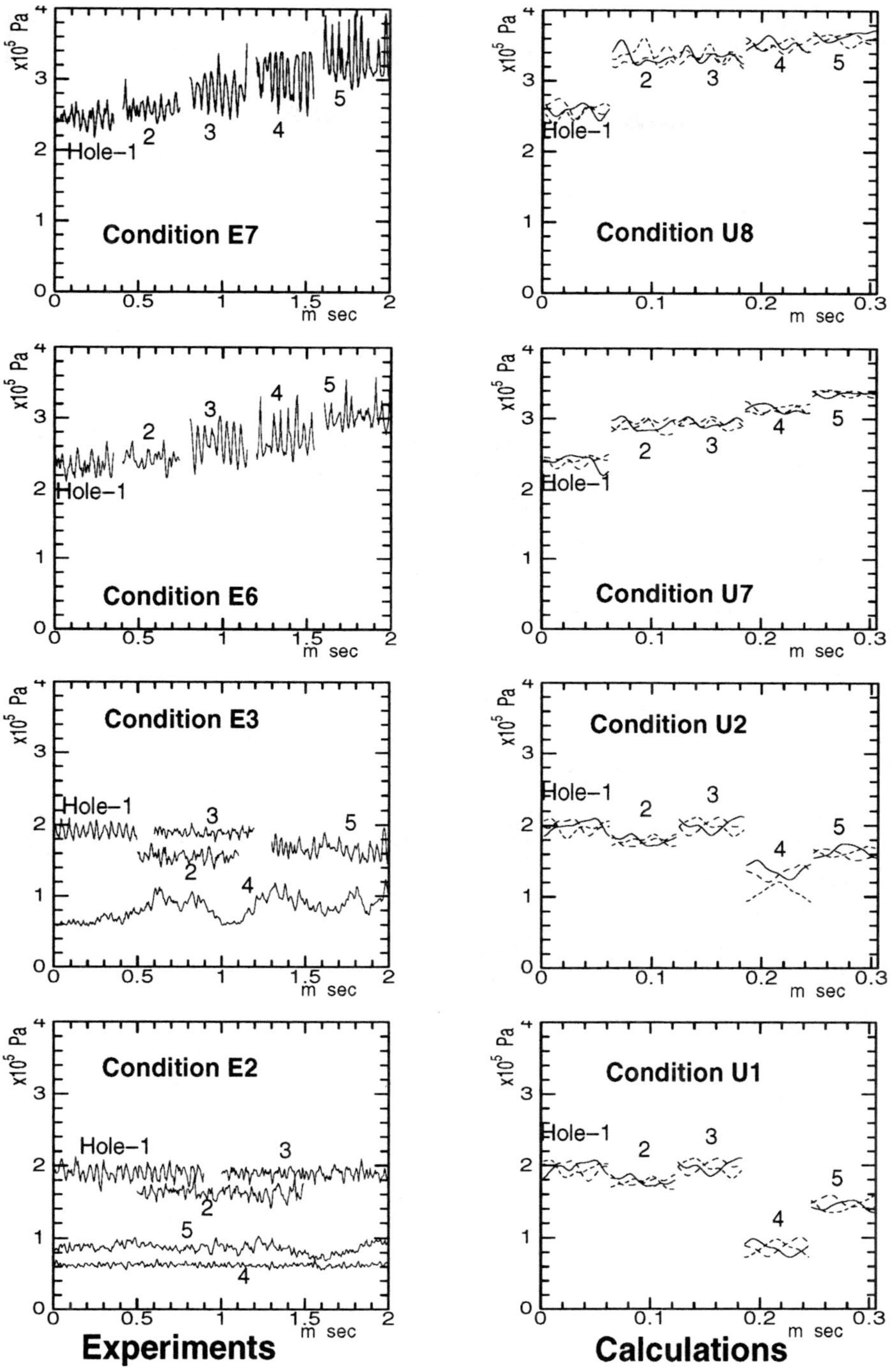

Figure 5. Instantaneous Pressure on Shroud Wall

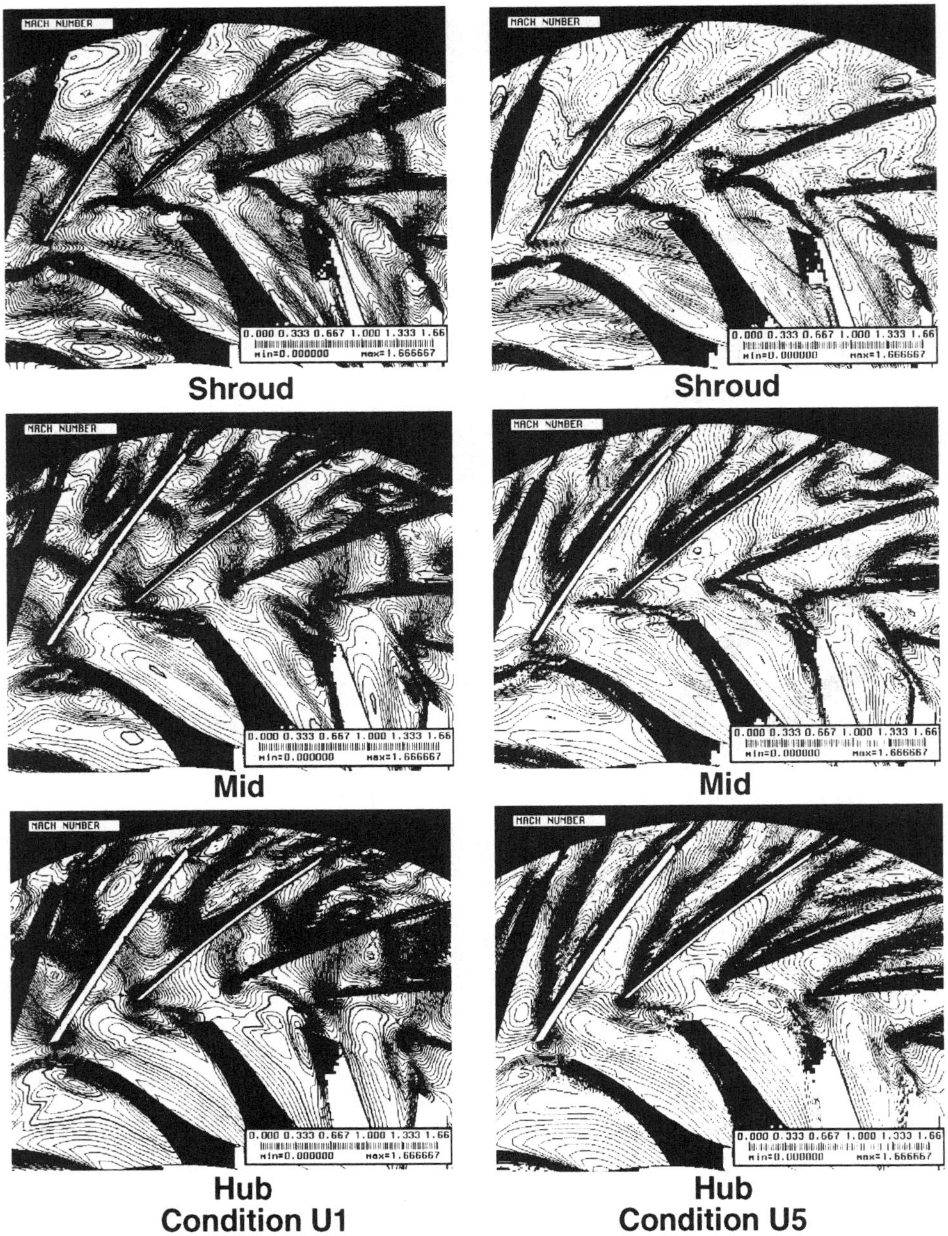

Figure 6. Mach Number Contours

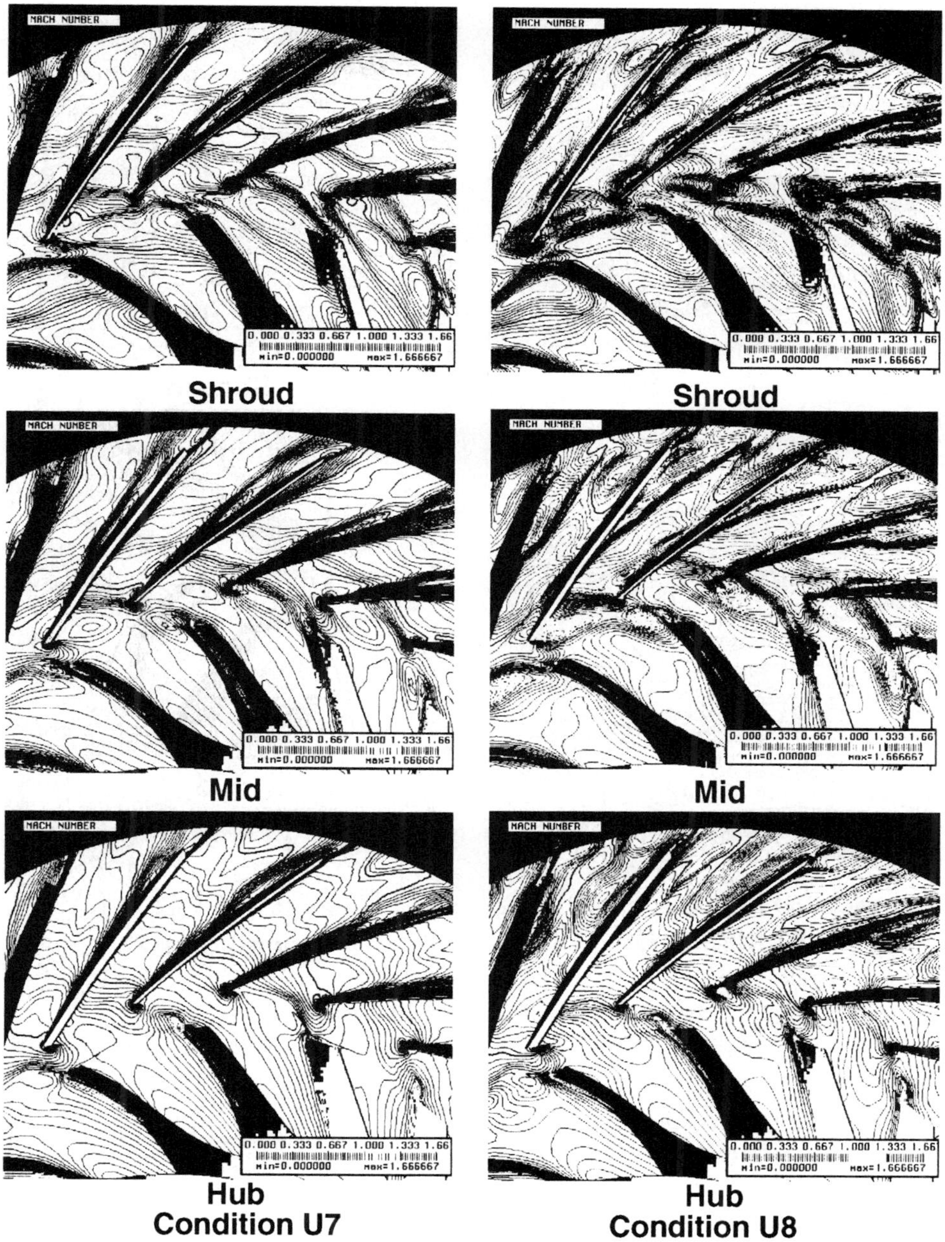

(Figure.6 Continued)

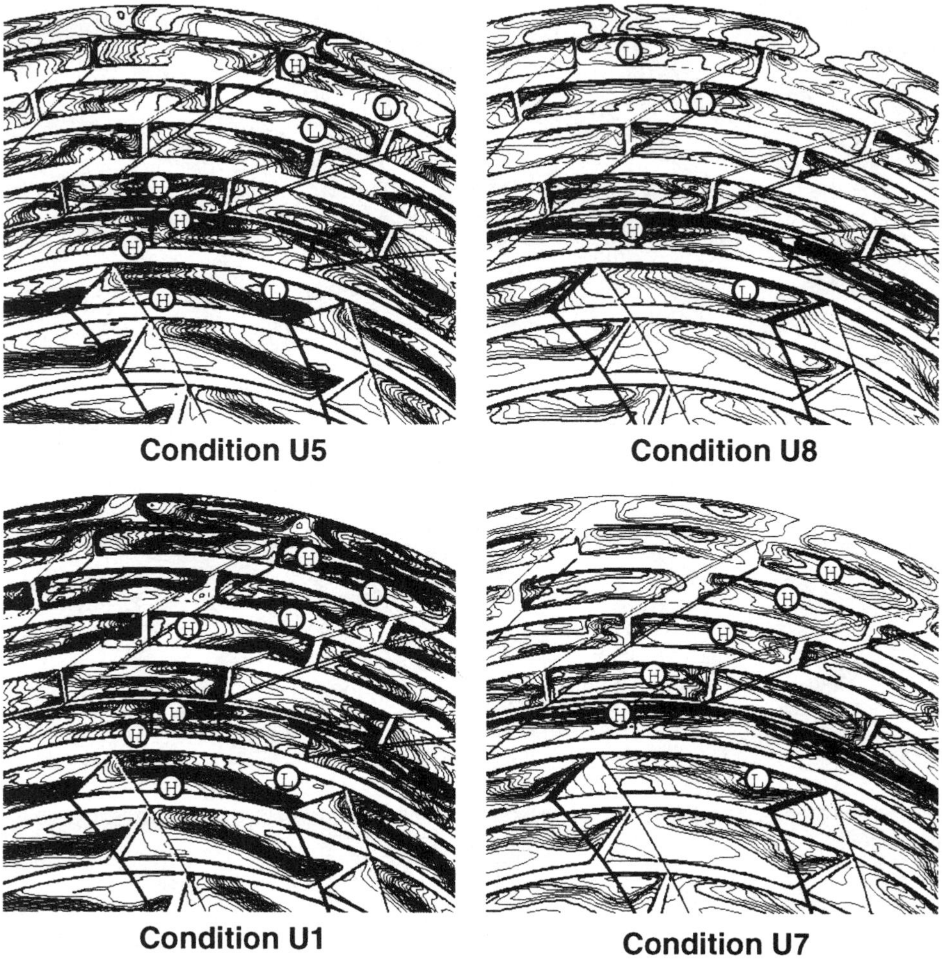

Figure 7. Normalized Total Pressure Contours

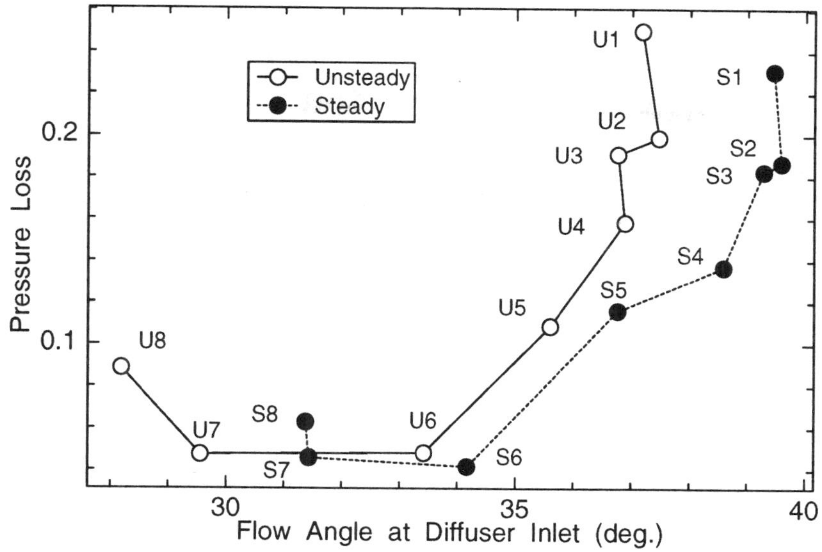

Figure 8. Overall Pressure Loss in Diffuser

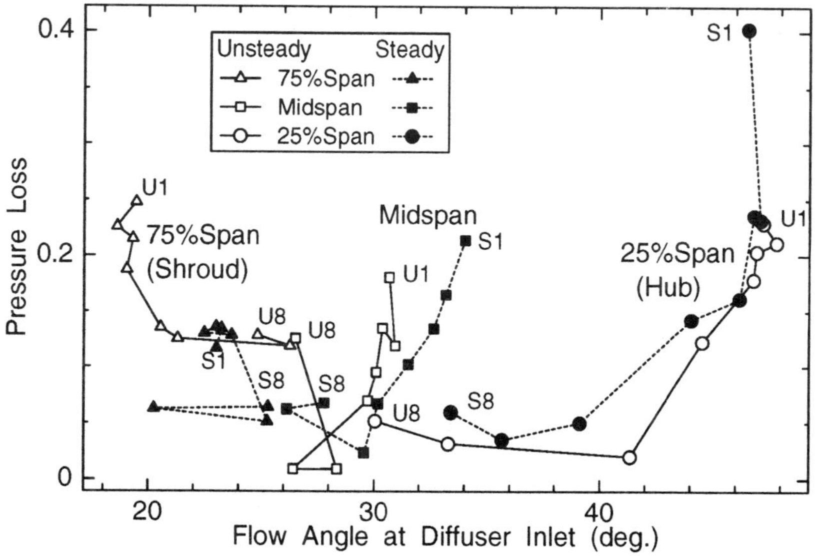

Figure 9. Pressure Loss on 3 Spanwise Planes

UNSTEADY FLOW AND TURBULENCE IN A LOW SPEED AXIAL COMPRESSOR

A. SENTKER AND W. RIESS
Institute for Turbomachinery
University of Hanover
Appelstr.9
D-30173 Hannover
Germany

1. Introduction

Flow in multistage turbomachines is necessarily unsteady. Traditionally this fact is excluded deliberately from theoretical and numerical treatment of turbomachinery flow by considering the flow relative to each blade row and circumferential averaging between blade rows. Even now 2D-meriodional flow or 3D calculations are using the simplification of circumferential averaging behind each blade row. The justification was -and still is to a large extent- the enormous volume of data and methodical problems of numerical calculations or considerable experimental difficulties arising in the treatment of unsteady flow.

Nowadays the engagement for the numerical and experimental research on unsteady flow phenomenon has grown (Lakshminarayana *et al.*, 1997; Jennions *et al.*, 1997; Adamczyk, 1996). The rotor-stator, stator-rotor interaction due to upstream wakes are a source of loss in efficiency. The mixing of the wakes dissipates energy. An information about the wake characteristics -decay and path- is essential for the prediction of the aerodynamic and mechanical performance.

Extensive experimental investigations on unsteady flow and turbulence have been conducted in the first stage of a low speed axial research compressor with emphasis on the rotor-stator, stator-rotor interaction. The flow field behind the IGV and the first rotor is presented. The IGV wake travels through the rotor and can be detected at a spatially fixed position. During the way through the rotor the wake becomes distorted because the region near the suction side moves faster than near the pressure side.

T.H. Fransson (ed.), Unsteady Aerodynamics and Aeroelasticity of Turbomachines, 273-286.

2. Experimental Technique

2.1. EXPERIMENTAL FACILITY

The two stage low speed experimental axial compressor (NGAV) of the Institute for Turbomachinery used for the extensive experimental investigation on the unsteady flow field is operating in a closed circuit non-pressurized facility with a water cooler, that guarantees constant temperature inlet and outlet conditions.
Typical data of the compressor are presented in table 1, for more detailed information see (Traulsen, 1989).

TABLE 1. Design Parameters

Outer diameter	d_a	$= 0.76\ m$
Blade Length	h	$= 140.0\ mm$
Hub/Tip ratio	ν	$= 0.63$
Rotational Speed (design)	n	$= 3000\ 1/min$
Total Pressure ratio	Π_0	$= 1.07$
Rotor Tip Mach number	Ma	$= 0.28$
Rotor Tip Reynolds number	Re	$= 455 \cdot 10^3$
Chord Lenght Rotor	C_R	$= 75.0\ mm$
Chord Lenght Stator	C_S	$= 90.0\ mm$
Chord Lenght IGV	C_{IGV}	$= 105.0\ mm$
Mass Flow Rate	$\dot{m}$	$= 16.5\ kg/s$

Probes can be attached to a mechanical traverse system, which can be mounted at different axial positions on the NGAV. With this system a positioning of the probes in circumferential and radial direction is possible in the axial gap between the blade rows and in four axial positions in the stator (Bigalk, 1995).

2.2. SPLIT FILM PROBES

For these investigations a split film probe (Type R57) (Dantec, 1995) with a radially oriented sensor operated in the constant temperature mode with an overheat ratio of 1.7 was used.
The sensor has an active length of 1.25 mm and consists of two 3 μm Nickel films deposited on a quartz fibre with a diameter of 200 μm. The relatively small dimension of the probe in circumferential direction offers a good opportunity for the measurement of the unsteady flow field in the $'S_1-Plane'$. Since the heated film is divided in radial direction it is possible

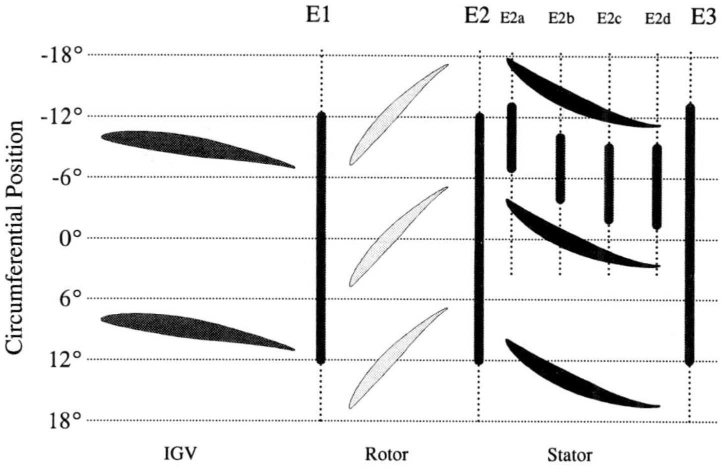

Figure 1. Measuring Planes

to acquire two components of the instantaneous flow velocity. The data are recorded by a 12 bit analogue/digital converter and stored on a personal computer (Riess *et al.*, 1996).

3. Measurement Procedure and Evaluation Method

All available measurement locations in the first stage of the compressor are indicated in figure 1. Measurements presented in this paper have been acquired in plane E1 behind the inlet guide vane and in E2 behind the first rotor from -12° to 12° in circumferential direction. The arc of 24° has been divided into 25 measurement stations with a distance between each station of 2° to 0.5^o depending on the local nonuniformity of the flow field. The distance between each station was chosen from earlier measurements. At locations with high unsteadiness i.e. blade wakes the stepping was 0.5^o whereas in the undisturbed flow field the stepping was 1° to 2°. This procedure helps to reduce the enormeous data volume.

The axial measurement locations are situated at 21.2% ($x_{ax}/C_r \cdot 100$) of the rotor axial chord upstream of the rotor and at 22.5% ($x_{ax}/C_r \cdot 100$) downstream of the rotor. The radial position of the probe for all measurements was at 50% blade height.

An operating point with a mass flow rate of $18.5kg/s$ and a pressure ratio of $\Pi = 1.07$ at a rotational speed of 2900 revolutions per minute was chosen for this investigations.

In each probe position the continuous velocity signal is averaged and digitized at a sampling rate of 200kHz. This procedure delivers 140 measured values per rotor blade pitch, each representing a time window of 4.92 μsec. Continuous measurements have been conducted for up to 126 rotor revolu-

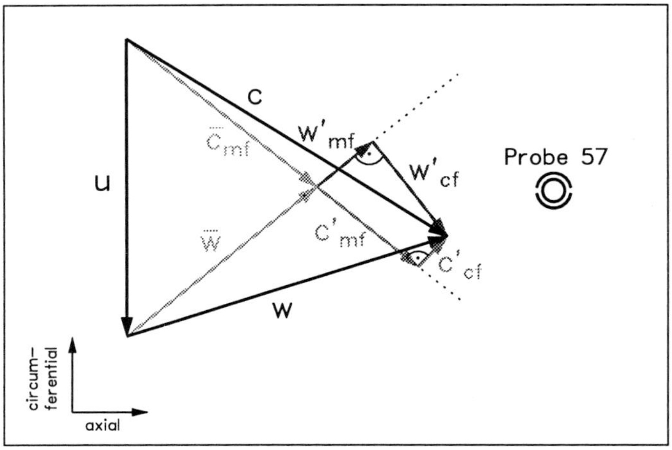

Figure 2. Instantaneous Velocity Components

tions, equivalent to a measuring time of 2.62 sec.
In the following evaluation the 126 measured velocity vector values in each of the 4200 time windows around the rotor circumference are ensemble averaged according to:

$$\overline{\vec{c}} = \frac{1}{n}\sum_{i=1}^{n} \vec{c_i} \qquad (1) \qquad\qquad \overline{\vec{w}} = \frac{1}{n}\sum_{i=1}^{n} \vec{w_i} \qquad (2)$$

for the absolute and relative velocity.
The subsequent determination of the fluctuation components of the velocities in the absolute resp. relative frame is made in a 'mainflow-crossflow' reference system acc. to fig. 2 (Riess *et al.*, 1996; Riess *et al.*, 1998), where

$$\overline{c_{mf}} = \overline{\vec{c}} \qquad \overline{c_{cf}} \equiv 0$$
$$\overline{w_{mf}} = \overline{\vec{w}} \qquad \overline{w_{cf}} \equiv 0$$

and the instantaneous fluctuation components are

$$c'_{mf,i} = c_{mf,i} - \overline{c_{mf}} \qquad (3) \qquad\qquad c'_{cf,i} = c_{cf,i} \qquad (4)$$
$$w'_{mf,i} = w_{mf,i} - \overline{w_{mf}} \qquad (5) \qquad\qquad w'_{cf,i} = w_{cf,i} \qquad (6)$$

The mean-square values of the velocity fluctuations are

$$\overline{(c'_{mf})^2} = \frac{1}{n}\sum_{i=1}^{n}(c'_{mf,i})^2 \qquad (7) \qquad\qquad \overline{(c'_{cf})^2} = \frac{1}{n}\sum_{i=1}^{n}(c'_{cf,i})^2 \qquad (8)$$

$$\overline{(w'_{mf})^2} = \frac{1}{n}\sum_{i=1}^{n}(w'_{mf,i})^2 \qquad (9) \qquad\qquad \overline{(w'_{cf})^2} = \frac{1}{n}\sum_{i=1}^{n}(w'_{cf,i})^2 \qquad (10)$$

Finally these values for corresponding points in the 30 rotor blade pitches are arithmetically averaged in a phase-locked mode.
It was found, that the differences of the mean square fluctuation values in different pitches are relatively small (Riess *et al.*, 1998). The 'rotor-averaged' values are shown in the following figures.
An 'overall relative unsteadiness value' can be calculated according to the formula generally applied for calculation of the turbulence level in macroscopically steady flow

$$Tu_c = \frac{1}{\bar{c}}\sqrt{\frac{1}{2}(\overline{(c'_{mf})^2} + \overline{(c'_{cf})^2})} \tag{11}$$

$$Tu_w = \frac{1}{\bar{w}}\sqrt{\frac{1}{2}(\overline{(w'_{mf})^2} + \overline{(w'_{cf})^2})} \tag{12}$$

These quantities are designated as 'unresolved unsteadiness' in (Lakshminarayana *et al.*, 1997). They are not necessarily identical with the familiar turbulence level Tu in macroscopically steady flow. First considerations concerning a deeper characterization have been presented in (Riess *et al.*, 1996).
With this multi window mode the periodic unsteady velocity component of the flow which results from the relative motion of the rotor with respect to the absolute frame can be suppressed. The fluctuation components characterize the superimposed random movement of the fluid.
For the measurement in plane E1 less measurements in each particular point have been acquired ($2 \cdot 10^{12}$ to $2 \cdot 10^{13}$) than in plane E2 with $2 \cdot 10^{19}$ single measurements.
In plane E1 behind the inlet guide vane the periodic component of the velocity is coordinate fixed whereas after the rotor we have the periodically oscillating velocity from the passing rotor blade. As a consequence more detailed information is needed to get a correct image of the unsteady velocity field.
The data measured behind the inlet guide vane have been evaluated only in the absolute system, whereas the data measured behind the rotor have been evaluated in both systems, absolute and relative.

4. Experimental Results

4.1. MEASUREMENTS BEHIND THE INLET GUIDE VANE

4.1.1. *Cycle-Averaged Properties*

In figure 3 the absolute velocity vector cycle averaged for each particular measurement position is presented. Cycle averaged means in this case, that the ensemble averaged and phase locked averaged properties are determined irrelevant of the momentary rotor position (Lakshminarayana *et al.*, 1997). In figure 3 the location of the measurement positions can be seen. At -6.5^o

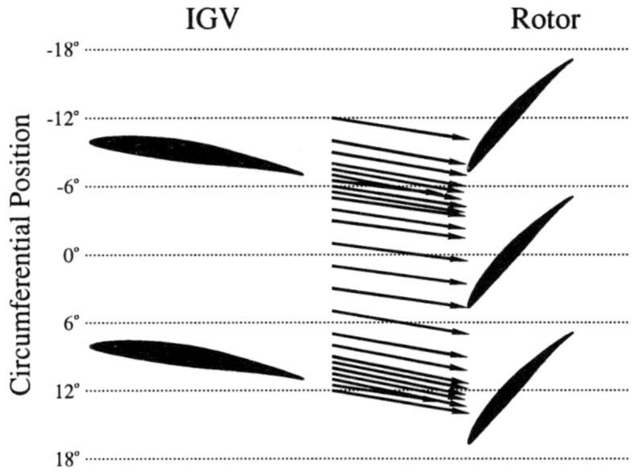

Figure 3. Absolute Velocity in Plane E1

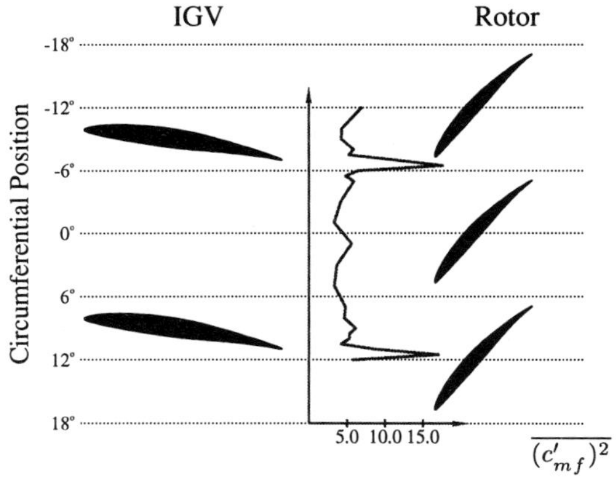

Figure 4. Mean Square Velocity Fluctuation in Mainflow Direction

and $+11.5^o$ the wake of the inlet guide vane causes a defect in the velocity field. The absolute velocity decreases up to 20 % and a deflection occurs, which leads to an increase of the absolute flow angle.

In figure 4 and 5 the variance of the unsteady flow component in mainflow direction and in crossflow direction (see figure 2) are presented as cycle averaged properties for each single measurement point.

The resulting plot shows that the fluctuations in mainflow direction are more than three times higher than the fluctuations in crossflow direction behind the inlet guide vane. In the wake of the inlet guide vane at measurement position -6.5^o and $+11.5^o$ the fluctuations increase sharply, in mainflow direction towards $17m^2/s^2$ and in crossflow direction towards $3m^2/s^2$. In the unsteady flow field between the blades the fluctuation in mainflow direction reaches values of 5 to $6m^2/s^2$ whereas the variance in crossflow

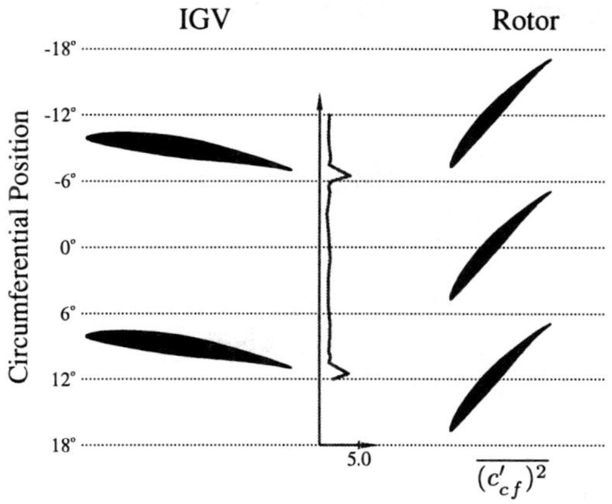

Figure 5. Mean Square Velocity Fluctuation in Crossflow Direction

direction is less than $1m^2/s^2$.

4.1.2. *Ensemble Averaged Properties*

In figure 6 the ensemble averaged absolute velocity in measuring plane E1 for variable rotor position is presented. The measured quantities have been collected into 50 time windows per pitch. The 3-D plot is a space and time resolved representation of the flow field, i.e. the coordinate rotor position is to be understood as a time scale.

At -6.5^o and $+11.5^o$ the decrease of the absolute flow velocity in the wakes of the inlet guide vane with $10m/s$ can be seen. The wake has a very small extension in circumferential direction. In addition the influence of the rotormovement behind the actual measurement position is obvious. At a circumferential distance of 12°, one rotor pitch, a decrease of the absolute velocity of $5m/s$ is visible, which changes the absolute position relative to the measurement position (IGV position). This velocity decrease is due to the local blockage the rotor blade exerts on the upstream flow field.

In figure 7 the fluctuations in mainflow direction $\overline{(c'_{mf})^2}$ are presented in the same way. Once again the space locked wakes of the inlet guide vanes can be seen as region where $\overline{(c'_{mf})^2}$ increases towards $20m^2/s^2$. No influence of the following rotor on the fluctuation profile of the flow field is visible. This fact is valuable for the fluctuations in crossflow direction, too. Since the upstream effect of the rotor blades is a 'potential' effect, it is evident that no increase of flow fluctuations is associated with it.

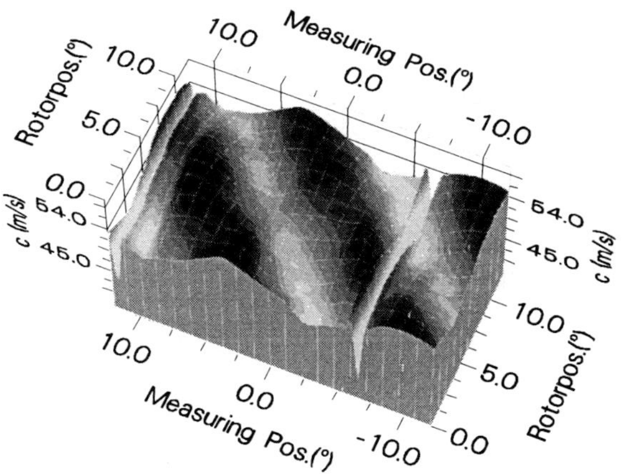

Figure 6. **Absolute Velocity in E1 Relative to the Rotor**

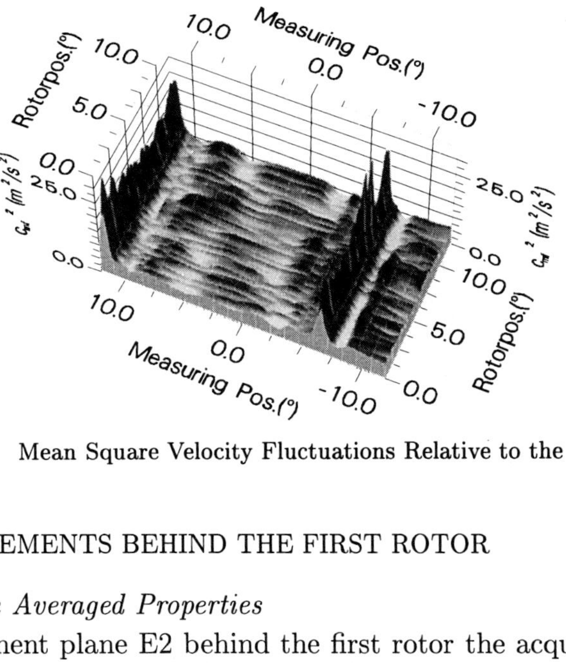

Figure 7. Mean Square Velocity Fluctuations Relative to the Rotor

4.2. MEASUREMENTS BEHIND THE FIRST ROTOR

4.2.1. *Ensemble Averaged Properties*

In the measurement plane E2 behind the first rotor the acquired data for one particular measurement point are ensemble averaged into 140 time windows per rotor pitch.

In figure 8 the relative velocity vector at rotor exit for the measurement position 5° is presented. Relative to the position of the inlet guide vane this position is situated in a region where the wake of the inlet guide vane exerts no influence on the flow field (see fig.1).

In the region between the rotor blades only the velocity vector for each tenth time window is presented to give an overview of the relative velocity

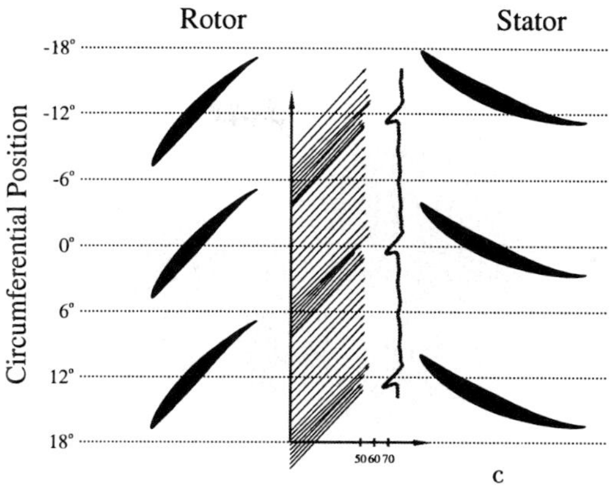

Figure 8. Relative Velocity in Plane E2

field. In the wake region the relative velocity for every fifth time window is presented. In addition the relative velocity is presented as a drawn line to show the contours of the relative velocity in the rotor wake.
The wake region in the relative system of the rotor is characterized by a decrease of the relative velocity and an increased deflection of the velocity towards the suction side of the rotor blade.

4.2.2. *Phase Locked Properties*

In the following part only spatially phase locked data for one representative rotor pitch are presented.
In the space-time resolved figures 9 and 10 the relative velocity field and the absolute velocity field behind the first rotor are presented. The axis 'Measuring Position' represents the probe positioning in the absolute system relative to the inlet guide vane and the first stator. The axis called 'Rotorpos.' represents the time resolved measurement data at the probe position relative to the rotor position.
In the relative field (fig. 9) the wake of the passing rotor blade is characterized by the the decrease of the relative velocity passing from the right to the left side of the figure as the rotor passes the measuring position. In the region from 0° to 5° the distance between the measuring positions was just 0.5^o whereas in the outer region the probe was traversed in 1° to 2° steps. This spacing has proven too large for the graphical evaluation. The 'holes' in the diagram result only from the insufficient volume of data for the interpolation process.
In the relative velocity field two regions situated at -3.5^o and 10° with a lower relative velocity can be seen. They are spatially fixed and have an extension of 3° in circumferential direction. The distance of 13.5^o between

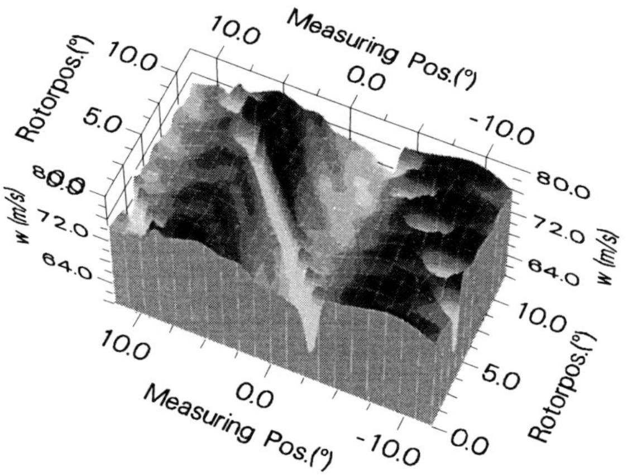

Figure 9. **Relative Velocity Field in Plane E2**

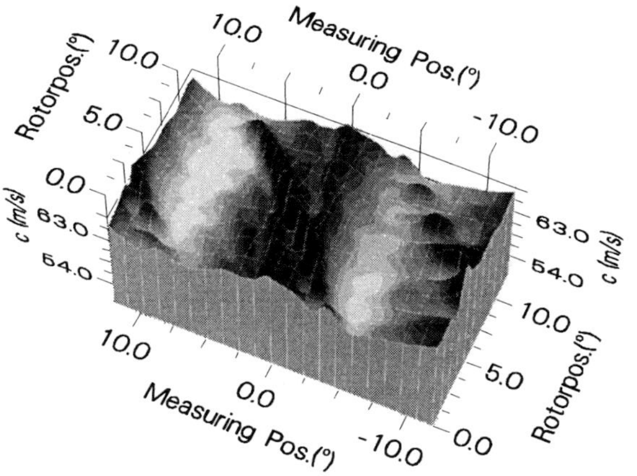

Figure 10. **Absolute Velocity Field in Plane E2**

these two regions corresponds to the distance between two stator rows. The leading edges of two stator blades are situated at -3.5^o and 10°. Their blockage effect is the reason for the decrease of the relative velocity.

In the absolute velocity field (fig. 10) the blockage effect of the following stator blades causes a decrease of the absolute velocity of 4 m/s at the measuring positions -5° and 8°. Compared to the relative flow field the blockage effect is shifted in direction of the absolute flow vector.

In the absolute flow velocity field the wakes of the passing rotor blades are mainly characterized by an increase of the absolute velocity on the suction side of the rotor blades. As a consequence the maxima which are passing from the right to the left side can be seen.

Remarkably in the absolute flow velocity field behind the first rotor, the

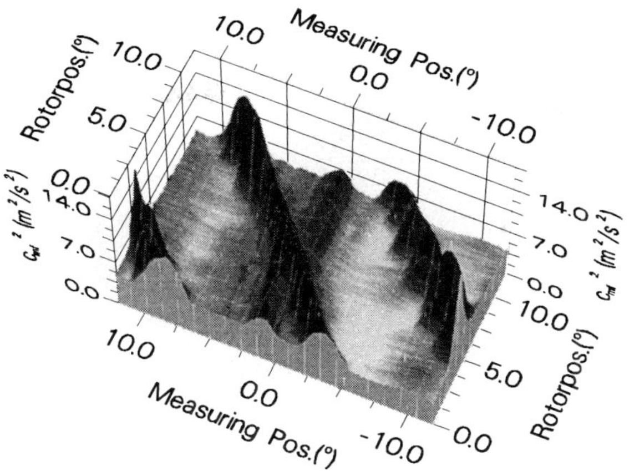

Figure 11. Mean Square Fluctuations in Mainflow Direction in E2

wake of the inlet guide vane can be seen spatially fixed as a region of increased absolute velocity. The wake extents from 0.5^o to 1.5^o at a relative rotor position of 0° to 5°. It is then chopped by the rotor wake and reappears from 6° to 12° rotor position at a spatially fixed measuring position of -1° to 0.5^o.

In figure 11, where the variance of the fluctuations in mainflow direction for the absolute system is presented, the wake of the inlet guide vane is clearly visible from 0.5^o to 1.5^o measuring position. The wake is characterized by an increase of the variance towards $7m^2/s^2$. In the undisturbed flow field the variance is lower with $3.5m^2/s^2$, in the wake of the rotor blades a variance of $13m^2/s^2$ is reached. In this figure the cutting effect of the rotor wake on the wake of the inlet guide vane is even more evident. The difference of the absolute position of the inlet guide vane wake is the result of the higher acceleration of the flow on the suction side of the wake in one rotor pitch where the wake of the inlet guide vane is transported as a separate part of the flow.

The same result can be observed if the variance in crossflow direction, which is not presented in the paper, is considered. With $3m^2/s^2$ the instantaneous fluctuations in this direction in the undisturbed flow field are generally less.

The turbulence intensity for the absolute flow field is presented in figure 12 in the same way. Crossing the rotor wake, where a turbulence intensity of 5.5% is reached, the transported wake of the inlet guide vane with 3.5% turbulence intensity is visible.

An influence of the following stator on the unsteady flow field could not be detected, as could be expected because of its 'potential' character.

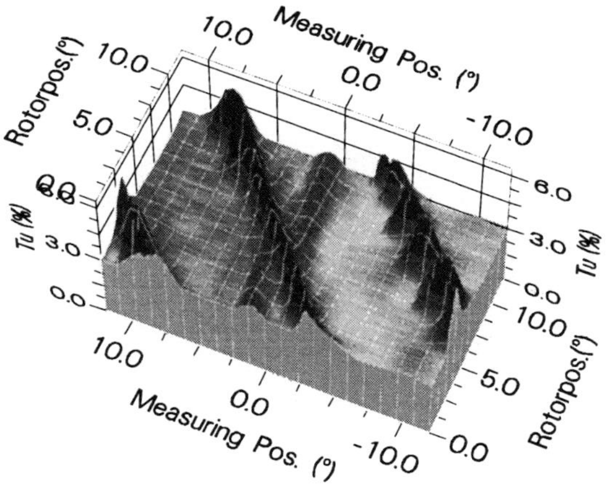

Figure 12. **Absolute Turbulence Intensity in E2**

4.3. COMPARISON OF ENSEMBLE AVERAGED PROPERTIES FOR DIFFERENT MEASURING POSITIONS

In order to provide a more detailed insight a comparison of the ensemble averaged unsteady flow components in the absolute system relative to the rotor position is presented. Three single measuring positions at 0.5^o, 1.0^o and 2.5^o are chosen to evaluate the influence the inlet guide vane wake exerts on the instantaneous flow field downstream of the first rotor. For a better comparison the rotor wakes have been positioned one over the other. In fig. 13 the variance of the fluctuations in main-flow direction over one representative rotor pitch are presented. The maximum of the variance characterizes the rotor wake with up to $14m^2/s^2$. Outside of the rotor wake the curves measured at 0.5^o and 1.0^o show a higher variance in the region between the wakes than at 2.5^o measuring position with $5m^2/s^2$. This is the effect of the inlet guide vane wake on the instantaneous flow field behind the first rotor. At 0.5^o the influence of the inlet guide vane wake is measurable only from 8° to 1° degrees of the relative rotor position. At 1.0^o the higher variance extents over the whole pitch. This fact underlines that the IGV-wake is chopped by the rotor blade. Additionally it can be observed, that the variance in main-flow direction in the rotor wake is $2m^2/s^2$ lower at the measuring position where the IGV-wake is visible.
The variance in crossflow direction has a value of 3 to $4m^2/s^2$ between the rotor wakes (see fig. 14) and is less high than the variance in mainflow direction. In the wake the fluctuations in crossflow direction reach values of $14m^2/s^2$. The wake of the inlet guide vane is responsible for the increase of the variance in crossflow direction between 5° and 12° degrees for the measuring position at 0.5^o and between 6° and 12° degrees for the measuring

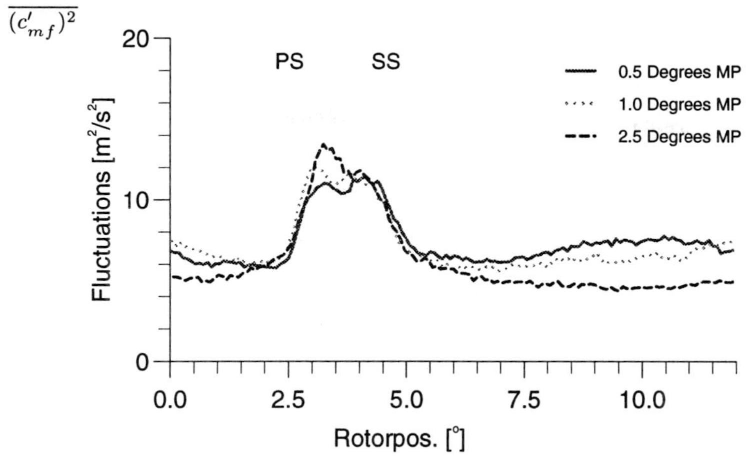

Figure 13. Fluctuations in Mainflow Direction for Different Measuring Points

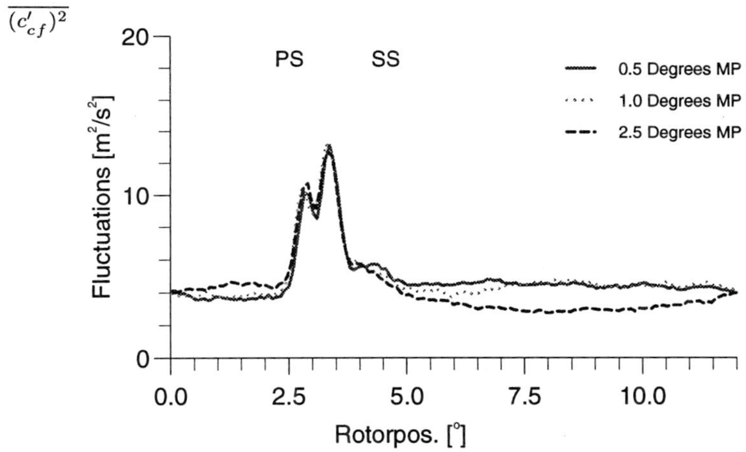

Figure 14. Fluctuations in Crossflow Direction for Different Measuring Points

position at 1.0^o. An influence of the inlet guide vane wake on the rotor wake concerning the unsteady velocity component in crossflow direction cannot be observed.

Comparing the rotor wake shape of both variances the double peak of the fluctuations in crossflow direction is characteristic for the unsteady flow field in the wake (Riess *et al.*, 1996). The peak on the suction side of the wake is higher than on the pressure side. At the boarder of the wakes the energy exchange with the 'undisturbed' flow field takes place. Due to the exchange the fluctuations in crossflow direction seem to rise. Mainly on the suction side because here the velocity difference and the friction between the two flow regions is higher. As a consequence the fluctuations in cross flow direction in the middle of the wake are lower.

Obviously the circumferential extension of the rotor wake region is bigger for the variance in mainflow direction with 2.5^o in contrast to 1.3^o in crossflow direction. This extension takes place mainly towards the suction side of the rotor blade.

5. Summary

Measurements of the unsteady flow have been made in the first stage of a low-speed axial compressor at 50% blade height. The use of split-film probes permits the determination of value and angle of the unsteady flow velocity, ensemble averaging and statistical evaluation of numerous sets of repeated measurements renders data for the velocity fluctuations, which can be differentiated in mainflow and crossflow direction.
The IGV-wakes can be identified in the flow behind the rotor. The rotor blade boundary layer is influenced by the interaction with the IGV-wake.
The data acquired can serve for the validation of the results of numerical simulations of unsteady flow in compressors, especially for comparison of the measured and calculated turbulence characteristics.

References

Traulsen, D. (1989) *Auslegung und Realisierung eines Axialverdichterprüfstandes zur Untersuchung von Randzonenströmungen*, Dissertation, VDI-Verlag, Reihe 7

Bigalk, J. (1995) *Sekundärströmung in der Seitenwandgrenzschicht mehrstufiger Axialverdichter*, Dissertation, VDI-Verlag, Reihe 7

Dantec (1995) *Dantec Measurement Technology*, Probe Catalogue

Sentker, A. and Riess, W. (1996) Turbulence Intensity Measurements in a Low Speed Axial Compressor, in W. Rodi and G.Bergeles (eds.), *Engineering Turbulence Modelling and Experiments 3*, Elsevier Science B.V., pp. 773-783.

Sentker, A. and Riess, W. (1996) High Resolution Measurement of Turbulence in Axial Compressors, *10. DGLR-Fach-Symposium der AG STAB*

Sentker, A. and Riess, W. (1998) Measurement of Unsteady Flow and Turbulence in a Low Speed Axial Compressor, *Experimental Thermal and Fluid Sciences*, **Special Issue**.

Zaccaria, M.A. and Lakshminarayana, B. (1997) Unsteady Flow Field Due to Nozzle Wake Interaction With the Rotor in an Axial Flow Turbine: Part I- Rotor Passage Flow Field, *Journal of Turbomachinery*, **Vol. 119** , pp. 201-213.

Riess, W. and Sentker, A. and Walbaum, M. (1996) Experimental Investigations of Unsteady Flow in Axial Compressors, in IMechE Seminar Publication, *Latest Advances in the Aerodynamics of Turbomachinery with Special Emphasis upon Unsteady Flows*, pp. 67-75.

Jennions, I.K. and Adamczyk, J. J. (1997) Evaluation of the Interaction Losses in a Transonic Turbine Rotor/LP Vane Configuration, *Journal of Turbomachinery*, **Vol. 119** , pp. 68-76.

Adamczyk, J. J. (1996) Wake Mixing in Axial Flow Compressors, *ASME Paper*, **96-GT-29**.

UNSTEADY INLET/COMPRESSOR INTERACTION EXPERIMENT TO SUPPORT THE MODELING OF COMPRESSOR-FACE BOUNDARY CONDITIONS

M. SAJBEN and D. D. FREUND
University of Cincinnati
Department of Aerospace Engineering and Engineering Mechanics
Cincinnati, OH 45221-0070
USA

1. Introduction

Atmospheric disturbances affecting high-speed, air-breathing propulsion systems may induce rapid flow transients. Such processes usually involve strong interactions between the inlet and the engine, and may lead to undesirable or even unacceptable consequences (e.g. inlet unstart, compressor stall or surge). Their reliable prediction requires, in principle, the simultaneous computation of the flow in both the inlet and the compressor. Such computations, while may be feasible in some simplified form [1,2], are too complex to fit into the cost and time constraint of industrial design processes. Current engineering practice is to perform computations for the two components separately, relying on the compressor face boundary conditions (CFBC's) to represent the engine for the inlet flow predictions, or the inlet for engine calculations.

The nature of CFBC's is far from well established. Current practice is to assume some simple condition, based on intuitive arguments and mathematical convenience. For inlet computations, the outflow condition is generally specified as constant pressure or constant velocity, although a number of other conditions have been proposed recently [3,4]. While these conditions apply to steady state situations, their validity for rapid transients is in serious doubt. A broad search uncovered no experimental information suitable for settling the problem.

This effort was originally motivated by concerns of high-speed inlet technology groups, who found that the results of unsteady inlet flow computations depend strongly on the postulated outflow conditions, and that the customarily used conditions allow a wide range of answers [5]. The authors are not aware of similar conclusions concerning engine computations but the upstream boundary conditions are likely to have a significant effect on at least the first stage during such transients.

The purpose of the experiment reported here is to provide empirical information to support the formulation and validation of boundary conditions that are capable of approximately describing the dynamics of a compressor-face during rapid transients of inlet/compressor systems.

T.H. Fransson (ed.), Unsteady Aerodynamics and Aeroelasticity of Turbomachines, 287-300.

2. Description of Experiment

2.1 CONCEPT

Investigations of joint inlet/engine dynamics have been reported that utilize externally introduced periodic disturbances of the inlet flow [6,7]. The difficulty with this approach is that periodic forcing creates a modal oscillation pattern involving and depending on the entire inlet, including the inflow boundary conditions (as may be created by a periodic disturbance generator). It is difficult, if not impossible, to extract information from such data that would specifically relate to and characterize the processes at the compressor face.

In the present work, the interface is explored by means of diagnostic pulses. A short-duration acoustic (isentropic) pulse is generated near the midpoint of a constant-area annular inlet and is allowed to propagate to and to be reflected from the compressor. The pulse and its reflection are tracked by fast-response pressure transducers. In possession of the detailed pulse profiles, it is possible to determine a frequency-resolved reflection coefficient that is a fairly complete dynamic characterization of the compressor face. This technique is an extension of methodologies developed for the characterization of various acoustic materials and duct system components [8,9,10].

It is known that small disturbances in compressible flows may be classified into three types: acoustic, vorticity and entropy [11]. If the mean flow is uniform, then these disturbances are non-interacting. The present work deals with acoustic disturbances only.

Figure 1 illustrates how pulse data can help to clarify compressor-face processes. The Figure shows one half of a sine wave, as a possible incident compression pulse. The Figure also shows the reflections predicted by three commonly employed CFBC's: the constant pressure ($p' = 0$), constant velocity ($u' = 0$) and the non-reflecting-boundary conditions (p' and u' are the time-dependent perturbations of the respective time-mean flow properties). The $p' = 0$ condition corresponds (in the long wave length limit) to an open ended pipe and produces an immediate, identical pulse with opposite sign (an expansion pulse). The $u' = 0$ condition describes a hard-wall termination moving at the flow velocity, and produces an immediate, like-sign reflection with an identical shape. The no-reflection condition, of course, produces no upstream-moving waves. It is clear that these practices produce conflicting results.

The inlet and the compressor configuration, the method of generating the required pulse, the diagnostic systems and the data reduction methodology have all been described in a series of four publications [12-15] and only a brief review is offered here.

2.2 COMPRESSOR

The compressor of a General Electric T58-3 helicopter engine is used. The engine has been modified for cold operation: the combustor has been removed and replaced with a plenum chamber that supplies the turbine with high pressure air from an external source. Since the inlet temperature of the turbine drive air is low (approximately ambient), the maximum value of the axial Mach number (M) achievable in this configuration is approximately 0.2 at 20,000 RPM, corresponding to 76% of design speed.

The compressor has ten stages, a 24.7 cm first stage tip diameter, and a variable inlet guide vane (VIGV) row with 38 vanes. The first stage rotor contains 30 blades, and the solidity and stagger angle at the mean diameter are 1.16 and 52°, respectively. The front rotor bearing is supported by four struts, located just upstream of the VIGV's. The struts represent a 9.2% blockage of the annulus area. Fig. 2 illustrates the configuration at the mean diameter. The location of the "compressor face" is defined arbitrarily as the midpoint of the first stage rotor.

2.3 INLET DUCT

Figure 3 shows the compressor rig mated with a constant-area annular inlet duct. The length of the duct is 180 cm and the hub/casing diameters of the annulus are 13.7/25.9 cm, respectively. Ambient air enters through a screened, 25:1 contraction ratio bellmouth intended to remove steady or unsteady distortions that might originate from the environment. The hub contains the mechanism for generating individual acoustic pulses.

2.4 PULSE GENERATION

The acoustic pulse must have a sufficiently large amplitude to be clearly separable from the compressor noise. This noise is dominated by a 7-9 kHz tone at the first rotor blade passing frequency, with an amplitude typically near 140 dB (re 2.0E -5 Pa). The pulse must also have a short duration, such that the reflection process at the compressor face can be completed before reflections from the upstream end of the inlet can interfere. In addition, the pulse must have a simple structure, be largely planar and free of concurrent entropy or vorticity variations.

Pulses conforming to these requirements are generated by an abrupt deformation of the hub contour. A 12.7 cm long section of the hub is replaced by a rubber tube (called a "boot") that is inflated slightly to form an smooth, axisymmetric bump. A low pressure is applied abruptly to the inside surface of the boot, to change the bump shape to a cylinder with a diameter equal to that of the hub. The shape change occurs in less than one ms, justifying the term "collapsing bump technique". The collapse generates two pulses: one moving in the upstream and one in the downstream direction.

The rapid reduction of internal pressure within the hub is achieved by configuring the hub interior as a shock tube and positioning the (helium-filled) driver section inside the collapsing boot. The process is triggered by bursting a diaphragm on command. The pulses are tracked by means of fast-response pressure transducers flush-mounted on the outer wall and the hub of the inlet. The signals are acquired at 100 kHz data rate.

Using this method, expansion pulses are routinely produced with approximately one ms duration, peak amplitudes up to 10 kPa and total energy content up to 52 joules.

Most runs were made with 4 kPa peak amplitude, which, at a mean axial Mach number of 0.17, corresponds to the rather sizable axial velocity disturbance of 16%.

Figure 4 illustrates the time history for a typical pulse. Data from four separate runs are superimposed in the Figure to illustrate the excellent repeatability. The 1.5 kHz ringing after the main pulse is a byproduct of the pulse generation process. It has been verified that the pulses are planar and axially symmetric. The pulses propagate at a constant speed equal to $a \pm u$, where a and u are the the speed of sound and flow speed, respectively, as determined from mean flow property measurements. The pulses do not appreciably change their shape while moving in the constant-area duct, which greatly simplified data interpretation. It has been shown that the struts in front of the engine do not generate significant reflections.

3. Results

3.1 TIME MEAN FLOW

The flow in the inlet duct is well-behaved, parallel, with a nearly zero-pressure-gradient. It is believed to be free of features that could obscure the interface reflection processes (e.g. high turbulence levels or separated regions). The Reynolds number based on annulus height varied in the range between 130,000 to 245,000. On the outer wall, transition to turbulence is thought to occur naturally on the bell mouth, near the beginning of the constant area section. The calculated displacement boundary layer thickness at the bump center is $\delta^* = 1.8$ mm (3 % of passage height). At the leading edge of the VIGV row, $\delta^* = 3.6$ mm (6 % of passage height). It is thus expected that an irrotational core flow persists along the entire length of the inlet duct.

3.2 TIME-DEPENDENT DATA

Figure 5 is an *x-t* diagram illustrating the pulses investigated (x is the streamwise coordinate, positive in the flow direction, origin at the midpoint of the first rotor). The pulse generation process, and also the pulses themselves, are of finite duration and finite axial length (approximately 1 ms and 40 cm, respectively), which explains the bands representing them in the Figure. Each band represents a pulse propagating with essentially unchanged shape. A reflection at either end will result in a pulse moving in the opposite direction, with a structure that differs significantly from the incident pulse. The pulses are labeled as Ux and Dx, designating upstream and downstream propagating pulses, respectively, where x is an identification number assigned in the temporal order of appearance.

Pressure-time histories obtained from fixed transducers correspond to time records obtained along vertical lines in the diagram. Inspection of the Figure shows that two pulses may overlap for some part of all time records. Having verified that the pulses propagate at accurately known speeds without changing their shape, it was possible to determine the reflected pulse by simply subtracting the known incident pulse (after an appropriate time shift) from the total signal. This type of data manipulation was feasible because the shortness of the pulses insured that a superimposition of three or more pulses did not occur during the data acquisition interval. Details are given in [15].

The reflected pulse could be obtained from any of several transducer signals, providing a degree of redundancy. The reflections determined in this fashion were in good agreement with each other, increasing confidence in the data.

The temperature of the ingested ambient air varied slightly from run to run. In order to remove the consequent minor variations from the data, the results are presented in terms of a nondimensional time or a nondimensional frequency, defined as $\tilde{t} = at/D$ and $\tilde{f} = fD/a$, respectively (a is the speed of sound, t is time, f is frequency and D is the outer diameter of the annular channel). Quantities normalized in this fashion are shown with a tilde. This normalization also provides the scaling of the results to larger systems. Pressure disturbances are normalized by the undisturbed (mean) static pressure, p_o.

Attention is focused on the first reflection that occurs at the engine face (D1/U2). This reflection is used to characterize the process, while the second reflection at the compressor face (D2/U3) is used to verify the method of characterization.

Figure 6 illustrates a composite of four time records from four different locations. The data shown have been low-pass filtered at 2 kHz ($\tilde{f}$ =1.523) to eliminate the blade passing noise. The initial D1 pulse is isolated and clearly identifiable at location 1. At other stations, it overlaps its own reflection (U2) to various degrees, making a clear recognition of either pulse less obvious. However, the main features are clear and are consistent with Figure 4. Pulses D2 and U3 are also recognizable at later times.

Figure 7 shows time records of pulse U2, as determined from four transducer locations. The traces were shifted in time to make the peaks coincide and facilitate the comparison. The U2 traces from the hub and the case agree well, indicating the absence of radial changes. U2 is seen to be an expansion pulse, just as its progenitor D1, but it has a much smaller amplitude and lasts about three times longer than the main part of D1. This seems to indicate that the pulse generates reflections not only from the first stage, but also from successive stages.

There are minor differences between the traces from locations 3 and 4 that appear as oscillations near $\tilde{f}$ =1.143 (~1.5 kHz). Higher order oscillatory modes were not investigated in this work but it is probably relevant that the duct has spinning modes with cut-on frequencies near $\tilde{f}$ =1.219 (~1.6 kHz). Thus it appears possible that the reflected pulse contains contributions from one or more spinning modes.

Similar data have been acquired at a number of conditions within the operating range of the facility and are reported in Ref. [15].

3.3 FREQUENCY DOMAIN ANALYSIS

Each reflection process involves an incident and a reflected pulse, both of which are specific to the pulse generation technique used in this experiment. By themselves, such data do not reveal what will happen if a different kind of pulse arrives to the compressor face. Fortunately, frequency domain analysis has been identified in acoustic technology [8-10] as a powerful method that can (a) provide a characterization of the reflection process that is independent of the pulse shape, and (b) provide a method to calculate the reflection to an arbitrarily shaped incident pulse.

The method assumes that the incident pulse (input) and the reflected pulse (output) are linearly related. The ratio of the respective Fourier transforms is defined as the transfer function that characterizes the process connecting input and output. Knowledge of the transfer function implies a prediction capability: multiplying the Fourier transform of an arbitrary incident disturbance with the transfer function yields the Fourier transform of the corresponding reflection. Retransforming into the time domain provides the time history of the reflection. The method is well established and commercial software is readily available to perform the required manipulations numerically (LabVIEW was used).

The transfer function is a complex quantity, given as a function of frequency. The absolute value (often called the "gain") is best viewed, in this context, as a frequency-resolved reflection coefficient. The argument of the number (ϕ) represents a phase shift. In the present context the phase shift does not have a direct physical meaning. However, it is easily shown that the phase shift divided by the angular frequency ($\tau = \phi / \omega$, where $\omega = 2\pi f$) represents a time delay, which is more easily associated with a reflection process. Accordingly, the argument of the transfer function is presented here as a time delay vs. frequency function. The time delay is presented in normalized form, as $\tilde{\tau} = \tau a/D$. The absolute value and the time delay jointly characterize the reflection process, independently of the pulse shape used in the experiment.

The transfer function does depend, of course, on the engine geometry and the flow conditions to which it applies. The data showed that the significant variables are the axial Mach number and the stagger angle of the VIGV row (α), while the effects of compressor pressure ratio were negligible. Pulse amplitude had no discernible effect on the transfer function, supporting the assumption of linear dependence.

Spectral analysis of the data indicated that the incident and reflected pulses contain only very minor contributions from frequencies over one kHz, corresponding to $\tilde{f} = 0.76$. Accordingly, the transfer functions and results of computations using them are presented only for this frequency band.

Figures 8 and 9 illustrate the transfer function for $M = 0.17$, and $\alpha = 25$ degrees. The absolute value curve (Fig. 8) displays a behavior similar to that of a low-pass filter, in that the reflection amplitudes steadily decline with frequency. Very little reflection occurs over a normalized frequency of $\tilde{f} = 0.76$. It follows that the amplitude of a reflected pulse depends on the shape and duration of the incoming pulse. Long duration pulses (low frequencies) produce high amplitude reflections, while short pulses (high frequencies) yield lower reflected amplitudes.

The normalized time delays may be given an approximate physical interpretation by assuming that the reflection occurs not at the compressor face, but at some plane located within the compressor. The time delay then corresponds to the time required by the wave to travel from the compressor face to this reflection plane and back. The dimensionless location of this "equivalent reflection plane" (designated by $\tilde{\xi}$) was computed by assuming that the speeds of the upstream and downstream waves inside the compressor are the same as in the duct. (The actual process within the compressor is far more complex than this simple view, involving much scattering among the stages.) The left side of the time delay plot in Fig. 9 is scaled to indicate the location of the reflection plane. A sketch of the blade rows has been added to facilitate the comparison. While the model is oversimplified, the distances calculated strongly suggest that (a) the involvement of interior stages depends on the shape (spectral content) of the disturbance and (b) that the reflection process may extend as far as the fourth stage.

Reference [15] includes a number of transfer functions, describing the behavior of the engine for axial Mach numbers from 0.1 to 0.18 and α values from 15° to 29.9°. The results show that the general character of the transfer function remains the same for all conditions, but the strength of the reflection increases with axial Mach number and also with the stagger angle. These trends agree with the numerical predictions of Paynter [16], made for linear cascade configurations.

3.4 PREDICTION

The transfer functions were determined using the D1/U2 reflection. The prediction method was then tested by applying this transfer function to predict the reflection to pulse D2, which induces the upstream pulse U3. Pulse D2 is significantly different from D1: it is compressive and also significantly longer. As Fig. 10 shows, the reflection obtained by the application of the transfer function method to pulse D2 agrees with the measured U3 quite well. The sign, magnitude and duration are correct. Higher order oscillations are evident in the measured data, which are probably related to higher order acoustic modes of the duct (not investigated in detail) and are not within the bandwidth of the prediction method. The conclusion is that the experimentally determined characterization of the compressor face, in conjunction with the frequency domain methods described, will correctly predict the response to an arbitrary incoming disturbance. Since we limited the transfer function to frequencies below $\tilde{f} = 0.76$ (1 kHz), the predicted pulses will not contain frequencies above this cutoff and will appear as smoothed shapes when compared to the data.

Figure 11 contains the responses to the original (D1) pulse, as predicted by three commonly used methods, together with the experimentally determined response. It is evident that none of the three customary CFBC's agree with the data. The constant pressure boundary condition does not predict even the sign of the response correctly. For completeness, Figure 11 also includes the "prediction" of U2 as obtained by applying the transfer function to pulse D1. Since the transfer function was obtained from this reflection in the first place, the agreement merely demonstrates that the data manipulations were performed consistently.

3.5 QUESTIONS REMAINING

The goal of the work was to obtain the experimental information necessary to formulate realistic boundary conditions for the inlet/compressor interface. Progress has been made in this direction, but the nature of the process dictated that the data be presented in frequency-domain terms. This format is not readily applicable to the time-domain computer codes used in propulsion practice. Clearly, a procedure is needed that is capable of using the empirically determined information (transfer function, or reflection coefficient) to impose an equivalent boundary condition in time-domain codes. The present effort did not include such a development.

Computational aeroacoustics (CAA) has essentially the identical problem of applying impedance boundary conditions for time-domain methods [17,18]. Current interest appears to be related primarily to solid boundaries of flow paths (e.g. specifying the characteristics of acoustic liners). Since there is a one-to-one relationship between acoustic impedance and acoustic pressure reflection coefficient, perhaps the solutions being developed in the CAA area can be extended to include outflow boundary conditions.

Another area needing additional work relates to the disturbance types. This study deals with acoustic disturbances only. In a real situation vorticity disturbances will be invariably present and entropy disturbances may be important in some cases. Unless these are also accounted for, the boundary conditions may not be fully representative of reality. A comparable exploration of entropy and vorticity disturbances is needed.

4. Summary

The compressor face boundary conditions (CFBC's) to be used in unsteady inlet or compressor computations represent a long-standing issue that remained unresolved for many years due to lack of applicable experimental information. The present effort employed a novel technique to obtain the data required. Large-amplitude, short-duration acoustic pulses were generated in a constant area inlet connected to an operating, ten-stage compressor. The pulses and their reflections from the compressor were recorded by means of wall-mounted pressure transducers. The wall pressure histories are appropriate for the determination of frequency-dependent reflection coefficients (transfer functions) that represent an empirical characterization of processes at the inlet/compressor interface. The availability of the reflection coefficients permits the prediction of the reflection to arbitrary incident disturbances, using well-developed frequency-domain methodologies. The prediction capability has been successfully demonstrated. We believe that the present study is the first experimental effort that explicitly targeted the CFBC issue.

The experimental results conflict sharply with predictions obtained by currently customary compressor-face boundary conditions. Also, currently used boundary conditions are tacitly assumed to be universally applicable to any engine, but the data show significant dependence on compressor geometry and axial Mach number (i.e.

reduced mass flow). The conclusion is that current practices concerning CFBC's are in need of review and revision.

Acknowledgments

The authors are indebted to Dr. Gerald C. Paynter of The Boeing Company and to Mr. Gary Cole of NASA Lewis Research Center, for encouragement and for numerous enlightening discussions on the subject. The authors are indebted to Mr. Russell DiMicco for invaluable technical support throughout the project. Initially the program was funded by the University of Cincinnati and by the Ohio Board of Regents. After January 1997, support has been provided by NASA Lewis Research Center, Mr. Gary Cole acting as Technical Officer.

References

1. Sugiyama, Y., Tabakoff, W., and Hamed., A., "J85 Surge Transient Simulation", (1989), *Journal of Propulsion and Power*, Vol. 5, No. 3, pp. 375-381.
2. Numbers, K., and Hamed, A., "Development of a Coupled Inlet-Engine Dynamic Analysis Method", (1997), AIAA paper 97-2880.
3. Decher, R., Mayer, D.W. and Paynter, G.C., "On Supersonic Inlet-Engine Stability," (1994), AIAA paper 94-3371.
4. Chung, J. and Cole, G.L., "Comparison of Compressor Face Boundary Conditions for Unsteady CFD Simulations of Supersonic Inlets," (1995), AIAA paper 95-2627.
5. Clark, L.T., "Dynamic Response Characteristics of a Mixed Compression Supersonic Inlet as a Part of a Larger System," (1995), AIAA paper 95-0036.
6. Peacock, R.E., "An Experimental Study of Pulsating Flow in a Three-Stage Axial Flow Compressor", (1978), ASME paper A79-2401 08-34.
7. Das, D.K., "Effects of Inlet Pressure Fluctuations on Axial Flow Compressors", (1989), *J. of Propulsion and Power*, Vol. 5, No. 1, P. 72-82.
8. "Standard Test Method for Impedance and Absorption of Acoustic Materials Using a Tube, Two Microphones, and a Digital Frequency Analysis System", (1986), ASTM Standard: E 1050 - 86.
9. Singh, R. and Katra, T., "Development of an Impulse Technique for Measurement of Muffler Characteristics", (1978), *Journal of Sound and Vibration*, Vol. 56, No. 2, pp. 279-298.
10. Chung, J.Y. and Blaser, D.A., "Transfer Function Method of Measuring In-Duct Acoustic Properties. Part II, Experiment", (1980), *Journal of the Acoustic Society of America*, Vol. 68, No. 3.
11. Kovasznay, L.S.G., "Turbulence in Supersonic Flow," (1953), *J. of the Aeronautical Sciences*, Vol. 20, No. 10, pp. 434-675.
12. Sajben, M. and Freund, D.D., "Experimental Exploration of Compressor-Face Boundary Conditions for Unsteady Inlet Flow Computations", (1995), AIAA paper 95-2886.
13. Freund, D.D. and Sajben, M., "Compressor-Face Boundary Condition Experiment: Generation of Acoustic Pulses in Annular Ducts", (1996), AIAA paper 96-2657.
14. Freund, D.D. and Sajben, M., "Experimental Investigation of Outflow Boundary Conditions Used in Unsteady Inlet Flow Computations", (1997), AIAA paper 97-0610.
15. Freund, D.D. and Sajben, M., "Reflection of Large Amplitude Acoustic Pulses from an Axial Flow Compressor," (1997), AIAA paper 97-2879.
16. Paynter, G.C. "Response of a Two-Dimensional Cascade to an Upstream Disturbance", (1997), *AIAA Journal*, Vol. 35, No. 3, pp. 434-440.
17. Tam, C.K.W., "Time-Domain Impedance Boundary Condition for Computational Aeroacoustics", (1996), AIAA paper 96-1662.
18. Ozyoruk, Y., and Long,. L.N., "Impedance Boundary Conditions for Time-Domain Computational Aeroacoustics Methods," (1997), AIAA paper 97-0021.

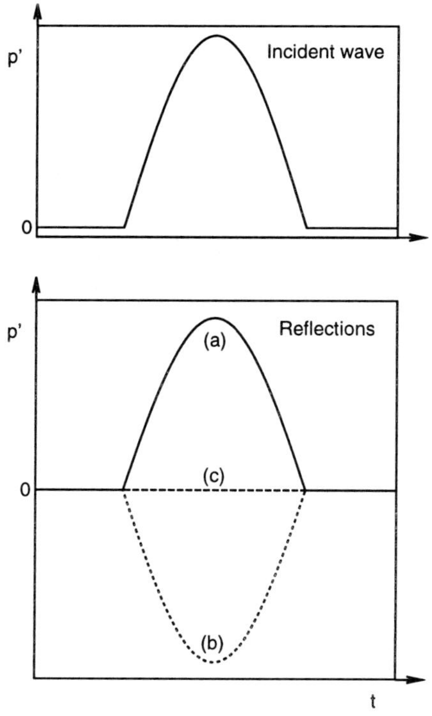

Fig. 1 - Reflections predicted for an incident, half-sine compression wave, using various common boundary conditions (no flow). (a) u'=0, (b) p'=0, (c) non-reflecting boundary.

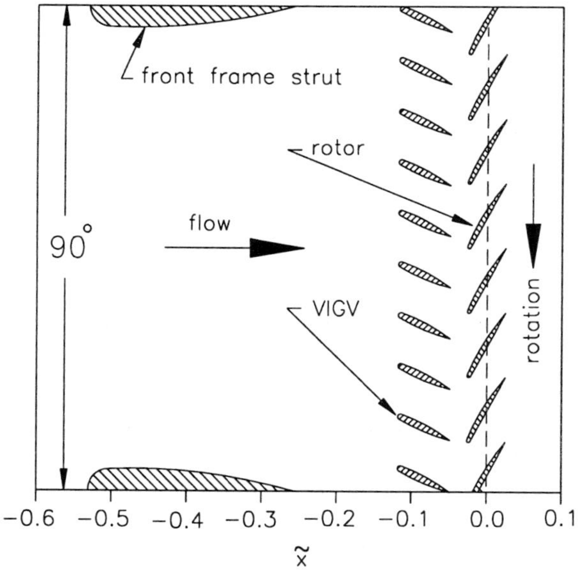

Fig. 2 - Geometry of compressor frame struts, VIGV's, and first stage rotor.

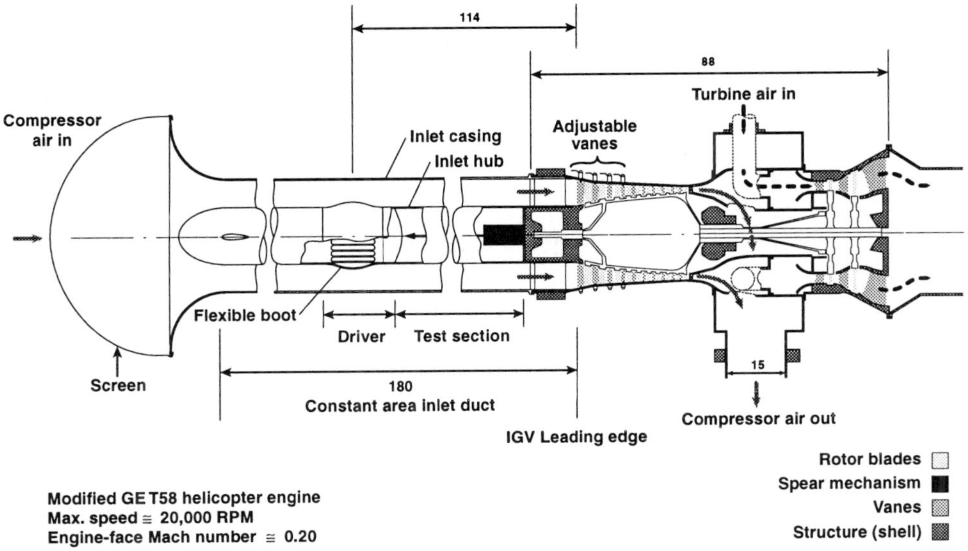

Fig. 3 - Inlet duct and compressor rig. All dimensions in cm.

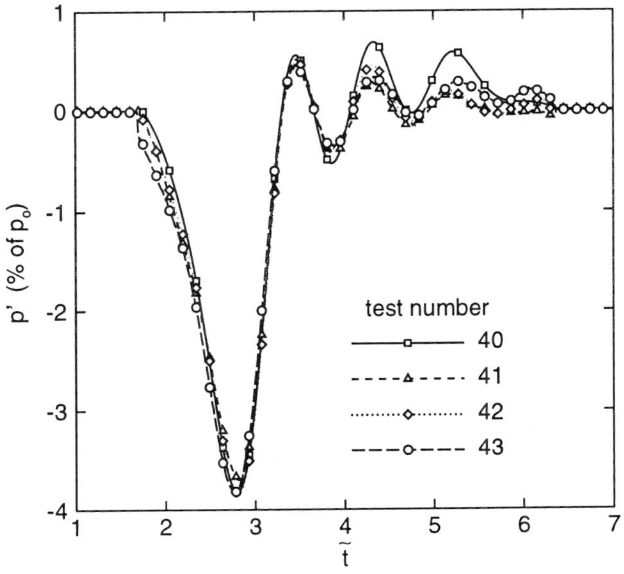

Fig. 4 - Initial downstream-moving pulse (D1) as determined in four runs. Most of the differences occur in the details of the ringing.

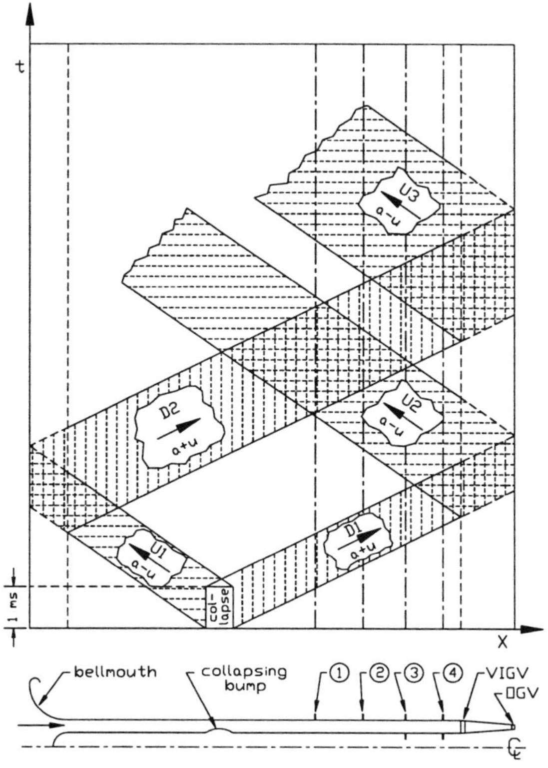

Fig. 5 - Distance-time (x-t) diagram of the pulses detected and analyzed. The finite duration of the pulses accounts for the width of the bands representing them. Pressure measurements are obtained along the vertical dash-dot lines corresponding to the transducers. Note some overlap of the pulses on all time records.

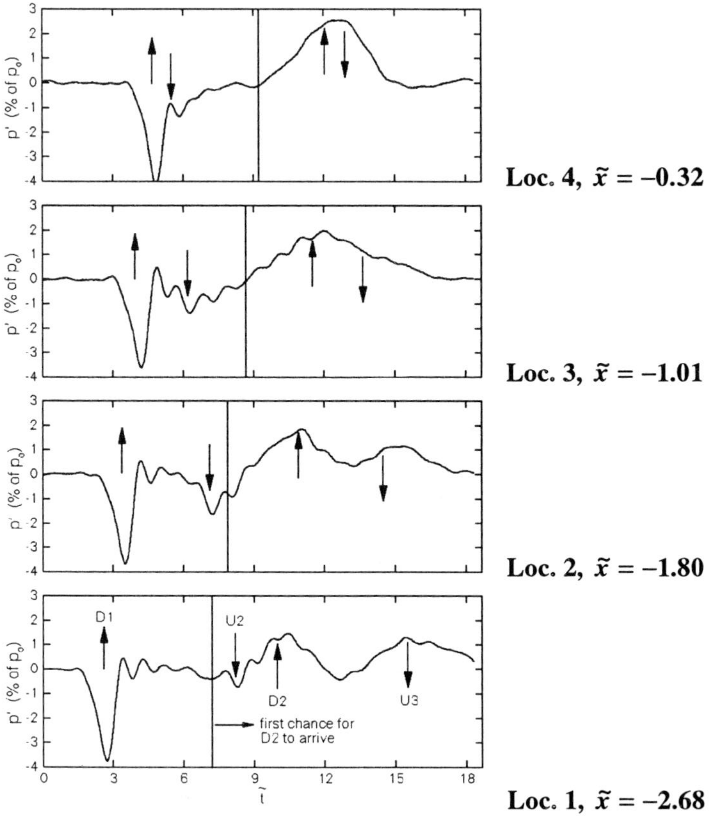

Fig. 6 - Time-pressure records from four locations. Arrows identify pulses. Vertical lines near the center of the time range indicate the first chance for pulses D2 and U2 to overlap.

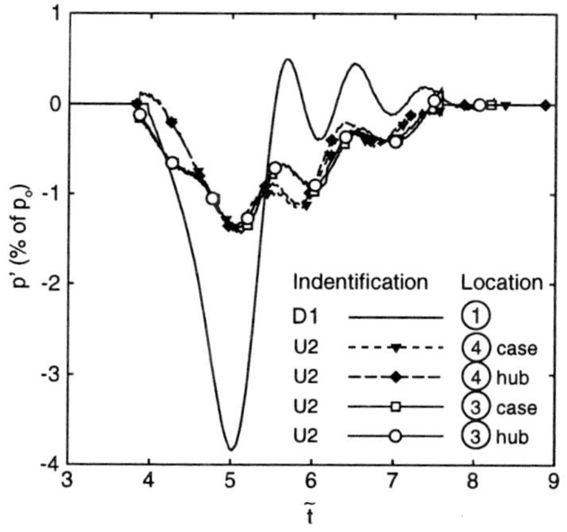

Fig. 7 - Measured pulse U2, as determined at four locations. Pulse D1 shown as reference. *M* = 0.17 and α = 24.8°.

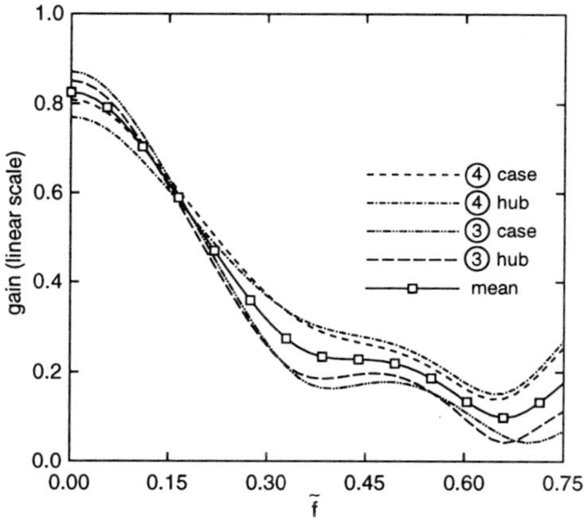

Fig. 8 - Transfer function absolute value from the D1/U2 transient, at *M* = 0.17 and α = 24.8°.

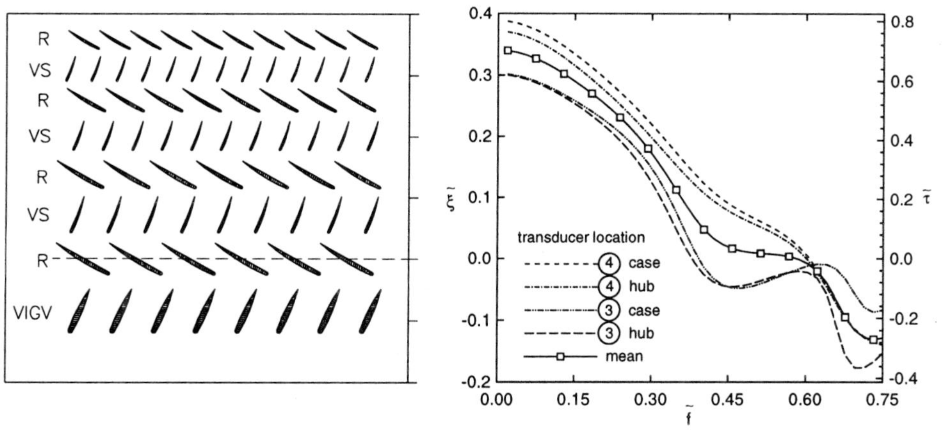

(a) **(b)**

Fig. 9 - Imaginary part of transfer function converted into a time delay. Determined from the D1/U2 transient, at *M* = 0.17 and α = 24.8°. (a) sketch of the first four compressor stages (axial length scaled to indicate reflection location directly) and (b) non-dimensional time delay and reflection location.

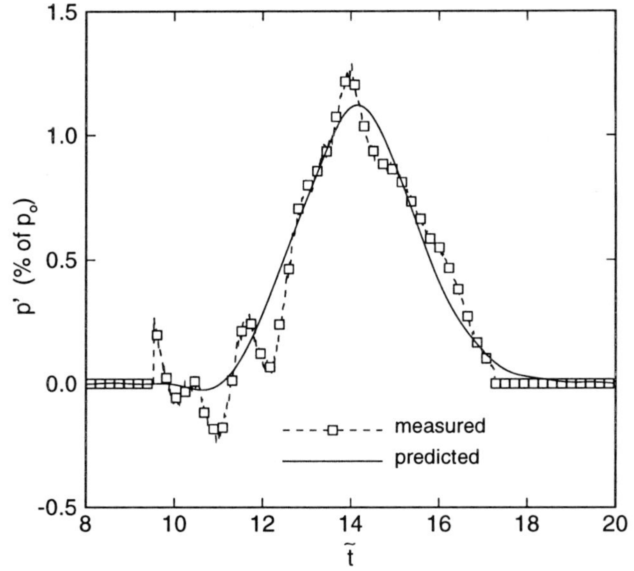

Fig. 10 - Shape of U3 pulse, as predicted by the transfer function method, compared with measurement, at $M = 0.17$ and $\alpha = 24.8^o$.

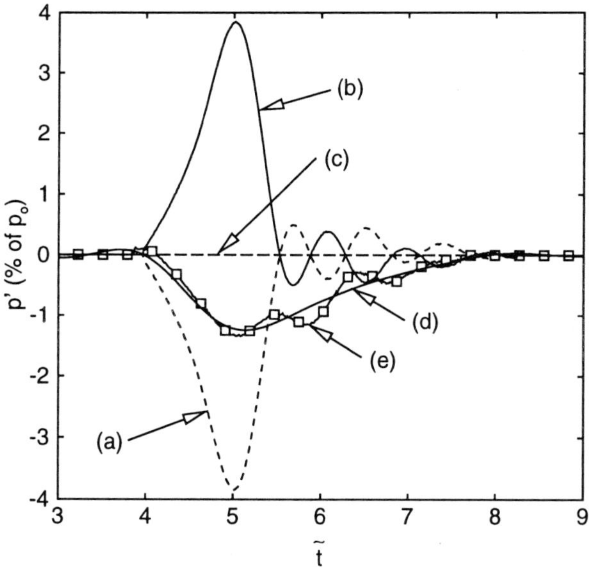

Fig. 11 - Shape of the pulse U2, as predicted by (a) *u'= 0*, (b) *p'= 0*, (c) non-reflecting boundary, and (d) transfer function method. Experimental data also shown as (e).

6: Aeroelastic Coupling - 1

THREE-DIMENSIONAL COUPLED MODEL FOR AEROELASTIC ANALYSIS OF TURBOMACHINE BLADE VIBRATIONS. ITS APPLICATION TO A HYDRAULIC TURBINE ROTOR

V.B. KURZIN, S.N. KOROBEINIKOV, V.P. RYABCHENKO AND L.A. TKACHEVA
M.A.Lavrent'ev Institute of Hydrodynamics SB RAS
Novosibirsk, Russia

1. Introduction

The analysis of aeroelastic vibrations in turbomachines is based on the unsteady flow theory and the structural dynamic theory. Each of these theories develops independently, as a rule. However to determine aerodynamic forces acting on virbating blades it is necessary to know frequencies and vibratory modes of the whole structure, which can depend on the aerodynamic blade interaction. For example it is known that the fluid/structure interaction has an influence on the eigenmodes of isolated blades when their partial natural frequencies nearly coincide, on the amplitudes and phases of vibrating blades in mistuned cascades and on the natural frequencies of hydraulic turbine blades [1 - 3]. So, it is necessary to develop the approaches of coupled fluid/structure models where elastic and aerodynamic problems should be solved simultaneously.

This report presents one of such approaches and its realization for the case of hydroturbine blade oscillations. It is based on the finite element method for the structural dynamic analysis. The unsteady aerodynamic forces acting on vibrating blades can be evaluated by any method of the unsteady flow theory. It is suggested that the cascade has a small mistuning.

2. Statement of the Problem

Harmonic vibrations of cascade blades in a fluid will be discussed in this paper. Assume that a cascade has some little geometric inhomogeneity and its blades interact only through a fluid.

T.H. Fransson (ed.), Unsteady Aerodynamics and Aeroelasticity of Turbomachines, 303-315.

In the linear approach the investigation of aeroelastic blade vibrations can be reduced then to the matrix equation system

$$(C_m - \lambda M_m)X_m = \sum_{n=0}^{N-1} A_{mn}X_m + F_m, \ \lambda = \omega^2 \ (m = 0,\ 1, ..., N-1). \quad (2.1)$$

Here N is the number of blades, X_m is the vector of generalized coordinates (harmonic amplitudes), C_m and M_m are stiffness and mass matrices, A_m is the matrix of aerodynamic response to structural vibrations, F_m is the vector of exciting aerodynamic loads, ω is the frequency of vibrations.

Let us introduce a small parameter ε, describing the inhomogeneity extent of the cascade. The matrices of eq. (2.1) can be written as

$$C_m = C + \varepsilon\tilde{C}_m, \ M_m = M + \varepsilon\tilde{M}_m, \ A_{mn} = A_{mn}^{(0)} + \varepsilon\tilde{A}_{mn}, \quad (2.2)$$

where C , M and $A_{mn}^{(0)}$ are matrices of an appropriate homogeneous cascade. In addition, assume that

$$max\left\{\frac{\omega_m^{(1)} - \omega^{(1)}}{\omega^{(1)}}\right\} = O(\varepsilon), \quad (2.3)$$

where $\omega_m^{(1)}$ is the natural frequency of the m blade in vacuum by the first form, $\omega^{(1)}$ is the appropriate natural frequency of homogeneous cascade blades.

The solution of this problem can be divided into three stages: 1) the solution of the eigenvalue problem for the proper homogeneous cascade; 2) the calculation of inhomogeneity influence on eigenvalues; 3) the evaluation of forced vibratory amplitudes and phases for mistuned cascades in the vicinity of resonance.

3. Natural Vibrations of a Homogeneous Cascade in Fluid at Rest.

It is well known that proper vibrations of the homogeneous cascade blades occur with the same amplitudes and a constant phase shift $\mu_j = 2\pi j/N$ ($j = 0, 1, 2, ...N-1$) between vibrations of neighbouring blades. Therefore, according to eqs.(2.1) and (2.2), the determination of these vibrations is reduced to the solving the eigenvalue problem for the equation

$$[C - \lambda_j(M + A_j)]X_j = 0, \quad (3.1)$$

where X_j is the vector of coefficients of approximation of the amplitude function of an initial blade vibrations ($m = 0$) over some system of base functions appropriate to the phase shift μ_j :

$$A_j = \sum_{s=0}^{N-1} L_s e^{i\mu_j s}, s = \begin{cases} n-m, & for\ n \geq m \\ N+(n-m), & for\ n < m \end{cases}$$

where L_s are matrices of coefficients of the generalized displacements of the n th blade on the generalized hydrodynamic forces acting on the m th blade. The finite element method is the most effective to describe elastic deformations of arbitrary shape. However, in this case there arise problems of constructing matrices A_j. The use of the iteration method to solve the problem (3.1) allows one to avoid the evaluation of matrices A_j. Indeed, the matrix A_j enters into eq.(3.1) only as a product by a vector X_j, ($P_j = A_j X_j$). At the n th step of the iteration process this vector is determined with the vibratory mode found at the prievious step by any one of the known unsteady flow theory of cascades methods.

The iteration process is performed as follows. Using the Cholesky factorization for the stiffness matrix $C = LL^{\mathrm{T}}$ (L is the lower triangular matrix) we perform a transformation of eq.(2.2):

$$[\nu_j J - L^{-1}(M + A_j)(L^{\mathrm{T}})^{-1}]Y_j = 0$$

(where $\nu_j = 1/\lambda_j$, $Y_j = L^{\mathrm{T}} X_j$, J is the unit matrix). As a result, we obtain the eigenvalue problem for the Hermitian matrix:

$$G_j = L^{-1}(M + A_j)(L^{\mathrm{T}})^{-1} \tag{3.2}$$

According to [5], the first eigenvalue ν_{1j} of the matrix G_j can be defined by the method of direct iterations of the form:

$$Y_j^{n+1} = G_j Y_j^n / ||Y_j^n||_\infty, \tag{3.3}$$

where $||\ ||_\infty$ is the norm in l_∞, i.e. the vector element with a maximum module. With allowance for (3.2), relation (3.3) is transformed into

$$Y_j^{n+1} = [L^{-1}M(L^{\mathrm{T}})^{-1}Y_j^n + L^{-1}P_j^n]/||Y_j^n||_\infty. \tag{3.4}$$

Here the matrix A_j is not contained. As the initial approximation of process (3.4) we use the eigenvector of blade vibrations in vacuum which is determined by the subspace iteration method [4].

According to [5], this iteration process converges so that

$$Y_j^n/||Y_j^n||_\infty \to Y_j/||Y_j||_\infty,\ ||Y_j^n||_\infty \to \nu_{1j}\ as\ n \to \infty.$$

The convergence speed is determined by the ratio ν_{1j}/ν_{2j}. Thus, if the eigenvalues ν_{2j} and ν_{1j} are close, the convergence speed is slow. It can be increased by means of a shift. In this case the matrix $F_j = G_j - \zeta J$ has eigenvalues $\nu_{1j} - \zeta$, $\nu_{2j} - \zeta$, ... and the same eigenvectors as the matrix G_j. A shift is chosen so that the eigenvalue $\nu_{1j}\zeta$ has a maximum module and, at the same time, ratio (3.3) for the matrix F_j is smaller. To determine the subsequent eigenvalues, we use the exhoustung method by means of unitary transformations [5].

In conclusion we notice that for cascade vibrations in a gas flow when defining the frequencies and blade vibration shapes one may generally restrict his consideration to the zero approximation of the above iterative process.

4. The Perturbation Method for the Solving the Eigenvalue Problem

Let us describe System (2.1) with the allowance of (2.2) in the form

$$(B - \lambda A)X = \varepsilon(\tilde{B} - \lambda\tilde{A})X, \tag{4.1}$$

where A, B are real block, symmetric matrices. According to [6], a general estimate of eigenvalue perturbations (4.1) takes the form:

$$|\lambda^{(s)} - \lambda_0^{(s)}| \leq \varepsilon||B|| \cdot ||A^{-1}||, \ (s = 1,\ 2, ..., N_1). \tag{4.2}$$

where $\lambda_0^{(s)}$ are eigenvalues (numbered in order of their increasing) of the problem on natural vibrations of an appropriate homogeneous cascade N_1 is the order of A and B matrices. From eq.(4.2) it does not follow a priori that the perturbation is small because the value $||B||$ may be large and the matrix A may be ill conditioned. However, considering eq. (2.3), we can show that for some set

$$\Lambda_0 = \{\lambda_0^{(r)}\}_{r=1}^{s_0}, \tag{4.3}$$

the following estimate is valid

$$\lambda^{(r)} = \lambda_0^{(r)}(1 + \tilde{\lambda}^{(r)}), \ \tilde{\lambda}^{(r)} = O(\varepsilon), \tag{4.4}$$

where the value of s_0 is dependent on the spectrum density and determined in the following way. Let N_r be the number of the elements of the subset S_r of numbers "s":

$$S_r = \{s\}, \tag{4.5}$$

which determine the values of $\lambda_0^{(s)}$ satisfying inequalities

$$|\lambda_0^{(s)} - \lambda_0^{(r)}|/\lambda_0^{(1)} \leq a = O(1). \tag{4.6}$$

Then

$$s_0 = s_1 + N_{s_1+1}, \; s_1 = \left[\frac{a}{\varepsilon}\right].$$

We represent the eigenvector $X^{(r)}$ of Eq. (4.1) as the linear combination of vector $X_0^{(s)}$, corresponding to $\lambda_0^{(s)}$

$$X^{(r)} = \sum_{s=1}^{N_1} \alpha_s^{(r)} X_0^{(s)}, \; ||X^{(r)}|| = 1 \tag{4.7}$$

and split it into two components

$$X_1^{(r)} = \sum_{s \notin s_r} \alpha_s^{(r)} X_0^{(s)}, \; X_2^{(r)} = \sum_{s \in s_r} \alpha_s^{(r)} X_0^{(s)}. \tag{4.8}$$

As the matrices A and B are real and symmetric we get with the account for (4.4) and (4.6)

$$||X_2^{(r)}||^2 = O(\varepsilon). \tag{4.9}$$

Substituting (4.4) and (4.7) into (4.1) and multiplying (4.1) from left by the rectangular matrix which lines are vectors $Y_0^{(p)} = [X_0^{(p)}]^{\mathrm{T}}$ $(p \in S_r)$, In view of (4.8) and (4.9), we find

$$min|\tilde{\lambda}^{(r)} - \lambda_t^{(r)}| = O(\varepsilon^{3/2}), \tag{4.10}$$

where $\lambda_t^{(r)}$ are eigenvalues of the matrix

$$T^{(r)} = \Delta^{(r)} + G^{(r)}, \tag{4.11}$$

$$\Delta^{(r)} = diag\left(\frac{\lambda_0^{(s)} - \lambda_0^{(r)}}{\lambda_0^{(r)}}\right), \; G^{(r)} = \{g_{ps}^{(r)}\}, \; p, s \in S^{(r)}$$

$$g_{ps}^{(r)} = \frac{\varepsilon}{Y_0^{(p)} B X_0^{(s)}} \cdot Y_0^{(p)} (\tilde{B} - \lambda_0^{(r)} \tilde{A}) X_0^{(s)}$$

5. On the Determination of Perturbed Components of Eigenvalues

To within the first order of the small value ε perturbed components of the eigenvalues in (4.1) are defined by the solving the eigenvalue problem for $T^{(r)}$ matrix. The appropriate matrix equation is a generalization of the known algorithm for the determination of the vibratory stability of a mistuned cascade in a gas flow when partial frequencies of blades are far apart.

Actually the aerodynamic forces in a gas flow are small as compared to elastic and inertial those. Consequently, for every vibratory mode of isolated blades there is the following subset of close eigenvalues

$$\Lambda^{(k)} = \{\lambda_0^{(kj)}\}_{j=0}^{N-1}, \tag{5.1}$$

where k is the number of vibratory mode caused by aerodynamic blade interaction. Let us now turn from the indices s and p in (4.11) to new those:

$$s = 1 + j + (k-1)N_0, \; p = 1 + q + (v-1)N_0, \tag{5.2}$$

$$j, \; q = 0, \; 1, ..., N-1$$

$$k, \; v = 1, \; 2, ..., N_1/N_0,$$

and describe the vector $X_0^{(s)}$ as follows

$$X_0^{(s)} = X_0^{(kj)} = \frac{1}{\sqrt{N}}\{X_{0n}^{(kj)}\}_{n=0}^{N-1}, \; X_{0n}^{(kj)} = X_{00}^{(k0)}e^{i\mu_j n}, \; ||X_{00}^{(k0)}|| = 1, \tag{5.3}$$

If partial blade frequencies are far apart, then the number of the subset (4. 5) members $N_2 = N$, that is, the matrix order $T^{(r)}$ in (4.11) will be equal to N. With the help of the matrix

$$E = \frac{1}{\sqrt{N}}\left\{exp\left(i\frac{2\pi}{N}mn\right)\right\}_{m, \; n=0}^{N-1},$$

we perform the similarity transformation of the matrix $T^{(r)}$. Based on (2.3), (4.11), (5.2) and (5.3), we derive

$$(L^{(k)} + diag(\tilde{r}_m^{(k)}) - \tilde{\lambda}^{k0}J)Z^{(k)} = 0, \tag{5.4}$$

where $L^{(k)}$ is the cyclic matrix whose members are equal to

$$l_{mn}^{(k)} = \frac{\varepsilon}{X_{00}^{(k0)}CX_{00}^{(k0)}}Y_{00}^{(k0)}\sum_{j=0}^{N-1} P_{00}^{(kj)}(1 - e^{-i\mu_j(n-m)}),$$

$$\tilde{r}_m^{(k)} = \frac{\varepsilon}{Y_{00}^{(k0)}CX_{00}^{(k0)}}Y_{00}^{(k0)}(\tilde{C}_m - \lambda_0^{(k0)}\tilde{M}_m)X_{00}^{(ko)},$$

$P_{00}^{(kj)}$ is the vector of the amplitude function of the unsteady hydrodynamic force acting on the initial blade of a phase shift μ_j.

According to (4.4), approximate eigenvalues (4.1) in this case will be defined as

$$\lambda^{(k0)} = \lambda_0^{(k0)}(1 + \tilde{\lambda}^{(k0)}) \tag{5.5}$$

where $\lambda_0^{(k0)}$ are eigenvalues of the problem on blade vibrations of an appropriate homogeneous cascade by the k th mode with a phase shift $\mu_0 = 0$; $\tilde{\lambda}^{k0}$ are eigenvalues of (5.4).

The case where we have K close of natural frequencies of isolated blade is of most interest in practice. Then, owing to the aerodynamic interaction, the number of close eigenvalues of a whole cascade will equal to $K \times N$. These values can be determined by the formula (5.5) where $\tilde{\lambda}^{(k0)}$ are eigenvalues of the matrix $T^{(r)}$ which here is transformed into a block matrix of the form:

$$\tilde{T}_r = [L^{(k)} + diag(\Delta^{(k)} + \tilde{R}_m^{(k)})], \; L^{(k)} = \{L_{mn}^{(k)}\}_{m,\; n=0}^{N-1},$$

where $L_{mn}^{(k)}$ are square matrices of the order K which involve the following members

$$l_{mn}^{(vu)} = \frac{\varepsilon}{Y_{00}^{(v0)} C X_{00}^{(u0)}} Y^{(v0)} \sum_{j=0}^{N-1} P_{00}^{(u_j)} (1 - e^{-i\mu_j (n-m)})$$

$$(u_1 v = k, \;\; k+1, ..., k+K-1)$$

and $\tilde{R}_m^{(k)}$ is the square matrix of the order K which involves the following members

$$\tilde{r}_m^{(vu)} = \frac{\varepsilon}{Y_{00}^{(v0)} C X_{00}^{(u0)}} Y_{00}^{(v0)} \left(\tilde{C}_m - \lambda_0^{(u0)} \tilde{M}_m \right) X_{00}^{(u0)}$$

6. The Examples of an Enclosing Media Influence on the Hydraulic Turbine Blade Vibrations.

Vibrations of hydraulic turbine blades in a liquid are the most prominent example as their description requires that the auxiliary problems on elastic blade vibrations and determination of unsteady hydrodynamic forces acting on these blades be solved simultaneously.

To solve this problem the following assumptions are introduced:

- blades are rather thin;
- a cascade is annular and set between two infinite circular cylinders;
- a liquid is ideal and incompressible.

For the building the stiffness and mass matrices the finite element method is used. For this purpose a shell finite element (incorporated into the library of the program PIONER [7]) is used to calculate thin-walled constructions (fig.1). Unsteady hydrodynamic forces acting on blades are determined in solving a three-dimensional problem on a liquid flow through a cascade by the vortex frame method [8].

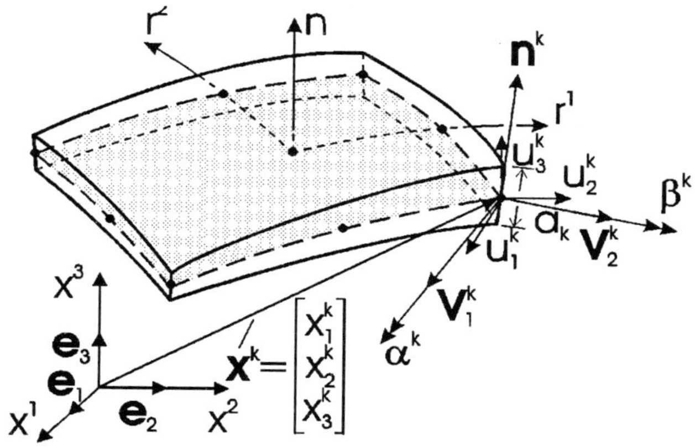

Figure 1. A shell finite element

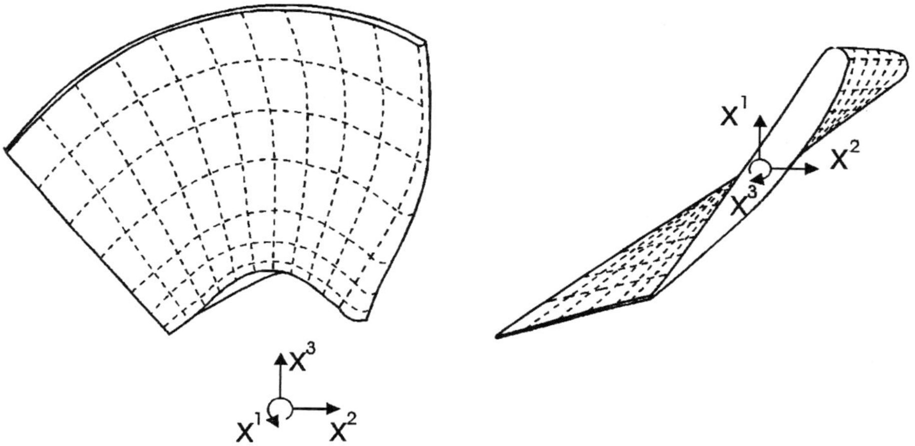

Figure 2. The general view of a blade

Calculations of natural frequencies and vibratory modes are performed for the homogeneous cascade model of PL 587V hydraulic turbine designed at the Leningradskii metallic factory" and tested by experiment (fig. 2). The numerical analysis has revealed a good convergence of the iterative process in calculating natural frequencies and vibratory modes. Within the limits of relative error equal to 0,001, the number of iteration was not over five.

Fig. 3 presents the first three vibratory modes of this blade.

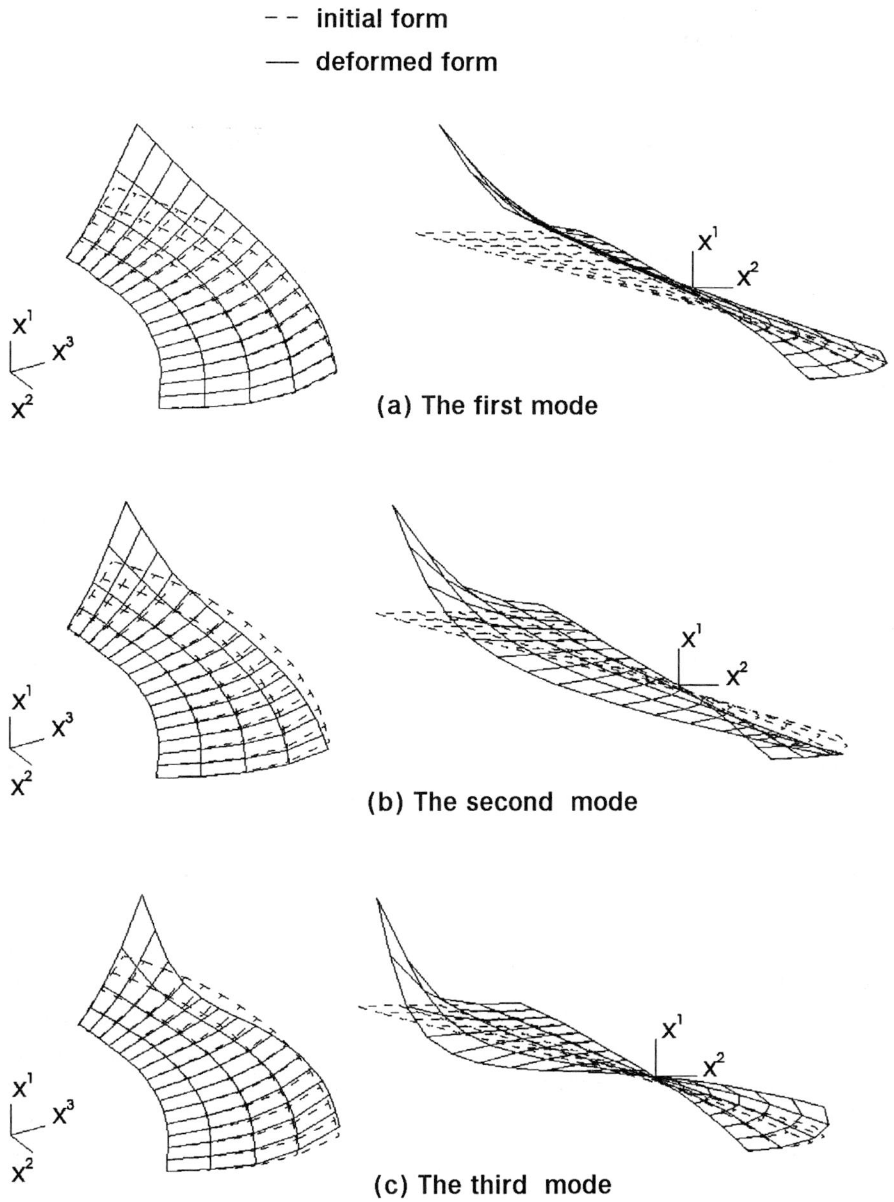

Figure 3. Eigenmodes of a blade

In Table 1 are given the calculation results and experimental data for natural oscillations of a blade clamped in a sleeve.

TABLE 1. Eigenfrequencies (cycle per second) of the blade vibrations in vacuum and water

Mode number	In vacuum Calcul	In vacuum Experiment	In water Calcul	In water Experiment
1	789	705-761	456	408-498
2	1065	880-1068	705	612-770
3	1525	1412-1496	1124	957-1066

The mistunity influence on the natural frequencies in a liquid are given in fig. 4.

The calculation has been performed by eq.(5.4) wherein an inhomegeneity parameter r estimating a mistunity of natural frequencies of blade oscillations

$$\tilde{r}_m^{(k)} = \frac{r\Delta\omega_m^{(k)}}{\omega^{(k)}}, \quad \Delta\omega_m^{(k)} = \omega_m^{(k)} - \omega^{(k)}, \quad \omega^{(k)} = \frac{\sum \omega_m^{(k)}}{N},$$

is specified as

$$\tilde{r}_m^{(k)} = \varepsilon \, cos \, \frac{2\pi}{N} ms, \quad (N = 6)$$

In fig. 4 the partial frequencies of blades without regard for their interaction are marked by dushed lines and solid lines corresponds to partial frequencies due to hydrodynamic interaction.

The mistunity of natural frequencies affects qualitatively the nature of excited cascade vibrations. Indeed, the exciting aerodynamic forces acting on blades are mainly determined by the circular nonuniformity of the flow. These forces represent the set of harmonics, acting on the adjacent blades of the cascade at the same frequencies and amplitudes as well as with a constant phase shift. Since natural vibrations of a homogeneous cascade likewise occur with the constant phase shift, to be in resonance, the forced cascade oscillations under the action of the above exciting forces, not only demand the frequency agreement of these with natural ones, but corresponding phase shifts also is required to be coincided. However, natural vibrations of inhomogeneous cascades are devoid of a generalized periodicity. So for each harmonic of an exciting force the resonance phenomenon would thus be expected to occur when its frequency coincides with any one of natural frequencies of cascade oscillations.

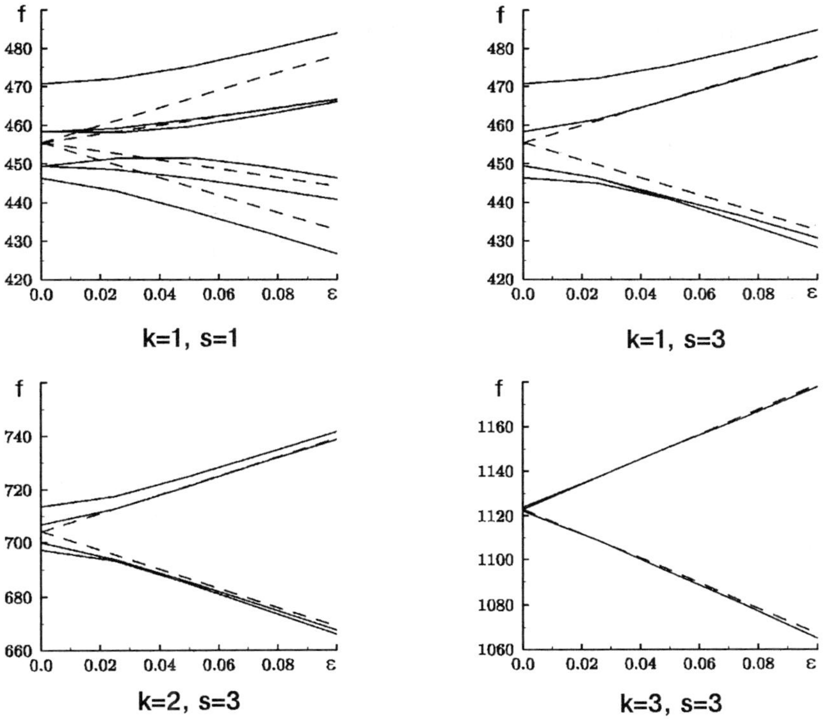

Figure 4. The dependence of blade eigenfrequencies on the mistunity

Fig. 5 shows amplitude-frequency characteristics of the above cascade by the action of exciting forces on blades having amplitudes as follows:

$$P_m = 1, \ (m = 1, \ 2, ..., N)$$

The indicated nature of exciting forces is taken for a qualitative comparison of calculation results with experimental data obtained in testing a cascade on a stand. Dushed/solid lines correspond to homogeneous/inhomogeneous cascade. Each solid line is due to some blade, their number is conditioned by the mistunity law.

The results obtained for inhomogeneous cascade are in qualitative agreement with experimental data. It is observed the stratification of eigenfrequencies for inhomogeneous cascade as compared to homogeneous cascade. The amplitude of forced vibrations for some blade of inhomogeneous cascade is always greater than that for homogeneous cascade.

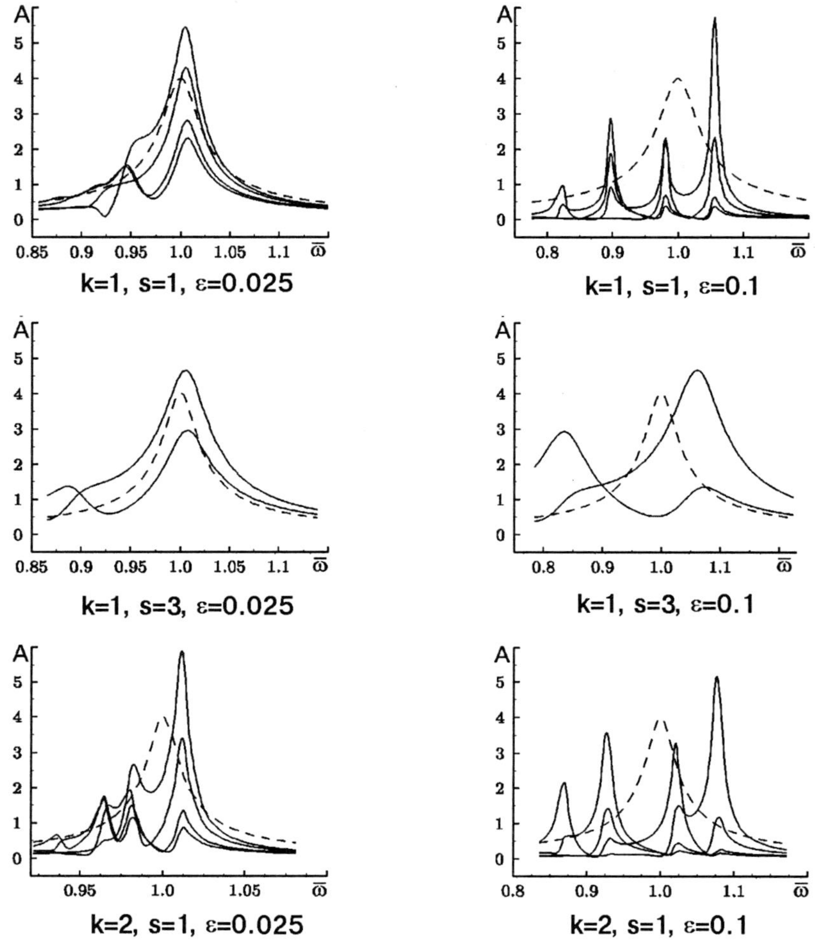

Figure 5. The dependence of the amplitude- frequency characteristics on the blade mistunity ($\bar{\omega} = \omega/\omega_0^{(j)}$)

References

1. Gorelov, D.N. and Guseva, L.A.: Blade vibration of axial hydraulic turbines in a fluid flow, in: G.S. Pisarenko *et al.* (eds.),*Aeroelasticity of Turbomachines* [in Russian], Naukova Dumka, Kiev, (1980), 81–89.
2. Naumenko, V.V, Satsuk, V. G, Strelnikova, E.A. and Simonenko, V.A.: Method of discrete singularities in problems on natural vibrations of hydraulic turbine blades in an ideal incompressible fluid, in: *Method of Discrete Singularities in Problems of Mathematical Physics and Its Role in the Development of a Numerical Computer Experiment* (Abstracts III All-Union Symposium), Izd. Khar'kov Univ., Khar'kov, (1987), 131–132.
3. Sundqvist, J.: An application of ADINA to the solution of fluid-structure interaction problems, *Computers & Structures*, **17**, (1983), 793–807.

4. Bathe, K.J.: *Finite Element Procedures in Engineering Analysis*, Prentice-Hall, Englewood Cliffs, N. J., 1982.
5. Wilkinson, J.: *The Algebraic Eigenvalue Problem*, Clarendon Press, Oxford, 1965.
6. Parlett, B.N.: *The Symmetric Eigenvalue Problem*, Prentice-Hall, Englewood Cliffs, N. J., 1980.
7. Korobeinikov, S.N., Agapov, V.P., Bondarenko, M.I. and Soldatkin, A.N.: The general purpose nonlinear finite element structural analysis program PIONER," in: B.Sendov *et al. Proc. Int. Conf. on Numerical Methods and Applications*, Publ. House Bulg. Acad. Sci., Sofia (1989), 228–233.
8. Ryabchenko, V.P.: Calculation of the associated mass of an annular blade cascade of arbitrary configuration by the discrete vortex method," in: *Proc. CIAM: Aeroelasticity of Turbomachine Blades* (1991), No. 1293, Issue 6, 79–86.

The 2D Flutter of a Bladed Disc in an Incompressible Flow

R. RZĄDKOWSKI
Institute of Fluid-Flow Machines, Polish Academy of Sciences, Gdańsk 80 − 952, *ul.Fiszera 14, Poland.*

V. GNESIN, A. KOVALYOV
Department of Hydromechanics of Power Machines,
Institute for Problems in Machinery
Ukrainian National Academy of Sciences
Pozarsky St. 2/10 Kharkov, 310046, Ukraine

1. Introduction

The traditional approach in flutter calculations of cascade has been to apply frequency domain analysis. In this approach, the blade motions are assumed to be harmonic functions of time with a constant phase lag between adjacent blades. The equations of motion then reduce to an eigenvalue problem; the eigenvalues determine stability. This analysis is restricted to linear structural equations and aerodynamic forces that are linearly dependant on the displacements. The unsteady aerodynamic forces used in such an analysis are typically based on classical linearized potential theory, but allows the use of efficient semianalytical solution methods.

In recent years the works of the coupled fluid -structure problems appeared. The computational method used to solve coupled fluid-structure problem was based on a time-marching algorithm, so it was natural to consider a time domain flutter analysis method. The time domain method

T.H. Fransson (ed.), Unsteady Aerodynamics and Aeroelasticity of Turbomachines, 317-334.

of flutter analysis is based on the simultaneous integration in time of the equation of motion for the structure and the fluid (Bakhale et al. [2]). The time domain method is not restricted to linear structural models and does not require the assumption of a linear relations between aerodynamic loads and blade displacements. The time domain approach allows an accurate representation of the onset of flutter in the cascades when the blades has the complex shape.

He [7], [8] used 2D Navier-Stokes equations, discretized spatially by using the cell-vertex finite volume scheme. The fluid-structure equations were solved in time using the multi-stage Runge-Kutta scheme. The motion of each blade is governed by a structural dynamic model with two degrees of freedom, so the influence of the disc was not taken into consideration.

In this work the aerodynamic forces and moments for a cascade of vibrating blades placed on the disc will be calculated. No restrictions on the motion of the blade were imposed. The model of coupled motion of the vibrating bladed disc and a fluid passing thorough the blades will be shown. In the flow models the unsteady 2D Euler equations are integrated using an explicit monotonous second-order accuracy Godunov's type difference scheme.

The unsteady equations of motion for the structure and the fluid are integrated simultaneously in time, starting with a steady flowfield and a small initial disturbances or the assumed unsteady forces, applied to the blades. Each blade is allowed to move independently, and the motion of the all blades is analyzed to determine the aeroelastic stability of the cascade. The results are compared with those from a separate frequency domain analysis.

2. The structure model

The blade model applied here is a one-dimensional beam described by an extended beam-theory including all important effects on a rotating blade. The model was described in detail in reference (see Rzadkowski [10]). The disc is modelled by a moderately thick plate theory. Assuming rigidly fixed blades on the disk rim, the displacements of any particle of the bladed disc are written in the form

$$\mathbf{u} = \mathbf{U}\mathbf{q}, \tag{1}$$

where

$$\mathbf{u} = col(\mathbf{u}_1, ..., \mathbf{u}_N, \mathbf{u}_d),$$

$$\mathbf{q} = col(\mathbf{q}_1, ..., \mathbf{q}_N, \mathbf{q}_d),$$

$$\mathbf{U} = \begin{bmatrix} \mathbf{U}_1 & 0 & \dots & 0 & \mathbf{U}_{sp1} \\ 0 & \mathbf{U}_2 & \dots & 0 & \mathbf{U}_{sp2} \\ \cdot & \cdot & \cdot & \cdot & \cdot \\ 0 & 0 & \dots & \mathbf{U}_N & \mathbf{U}_{spN} \\ \mathbf{U}_{sp1}^T & \mathbf{U}_{sp2}^T & \dots & \mathbf{U}_{spN}^T & \mathbf{U}_d \end{bmatrix},$$

$\mathbf{u}_i$ is the vector of generalized displacements of each point of the ith blade, $\mathbf{u}_d$ is the vector of generalized displacements of each point of the disc, $\mathbf{q}_i$ are the unknown functions of the ith blade and the $\mathbf{q}_d$ are the unknown functions of the disc, both describing the displacement field. The matrices $\mathbf{u}, \mathbf{q}, \mathbf{U}_i, \mathbf{U}_d, \mathbf{U}_{spi}$, i=1, ... , N are shown in (Rzadkowski [10], [11], [12]).

The blades and disc material is modeled as a Hooke's material.

In order to find an approximate solution of the forced vibrations of the mistuned bladed disc, the Hamilton principle can be used.

The equation of motion of the bladed disc for the forced vibration is represented by the equation

$$\mathbf{M}\ddot{\mathbf{u}} + \mathbf{C}\dot{\mathbf{u}} + \mathbf{K}\mathbf{u} = \mathbf{F}, \tag{2}$$

where $\mathbf{M}$, $\mathbf{C}$, $\mathbf{K}$ are the mass, damping and stiffness matrix of the bladed disc, respectively, and $\mathbf{F}$ is the vector of external forces.

The aeroelatic equation (see equation (2)):

$$\mathbf{M}\ddot{\mathbf{u}}_{t_m} + \mathbf{C}\dot{\mathbf{u}}_{t_m} + \mathbf{K}\mathbf{u}_{t_m} = \mathbf{F}_{t_m} \tag{3}$$

is integrated at a considered point of time $t_m = t_o + m\Delta t$, by a method of constant average acceleration.

For each point of time t_m, the excitation generalized forces $\mathbf{F}_{t_m}$ must be calculated from the flow model, for the position of the blades in cascade, calculated by the structure model.

To obtain the solution for the general displacements at time t + Δt, the equilibrium conditions are considered at time t+ $\theta\Delta$t. However, because the acceleration is assumed to vary linearly, a linearly projected load vector is used; i.e., the equation employed is

$$\mathbf{M}\ddot{\mathbf{y}}_{t+\theta\Delta t} + \mathbf{C}\dot{\mathbf{y}}_{t+\theta\Delta t} + \mathbf{K}\mathbf{y}_{t+\theta\Delta t} = \mathbf{F}_{t+\theta\Delta t}, \tag{4}$$

where

$$\mathbf{F}_{t+\theta\Delta t} = \mathbf{F}_t + \theta(\mathbf{F}_{t+\Delta t} - \mathbf{F}_t). \tag{5}$$

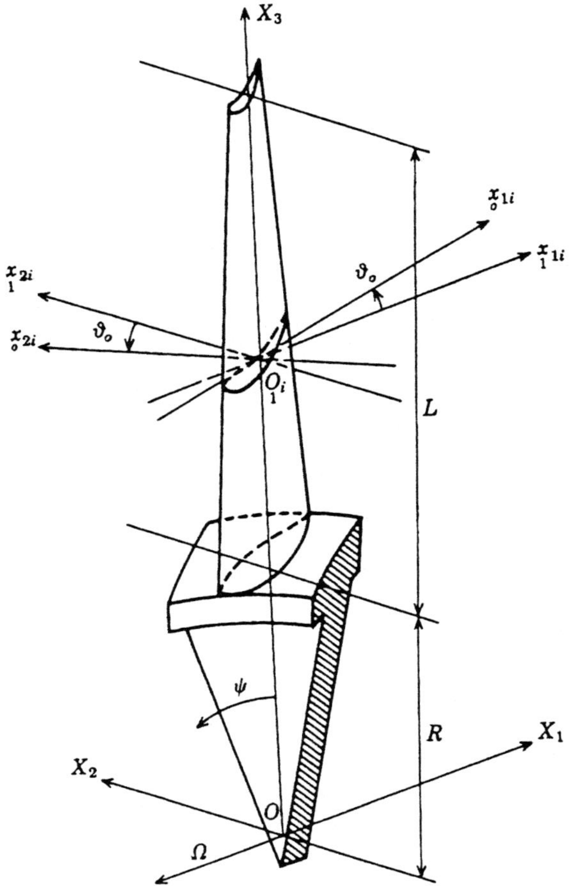

Figure 1: Model of the bladed disc.

The Wilson θ method (see Bathe et al. [3]), in which a linear variation of acceleration from time t to time t + Δt is assumed. For the unconditional stability we employ $\theta = 1.4$.

Let τ denote the increase in time, where $0 < \tau < \theta\Delta t$; then for the time interval t to t + $\theta\Delta$ t, it is assumed that

$$\ddot{\mathbf{y}}_{t+\tau} = \ddot{\mathbf{y}}_t + \frac{\tau}{\theta\Delta t}(\ddot{\mathbf{y}}_{t+\theta\Delta t} - \ddot{\mathbf{y}}_t). \tag{6}$$

Integrating equation (6), we obtain

$$\dot{\mathbf{y}}_{t+\tau} = \dot{\mathbf{y}}_t + \ddot{\mathbf{y}}_t\tau + \frac{\tau^2}{2\theta\Delta t}(\ddot{\mathbf{y}}_{t+\theta\Delta t} - \ddot{\mathbf{y}}_t), \tag{7}$$

and

$$\mathbf{y}_{t+\tau} = \mathbf{y}_t + \dot{\mathbf{y}}_t\tau + 0.5\ddot{\mathbf{y}}_t\tau^2 + \frac{1}{6\theta\Delta t}\tau^3(\ddot{\mathbf{y}}_{t+\theta\Delta t} - \ddot{\mathbf{y}}_t). \tag{8}$$

Using (7) and (8), we have at time t + $\theta\Delta$ t,

$$\dot{\mathbf{y}}_{t+\theta\Delta t} = \dot{\mathbf{y}}_t + \frac{\theta\Delta t}{2}(\dot{\mathbf{y}}_{t+\theta\Delta t} + \dot{\mathbf{y}}_t), \tag{9}$$

$$\mathbf{y}_{t+\theta\Delta t} = \mathbf{y}_t + \theta\Delta t\dot{\mathbf{y}}_t + \frac{\theta^2\Delta t^2}{6}(\ddot{\mathbf{y}}_{t+\theta\Delta t} + 2\ddot{\mathbf{y}}_t), \tag{10}$$

from which we can solve for $\ddot{\mathbf{y}}_{t+\theta\Delta t}$ and $\dot{\mathbf{y}}_{t+\theta\Delta t}$ in terms of $\mathbf{y}_{t+\theta\Delta t}$

$$\ddot{\mathbf{y}}_{t+\theta\Delta t} = \frac{6}{\theta^2\Delta t^2}(\mathbf{y}_{t+\theta\Delta t} - \mathbf{y}_t) + \frac{6}{\theta\Delta t}\dot{\mathbf{y}}_t - 2\ddot{\mathbf{y}}_t, \tag{11}$$

and

$$\dot{\mathbf{y}}_{t+\theta\Delta t} = \frac{3}{\theta\Delta t}(\mathbf{y}_{t+\theta\Delta t} - \mathbf{y}_t) - 2\dot{\mathbf{y}}_t - \frac{\theta\Delta t}{2}\ddot{\mathbf{y}}_t. \tag{12}$$

3. Aerodynamical models

The aerodynamic model developed by Gnesin et al. [1] is capable of representing 2D-flows over a wide Mach number range from low subsonic to supersonic, including transonic flows. The aerodynamic model fully accounts for blade thickness and camber and the angle-of-attack effects. The unsteady Euler 2D equations are integrated using an explicit monotonous second-order accuracy Godunov's type difference scheme.

The transonic flow, including in general case the strong discontinuity of parameters, is described in relative coordinates frame, by unsteady Euler equations, represented as integral conservation laws:

$$\frac{\partial}{\partial t}\mathbf{U} + \frac{\partial}{\partial x}\mathbf{F}_1 + \frac{\partial}{\partial y}\mathbf{F}_2 = 0, \tag{13}$$

where

$$\mathbf{U} = \begin{bmatrix} \rho \\ \rho v_1 \\ \rho v_2 \\ E \end{bmatrix}, \quad \mathbf{F}_1 = \begin{bmatrix} \rho v_1 \\ P + \rho v_1^2 \\ \rho v_1 v_2 \\ (P+E)v_1 \end{bmatrix}, \quad \mathbf{F}_2 = \begin{bmatrix} \rho v_2 \\ \rho v_1 v_2 \\ P + \rho v_2^2 \\ (P+E)v_2 \end{bmatrix}.$$

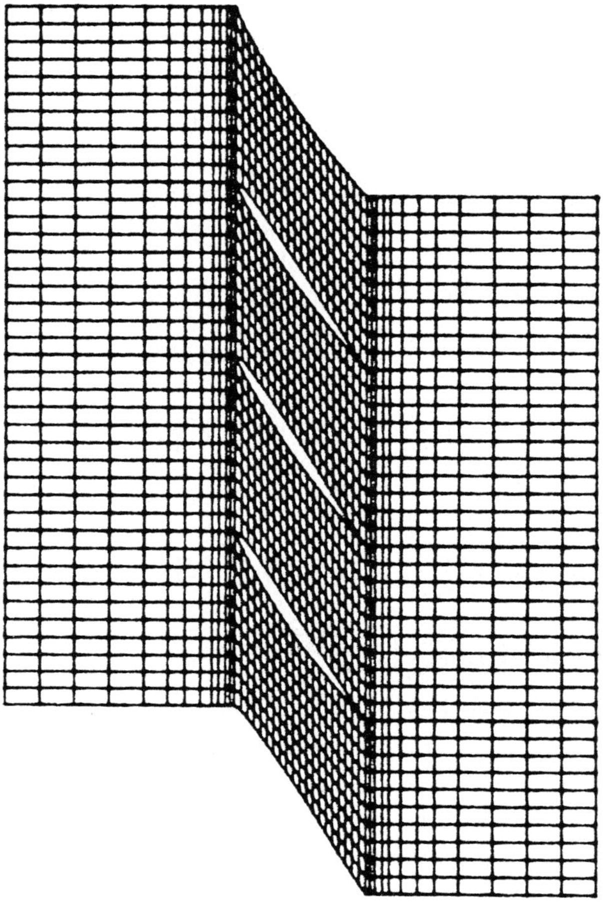

Figure 2: Calculation domain.

The state equation is added to equation (13)

$$\varepsilon = \frac{1}{\kappa - 1}\frac{p}{\rho}. \tag{14}$$

Usually the axial velocity in the turbine stage is subsonic, so the boundary conditions are imposed:

at the inlet of calculation domain

$$P_o = P_o(x_1, x_2), \qquad T_o = T_o(x_1, x_2), \tag{15}$$

$$\alpha = \alpha(x_1, x_2), \qquad \gamma = \gamma(x_1, x_2), \tag{16}$$

where P_o the total pressure, T_o the total temperature and two angles in tangential and meridian planes;

at exit behind the rotor wheel

$$P = P(x_1, x_2), \tag{17}$$

where P static pressure.

At the boundaries which coincide with the blade surface, casing and hub, the following condition is used

$$\mathbf{vn} = 0. \tag{18}$$

The time -spacing generalized conditions of periodicity is imposed at the boundary between neighboring blade passage.

The unsteady Euler equations are integrated on H-type difference grid, by using the explicit monotonous second-order accuracy Godunov's scheme written in form of difference conservation laws.

$$\mathbf{f}^{t+\Delta t}_{m_i+0.5} = \mathbf{f}^{t}_{m_i+0.5} + \frac{\Delta t}{\Omega_{m_i+0.5}}[\sum_{i=1}^{3}(\mathbf{F}_i\mathbf{n}^o\sigma_{m_i+0.5} -$$

$$\sum_{i=1}^{3}\mathbf{F}_i\mathbf{n}^o\sigma_{m_i}] - \Delta t\mathbf{H}, \tag{19}$$

where $\mathbf{f} = \{\rho, \rho v_1, \rho v_2, E\}$, $\mathbf{F} = \{\rho\mathbf{v}, \rho v_1\mathbf{v}_i + \delta_{1i}p, \rho v_2\mathbf{v}_i + \delta_{2i}p, (E+p)\mathbf{v}\}$, $\mathbf{v}_i$ is the normal vector of the ith surface, δ_{1i} is the Kroneker delta, ρ is the flow density, $\rho\mathbf{v}$ is the impulse and E is the energy through lateral sides of cell.

This model can be used as a basis for prediction of both the unsteady aerodynamic forces acting on turbine blades and the time - averaged flow field in a turbine stage under different regime conditions and geometrical characteristics.

4. Numerical results

In order to verified the numerical program, obtained results were compared to experimental and theoretical results presented by Bölcs and Fransson [4] and [5]. The First Standard Configuration (compressor cascade in low subsonic flow) were used.

The cascade configuration consists of eleven vibrating NACA 65-series blades, each having a chord 2c=0.1624 m and length L= 0.254 m, with a 10° circular arc camber and a thickness-to-chord ratio of 0.06. The pitch-to-chord ratio is 0.750 and the stagger angle for the experiments presented here is 55 °. The cascade geometry and profile coordinates are given in [5].

Table 1: The natural frequencies of a non-rotating blade and bladed disc.

Group		Natural Frequency [Hz] for rotational speed 0 rpm						blade
		number of nodal diameters						
		0	1	2	3	4	5	
1 bending	ho/R=0.16	97.70	97.16	98.92	100.25	100.62	100.77	101.091
	ho/R=0.11	92.58	89.76	94.70	98.67	99.78	100.25	
	ho/R=0.06	66.14	57.80	70.59	87.83	93.96	96.65	
2 torsion	ho/R=0.16	327.63	337.81	384.76	392.55	393.72	394.22	395.67
	ho/R=0.11	234.81	230.67	295.02	380.85	389.04	391.21	
	ho/R=0.06	146.77	158.15	175.67	235.26	296.75	342.34	

For such a data the free vibration of the singular blade and the palisade with eleven blades were calculated. The diameter of disc was found from the pitch-to-chord ratio. The palisade has the following physical properties; disc r_o = 0.02 m, R =0.2 m; ρ =0.78 10^4 kg m^{-3}, E = 0.207 10^{12} N/m^2, ν=0.3. The thickness of the disc is h_o=0.1245 m, so h_o/R= 0.62. The theory of the free vibration of disc presented in [10] has limitation to h_o/R $<$ 0.16, so the thickness of the disc was assumed to be h_o=0.0325 m. Table 1 presents the values of natural frequencies of the rigid blade and bladed disc for different values of h_o/R.

These results show the influence of the disc thickness on the natural frequencies of the system. When the disc thickness increase the natural frequencies increase as well.

As a first step of verification of the numerical program the steady flow was calculated for a cascade of 11 blades. The results of the blade surface pressure coefficient (see Bölcs and Fransson [4] pp.21) for angle of incidence equal to 6° was presented in Figure 3. The solid line presents the numerical values obtained for the upper surface (suction surface), the dashed lines for the lower surface (pressure surface). Figure 4 presents the damping coefficient versus the interblade phase angle. The points $(\times, \times, \diamond, \diamond)$ are experimental results presented by Bölcs and Fransson [4].

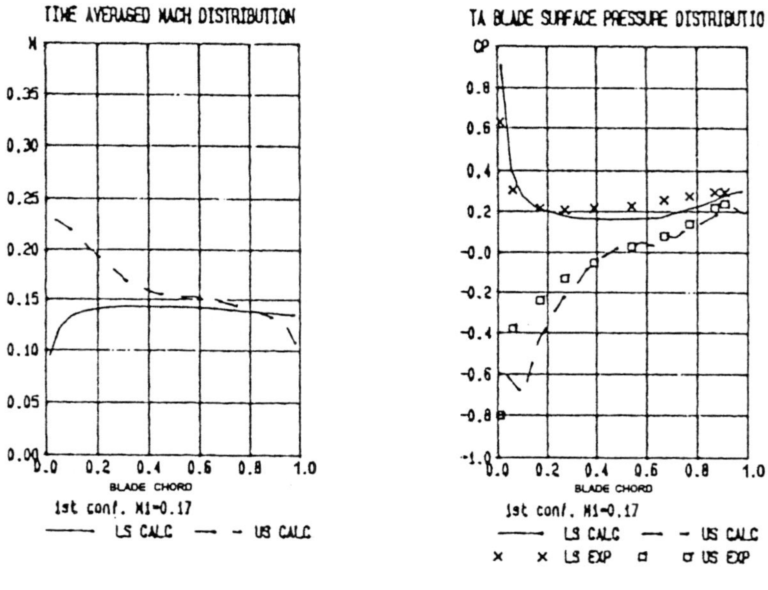

Figure 3: Time averaged blade surface pressure coefficient.

The two-dimensional differential work, per unit span, done on a rigid system by the aerodynamic forces and moments is conventionally expressed by the product of the in-phase components of force and differential translation, as well as moment and differential torsion. The aerodynamic work can be expressed in normalized form as the aerodynamic damping coefficient Ξ (see Bölcs and Fransson [4] pp. 26 and Carta [6]). This coefficient is positive for a stable motion.

For a cascade with eleven blades consider here, the values of interblade phase angles, from open region (0,180°) [4] (see Figure 4), are critical values (unstable conditions) obtained in the frequency domain analysis for a torsional vibration (see Bölcs and Fransson [4] pp.66).

The interblade phase angle are used in flutter analysis (see Verdon [13]) in order to modelled the movement of the blades of the palisade. The displacement of jth blade in cascade is presented by equation

$$u_j = u_o e^{i(\omega t + j\sigma)} \tag{20}$$

where u_o amplitude of the displacement, ω the oscillation frequency, $\sigma = 2\pi n1/N$ the phase angle, N is the number of rotor blades, n1=0, ... , N-1.

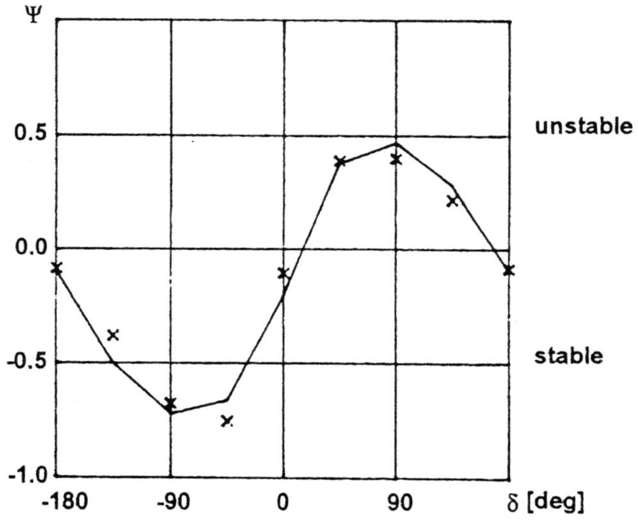

Figure 4: Aerodynamic damping coefficient versus interblade phase angle.

When the blades are mounted on the disc the interblade angle can not be chosen properly. The interblade angle is connected with the number of nodal diameters in the vibration mode. From the forced analysis of the bladed disc the nodal diameters in the vibration mode are connected with the number of stator blade. So, the phase angle of the jth blade presented in flutter analysis of cascade is connected with the number of stator blade $N1$:

$$u_j = u_o e^{i(\omega t + j2\pi N1/N)} \tag{21}$$

where N is the number of rotor blades.

For a considered bladed disc with eleven blade, zero nodal diameter mode are adequate to zero phase angle and for example N1=11 (see equation (21)), one nodal diameter mode with the phase angle equal to 32.7° (N1=12) or - 32.7° (N1=10), two nodal diameter modes is adequate to the phase angle equal to 65.4° (N1=13) or - 65.4 ° (N1=9), three nodal diameter modes to the phase angle equal to 98.1° (N1=14) or -98.1° (N1=8), four nodal diameter modes to the phase angle equal to 130.8° or - 130.8° and five nodal diameter modes to the phase angle equal to 163.5° or -163.5°.

Next the time-dependant results are considered. The excitation torsional frequency is equal to 15.5 Hz, so this is not the natural frequency of the blade. This type of vibration is called "the control vibration" . In the aerodynamic test the blades vibrations during experiments are kept constants, either in time or between the different blades. The experimentally determined time-averaged blade surface pressure distributions are used for calculation of the aerodynamic work.

From the structure point of view it is interesting to analyse the flutter near the natural frequency of system (see Henry et al. [9]). The material parameters of the blade and bladed disc were changed in order to obtained the natural frequencies close to 15.5 Hz.

The bladed disc has the following physical properties; disc r_o = 0.02 m, R =0.2 m; ρ =0.78 10^4 kg m^{-3}, E = 0.344 10^8 N/m^2, ν=0.3.

Table 2: The natural frequencies of a non-rotating blade and bladed disc.

Group		Natural Frequency [Hz] for rotational speed 0 rpm						blade
		number of nodal diameters						
		0	1	2	3	4	5	
2 torsion	ho/R=0.16	14.13	15.4167	15.4187	15.4193	15.4199	15.421	15.4383

The calculation of motion of the cascade were done for different interblade phase angles. The blades oscillated in a sinusoidal motion with the specified interblade phase angle. The blade displacements and velocities and the flow variables in each period of time are used as the initial conditions for the time-integration procedure. The amplitude of oscillations grows or decays. In this procedure, the blade motion initially confined to the specific phase angle selected in the initial condition.

Consequently, the stability of the cascade can be determined for the number of stator blade selected in the starting procedure.

The blade motion was initially forced with a phase angle of 98.1 deg. After being released, the blade motion shows the same phase angle for some time: since this phase angle corresponds to unstable motion, the amplitude grows.

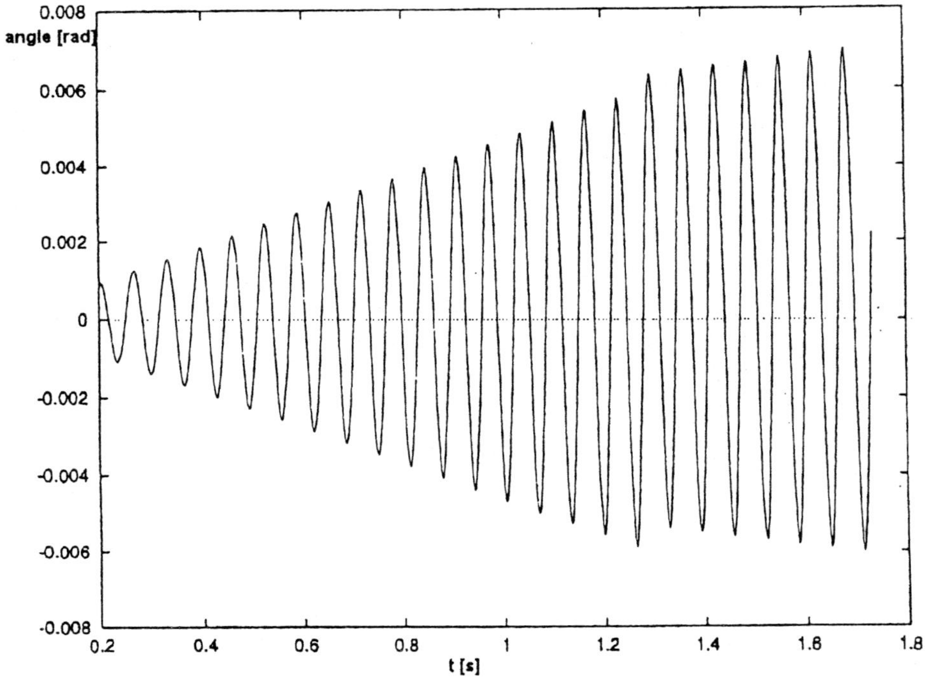

Figure 5: Torsional blade motion for the excitation frequency f=15.44 Hz and the phase angle 98.1 deg.

Figure 5 illustrates the starting procedure for torsional vibration for the excitation frequency f=15.44 Hz and the phase angle 98.1°. This phase angle is the critical values obtained in the frequency domain analysis (see Figure 4). Only the motion of one blade is shown. The motions of the remaining blades are similar, except that adjacent blades are displaced along the time axis by a time interval corresponding to 98.1°. It can be seen that the phase angle 98.1° leads to growing oscillations. The similar situation is for the phase angles 65.4°, 98.1° and 130.8°.

To investigate the stability and behaviour at some other interblade phase angle, calculations were started from a forced torsional motion with the phase angle -98.2°. Figure 6 illustrates the starting procedure for the phase angle equal to -98.2°. It can be seen that the phase angle equal to -98.2° results in a decay in the amplitude of blade oscillations. A frequency domain analysis has shown that this phase angle corresponds to a stable condition.

From the above results the good agreement between the experimental and numerical results is observed. In the numerical calculations the material damping (the matrix **C** in equation (2)) was neglected.

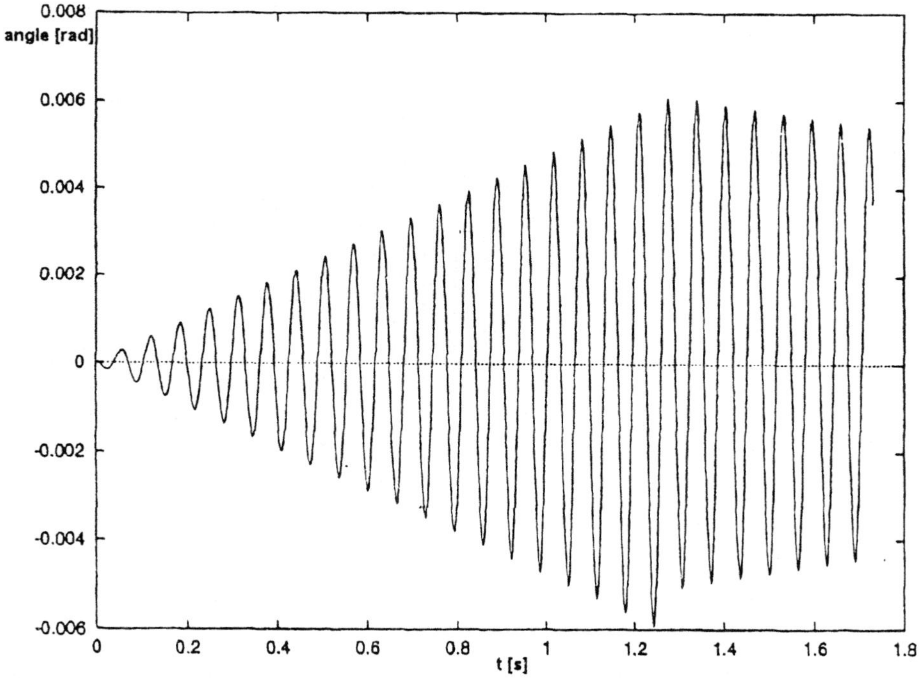

Figure 6: Torsional blade motion for the excitation frequency f=15.44 Hz and the phase angle -98.1 deg.

In the above calculations the small number of time periods (equal to ten) were used in the stability calculations. It is called the short time response of the blades. When the number of periods were increased to 47 the response picture were changed in the stable motion (see Figure 7).

A careful study of the results revels the following. The blade motion was initially forced with a phase angle of -98.1 deg. After being released, the blade motion shows the same phase angle for some time. Since this phase angle corresponds to stable motion, the amplitude decays. The phase angle was changing, for all time, because of the aerodynamic damping. Then the transient behaviour is observed near t = 3 s, when the phase angles is changing to unstable conditions and amplitude is starting to grows.

This clearly shows that the short time response of a blades is dominated by the phase angle used in the initial conditions. The long time response, on the other hand, is dominated by the most unstable phase angle (see Figure 7, 8).

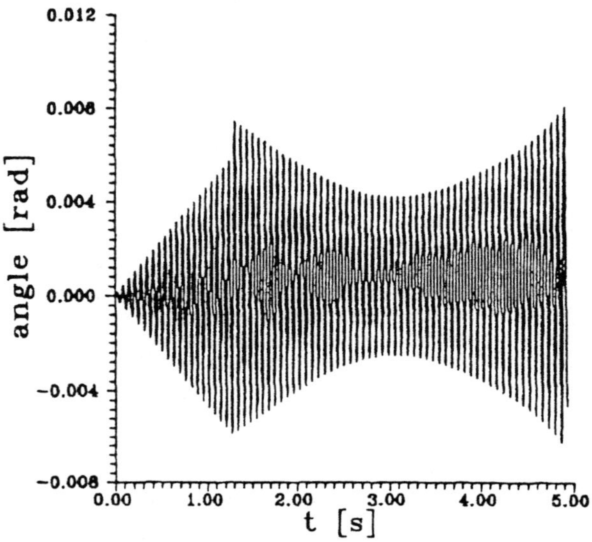

Figure 7: Torsional blade motion for the excitation frequency f=15.44 Hz and the phase angle -98.1 deg. in a long time response.

Next the calculation of motion of the bladed disc were done for different interblade phase angles and the different frequencies of excitations presented in Table 2, corresponding to the nodal diameter modes. It can be seen that the phase angle equal to -98.2°(the number of nodal diameters is equal to 3) and frequencies of excitations 15.4187, 15.4193, 15.4199, 15.421 [Hz] results in a decay in the amplitude of blade oscillations and that the phase angle equal to 98.2° (the mode shape of the disc with the number of nodal diameters equal to 3) and frequencies of excitations 15.4187, 15.4193, 15.4199, 15.421 [Hz] lead to growing oscillations. For the phase angles equal to 98.2° and -98.2°, and the frequency of excitation 14.13 [Hz] the flutter is not observed (the short time response).

A second starting procedure was developed to take advantage of the observation that the least stable phase angle mode dominates the long time response, irrespective of which phase angle is used in the initial conditions.

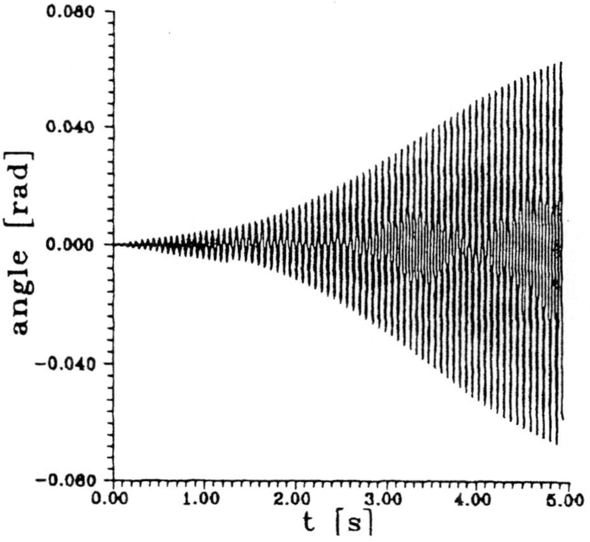

Figure 8: Torsional blade motion for the excitation frequency f=15.44 Hz and the phase angle 98.1 deg. in a long time response.

In this second procedure, the blades are not oscillated in a forced motion. Instead, initial displacement and velocities are zero for all but one blade. This blade is given a torsion moment. Unlike the first procedure, these initial conditions do not correspond to a sinusoidal motion with a specific interblade phase angle. Consequently, the blade displacements obtained from this procedure can be used to determine the stability of the cascade for all possible interblade phase angles since no specific phase angle is used in the starting procedures. This procedure eliminates the need to check the stability for each phase angle separately and is therefore preferable to the first procedure.

To demonstrate the second starting procedure. Only one of eleven blades is given a small initial torsion moment. All the other initial blade displacements and velocity are zero. The response of the blade with the nonzero initial velocity is shown in Figure 9. The motion of the other ten blades in not shown due to space limitations.

Thus it is clear that the time domain method predict the same value of interblade phase angle as the frequency domain method. It is interesting to observe that the time domain method predicts the phase angle corresponding to the most unstable motions for both stable and unstable

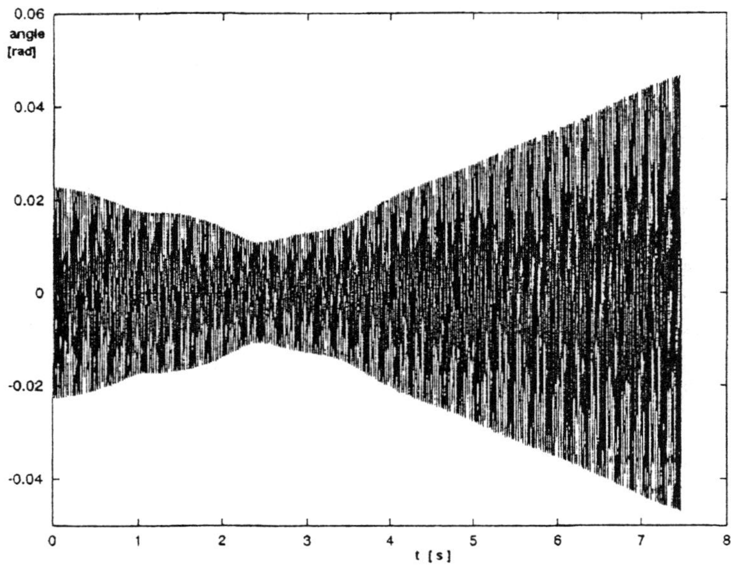

Figure 9: Torsional blade motion for a second starting procedure.

motions.

Figure 10 presents the areodynamic work coefficient of the palisade versus the interblade phase angle as a function of the natural frequency of the blades (see Tables 1,2). When the natural frequency of the blade increase the stability of the palisade increase as well.

5. Conclusions

In the present study, the simultaneous time domain method is used to determine the aeroelastic stability of the cascade. The unsteady equations of motion for the structure and the fluid are integrated simultaneously in time starting with a steady flowfield and an assumed unsteady forces. Each blade is allowed to move independently, and the motion of all blades is analyzed to determine the aeroelastic stability of the bladed disc. Good agreement between the results is observed.

The time of numerical computation of eleven bladed disc and for one frequency is 2880 min. CPU of the Power Station Indygo.

The presented above model was the first attempt to analyze the self-excited vibration of a bladed disc,in the time domain method, where the blades were modelled by the continuous models. The numerical program was tested for a low subsonic flow. The presented time domain method

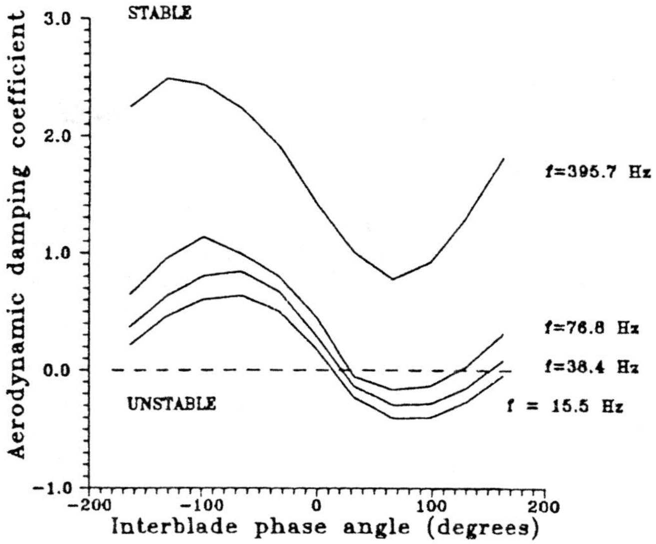

Figure 10: The influence of the natural frequency of the blade on the aerodynamic damping coefficient.

more realistic simulation of the motion of the fluid and the cascade blades that should lead to a better physical understanding. The next step is the verification of the presented model for a subsonic, transonic and supersonic flow.

All numerical calculations were made in the Academic Computer Center TASK, in Gdańsk.

Acknowledgement

The authors wish to gratefully acknowledge the Józef Mianowski Fund and Foundation for Polish Science, Warsaw for a research fellowship to the second author, which enable to carry out this work.

References

[1] Arkadyev A.B., Gnesin V.I., Vanin V.A., Yershov S.V.: Unsteady Blade Forces Generated by Rotor-Stator Aerodynamic Interaction and Cascade Airfoils Oscillations, *International Conference Engineering Aerohydroelasticity EAHA*, December 5-8, Prague, Czech Republic, pp.366-371, 1989.

[2] Bakhale M.A., Reddy T.S.R., Keith T.G.: Time Domain Flutter Analysis of Cascade Using Full-Potential Solver, *AIAA Journal*, Vol. 30, No.1, pp.163-170, January 1992.

[3] Bathe K., Wilson E.: *Numerical Methods in Finite Element Analysis*, Prentice-Hall, Inc., Englewood Cliffs, New Jersey, 1976.

[4] Bölsc A., Fransson T.H.: Aeroelasticity in Turbomachines Comparison of Theoretical and Experimental Cascade Results, *Communication du Laboratoire de Thermique Appliquée et Turbomachines*, Nr.13. Lusanne, Epfel, 1986.

[5] Bölsc A., Fransson T.H.: Aeroelasticity in Turbomachines Comparison of Theoretical and Experimental Cascade Results, *Communication du Laboratoire de Thermique Appliquée et Turbomachines*, Nr.13, Appendix A5 All Experimental and Theoretical Results for the 9 Standard Configurations,

[6] Carta F.O.: Coupled Bladed-Disk-Shroud Flutter Instabilities in Turbojet Engine Rotors, *Transactions of the ASME, Journal of Engineering for Power*, Vol. 89, No. 3, pp. 419-426, 1967.

[7] He L., Denton J.D.: Three Dimensional Time -Marching Inviscid and Viscous Solutions for Unsteady Flows Around Vibrating Blades, *ASME Paper, 93-GT-92*, 1993.

[8] He L.: Rotating-Stall / Stall-Flutter Prediction Using A Fluid / Structure Coupled Method, *Proceedings of 7th International Symposium on Unsteady Aerodynamics and Aeroelasticity of Turbomachines*, September 25-29, Fukuoka, Japan, pp. 597-608, 1994.

[9] Henry R., Jacquet-Richardet G.: Flutter Analysis in Blade Dynamics, *8th IFToMM World Congress, Theory of Machines and Mechanisms, Forced Vibrations of Turbomachine Blades*, Prague, August 26-31, 1991.

[10] Rzadkowski R.: The General Model of Free Vibrations of Mistuned Bladed Discs. Part I. Theoretical Model. Part II. Numerical Results, *Journal of Sound and Vibration*, 173(3), pp.395-413, 1994.

[11] Rzadkowski R.: *Dynamics of Steam Turbine Blading, Part II, Bladed Discs* (books accepted for publication to Ossolineum in 1997)

[12] Rzadkowski R.: Numerical Analysis of Free and Forced Vibration of Tuned and Mistuned Bladed Discs, *Zeszyty Naukowe IMP PAN* 483/1438/97 (Doctor Dissertation).

[13] Verdon J.K.: Linearized Unsteady Aerodynamic Theory, *AGARD Manual on Aeroelasticity in Axial Flow Turbomachines*, Vol.I, pp.(2-1,2-31), 1987.

STRUCTURAL MISTUNING AND AERODYNAMIC COUPLING IN TURBOMACHINERY BLADINGS

G. KAHL
MTU Motoren- und Turbinen-Union München GmbH
P.O. Box 50 06 40
D - 80976 München
Germany

1. Abstract

Most calculation methods for flutter and forced response in turbomachines rely on ideally tuned modes in the blading. In reality, small deviations in blade geometry lead to a scatter of the blade eigenfrequencies. The mechanical damping is usually low, while oscillation-induced unsteady aerodynamic forces can be very significant. In this situation, small mistuning of the blades can lead to localization effects, where large variations in the resonant amplitudes in a cascade occur.

The present paper describes a method to evaluate the effect of structural mistuning on the unsteady aerodynamics, aeroelastic stability and the resonant amplitudes of oscillating cascades.

To evaluate the unsteady aerodynamics of a mistuned oscillating cascade, 3D time-linearized Euler-calculations of tuned modes with different interblade phase angles are used. The Euler code is the 3D version of a Quasi-3D linearized Euler code previously presented by the author. It is capable of determining the unsteady aerodynamics in oscillating cascades of arbitrary tuned eigenmodes. The resulting aerodynamic forces are superimposed to yield the influence coefficients of single oscillating blades, using the eigenmodes as the generalized coordinates.

Provided that the blade eigenfrequencies are well separated and that the mass parameter is fairly high, coupling between the blade eigenmodes can be neglected. Depending on the problem under consideration - forced response or flutter calculation - either a linear system of equations results for the resonant amplitudes and phase lags of all the blades in the cascade, or an eigenvalue problem, yielding the eigenfrequencies and eigenmodes of the aeroelastic system can then be set up and solved. Once the tuned mode coefficients have been determined, the solution requires very little computer time, as the number of degrees of freedom is reduced to one for each blade in the cascade. Consequently, the method is very well suited for parametric studies.

Sample calculations for a subsonic LP turbine and a transonic compressor rotor are shown. The comparison of the current results and tuned mode analysis indicates the necessity to include mistuning effects in flutter and forced response calculations of turbomachinery bladings.

T.H. Fransson (ed.), Unsteady Aerodynamics and Aeroelasticity of Turbomachines, 335-346.

2. Introduction

Design requirements for modern turbomachinery used in aeropropulsion include low weight, high power output, wide operating ranges and high reliability. These conflicting goals lead to designs where blade resonances cannot be avoided over the whole operating range. To assure structural integrity of the blades in this situation, the designer needs computational tools to predict the resonant amplitudes and stresses. Most of the currently used methods for predicting forced response amplitudes rely on the assumption of "tuned modes", all blades in a cascade being identical and consequently having identical resonant amplitudes. In reality, significant variations of the resonant amplitudes of the individual blades in a cascade are often observed. The reason for this is believed to lie in small deviations in blade geometry or root fixture, leading to a scatter in the eigenfrequencies.

Such mistuned systems have been studied by various authors, e.g. Kielb and Kaza (1983), Pierre et al. (1994), Srinivasan and Tavares (1994) and others. In the present analysis, a fully 3D aerodynamic model is used to predict the coupling effects introduced by the unsteady pressures induced by the oscillating blades.

3. Structural and Aerodynamic Model

The system under consideration in this paper consists of an assembly of n_{blades} engine blades that are slightly different from each other and coupled only through aerodynamic forces, as in the previous work of Pierre et al. (1994). The differences among the blades may arise from manufacturing tolerances, repair of single blades, or similar sources and are assumed to result in a small deviation in the eigenfrequencies of the blades. The effect on the blade eigenmodes is assumed to be negligible, hence all blades have identical eigenmodes. Furthermore, the eigenmodes of each blade are assumed to be well separated, consequently coupling between different eigenmodes is neglected.

Under these assumptions, each blade can be modeled as a single-degree of freedom-oscillator. The generalized amplitude $\tilde{a}_k$ (the magnitude of movement in the direction of the eigenmode Φ under consideration for blade k) is then sufficient to describe the displacement $\Delta \boldsymbol{x}_k$ of the blade k at each location $(z, r\theta, r)$:

$$\Delta x_k (z, r\theta, r) = \tilde{a}_k \Phi (z, r\theta, r) \tag{1}$$

The generalized unsteady forces F_k in the direction of the eigenmode Φ acting on a blade k can be calculated from the surface pressure disturbance distribution p'_k using

$$F_k = \oint_A -p'_k(z,r\theta,r)\, n\, \Phi(z,r\theta,r)\, dA \tag{2}$$

where $\boldsymbol{n}$ is the unit normal to the surface element dA. This pressure disturbance p'_k may arise either from the oscillation of the blades or from inlet disturbances, wakes or external pressure fields acting upon the cascade. We will therefore split F into the two parts F_{ind} for motion induced forces and F_{ext} for "external" forces. For oscillation induced unsteady pressures, the pressure disturbance p'_k on blade k is assumed to be a linear function of the blade displacements $\tilde{a}_l$ of all blades in the cascade, including blade k itself. Consequently, this hold true for the motion induced generalized forces F_{ind} too, so that

$$F_{ind,k} = \sum_{l=1}^{nblades} c_{k,l}\, \tilde{a}_l \tag{3}$$

while the external forces F_{ext} are assumed to be independent of blade motions. We can now set up the basic matrix equation for the coupled system as

$$M\ddot{\tilde{a}} + D\dot{\tilde{a}} + (K + C)\,\tilde{a} = \tilde{F}_{ext} \tag{4}$$

Here, **M**, **D** and **K** are the diagonal mass, damping and stiffness matrices, respectively, while **C** holds the aerodynamic coefficients $c_{k,l}$.

Let us first take a look at the situation in the absence of external forces $\mathbf{F}_{ext}$. For convenience, it is assumed that all blades have identical mass m and mechanical damping d, while the stiffness terms in **K** differ slightly, thus introducing the mistuning in the system. Using a common exponential approach for $\tilde{a}_k$

$$\tilde{a}_k = a_k\, e^{\left(i(\omega + i\gamma) - \frac{d}{2m}\right)t} \tag{5}$$

we arrive at the following equation, where **E** denotes the unit matrix :

$$\left\{M^{-1}(K+C) - E\left(\tfrac{d^2}{4m^2}\right) - E\left(\omega + i\gamma\right)^2\right\}a = 0 \tag{6}$$

Solving this eigenvalue problem of the order n_{blades} yields the aeroelastic eigenfrequencies ω with their corresponding damping γ, along with the aeroelastic eigenmodes $\boldsymbol{a}$. In this context, it is important not to get mixed up with the eigenmodes of each blade Φ, for example a first torsion mode, and the eigenmode of the cascade $\boldsymbol{a}$, i.e. the distribution of the complex amplitudes in the direction of Φ for all blades of the cascade.

If the amplitudes of the blades under the influence of external forces $\mathbf{F}_{ext}$ (e.g. induced by wakes of upstream blade rows or pressure fields of downstream struts) are sought, the oscillation frequency ω is known beforehand. As only the steady state amplitudes (after transients have decayed) are of interest here, it is thus sufficient to set

$$\tilde{a}_k = a_k\, e^{i\omega t} \tag{7}$$

In this case we have to solve a set of linear equations for the blade amplitudes a_k:

$$\left(K + C + D\omega - M\omega^2\right)a = F_{ext} \tag{8}$$

Here, the assumption of identical mass and damping for all blades is not necessary. Both equations (6) and (8) remain valid as long as only frequencies in the vicinity of the blade eigenfrequency for the mode Φ are considered.

The eigenmodes and eigenfrequencies used in the present study were computed by the finite-element method MSC-NASTRAN, where no restrictions are placed on the complexity of the eigenmodes. Simple bending or torsional modes can be handled in exactly the same way as higher modes with components of bending, torsion and chordwise bending.

4. Linearized Euler Method

The tool used to predict the unsteady aerodynamic pressures on the vibrating blades used in this study is a time linearized 3D Euler method. This method is the 3D extension of a quasi-3D method presented earlier by the author (Kahl 1995, Kahl and Klose 1993). It is based on a finite-volume Runge-Kutta solver using a structured H-type mesh and can handle sub- and transonic flows in axial compressors and turbines. Arbitrary tuned mode eigenforms can be prescribed for the airfoils, not limited to rigid airfoil sections, but restricted to constant amplitude and constant interblade phase angle for all blades in the cascade. Alternatively, up- or downstream pressure disturbances or upstream vorticity disturbances can be prescribed to model the pressure field or wakes of neighboring blade rows.

Due to the assumption of tuned modes (i.e. identical eigenmodes for all blades and constant interblade phase angle between the motions of adjacent blades) the computational domain can be restricted to a single blade passage with appropriate phase-shifted periodic boundary conditions at the circumferential boundaries. The far-field (inlet and outlet) boundaries are modeled by nonreflecting boundary conditions as used by Hall and Lorence (1992), based on the work of Saxer and Giles (1990).

The method is currently run on SGI Challenge workstations. Calculation times are approximately 2h per interblade phase angle on a single R8000 processor for a 117x41x25 (axial x tangential x radial) mesh.

For any given eigenform and frequency, the linearized Euler method can calculate the generalized forces F_{tuned} (φ) for tuned modes with arbitrary interblade phase angles, where

$$F_k = F_{tuned}(\varphi)\, e^{-i\varphi(k-1)} = a_k\, c_{tuned}(\varphi) \tag{9}$$

As was pointed out in section 3, the influence coefficients $c_{k,l}$ are required instead of the tuned mode coefficients c_{tuned} to set up the equations (6) or (8). Following the work of Srinivasan and Tavares (1994), the values of $c_{k,l}$ can be calculated from a complete set of c_{tuned} $(n\varphi)$ with $n=0 \ldots n_{blades} - 1$ and $\varphi = 2\pi / n_{blades}$:

$$c_{k,l} = 1 / n_{blades} \sum_{n=0}^{nblades-1} c_{tuned}(n\varphi)\, e^{i(l-k)\, n\varphi} \tag{10}$$

These influence coefficients form the elements of the matrix $\boldsymbol{C}$ which is used in eqs. (6) and (8).

For cascades with a large number of airfoils (say, more than 50 airfoils), the computation of tuned mode coefficients for all possible interblade phase angles may require a lot of computer time. In these cases, it is possible either to restrict the number of airfoils by assuming a number of equally mistuned segments around the circumference, thus reducing the number of required calculation runs, or by calculating the coefficients for a restricted set of interblade phase angles and then analytically interpolating for the missing values.

The aerodynamic coefficients are strictly valid only for the frequency at which they were calculated, usually the in-vacuo eigenfrequency of the blades. When doing the eigenvalue analysis, the resulting frequencies will differ from the in-vacuo-frequencies. If this difference is large, a new run of the linearized Euler method will be necessary for this frequency, thus forming an iterative procedure converging to the true aeroelastic eigenfrequencies. For the cases studied here, the frequency shift was found to be very small and the aerodynamic coefficients were found to be nearly constant with respect to the small change in frequency. Consequently, this variation is neglected. The same holds true for the forced response calculations, where the mistuning of individual blades that is studied here is on the same order of magnitude as the shift in eigenfrequencies due to aerodynamic forces.

5. Sample Results

5.1. TRANSONIC COMPRESSOR ROTOR

The first example under consideration is the rotor of an experimental single stage compressor. It consists of 16 CFRP blades. Relative inlet Mach number ranges from 0.7 at the hub to 1.3 at the tip. The rotor is shown in *Fig. 1* along with the mesh used for the linearized Euler method (every 2nd mesh line shown). The 2nd bending eigenform at approximately 2 kHz is taken for this example, yielding reduced frequencies based on full chord in the range of 2.9 (hub) to 2.3 (tip).

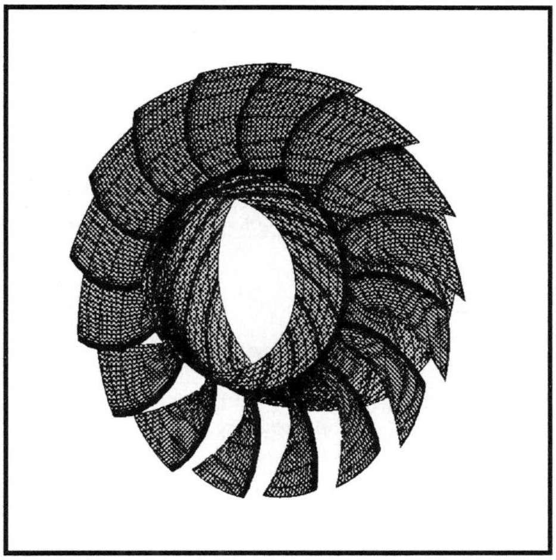

Fig. 1. Experimental transonic compressor rotor with aero mesh, every 2nd grid line shown.

In *Fig. 2,* the tuned rotor eigenvalues (squares) are shown as frequency / damping pairs, where the frequency is normalized by the in vacuo eigenfrequency of the blades and damping is shown as the logarithmic decrement. The variation in eigenfrequency is in the range of -0.6% and +0.2%, corresponding to approx. -12 to +4 Hz. This very small change supports the assumption that the changes in the blade eigenmode should be negligible. The resulting eigenmodes of the cascade (the distribution of amplitude and phase around the circumference) are, not surprisingly, the tuned modes with constant amplitude and constant interblade phase angle for all blades in the rotor.

Subjecting the tuned rotor to a 4th engine order (EO) excitation, corresponding to an interblade phase angle φ=-90 deg., and varying the excitation frequency from 0.97 to 1.03 times the in-vacuo eigenfrequency yields the blade response shown as the solid line in *Fig. 2*. The response amplitude is normalized by the maximum value for this case, this reference value will be used in all further plots for the transonic compressor rotor. Of course, all the blades show identical responses for this tuned configuration.

As an example for the effects of mistuning on the eigenvalues, consider a configuration where all the even numbered blades are mistuned by +0.5% while the odd numbered ones are mistuned by -0.5%. The eigenvalues (the triangles in *Fig. 2*) now form two separate clusters, centered about the eigenfrequencies of the blades at 0.995 and 1.005. The range of the aerodynamic damping for these eigenvalues is much reduced compared to the tuned configuration, with a considerably higher minimum value of aerodynamic damping. This is in line with findings from other authors that mistuning is beneficial in terms of aeroelastic stability (e.g. Pierre et al, 1994).

For higher magnitudes of the mistuning (not shown here), the aerodynamic damping converges to values close to the average value of the tuned configuration. This is similar to a completely decoupled configuration, where only the self-induced aerodynamic forces contribute to the aerodynamic damping. Looking at eq. (10), it can be seen that the self induced forces (index l=k) are the average of the tuned mode coefficients for all interblade phase angles because the exponential term vanishes with l-k=0.

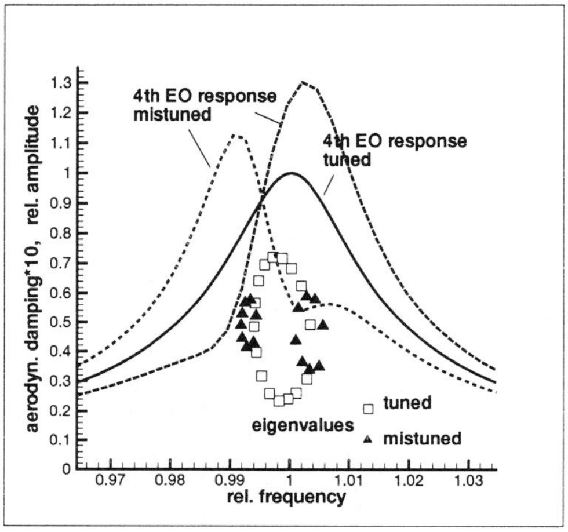

Fig. 2. Aeroelastic eigenvalues and response to 4th EO excitation for transonic compressor rotor

Subjecting this mistuned rotor to the same 4th EO excitation frequency sweep as above now yields two sets of response curves (dashed lines in *Fig. 2*). The even numbered blades (positive mistuning) show the highest response amplitudes, with maximum values up by approximately 30% compared to the tuned configuration. Again, this confirms the common belief that mistuning generally leads to higher resonant amplitudes. As this can pose a serious problem for the designer, the following examples focus on the effects of mistuning on the forced response amplitudes.

The simplest case for mistuning is a single mistuned blade in an otherwise perfectly tuned rotor. *Fig. 3* depicts a typical result for such a case, where blade no. 8 is mistuned by -1%. The graph shows the amplitudes of blades 4 through 11 as a function of the forcing frequency for a 4th engine order excitation. The excitation frequency is varied from 0.975 to 1.025 times the tuned mode frequency in 150 steps, hence the picture contains the results of 150 computation runs.

The amplitudes are normalized by the "tuned mode" resonant response for this case. The bars under the ribbons mark the in vacuo eigenfrequencies of each blade. The blade next to the mistuned blade (no. 7) shows an increase in amplitude of 40% over the tuned mode maximum.

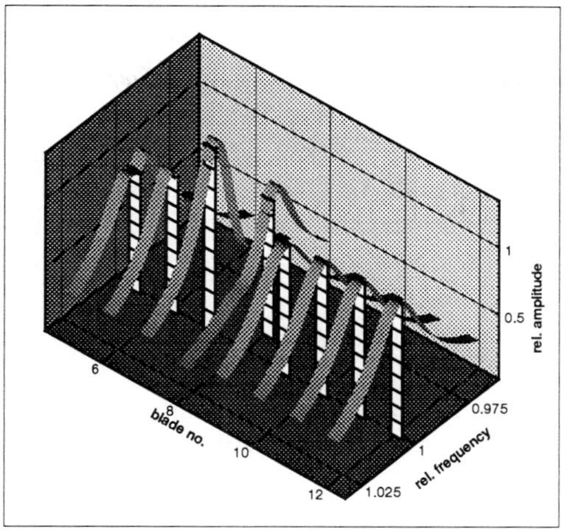

Fig. 3. Transonic compressor rotor amplitudes, blade 8 mistuned by -1%, 4th EO excitation..

The amplitude of the mistuned blade itself is approx. 10% higher than for the tuned mode. The rest of the blades show small deviations from their "tuned mode" behavior, this also holds true for the blades 1-3 and 13-16 which are not shown in this picture.

The speed of the current method makes it well suited for parameter variations. First, we can vary the amount of mistuning introduced to blade no.8. *Fig. 4* shows the result of such a variation, the mistuning varying from -5% to +5%. Although the assumption of constant aerodynamic coefficients becomes questionable for such a large variation in frequency, the results should nevertheless show the correct trends. The picture shows the maximum amplitude occurring on any of the rotor blades when doing frequency sweeps with varying blade 8 mistuning.

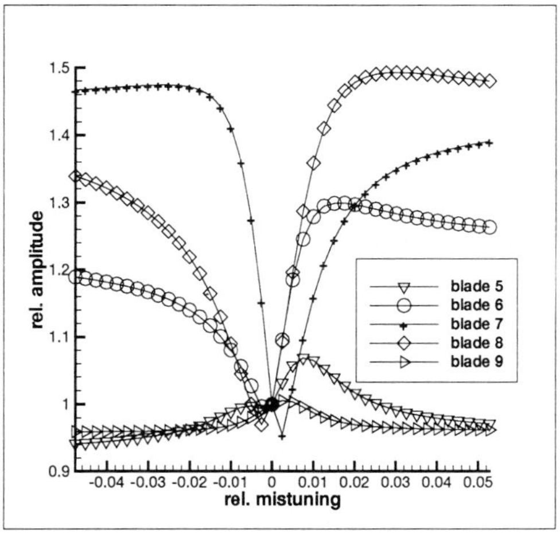

Fig. 4. Transonic Compressor rotor max. amplitudes, blade 8 mistuned by -5% through +5%, 4th EO excitation.

We can see a large variation in the behavior of the mistuned and its neighboring blades as a function of the magnitude of mistuning. For negative values of mistuning, we see a similar behavior to what was shown in *Fig. 3*. For positive values, the location of max. amplitude switches to the mistuned blade itself, while blade 7 max. amplitude drops to considerably lower values. Blade 6 behaves very similar to blade 7, while yielding smaller changes in the amplitudes. All the other blades, including blades 1-4 and 10-16 which are not shown, are only marginally affected by the blade 8 mistuning.

In this case, the maximum amplitude of the mistune blade itself is increased. This is not necessarily so, as can be seen from the next example, *Fig. 5*. Here, the conditions are the same as for *Fig. 3*, but the interblade phase angle of the excitation has been changed to -22.5°. This corresponds to a 1st or 17th EO excitation. The amplitudes are still normalized by the tuned mode amplitude of the 4th EO excitation. The aerodynamic damping of the tuned mode is lower for this interblade phase angle, resulting in an almost double amplitude for this case.

The effect of the mistuning leads to smaller max. amplitudes for blade 8 and most of its neighbors, except for a minor increase on blade 8 for small positive values of mistuning. Nevertheless, for all magnitudes of mistuning, the maximum amplitude in the complete cascade never drops below the "tuned mode" value.

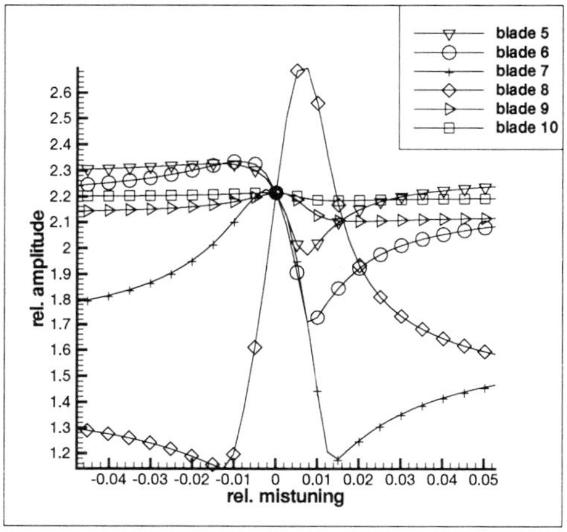

Fig. 5. Transonic Compressor rotor max. amplitudes, blade 8 mistuned by -5% through +5%, 1st EO excitation.

Varying the interblade phase angle over the whole range of -180 to 180 deg and taking the maximum amplitude of all the blade in the cascade yields the results shown in *Fig. 6*. A number of points are interesting to note about this figure.

First, there is no case, where the overall maximum amplitude is lower for a mistuned configuration than for the corresponding tuned configuration. Hence the mistuning of a single blade always has a negative effect for the rotor under consideration. Second, the mistuning has a small effect on the overall maximum amplitude for the cases with high tuned mode amplitudes (e.g. φ = 45 deg.) and a large effect for some other cases, e.g. φ = -45 deg.

These results for a very simple structure of the mistuning (single mistuned blade) already show a wide variety of effects, strongly depending on the interblade phase angle and the magnitude of mistuning. In real turbomachinery cascades, there may be combinations of deterministic mistuning (e.g. repaired blades) and stochastic mistuning (due to manufacturing tolerances), leading to a considerably more complicated situation.

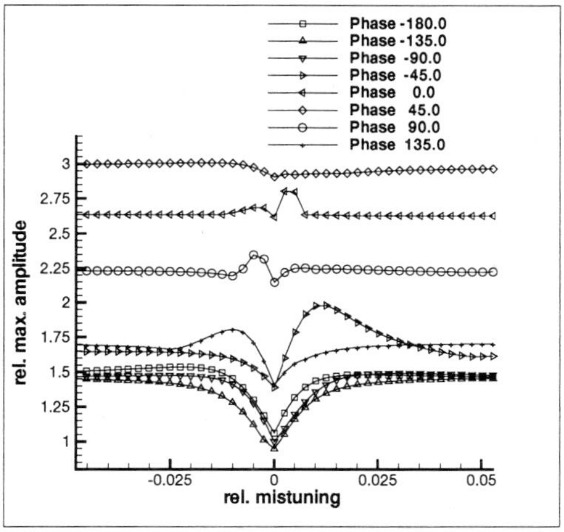

Fig. 6. Transonic Compressor rotor max. amplitudes, blade 8 mistuned by -5% through +5%, variation of excitation EO.

Fortunately, the speed of the current method makes it possible to address these problems. As it can to examine a large number of these configurations analytically in a short time, statistical or "worst case" information about the amplitudes that have to be expected can be gained with reasonable effort.

5.2. LP TURBINE

The second example considered here is a LP turbine vane stage with 120 airfoils in high subsonic flow conditions. The eigenform under consideration is 1st torsion, with reduced frequencies in the range of 0.5 (hub) to 0.95 (tip). Due to the large number of vanes, only a fifth segment of the whole cascade is considered here, resulting in 24 vanes per segment.

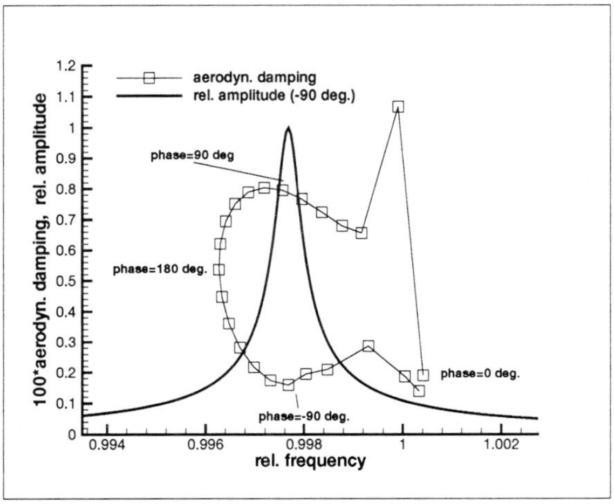

Fig. 7. Aerodynamic damping and response to -90 deg. excitation for tuned LP turbine vane stage.

Fig. 7 depicts the aerodynamic damping and response amplitude to a φ = -90 deg excitation in the same format as in *Fig. 2* (note the difference in scale for the aerodynamic damping and the relative frequency).

Obviously, the aerodynamic damping is much lower (about 1/10) than for the compressor example. This of course leads to a much sharper resonance peak, and also to a lower scatter of the aeroelastic eigenfrequencies of the cascade. This behavior can be attributed to a much higher mass parameter for this case, or, equivalently, comparably low unsteady aerodynamic forces. As in the compressor example, the resonant amplitude of the tuned configuration at φ =-90 deg is taken as the reference value for all further configurations.

Due to the sharp resonance peaks, the mistuning effects of interest occur at very small mistuning magnitudes. The first example for this configuration consists of two vanes (nos. 10 and 11) mistuned by only -0.25%. *Fig. 8* shows the resulting amplitude distributions for vanes 5 through 16 for this configuration with a φ = -90 deg excitation. The layout of the figure is equivalent to that of *Fig. 3* for the compressor example.

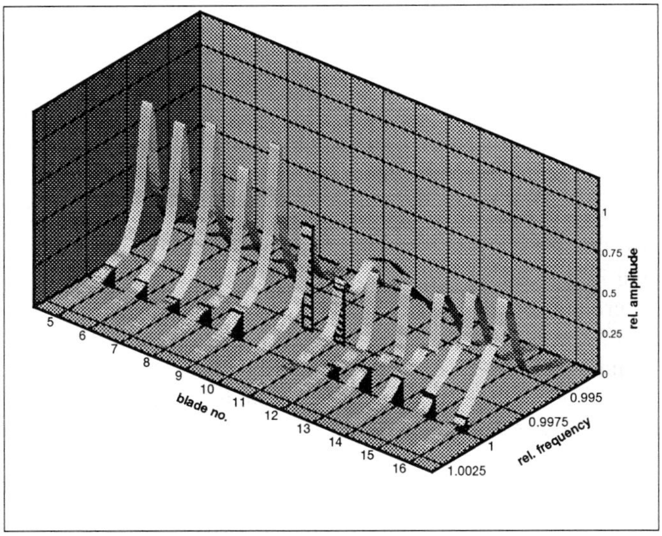

Fig. 8. Low pressure turbine vane amplitudes, vanes 10 and 11 mistuned by -0.25%, φ = -90 deg.

The mistuned vanes 10 and 11 as well as their pressure side neighbors 12 though 14 show strongly reduced amplitudes, while the neighboring vanes 6 through 9 show increased maximum amplitudes. All vanes except vane 11 reach their maximum amplitudes at approximately the same excitation frequency, which is very close to the aeroelastic eigenfrequency of the tuned configuration for φ = -90 deg.

A totally different situation arises in *Fig. 9*, where the excitation has been changed to φ = 180 deg. In this case, the highest amplitudes are observed on the mistuned vanes themselves, while the neighbors on both sides stay at very low levels for all excitation frequencies. Also interesting to note is the form of the resonance curve for vane 12, which shows two distinct peaks, both of which are significantly off the in-vacuo eigenfrequency.

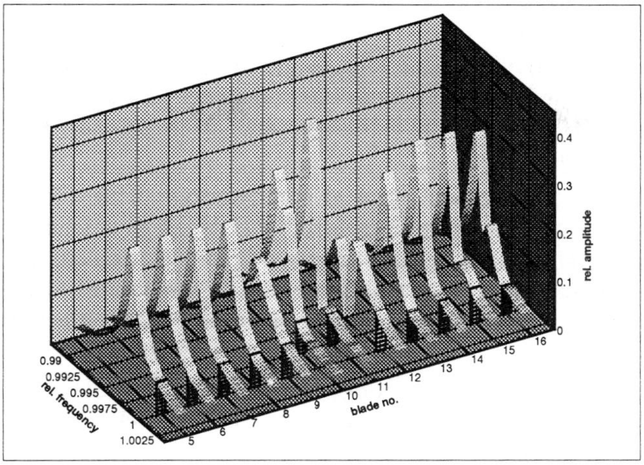

Fig. 9. Low pressure turbine vane amplitudes, vanes 10 and 11 mistuned by -0.25%, φ = 180 deg.

Fig. 10 depicts the results of a variation of excitation EO over the whole range of possible interblade phase angles. The layout of the figure is identical to *Fig. 6* for the compressor rotor, hence only the maximum amplitude for each magnitude of mistuning is plotted. As for the compressor rotor, the mistuning in no case reduces the maximum amplitude arising in the cascade. However, the increase in maximum amplitudes varies strongly with interblade phase angle. Another point worth noting is that the maximum amplitudes remain almost constant for mistuning magnitudes below -0.5% and above +0.5%. This range corresponds approximately to the spread of the tuned mode aeroelastic eigenfrequencies shown in *Fig. 7*. Looking back on *Fig. 2* and *Fig. 6*, a similar behavior can be seen for the compressor rotor. It could be suspected that the influence of mistuning and aerodynamic coupling is strongest, when the mistuning magnitude remains in the range that is spanned by the aeroelastic eigenfrequencies of the tuned cascade.

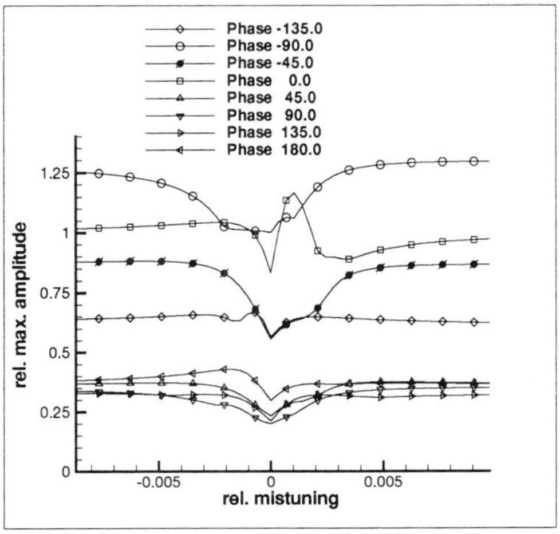

Fig. 10. Low pressure turbine vane amplitudes, 2 vanes mistuned by -1% through +1%, variation of excitation EO.

6. Concluding Remarks

The method described in this paper allows the designer to assess the effect of mistuning on the resonant amplitudes of turbomachinery bladings. Under some simplifying, but realistic assumptions, the amplitude distributions in a cascade with given, small mistuning can be calculated very quickly.

Once the aerodynamic coefficients for a given eigenmode of the blades are evaluated, many mistuned configurations can be analyzed at little computational cost. This makes the method applicable for parametric variations or even for Monte-Carlo simulations of stochastically mistuned bladings. For bladings with closely spaced eigenfrequencies, an extension of the method to account for multiple blade eigenmodes is fairly straightforward.

The examples shown for a transonic compressor and a low pressure turbine show large deviations from the tuned mode behavior even if only a single blade is mistuned. The influence of mistuning and aerodynamic coupling seems to be strongest, when the mistuning magnitude remains in the range that is spanned by the aeroelastic eigenfrequencies of the tuned cascade.

As real turbomachinery bladings will always be mistuned, these results show the necessity to include mistuning effects in a forced response analysis to obtain realistic estimates of the resonant amplitudes and stresses in the design phase. In practical applications of the method, it may turn out to be a bigger problem to obtain a realistic estimate of the distribution and magnitude of the mistuning than to calculate the resonant response from this data.

7. References

Hall, K. C., Lorence, C. B, 1992, "Calculation of Three-Dimensional Unsteady Flows in Turbomachinery Using the Linearized Harmonic Euler Equations", *ASME paper 92-GT-136.*

Kahl, G., Klose A., 1993, "Computation of Time Linearized Transonic Flow in Oscillating Cascades", *ASME paper 93-GT-269.*

Kahl, G. , 1995, "Application of a Time Linearized Euler Solver to Aeroelastic Problems in Turbomachinery", *ASME paper 95-GT-123.*

Kielb, R. E.; Kaza, K. R. V, 1983, "Aeroelastic Characteristics of a Cascade of Mistuned Blades in Subsonic and Supersonic Flow", *Trans. ASME J. Vibration, Acoustics, Stress and Reliability in Design,* Vol. 105, pp. 425 - 433.

Kielb, R. E.; Chiang, H. D., 1992, "Recent Advancements in Turbomachinery Forced Response Analyses" *AIAA paper 92-001.*

Pierre, C. et al., 1994:, "Localization of Aeroelastic Modes in Mistuned High-Energy Turbines", *Journal of Propulsion and Power*, Vol. 10, No. 3 pp. 318 - 328.

Saxer, A. P., Giles, M. B., 1990, "Inlet Radial Temperature Redistribution in a Transonic Turbine Stage", *AIAA paper 90-1543.*

Srinivasan, A.V., Tavares, G. G., 1994, Direct Use of Unsteady Aerodynamic Pressures in the Flutter Analysis on Mistuned Blades, *ASME paper 94-GT-347.*

WHOLE-ASSEMBLY FLUTTER ANALYSIS OF A LOW PRESSURE TURBINE BLADE

A.I. SAYMA, M. VAHDATI, J.S. GREEN* AND M. IMREGUN

Imperial College of Science Technology and Medicine
Mechanical Engineering Department
Exhibition Road, London SW7 2BX, UK
**Rolls-Royce plc, PO Box 31, Derby DE2 8BJ, UK*

Abstract

This paper reports the findings of a flutter investigation on a low pressure turbine blade using a 3D non-linear integrated aeroelasticity method. The approach has two important features: (i) the calculations are performed in a time-accurate and integrated fashion whereby the structural and fluid domains are linked via an exchange of boundary conditions at each time step, and (ii) the analysis is performed on the entire bladed-disk assembly, thus removing the need to assume a critical interblade phase angle. Although such calculations are both CPU and in-core memory intensive, they do not require pre-knowledge of the flutter mode and hence they allow a better understanding of the aeroelasticity phenomena involved.

The flow is modelled inviscidly but the steady-state viscous effects are accounted for using a distributed loss model. The structural model was obtained from a standard FE eigensolution and a large number of assembly modes were included in the calculations. The study was focused on a part-speed condition at which a number of unstable modes were known to exist from the available experimental data. The whole assembly was modelled using about 650,000 nodal points and predictions were made of aeroelastic modal time histories. From those time histories, it was possible to identify the forward and backward traveling waves and to deduce the unstable modes of vibration. The theoretical predictions were found to be in very good agreement with the experimental findings.

T.H. Fransson (ed.), Unsteady Aerodynamics and Aeroelasticity of Turbomachines, 347-359.

1. Introduction

Modern turbomachines operate under very complex flow regimes where a mixture of subsonic, transonic and supersonic regions coexist. Shock wave interactions and shock wave boundary layer interactions lead to a complex flow structure, the modelling of which needs to be both computationally-efficient and numerically accurate. The accuracy requirement becomes even more important in flutter calculations, where the position of the shock is the most single important factor[1, 2]. With recent advances in internal compressible flow modelling and increased computational power, it is now possible to undertake both steady and unsteady flow analyses of very complex turbomachinery geometries[10]. Although the use of unstructured meshes is becoming increasingly common for turbomachinery applications, most of the past impetus has come from external compressible flows[11, 12].

A literature survey on flutter prediction methods is beyond the scope of this paper and the interested reader should consult [13]. However, a very brief overview will be given for sake of completeness. Most flutter computations consider a typical sector vibrating in some given assembly mode (or interblade phase angle) for which flutter is expected to occur. In other words, the flutter mode must be known before the analysis, though it is also possible to consider the individual stability of each mode in turn. In such analyses, the interblade phase angle must be prescribed at the periodic boundaries. Although such an approach has relatively modest CPU and in-core memory requirements, it also has a number of serious shortcomings:

- The flutter frequency of the aeroelastic system must be assumed to be identical to the natural frequency of the structural system.
- No interaction can take place between the various assembly modes.
- The approach is valid only when the aeroelastic system has reached a periodic steady-state condition. In most cases, extra computations are needed to impose the required interblade phase angle pattern over several cycles, a feature that increases the computational cost.
- If the flutter mode is not known in advance, all possible interblade phase angles need to be tried as a separate computation.

In the present work, it is proposed to undertake a whole-assembly analysis in order to overcome the problems above. The non-linear aeroelasticity method described in [3] will also be used here. Such an approach allows the inclusion of all assembly modes and hence removes the need to assume a critical interblade phase angle. In other words, the unstable modes are found as a result of the computation and not provided as part of the input data.

2. The flow model

Although the flow model is described in detail in ref [3], a brief summary will be given here for the sake of completeness.

2.1. GOVERNING EQUATIONS

The flow equations will be given for a three dimensional Cartesian coordinate system which is fixed to a rotating blade. In this case, the conservation form of the unsteady, Reynolds-averaged Navier-Stokes equations is given by:

$$\frac{\partial \mathbf{U}}{\partial t} + \frac{\partial \mathbf{F}_i}{\partial x_i} = \frac{\partial \mathbf{G}_i}{\partial x_i} + \mathbf{I}; \tag{1}$$

where $\mathbf{U}$ is the solution vector, $\mathbf{F}_i$ and $\mathbf{G}_i$ are the inviscid and viscous fluxes respectively and $\mathbf{I}$ contains the terms due to the rotation of the coordinate system. These terms can be written as:

$$\mathbf{U} = \begin{bmatrix} \rho \\ \rho u_i \\ \rho \epsilon \end{bmatrix}; \quad \mathbf{F}_i = \begin{bmatrix} \rho(u_j - w_j) \\ \rho u_i(u_j - w_j) + p\delta_{ij} \\ \rho E(u_j - w_j) + P u_j \end{bmatrix}; \quad \mathbf{G}_i = \begin{bmatrix} 0 \\ \sigma_{ij} \\ u_k \sigma_{ik} + \kappa \dfrac{\partial T}{\partial x_i} \end{bmatrix}; \tag{2}$$

were ρ is the density, ϵ is the specific total energy, u_i is the absolute velocity vector, w_i is the contra variant velocity vector, κ is the summation of the molecular and eddy diffusivity. The viscous stress tensor σ_{ij} is given by:

$$\sigma_{ij} = \mu \left(\frac{\partial u_i}{\partial x_j} + \frac{\partial u_j}{\partial x_i} \right) - \frac{2}{3} \delta_{ij} \mu \frac{\partial u_k}{\partial x_k} \tag{3}$$

where μ is the summation of the molecular and eddy viscosities.

The term $\mathbf{I}$ of (1) is given by:

$$\mathbf{I} = \begin{bmatrix} 0 \\ 0 \\ \rho \Omega u_3 \\ \rho \Omega u_2 \\ 0 \end{bmatrix} \tag{4}$$

where Ω is the rotational speed. The above system of equations is complemented by the perfect gas equation :

$$p = (\gamma - 1)\rho \left(\epsilon - \frac{u^2}{2} \right) \tag{5}$$

where γ denotes the ratio of the fluid specific heats.

The Eddy viscosity is calculated using the one-equation turbulence model of Baldwin and Barth [5].

2.2. NUMERICAL METHODOLOGY

The spatial solution domain is discretized using linear tetrahedral elements. The equations are discretized by first obtaining a variational weak formulation. Then the variables are represented over the discrete domain via piecewise linear interpolation shape functions. The resulting central difference scheme is numerically stabilised using a fourth order dissipation scheme for smooth flows. The matrix dissipation scheme used follows that proposed by Swanson and Turkel [6]. A more dissipative scheme is required near discontinuities such that unwanted numerical oscillations can be prevented. This is achieved by switching to a second-order dissipation scheme in such situations. The switching between the two schemes is determined using a pressure-based sensor[3].

The semi-discrete system of equations is advanced in time using an implicit time integration. The resulting system of non-linear equations is solved using Jacobi iterations and their final form is given by:

$$\mathbf{K}(\mathbf{U})\mathbf{\Delta U} = \mathbf{RHS_i} + \mathbf{RHS_v} \tag{6}$$

where $\mathbf{K}(\mathbf{U})$ is given in [7], **RHS** is the right hand side evaluated using the latest values of the solution vector, subscripts i and v referring to the inviscid and viscous contributions respectively.

2.3. THE LOSS MODEL

A viscous simulation of a full assembly will have very substantial computational requirements so that detailed parametric studies are not yet feasible. Consequently, a distributed loss model is used to represent the steady-state viscous fluxes and the unsteady viscous effects are inherently assumed to be negligible. Despite this limitation, the method is expected to be applicable to a wide range of turbomachinery operating conditions.

In general, a loss model assumes that the effect of the shear stresses on the motion can be represented by a distributed friction force, the resulting flow representation being still of inviscid nature, but not isentropic [8]. The success of the loss model in approximating the true viscous flow will largely depend on the definition of the distributed friction force. The simplest representation consists of an averaged shear stress formulation using semi-empirical data, such models being commonly used for multi-stage turbomachinery flows in connection with quasi-three dimensional approximations [9]. In the present work, the loss model is defined numerically using data from a previous viscous analysis of a single blade-passage on a much finer mesh. The viscous solution is then interpolated onto the coarser mesh which is used for an inviscid analysis of the whole-assembly. The viscous stresses of the first solution are thus saved on the second mesh and used throughout the unsteady inviscid solution as a source terms.

3. The structural model

The structural model is based on a linear modal model, the mode shapes and natural frequencies being obtained via standard FE analysis techniques. The mode shapes are interpolated from the structural analysis mesh onto the aerodynamic mesh as the two discretization levels are unlikely to be coincident.

The structural part of the aeroelastic equations of motion are uncoupled by using the mode shape matrix ϕ.

$$\frac{d^2\eta_r}{dt^2} + \omega_r^2\eta_r = \sum_{i=1}^{N} \phi_{i,r}.F_i \tag{7}$$

where r is the mode index, i is the node number, η is the modal deflection, N is the number of surface nodes and ω is the natural frequency. F is the aerodynamic force which is obtained using:

$$F_i = p_i \delta A_i \mathbf{n}_i \tag{8}$$

where p denotes the static pressure, δA denotes the elemental area and $\mathbf{n}$ denotes the unit normal vector, all at node i.

Equation (7) is advanced in time using the Newmark-β method. Boundary conditions from the structural and aerodynamic domains are exchanged at each time step and the aerodynamic mesh is moved to follow the structural motion.

4. Numerical results

The above method was used for the flutter analysis of an early-development-standard low-pressure turbine for which some experimental observations were available at 88%, 94% and 99% of maximum shaft speed conditions. As discussed previously, a steady-state viscous solution was obtained first for a single blade passage. Unstructured triangular-element meshes were generated at several radial stations and these were connected to produce the final mesh of tetrahedral elements with 85,275 points. Special attention was paid to have a good distribution of the mesh points in the axial and tangential directions while keeping the resolution in the radial direction to a minimum. The boundary layer mesh was generated using an advancing-normal method. A view of the surface mesh is shown in Fig. 1 while a zoom view of the hub section is given in Fig. 2. A viscous steady-state solution was obtained for the 99% speed case and the results are given in Figs 3 & 4 as Mach number and pressure contours at mid-section. The predicted flow results were compared to available measured data and good agreement was observed in all cases.

The viscous solution obtained on the fine mesh was interpolated onto an inviscid single-blade-passage mesh of about 8,000 points. The single passage mesh was then expanded to a whole-assembly mesh of about 664,000 points. A view of a sector of this mesh is shown in Fig. 5.

A total of 15 assembly modes, namely nodal diameters 3 to 12, 15, 20,25, 33 and 42 for the first flap mode, were included in the flutter calculations. These are listed in Table 1 for the three speeds considered. Because of computational considerations, the aeroelastic time histories were computed for 0.32 seconds for the 99% speed, for 0.1 seconds for 94% speed and 0.2 seconds for 84% speed. The calculations were started by giving initial modal velocities to some of the assembly modes, analogous to hitting a blade and monitoring whether the response is decaying or growing in time. Typical unsteady pressure contours are shown in figure 6 and examples of stable and unstable traveling wave modal time histories given in figures 7 and 8.

For each time history, it is possible to calculate a logarithmic decrement (logdec) value which is a linear measure of the decay rate between two given cycles in the time history. In general, the time histories do not show linear behaviour and care should be taken in interpreting the results obtained using the linear logdec tool. Two important observations can be made.

1. Significantly different logdec values may be obtained by using different sections of the time histories. Therefore, it may not always be possible to characterise the aeroelastic behaviour with a single equivalent damping value.
2. A study of the individual modal time histories seems to indicate that interactions take place between the different assembly modes.

In any case, logdec values were computed for the available time histories for all three speeds and the results are summarised in figure Fig. 10. The instability is seen to be in the 6 to 12 nodal diameter range, the 94% speed case being worst, at which the 8ND mode is the least stable. These findings are in very good agreement with experimental observations which indicate instability at the same speed condition and for the same modes. Finally, the actual displacement time history at a point near the blade's tip is shown in

Nodal Diameter	Frequency(Hz)		
	88%	94%	99%
3	171.095	171.096	174.774
4	190.831	187.528	191.253
5	220.147	213.492	217.241
6	250.463	240.562	244.393
7	281.761	268.919	272.861
8	316.007	300.771	304.807
9	354.185	337.317	341.404
10	396.407	378.855	382.942
11	441.981	424.893	428.927
12	489.259	474.066	478.001
15	607.398	606.689	610.314
20	670.106	672.203	676.120
25	690.428	689.416	693.351
33	706.216	703.028	706.899
42	711.670	707.913	711.749

TABLE 1. Structural frequencies at % of maximum shaft speed

Fig. 11. The instability can easily be inferred from the beating-like motion whose time history is showing an overall growing trend.

The findings are in good agreement with the experimental observations since flutter was measured in 7, 8 and 9 nodal diameters. Furthermore, mechanical damping was not included in the calculations, a feature that explains why more modes were predicted to be unstable by the numerical model.

5. Concluding Remarks

1. A 3D integrated non-linear flutter prediction method has been presented and applied to a low-pressure turbine. Numerical results have been found to be in very good agreement with experimental observations.
2. All modes of interest were included in one computation and no assumptions about critical interblade phase angles were necessary. The results showed non-linear time histories, highlighting the difficulties of using equivalent linear damping for stability purposes. Interactions between the various assembly modes were also observed.
3. The computational cost for such 3D non-linear calculations is high, both in CPU seconds and memory requirements. Therefore, standard

whole-assembly calculations are likely to be limited to inviscid flow modelling for the next few years, though the first whole-assembly viscous calculations should appear well before the end of the Century.

Acknowledgments

The authors would like to thank Rolls-Royce plc for both sponsoring the work and allowing the publication of this paper.

References

1. M. E. Goldstein, W. Braun and J. J. Adamczyk(1977), Unsteady flow in supersonic cascade with strong in-passage shocks, *J. Fluid Mech.* **83**, 569-604.
2. O. O. Bendiksen(1986), Role of shocks in transonic/supersonic compressor rotor flutter, *AIAA Journal* **24**, 1179-1186.
3. M. Vahdati and M. Imregun(1995), Non-linear aeroelasticity analyses using unstructured dynamic meshes, In Y. Tanida and M. Namba (eds.), *Unsteady Aerodynamics and Aeroelasticity of turbomachines*, Elsevier Pubbblishers, pp. 177-190.
4. K. Morgan, J. Peraire and J. Peiró, Unstructured grid methods for compressible flow, *AGARD Report 787*, Special Course on Unstructured Grid Methods for Advection Dominated Flows.
5. B. S. Baldwin and . J. Barth(1991), A one equation turbulence transport model for high Reynolds number wall-bounded flows, *AIAA Paper* 91-0610.
6. R. C. Swanson and E. Turkel(1992), On central difference and upwind schemes, *J. Comp. Phys.* **101**, 292-306.
7. M. Vahdati, K. Morgan and J. Peraire(1993), The computations of viscous compressible flows using an upwind algorithm and unstructured meshes, *Computational Non-linear Mechanics in Aerospace engineering*, AIAA Progress in Aeronautics and Astronautics series.
8. Ch. Hirsch(1995), *Numerical computation of Internal and external flows Volume I* John Wiley & Sons.
9. Ch. Hirsch and H. Deconincki(1985), Through flow models in turbomachines: stream surface and passage averaged representations, In S. A. Ucer, P. Stow and Ch. Hirsch (eds.), *Thermodynamics and fluid Mechanics of turbomachinery, Vol I*, NATO ASI Series.
10. W. N. Dawes(1988),Development of a three dimensional Navier-Stokes solver for application to all types of turbomachinery, ASME Paper No. 88-GT-70,
11. A. Jameson, T. J. Baker and N. P. Weatherill(1986), Calculation of inviscid transonic flow over a complete aircraft, AIAA Paper 85-0103.
12. Peraire, J. Peiró, and K. Morgan(1993), Finite Element multigrid solution of Euler flows past installed aero-engines, *Computational Mechanics* **11**,433-451.
13. Marshall, J. G. & Imregun, M.(1996), A survey of aeroelasticity methods with emphasis on turbomachinery, *J. of Fluids and Structures.*

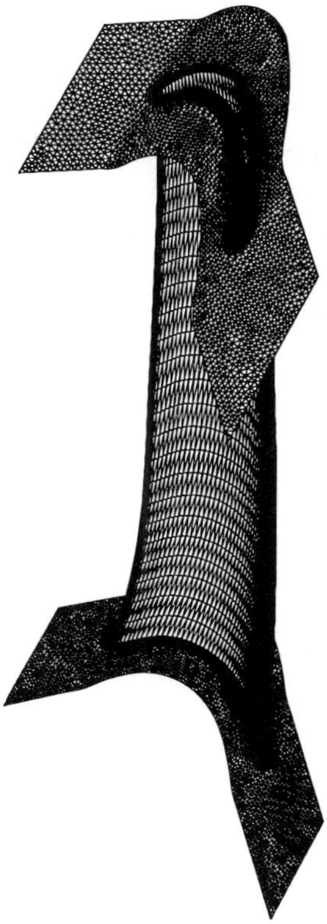

Figure 1. View of the viscous flow mesh used

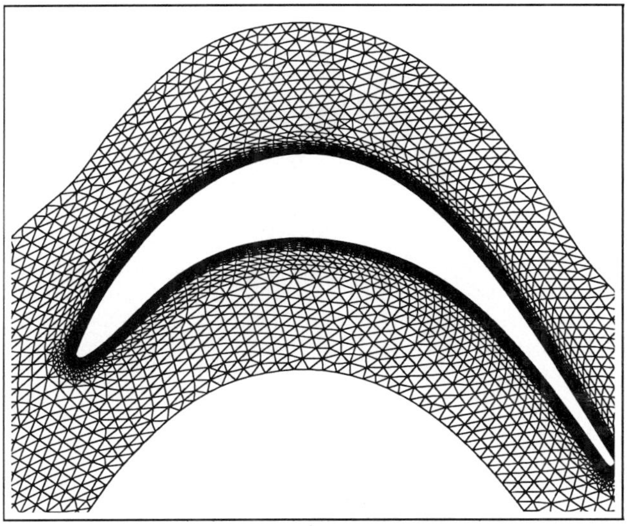

Figure 2. Close up of the viscous mesh at the hub section

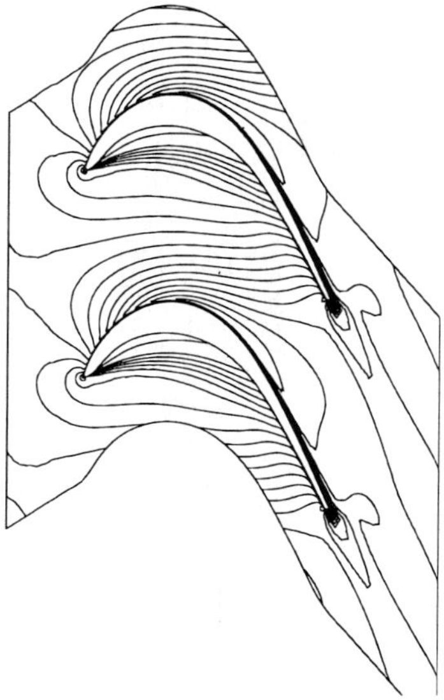

Figure 3. Steady-state Mach number contours at mid-section

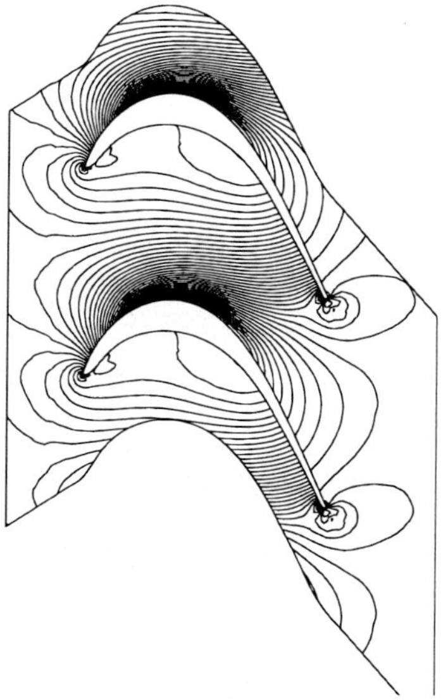

Figure 4. Steady-state pressure contours at mid-section

Figure 5. A view of a sector of the whole assembly mesh

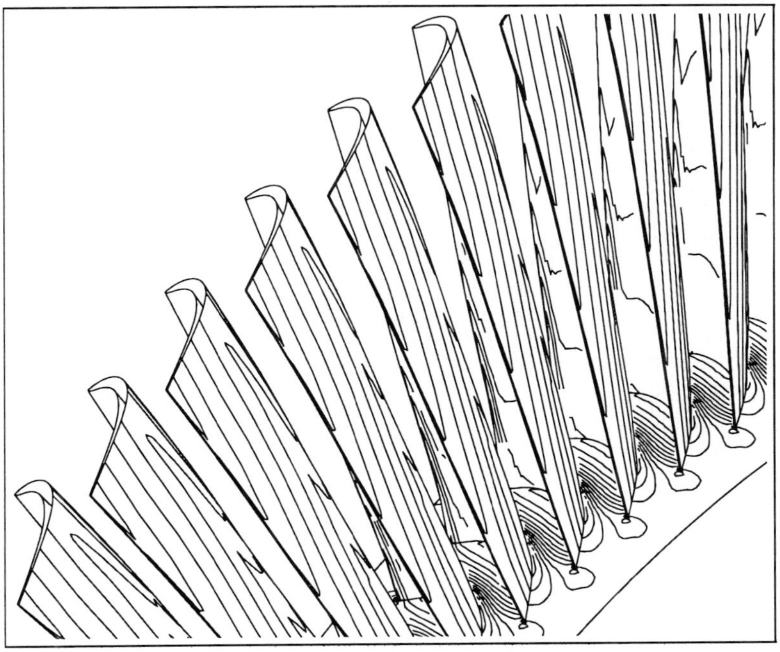

Figure 6. Unsteady pressure contours on the turbine blades

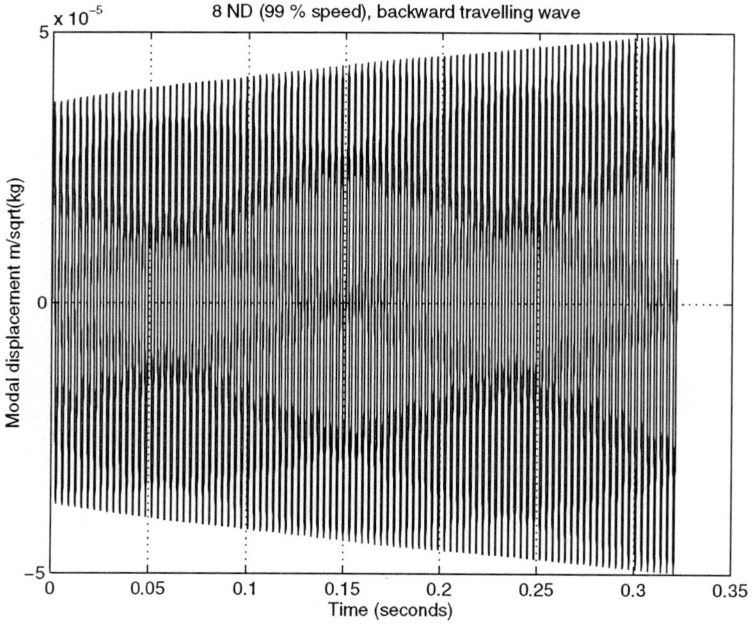

Figure 7. Modal displacement for an unstable mode

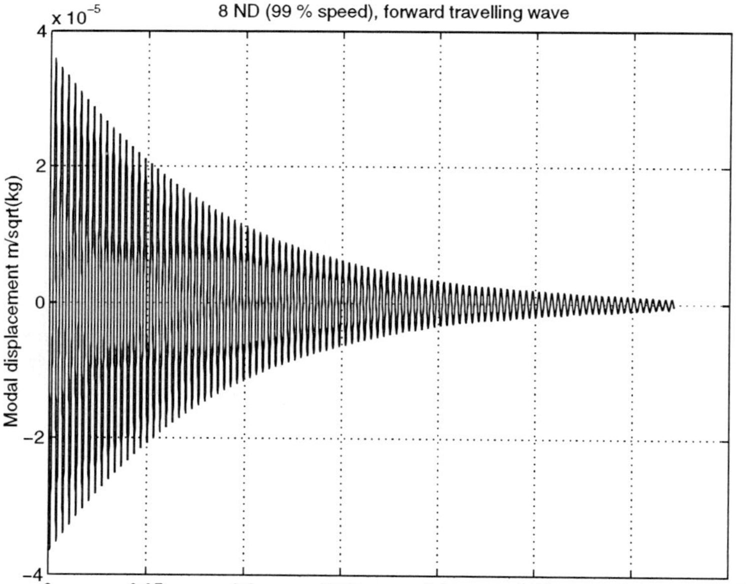

Figure 8. Modal displacement for a stable mode

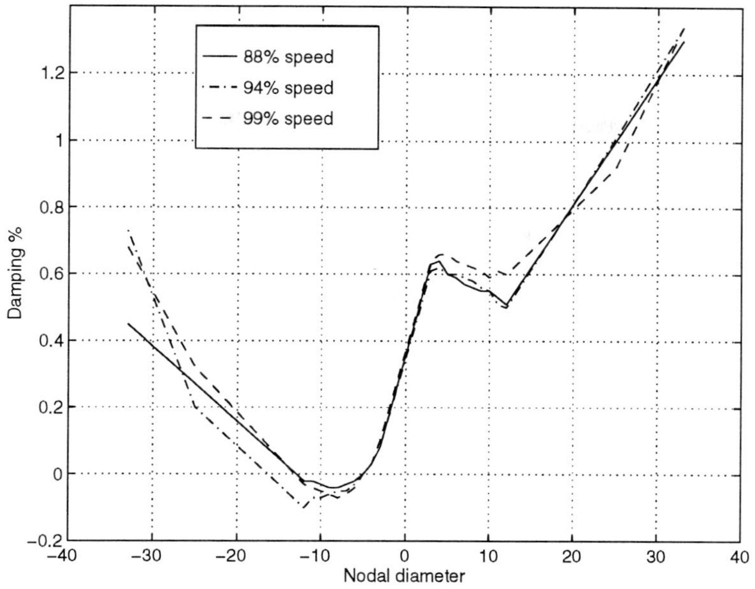

Figure 9. Damping vs. nodal diameter for the three speeds

7: Experimental Blade Row Interactions

PASSAGES OF PRECEDING BLADE WAKES AND THEIR EFFECTS IN THE ROTOR PASSAGES OF A SINGLE STAGE AXIAL-FLOW FAN

Tsutomu ADACHI
Department of Mechanical Engineering
for Transportation, Faculty of Engineering,
Osaka Sangyo University,
3-1-1, Nakagaito, Daito, Osaka , JAPAN(574)

Yutaka YAMASHITA
Hitachi LTD
Hitachi Works

1.INTRODUCTION

In this paper, three dimensional unsteady velocity were measured in the moving blade passage of a single stage axial-flow fan with a stationary blade in front of fan wheel. In order to measure three dimensional unsteady velocity distributions in the moving blade row, a specially designed hot wire traversing apparatus was used. A hot wire probe which was 45° inclined to the stem was used. In the moving blade passage, measurements were made in 6 axial positions, in 11 radial positions and in 20 circumferential positions. Rotational speeds were 600, 1500 and 2400 rpm and the opereating conditions were three points on the decreasing inclination zone on the pressure-discharge curve. Effects of secondary flow, leakage through the tip clearance on the unsteady flow phenomena in the rotor passage were discussed.

NOMENCLATURES

R: Radial direction
Θ:Circumferential direction
Z: Axial direction
FP1: Maximum discharge
FP2: Design point
FP3: Maximum pressure

2. EXPERIMENTAL APPARATUS AND METHOD OF MEASUREMENT

2.1 EXPERIMENTAL APPARATUS

Table 1 and Fig.1 show the specifications and main dimensions of a

T.H. Fransson (ed.), Unsteady Aerodynamics and Aeroelasticity of Turbomachines, 363-377.

single-stage axial flow fan used for this research. It has a comparatively small type number of K=2.06 for an axial-flow fan. The flow pattern of this fan is a free vortex type. The moving and stationary blades are made of duralmin and machined precisely by a numerically controlled milling machine. It can be operated in a wide range of 300 to 3000 rpm.

For measurements, a hot wire probe with a thickness of 5 μ m tungsten wire which was 45° inclined to the stem, as shown in Fig.2, was used. Measurements were made 4 times at the same measuring positions.

In order to traverse a hot wire probe in the circumferential direction during rotation, a traversing apparatus as shown in Fig. 3 was used. This apparatus was composed of a main gear, a main body, two stepping motors with pinions, two hot wire supports and a supporting plate. It rotated together with the fan wheel. If the stepping motor is rotated during the rotation of the fan wheel, then the main body shifts to a position relative to the shaft. The angle of rotation was 0.135° per each step of the stepping motor. The radial position of the probe was adjusted in the static state. The feed current to the stepping motor and the output signals from the hotwire probe were sent through the slip-ring that was attached at one end of the shaft.

2.2 MEASUREMENTAL METHOD

Figure 4a,b shows the measuring positions in the blade passage. Figure 4a shows the positions in the axial and radial directions. In the axial direction the leading and the trailing edges at the hub are designated by Z=0 and Z=1.0 respectively. Dividing the axial distance from Z=0 and Z=1.0 by 5, we get Z=0.2 , 0.4, 0.6, 0.8. To show the radial positions, the values of the radial distance from the hub was divided by the distance from the hub and the tip, and then was multiplied by 100. Figure 4b shows positions in the radial and circumferential directions at Z=0.4 and Z=1.0. The circumferential positions were taken at every 2° from the trailing edge of the hub in the counter rotating direction at the outlet from the moving blade row.

Measurement begins at the point where the leading edge of the one moving blade and the trailing edge of the stationary blade met at the Z-direction at the tip (Fig.5). The sampling period was 0.5ms for 600 rpm, 0.2 ms for 1500 rpm, and 0.12 ms for 2400 rpm. 60 samplings were taken and were summed up 2^9 times, therefore random components were excluded. Measurements were done 4 times at the same measuring positions with the same hot wire probe in the 4 different directions as shown in Fig. 2. The magnitude and the direction of velocity were calculated from 4 signals taken at the same positions. The calculating method was shown in the reference.

The operating conditions were chosen at the maximum discharge (FP1), the design point(FP2), and the maximum pressure point(FP3), as shown in Fig.6.

Rotational speeds were at 600, 1500 and 2400 rpm.
Experimental results on the unsteady phenomena in the moving blade passage, that was synchronized with the movement of moving blade were shown and unsteady flow were discussed.

3. UNSTEADY FLOW IN THE MOVING BLADE PASSAGE

3.1 FLUCTUATING FLOW IN THE Θ-ZPLANE

Figure 7 and 8 show fluctuating velocity distributions in the circumferential and axial-plane(Θ-Zplane) in the progress of time. In these figures, arrows show fluctuating velocity components in the Θ-Z plane and • makers show those in the radial direction, positive in the upward direction from the measuring positions. The ⊘ makers show the minimum axial velocity point and these correspond to the wake center. In the plane near the tip, leakage from the tip clearance arises. The interference of leakage flow and wake show unperiodical disturbances. In the plane midway between the tip and hub and near the hub, the passages of wake can clearly be recognized. The radial outward flow in the wake and larger axial flow arise in the passage.

3.2 FLUCTUATING FLOW IN THE RADIAL AND CIRCUMFERENTIAL-PLANE (R-Θ PLANE).

Figure 9 and 10 show fluctuating velocity distributions in the radial and circumferential directions. In these figures arrows show fluctuating velocity components. The ⊘ makers show the positions of the minimum axial velocity, and they correspond to the center of the wakes. The radial outward flow across the wake surface can be seen. The effect of tip leakage flow have some unsteady effects. This effect is more remarkable at FP3 than at FP2 and FP1. Circulating flow arise in the rotor passage as a result of these flow. The wake surfaces are distored as they flow downstream. The effect from these distortions is more remarkable with an increase in the rotational speed, i.e., Reynolds number.

3.3 FLUCTUATING FLOW VELOCITY AND ANGLE NEAR THE OUTLET OF MOVING BLADE ROW

Figures 11 and 12 show fluctuating flow angle in the Θ-Z plane near the trailing edge of the rotor blade. In these figures, abscissa show rotating angle of the rotor and ordinate shows flow outlet angle. Figure 11 was taken midway between hub and tip and near the trailing edge of the moving blade. The fluctuating velocity and flow angle show periodical variation with the period of passing of a stationary blade pitch (27.7°). Figure 12 was taken near the tip. Some irregular components are contained in this figure. These may come from leakage from the tip clearance.

Figures 13a,b,c show the contours of amplitude of fluctuating flow outlet angle in the R-Θ plane near the outlet of the moving blade. In these figures ordinate shows radial positions between the hub and tip, and abscissa show circumferential positions of one pitch of moving blade, as shown in Fig. 4b.

Figures 14a,b,c show the contours of amplitude of fluctuating flow outlet angle in the Θ-Z plane near the outlet of the moving blade. The tendency of fluctuating outlet angle in the R-Θ and in the Θ-Z plane are similar. Figures 15a,b,c show the contours of amplitude of fluctuating velocity (Δ V/V) in the Θ -Z plane near the outlet of the moving blade. The tendency of these fluctuatig values i.e., fluctuating flow outlet angle and fluctuating flow velocity show some similarity and it also similar to the steady loss distribution across the rotor passage. The fluctuating amplitudes take the largest values in the corner of suction surface and hub. The fluctuating values also take large values near the tip midway between moving blades. It corresponds to the effect of the tip leakage. They correspond to domain of large secondary loss. The variations of these values with rotational speeds may be considered to be smaller with an increase in rotational speed. For the operating condition FP3 the domain of the maximum loss shifts to some other place because of the effect of flow disturbance.

4. CONCLUSIONS

In this paper, measurements of three dimensional unsteady velocity were made for various operating conditions, rotational speeds in the various positions of rotor passage. Wake passage of the preceding stationary blade were distored by the effect of rotor blade. The distribution is also effected by leakage flow from the blade tip. The outlet velocity and flow angle of rotor blade also fluctuate by the influence of preceding stationary blade row. The amplitude of fluctuation takes a large value at the corner of suction and hub surfaces and midway between moving blades near the hub.

REFERENCES

1. Adachi, T., Tomisawa, M., Hino, T., Yamashita, Y., Kawai, T., and Matsuuchi, K., Study on the Characteristics and Flow in an Axial-Flow Fan- Effect of Reynolds - Number(in Japanese), Turbomachinery, 6(1995), 330-3 47.

Table1 Specification and main dimensions of the fan used in this research

Discharge coefficient	0.507
Pressure coefficient	0.550
Type number	2.06
No. of revolution	300~3000 rpm
No. of stationary blades	13
No. of moving blades	10
Outer diameter	0.5 m
Hub ratio	0.6
Flow pattern	Free vortex Type

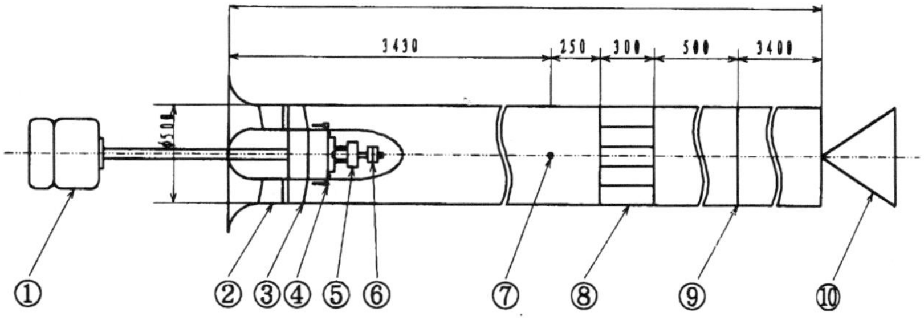

① Motor ② Stationary blade ③ Moving blade
④ Hot wire probe ⑤ Traversing apparatus
⑥ Slip ring ⑦ Static pressure hole
⑧ Honey comb ⑨ Pitot tube ⑩ Damper

Fig. 1 Single stage axiall flow fan

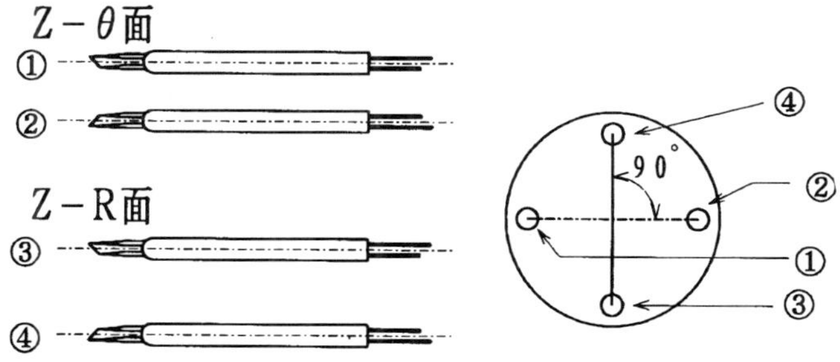

Fig. 2 Hot wire probe

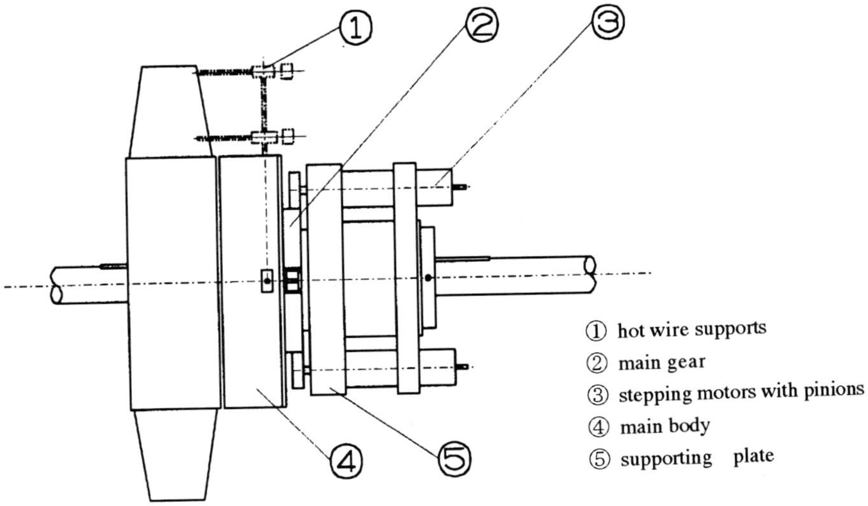

Fig. 3 Traversing apparatus

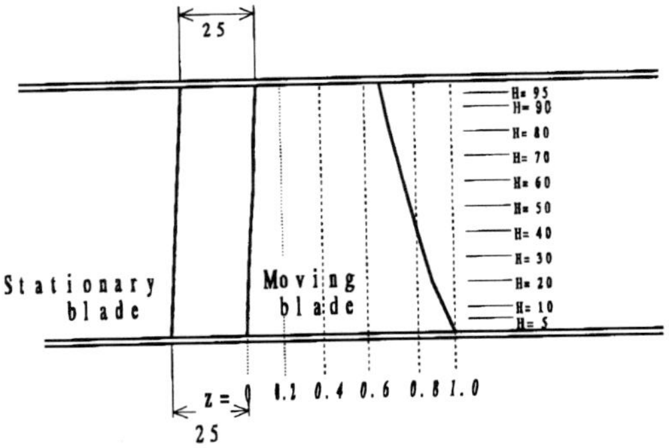

Fig. 4a Measuring positions in the blade passage in the axial and radial direction

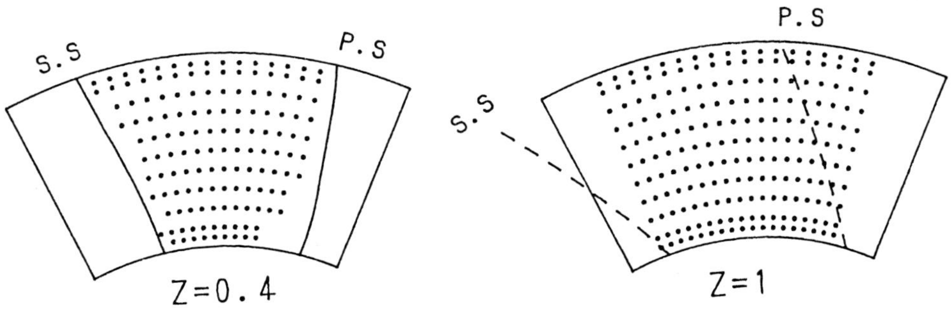

Fig. 4b Measuring positions in the blade passage in the radial and circumferential direction

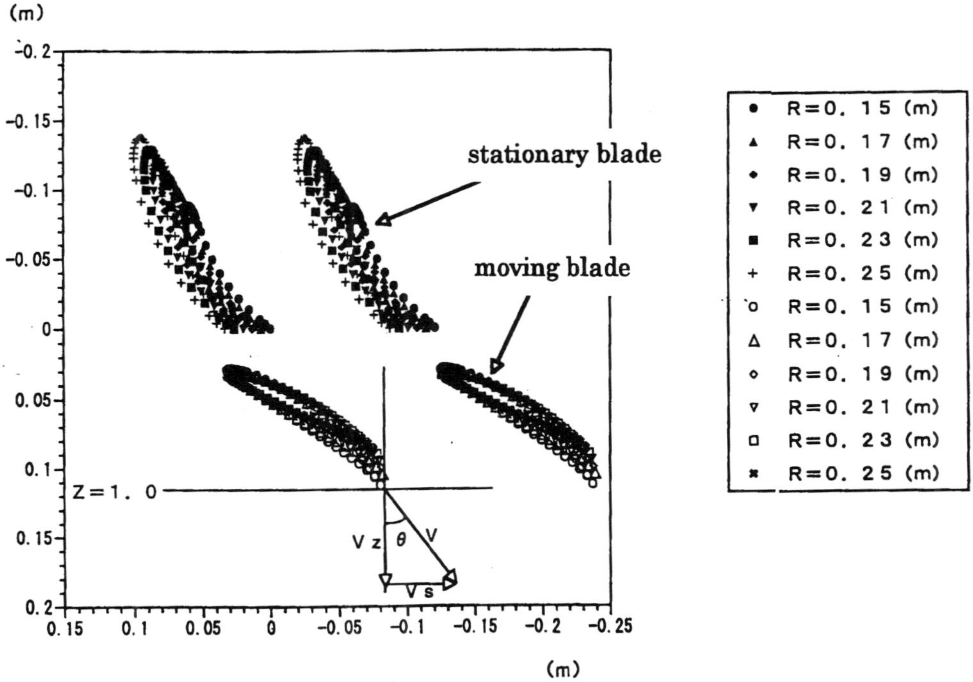

Fig. 5 The stationary and moving blade row at the start of measurement (t=0) viewed in the radial direction

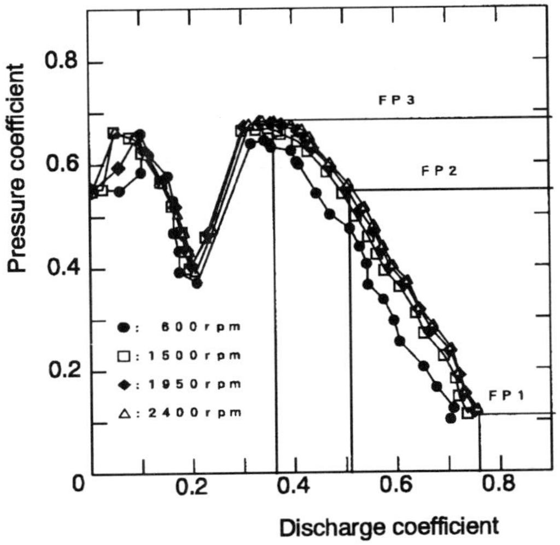

Fig. 6 Operating conditions

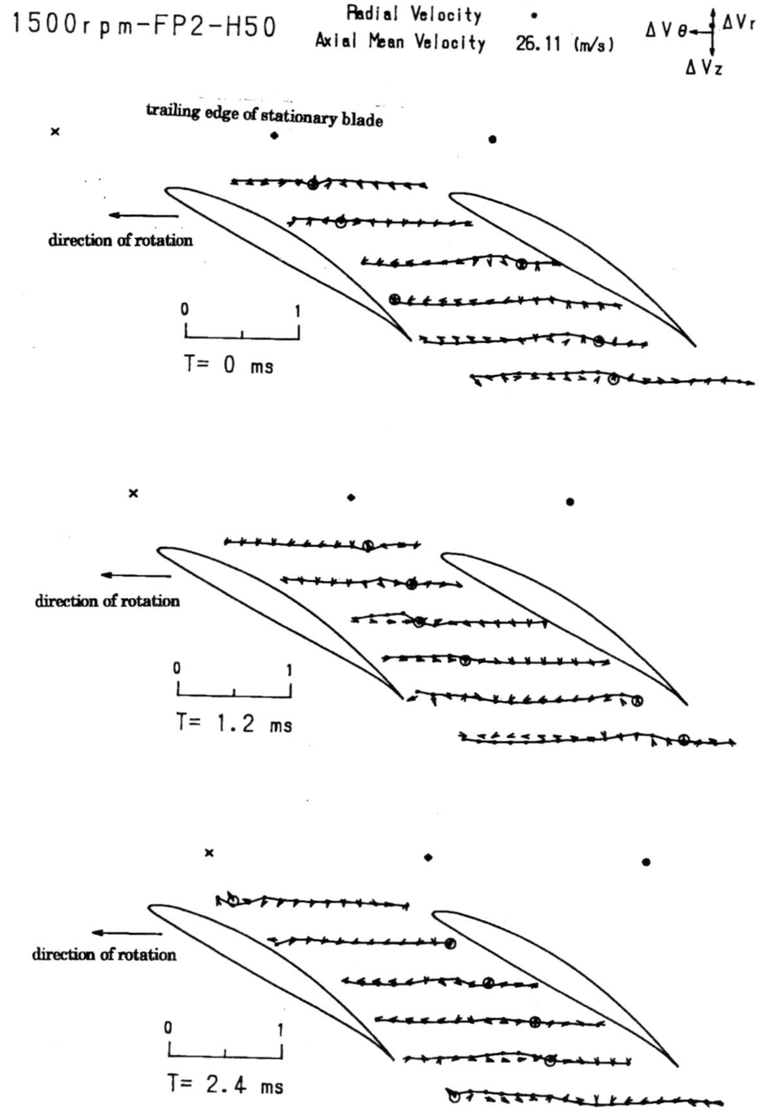

Fig. 7 Fluctuating velocity distributions in the circumferential and axial plane (Θ-Z planme) in the progress of time, 1500 rpm, FP2, H50

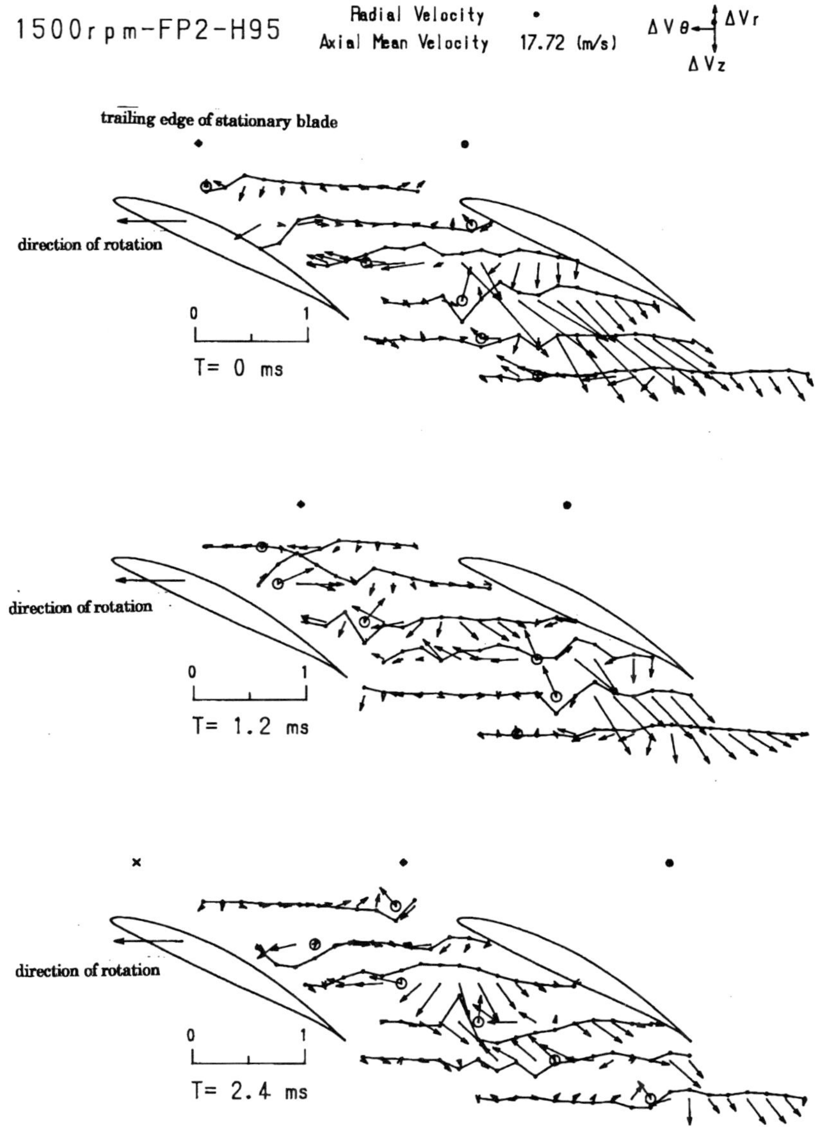

Fig. 8 Fluctuating velocity distributions in the circumferential and axial plane (Θ-Zplane), in the progress of time, 1500 rpm, FP2, H95

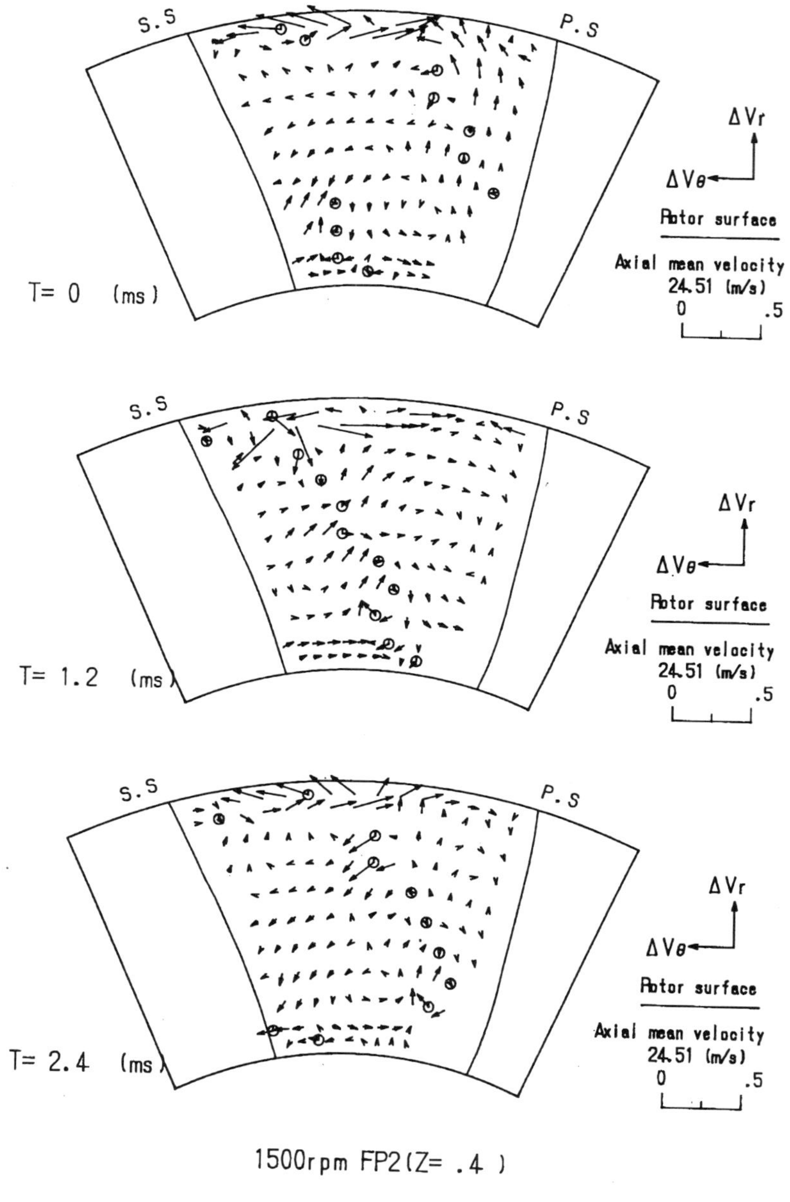

Fig. 9 Fluctuating velocity distributions in the radial and circumferential plane (R-Θ plane), in the progress of time, 1500 rpm, FP2, Z=0.4

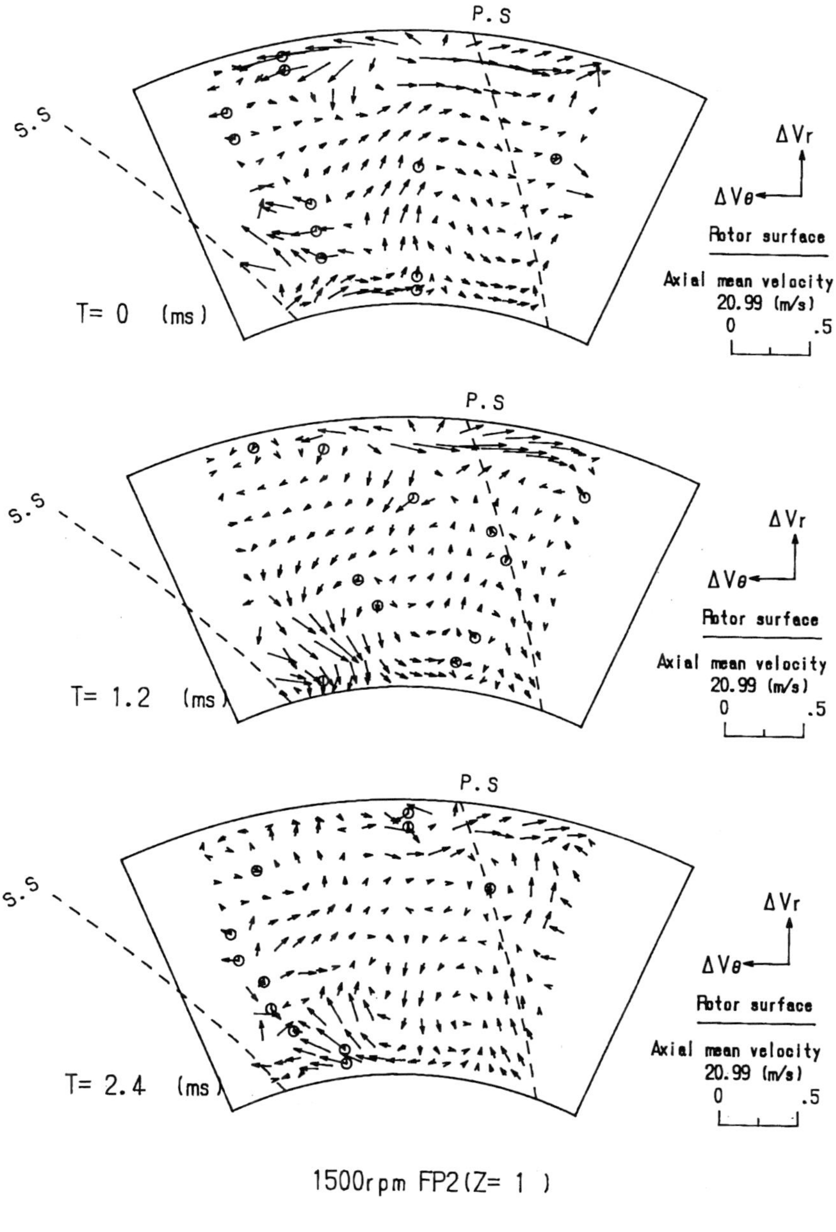

Fig. 10 Fluctuating velocity distributions in the radial and circumferential plane (R-Θ plane), in the progress of time, 1500 rpm, FP2, Z=1.0

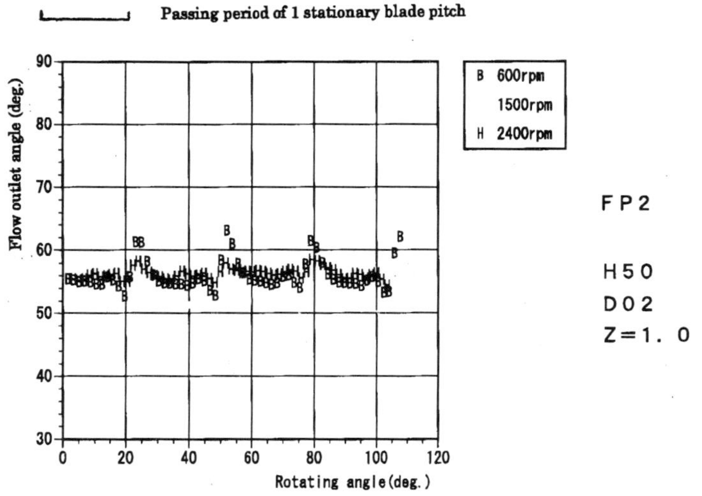

Fig. 11 Fluctuating flow angle in the Θ-Z plane near the trailing edge of moving blade H50

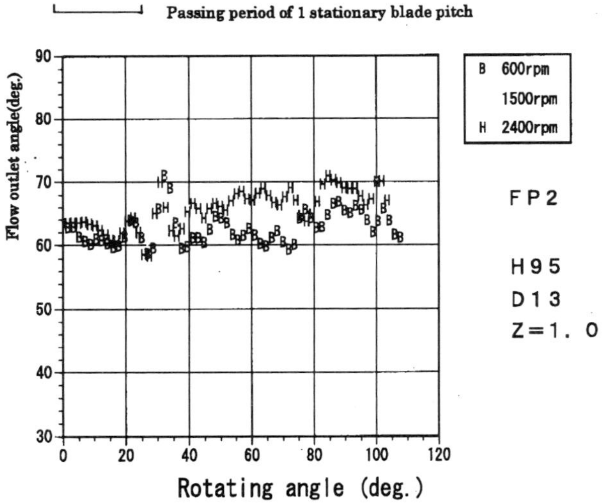

Fig. 12 Fluctuating flow angle in the Θ-Z plane at outlet of moving blade Z=1.0, H95

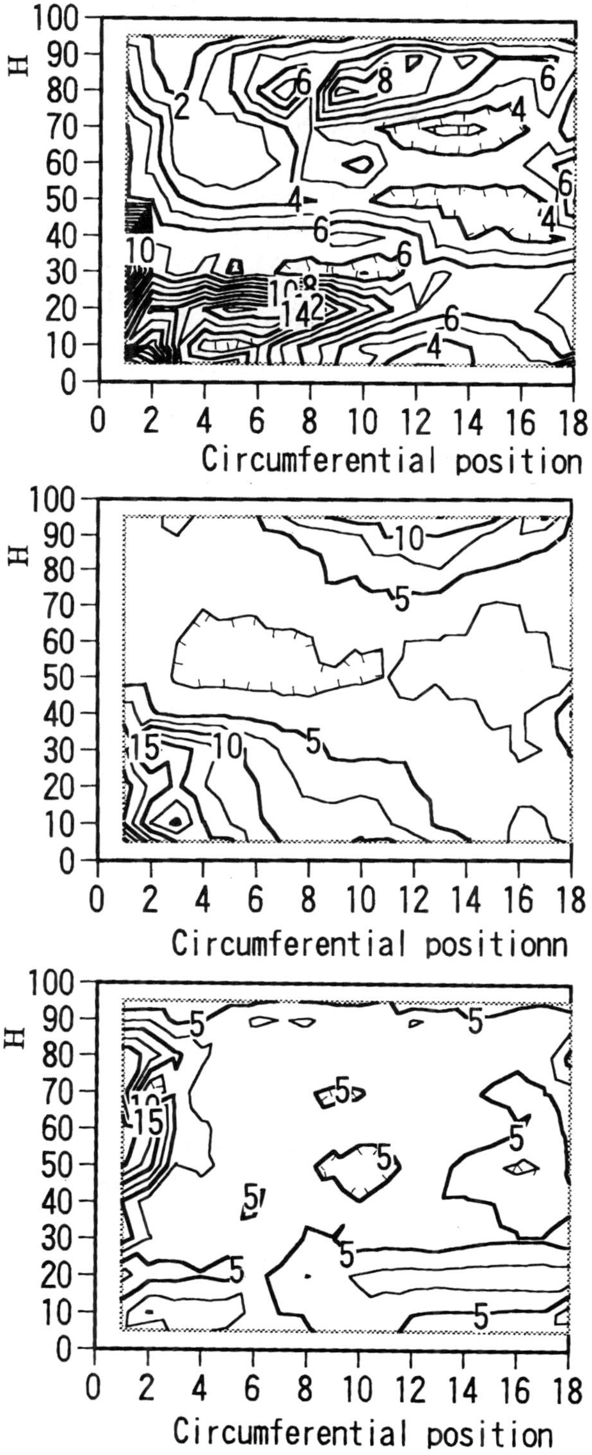

a. 1500 rpm, FP2

b. 2400 rpm, FP2

c. 1500 rpm, FP3

Fig.13 Contours of amplitude of fluctuating flow outlet angle of moving blade in R-Θ plane

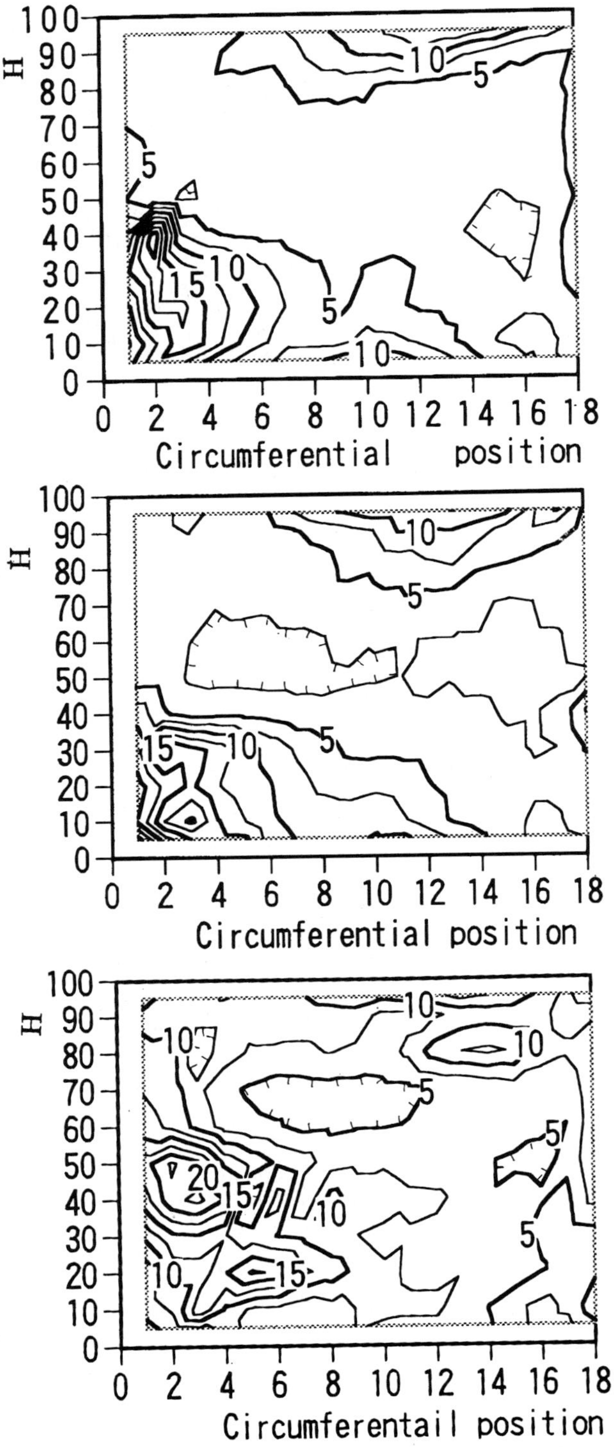

a. 1500 rpm, FP2

b. 2400 rpm, FP2

c. 1500 rpm, FP3

Fig.14 Contours of amplitude of fluctuating flow outlet angle of moving blade in Θ-Z plane,

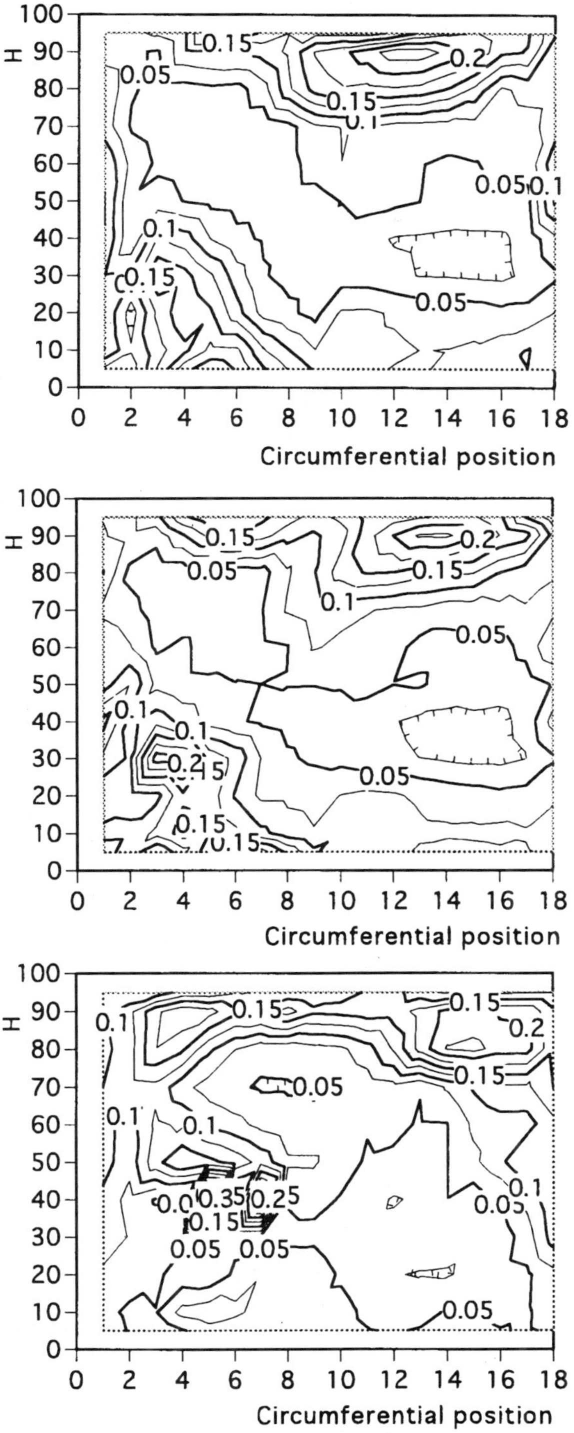

a. 1500 rpm, FP2

b. 2400 rpm, FP2

c. 1500 rpm, FP3

Fig. 15 Contours of amplitude of fluctuatiing out let velocity (Δ V/V) of moving blade in Θ-Z plane,

Forced Response Vibrations of a Low Pressure Turbine Due to Circumferential Temperature Distortions

Steven R. Manwaring
Kevin L. Kirkeng
GE Aircraft Engines
Cincinnati, OH USA 45215

Abstract

This paper presents the results of an experimental and analysis program which investigated the forced response of a low pressure turbine (LPT) blade due to circumferential temperature distortions which emanated from the combustor and propagated through the entire turbine system. Two–per–rev temperature variations of 350°C were measured at the LPT inlet and 300°C at the LPT exit. Total pressure fluctuations of 4% of the steady value were generated in the turbine system due to these temperature distortions. These distortions generated entropic and vortical waves which gave excitations large enough to give unsteady loadings and stresses on the 5th stage LPT blade sufficient to cause HCF failures for a disk/blade/tip–shroud system mode crossing. Using the measured temperature and pressure variations at the aft of the LPT, an analysis system was developed for predicting the stresses on the LPT blade. The predicted stress at the strain gage locations were within 20%–30% of the measured values, thus demonstrating the capability of determining the impact of circumferential staging of combustor fuel nozzles on turbine system blading vibrations. Therefore, in designing turbine blading, mode crossings of counts from combustor staging should be avoided or analysis similar to the above should be performed to ensure acceptable tolerances on blade vibrations.

1.0 Introduction

Significant circumferential temperature distortions in turbine blading are becoming increasingly prevalent due to the use of multi–dome combustors to reduce nitric–oxide (NOX) emissions. During certain operations of the gas–turbine engine, one or two of the domes will have portions of fuel nozzles around the circumference turned off, giving circumferential variation of total temperatures of nearly 300° C. This variation persists through the many rows of the high and low pressure turbine blading causing distortions in the flow and thus excitation of low order system modes of vibration.

Over the past decade or two measurements and analyses of blade unsteady loading or vibratory responses due to many different sources of excitation have been performed. Fleeter and his colleagues (1978–1997) have performed numerous measurements of unsteady pressure loadings on compressor and turbine bladerows due to wakes from upstream airfoils and struts. Gallus and his colleagues (1980) have also extensively measured the unsteady loadings in compressor blading due to upstream airfoil wakes. Manwaring and Wisler (1992) measured and analyzed unsteady loads on compressor and turbine blades due to wakes and potential fields. Fleeter and Manwaring (1990) and Manwaring, et.al. (1996) determined from measurements and analyses the impact of inlet distortions on first stage compressor blading. O'Brien, et.al. (1983), Chiang and Turner (1992) and Markland, et.al., (1996) have measured and analyzed the unsteady loading and vibratory responses of fan blades due to downstream strut potential flows. Dunn has

T.H. Fransson (ed.), Unsteady Aerodynamics and Aeroelasticity of Turbomachines, 379-392.

worked closely with Delaney (1992–1994) to measure and analyze the unsteady pressures on turbine nozzles and blades resulting from shocks.

There has been, however, no publications documenting measurements or analysis of the unsteady loading or vibratory response where the excitation source is circumferential temperature variations. Therefore, the objective of this work is to determine if temperature variations would cause excitations strong enough to result in vibratory stresses which would result in blade failures. A parallel objective is to be able to predict the blade vibratory levels due to these temperature distortions. The approach is to measure the circumferential distortions obtained in a turbine system of a gas turbine engine with and without combustor indexing of fuel nozzle operation and then, measure the resulting vibratory stresses on a downstream blade. The indexing for this case is to have the nozzles operating to form a two–per–rev pattern around the circumference. The distortion is then analyzed to determine boundary conditions for blade row unsteady loading and vibratory stress predictions.

2.0 Test and Analysis Program

A testing and analysis program is described for determining the impact of temperature variations emanating from the combustor on the fifth stage blade of the low–pressure turbine in the gas turbine engine shown in Figure 1.

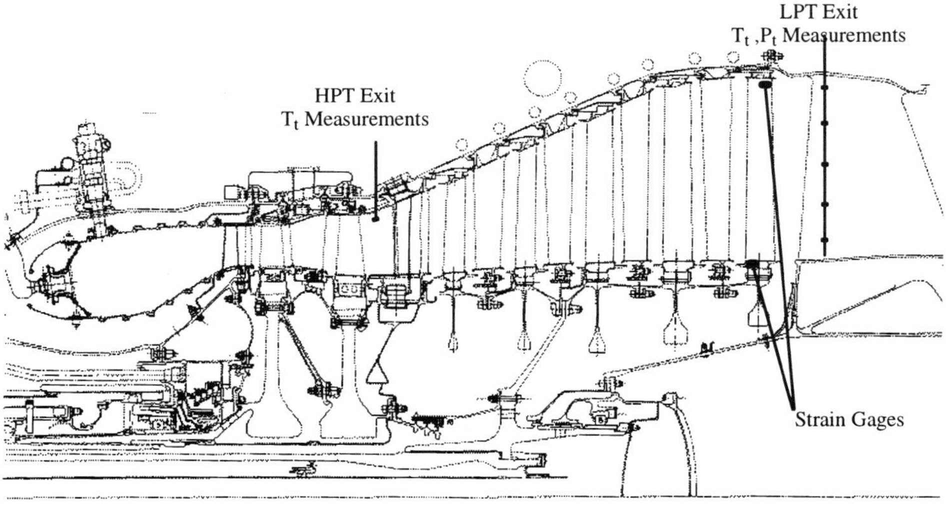

Figure 1. Schematic of Combustor and Turbine System for Engine Tests

2.1 Test Program

A test series was performed to obtain back–to–back measurements of the circumferential distortions existing in turbine system and the resulting 5th stage LP blade vibrations. The two tests were with fuel nozzles operating with normal and circumferentially distorted flow patterns as shown in Figure 2. The latter of these two conditions gives a two–per–rev circumferential distortion pattern in total temperature and pressure. Figure 3 schematically shows the Campbell diagram which highlights the bladed–disk system mode crossings with the 2/rev occurring near maximum speed.

The circumferential variations in temperature are measured at the exit of the high–pressure turbine (inlet to the low pressure turbine) as shown in Figure 1 using eight temperature probes located around the circumference at 95.8% span. Circumferential variations of total temperature as well as total pressure are measured at the LP turbine exit using radial five–element rakes, as shown in Figure 1. The rakes are located in four locations over half the circumference, such that one period of the two nodal diameter distortion is measured.

The fifth stage of the LP blade was instrumented with strain gages to determine blade vibratory stresses and deflections. 2 gages were located on the blade and its shank for five different blades. The finite element model (FEM) mesh shown in the analysis section of the paper gives the locations of the strain gages. The blade has a double tang dovetail for attachment to the disk and a shroud at the tip. Table 1 below describes the blading shape parameters.

Table 1. LPT Stage 5 Blading Design Parameters

Midspan Solidity	1.35
Midspan Camber	74.9°
Midspan Chord	4.52 cm
Midspan Stagger	26.1°
Aspect Ratio	6.32
No. of Airfoils	98

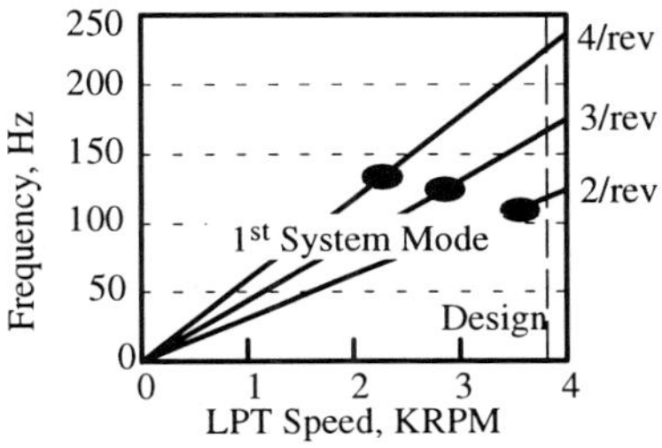

Figure 3. Campbell Diagram of Crossings

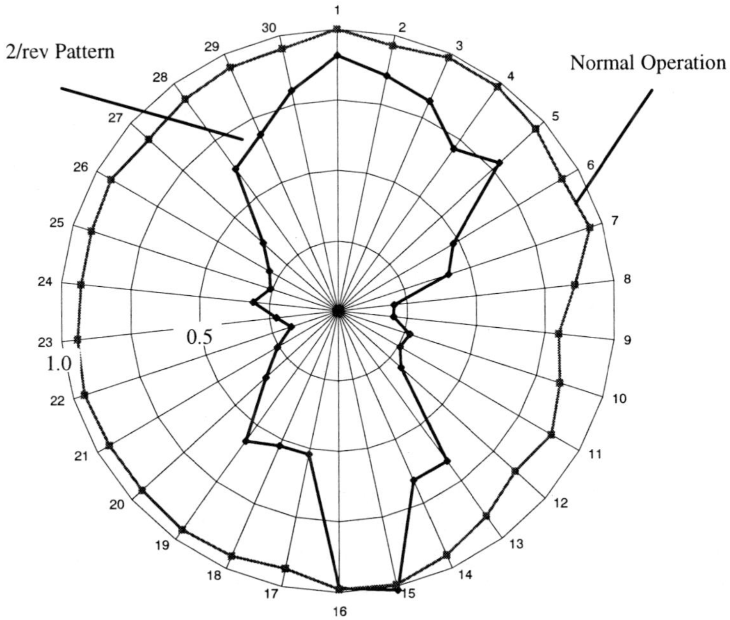

Figure 2. Normalized Fuel Flows of Combuster Nozzles

2.2 Analysis Program

A predictive system was developed to analyze the measured circumferential distortions due to temperature variations in order to set the boundary conditions for a forced response prediction system. The forced response system consists of unsteady aerodynamics calculations, cyclic–symmetry finite element modal analysis, mapping of FEM predicted deflections to the aerodynamics analysis and modal integration, and finally single–degree–of–freedom response calculations to obtain blade stresses.

The following diagram describes the process developed to generate unsteady flow conditions for the fifth stage blade from the measurements of circumferential total pressure and temperature obtained at the LPT exit.

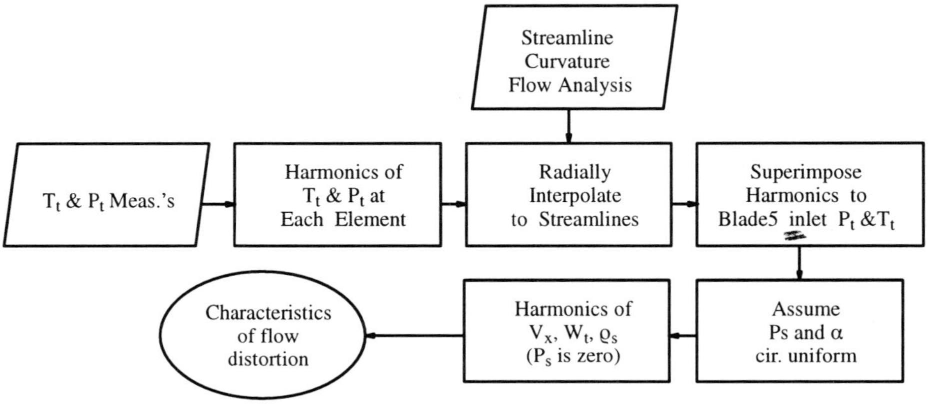

Figure 4. Analysis Process of Distortion Measurements at LPT Exit

A through–flow analysis was used to determine the steady flow conditions at the inlet to stage five of the LPT. The harmonic fluctuations about the mean values of total temperature and pressure measured at the blade exit (LPT exit) were assumed to remain constant across the blade. The static pressure and the absolute gas angle aft of the fifth stage nozzle (blade inlet) were assumed to be uniform circumferentially. Thus, at the blade inlet, by using the through–flow predicted total and static pressure, gas angles and total temperature, plus the harmonic fluctuations of P_t and T_t, harmonic variations of axial and tangential velocities and static density were calculated. The harmonic unsteadiness in primitive flow variables (density, velocities and static pressures) are then used to define the unsteady flow characteristics per the work of Giles (1988) and Hall and Crawley (1989). These characteristics were then used as the boundary conditions for the linearized unsteady Euler analyses described below.

The unsteady aerodynamics due to the distortions (gust response) and the blade motion (aerodynamic damping) were determined using a linearized unsteady Euler solver developed by Holmes, Mitchell and Lorence (1997). The grid shown in Figure 5 defines the inviscid mesh used to perform the non–linear steady and linear unsteady calculations. 113 grid lines were used in the flowwise direction, 33 in the blade–to–blade direction and 33 in the spanwise direction. The trailing edge was modeled as a cusp to reduce errors in the inviscid calculations about the actual rounded geometry.

Predictions of the modeshapes and were obtained using NASTRAN cyclic symmetry solution with a blade/shank/shroud finite element model defining the airfoil and attachments; Figure 6 shows the mesh. A total of 6560 elements were used, 20 across the span of the airfoil, 10 from leading to trailing edge and one through the blade; roughly 6000 were used in the blades shank definition. The modeshapes are mapped to the Euler grid defined above in order to modally integrate the blade surface unsteady pressures to obtain the modal force, $\overline{F}$. Also the mapped deflections are used to perform an aerodynamic damping calculation using the linearized Euler solver.

The vibratory stresses were obtained by applying the single degree of freedom equation for a spring–mass–damper modal system (Vierck, 1979).

$$\overline{Q} = \frac{\overline{F}/K}{2\zeta} \qquad (1)$$

where $\overline{Q}$ is the modal amplitude, K is the blade stiffness and ζ is the blade damping. The stiffness is obtained from $K{=}\omega_n^2{*}M$, where the modal mass, M, was set to unity for the finite element analysis. The blade damping was estimated by applying the half power rule to the engine strain gage measurements: $\zeta = 2\Delta f/f_o$, where Δf is the frequency delta at the blossom half power ($1/\sqrt{2}$ times peak amplitude) and f_o is the natural frequency at the peak amplitude. The predicted stresses on the blade are obtained by multiplying the FEM stresses by the modal amplitude, $\overline{Q}$.

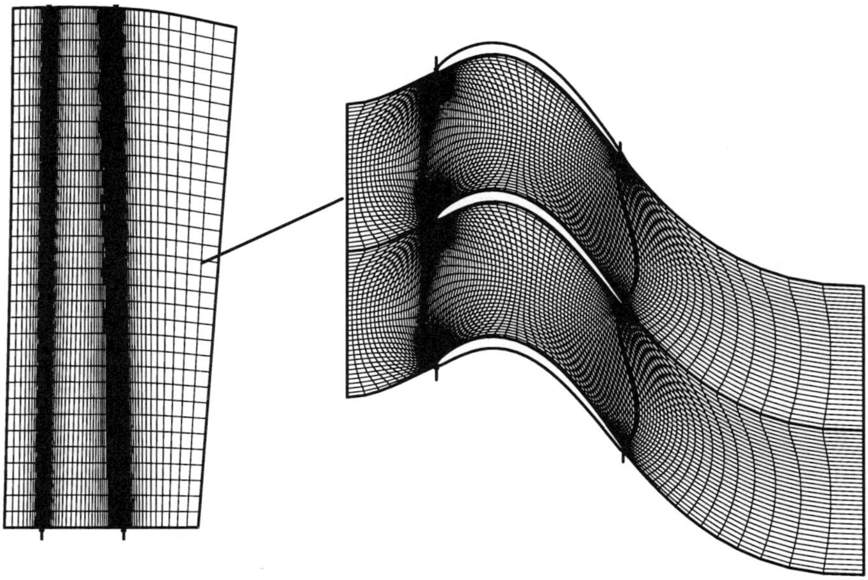

Figure 5. Elliptic H–Grid used for Euler CFD Calculations

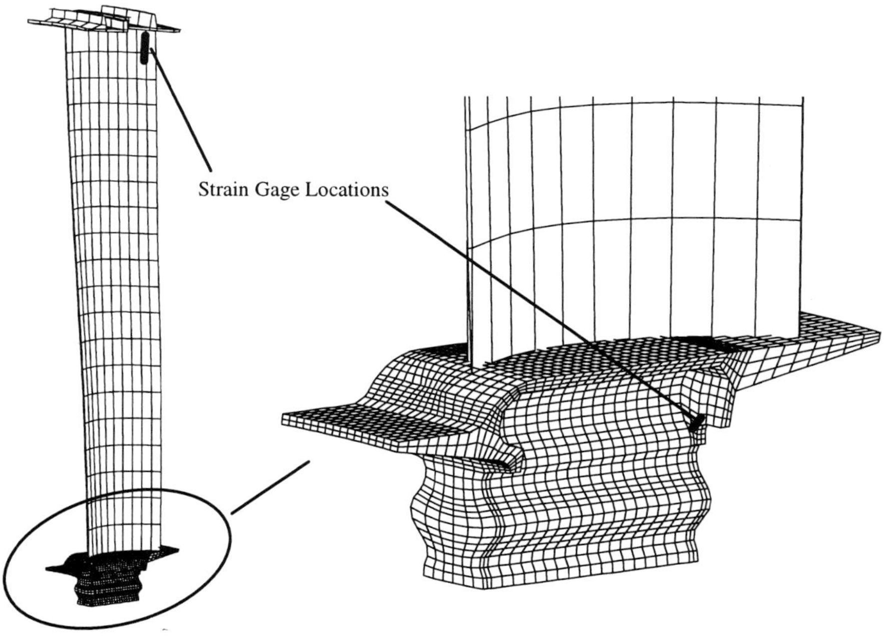

Figure 6. Mesh for Finite Element Model of Airfoil and Attachment Region

3.0 Data and Analysis Results

This section presents the results of the data and analysis of the circumferential distortion measurements due to the temperature variations and the resulting strain gage responses on the 5th stage of the LP turbine blade. The definitions of the distortions will be given first, followed by the results of the modal analysis using the finite element model. Next the unsteady aerodynamic response due to the distortions will be given. Lastly, the comparison of the predicted and measured strains on the 5th stage of the LP turbine blade will be reported.

3.1 Circumferential Distortions Due to Temperature Variations

Measurements of the circumferential variation in total temperature into and out of the LP turbine were obtained. Also, at the exit of the LP turbine the total pressures were measured. Figure 7 shows the 2/rev temperature variation at the LPT inlet and exit. Peak–to–peak variations of up to nearly 350° C (600° F) were measured at approximately 95% span at the LP inlet. This temperature variation remains relatively unchanged through the five stages of the LP blading as shown by the nearly 300° C (500° F) variation measured across the span at the LP exit.

The variation in absolute total pressure due to the temperature distortions generated at the combustor is shown in Figure 8, which plots the variation in total pressure from the average and then normalized by the average. Peak–to–peak variations of up to 4% of the steady total pressure occur in the outer half of the annulus span. In the inner half of the span the variations are minimal. The assumed cause of the total pressure fluctuations is from the rotor due different amounts of work depending on the inlet temperature conditions around the circumference.

The circumferential variations in total temperatures and pressures at the LPT exit are Fourier decomposed to obtain conditions for the linearized unsteady analyses. Figure 9 shows the spanwise profiles of the amplitude of the first four harmonics for both the total temperatures and pressures. For the temperature, the circumferential variation is predominated by the first harmonic (fundamental, 2/rev) while for the pressures, both the first and second harmonics are significant. The harmonic amplitude show once again that the temperature variations are nearly constant across the span while the pressure fluctuations are stronger in the OD region and decrease with decreasing span.

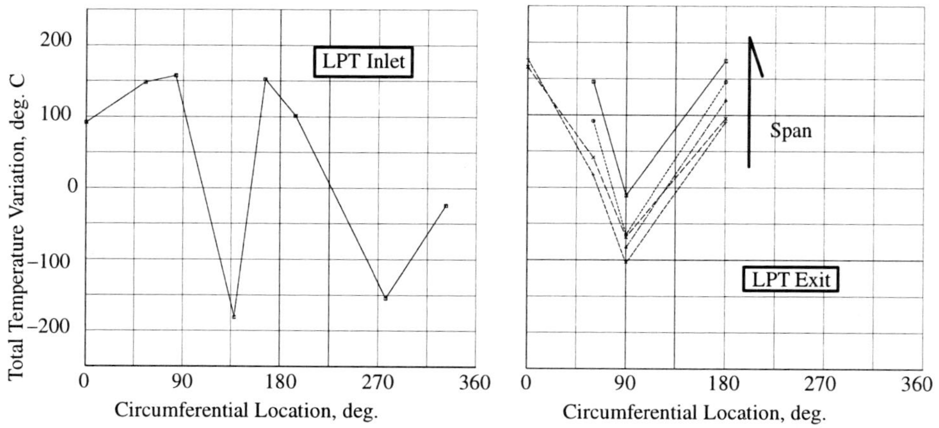

Figure 7. Temperature Variations at the 5–stage LPT inlet and exit

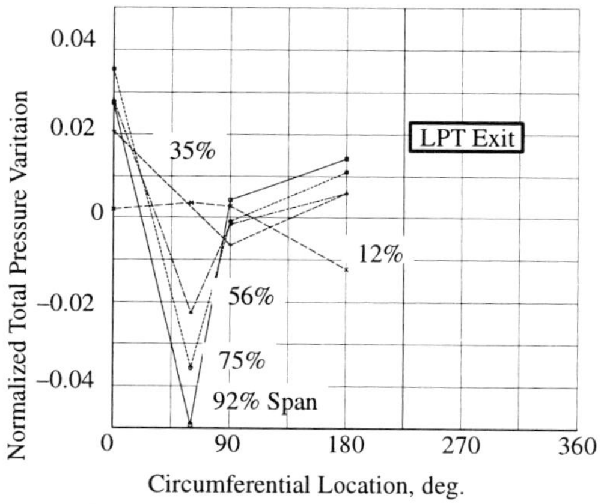

Figure 8. Total Pressure Variations at the 5–stage LPT exit

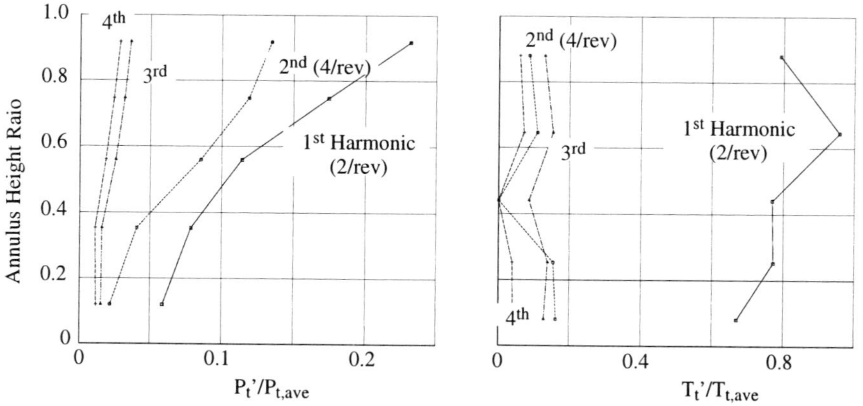

Figure 9. Total Pressure and Temperature Harmonics at the 5–stage LPT exit

The 2D characteristic definitions of the distortion excitation of the 5th stage LPT blade is shown in Figure 10. The flow defect is composed of two types: an entropic wave generated largely by the temperature variations and a vortical wave generated predominantly by the total pressure variations. The entropy wave is generally 30% stronger than the vortical wave. The acoustic wave is nearly zero due to the assumption in the data analysis of zero circumferential variation in static pressure.

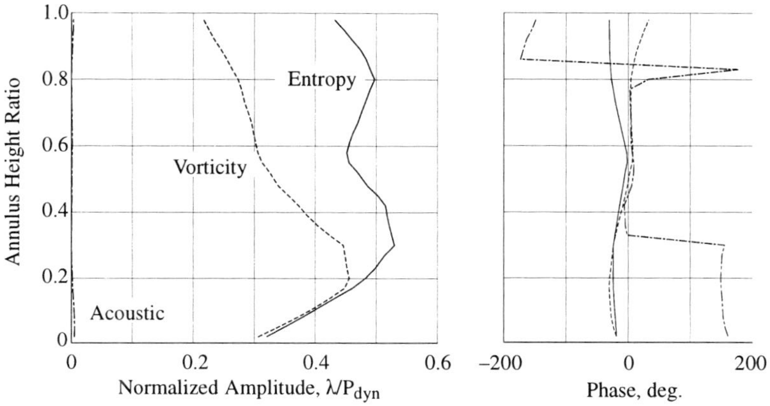

Figure 10. 2D characteristic definitions of distortion defect on the 5th Stage Blade

3.2 Modal Analysis

The results of the cyclic symmetry finite element model were used to determine the modeshapes and stress distributions for the 5th stage LPT blade. The two–nodal diameter modeshape of the shank/airfoil/tip–shroud model can be seen by the contour plots of tangential deflections shown in Figure 11. The deflections are predominantly in the outer half of the blade with the mode being a combination of torsional and flexural components. The torsional mode is generally in the outer third of the blade and the bending mode is in the middle third of the blade.

3.3 Unsteady Aerodynamic Forcing Function

The predicted unsteady loading due to the entropic and vortical waves (temperature and pressure distortions) is shown in Figure 12. The unsteady pressures, Figure 12a, are largest in the outer 20%–30% of both surfaces and along the leading edge. After weighting the blades normal deflections with the pressures to obtain the modal force distributions, Figure 12b, the leading edge is seen to be still an important area of high unsteady loading. However, due to the modeshape node line being located at approximately 15–20% chord the area of high unsteady loading in the outer quarter of the blade is only on the back half of the blade. The modal force distribution is integrated to obtain the force used in the single degree of freedom calculation.

To determine which component of the excitation, entropic or vortical, if either, is dominant over the other in generating unsteady forces, the linearized analysis was performed with entropic–only and vortical–only boundary conditions. The predicted modal

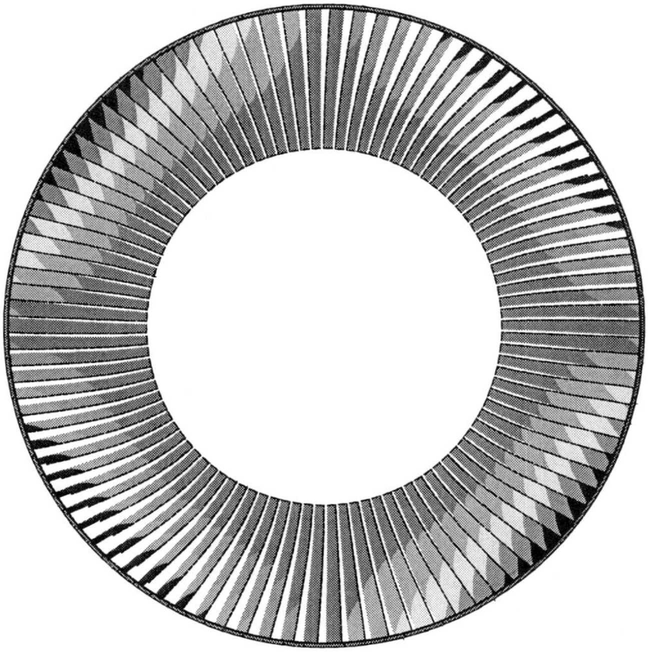

Figure11.Contours of tangential deflections for 2–nodal diameter system mode

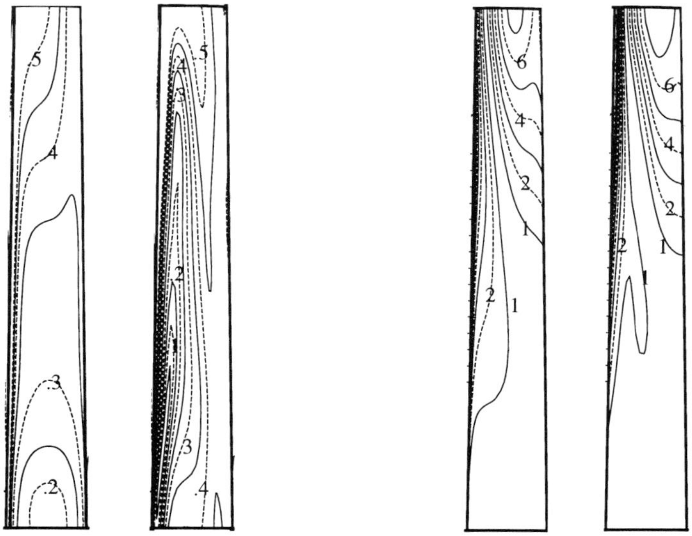

a.) Blade surface unsteady pressures

b.) Blade modal force

Figure 12. Predicted unsteady loading on LPT blade due to temperature distortions force for the vortical wave was approximately 35% larger than that of the entropic wave. Therefore, both the temperature variations, entropic waves, and the total pressure distortion, vortical waves, are of nearly equal importance for determining the unsteady loading in the LPT blade.

3.4 LPT Blade Vibratory Stresses

The measured and predicted stresses on the LPT 5th stage blade are presented in this section. Figure 13 shows the measured campbell diagrams for the strain gage located at the shank of the blade. The vibratory levels for the 2/rev crossing at the peak of the resonance with the temperature distortion are a significant percentage of the materials fatigue capability. Therefore, in designing turbine blading, the stimulus from circumferential staging of combustor nozzles must be accounted for by adjusting the stimulus count (#/rev) to avoid blade or system modes or performing analyses similar to this to assure that the forcing functions are weak and will not give stresses above tolerable levels.

To obtain predictions of the blade stresses using the single degree of freedom (SDOF) equation and the FEM stress distribution, the overall system damping is required. Predictive capabilities of this quantity are immature and therefore, the measured value obtained from applying the half power rule to the resonance blossom of the 2/rev crossing in Figure 11b, $\zeta = 2.8\%$, was used in the SDOF calculation. A similar calculation performed for the 3/rev resonance blossom, gave a value of $\zeta = 2.0\%$, which verifies the level of overall damping. To help determine the parts of the damping which give this total damping, a blade motion, unsteady aerodynamic calculation was obtained using the 3D, linearized unsteady Euler solver. The aerodynamic damping was predicted to be $\zeta = -0.4\%$, and, thus, the mechanical and material damping should be 3.2%. It is noted that the majority of this damping is coming from the frictional damping caused by the tip shroud interlock.

Comparison of the predicted stresses at the two strain gage locations with those measured is shown in Table 2 below.

Table 2. Comparison of Predicted Stress with Measurements

Gage Location	%Δ of Prediction From Measurement
Shank	19
Blade Tip	28

The predictions show very good comparisons with the measurements which demonstrates the abilities to predict the impact on blade vibrations due to circumferential variations in temperature from staging of combustor fuel nozzles.

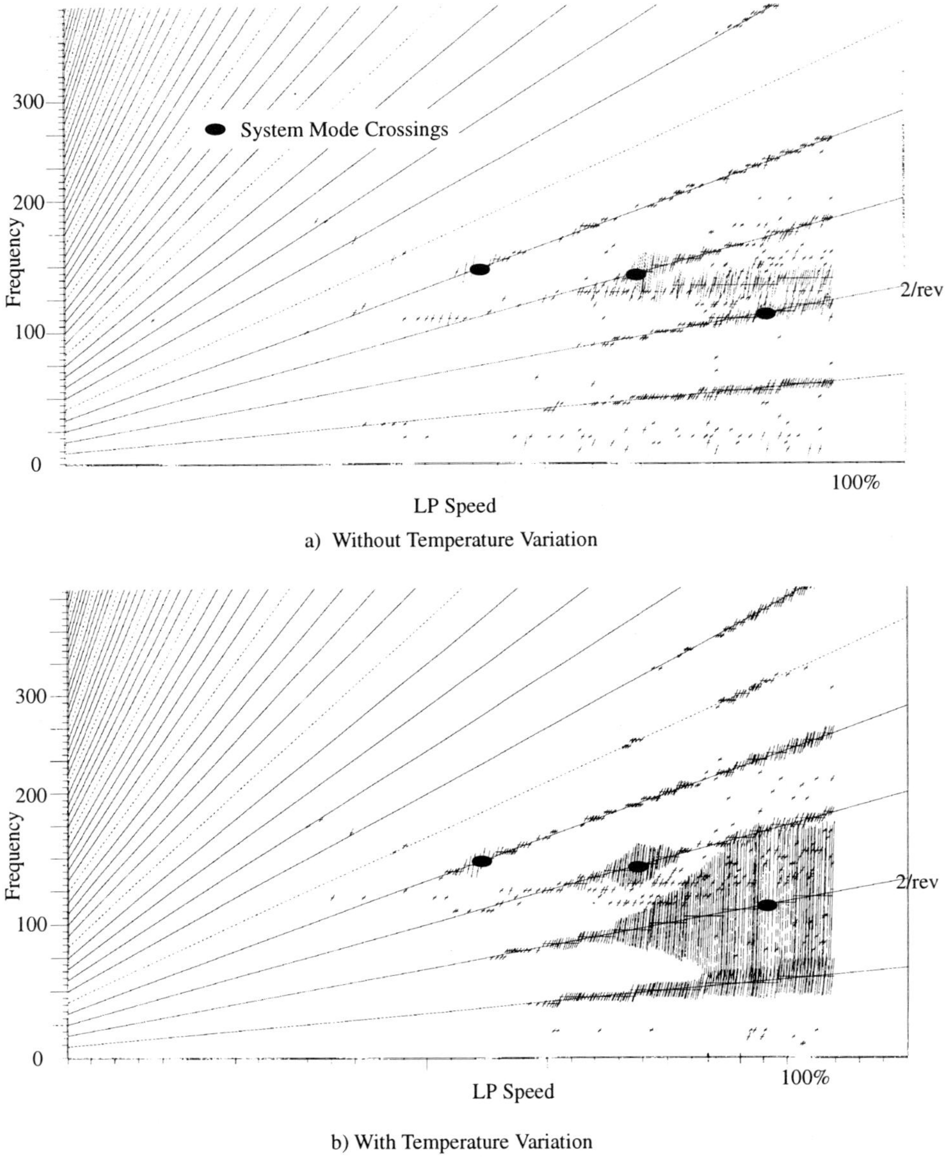

b) With Temperature Variation

Figure 13. Measured strains on the fifth stage LP blade

4.0 Summary and Conclusions

An experimental and analysis program was performed to determine the impact of temperature distortions from the combustor on the potential vibrations of turbine system blading. In particular, the LPT 5th stage blade was instrumented with strain gages and measured vibrations large enough to cause HCF failures. Measurements of circumferential variations in temperature of 300°C were seen throughout the HP and LP turbine. These temperature distortions generated total pressure variations, which were measured to be 4% of the average value at the LPT exit. The total temperature variations generated entropic type excitations and the total pressure variations generated vortical type excitations on the last stage LPT blade. The unsteady loading due to vortical disturbance (total pressure variations) was shown to be dominant over those of entropic disturbances (temperature variations). The predicted stresses using a 3D, linearized unsteady Euler solver compared within 20%–30% of the measured stresses, thus demonstrating the capability to analyze the impact of staging of combustor fuel nozzles on turbine system blading. Therefore, in designing turbine blading proper accounting of the stimulus from combustor staging of fuel nozzles is needed.

Acknowledgements

The authors would like to thank members of GE Aircraft Engines aeromechanics group for their guidance and insight throughout this program. In particular, we would like to thank Dr. Robert E. Kielb for his excellent comments and suggestions during the writing and presentation preparation of this work. The authors also would like to thank GE Aircraft Engines for permission to publish this paper.

References

Capece, V.R., Manwaring, S.R., and Fleeter, S., 1986, "Unsteady Bladerow Interactions in a Multi–Stage Compressor," *AIAA Journal of Propulsion*, Vol. 2, No. 2, pp. 168–174.

Capece, V.R., and Fleeter, S., 1987, "Unsteady Aerodynamics Interactions in a Multi–Stage Compressor," ASME Paper 87–GT–171.

Chiang, H.–W.D., and Turner, M.G., 1996, "Compressor Blade Forced Response Due to Downstream Vane–Strut Potential Interaction," *ASME Journal of Turbomahinery*, Vol. 118 No. 1, pp. 134–142.

Dunn, M.G., Bennett, W.A., Delaney, R.A., Rao, K.V., 1992, "Investigation of Unsteady Flow Through a Transonic Turbine Stage: Data/Prediction Comparison for Time–Averaged and Phase–Resolved Pressure Data," *ASME Journal of Turbomachinery*, Vol. 114.

Rao, K.V., Delaney, R.A., and Dunn, M.G., 1994, "Vane–Blade Interaction in a Transonic Turbine, Part I: Aerodynamics," *AIAA Journal of Propulsion and Power*, Vol. 10, No. 3, pp. 305–311.

Fleeter, S., Jay, R.L., and Bennett, W.A., 1978, "Rotor Wake Generated Unsteady Aerodynamic Response of a Compressor Stator," *ASME Journal of Engineering for Power*, Vol. 100, pp. 664–675.

Fleeter, S., Bennett, W.A., and Jay, R.L., 1980, "The Time–Variant Aerodynamic Response of a Stator Row Including the Effects of Airfoil Camber," *ASME Journal of Engineering for Power*, Vol. 102, pp. 334–343.

Gallus, H.E., Lambertz, J., and Wallman, Th., 1980, "Blade–row Interaction in an Axial–Flow Subsonic Compressor Stage," *ASME Journal of Engineering for Power*, Vol. 102, pp. 169–177.

Giles, M.B, "Non–reflecting Boundary Conditions for the Euler Equations," Computational Fluid Dynamics Laboratory Report CFDL–TR–88–1, Feb 1988.

Hall, K.C., and Crawley, E.F., "Calculation of Unsteady Flows in Turbomachinery Using the Linearized Euler Equations," AIAA Journal, Vol. 27, June 1989, pp. 777–787.

Henderson, G.H., and Fleeter, S., 1992, "Forcing Function Effects on Unsteady Aerodynamic Gust Response Part 2: Low Solidity Airfoil Row Response," ASME Paper No. 92–GT–175.

Holmes, D.G., Mitchell, B.E., and Lorence, C.B., 1997, "Three Dimensional Linearized Navier–Stokes Calculations for Flutter and Forced Response," *Proceedings of the 8th International Symposium on Unsteady Aerodynamics and Aeroelasticity of Turbomachines*.

Johnston, R.T., Feiereisen, J.M., and Fleeter, S.,1994, "Rotor Wake and Potential Forcing Functions, Including Blade Row Interactions," AIAA Paper 94–2975

Manwaring, S.R., and Fleeter, S., 1990, "Inlet Distortion Generated Periodic Aerodynamic Rotor Response," *ASME Journal of Engineering for Power*, Vol. 112, pp. 298–307.

Manwaring, S.R., and Fleeter, S., 1991, "Forcing Function Effects on Rotor Periodic Aerodynamic Rotor Response," *ASME Journal of Engineering for Power*, Vol. 112, pp. 312–319.

Manwaring, S.R., and Wisler, D.C., 1993, "Unsteady Aerodynamics and Gust Response in Compressors and Turbines," *ASME Journal of Engineering for Power*, Vol. 115, pp. 724–740.

Manwaring, S.R., Rabe, D.C., Lorence, C.B., and Wadia, A.R., "Inlet Distortion Generated Forced Response of a Low–Aspect–Ratio Transonic Fan," ASME Paper No. 96–GT–376.

Markland, G.J., Barton, M.T., and O'Brien, W.F., "The Use of Circumferentially Nonuniform Stators to Attenuate LP Compressor Rotor–Stator–Strut Aerodynamic and Mechanical Interactions," ASME Paper 96–GT–154.

O'Brien, W.F., Reimers, S.L., and Richardson, S.W., 1983, "Interaction of Fan Rotor Flow with Downstream Struts," AIAA Paper No. 83–0682.

Sanders, A., and Fleeter, S., 1997, "Rotor Blade–to–Blade Wake Variability and Its Effect on Downstream Vane Response," AIAA Paper No. 97–2753.

Vierck, R.K., *Vibration Analysis*, Harper & Row, 1979.

INVESTIGATION OF THE BOUNDARY LAYER DEVELOPMENT ON A HIGHLY LOADED LOW PRESSURE TURBINE CASCADE UNDER THE INFLUENCE OF UNSTEADY FLOW CONDITIONS

PAOLA ACTON, LEONHARD FOTTNER
Institut für Strahlantriebe
Universität der Bundeswehr München
D-85577 Neubiberg
Germany

Abstract

The flow in turbomachines is characterised by high turbulence levels as well as by strong non uniformity and non stationarity. Each blade row generates wakes with lowered velocities and higher turbulence levels. The relative motion between rotor and stator rows gives rise to a non-uniform and unsteady flow impinging on the downstream blade rows. This phenomenon strongly influences the boundary layer behaviour on the blade surfaces and hence the losses and the loading of the profile. Measurements of profiles in cascade wind tunnels with homogeneous inlet conditions have often revealed inferior aerodynamic performance compared with that obtained in a real turbomachine.

The High-Speed Cascade Wind Tunnel of the Universität der Bundeswehr München has been equipped with a wake generator that can simulate a moving blade row in front of the fixed linear cascade mounted in the test section.

Measurements were performed on a highly loaded low pressure turbine cascade varying the velocity and the pitch and hence the redzced frequency of the wake-generating bars. The loss measurements showed an optimal reduced frequency, at which the profile losses were reduced to 77% of those observed under stationary inlet flow conditions. Hot-film measurements on the other hand were able to give a deeper insight into the physical basis of the complex phenomenon of unsteady boundary layer transition.

Nomenclature

c_{ax}	[m/s]	axial velocity
E	[Volt]	anemometer output tension
f	[Hz]	frequency
l	[m]	chord length
s	[m]	position along the surface

T.H. Fransson (ed.), Unsteady Aerodynamics and Aeroelasticity of Turbomachines, 393-406.

t	[m]	pitch
t	[s]	time
T	[s]	time period
U	[m/s]	velocity
β	[°]	flow angle
μ_3	[-]	skewness
ω	[-]	total pressure losses

Abbreviations

Ma	**Ma**ch-number
Re	**Re**ynolds-number
RMS	**R**oot-**M**ean-**S**quare
So	**Str**ouhal-number

Subscripts

2th	downstream conditions for isentropic flow
tot	total
b	referring to the bars
~	ensamble averaged
∞	referring to undistrubed flow conditions

1. Introduction

An enhanced knowledge of the boundary layer behaviour on turbomachine blade rows under different loading conditions leads to an optimised design of the blade profiles that allows for a reduction in the number of blades and a increas in the efficiency of the whole component. Until now blade design has been based mainly on correlations established from the manifacturer's experience or on results of stationary numerical codes. The quality of the design has thereafter been tested in cascade wind tunnels with stationary inlet flow conditions. Recently however, the performance of the bladings under real engine conditions has been found to be distinctly different from that observed during stationary cascade testing. This has been attributed to differences in the transitional behaviour of the boundary layer exposed to non-homogeneous and unsteady inlet flow conditions. Engine and low-speed wind tunnel tests have helped to analyse this phenomenon. The present paper introduces results obtained in a high speed cascade wind tunnel maintaining most of the similarity parameters typical of real engine conditions, with the aim of establishing an improved understanding of the boundary layer behaviour under the influence of unsteady inlet flow conditions.

2. Test Facility and Measurement Techniques

2.1 THE HIGH SPEED CASCADE WIND TUNNEL

The High Speed Cascade Wind Tunnel (HGK) of the University of the Federal Armed Forces in Munich is an open loop facility which can operate continuously and reach Mach numbers up to Ma = 1.05 in the test section (Fig. 1). As the wind tunnel is built inside a large pressure tank it is possible to vary the Mach number and the Reynolds number in the test section independently (see *Sturm et al.* (1985)) in order to simulate the flow conditions inside real turbomachines correctly.

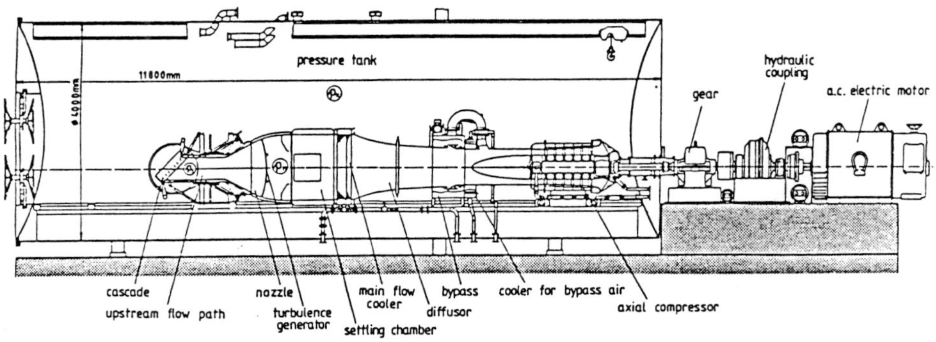

Figure 1 The High-Speed Cascade Wind-Tunnel (HGK)

The air is supplied by a six-stage axial compressor, driven by a 1.3 MW a.c. electric motor which is situated outside the pressure tank. The air flows through a diffuser into a settling chamber, where it is cooled down to an adjustable constant temperature (30°C to 60°C), and then through a turbulence generator; finally it is accelerated in the nozzle before reaching the 300 mm wide, variable height test section.

2.2 THE MOVING-BAR FACILITY FOR THE GENERATION OF UNSTEADY FLOW CONDITIONS

The rotor-stator interaction is simulated in the HGK by cylindrical rods moving in front of the cascade. As was shown by *Pfeil et al.* (1975) the far wake of a cylinder is similar to that of an airfoil given that both cause the same total pressure losses.

In order to simulate the wakes of turbomachine bladings with the help of cylindrical bars in a wind tunnel several similarity parameters have to be met correctly. The width, the total pressure loss (i.e. the velocity deficit) and the maximum turbulence level in the wake have to be matched. This can be done by adjusting the diameter of the bars and the distance between the bars and the inlet plane of the cascade.

Furthermore the Strouhal number similarity also has to be taken into account. The Strouhal number is defined as: $So = (f \cdot l)/c_{ax}$, where f is a frequency, l a typical length (here the chord length) and c_{ax} the axial velocity. In modern turbines values of around So = 0.8 are reached. The wake generating facility has been designed in order to stand velocities up to 60 m/s and to allow adjustment of the velocity for simulating different Strouhal-numbers.

The wake generator has been designed with emphasis on modularity in order to allow alteration of the different wake parameters with reasonably little effort. The working principle of the wake generator is based on cylindrical bars fastened to two rubber timing belts moving in front of and behind the cascade. The timing belts, having a length of 4000 mm, span over two main pulleys of diameter 400 mm positioned

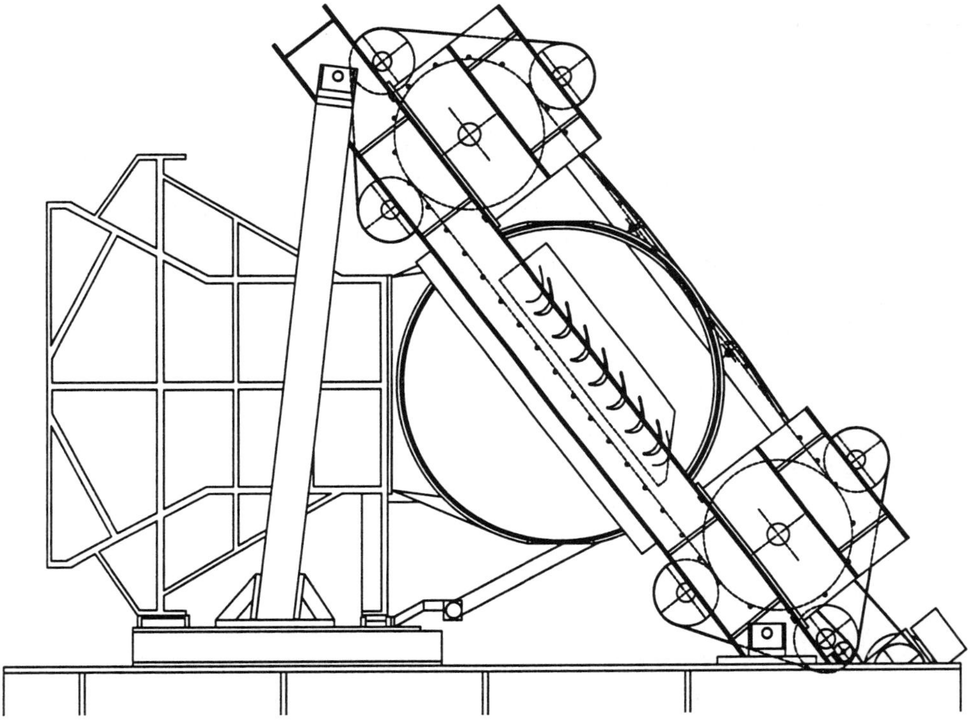

Figure 2 The generator of unsteady inlet flow conditions (EIZ)

above and below the cascade (Fig. 2). The distance between the trailing edge of the cascade and the returning bars is big enough not to disturb the flow in the blade passages and to allow the traverse of all probe types usually employed during the tests. Pulleys of big dimensions were chosen in order to reduce the high centrifugal forces due to the deviation of the bars. At the maximum velocity the centrifugal force on steel bars of 2 mm in diameter amounts to F = 136 N, which may cause a plastic deformation of the bars. In order to reduce the loading on the bars they are guided by two additional flat belts and two pulleys as is shown in Fig. 3. The toothed disks and the pulleys for flat belts are fastened to two main axes, which are fixed to a rigid frame, ensuring the stability of the whole wake generator. The mounting supports of the gear boxes are designed in order to enable control of the belt tension and the parallelism of the main shaft.

The driving pulley is positioned on the lower main shaft and is connected by a timing belt to the a.c. electric motor using a 1:1 transmission ratio. The water-cooled motor has a nominal maximum power of 17 kW at 3000 rpm. The motor can turn in both directions for simulating the rotor-stator interaction of compressor as well as of turbine stages. The velocity of the motor can be digitally regulated with an accuracy of 0.1 rpm from 100 rpm to 3000 rpm.

Because one part of the timing belts moves inside the wind tunnel nozzle upstream of the cascade it was necessary to reduce the width of the test section in order to avoid disturbances of the inlet flow to the cascade. Two wooden side walls were designed for this purpose according to the concepts of *Cohen et al.* (1962); the width of the test section was reduced to 170 mm, thus setting the blade aspect ratio for the usually employed cascades to 1.7.

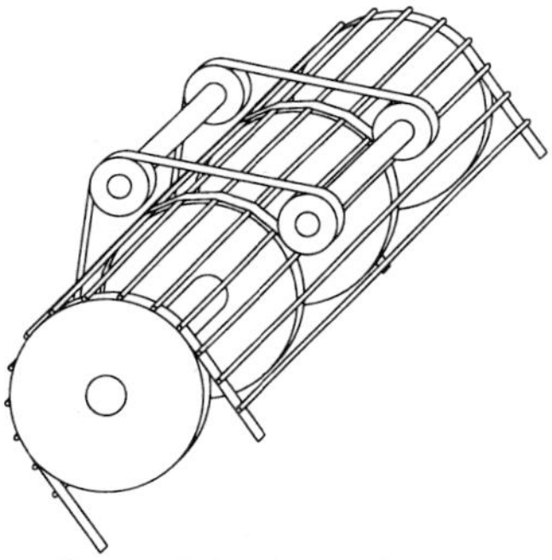

Figure 3 Guidance facitlity

For performing an ensemble averaged analysis of the high frequency signals of the fast response sensors a trigger module was fixed to the main frame in order to detect the passing bars.

As the measurements are very time consuming because of the large amounts of data which have to be stored, the two halves of the timing belts were equipped with bars of different pitch, allowing for the observation of two flow conditions in one measurement cycle.

2.3 MEASUREMENT TECHNIQUES

2.3.1 *Pneumatic Measurements*

In the first evolutionary step of the new test facility total pressure losses and surface pressure distribution were measured as mean values. For the loss evaluation the total pressure p_{t2} was measured by a wedge probe traversed over a whole pitch at 40% of the chord length downstream of the trailing edge plane of the cascade.

2.3.2 *Glue-On Hot-Film Sensors*

Surface-mounted hot-film gauges were glued onto the suction and the pressure sides of the blades in order to analyse the development of the shear stress and of the boundary layer conditions along the profile. Although it is theoretically possible to calibrate the sensors with respect to the wall shear stresses no calibration was performed as it is a rather laborious procedure. Furthermore no calibration is necessary in order to obtain the desired information about transition and separation phenomena of the boundary layer, as was shown by *Schröder* (1991). The location and the form of the transition can be efficiently detected with the help of the normalised RMS value of the anemometer signal.

For the present measurements, hot-film arrays designed by MTU Munich were glued onto the whole suction and pressure sides of two adjacent blades. Each sensor consists of a 3 mm long, ultra-thin nickel film (0.4 μ) deposited on a 0.1 mm poly-imide substrate. One array consists of up to 80 sensors spaced 3 mm apart. During the first test measurements, 28 sensors were used on the suction side and 24 on the pressure side.

2.3.3 *Hot-Wire Measurements*

The velocity and turbulence level distributions downstream of the wake generator and of the cascade, respectively, were measured by a single wired DANTEC subminiature BL probe of the type 55P15 connected to a constant-temperature-bridge of the type 55M01. The traverse of the probe can be automatically controlled by two computer driven step motors. The sensor calibration is performed in the open test section of the HGK and a 4th degree polynomial is used for the approximation of the calibration curve. The polynomial calibration was chosen because it is more suitable for the high velocity conditions of the HGK than the application of King's law.

2.4 DATA ACQUISITION AND PROCESSING

The data acquisition and the processing, as well as the functioning of the stepper motors, are fully automated under the control of two separated computer systems, one for the pneumatic measurements and one for the hot-sensor measurements.

The stationary pressure measurements are performed by means of the PC-driven software WINPANDA, developed by *Ganzert et al.* (1996).

The hot-sensor measurement techniques are controlled by a HP 3565 work station, which is able to acquire up to three channels simultaneously with a maximum sample frequency for each channel of 260 kHz. The analog data are first filtered by an anti-aliasing filter and then digitised by a 14-bit A/D-converter.

The high frequency data from the thermal anemometry measurements are analysed in the time domain as raw data, as well as in the frequency domain with the help of an FFT routine. Furthermore the well established ensemble averaging technique has been applied, using 300 ensembles, each consisting of 3146 samples corresponding to 6 wake passages. The ensemble average of a time-dependent quantity b is given by:

$$\tilde{b}(t) = \frac{1}{N} \cdot \sum_{j=1}^{N} b_j(t) \tag{1}$$

where N is the number of ensembles and the ensemble averaged RMS is given by:

$$\sqrt{\tilde{b}'^2(t)} = \sqrt{\frac{1}{N-1} \cdot \sum_{j=1}^{N} \left(b_j(t) - \tilde{b}(t) \right)^2} \tag{2}$$

The skewness (third-order moment)

$$\tilde{\mu}_3 = \frac{1}{\left(\tilde{b}(t)\right)^3} \cdot \frac{1}{N-1} \cdot \sum_{j=1}^{N} \left(b_j(t) - \tilde{b}(t)\right)^3 \tag{3}$$

has also been calculated in order to gain a deeper insight into the boundary layer characteristics.

In the case of the hot-film data, b can be substituted by the shear stress τ_w, and for the hot-wire measurements by the velocity.

3. Experimental Results

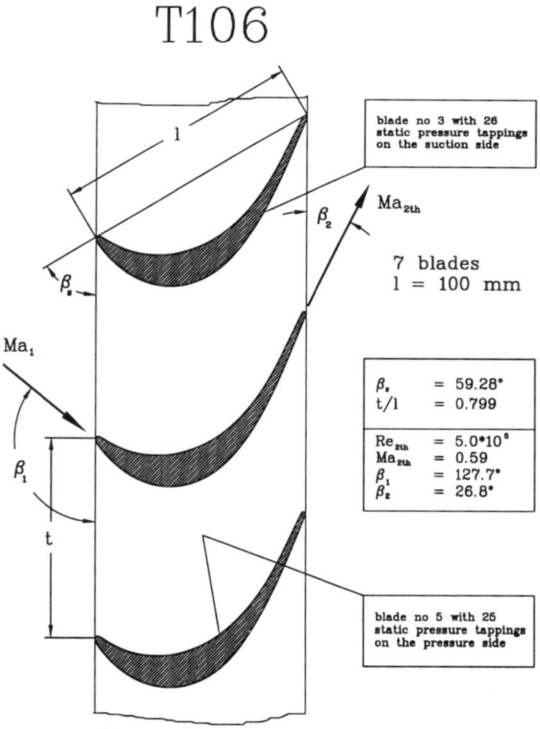

Figure 4 The turbine cascade T106

The measurements were performed on the high pressure turbine cascade T106 (Fig. 4) for one combination of Reynolds- and Mach-number: $Re_{2th} = 2 \cdot 10^5$ and $Ma_{2th} = 0.40$.

The aim of the investigation was to analyse the boundary layer behaviour under the influence of different unsteady inlet flow conditions. Therefore the Strouhal-number was varied by changing the velocity and frequency of the wakes independently.

The wakes were generated with the help of bars of diameter 2 mm positioned 70 mm upstream of the cascade inlet plane. Measurements were performed with two different bar pitches and four bar velocities, thereby achieving the Strouhal-numbers varying between So = 0.21 and So = 1.26.

3.1 MEASUREMENTS DOWNSTREAM OF THE WAKE-GENERATOR

Prior to the cascade tests, single wire probe measurements were performed in the free stream behind the wake generator in order to analyse the shape and intensity of the generated wakes. The probe was positioned in the centre of the test section and the bars were moved with different velocities in front of it. The ensemble averaged results are shown in Fig. 5 for the two bar-pitches: $t_b = 40$ mm and $t_b = 80$ mm respectively. The inlet flow conditions correspond to the theoretical outlet flow conditions $Ma_{2th} = 0.40$ and $Re_{2th} = 2 \cdot 10^5$ for the turbine cascade T106. In each column the wakes for the two different bar velocities U = 20 m/s and U = 30 m/s are represented in separate diagrams in the form of ensemble averaged quantities. On the abscissa the time is normalised with respect to the wake passing period. On the ordinate the ensemble averaged RMS value of the velocity fluctuation and the ensemble averaged velocity are normalised with the corresponding free stream values.

The wakes behind the bars are very similar at all analysed bar velocities. For the greater bar pitch a distinct range of non-disturbed flow is visible between the wakes. At the higher bar velocity a stronger velocity deficit corresponding with increased turbulence levels is apparent. The distribution of the turbulence level is non-symmetric because the bars do not move perpendicularly to the main flow.

The shape of the wakes generated by the bar arrangement with the smaller pitch is similar at both bar velocities whilst the intensity of the wake is higher at higher bar velocity. The undisturbed free stream flow almost vanishes because the wakes tend to join in the cascade inlet plane.

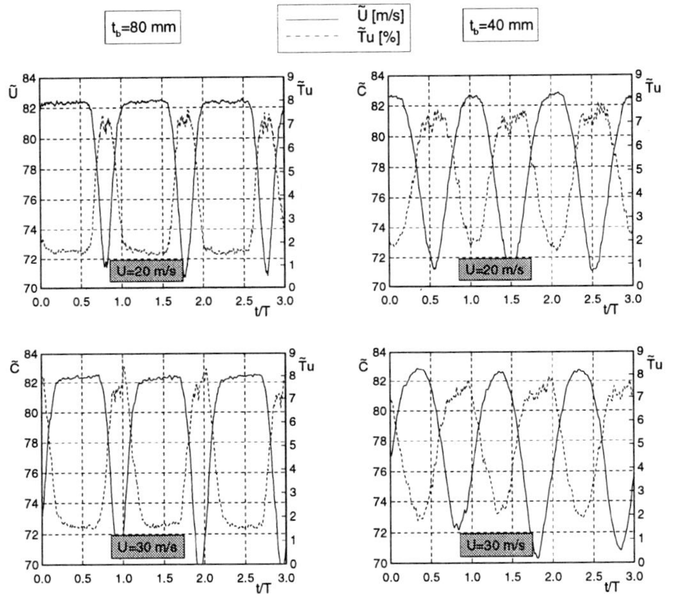

Figure 5 Velocity and turbulence level distribution behind the wake-generator

3.2 HOT-FILM MEASUREMENTS ON THE PROFILE

The signals of the hot-films glued onto the suction side of the profile have been analysed with the help of the abovementioned ensemble averaged technique in order to separate the random from the periodic effects. The results can best be described in the form of distance-time diagrams, where the position along the profile s is displayed on the abscissa and the time t on the ordinate; the two quantities are made nondimensional with respect to the profile's length s_{tot} and the bar period T, respectively.

Measurements have been performed for one combination of Mach-number $Ma_{2th} = 0.40$ and Reynolds-number $Re_{2th} = 2 \cdot 10^5$, at which the pitch and velocity of the bars were varied in order to achieve a Strouhal-number between So = 0.21 and 1.26. The general physical phenomena occurring along the profile in the case of unsteady inlet flow conditions will be described in detail for a bar pitch of t_b = 80 mm and a velocity of U = 10 m/s. Fig. 6 shows the distance-time diagrams of the ensemble averaged signals (a), of the ensemble averaged RMS values (b) and of the ensemble averaged skewness μ_3 (c) for this given flow condition.

Due to the very good resolution of the hot-films the influence of the passing wakes can be analysed near to the leading edge as well as in the transition region further downstream. In the region between two wakes **a** (t/T = 0.8 to 1.1) a small separation bubble occurs at $s/s_{tot} = 0.06$, which is characterised by a local maximum of the RMS-values. As the wake approaches the suction side, the flow field between the profile surface and the wake is accelerated, and the separation bubble disappears since there is no longer an adverse pressure gradient in this region. This results in a reduction in the intensity of the RMS-values in region **b** for t/T = 1.1 to 1.5. The increased turbulence level in the wake influences the flow around the leading edge only later on in **c**: the high random fluctua-

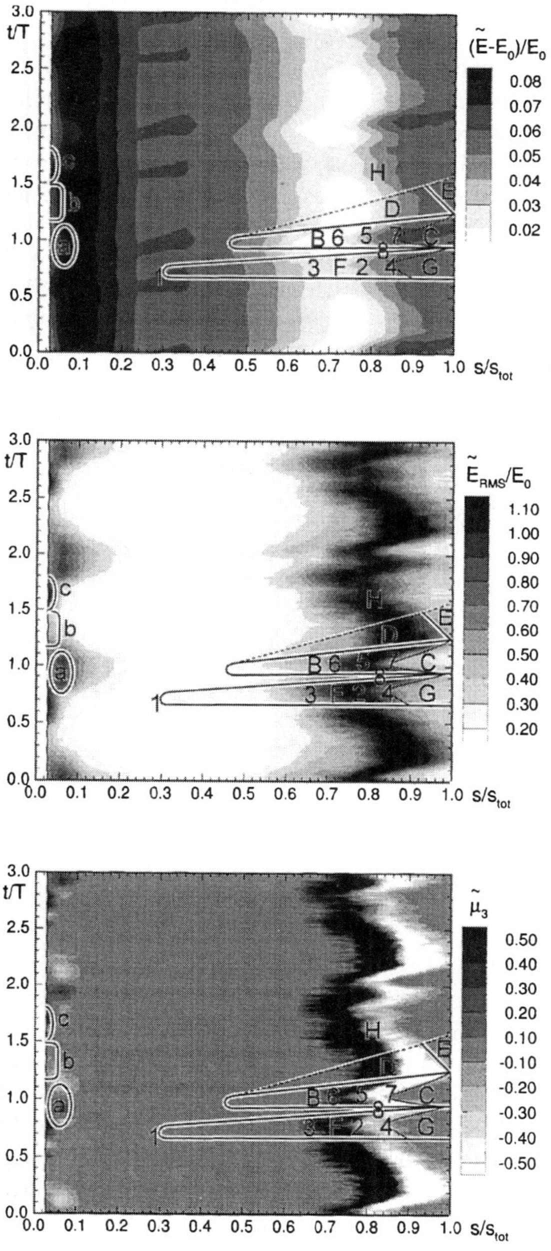

Figure 6 Hot film results, U=10 m/s, t_b=80 mm,

tions in the middle of the wake generate a short turbulent boundary layer, which again suppresses the separation bubble. The strong acceleration which characterises the first part of the suction side causes a relaminarisation of the boundary layer independently of the wake passing. Therefore the first third of the suction side behaves as an undisturbed laminar boundary layer, where the disturbances from outside are suppressed.

Beginning at $s/s_{tot} = 0.3$ wake-influenced areas and regions between two wakes can be detected again. The higher RMS-values at point **1** show the first instability caused by the oncoming wake. Downstream of this point a wake-induced transition takes place in the regions **B** and **F** and a wake-induced turbulent boundary layer can be found in **C** and **G**. Later on, i.e. for higher time values, the boundary layer is not longer influenced by the wakes and transition takes place in the regions **D** and **H**.

The oncoming wakes generate a shift of the transition point along the blade surface, which is normally in phase with the wake-period (*Cumpsty et al.*, 1995; *Schulte et al.*, 1996). In the present case, however, the transition point moves twice upstream (points **2** and **5**) and downstream during one bar-period. To the authors' knowledge such behaviour has not yet been reported in the literature.

The two transition regions which are located further upstream differ strongly from each other in their dimensions as well as in the effects on the following boundary layers. The first instability point, which will lead to transition point **2** (at $s/s_{tot} = 0.76$) can be detected at point **1**. Upstream of point **2** the shear stress decreases gradually (smaller nondimensional mean values $(E-E_0)/E_0$). At $s/s_{tot} = 0.63$ the skewness shows a positive value for the first time and marks the onset of transition. Behind the transition point the skewness becomes negative and reaches the value zero at **4** (at $s/s_{tot} = 0.86$), which denotes the end of transition. A full turbulent boundary layer (**G**) develops downstream thereof and reaches the trailing edge. This region is characterised by relatively small shear stresses. The transition point **2** is therefore identified by a long transition region, which extends along 56% of the suction side.

The first instability point upstream of transition point **5**, however, lies further downstream at $s/s_{tot} = 0.46$. Transition begins at **6** ($s/s_{tot} = 0.71$) and ends at **7** ($s/s_{tot} = 0.84$). Upstream of point **5** the shear stress decreases abruptly in contrast to the behaviour upstream of point **2**. The transition region is characterised by a shorter extent (only 38% of the profile length) and by higher shear stress values in the following full turbulent region.

Because of the markedly different behaviour of the two wake-induced transition regions it can be assumed that they are due to two different physical phenomena. When the wake approaches the suction side, the flow field accelerates not only in the vicinity of the leading edge but also along the whole first part of the suction side, as can be confirmed by numerical calculation. For given flow conditions downstream of the cascade this acceleration will cause a stronger deceleration in the rear part of the profile, which will move the transition point further upstream to point **2**. As the wake further approaches the suction side this acceleration effect is reduced and the transition region moves downstream to point **8**. As the distance between wake and leading edge decreases further, the higher turbulence level in the wake influences the boundary layer and causes a second upstream shift of the transition region to point **5**.

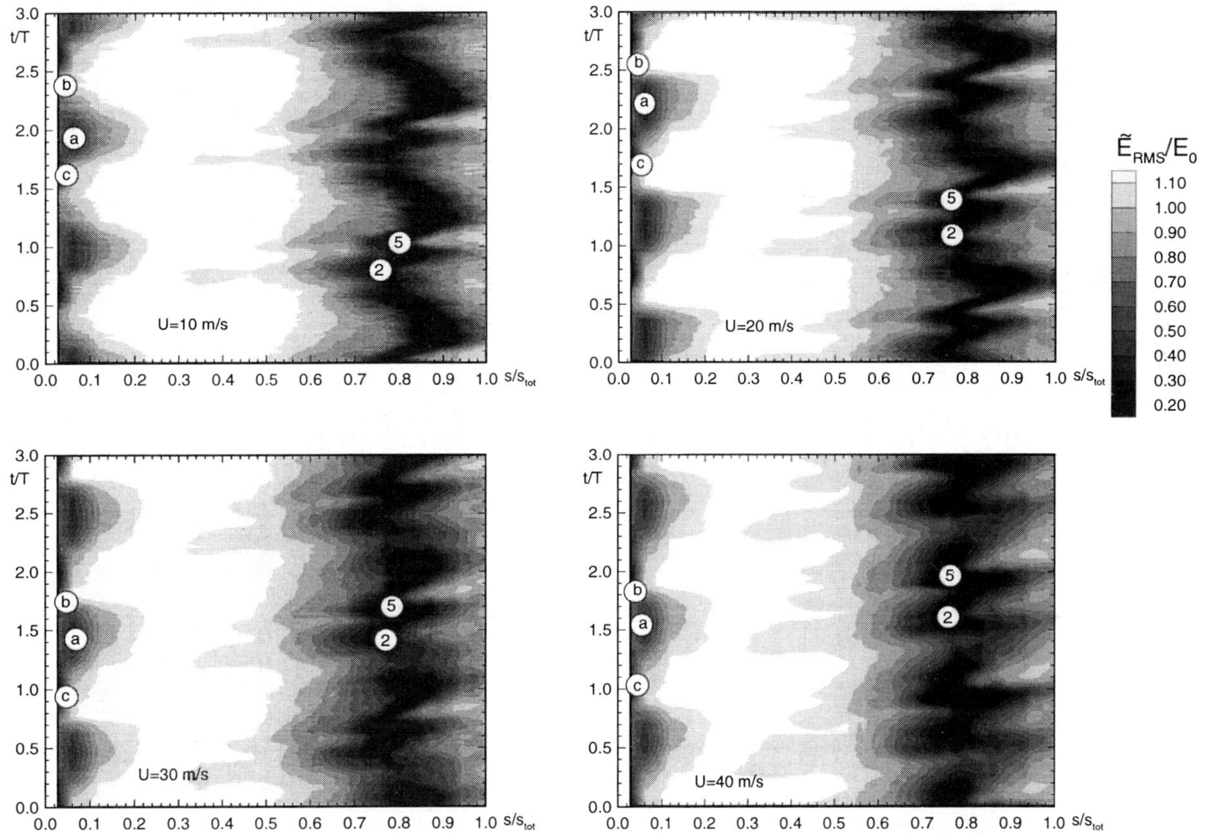

Figure 7 Hot film results for different bar velocities ($t_b = 80$ mm)

Thereafter the transition moves downstream to $s/s_{tot} = 0.87$. This point lies further downstream than the transition point with stationary inlet flow conditions ($s/s_{tot} = 0.83$). The region **D** is the so-called becalmed region, which arises from the different velocities of the upstream and downstream parts of the turbulent spots generated by the incoming wakes. This region is characterised by small random fluctuations and by skewness around zero. Nevertheless it should not be confused with a laminar boundary layer. The boundary layer in the becalmed region is particularly thin and therefore generates higher shear stresses; at the same time it is not susceptible to disturbances, so it causes a downstream shift of the transition region. These peculiarities of the becalmed region attenuate asymptotically with increasing distance from the turbulent spot. Hence it is not possible to locate an exact end to this region.

When the influence of the becalmed region on boundary layer development decreases, the transition region (**H**) moves upstream again, to the position it would have under undisturbed inlet flow conditions. Before the transition region can stabilise in this position, however, the next wake arrives and causes the already explained upstream shift of the transition point.

The influence of the bar velocity on the boundary layer development along the suction side can be analysed with the help of the four distance-time diagrams shown in Fig. 7 for the bar pitch of $t_b = 80$ mm. At all bar velocities wake-influenced regions and regions between wakes can be detected. A short separation bubble (**a**) near to the leading edge ($s/s_{tot} = 0.06$) can be seen between two wakes at all bar velocities. An area influenced by the high turbulence level in the wake (**c**) can be noticed through the higher fluctuation intensities, while the region (**b**) which is influenced by the accelerating flow becomes smaller with increasing bar velocity, tending to vanish at the highest bar velocity $U = 40$ m/s.

The transition point in the rear part of the profile moves upstream and downstream twice during a wake period. The distance between the two upstream transition points, however, becomes bigger with increasing bar velocity. At the same time the extension of the becalmed region decreases because it doesn't have enough time to develop completely before the next wake influences the boundary layer again. The position of the first instability point as well as the position of the most upstream transition point is not influenced by the bar velocity. The bar velocity has its strongest influence on the transition point lying furthest downstream.

For conciseness the results of the measurements performed with the smaller bar pitch of $t_b = 40$ mm will not be described. However in Fig. 8 the positions of the transition points for all bar velocities and both bar pitches are summarised together with data on the extension of the transition region. In the following description the most upstream transition point will be called the "first transition point" and the most downstream the "last transition point". It can be noticed that the first transition point moves only slightly with changing Strouhal-number and that its position is the same whether the Strouhal-number is due to a high bar velocity and a large bar pitch or to a low bar velocity and a small bar pitch. The position of the last transition point, however, changes noticeably with the Strouhal-number and it should be remarked that the position is different according to the bar pitch. The distance between the first and the last transition point is shown below. For the lower Strouhal-numbers it differs

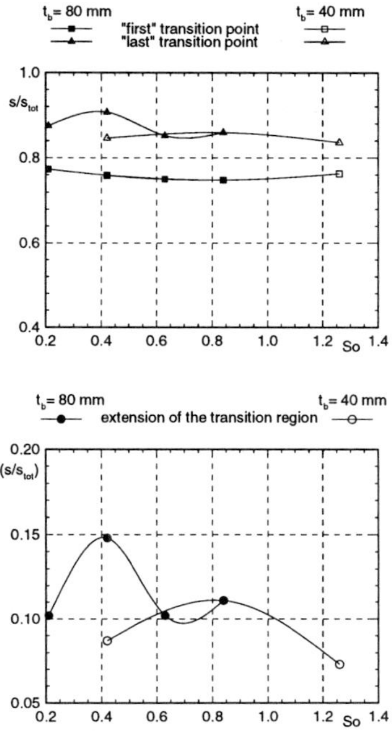

Figure 8 Location of transition points for different Strouhal-numbers

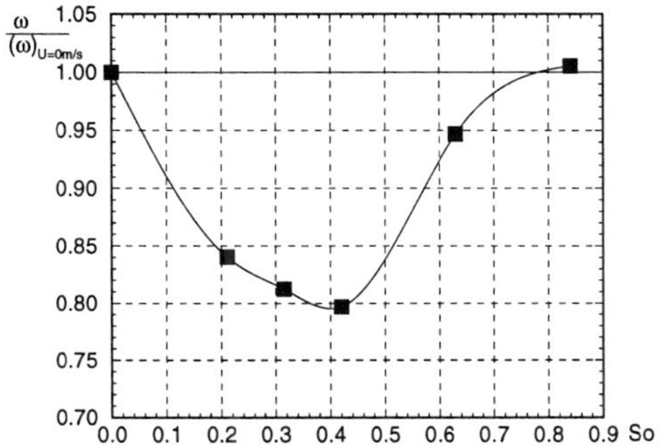

Figure 9 Cascade losses for different Strouhal-numbers

substantially between the two different bar pitches, while for higher Strouhal-numbers this difference becomes negligible.

3.3 LOSS MEASUREMENTS

Loss measurements were performed downstream of the cascade with the higher bar pitch and at varying bar velocity. These are summarised in Fig. 9, where the losses ω have been made nondimensional with respect to the cascade losses occurring for stationary inlet flow conditions $\omega_{U=0\ m/s}$. Up to So = 0.42 the total pressure losses decrease down to a value which is only 77% of that reached by stationary inlet flow conditions. This Strouhal-number corresponds to that at which the becalmed region has its strongest influence on the boundary layer and the "last transition point" is positioned furthest downstream (see Fig. 8). After this minimum the losses increase again to the value reached under stationary inlet flow conditions.

A profile design which takes into account the rotor-stator interaction could therefore lead to lower profile losses and hence to higher efficiencies of the complete turbomachine.

4. Summary

An experimental analysis of the influence of the rotor-stator interaction on the boundary layer behaviour of a high loaded turbine profile was performed on a large scale cascade in the High-Speed Cascade Wind Tunnel of the Universität der Bundeswehr München. For one given flow condition the frequency and the velocity of the oncoming wakes were varied, in order to analyse separately the different parameters influencing this complex physical phenomenon. High response sensors were used for the investigation of the boundary layer, while traditional pressure probes were used for the total pressure losses.

The importance of both the turbulence level and the total pressure loss in the impinging wakes for the boundary layer development along the suction side of the profile were shown by the measurements. A reduction of the total pressure losses was achievable for a particular Strouhal-number, which demonstrates that taking into account the rotor-stator interaction during the design of new profiles can lead to enhanced efficiencies of the whole component.

5. References

Bellhouse, B.J., Schultz, D.L. (1966), *"Determination of Mean and Dynamic Skin Friction, Separation and Transition in Low-Speed Flow with a Thin-film heated element"*, J. of Fluid M. Vol. 24, part 2, pp.:379-400

Cohen, M.J., Ritchie, N.J.B. (1962), *"Low-Speed Three-Dimensional Contraction Design"*, J. of the Royal Aeronautical Society, Vol. 66, pp.:231-236

Cumpsty, N.A., Dong, Y., Li, Y.S., (1995), *"Compressor Blade Boundary Layers in the Presence of Wakes"*, ASME Paper 95-GT-443

Ganzert, W., Fottner, L. (1996), *"WINPANDA - an Enhanced PC-Based Data Acquisition System for Wake and Profile Distribution "*, 13th Symp. on Meas. Tech. for Transonic and Supersonic Flows in Cascades and Turbomachines, Zürich

Ludwieg, H. (1949), *"Ein Gerät zur Messung der Wandschubspannung turbulenter Reibungsschichten"*, Ingenieur Archiv, Band 17, pp.:207-218

Pfeil, H., Eifler, J. (1975), *"Messungen im turbulenten Nachlauf des Einzelzylinders"*, Forschung im Ingenieurwesen, Vol. 41, Nr.5, pp.:137-168

Schröder, Th., (1991), *"Investigation of the Blade Interaction and Boundary Layer Transition Phenomena in a Multistage Aero Engine Low-Pressure Turbine by Measurements with Hot-Film Probes and Surface-Mounted Hot-Film Gauges"*, Boundary Layers in Turbomachines, VKI Lecture Series 1991-6

Sturm, W., Fottner, L. (1985), *"The High-Speed Cascade Wind Tunnel of the German Armed Forces University Munich"*, 8th Symp. on Meas. Tech. for Transonic and Supersonic Flows in Cascades and Turbomachines, Genoa

Schulte, V., Hodson, H.P., (1996), *"Unsteady Wake-Induced Boundary Layer Transition in High Lift LP Turbines"*, ASME Paper 96-GT-486

Experimental and Numerical Investigation of Unsteady Rotor-Stator Interaction on Axial Compressor Stage (with IGV) Performance

V.E. Saren, N.M. Savin [1]
Central Institute of Aviation Motors
Moscow, Russia
D.J. Dorney
GMI Engineering & Management Institute
Flint, MI, USA
R.M. Zacharias
Pratt & Whitney
East Hartford, CT, USA

ABSTRACT The unsteady flowfield in an axial compressor stage (with an inlet guide vane) has been examined experimentally and computationally. In particular, the effects produced by mutually shifting (clocking) the airfoils in the stationary blade rows has been investigated. The experimental data and computational results indicate that airfoil clocking can be used to increase the efficiency and reduce the unsteady rotor loadings in axial compressors.

1 Introduction

The physical mechanisms governed by the interaction of rotors and stators in axial-flow turbomachines are not completely understood; in particular there is little published data on experiments for three or more interacting blade rows. For the most part, the published literature is concerned with the unsteady blade loads associated with the aerodynamic interaction of two adjacent blade rows, where one blade row is rotating. This research, which has been both theoretical and experimental, has enabled a better understanding of the roles of axial gap and blade-count ratio on the behavior of unsteady flow fields. It should be noted, however, that the majority of the published literature has not addressed the influence of the unsteadiness on the time-averaged performance characteristics of turbomachinery, such as efficiency and work. It is reasonable to expect that temporal changes in the flow field

[1] Financial support provided under RFFI Grant 96-01-01847 and CRDF 1-195

T.H. Fransson (ed.), Unsteady Aerodynamics and Aeroelasticity of Turbomachines, 407-424.

due to interacting blade rows can increase/decrease the losses; developing methods to advantageously control the unsteadiness in the flow will result in a significant advancement in the design of modern turbomachines.

One example of manipulating flow unsteadiness to benefit performance arises when the mutual circumferential (clocking) position of two stators (in a multiple blade row environment) are varied. Traditional aerodynamic design systems based on steady flow procedures would predict that clocking the stator airfoils would not affect the performance of the turbomachine. In actual turbomachines, however, performance variations occur with airfoil clocking for at least two reasons: 1) the changes in unsteady (periodic) disturbances experienced by the rotor blades because of mutual shifts in the nonuniform flow fields which occur in the stator rows (IGV and stator) as the clocking position is varied, and 2) the changes in the stator outlet flow because of the circumferential shift in the wakes from the inlet guide vane (IGV) or upstream stator.

Several experimental, theoretical and numerical studies relating to airfoil clocking have been published during the past few years. The effects of airfoil clocking (or indexing) on turbine performance has been investigated by Huber *et al.* [1], Griffin *et al.* [2] and Dorney and Sharma [3]. The studies showed changes in efficiency on the order of 0.5% as the stators were clocked. The highest efficiencies occurred when the first-stage stator wake impinged on the second-stage stator while the lowest occurred when the first-stage stator wake convected through the middle of the second-stage stator passage. The effects of airfoil clocking on compressor performance have also received attention during the last few years. In the works of Saren, experiments were used to reveal the influence of stator clocking on the flow in the rotor passage, and theoretical explanations of the phenomena were proposed [4, 5]. Hsu and Wo experimentally demonstrated that rotor clocking can be used to reduce the unsteady loads experienced by downstream blade rows [6], while in another experimental investigation Walker *et al.* studied the effects of inlet guide vane clocking on the boundary layer quantities and losses of a downstream stator blade row [7]. Barankiewicz and Hathaway [8] also performed stator clocking experiments, but their data showed only a 0.2% performance change due to clocking. Gundy-Burlet and Dorney performed computational studies of stator clocking in a 2-1/2 stage compressor, with the results showing efficiency variations on the order of 0.5% [9]. It should be noted that all the studies of airfoil clocking discussed above were performed for low-speed machines.

The magnitude of the variations in the performance parameters in previous studies emphasizes the importance of airfoil clocking and justifies further research. The current research is devoted to the experimental and numerical investigation of time-averaged and unsteady flow parameters in a subsonic 1-1/2 stage axial compressor with clocking of the IGV and stator blade rows. The main goal of the work is to establish the benefits of airfoil clocking and to reveal the physical mechanisms inducing the effects.

2 Theoretical Background

The problem of interest involves the unsteady flow in a stage of an axial compressor with an IGV. It is assumed that the flow at the inlet of the IGV is uniform. According to the theory of gasdynamics for interacting blade rows, there will be temporal (periodic) variations in the relative speed of the gas, causing variable circulation and loads on the blades. A portion of the unsteadiness in interacting blade rows is associated with potential field non-uniformities, while part is connected with the system of wakes generated by upstream blade rows. Clearly, the level of the unsteadiness is related to the distance (axial gaps) between blade rows, both in terms of the strength of the potential field and the intensity of the wakes. It has been shown in calculations and experiments that potential disturbances can remain significant for up to one pitch upstream of the blade row generating the disturbance. Wakes can require much larger axial distances to spread and decay.

2.1 Experimental Effort

Based on the knowledge that flow unsteadiness can be transmitted both upstream and downstream of a given blade row, theoretical and experimental investigations should include all the blade rows in a multi-stage environment. The total influence of the IGV and stator on the rotor is comprised of the IGV wakes and the potential disturbances generated by the IGV and stator. A mutual circumferential shifting (clocking) of the IGV and stator blades will alter the interferential position of their flow fields, which in turn can influence the unsteady loads and wakes generated by the rotor blades. Experimentally these effects can be quantified by measuring the (time-averaged and unsteady) static and total pressures at the rotor blade-passing frequency using high-frequency transducers. Note that the time-averaged aerodynamic load acting on the rotor blade is not dependent on the clocking position, implying that the IGV and stator blades act on the rotor independently. Measuring the moment on the compressor shaft should confirm (or refute) this assumption. Measuring the unsteady static pressure on the rotor casing will also quantify the influence of stator clocking on the unsteady rotor loading.

As the stator blades are clocked there is also a shift in the location where the IGV wakes impact the stators. Such a shift can alter the character of the flow in the stator passage. Previous studies have shown that the highest efficiencies are obtained when the IGV wakes impact the leading edges of the stator blades, while the lowest efficiencies are obtained when the IGV wakes are aligned with the mid-passage region of the stator passages [1]. The alignment of the IGV wakes in the stator passage can be classified as a steady flow mechanism.

2.2 Computational Effort

Numerical simulations of stator clocking have been useful in highlighting the physical mechanisms associated with airfoil clocking because it is possible to isolate wakes and interrogate boundary layer quantities [2, 3, 9]. In the current study, a series

Quantity	Symbol	Value
Rotor Tip Speed	U_{tip}	300 m/sec
Loading Coefficient	$\bar{H}_t$	0.29
Flow Coefficient	ϕ	0.50
Stage Total Pressure Ratio	PR_t	1.31
Stage Adiabatic Efficiency	η_{ad}	0.88
IGV, Rotor, Stator Exit Flow Angles	$\alpha_1, \beta_2, \alpha_3 (see Fig.1)$	30°
Stage Outside Diameter	D_c	0.32 m
Rotor Inlet Hub-to-Tip Ratio	$\bar{d}_1$	0.75

Table 1: Compressor design parameters.

of unsteady two-dimensional Navier-Stokes simulations have been performed for the IGV-rotor-stator geometry.

The numerical procedure discretizes the flow field into two types of zones. Inner "O" grids are used to resolve the flowfield near the airfoils. These "O" grids are overlaid on outer "H" grids which are used to resolve the flowfield in the passages between airfoils. The "H" grids are allowed to slip relative to one another to simulate the relative motion between rotors and stators. The governing equations are cast in the strong conservation form. A fully implicit, finite-difference method is used to advance the solution of the governing equations in time. A Newton-Raphson subiteration scheme is used to reduce the linearization and factorization errors at each time step. The convective terms are evaluated using a third-order-accurate upwind-biased scheme. The viscous terms are evaluated using second-order accurate central differences. An algebraic two-layer turbulence model is used to compute the turbulent eddy viscosity. More details on the numerical procedure can be found in Refs. [9, 10].

3 Experimental Geometry and Measurements

The experimental test article studied in this investigation is an axial compressor with the design parameters given in Table 1. Note, both the loading and flow coefficients in Table 1 are referenced to the tip speed of the rotor. The flow path and measurement stations are shown in Fig. 1. The IGV and stator rows each contain $Z_{IGV} = Z_{st} = 34$ blades, while the rotor row contains $Z_r = 35$ blades.

The radial clearance between the rotor blade tips and the outer casing is on average equal to 0.5 mm at cold conditions, with this clearance reducing to approximately 0.3 mm due to rotor rotation (during compressor operation). The radial clearance near the hub of cantilevered stators is approximately 0.3 mm, and this value does not vary during compressor operation. The axial gap between the IGV and rotor blade rows for the baseline compressor configuration is 70% as referenced to the chord length of the rotor ($b = 33$ mm). The axial gap between the rotor and stator blade rows is 66%, again referenced to the rotor.

The construction of the compressor stage allows the IGV blades (and disk) to be moved 30° in the circumferential direction (about three IGV pitches). The stator blades can slowly rotate at approximately 2 rev/min. The measurement ring at the exit of the stage can also be rotated approximately 30°. Thus, for all the relative clocking positions of the IGV and stator blades the stage characteristics can be measured by rotating the measuring ring at the exit of the stage.

The main geometric characteristics of the compressor stage are illustrated in Fig. 1. The measurements taken within the stage can be classified into two groups:

1. Time-averaged (steady) flow parameters - The time-averaged measurements are obtained using pneumatic probes and inertial probes. There are static pressure probes on the casing and hub, three 9-point radial rake probes to determine the total pressure at the rotor inlet, six 8-point radial rakes to determine the total pressure at the stage exit and two 4-point radial thermocouple rakes to determine the total temperature at the exit of the stage.

2. Unsteady flow parameters - The unsteady measurements are obtained using high-frequency Endevco transducers located directly in the flow. Total pressure measurements are taken at the stage exit with the help of a movable radial probe and unsteady static pressure measurements are obtained on the rotor casing at six points.

The rotation frequency and moment on the rotor shaft are determined using a torquemeter, while the mass flow rate is obtained using a collector at the entrance to the facility. In calculating the work and efficiency of the stage, the effects of the friction in the bearings and the disk were subtracted out. The average errors in the measurements were 0.2% for the pressure, 0.5^o for the temperature and 0.3% for the torque. The total error in the determination of the efficiency was 0.5%.

The experimental tests were performed without insert probes (flow path obstructions), except for the stage exit. The total pressure upstream of the rotor was determined by preliminary calibration of the losses at the inlet part of the stage using three 9-point radial rake probes. During the tests involving clocking the radial probes were removed from the flow path. During some of the tests the Endevco transducers were inserted after the rotor and at the stage exit.

The experimental tests were conducted using the following procedure:

1. For a given blade rotation speed and back pressure (exit throttle position) all the flow parameters were recorded at 10 divisible frequencies at 10 positions of the measuring ring over one stator pitch at the exit of the stage. Thus, the flow characteristics were obtained for some initial clocking position between the IGV and stator.

2. At the design operating conditions the clocking of the IGV and stator rows was varied by rotating the IGV by an amount Δt_{IGV}. Then, the flow parameters were again measured at the 10 positions of the measuring ring at the stage exit. For certain clocking positions the output of the high-frequency Endevco transducers was recorded on magnetic tape.

3. The collection and processing of the time-averaged flow parameters were made using an automated system which periodically prints results, while the unsteady parameters were processed using a spectrum analyzer.

4 Experimental Results

All the measurements discussed below were obtained at the design operating conditions given in Table 1.

4.1 Time-Averaged Flow Parameters

The stage total pressure ratio, PR_t, was determined as the ratio of the time-averaged total pressure downstream of the stator over the time-averaged total pressure just upstream of the rotor. Figure 2 illustrates the total pressure ratios at the tip, midspan and hub radii of the stator for two IGV positions separated by one-half the pitch ($\Delta t_{IGV} = 0.9$ and $\Delta t_{IGV} = 1.5$). It is clear from Fig. 2 that the pressure ratios are appreciably different for different clockings of the IGV and stator rows.

Figure 3 contains stator pitch-wise averaged total pressure ratios at the same radii (as in Fig 2) for IGV displacements of 10% of the IGV pitch, and for a total distance of more than 2 IGV pitches. Figure 3 shows the significant influence of airfoil clocking in a compressor, and that the effects are periodic over the pitch of the IGV (or stator) airfoils for similar blade count ratios. Variations in the term $({PR_t}^{(k-1)/k} - 1)$, which is proportional to the adiabatic (useful) work and efficiency of a stage, reaches 7-8% in the endwall regions and 3% near midspan. The main cause of the larger efficiency variations in the endwall regions is that the wakes from the IGV's are stronger in the endwall regions than near midspan (as determined from total pressure measurements).

Changes in the overall stage parameters as a function of clocking are presented in Fig. 4. In Fig. 4, the values of the pressure ratio were obtained by averaging over the entire stage exit (radially and circumferentially). The stage loading coefficient, $\bar{H}_t$, was obtained using measurements of the torque on the shaft of the compressor. As the IGV's are rotated circumferentially the loading coefficient displays little variation, while the efficiency $({PR_t}^{(k-1)/k} - 1)$ changes about 1.4%. The smaller changes in the overall total pressure ratio, compared to the values in Fig. 3, are due to the fact that the extrema of the total pressure ratios in Fig. 3 occur at different circumferential locations of the IGV's. Thus, averaging the total pressure and efficiency over the circumferential and radial directions reduces the overall effects of airfoil clocking.

4.2 Unsteady Total Pressure at the Stage Exit

The unsteady total pressure at the stage exit was measured using a radial probe of Endevco transducers mounted on a measuring ring. The measurements were obtained for five clocking positions of the IGV and stator (over one blade pitch) and three radii (tip, midspan and hub). At each of the five IGV positions (Δt_{IGV}) and the three radii, measurements were taken at 10 locations over a stator pitch. Using the recorded data and a spectrum analyzer the following integrated parameters were determined:

- $\bar{P}_t$ - time-averaged total pressure
- $\epsilon^*{}_T$ (in % of $\bar{P}_t$) - intensity of the turbulent component of the unsteady total pressure in the frequency range of 0-10 kHz
- $\epsilon^*{}_{Pk}$ - intensity of the periodic component of the unsteady total pressure for the $k-th$ $(k = 1, 2, 3, 4)$ harmonic of rotor blade-passing frequency
- $\epsilon^*{}_P = \sqrt{(\epsilon^*{}_{P1})^2 + (\epsilon^*{}_{P2})^2 + (\epsilon^*{}_{P3})^2 + (\epsilon^*{}_{P4})^2}$ - Total intensity of the periodic fluctuations
- $\epsilon^*{}_\Sigma = \epsilon^*{}_T + \epsilon^*{}_P$ - Intensity of summed fluctuations in the range of $0 - 40\ kHz$

From the resulting distributions it was determined that the intensity of the turbulent fluctuations, $\epsilon^*{}_T$, was largest in the region of the stator blade wakes, whereas the periodic fluctuations were associated with the rotor wakes. As the IGV and stators were clocked, the pitch-wise averages of the total pressure, $\bar{P}_t$, and total intensity of the periodic fluctuations, $\epsilon^*{}_P$, showed significant changes, while the values of $\epsilon^*{}_T$ displayed minimal changes.

The pitch-wise averages of $\epsilon^*{}_P$ and $\epsilon^*{}_T$ are shown as a function of IGV location in Fig. 5. The values of $\epsilon^*{}_P$ show much larger changes than $\epsilon^*{}_T$ at all three radii, implying a change in the intensity of the wakes from the rotor blades. This trend is shown more clearly in Fig. 6 which illustrates changes in the radial averages of $\epsilon^*{}_P$, $\epsilon^*{}_T$ and $\epsilon^*{}_\Sigma$ as function of IGV position. Time-averaged values of the unsteady total pressure obtained with the transducers display excellent with the values obtained with the pneumatic measurements.

4.3 Unsteady Pressure on the Rotor Casing

Fluctuations in the pressure on the rotor casing were measured at six points along the axis of the stage in the interblade regions before and after the rotor (transducers 1 and 6), on the leading and trailing edges of the rotor (transducers 2 and 5) and in the rotor passage (transducers 3 and 4). The measurements were made for IGV circumferential location variations of 10% of a pitch, over one complete pitch.

The results of the spectral analysis are shown in Fig. 7 for changes along the flow path of the pressure $\bar{P}$ (referenced to atmospheric pressure) and intensities of $\epsilon^*{}_P$ and $\epsilon^*{}_T$. In Fig. 7 the values of $\epsilon^*{}_P$ (which characterize the difference in the pressure between the pressure and suction surfaces of the rotor) display the largest changes. It is reasonable to conclude that changing the location of the IGV blades influences the loading and character of the flow in the rotor passage. Figure 8, which shows the unsteady pressure recorded with transducers 1,4 and 6, clearly shows the effects of clocking.

5 Computational Results

Two sets of numerical simulations have been performed. In both sets of simulations the blade counts were set at $Z_{IGV} = Z_{st} = Z_r = 34$ airfoils, and the rotor airfoils were scaled to keep the correct pitch-to-chord ratio. One set of simulations was performed at the design conditions ($\phi = 0.50$), and one set was performed at low-flow conditions ($\phi = 0.40$, near stall). The computational grid used to discretize the flow field contained 73,251 grid points. Eight equally-spaced increments were used in clocking the airfoils over one IGV/stator pitch. The average value of y^+, the non-dimensional distance of the first grid point above the surface, was approximately 0.75 for all three blade rows, and the surface boundary layers contained (on average) 25 grid points. All flow parameters were time-averaged over 12 rotor blade-passing cycles to allow for low-frequency unsteadiness.

Figure 9 illustrates the time- and stator pitch-averaged efficiency (as a function of airfoil clocking) computed in the simulations, along with the experimental data at midspan. The computed results exhibit lower efficiencies than the experimental data at the design condition, while the computed low-flow solution displays higher efficiencies. All three curves, however, display similar trends, with the highest efficiencies being observed when $\Delta t \approx 0.20$ and the lowest efficiencies occurring when $\Delta t \approx 0.65$. The experimental data show that $\Delta\eta \approx 2.5\%$, while the computed results indicate $\Delta\eta \approx 1.75\%$. It is significant that the computed results show that airfoil clocking is effective at off-design flow conditions.

The time-averaged results were similar at all clocking positions for the IGV and rotor, so the remainder of the discussion of the computational results will focus on the stator passage. Figures 10 and 11 illustrate the computed time-averaged vorticity contours in the stator passage at the design flow coefficient for the clocking positions corresponding to the highest ($\Delta t \approx 0.25$) and lowest ($\Delta t \approx 0.625$) efficiencies, respectively. Time-averaging the vorticity in the stator passage is a useful method for tracking the IGV wakes. These two figures show that the highest efficiency is obtained when the IGV wakes impact the leading edge of the stator airfoils, while the lowest efficiency occurs when the IGV wakes reside in the mid-passage region between the stator airfoils. These results agree with the results of previous investigations of airfoil clocking in turbines and compressors (*e.g.,* Refs. [1, 3, 9]). Figure 12 contains instantaneous entropy contours in the stage for $\Delta t = 0.25$. The entropy is another flow variable which is well-suited to wake-tracking. These contours highlight the path of the IGV wake at the clocking position corresponding to the peak efficiency, again showing that the wakes impact the leading edge of the stator. It is also evident from Fig. 12 that the rotor induces periodic shedding of the stator suction surface boundary layer.

Figure 13 shows the computed time-averaged pressure distributions on the surface of the stator at the design-flow operating conditions. The pressure is slightly lower at the clocking position corresponding to the highest efficiency, which is analogous to the results of several turbine airfoil clocking studies [2, 3]. Figure 14 contains the pressure envelopes for the clocking positions corresponding to the highest and lowest efficiencies. In this investigation, the term *envelope* refers to the minimum and maximum pressures experienced by an airfoil over the course of one or more rotor-passing cycles. Figure 14 shows the stator experiences more unsteadiness at the peak efficiency clocking position than at the lowest efficiency clocking position (most notably on the suction surface). This is in contrast to the rotor, which displayed less unsteadiness at the peak efficiency clocking position.

Figure 15 illustrates the time-averaged skin friction coefficient distributions at the extrema of the efficiency. The time-averaged skin friction (see Fig. 15) is greater at the highest efficiency, and the extent of flow separation (*i.e.*, $C_f < 0$) is reduced. This suggests that the process of the IGV wake (and associated vorticity) impacting the stator surface may enable the stator boundary layer to better withstand flow separation. The displacement thickness and momentum thickness distributions displayed a more gradual increase at the clocking location corresponding to the highest efficiency than at the clocking location corresponding to the lowest efficiency. In addition, the maximum thickness locations were located further aft on the airfoil at the highest efficiency, suggesting a healthier boundary layer.

6 Conclusions

The experimental and computational work performed in this investigation has led to the following conclusions concerning airfoil clocking in compressors:

- Mutual clocking of the IGV and stator significantly changes the stage total pressure ratio and efficiency. The experimental data showed changes in the stage total pressure ratio of 0.5%, and changes in the stage efficiency of approximately 1% (7-8% in the endwall regions).

- Changes in the stage aerodynamic parameters due to stator clocking are accompanied by changes in the rotor passage flow field. According to measurements of unsteady static pressure on the rotor casing, and unsteady total pressure measurements at the stage exit, the aerodynamic load on the rotor varies by up to 20% and the intensity of the rotor wakes changes by up to 50%.

- Computational simulations, which were conducted using an unsteady two-dimensional Navier-Stokes analysis (with an equal number of airfoils modelled in each blade row), showed that the highest stage efficiencies are observed when the airfoils are clocked such that the IGV wakes directly impact the surface of the stator. The lowest efficiencies were obtained when the IGV wakes are located mid-passage between stators. The action of the IGV wakes impacting the stator airfoils (mixing/vorticity) appears to enable the stator to better withstand flow separation.

References

[1] Huber, F. W., Johnson, P. D., Sharma, O. P., Staubach, J. B., and Gaddis, S. W., "Performance Improvement Through Indexing of Turbine Airfoils: Part I - Experimental Investigation," ASME Paper 95-GT-27, Houston, TX, 1995, also to appear in the *ASME Journal of Turbomachinery*.

[2] Griffin, L. W., Huber, F. W., and Sharma, O. P., "Performance Improvement Through Indexing of Turbine Airfoils: Part II - Numerical Simulation," ASME Paper 95-GT-28, Houston, TX, 1995, also to appear in the *ASME Journal of Turbomachinery*

[3] Dorney, D. J., Sharma, O. P., "A Study of Turbine Performance Increases Through Airfoil Clocking", AIAA Paper 96-2816, Lake Buena Vista, FL, 1996.

[4] Saren, V. E., "Relative Position of Two Rows of Axial Turbomachine Effect on Aerodynamics of Row Placed Between Them," *Unsteady Aerodynamics and Aeroelasticity of Turbomachines*, Elsevier, 1995, pp. 421-425.

[5] Saren, V. E., "Some Ways of Reducing Unsteady Loads Due to Blade Row Hydrodynamic Interaction in Axial Flow Turbomachines," Second International Conference EAHE, Pilsen, Czech Republic, 1994, pp. 160-165.

[6] Hsu, S. T., and Wo, A. M., "Reduction of Unsteady Blade Loading by Beneficial Use of Vortical and Potential Disturbances in an Axial Compressor with Rotor Clocking," ASME Paper 97-GT-86, Orlando, FL, 1997, also to appear in the *ASME Journal of Turbomachinery.*

[7] Walker, G. J., Hughes, J. D., Kohler, I., and Solomon, W. J., "The Influence of Wake-Wake Interactions on Loss Fluctuations of a Downstream Axial Compressor Blade Row," ASME Paper 97-GT-469, Orlando, FL, 1997.

[8] Barankiewicz, W. S., and Hathaway, M. D., "Effects of Stator Indexing on Performance in a Low Speed Multistage Axial Compressor," ASME Paper 97-GT-496, Orlando, FL, 1997.

[9] Gundy-Burlet, K. L., and Dorney, D. J., "Investigation of Airfoil Clocking and Inter-Blade Row Gaps in Axial Compressors," AIAA 97-3008, Seattle, WA, 1997.

[10] Dorney, D. J., and Verdon, J. M., "Numerical Simulations of Unsteady Cascade Flow," *ASME Journal of Turbomachinery*, Vol. 116, No. 4, October 1994, pp. 665-675.

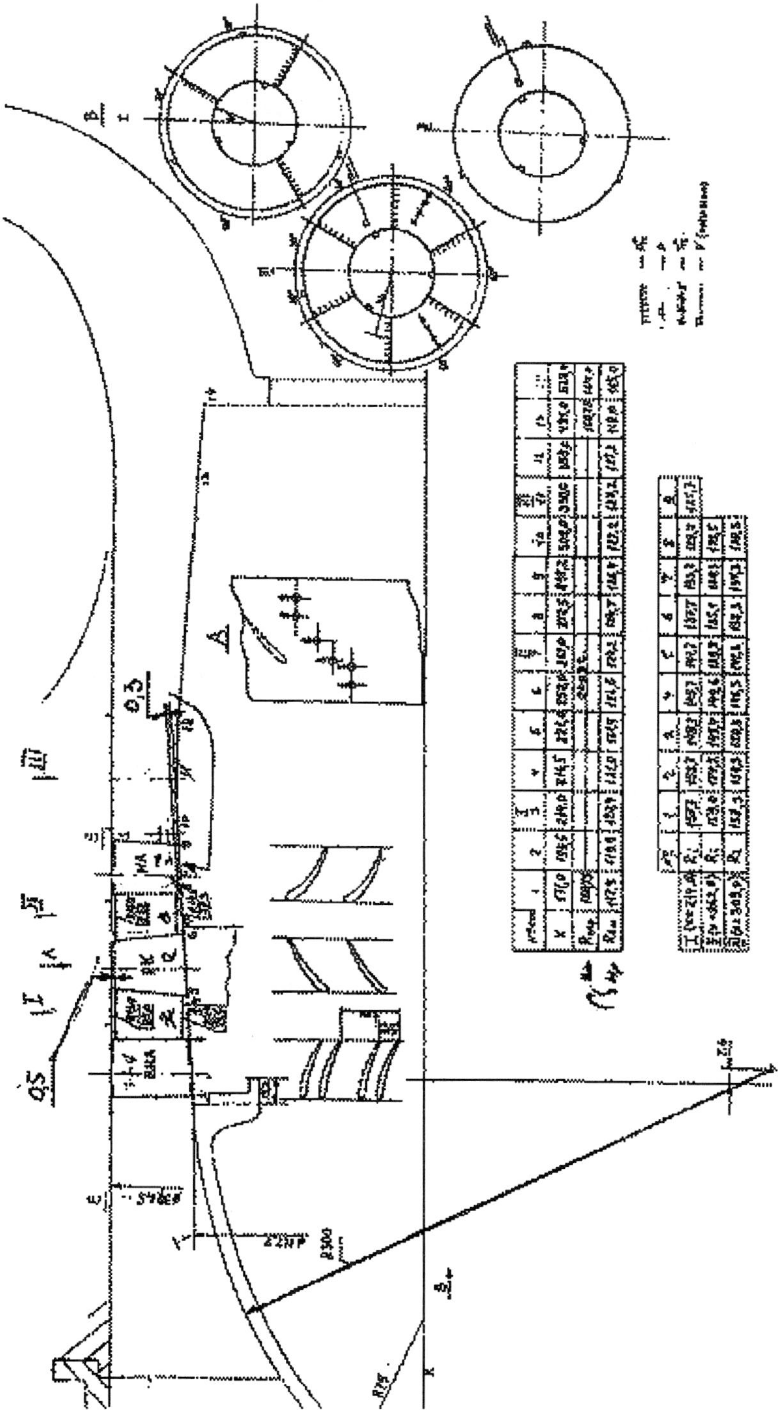

Figure 1: Compressor flow path and measurements stations

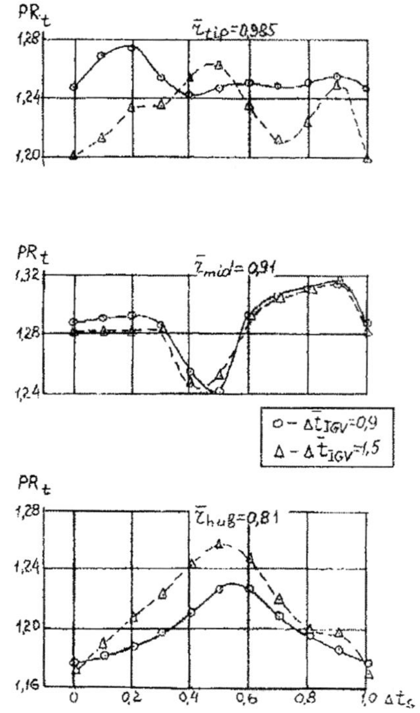

Figure 2: Stage total pressure ratios at the tip, midspan and hub radii of the stator for two clocking positions

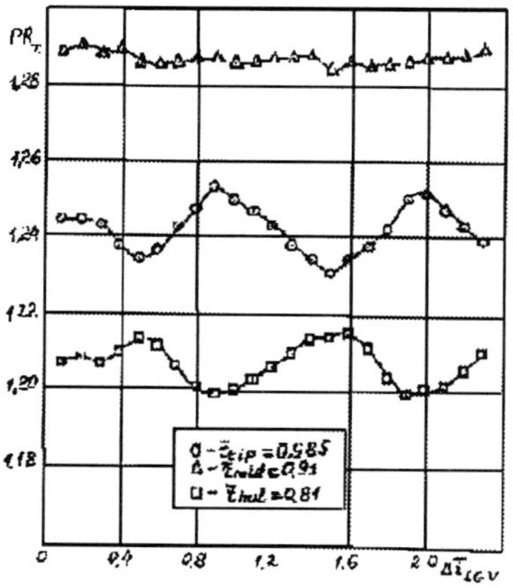

Figure 3: Stator pitch-wise averaged total pressure ratios at the tip, midspan and hub radii of the stator for clocking over more than 2 IGV pitches

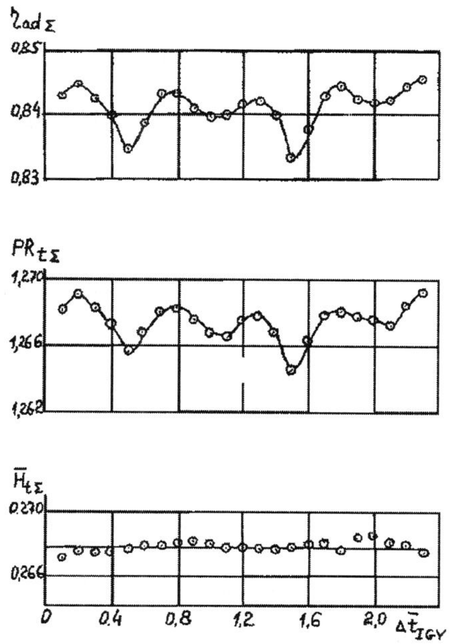

Figure 4: Variation of overall stage flow parameters as a function of IGV location

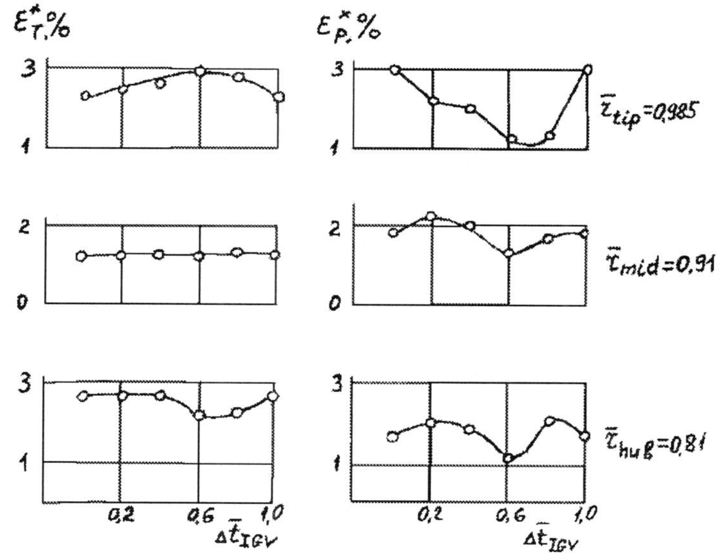

Figure 5: Variation of $\epsilon^*{}_P$ and $\epsilon^*{}_T$ as a function of IGV location

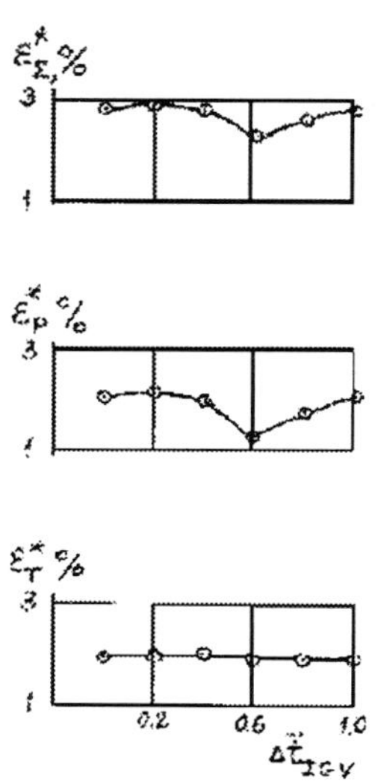

Figure 6: Variation of radial averages of $\epsilon^*{}_P$, $\epsilon^*{}_T$ and $\epsilon^*{}_\Sigma$ as a function of IGV location

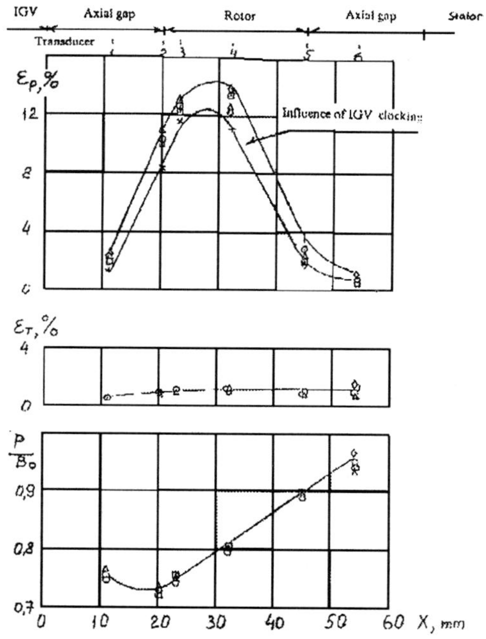

Figure 7: Streamwise variations in $\bar{P}$, $\epsilon^*{}_P$ and $\epsilon^*{}_T$ as a function of IGV location

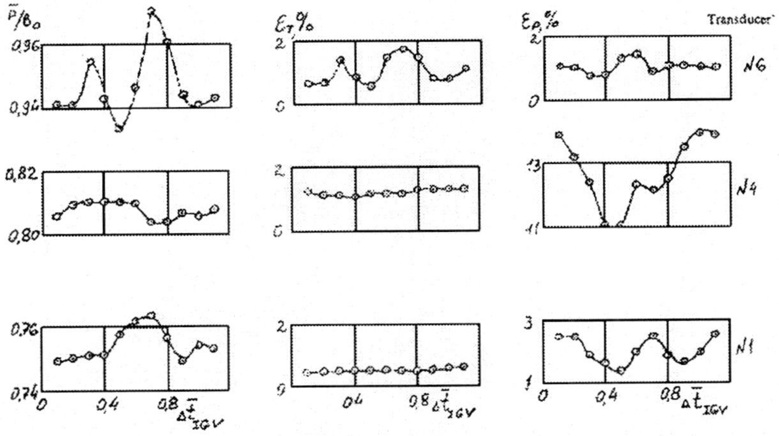

Figure 8: Variations in $\bar{P}$ at transducers 1,4 and 6 as a function of IGV location

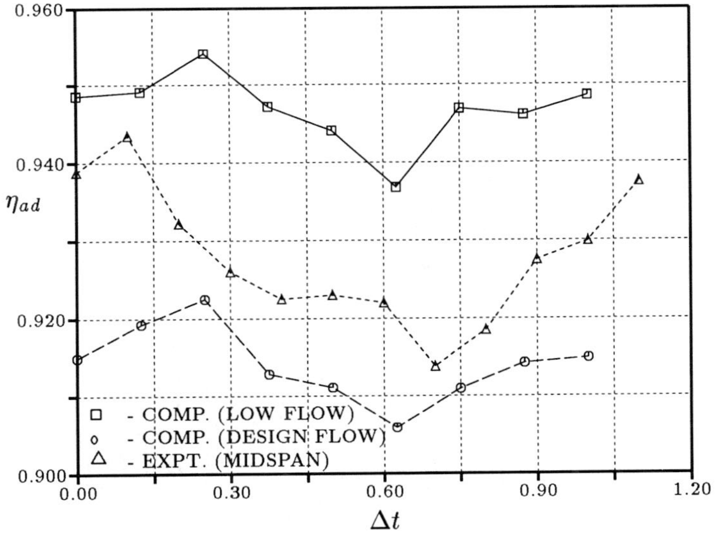

Figure 9: Computed and experimental efficiencies as a function of airfoil clocking

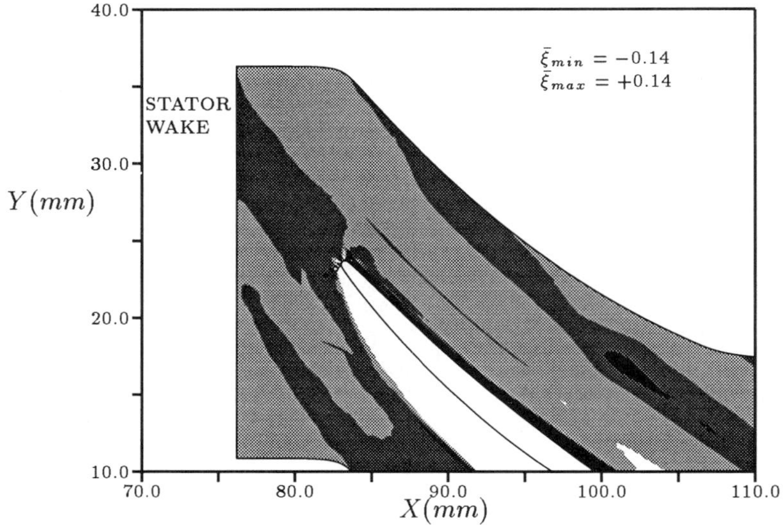

Figure 10: Computed time-averaged vorticity contours at the highest efficiency - $\Delta t = 0.25$

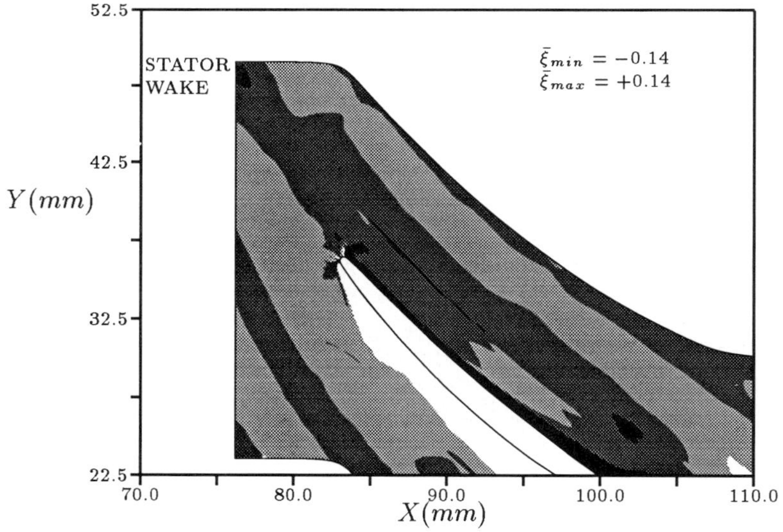

Figure 11: Computed time-averaged vorticity contours at the lowest efficiency - $\Delta t = 0.625$

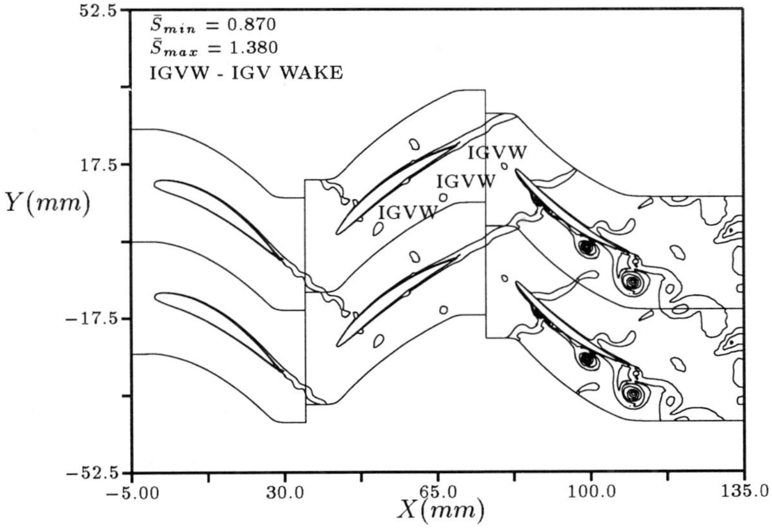

Figure 12: Computed instantaneous entropy contours at the highest efficiency - $\Delta t = 0.25$

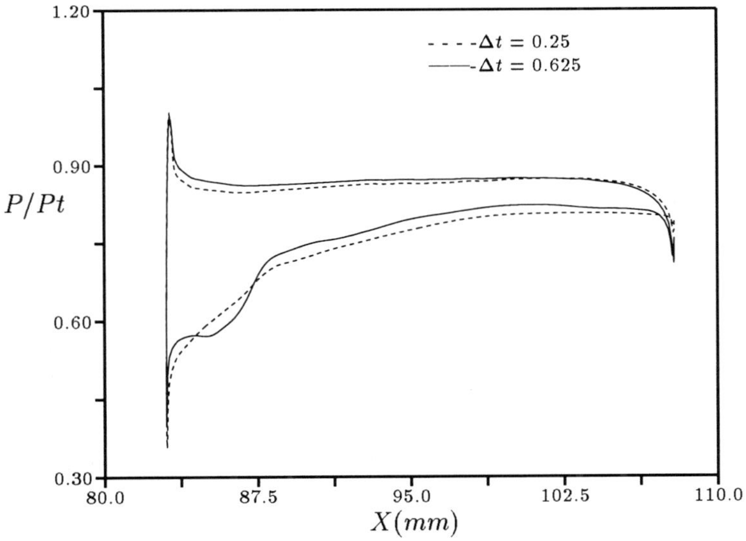

Figure 13: Computed time-averaged stator surface pressure at the highest ($\Delta t = 0.25$) and lowest ($\Delta t = 0.625$) efficiencies

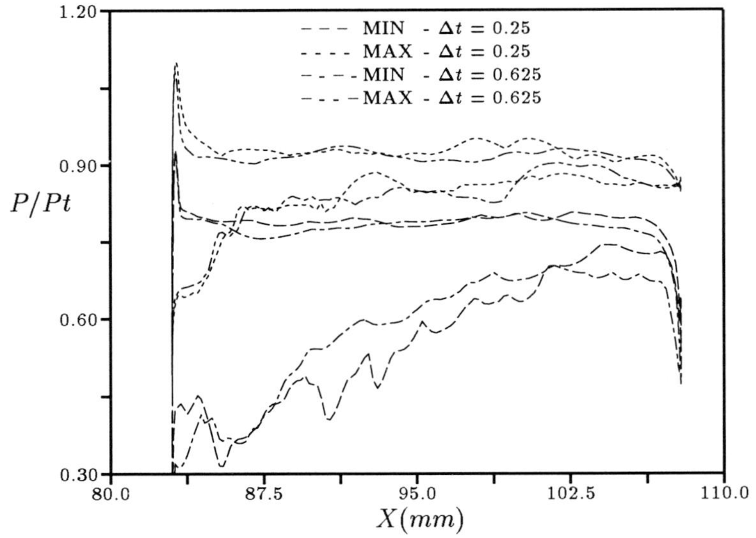

Figure 14: Computed unsteady stator pressure envelopes at the highest ($\Delta t = 0.25$) and lowest ($\Delta t = 0.625$) efficiencies

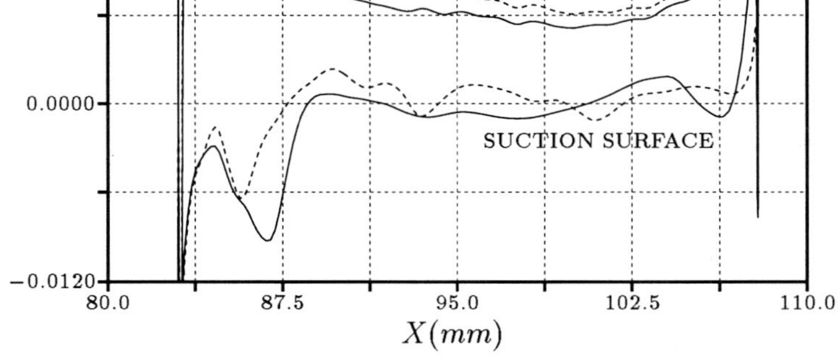

Figure 15: Computed time-averaged skin friction coefficient at the highest ($\Delta t = 0.25$) and lowest ($\Delta t = 0.625$) efficiencies

8: Theoretical Studies of Unsteady Flow Around Oscillating Blades - 3

A 3D LINEARIZED EULER ANALYSIS FOR BLADE ROWS PART 1: AERODYNAMIC AND NUMERICAL FORMULATIONS

MATTHEW D. MONTGOMERY AND JOSEPH M. VERDON
United Technologies Research Center
East Hartford, CT 06108, USA

Abstract. A three-dimensional, linearized, unsteady, Euler analysis is being developed to efficiently predict the aeroelastic and aeroacoustic behaviors of axial-flow turbomachinery blading. The aerodynamic and numerical formulations are presented in Part 1 of this paper. In particular, the field equations and boundary conditions governing nonlinear and linearized, unsteady, inviscid flows through blade rows in cylindrical annular ducts are derived. A linearized numerical formulation, coupling a near-field, wave-split, finite-volume analysis to far-field eigenanalyses, is also described. The linearized aerodynamic and numerical models have been implemented into a three-dimensional, unsteady flow code, LINFLUX. This code is applied in Part 2 to benchmark, unsteady, subsonic flows to demonstrate its accuracy and prediction capabilities.

1. Introduction

Predictions of unsteady flows through turbomachinery blade rows are motivated primarily by the need to determine the aeroelastic (flutter and forced vibration) and aeroacoustic (sound generation and propagation) characteristics of the blading. Accurate and efficient aerodynamic analyses are needed to determine the unsteady blade loads and the acoustic responses of the blade row, for various sources of excitation. The excitations are blade vibrations and aerodynamic disturbances at the inlet and exit that carry energy towards the blade row. The computational resources required by simulations of nonlinear and/or viscous unsteady flows prohibit their use in aeroelastic or aeroacoustic design studies. Therefore, linearized, inviscid analyses are needed for efficient aerodynamic response predictions.

T.H. Fransson (ed.), Unsteady Aerodynamics and Aeroelasticity of Turbomachines, 427-444.

The first linearized analyses available for turbomachinery aeroelastic and aeroacoustic applications were restricted to shock-free flows through lightly-loaded blade rows in two- and three-dimensions. Because of these restrictions, two-dimensional unsteady linearizations, relative to nonuniform, potential, mean flows, were developed. Such linearized analyses account for the effects of real blade geometry, mean blade loading, and weak shocks on the unsteady aerodynamic response.

More comprehensive linearizations are needed to predict two- and three-dimensional flows in which strong shocks occur, and three-dimensional flows with mean swirl. For such flows, the nonlinear Euler equations are required to model the nonisentropic and rotational steady background flows, and the linearized Euler equations are required to model the unsteady perturbations. Thus, two- [1, 2, 3] and three-dimensional [4, 5] linearized Euler analyses are currently being developed. The present authors have developed the two-dimensional LINFLUX analysis, and applied this analysis to subsonic and transonic unsteady flows excited by prescribed blade motions and external aerodynamic disturbances [6, 7].

The goals of the present effort were to develop, demonstrate, and validate a three-dimensional LINFLUX analysis for unsteady flows through blade rows in cylindrical annular ducts. A 3D code has been constructed, which couples a near-field, wave-split, finite-volume analysis for the unsteady perturbations of general mean flows to far-field, semi-analytic eigenanalyses for the perturbations of fully-developed, axisymmetric mean flows. The near-field analysis is based upon the numerical schemes used in the nonlinear unsteady analysis, TURBO [8]. TURBO is an implicit, flux-split, cell-centered, finite-volume analysis used herein to predict three-dimensional steady background flows. The flux or wave splitting allows a sharp resolution of nonlinear shock phenomena — a feature which should facilitate the accurate prediction of impulsive unsteady shock loads with future linearized analyses. The far-field analyses, which are coupled to the linearized near-field analysis at the computational inflow and outflow boundaries, allow incoming aerodynamic excitations to be prescribed, and outgoing disturbances to pass through the computational boundaries without reflection.

The LINFLUX analysis is described in Part 1, and applied in Part 2 to predict unsteady flows through relatively simple blade rows; namely, a helical fan and a real-blade configuration based on the Tenth Standard Cascade [9]. We considered unsteady, subsonic flows excited by blade vibrations or acoustic excitations. Comparisons with 2D analyses indicate that the 3D LINFLUX analysis provides accurate unsteady aerodynamic response information, provided that meshes of sufficient density and clustering are employed. However, additional work is needed to enhance the computational efficiency of LINFLUX and extend its range of application.

2. Unsteady Flow through a Blade Row

We consider time-dependent flow of an inviscid, non-heat conducting, perfect gas, with negligible body forces, through a rotating and vibrating blade row in a stationary annular duct, see Figure 1. The duct is of infinite axial extent and has hub and duct radii of $r = r_H$ and $r = r_D$. The blade row consists of N_B blades, which rotate at constant angular velocity $\boldsymbol{\Omega} = \Omega \mathbf{e}_\xi$. In the absence of vibrational motion, the blades are identical in shape and orientation, and equally spaced around the rotor.

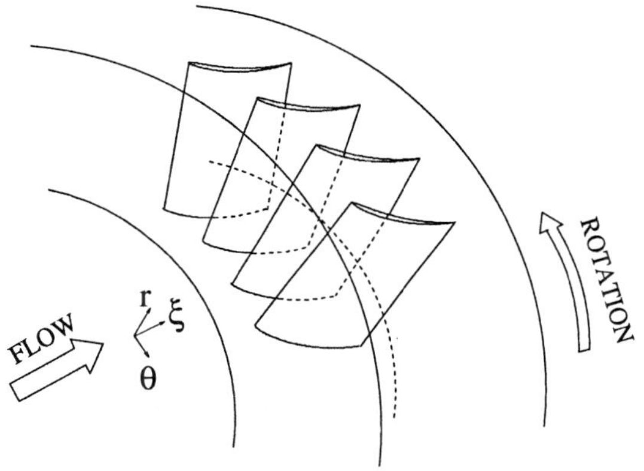

Figure 1. Rotating axial compressor blade row operating within an annular duct.

We will analyze unsteady flows in a relative reference frame that rotates with the blade row, using right-handed cylindrical $(r,\ \theta,\ \xi)$ and Cartesian $(x_1, x_2, x_3) = (\xi,\ r\sin\theta,\ -r\cos\theta)$ coordinates. Here ξ and r measure distance along and radially outward from the duct axis, respectively, and θ measures angular distance opposite to the rotation direction. The rotation is assumed to be counterclockwise relative to an observer looking in the axial flow direction, so that $\Omega < 0$. We will also examine the flow in a stationary, or absolute, reference frame, and will use the superscript "abs" to indicate an absolute physical quantity; e.g., $\theta^{\text{abs}} = \theta + \Omega t$.

To describe flows in which the fluid domain deforms with time, we consider two sets of independent variables, say $(\mathbf{x}, t)$ and $(\bar{\mathbf{x}}, t)$. The position vector $\mathbf{x}(\bar{\mathbf{x}}, t) = \bar{\mathbf{x}} + \mathcal{R}(\bar{\mathbf{x}}, t)$ describes the instantaneous location, in the relative frame, of a moving field point, $\bar{\mathbf{x}}$ refers to the mean position of this point, and $\mathcal{R}(\bar{\mathbf{x}}, t)$ is its displacement. The displacement field, $\mathcal{R}$, is prescribed so that the solution domain deforms with the vibrational blade motions, but is rigid far from the blade row.

For aeroelastic and aeroacoustic applications, we are usually interested in unsteady flows that can be regarded as perturbations of flows that are steady in the relative frame. Moreover, the steady background flows far from the blade row are assumed to be at most a small steady perturbation from a fully-developed, axisymmetric flow. The unsteady fluctuations arise from temporally and circumferentially periodic excitations of small amplitude; i.e., prescribed vibrational blade motions and aerodynamic disturbances at inlet and exit that carry energy towards the blade row.

For example, if the blades vibrate at reduced frequency ω, in the rotating frame, with a constant interblade phase angle, σ, we can write

$$\mathcal{R}_{\mathrm{B}_n}(\bar{r},\ \bar{\theta}+n\bar{\theta}_G,\ \bar{\xi},\ t) = \mathbf{T}_n Re\{\mathbf{R}_{\mathrm{B}}(\bar{r},\ \bar{\theta},\ \bar{\xi})\exp[i(\omega t + n\sigma)]\}. \quad (1)$$

Here, $\mathcal{R}_{\mathrm{B}_n}$ is the displacement of a point on the nth blade; $\bar{\theta}_G = 2\pi/N_B$ is the angular pitch of a blade passage; $\mathbf{T}_n$ is a rotation matrix, which rotates vectors through n passages; $n = 0, 1, 2, \ldots, N_B - 1$; $Re\{\ \}$ denotes the real part of $\{\ \}$; and $\mathbf{R}_{\mathrm{B}}$ is the complex amplitude of the reference ($n = 0$) blade displacement. The interblade phase angle, σ, is determined by the nodal diameter pattern of the vibration, i.e., $\sigma = 2\pi N_{\mathrm{D}}/N_B$, where $|N_{\mathrm{D}}|$ is the number of times the disturbance pattern repeats around the duct. The sign of N_{D} depends upon the direction of rotation of the disturbance pattern; if this pattern moves in the direction of blade rotation, $N_{\mathrm{D}} > 0$.

Typically, an external unsteady aerodynamic excitation is represented as a superposition of fundamental disturbances. The fundamental disturbances are harmonic in time, at temporal frequency ω, and in the circumferential direction, at angular wave number $\bar{m} = N_{\mathrm{D}} + mN_{\mathrm{B}}$, where $m = 0, \mp 1, \mp 2, \ldots$. The quantities ω, N_{D}, and the disturbance amplitudes are prescribed. The radial and axial behaviors are determined from the equations that govern the unsteady flow in the far field.

3. Unsteady Aerodynamic Formulations

In this discussion, all variables are dimensionless. To allow convenient comparisons between 2D and 3D solutions at midspan, the reference length is the blade chord at midspan, and the reference density and flow speed are the inlet mean density and mean relative flow speed at midspan.

Nonlinear equations that govern the unsteady flow are derived from the conservation laws for mass, momentum, and energy and the thermodynamic relations for a perfect gas. Consider an arbitrary moving control volume, $\mathcal{V}(t)$, bounded by the surface $\mathcal{A}(\mathbf{x}, t)$. The conservation laws for the fluid in $\mathcal{V}$, in a reference frame rotating with the blade row, can be written as

$$\frac{d}{dt}\int_{\mathcal{V}} \tilde{\mathbf{U}}d\mathcal{V} + \int_{\mathcal{A}} [\tilde{\mathbf{F}}_{x_j} - \tilde{\mathbf{U}}\dot{\mathcal{R}}_{x_j}]n_{x_j}d\mathcal{A} = \int_{\mathcal{V}} \tilde{\mathbf{S}}d\mathcal{V}\ . \quad (2)$$

Here, the symbol $\tilde{}$ indicates an unsteady flow quantity; $\dot{\mathcal{R}}$ is the velocity of a moving field point; $\mathbf{n}$ is a unit normal vector pointing outward from $\mathcal{A}$; and a summation over repeated j indices is implied.

The state, $\tilde{\mathbf{U}}$, flux, $\tilde{\mathbf{F}}_{x_j}$, and source term, $\tilde{\mathbf{S}}$, vectors in equation (2) are defined in Cartesian coordinates by $\tilde{\mathbf{U}}^T = \left[\tilde{\bar{\rho}},\ \tilde{\bar{\rho}}\tilde{V}_{x_1},\ \tilde{\bar{\rho}}\tilde{V}_{x_2},\ \tilde{\bar{\rho}}\tilde{V}_{x_3}, \tilde{\bar{\rho}}\tilde{E}_T\right]$,

$$\tilde{\mathbf{F}}_{x_j} = \begin{Bmatrix} \tilde{U}_{j+1} \\ \tilde{U}_{j+1}\tilde{U}_2/\tilde{U}_1 + \tilde{P}\delta_{1j} \\ \tilde{U}_{j+1}\tilde{U}_3/\tilde{U}_1 + \tilde{P}\delta_{2j} \\ \tilde{U}_{j+1}\tilde{U}_4/\tilde{U}_1 + \tilde{P}\delta_{3j} \\ \tilde{U}_{j+1}(\tilde{U}_5 + \tilde{P})/\tilde{U}_1 \end{Bmatrix}, \quad \tilde{\mathbf{S}}(\tilde{\mathbf{U}}, \mathbf{x}) = \Omega^2 \begin{Bmatrix} 0 \\ 0 \\ \tilde{U}_1 x_2 + 2\tilde{U}_4/\Omega \\ \tilde{U}_1 x_3 - 2\tilde{U}_3/\Omega \\ \tilde{U}_3 x_2 + \tilde{U}_4 x_3 \end{Bmatrix}, \tag{3}$$

where $\tilde{\bar{\rho}}$, $\tilde{\mathbf{V}}$, $\tilde{E}_T = \tilde{E} + \tilde{V}^2/2$ and $\tilde{P} = (\gamma - 1)\tilde{\bar{\rho}}\tilde{E} = (\gamma-1)[\tilde{U}_5 - \tilde{U}_1^{-1}(\tilde{U}_2^2 + \tilde{U}_3^2 + \tilde{U}_4^2)/2]$ are the time-dependent fluid density, relative velocity, relative specific total internal energy, and pressure, respectively.

We will also require a differential form of equation (2), that applies at fixed locations ($\dot{\mathcal{R}} = \mathbf{0}$) in the rotating frame, to describe the flow far from the blade row. In terms of cylindrical coordinates, this equation is

$$\left.\frac{\partial \tilde{\mathbf{U}}}{\partial t}\right|_{\mathbf{x}} + r^{-1}\frac{\partial r\tilde{\mathbf{F}}_r}{\partial r} + r^{-1}\frac{\partial \tilde{\mathbf{F}}_\theta}{\partial \theta} + \frac{\partial \tilde{\mathbf{F}}_\xi}{\partial \xi} = \tilde{\mathbf{S}}\ , \tag{4}$$

where $\tilde{\mathbf{U}}^T = \left[\tilde{\bar{\rho}},\ \tilde{\bar{\rho}}\tilde{V}_r,\ \tilde{\bar{\rho}}\tilde{V}_\theta,\ \tilde{\bar{\rho}}\tilde{V}_\xi,\ \tilde{\bar{\rho}}\tilde{E}_T\right]$,

$$\tilde{\mathbf{S}} = r^{-1}\begin{Bmatrix} 0 \\ (\tilde{U}_3)^2/\tilde{U}_1 + \tilde{P} \\ -\tilde{U}_2\tilde{U}_3/\tilde{U}_1 \\ 0 \\ 0 \end{Bmatrix} + \Omega \begin{Bmatrix} 0 \\ 2\tilde{U}_3 + \tilde{U}_1\Omega r \\ -2\tilde{U}_2 \\ 0 \\ \tilde{U}_2\Omega r \end{Bmatrix}, \tag{5}$$

and the flux vectors $\tilde{\mathbf{F}}_r(\tilde{\mathbf{U}})$, $\tilde{\mathbf{F}}_\theta(\tilde{\mathbf{U}})$ and $\tilde{\mathbf{F}}_\xi(\tilde{\mathbf{U}})$ have functional forms similar to those given for $\tilde{\mathbf{F}}_{x_j}(\tilde{\mathbf{U}})$, $j = 1, 2, 3$, in equation (3).

In addition to the foregoing field equations, conditions on the unsteady flow at the blade surfaces, the duct walls, and at the inflow and outflow boundaries of the computational domain are required. Flow tangency conditions, i.e., $(\tilde{\mathbf{V}} - \dot{\mathcal{R}}) \cdot \mathbf{n} = 0$, apply at the moving blade surfaces, $\mathcal{B}_n$, and at the stationary ($\dot{\mathcal{R}} = \mathbf{0}$) duct walls. At the inflow boundary, temporally- and circumferentially-averaged values of the total pressure, total temperature and flow angle are specified as functions of radius. At the outflow boundary, the temporally- and circumferentially-averaged pressure is specified, consistent with radial equilibrium. The unsteady fluctuations at inlet and exit that carry energy towards the blade row must also be specified; those that carry energy away from the blade row must be determined as part of the nonlinear unsteady solution.

3.1. LINEARIZED UNSTEADY AERODYNAMIC MODEL

If the unsteady excitations are assumed to be small (e.g., $|\mathcal{R}_{B_n}| \sim \mathcal{O}(\epsilon) << 1$) and to occur at a single temporal frequency, ω, then the unsteady, inviscid flow can be approximated as a first-order (in ϵ), harmonic perturbation of an underlying zeroth-order, background flow that is steady in the rotating frame. This approximation offers several computational advantages. First, physical time dependence can be removed from the first-order unsteady boundary value problem. Second, because of the assumed periodic nature of the rotor geometry and the flow conditions, the computational domain for the steady and linearized unsteady flows can be limited to a single blade passage. Finally, eigensolutions for the perturbations of fully-developed, axisymmetric, mean flows in the far-field can be constructed, and matched to computational near-field solutions to reduce the axial extent of the computational domain.

To determine the linearized unsteady aerodynamic equations, we expand the state vector, $\tilde{\mathbf{U}}$, into an asymptotic series of the form [1]

$$\tilde{\mathbf{U}}[\mathbf{x}(\bar{\mathbf{x}},t),t] = \mathbf{U}(\bar{\mathbf{x}}) + \tilde{\mathbf{u}}[\mathbf{x}(\bar{\mathbf{x}},t),t] + \dots \;, \tag{6}$$

where the column vector $\mathbf{U}(\bar{\mathbf{x}})$ contains the conservation variables for the steady background flow at $\bar{\mathbf{x}}$ and $\tilde{\mathbf{u}}[\mathbf{x}(\bar{\mathbf{x}},t),t] = Re\{\mathbf{u}(\bar{\mathbf{x}})\exp(i\omega t)\}$ contains the first-order unsteady variables at the instantaneous position, $\mathbf{x}$. The components of $\mathbf{u}$ are the complex amplitudes of the first-order, conservation variables, i.e., $\mathbf{u}^T = [\rho \;,\; \bar{\rho}v_{x_1} + \rho V_{x_1} \;,\; \bar{\rho}v_{x_2} + \rho V_{x_2} \;,\; \bar{\rho}v_{x_3} + \rho V_{x_3} \;,\; \bar{\rho}e_T + \rho E_T]$, where $\bar{\rho}$, $\mathbf{V}$, and E_T are the primitive steady flow variables and ρ, $\mathbf{v}$, and e_T are the complex amplitudes of the primitive, first-order, unsteady, flow variables. The unsteady flux, $\tilde{\mathbf{F}}_{x_j}$, and source term, $\tilde{\mathbf{S}}$, vectors in equation (2) are approximated using Taylor series expansions about the mean flow state, $\mathbf{U}$, and the reference spatial location, $\bar{\mathbf{x}}$, i.e.,

$$\tilde{\mathbf{F}}_{x_j}(\tilde{\mathbf{U}}) \approx \mathbf{F}_{x_j}(\mathbf{U}) + \frac{\partial \mathbf{F}_{x_j}}{\partial \mathbf{U}}\tilde{\mathbf{u}} \text{ and } \tilde{\mathbf{S}}(\tilde{\mathbf{U}},\mathbf{x}) \approx \mathbf{S}(\mathbf{U},\bar{\mathbf{x}}) + \frac{\partial \mathbf{S}}{\partial \mathbf{U}}\tilde{\mathbf{u}} + (\mathcal{R}\cdot\boldsymbol{\nabla})\mathbf{S} \;. \tag{7}$$

Finally, the nonlinear field equation (2) has been expressed in terms of integrals over the moving volume, $\mathcal{V}$, and the moving surface, $\mathcal{A}$. However, the steady and linearized unsteady equations will be expressed in terms of integrals over the mean volume, $\bar{V}$, and mean area, $\bar{A}$. To within first-order in ϵ, $d\mathcal{V} = (1 + \boldsymbol{\nabla}_{\bar{\mathbf{x}}}\cdot\mathcal{R})d\bar{V}$ and $\mathbf{n}d\mathcal{A} = \bar{\mathbf{n}}d\bar{A} + Re\{\Delta(\bar{\mathbf{n}}d\bar{A})\exp(i\omega t)\}$, where

$\bar{\mathbf{n}}$ is the unit outward normal to the surface $\bar{A}$, and $\Delta(\bar{\mathbf{n}}d\bar{A})$ is the complex amplitude of the first-harmonic component of $\mathbf{n}d\mathcal{A} - \bar{\mathbf{n}}d\bar{A}$.

The zeroth-order steady and the first-order unsteady equations are obtained by substituting the foregoing expansions into the nonlinear governing equations, equating terms of like power in ϵ, and neglecting terms of higher order in ϵ. This procedure leads to nonlinear and linear, variable-coefficient equations, respectively, for the zeroth- and first-order flows. The coefficients in the first-order equations depend on the underlying steady flow.

The integral field equation governing the zeroth-order, steady flow is

$$\int_{\bar{A}} \mathbf{F}_{x_j} n_{x_j} d\bar{A} = \int_{\bar{V}} \mathbf{S} d\bar{V} \ . \tag{8}$$

This equation is solved subject to the flow tangency conditions, i.e., $\mathbf{V} \cdot \mathbf{n} = 0$, at the blade surfaces and the duct walls, and periodic conditions, e.g., $\mathbf{V}(\bar{r},\ \bar{\theta} + n\bar{\theta}_G,\ \bar{r},\ \xi) = \mathbf{T}_n \mathbf{V}(\bar{r},\ \bar{\theta},\ \bar{\xi})$, at the passage boundaries. At the computational inflow and outflow boundaries, averaged values of the appropriate steady flow variables are specified, as described above. Small departures from these prescribed conditions, due to outgoing disturbances, are determined as part of the steady solution.

The field equation governing the first-order, unsteady flow is

$$i\omega \int_{\bar{V}} \mathbf{u} d\bar{V} + \int_{\bar{A}} \left(\frac{\partial \mathbf{F}_{x_j}}{\partial \mathbf{U}} \mathbf{u} \right) \bar{n}_{x_j} d\bar{A} - \int_{\bar{V}} \frac{\partial \mathbf{S}}{\partial \mathbf{U}} \mathbf{u} d\bar{V} = - \int_{\bar{V}} i\omega \left(\frac{\partial R_{x_m}}{\partial \bar{x}_m} \right) \mathbf{U} d\bar{V}$$
$$+ \int_{\bar{A}} \left[i\omega R_{x_j} \mathbf{U} \bar{n}_{x_j} d\bar{A} - \mathbf{F}_j \Delta(\bar{\mathbf{n}} d\bar{A}) \right] + \int_{\bar{V}} \frac{\partial}{\partial \bar{x}_j} (R_{x_j} \mathbf{S}) d\bar{V} \ , \tag{9}$$

where the terms on the right-hand side, which depend only on steady flow quantities and the prescribed displacement field, $\mathbf{R}$, are regarded as known source terms. Linearized flow tangency conditions, i.e., $\mathbf{v} \cdot \bar{\mathbf{n}} = i\omega \mathbf{R} \cdot \bar{\mathbf{n}} + \mathbf{V} \cdot \boldsymbol{\nabla}(\mathbf{R} \cdot \bar{\mathbf{n}})$, apply at the mean blade surfaces and duct walls. Phase-shifted periodicity conditions, e.g., $\mathbf{v}(\bar{r},\ \bar{\theta} + n\bar{\theta}_G,\ \bar{\xi}) = \mathbf{T}_n \mathbf{v}(\bar{r},\ \bar{\theta},\ \bar{\xi}) \exp(in\sigma)$, apply on the upstream and downstream passage boundaries. Finally, unsteady far-field solutions are matched to a numerical near-field solution at the computational inflow and outflow boundaries.

3.2. SOLUTION STRATEGY

We use the nonlinear analysis, TURBO [8], to determine the steady background flow, and then solve the linearized, unsteady problem using the LINFLUX analysis. For the near-field unsteady analysis, it is advantageous to regard the state vector $\mathbf{u}$ as pseudo-time dependent. Thus, we set $\mathbf{u} = \mathbf{u}(\bar{\mathbf{x}}, \tau)$, where τ is a pseudo-time variable, and add the term

$\partial(\int_{\bar{V}} \mathbf{u} d\bar{V})/\partial\tau$ to the left-hand side of equation (9). This allows the use of a conventional time-marching algorithm to converge the solution for the complex amplitude of the unsteady state vector.

The complex-amplitude of the displacement field, $\mathbf{R}(\bar{\mathbf{x}})$, must be prescribed over the entire solution domain. This field is defined so that the solution domain deforms with the blade motion, i.e., $\mathbf{R} = \mathbf{R}_{\mathrm{B}_n}$ for $\bar{\mathbf{x}}$ on B_n, slides along the stationary hub and duct walls ($\mathbf{R} \cdot \bar{\mathbf{n}} = 0$ for $\bar{r} = r_H, r_D$), and is rigid far from the blade row ($\mathbf{R} = \mathbf{0}$ for $\bar{\xi} \lessgtr \bar{\xi}_{\mp}$). In addition, $\mathbf{R}$ is prescribed along the passage boundaries to satisfy phase-shifted periodicity. In the near field, $\mathbf{R}$ is first determined along the hub and duct walls as solutions of Laplace's equation, $\nabla^2_{\bar{\mathbf{x}}}\mathbf{R} = \mathbf{0}$, in two dimensions. It is then determined in the interior of the computational domain as a solution of Laplace's equation, in three dimensions, subject to Dirichlet boundary conditions. If the blades do not vibrate, we set $\mathbf{R} = \mathbf{0}$.

4. Unsteady Perturbations in the Far-Field

Far-field solutions, based on reduced sets of governing equations, can be applied to restrict the axial extent of the near-field computational domain. Such solutions allow incoming unsteady excitations to be prescribed, and render the computational inflow and outflow boundaries transparent to outgoing disturbances. The approach taken here is a similar, but expanded, version of that adopted in [10]. In particular, approximate representations for the unsteady perturbations of fully-developed, axisymmetric, mean flows are used to describe the behaviors of the convected and the modal type unsteady disturbances that can exist far from a blade row.

To develop the far-field analysis, we require differential forms of the steady and linearized unsteady field equations, which are derived from equation (4). We assume that the steady background flow in the far field is dependent only on radial position, and that the steady radial velocity is negligible, i.e., $\bar{\rho} = \bar{\rho}(r)$, $P = P(r)$, and $\mathbf{V} = V_\theta(r)\mathbf{e}_\theta + V_\xi(r)\mathbf{e}_\xi$. Under these conditions, the steady field equation reduces to $dP/dr = \bar{\rho} r^{-1}(V_\theta^{\mathrm{abs}})^2$. For the mean flow conditions just described, the linearized unsteady field equation has the form

$$i\omega\mathbf{u} + r^{-1}\frac{\partial(r\mathbf{A}_2\mathbf{u})}{\partial r} + r^{-1}\mathbf{B}_2\frac{\partial\mathbf{u}}{\partial\theta} + \mathbf{C}_2\frac{\partial\mathbf{u}}{\partial\xi} - \mathbf{D}\mathbf{u} = 0\ , \tag{10}$$

where $\mathbf{A} = \partial\mathbf{F}_r\ /\partial\mathbf{U}$, $\mathbf{B} = \partial\mathbf{F}_\theta\ /\partial\mathbf{U}$ and $\mathbf{C} = \partial\mathbf{F}_\xi\ /\partial\mathbf{U}$ are flux Jacobian matrices and $\mathbf{D} = \partial\mathbf{S}/\partial\mathbf{U}$ is the source-term Jacobian. The subscript 2 on the flux Jacobians in (10) indicates that they are evaluated at $U_2 = \bar{\rho}V_r = 0$.

4.1. UNIFORM MEAN FLOW

For the special case of a uniform mean flow, in the absolute frame, exact solutions to equation (10) can be determined [7]. Such solutions indicate that an arbitrary unsteady perturbation can be described as the sum of independent entropic, vortical, and irrotational acoustic disturbances. The entropic and vortical disturbances are convected by the mean flow. Thus, for an unsteady flow occurring at frequency ω, a state vector, $\mathbf{u}_C$, describing such disturbances is a solution of the convection equation $\bar{D}\mathbf{u}_C/Dt = i\omega\mathbf{u}_C + \mathbf{V}\cdot\nabla\mathbf{u}_C = 0$, of the form

$$\mathbf{u}_C = \sum_{m=-\infty}^{\infty} \mathbf{u}_m(r)\exp[i(\kappa_{\xi,m}\xi + \bar{m}\theta)] \ . \tag{11}$$

Here, the $\mathbf{u}_m$ are arbitrary functions of radius, $\kappa_{\xi,m} = -(\omega - \bar{m}\Omega)V_\xi^{-1} = -\omega^{\mathrm{abs}}V_\xi^{-1}$ and $\bar{m} = N_D + mN_B$.

The acoustic disturbances are governed by a convected wave equation for the unsteady pressure, which can be solved analytically [11, 12]. The resulting solution for the complex-amplitude of the unsteady pressure is

$$p = \sum_{m=-\infty}^{\infty} e^{i\bar{m}\theta} \sum_{\mu=0}^{\infty} \left[a_{m\mu}^{-}\exp(\chi_{m\mu}^{-}\xi) + a_{m\mu}^{+}\exp(\chi_{m\mu}^{+}\xi)\right] E_{m\mu}(r) \ . \tag{12}$$

Here $a_{m\mu}^{-}$ and $a_{m\mu}^{+}$ are the amplitudes of the downstream and upstream traveling (in a subsonic axial mean flow) pressure waves, and $E_{m\mu}(r) = J_{\bar{m}}(k_{m\mu}r) + Q_{m\mu}Y_{\bar{m}}(k_{m\mu}r)$ are the "characteristic E-functions" of [11]. The E-functions are combinations of Bessel functions, of order $\bar{m}$, of the first and second kinds. The constants $k_{m\mu}$ and $Q_{m\mu}$ in (12) are determined by the duct-wall boundary conditions, and the index $\mu = 0, 1, 2, \ \ldots$ indicates the number of zero crossings or nodes in the μth radial mode.

The axial exponential coefficients, $\chi_{m\mu}^{\mp} = \beta_{m\mu}^{\mp} + i\kappa_{\xi,m\mu}^{\mp}$, are given by

$$\chi_{m\mu}^{\mp} = \left(1 - M_\xi^2\right)^{-1} A^{-1}\left[iM_\xi\omega_m^{\mathrm{abs}} \mp \sqrt{(1-M_\xi^2)A^2k_{m\mu}^2 - (\omega_m^{\mathrm{abs}})^2}\ \right] , \tag{13}$$

where $M_\xi = V_\xi/A < 1$ is the mean axial Mach number. If $(\omega_m^{\mathrm{abs}})^2 > (1-M_\xi^2)A^2k_{m\mu}^2$, then the $\chi_{m\mu}^{\mp}$ are purely imaginary, and the $m\mu^{\mathrm{th}}$ pressure patterns propagate. If the $\chi_{m\mu}^{\mp}$ are complex, then one pattern attenuates, and the other grows exponentially, with axial distance. Each upstream and downstream traveling acoustic mode has a unique axial exponential coefficient, $\chi_{m\mu}^{\mp}$, except at an acoustic resonance condition, and thus a unique propagation/decay behavior and state vector.

The perturbation state vector for acoustic disturbances is given by

$$\mathbf{u}_A = \sum_{m=-\infty}^{\infty} e^{i\bar{m}\theta} \sum_{\mu=0}^{\infty} \left[a_{m\mu}^- \exp\left(\chi_{m\mu}^- \xi\right) \mathbf{u}_{m\mu}^- + a_{m\mu}^+ \exp\left(\chi_{m\mu}^+ \xi\right) \mathbf{u}_{m\mu}^+\right] . \tag{14}$$

where the modal state vectors, $\mathbf{u}_{m\mu}^{\mp}(r)$, are determined, in terms of the pressure, from the linearized unsteady field equations. Note that, in addition to different axial behaviors, the acoustic disturbances in (14) and the convected disturbances in (11) have different radial behaviors. The former occur in radial modes that are determined by the unsteady field equations, whereas the radial dependence of the latter must be specified.

4.2. NONUNIFORM MEAN FLOW

Guided by the exact solutions for uniform mean flows, we will construct approximate solutions to equation (10) for nonuniform, but fully-developed and axisymmetric, mean flows. For this purpose, we set $\mathbf{u} = \mathbf{u}_C + \mathbf{u}_W$, where $\mathbf{u}_C$ describes the convected disturbance field, and $\mathbf{u}_W$ describes a series of wave, or modal, type disturbances, i.e.,

$$\mathbf{u}_W = \sum_{m=-\infty}^{\infty} \sum_{n=0}^{\infty} a_{mn} \mathbf{u}_{mn}^R(r) \exp(\chi_{mn}\xi + i\bar{m}\theta) . \tag{15}$$

The index n in (15) refers to different possible types of modes, such as upstream- and downstream-traveling modes and modes with different numbers of radial nodes. The classification of these modes is discussed below.

The convected disturbance is a solution to equation (10) that satisfies the convection equation, $\bar{D}\mathbf{u}_C/Dt = 0$, and therefore, has a solution of the form (11), but with $\kappa_{\xi,m} = -[\omega + mr^{-1}V_\theta(r)]/V_\xi(r)$. The convected field may contain entropic and/or vortical disturbances, depending on the properties of the mean flow. However, for general nonisentropic, rotational mean flows no convected disturbance field will exist, because entropic and vortical disturbances are coupled to the modal disturbances [13].

The modal disturbances are determined by substituting the assumed form of the solution, (15), into the field equation (10), yielding the system

$$i\omega \mathbf{I}\mathbf{u}_{mn}^R + \frac{1}{r}\frac{\partial}{\partial r}\left(r\mathbf{A}_2\mathbf{u}_{mn}^R\right) + \frac{i\bar{m}}{r}\mathbf{B}_2\mathbf{u}_{mn}^R + \chi_{mn}\mathbf{C}_2\mathbf{u}_{mn}^R - \mathbf{D}_2\mathbf{u}_{mn}^R = \mathbf{0} , \tag{16}$$

which must be solved numerically. After discretizing (16) and applying the duct-wall boundary condition, $v_r = 0$, we obtain the matrix equation

$$\left(\mathbf{P} - \chi_{mn}\mathbf{C}_2\right) \mathbf{u}_{mn}^R = \mathbf{0}, \tag{17}$$

where $\mathbf{P} = -i\omega\mathbf{I} - \mathbf{L}(r, \mathbf{A}_2) - i\bar{m}r^{-1}\,\mathbf{B}_2 + \mathbf{D}_2$ and $\mathbf{L}(r, \mathbf{A}_2)\mathbf{u}^R_{mn}$ is a finite difference approximation to $r^{-1}\partial(r\mathbf{A}_2\mathbf{u}^R_{mn})/\partial r$. The column vector $\mathbf{u}^R_{mn}$ in equation (17) contains an entry for each of the five conservation variables at each radial discretization point.

Equation (17) can be solved, using a standard linear algebra routine, to determine the axial eigenvalues, χ_{mn}, and the associated right eigenvectors, $\mathbf{u}^R_{mn}(r)$, of the modal, far-field, unsteady perturbations. The left eigenvectors are determined by solving the equation $(\mathbf{P} - \chi_{mn}\mathbf{C}_2)^H\mathbf{u}^L_{mn} = \mathbf{0}$, where the superscript H denotes the conjugate transpose. An orthonormal set of left eigenvectors can then be obtained by setting $(\mathbf{v}^L_{mn})^H = (\mathbf{u}^L_{mn})^H\mathbf{C}_2/[(\mathbf{u}^L_{mn})^H\mathbf{C}_2\mathbf{u}^R_{mn}]$. By invoking the orthogonality of left and right eigenvectors, the complex amplitudes of the modal disturbances, a_{mn}, can be determined by taking inner products of $\mathbf{v}^L_{mn}$ with $\mathbf{u}_W$, i.e.,

$$a_{mn} = \langle \mathbf{v}^L_{mn}, \frac{N_B}{2\pi}\int_\theta^{\theta+2\pi/N_B} \mathbf{u}_W \exp(-i\bar{m}\theta)\, d\theta\rangle = \int_{r_H}^{r_D} \frac{(\mathbf{v}^L_{mn})^H \cdot \mathbf{u}_{W,m}}{r_D - r_H} dr\ . \tag{18}$$

Sample numerical solutions to equation (17) for an acoustic perturbation from Part 2 of this paper are presented in Figures 2 and 3. The computed axial eigenvalues, $\chi^{\mp}_{m\mu}$, are shown in Figure 2, where the numbers m, μ above each symbol indicate the circumferential (m) and radial (μ) mode order. The pressure, $p^R_{m\mu}(r)$, in the eigenmodes $m = -1, 0, 1$ and $\mu = 0, 1, 2$ are shown in Figure 3. Only the acoustic disturbances in the (0,0) mode, which show little variation with radius, are of propagating type. The numerical predictions in Figures 2 and 3 are in excellent agreement with the corresponding analytical predictions based on equations (12) and (13).

4.3. DISCUSSION

As described above, unsteady perturbations of uniform mean flows can be represented as superpositions of convected (entropic and vortical) disturbances, and upstream- and downstream-traveling irrotational pressure disturbances. For nonuniform mean flows, the situation is more complicated [14]. In particular, for rotational, but isentropic, mean flows, the unsteady entropy is an independent convected disturbance. However, because of coupling between vortical and acoustic disturbances, due to mean-flow vorticity, neither convected vortical nor irrotational acoustic disturbances exist. Instead, nearly-convected or vorticity-dominated modal disturbances, that contain pressure, and acoustic or pressure-dominated modal disturbances, that contain vorticity, occur [15]. These two types of disturbances emerge as solutions of the eigenvalue problem (16).

The group velocity,

$$V_{g,mn} = \partial\omega/\partial\chi_{mn} = \langle \mathbf{v}^L_{mn}, \mathbf{C}\mathbf{u}^R_{mn}\rangle / \langle \mathbf{v}^L_{mn}, (\partial\mathbf{P}/\partial\omega)\,\mathbf{u}^R_{mn}\rangle\ , \tag{19}$$

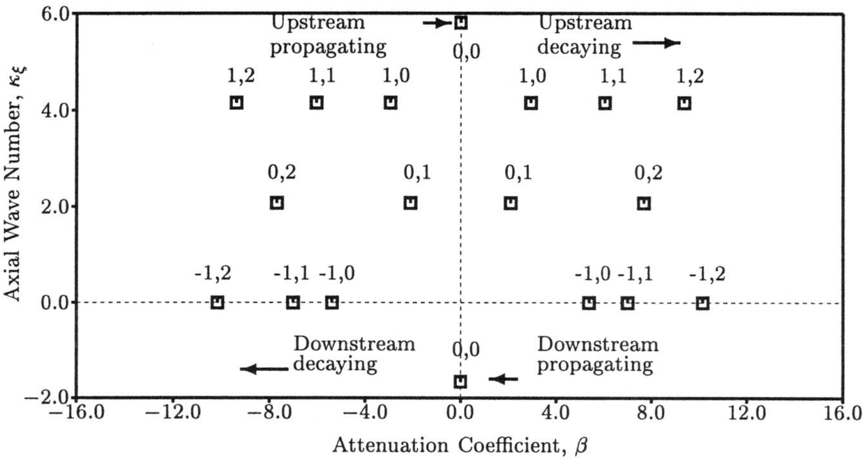

Figure 2. Axial eigenvalues, $\chi = \beta + i\kappa_\xi$, for three $(m = -1, 0, 1)$ circumferential harmonics and three $(\mu = 0, 1, 2)$ radial modes of an acoustic perturbation, at $\omega = 3.332$, $N_\mathrm{B} = 24$ and $N_\mathrm{D} = 6$, of a uniform absolute mean flow, at $M_\xi = 0.495$, within a cylindrical annular duct with $r_D = 4.244$ and $r_H/r_D = 0.8$.

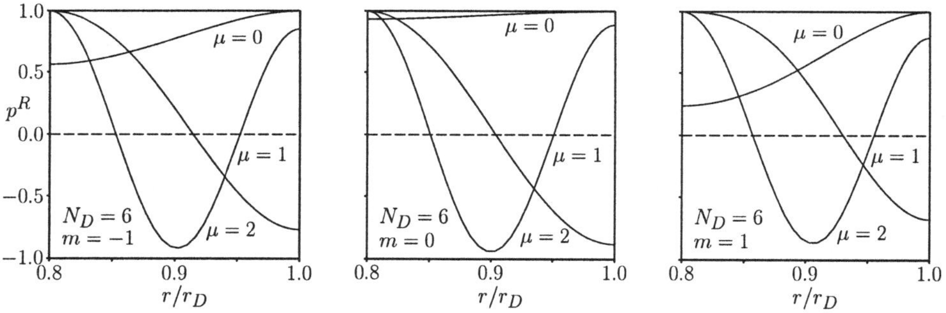

Figure 3. Radial pressure modes, $p^R_{m\mu}(r)$, $m = -1, 0, 1$, $(\bar{m} = -18, 6, 30)$, $\mu = 0, 1, 2$, for an acoustic perturbation, at $N_\mathrm{B} = 24$ and $N_\mathrm{D} = 6$, of a uniform axial flow, at $M_\xi = 0.495$, within a cylindrical duct.

i.e., the axial velocity at which an mnth modal disturbance carries energy, is used to classify disturbances. Nearly-convected disturbances travel downstream, without attenuation, at axial speeds slightly less than and slightly greater than the mean flow speed. If the mean axial flow is subsonic, then acoustic disturbances travel both upstream and downstream.

For nonuniform flows, we can further decompose the unsteady state vector to account for the two types of modal disturbances. Thus, we set $\mathbf{u} = \mathbf{u}_C + \mathbf{u}_W = \mathbf{u}_C + \mathbf{u}_A + \mathbf{u}_N$, where $\mathbf{u}_C$, $\mathbf{u}_A$ and $\mathbf{u}_N$ are the complex amplitudes of the convected, the acoustic and the nearly-convected unsteady disturbances, respectively. The functional form of the state vector for the convected disturbances has been given above. The state vector for acoustic disturbances is

$$\mathbf{u}_A = \sum_{m=-\infty}^{\infty} e^{i\bar{m}\theta} \sum_{\mu=0}^{\infty} \left[a^-_{m\mu,A} e^{\chi^-_{m\mu,A}\xi} \mathbf{u}^{R,-}_{m\mu,A}(r) + a^+_{m\mu,A} e^{\chi^+_{m\mu,A}\xi} \mathbf{u}^{R,+}_{m\mu,A}(r) \right] , \tag{20}$$

where μ indicates the number of radial nodes, and the $-$ and $+$ superscripts refer to the downstream- and upstream-traveling acoustic disturbances. The nearly-convected disturbances, i.e.,

$$\mathbf{u}_N = \sum_{m=-\infty}^{\infty} e^{i\bar{m}\theta} \sum_{\mu=1}^{\infty} \left[a^-_{m\mu,N} e^{i\kappa^-_{m\mu,N}\xi} \mathbf{u}^{R,-}_{m\mu,N}(r) + a^+_{m\mu,N} e^{i\kappa^+_{m\mu,N}\xi} \mathbf{u}^{R,+}_{m\mu,N}(r) \right] , \tag{21}$$

are also ordered by the number of radial nodes, but in this case starting with $\mu = 1$, and the $-$ and $+$ superscripts in (21) refer to disturbances that travel slightly slower and slightly faster than the convection speed.

In numerical calculations, the series in equations (20) and (21) must be truncated, since only a finite number of circumferential and radial modes can be accommodated. Also, the numerical solutions will yield spurious modes, i.e., modes that satisfy the difference equation (17), but not the differential equation (16). These modes must be filtered out to yield a valid solution set. Here, the filtering is based on the number of radial nodes and the oscillations of a computed radial mode. To prevent aliasing errors, the number of nodes, N_Z, in a given mode is limited by $N_Z \leq (r_D - r_H)/(4\Delta r_{\max})$, where $\Delta r_{\max}$ is the maximum radial grid spacing. To further eliminate spurious modes, the number of crossings about the mean value of the mode is also limited. These filtering criteria have been usually found to yield only genuine modes, but the filtering algorithm is still under development.

Since only a finite number of modes are retained after the truncation and filtering processes, the far-field modal description may be incomplete. This caveat applies to both the circumferential and radial modes. The number of circumferential modes required for acoustic disturbances is determined by the number of cut-on modes. For subsonic tip relative Mach number typically only a few, if any, acoustic modes are cut-on, but for supersonic tip relative Mach number an infinite number of acoustic modes may be cut-on. The number of radial modes required for the nearly-convected waves is unknown at this time.

Since the acoustic and nearly-convected modes are ordered by the number of radial nodes, the number of nodes in each kept numerical mode can be counted to determine if any genuine modes are missing, or if spurious modes are being kept. If the number of kept modes with a given number of nodes is one for the upstream- and downstream-traveling acoustic modes, or two for the nearly-convected modes, then it is likely that only the genuine modes have been retained. Based on our numerical results, the exclusion of genuine modes, or the inclusion of spurious modes, is detrimental to both the accuracy and convergence properties of the LINFLUX analysis.

4.4. NEAR-FIELD/FAR-FIELD MATCHING PROCEDURE

The far-field solutions must be applied in conjunction with a numerical near-field solution to determine the unsteady flow. Incoming disturbances are prescribed, and outgoing disturbances are determined by matching the near- and far-field solutions. The convected, nearly-convected, and acoustic disturbances that travel downstream are incoming disturbances at the inflow boundary ($\xi = \xi_-$) and outgoing disturbances at the outflow ($\xi = \xi_+$) boundary. Acoustic waves that travel upstream are incoming disturbances at $\xi = \xi_+$ and outgoing disturbances at $\xi = \xi_-$.

The amplitudes of the outgoing modal disturbances are determined by taking inner products, cf. (18), using the near-field state vector, $\mathbf{u}$, in lieu of $\mathbf{u}_W$. This requires invoking the assumption that $\langle \mathbf{v}^L_{mn}, \mathbf{u} \rangle = \langle \mathbf{v}^L_{m,n}, \mathbf{u}_W \rangle$, i.e., that the left eigenmodes of the modal disturbances are orthogonal to the convected disturbances. Once the amplitudes of the outgoing modes are determined, the wave-type modes are superposed to provide solutions for $\mathbf{u}_W = \mathbf{u}_A + \mathbf{u}_N$ for a finite number of modes.

At the upstream far-field boundary, the convected disturbance is set to describe any incident convected gust. At the downstream boundary, the wave-type modes are subtracted from the total unsteady disturbance and the remainder, $\mathbf{u}_C = \mathbf{u} - \mathbf{u}_W$, is treated as a convected disturbance. The convected disturbance in the region $\xi > \xi_+$ is computed, by the method of characteristics, as a solution of $\bar{D}\mathbf{u}_C/Dt = 0$. Since the mean radial velocity has been assumed to be negligible, $\mathbf{u}_C(r, \theta, \xi) = \mathbf{u}_C(r, \theta, \xi_+) \exp[-i\omega(\xi - \xi_+)/V_\xi]$ along the constant-radius characteristics.

In the near-field, the linearized unsteady equations are solved using a pseudo-time marching technique. After each update of the near-field solution, the amplitudes of the wave-type modes, i.e., $a^{\mp}_{m\mu,A}$ and $a^{\mp}_{m\mu,N}$, and the far-downstream convected disturbance, $\mathbf{u}_C(r, \theta, \xi_+)$, are updated. The far-field solutions, which are the sums of wave-type and convected disturbances, are then updated, and used to supply the far-field boundary information needed for the next near-field update.

5. Near-Field Numerical Model

The numerical procedures used in LINFLUX to resolve the linearized unsteady flow in the near field are based on those employed in the nonlinear analysis, TURBO [8]. Both analyses use sheared H-meshes, which define a curvilinear coordinate system, such that, there is a one-to-one correspondence between points, $\bar{\mathbf{x}}$, in the physical domain and points, $\boldsymbol{\alpha}$, in a computational domain, where the grid is uniform and orthogonal. The α_1, α_2 and α_3 computational coordinates, or the I, J, K mesh indices, correspond, generally, to the axial, radial and circumferential directions. Cell faces are surfaces of constant computational coordinate, with each cell bounded by six surfaces: $\alpha_1 = I \mp 1/2$, $\alpha_2 = J \mp 1/2$, and $\alpha_3 = K \mp 1/2$.

5.1. FINITE VOLUME EQUATIONS

If we let the symbol ˆ denote a quantity expressed in terms of cell geometry, then a finite-volume discretization of the pseudo-time dependent form of the linearized unsteady equation (9) can be written as

$$\frac{\partial \hat{\mathbf{u}}}{\partial \tau} = -i\omega\hat{\mathbf{u}} - \delta_j\hat{\mathbf{f}}_j + \frac{\partial \mathbf{S}}{\partial \mathbf{U}}\hat{\mathbf{u}} - i\omega(\Delta\vartheta)\mathbf{U} - \delta_j\hat{\mathbf{f}}_j^d + (\Delta\vartheta)\mathbf{S} + \bar{\vartheta}(\mathbf{R}\cdot\nabla_{\bar{\mathbf{x}}})\mathbf{S} = -\hat{\mathbf{r}} \quad (22)$$

where $\hat{\mathbf{u}} = \bar{\vartheta}\mathbf{u}$, $\hat{\mathbf{f}}_j = (\partial\hat{\mathbf{F}}_j/\partial\mathbf{U})\mathbf{u} = \bar{A}_{jk}(\partial\mathbf{F}_k/\partial\mathbf{U})\mathbf{u} = \bar{A}_{jk}\mathbf{f}_k$, and $\hat{\mathbf{f}}_j^d = -\dot{\vartheta}_j\mathbf{U} + a_{jk}\mathbf{F}_k$. The quantities $\mathbf{u}$, $\mathbf{U}$, and $\mathbf{S}$ represent average values of the physical state and source term vectors over a mean cell volume; $\bar{\vartheta}$ is the mean cell volume; $\bar{A}_{jk}$ is the mean area of a constant α_j cell face projected in the $\bar{x}_k$ direction; the vectors $\hat{\mathbf{F}}_j$ and $\hat{\mathbf{f}}_j$ are the steady and unsteady fluxes, respectively, across a constant α_j cell face; $\hat{\mathbf{f}}_j^d$ is the unsteady flux associated with the displacement field, $\mathbf{R}$; and $\hat{\mathbf{r}}$ is the residual of the first-harmonic unsteady equation. The steady quantities $\bar{\vartheta}$, $\bar{A}_{jk}$, $\mathbf{U}$, $\mathbf{F}_j(\mathbf{U})$ and $\mathbf{S}(\mathbf{U}, \bar{\mathbf{x}})$ and the unsteady displacement field, $\mathbf{R}$, are regarded as known for the linearized unsteady analysis.

The operator δ_j in equation (22) denotes the difference in the j-direction across adjacent cell interfaces, and a repeated j index implies summation over all computational coordinate directions. Thus, the terms $\delta_j\hat{\mathbf{f}}_j$ and $\delta_j\hat{\mathbf{f}}_j^d$ are the net unsteady fluxes through a cell due to the unsteady fluid motion and the grid motion, respectively. The linearized perturbation equation contains source terms associated with changes in cell volume, cell face area, and cell radial location. In evaluating the volume terms, $\Delta\vartheta = \delta_j(\bar{A}_{jk}R_k)$ and $\dot{\vartheta}_j = i\omega\bar{A}_{jk}R_k$, the R_k are taken to be the average displacements over a cell face. The perturbations in cell face areas, a_{jk}, and in cell location are computed using the displacements of the cell vertices.

The boundary conditions in the near-field are flow tangency and phase-shifted periodicity. The flow tangency conditions are implemented using phantom cells inside a solid surface. The density and pressure in a phantom cell are defined by a first-order accurate reflection condition. The phantom cell velocity is defined such that the velocity at the solid surface, which is the average of the velocities in the phantom and the interior cells, satisfies the flow tangency condition. The phase-shifted periodicity condition is imposed along a passage boundary by taking advantage of grid periodicity in the blade-to-blade direction. Thus, we set $\mathbf{u}|_{I,J,K} = \mathbf{T}_{\pm 1}\mathbf{u}_{I,J,K\pm(NK-1)}\exp(\pm i\sigma)$, where NK denotes the number of blade-to-blade grid points, and the $\mathbf{T}_n$ matrix rotates the x_2 and x_3 components of the momentum vector through n blade passages. To complete the linearized unsteady boundary value problem, analytic/numeric far-field solutions are matched to the numerical near-field solution at the computational inflow and outflow boundaries, as described above.

5.2. EVALUATION OF FLUX TERMS AND PSEUDO-TIME INTEGRATION

To perform the pseudo-time marching, we replace the τ-derivative in (22) by a two-point, backward difference expression, and expand the residual about the nth time level, leading to the following discretized field equation

$$\left(\hat{\Theta}\mathbf{I} + \frac{\partial\hat{\mathbf{r}}}{\partial\mathbf{u}}\right)\Delta\mathbf{u}^n = \left(\hat{\Theta} + i\omega\bar{\vartheta} - \bar{\vartheta}\frac{\partial\mathbf{S}}{\partial\mathbf{U}}\right)\Delta\mathbf{u}^n + \delta_j\left(\frac{\partial\hat{\mathbf{F}}_j}{\partial\mathbf{U}}\Delta\mathbf{u}^n\right) = -\hat{\mathbf{r}}^n . \tag{23}$$

In equation (23), the superscript n refers to the nth pseudo-time level, $\hat{\Theta} = \bar{\vartheta}/\Delta\tau$, $\Delta\mathbf{u}^n = \mathbf{u}^{n+1} - \mathbf{u}^n$, and $\hat{\mathbf{r}}$ is as defined in (22).

Flux difference splitting is applied to evaluate the flux terms in equation (23). In particular, we evaluate the flux, say $\hat{\mathbf{f}}_{J+1/2}$ at the $J+1/2$ cell interface, in terms of the flux in the cell to the left (J) of the interface and the flux due to waves approaching the interface from the right; i.e.,

$$\hat{\mathbf{f}}_{J+\frac{1}{2}} = \left(\frac{\partial\hat{\mathbf{F}}}{\partial\mathbf{U}}\mathbf{u}\right)_{J+\frac{1}{2}} \approx \hat{\mathbf{f}}(\mathbf{u}_J, \bar{A}_{J+\frac{1}{2}}) + \left.\frac{\partial\hat{\mathbf{F}}}{\partial\mathbf{U}}\right|^{-}_{\mathbf{U}^{\mathrm{Roe}},\bar{A}_{J+\frac{1}{2}}} (\mathbf{u}_{J+1} - \mathbf{u}_J) , \tag{24}$$

where $\hat{\mathbf{f}}(\mathbf{u}_J, \bar{A}_{J+1/2})$ is a flux based on the state vector in the Jth cell and the mean area of the $J+1/2$ cell interface. The flux Jacobian matrix $\partial\hat{\mathbf{F}}/\partial\mathbf{U}|^{-}_{\mathbf{U}^{\mathrm{Roe}},\bar{A}_{J+1/2}}$ is evaluated in terms of the Roe-averaged [16] intermediate state vector, $\mathbf{U}^{\mathrm{Roe}}$, and the area $\bar{A}_{J+1/2}$. The $(-)$ superscript on $\partial\hat{\mathbf{F}}/\partial\mathbf{U}$ indicates a summation over left traveling disturbances, i.e., those corresponding to negative eigenvalues of the flux Jacobian matrix. The discrete flux approximation (24) is first-order accurate, since the interfacial fluxes are based only upon information from adjacent cells. Second order spatial accuracy is achieved by adding corrective fluxes, which bring in information from additional neighboring cells.

The linearized unsteady equation (23) is discretized, as outlined above, leading to a system of linear algebraic equations. This system is solved, at each pseudo-time step, using a symmetric Gauss-Seidel (SGS) iteration procedure, in which the left-hand side matrix is decomposed into diagonal and off-diagonal submatrices. Thus, we set

$$\left(\hat{\Theta}_J \mathbf{I} + \frac{\partial \hat{\mathbf{r}}}{\partial \mathbf{u}}\right) \Delta \mathbf{u}_J^n = \mathbf{D}_J \Delta \mathbf{u}_J^n - \mathbf{M}_{J-1}^{+} \Delta \mathbf{u}_{J-1}^n + \mathbf{M}_{J+1}^{-} \Delta \mathbf{u}_{J+1}^n = -\hat{\mathbf{r}}_J^n \,, \quad (25)$$

where the J subscript indicates the cell index, the $\mathbf{D}$ submatrix contains the diagonal elements of the original matrix, and the $\mathbf{M}^+$ and $\mathbf{M}^-$ submatrices contain the off-diagonal elements in the negative and positive computational coordinate directions, respectively. Equation (25) is marched in pseudo-time until a converged solution is achieved. The current LINFLUX implementation uses explicit boundary conditions. These conditions are incorporated into the SGS iteration procedure, so that they are imposed in a semi-implicit manner. This treatment has been found to yield better convergence properties than a purely explicit implementation.

6. Concluding Remarks

LINFLUX, a three-dimensional linearized unsteady aerodynamic analysis, is being developed for turbomachinery aeroelastic and aeroacoustic design applications. This analysis is based on the linearized Euler equations of fluid motion; a near-field, implicit, flux-split, finite-volume analysis for the unsteady perturbations of arbitrary mean flows; and far-field eigenanalyses for the unsteady perturbations of fully-developed, axisymmetric mean flows. The near-field numerical model is based on the scheme used in the nonlinear analysis, TURBO [8]. The far-field eigenanalyses, which couple to the near-field finite-volume analysis at computational inflow and outflow boundaries, allow incoming aerodynamic excitations to be prescribed, and response disturbances to pass through these computational boundaries without spurious reflection. To date, the theoretical effort has been focused on formulating the linearized, inviscid, unsteady aerodynamic equations and the near- and far-field solution procedures, and on implementing these models.

In Part 2 of this paper, we apply the three-dimensional LINFLUX code to predict unsteady subsonic flows through a helical fan and through a "real" blade row, which is a three-dimensional version of the Tenth Standard Cascade Configuration [9]. We consider unsteady flows excited by prescribed blade vibrations or incoming acoustic disturbances at inlet or exit. In each case a nearly two-dimensional excitation is imposed, so that the LINFLUX predictions can be compared and validated against the predictions of proven two-dimensional analyses.

References

1. K. C. Hall and W. S. Clark. Linearized Euler Predictions of Unsteady Aerodynamic Loads in Cascades. *AIAA Journal*, 31(3):540–550, March 1993.
2. D. G. Holmes and H. A. Chuang. 2D Linearized Harmonic Euler Flow Analysis for Flutter and Forced Response. In *Unsteady Aerodynamics, Aeroacoustics, and Aeroelasticity of Turbomachines and Propellers*, pages 213–230. Edited by H. M. Atassi, Springer-Verlag, New York, 1993.
3. G. Kahl and A. Klose. Computation of Time Linearized Transonic Flow in Oscillating Cascades. ASME Paper 93–GT–269, 38th International Gas Turbine and Aeroengine Congress and Exposition, Cincinnati, Ohio, May 24–27 1993.
4. K. C. Hall and C. B. Lorence. Calculation of Three-Dimensional Unsteady Flows in Turbomachinery Using the Linearized Harmonic Euler Equations. *Transactions of the ASME: Journal of Turbomachinery*, 115(4):800–809, October 1993.
5. K. C. Hall, W. S. Clark, and C. B. Lorence. A Linearized Euler Analysis of Unsteady Transonic Flows in Turbomachinery. *Transactions of the ASME: Journal of Turbomachinery*, 116(3):477–488, July 1994.
6. M. D. Montgomery and J. M. Verdon. A Linearized Unsteady Euler Analysis for Turbomachinery Blade Rows Using an Implicit Wave-Split Scheme. In *Unsteady Aerodynamics and Aeroelasticity of Turbomachines*, pages 143–160. Edited by Y. Tanida and M. Namba, Elsevier, Amsterdam, 1995.
7. J. M. Verdon, M. D. Montgomery, and K. A. Kousen. Development of a Linearized Unsteady Euler Analysis for Turbomachinery Blade Rows. NASA CR 4677, June 1995.
8. J. M. Janus, H. Z. Horstman, and D. L. Whitfield. Unsteady Flowfield Simulation of Ducted Prop-fan Configurations. Paper 92–0521, AIAA 30th Aerospace Sciences Meeting and Exhibit, Reno, Nevada, January 6-9 1992.
9. T. H. Fransson and J. M. Verdon. Standard Configurations for Unsteady Flow through Vibrating Axial-Flow Turbomachine Cascades. In *Unsteady Aerodynamics, Aeroacoustics and Aeroelasticity of Turbomachines and Propellers*, pages 859–889. Edited by H. M. Atassi, Springer-Verlag, New York, 1993.
10. K. C. Hall, C. B. Lorence, and W. S. Clark. Nonreflecting Boundary Conditions for Linearized Unsteady Aerodynamic Calculations. AIAA Paper 93–0882, 31st Aerospace Sciences Meeting & Exhibit, Reno, Nevada, Jan 11–14 1993.
11. J. M. Tyler and T. G. Sofrin. Axial Flow Compressor Noise Studies. *SAE Transactions*, 70:309–332, 1962.
12. K. Imanari and H. Kodama. Application of Non-Reflecting Boundary Conditions to Three-Dimensional Euler Equation Calculations for Thick Strut Cascades. Paper 92–3045, AIAA 28th Joint Propulsion Conference and Exhibit, Nashville, Tennessee, July 6-8, 1992.
13. J. L. Kerrebrock. Small Disturbances in Turbomachine Annuli with Swirl. *AIAA Journal*, 15(6):794–803, June 1977.
14. K. A. Kousen. Eigenmode Analysis of Ducted Flows with Radially Dependent Axial and Swirl Components. In *Proceedings of the First Joint CEAS/AIAA Aeroacoustics Conference,* Vol, II, pages 1085–1094, Munich, Germany, June 12–15 1995.
15. V. V. Golubev and H. M. Atassi. Acoustic-Vorticity Waves in Swirling Flows. *Journal of Sound and Vibration*, to be published in 1997.
16. P. L. Roe. Approximate Riemann Solvers, Parameter Vectors and Difference Schemes. *Journal of Computational Physics*, 43:357–372, 1981.

A 3D LINEARIZED EULER ANALYSIS FOR BLADE ROWS PART 2: UNSTEADY AERODYNAMIC RESPONSE PREDICTIONS

MATTHEW D. MONTGOMERY AND JOSEPH M. VERDON
United Technologies Research Center
East Hartford, CT 06108, USA

Abstract. The linearized aerodynamic and numerical models, described in Part 1, have been implemented into a three-dimensional unsteady flow code, LINFLUX. This code is applied in Part 2 to benchmark, unsteady, subsonic flows to demonstrate its accuracy and prediction capabilities. Numerical results are presented for a helical fan and a three-dimensional version of the Tenth Standard Cascade. These results indicate that important progress has been made in developing a reliable and useful three-dimensional prediction capability for aeroelastic and aeroacoustic design studies.

1. Introduction

The aerodynamic and numerical formulations underlying the linearized unsteady aerodynamic analysis, LINFLUX, are presented in Part 1 of this paper [1]. A three-dimensional version of the LINFLUX code [2] has been developed and applied to predict unsteady flows through relatively simple three-dimensional blade rows rotating within annular ducts; namely, a helical fan and a real-blade configuration based on the Tenth Standard Cascade [3]. We have considered unsteady subsonic flows excited by blade vibrations or by acoustic excitations at inlet or exit.

For validation purposes, we selected nearly two-dimensional unsteady excitations and compared the 3D LINFLUX results with those of 2D analyses; the classical analysis of Smith [4] and a potential-based linearization, LINFLO [5]. The full-potential analysis CASPOF [6] provided the steady background flow for the LINFLO calculations, and the TURBO [7] analysis provided this information for the LINFLUX calculations. Our predictions indicate that the 3D LINFLUX analysis yields accurate unsteady aerodynamic response information, if suitable meshes are employed.

T.H. Fransson (ed.), Unsteady Aerodynamics and Aeroelasticity of Turbomachines, 445-464.

2. Unsteady Excitations

The unsteady fluctuations are excited by prescribed blade motions or acoustic excitations at inlet or exit. The blade vibrations occur at unit frequency, $\omega = 1$, and at constant interblade phase angle, $\sigma = 2\pi N_{\rm D}/N_{\rm B}$, where $N_{\rm D}$ is the number of nodal diameters and $N_{\rm B}$ is the number of blades. These motions consist of pure translations normal to the sectional blade chords (bending) or pure rotations about axes at the blade midchords (torsion). For convenient comparison between 3D and 2D predictions, the amplitudes of the blade motions are taken to be constant along the span.

The linearized analyses are applied to predict the local (w_C) and global (W_C) work per cycle responses; i.e.,

$$w_C(\bar{\mathbf{x}}_{\rm B}) = -\pi Im\{(P_{\rm B}\Delta(\bar{\mathbf{n}}_{\rm B}d\bar{A}_{\rm B})/\ d\bar{A}_{\rm B} + p_{\mathcal{B}}\bar{\mathbf{n}}_{\rm B}) \cdot \mathbf{R}^*_{\rm B}\}$$

and (1)

$$W_C = \textstyle\int_{\bar{A}_{\rm B}} w_C(\bar{\mathbf{x}}_{\rm B})d\bar{A}_{\rm B}\ ,$$

to prescribed blade vibrations. Here, $Im\{\ \}$ denotes the imaginary part of $\{\ \}$, P is the steady pressure, p is the complex amplitude of the unsteady pressure, the subscripts $\mathcal{B}$ and B refer to the instantaneous and mean blade surfaces, respectively, and the superscript $*$ denotes the complex conjugate.

We also consider unsteady flows excited by acoustic disturbances at inlet or exit that travel towards the rotor blade row. These acoustic excitations model one circumferential mode of the aerodynamic coupling between the rotor and an upstream or downstream stator, consisting of N_V vanes. In particular, we consider acoustic excitations of the rotor having zero radial crossings ($\mu = 0$) and occurring at the vane passing frequency, $\omega = -N_{\rm V}\Omega$, and at interblade phase angle $\sigma = 2\pi(N_{\rm B} - N_{\rm V})/N_{\rm B}$. This phase angle was chosen so that the excitation waves are of propagating type.

The state vector for acoustic excitations, cf. [1] equation (20), is

$$\mathbf{u}_A^{\mp}(r,\theta,\xi) = a_A^{\mp}\mathbf{u}_A^{R,\mp}(r)\exp(\chi^{\mp}\xi + iN_{\rm D}\theta)\ , \quad \xi \lessgtr \xi_{\mp}\ . \qquad (2)$$

where $\chi^{\mp}$ and $\mathbf{u}_A^{R,\mp}(r)$ are solutions of a far-field eigenvalue problem. The eigenmodes in equation (2) are scaled such that if $a_A = (1,0)$, then the amplitude of the pressure associated with the acoustic excitation, i.e.,

$$p_A^{\mp}(r,\theta,\xi) = a_A^{\mp}p^{R,\mp}(r)\exp(\chi^{\mp}\xi + iN_{\rm D}\theta)\ , \qquad (3)$$

is (1,0) at the location of maximum magnitude on the reference radial line. This reference line passes through the leading edge of the reference blade at midspan. See Figure 3 in Part 1 and Figure 13 for examples of $p^R(r)$, the pressure component of an eigenmode.

3. Geometry and Meshes

We consider unsteady subsonic flows through a helical fan with no mean loading and a three-dimensional version of the Tenth Standard Cascade Configuration [3]. Both blade rows consist of $N_B = 24$ blades, operate within a cylindrical annular duct of hub radius $r_H = 3.395$ and duct radius $r_D = 4.244$, and rotate at constant angular velocity $\boldsymbol{\Omega}$. There is no clearance between the blades and the outer duct wall. For both blade rows, the steady flow at inlet is axial and uniform in the absolute frame.

The TURBO nonlinear steady and the LINFLUX linearized unsteady solutions were determined on the same H-type grid, which extends over a single passage. Different grids were used for the helical fan [8] and the 3D Tenth Standard Cascade [9], but both consist of 141 axial, 41 tangential, and 11 radial surfaces (56,000 cells), with 81 axial lines intersecting the blade surfaces. Both grids extend one axial chord, at midspan, upstream and downstream from the blade row. Axial grid lines are clustered near leading and trailing edges; circumferential lines, near blade surfaces; and the radial grid is uniform. A finer mesh of 161 axial, 51 tangential, and 17 radial surfaces was also used for a mesh resolution study.

The LINFLUX near-field, finite-volume solutions are coupled to far-field eigensolutions, which were determined on a radial grid of 24 points clustered near the hub and duct walls. For the acoustically dominated unsteady flows under consideration, the convected and nearly-convected disturbances occurring downstream of the blade row were numerically convected through the computational outflow boundary.

The CASPOF full-potential steady and the LINFLO linearized unsteady solutions were determined using local C-meshes embedded in global H-meshes, which extend one axial chord upstream and downstream from the blade row. The LINFLO H- and C-meshes consist of 155 axial and 41 tangential lines and 101 radial and 21 circumferential lines, respectively. Coarser H- and C-meshes were used for the CASPOF calculations.

The 3D numerical solutions reported herein were determined on an IBM-3CT workstation. The LINFLUX calculations required from 1,000 to 3,500 pseudo-time steps to converge to an 0.1% tolerance for global unsteady aerodynamic response quantities. No convergence acceleration techniques are currently implemented in the LINFLUX code. Using two SGS subiterations, approximately 600 time steps could be completed per CPU hour on an IBM-3CT, corresponding to 110 μsec/time-step/cell or 1.5 to 5 CPU hrs. for a converged solution. The computing resources required by LINFLUX are much less than those required to determine a nonlinear unsteady solution using TURBO, especially for excitations at nodal diameters that require nonlinear solutions over many blade passages.

4. Helical Fan

The helical fan operates at Mach number $M^{\text{abs}} = M_\xi = 0.495$, and rotates at angular speed $|\mathbf{\Omega}| = 0.185$. The blade mean positions, in the rotating frame, are defined by $r\theta = \xi \tan\Theta(r) + nG(r)$, $2|\xi| < \cos\Theta(r)$, $n = 0, \ldots, N_{\text{B}} - 1$, where $G(r) = 2\pi r/N_{\text{B}}$ is the circumferential blade spacing, and $\Theta(r) = \tan^{-1}(\Omega r/V_\xi)$ is the blade stagger angle (see Figure 1 of Part 1). The blade chord, $c(r)$, is taken to be unity along the span; therefore, because of twist, the axial chord, $c_{ax}(r) = \cos\Theta(r)$, varies from 0.748 at the hub to 0.669 at the tip. Since the blade mean positions are aligned with the inlet relative mean flow velocity, the steady relative velocity is $\mathbf{V} = \mathbf{V}^{\text{abs}}_{-\infty} - \mathbf{\Omega} \times \mathbf{r}$ throughout the entire fluid domain, and the steady thermodynamic properties of the fluid are constant. Hence, no steady loads act on the blades, and the inlet and exit conditions are identical.

The helical fan geometry was chosen to provide a relatively simple 3D test case, and to allow meaningful comparisons between the 3D LINFLUX predictions at midspan and those of 2D analyses. The 2D results apply to a flat plate cascade, with $G = 1$ and $\Theta = 45\,\text{deg}$, operating in a uniform relative steady flow at $M = 0.7$ and $\Omega = 45\,\text{deg}$, where Ω is the flow angle.

4.1. BLADE VIBRATION

The 3D LINFLUX and the 2D Smith analyses were applied to predict the unsteady aerodynamic responses of the helical fan undergoing pure torsional vibrations about midchord and pure bending vibrations. Unsteady pressure difference predictions, i.e., $[\![p]\!] = p_{\text{lower}} - p_{\text{upper}}$, at midspan ($r/r_{\text{D}} = 0.9$) are shown in Figure 1 for motions at $\omega = 1$ and $\sigma = 90\,\text{deg}$. The results are in very good agreement, with small differences in the blade leading edge region, which are probably due to the small radial pressure variations present in the helical fan flows. The radial variation in the unsteady pressure response at the helical blade surfaces (not shown) is small, indicating that the response is nearly two-dimensional.

Global work per cycle predictions for the helical fan and corresponding flat-plate cascade undergoing vibrational motions are shown in Figure 2. The 3D LINFLUX results, indicated by the symbols, were determined for $N_D = -6, -5, \ldots, 18$; the Smith code results in increments of one degree. We have multiplied the 2D Smith work per cycle predictions by the blade span, $r_D - r_H = 0.849$, for comparison with the 3D LINFLUX predictions. The agreement is quite good over the entire interblade phase angle range, even near acoustic resonance conditions.

The vertical lines above the curves in Figure 2 indicate the 2D cut-off or resonance conditions, i.e., $\sigma^-_{\text{2D}} = -29.4\,\text{deg}$ and $\sigma^+_{\text{2D}} = 107.3\,\text{deg}$, and approximate resonance conditions for the helical fan. If all acous-

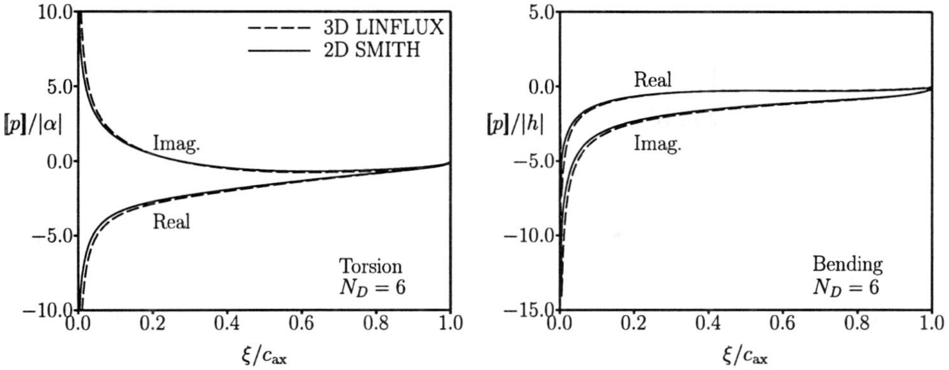

Figure 1. Unsteady pressure difference distributions at midspan ($r/r_D = 0.9$), as predicted using the 3D LINFLUX and the 2D Smith analyses, for the helical fan undergoing torsional blade vibrations about midchord and bending vibrations $\omega = 1$, $\sigma = 90$ deg.

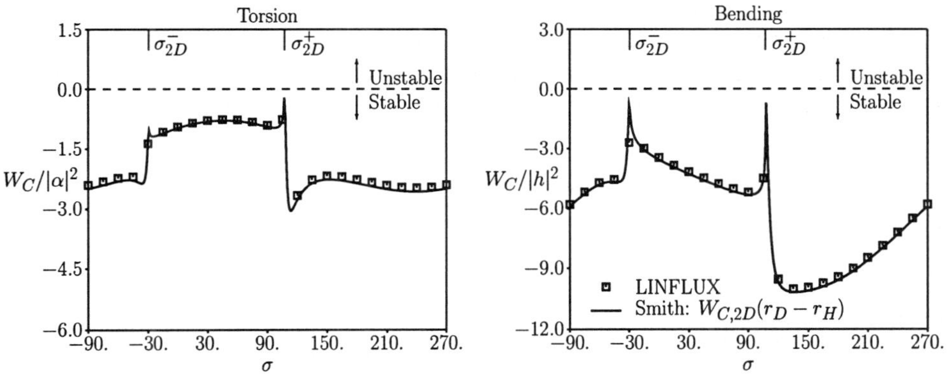

Figure 2. Work per cycle versus interblade phase angle for the helical fan undergoing pure torsional vibrations about midchord and pure bending vibrations at $\omega = 1$.

tic disturbances associated with the blade motion attenuate with increasing distance from the blade row, we refer to the blade motion as being sub-resonant; if any acoustic disturbances propagate, the motion is super-resonant. The blade motions at $\sigma_{2D}^- < \sigma < \sigma_{2D}^+$ are super-resonant; those at $-90\,\text{deg} \leq \sigma < \sigma_{2D}^-$ and $\sigma_{2D}^+ < \sigma \leq 270\,\text{deg}$ are sub-resonant.

4.2. ACOUSTIC EXCITATION

We also considered the unsteady responses of the helical fan to acoustic excitations, which model acoustic interactions with an adjacent upstream or downstream stator row with N_V vanes. In particular, we considered stators with $N_V = 18$ and $N_V = 21$; resulting in acoustic excitations at $\omega = 3.332$ and $\sigma = 90\,\text{deg}$ ($N_D = 6$) and $\omega = 3.887$ and $\sigma = 45\,\text{deg}$ ($N_D = 3$), respectively. In this discussion, an acoustic excitation from upstream refers to a downstream-propagating $(m, \mu) = (0, 0)$ disturbance imposed at the computational inflow boundary. Here m is the circumferential harmonic index, so the number of circumferential lobes is $\bar{m} = N_D + mN_B$. An acoustic excitation from downstream refers to an upstream-propagating (0,0) disturbance imposed at the computational outflow boundary.

The computed axial eigenvalues, $\chi_{m\mu}$, $m = -1, 0, 1$ ($\bar{m} = -18, 6, 30$) and $\mu = 0, 1, 2$, and radial pressure modes, $p^R_{m\mu}(r)$, in the far field of the helical fan are shown in Figures 2 and 3, respectively, in Part 1, for disturbances at $\omega = 3.332$. Only the disturbances in the (0,0) mode, for which the pressure shows little variation with radius, are of propagating type for this frequency. The downstream-propagating (0,0) disturbance has an axial wave number, κ_ξ, of -1.662; the upstream-propagating (0,0) disturbance has an axial wave number of 5.740.

Unsteady surface pressure distributions for the helical fan subjected to an acoustic excitation from upstream are shown in Figure 3. Here, the real and imaginary pressure components are in- and out-of-phase, respectively, with the excitation pressure at the midspan leading-edge point. The LINFLUX predictions indicate that the radial surface pressure variations are relatively small. As a result, the 3D LINFLUX and 2D LINFLO surface pressures at midspan are in very good agreement. According to the LINFLUX solution, the unit-amplitude pressure excitation from upstream gives rise to upstream and downstream-propagating, (0,0), acoustic response waves at amplitudes of 0.617 and 0.519, respectively. The corresponding LINFLO predictions are 0.694 and 0.517. Both analyses indicate that the attenuating acoustic response waves are small at the far-field boundaries.

Contours of the in-phase component of the unsteady pressure, $Re\{p\}$, at midspan for the acoustic excitation from upstream, are shown in Figure 4. The results are in good agreement, but differ slightly upstream due to the

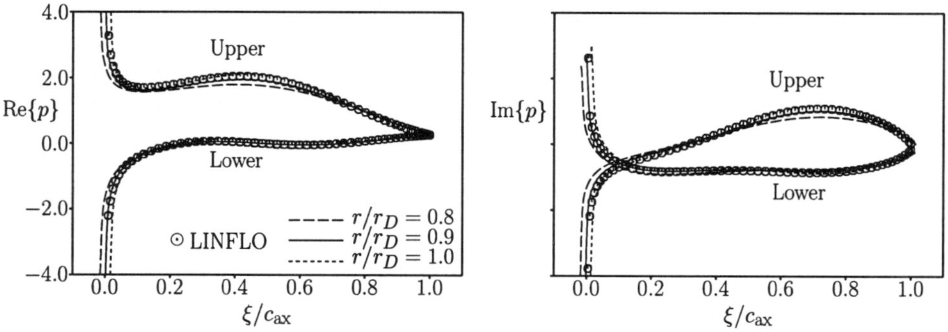

Figure 3. Unsteady surface pressure distributions due to an acoustic excitation from upstream [$p^-_{A,\text{Ref}} = (1, 0)$, $\omega = 3.332$ and $\sigma = 90$ deg] interacting with the helical fan.

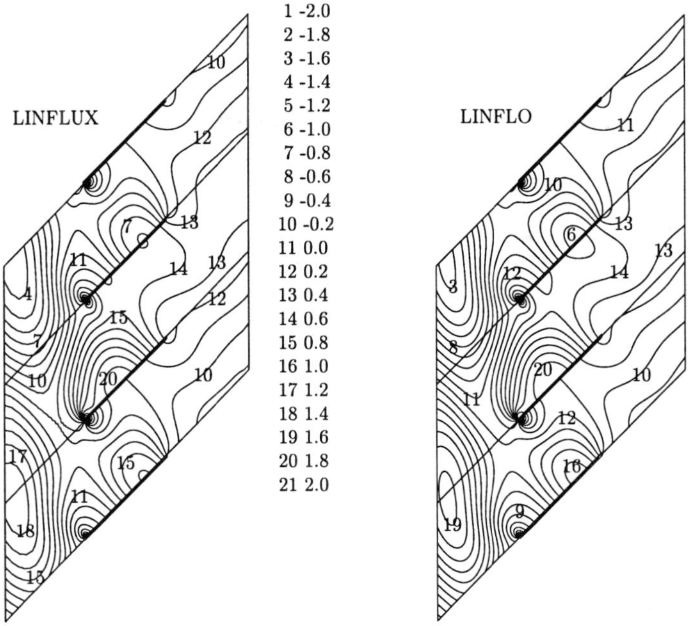

Figure 4. Contours of the in-phase component of the unsteady pressure at midspan due to the interaction of an acoustic excitation from upstream $[p^{-}_{A,\text{Ref}} = (1,0),\ \omega = 3.332,\ \sigma = 90\,\text{deg}]$ with the helical fan.

different upstream-propagating acoustic responses mentioned above. The LINFLUX results show no acoustic reflections at the computational inflow and outflow boundaries, indicating a successful coupling of the LINFLUX near-field, finite-volume analysis and the far-field eigenanalyses.

For a unit-amplitude, (0,0), acoustic excitation from downstream, LINFLUX predicts propagating, (0,0), acoustic responses at amplitudes of 0.560 upstream and 0.266 downstream. This downstream excitation also produces attenuating, (1,0), acoustic responses at amplitudes of 0.111 at both computational boundaries, indicating some three-dimensional content (see Figure 3 of Part 1) in the unsteady flow. The LINFLO calculations indicate propagating acoustic responses at amplitudes of 0.686 upstream and 0.266 downstream. The amplitudes of the higher order response modes are small.

The three-dimensional content of this unsteady flow is illustrated by the radial variation in the unsteady surface pressure distributions, shown in Figure 5. Because of such three-dimensional variations, there are small differences between the 3D LINFLUX and 2D LINFLO surface pressure predictions at midspan, but the two predictions are in good agreement. The unsteady pressure contours at midspan (not shown) are also in good agreement. To investigate whether the 2D vs. 3D differences were due to mesh resolution, the LINFLUX calculation for the acoustic excitation from

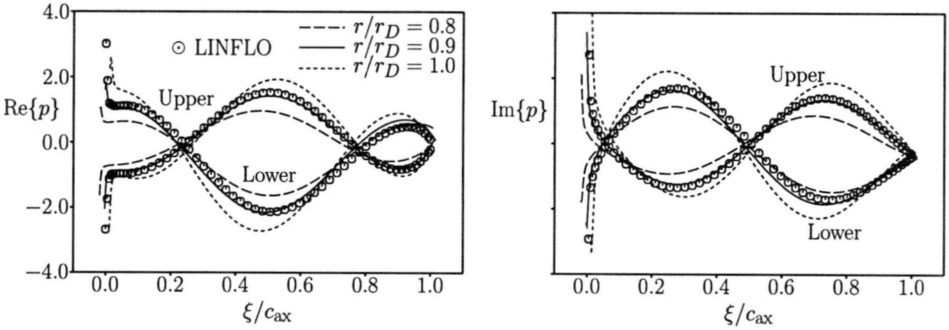

Figure 5. Unsteady surface pressure distributions due to the interaction of an acoustic excitation from downstream $[p^{+}_{A,\text{Ref}} = (1, 0),\ \omega = 3.332,\ \sigma = 90\,\text{deg}]$ with the helical fan.

downstream was also performed on a 161 × 51 × 17 mesh, with only slight changes in the 3D response predictions.

Unsteady surface pressure responses have also been determined for the helical fan subjected to a unit-amplitude, (0,0), acoustic excitation from a downstream stator consisting of 21 vanes; i.e., an excitation from downstream at $\omega = 3.887$, and $\sigma = 45\,\text{deg}$. In contrast to the $N_V = 18$ excitation, the $N_V = 21$ acoustic excitation produces two propagating acoustic response disturbances, at $(m, \mu) = (0, 0)$ and (1,0), in each far-field region. The (1,0) disturbances, which vary significantly with radius, are associated with strong radial variations in the near-field unsteady response. The predicted surface pressure responses to the $N_V = 21$ downstream excitation, see Figure 6, show a stronger three-dimensional content than the $N_V = 18$ downstream excitation, and, therefore, more significant differences between the 3D and 2D predictions at midspan. The $N_V = 21$ downstream excitation produces (0,0) and (1,0) propagating pressure responses at amplitudes of 0.167 and 0.172 far upstream and at 0.178 and 1.350 far downstream. It is interesting to note that the downstream excitation in the (0,0) mode produces a much stronger downstream response in the (1,0) mode than in the (0,0) mode.

In summary, the LINFLUX predictions for the helical fan undergoing vibrational motions, Figures 1 and 2, are in very good agreement with those of the classical analysis of Smith. The LINFLUX predictions for the helical fan subjected to acoustic excitations from upstream and downstream at $\omega = 3.332$ and $\sigma = 90\,\text{deg}$ ($N_D = 6$) are in good agreement with the 2D LINFLO predictions at midspan. The predictions for an acoustic excitation from downstream, at $\omega = 3.887$ and $\sigma = 45\,\text{deg}$ ($N_D = 3$), show stronger radial variations, due to the excitation of an acoustic mode with significant radial variation. Therefore, as might be expected, the 3D LINFLUX and 2D LINFLO results at midspan differ for this excitation.

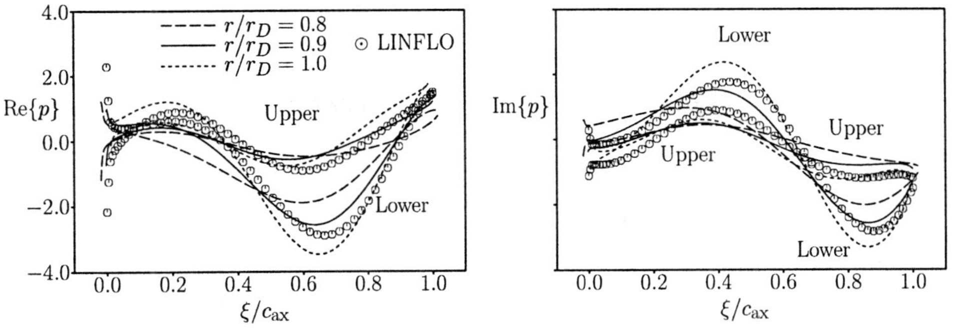

Figure 6. Unsteady surface pressure distributions due to the interaction of an acoustic excitation from downstream $[p^+_{A,\mathrm{Ref}} = (1,0),\ \omega = 3.887,\ \sigma = 45\,\mathrm{deg}]$ with the helical fan.

5. 3D Tenth Standard Configuration

The 3D Tenth Standard Cascade consists of 24 blades, which are twisted to reduce the variation in mean incidence due to blade rotation. At midspan ($r = r_{\mathrm{mid}}$), the blades are staggered at $\Theta = 45\,\mathrm{deg}$ with a circumferential spacing, $G = 2\pi r/N_{\mathrm{B}}$, of unity, and the blade sections are NACA 5506 airfoils, modified [10] to have wedge-shaped trailing edges. The mean chord lines are located at $r\theta = \xi \tan\Theta(r) + nG(r)$, $0 \le \xi \le \cos\Theta$, $n = 0, \ldots, N_{\mathrm{B}} - 1$, where $\Theta(r) = \tan^{-1}(r/r_{\mathrm{mid}})$, and the axial chord is constant. Therefore, the leading and trailing edge ξ and θ coordinates are constant along the entire span. The airfoil chord varies from 0.946 at the hub to 1.057 at the tip, because of twist, and the local thickness to chord ratio varies to maintain constant thickness. The cascade operates in a uniform axial inlet flow, which occurs at $M^{\mathrm{abs}}_{-\infty} = 0.4015$, and rotates at an angular speed of $|\mathbf{\Omega}| = 0.2145$. This 3D configuration was chosen to match the subsonic Tenth Standard Configuration [3] at midspan, where the relative inlet Mach number, $M_{-\infty}$, is 0.7 and the relative inlet flow angle, $\Omega_{-\infty}$, is 55 deg.

The TURBO steady-flow predictions, shown in Figure 7, indicate that the isentropic Mach numbers on the blade surfaces, based, here, on the pressure and the relative total pressure at a given radius, vary moderately with radius. Also, the 3D TURBO predictions at midspan are in close agreement with the 2D CASPOF predictions. The TURBO results indicate that the maximum Mach numbers on the suction surface of a blade are 0.849 at the hub, 0.906 at midspan, and 0.961 at the tip. These values occur at $\xi/c_{\mathrm{ax}} = 0.053$, 0.073, and 0.085, respectively. Thus, the flow on the suction surface is nearly sonic in the tip region just aft of the leading edge. The CASPOF predictions for the 2D cascade indicate a maximum Mach number of 0.916 at $\xi = 0.065$.

The mean inlet flow to the 3D Tenth Standard Cascade is uniform

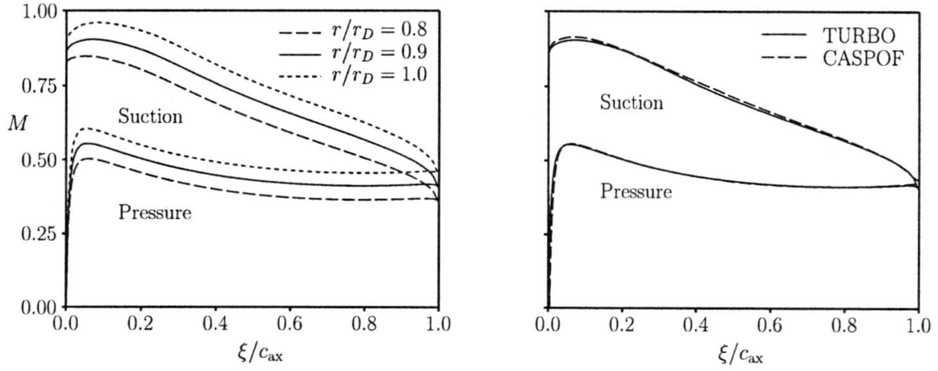

Figure 7. Relative isentropic Mach number distributions for the 3D 10th Standard Cascade ($M_{-\infty}^{\rm abs} = 0.4015$, $|\mathbf{\Omega}| = 0.2145$): (a) TURBO predictions; (b) TURBO and CASPOF predictions at midspan, $r/r_D = 0.9$.

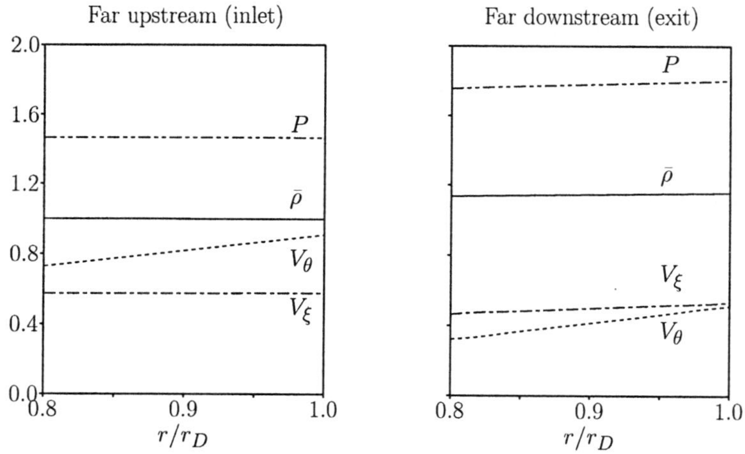

Figure 8. Relative frame steady flow properties in the far field of the 3D 10th Standard Cascade ($M_{-\infty}^{\rm abs} = 0.4015$, $|\mathbf{\Omega}| = 0.2145$).

and axial in the absolute frame, but, because of steady blade loading, the mean exit flow has swirl and axial shear, see Figure 8. The steady pressure ($P = 1.458$), density ($\bar{\rho} = 1.0$), and axial velocity ($V_\xi = 0.574$) are constant at inlet, and the relative circumferential velocity, $V_\theta = -\Omega r$ varies linearly from 0.728 at the hub to 0.910 at the tip. At the exit boundary, the steady pressure, density, and axial velocity vary with radius (mean shear), and $V_\theta \neq -\Omega r$ (mean swirl). The steady blade loading increases the pressure and density, and decreases the axial and circumferential velocities, especially the latter. The change in V_θ from inlet to exit is nearly constant, which is consistent with a small radial variation in mean loading.

5.1. BLADE VIBRATION

The 3D LINFLUX and the 2D LINFLO analyses were applied to predict the unsteady aerodynamic responses of the 3D and 2D Tenth Standard Cascades to pure torsional vibrations about midchord and pure bending vibrations, both at unit frequency. Because of mean blade loading, the steady inlet and exit conditions differ for the Tenth Standard Cascade. As a result, the acoustic properties differ upstream and downstream. For example, for an unsteady excitation at $\omega = 1$ and $\sigma = 90$ deg, the acoustic response in the (0,0) mode propagates far upstream, but attenuates far downstream. Since a propagating acoustic response exists, this unsteady excitation is classified as super-resonant.

Local (w_C) and global (W_C) work per cycle predictions for the 3D Tenth Standard Cascade vibrating in torsion and bending are shown in Figure 9. These results indicate that the local work per cycle responses do not vary significantly along the span, and thus are expected to be in good agreement with 2D predictions at each spanwise station.

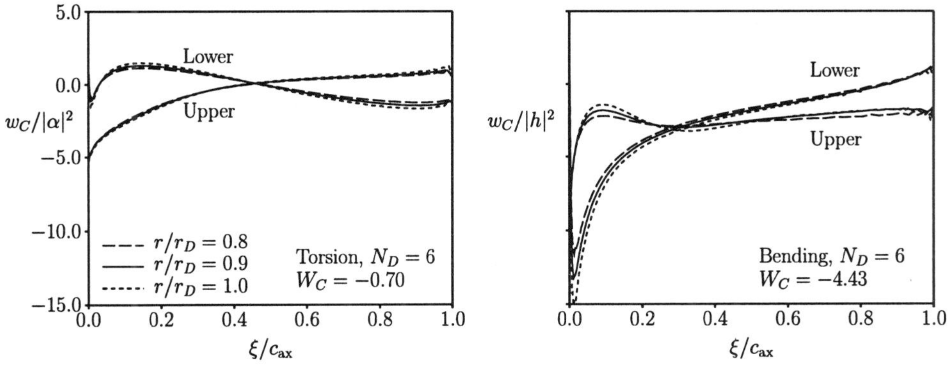

Figure 9. **Local and global work per cycle predictions for the 3D 10th Standard Cascade undergoing a torsional vibration about midchord and a bending vibration, at $\omega = 1$ and $\sigma = 90$ deg ($N_D = 6$).**

Local work per cycle predictions at midspan for the Tenth Standard Cascade vibrating in torsion and bending at interblade phase angles of ∓ 90 deg ($N_D = \mp 6$) are shown in Figure 10. The 3D LINFLUX and the 2D LINFLO predictions for the torsional blade vibrations are in very good agreement. Those for the bending vibration at $\sigma = -90$ show small differences over the entire blade. The reasons for these differences are not understood at present, but similar differences were reported in earlier work comparing the predictions of 2D nonlinear [11] and 2D linearized [12, 13] Euler analyses with LINFLO results. The local work per cycle predictions for the bending vibration at $\sigma = 90$ deg show similar small differences along the pressure surface, but large differences on the suction surface over the forward half of the blade. The reasons for the large discrepancies have not been established, but we suspect that the short wavelength acoustic response, in regions of high subsonic steady Mach number, is not adequately resolved on the 141 × 41 × 11 H-mesh used for the LINFLUX calculations. This hypothesis is consistent with the differences in the upstream-propagating acoustic response described below.

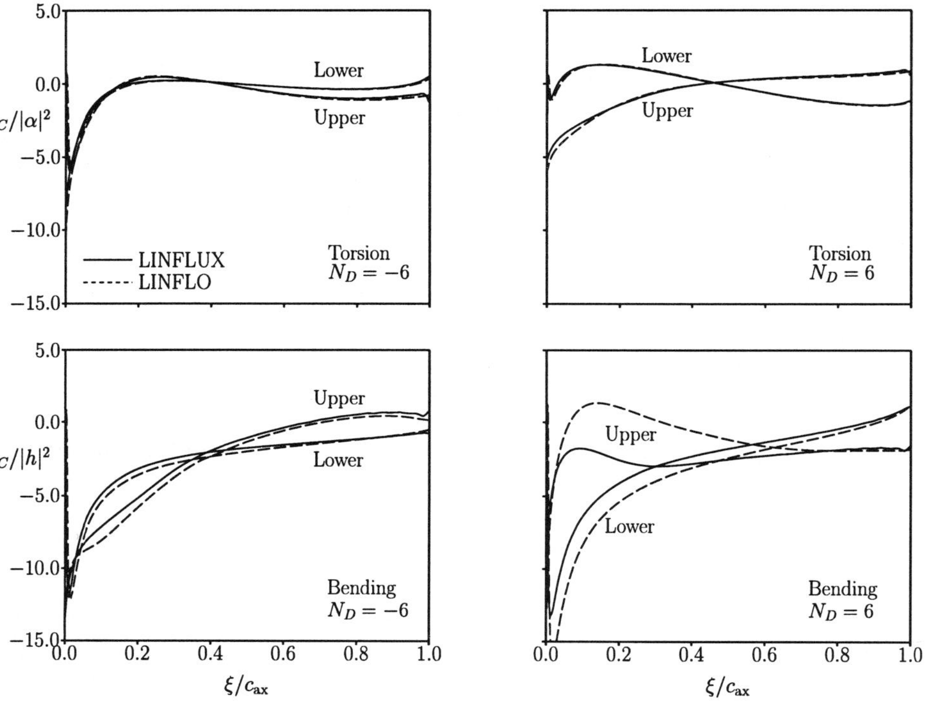

Figure 10. Local work per cycle distributions at midspan, as predicted using the 3D LINFLUX and the 2D LINFLO analyses, for the 10th Standard Cascade undergoing torsional blade vibrations about midchord and bending vibrations, at $\omega = 1$ and $N_D = \mp 6$.

The LINFLUX calculations for both the sub-resonant, $\sigma = -90$ deg, and the super-resonant, $\sigma = 90$ deg, blade motions reveal that the attenuating acoustic responses are of small amplitude at the computational far-field boundaries. The super-resonant torsional and bending vibrations produce upstream-propagating acoustic response disturbances at $\kappa_\xi = 1.583$ with amplitudes, a_A, of 1.352 and 1.540, respectively. The corresponding LINFLO predictions are $\kappa_\xi = 1.603$ with amplitudes of 1.529 and 2.822. Thus, there is a substantial difference between the LINFLUX and LINFLO predictions for the amplitude of the upstream-propagating acoustic response wave caused by the super-resonant bending vibration.

Global work per cycle versus interblade phase angle predictions for the 2D and 3D Tenth Standard Cascade cascades are shown in Figure 11. The 3D LINFLUX results, indicated by the discrete symbols, were determined for $N_D = -6, -5, \ldots, 18$; the 2D LINFLO results in increments of one degree. As in the case of the helical fan, we have multiplied the 2D work per cycle predictions at midspan by the blade span to allow a convenient comparison between the 3D LINFLUX and the 2D LINFLO predictions.

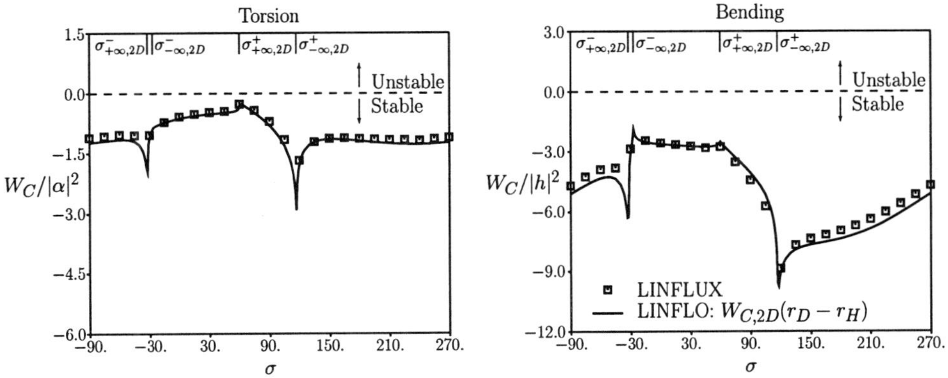

Figure 11. Work per cycle versus interblade phase angle for the 3D 10th Standard Cascade undergoing torsional vibrations about midchord and bending vibrations at $\omega = 1$.

The vertical lines above the curves in Figure 11 indicate the resonance conditions, at $\omega = 1$, for the 2D cascade. Because of the high hub-to-duct radius ratio, the 2D conditions approximate the 3D resonance conditions. The 2D resonance conditions are $\sigma_{-\infty}^{-} = -26.93$ deg and $\sigma_{-\infty}^{+} = 117.12$ deg in the far upstream region and $\sigma_{+\infty}^{-} = -31.80$ deg and $\sigma_{+\infty}^{+} = 59.79$ deg in the far downstream region. The super-resonant blade motions, which occur for $\sigma_{+\infty}^{-} < \sigma < \sigma_{-\infty}^{+}$, send a propagating wave into the upstream and/or downstream regions of the flow. The blade motions at $-90\,\text{deg} \leq \sigma < -31.80\,\text{deg}$ and $117.12\,\text{deg} < \sigma \leq 270\,\text{deg}$ are sub-resonant.

The results in Figure 11 indicate good agreement between the 3D LINFLUX and the 2D LINFLO global response predictions over the entire interblade phase angle range. However, the results for the bending vibrations must be interpreted with some caution. In particular, for the super-resonant bending vibrations at $\sigma = 75$, 90 and 105 deg, in which a single propagating acoustic response disturbance occurs far upstream, the LINFLUX and LINFLO local responses show large differences along the blade suction surfaces, similar to those revealed by the local work per cycle predictions in Figure 10 for the bending vibration at $N_D = 6$ ($\sigma = 90\,\text{deg}$).

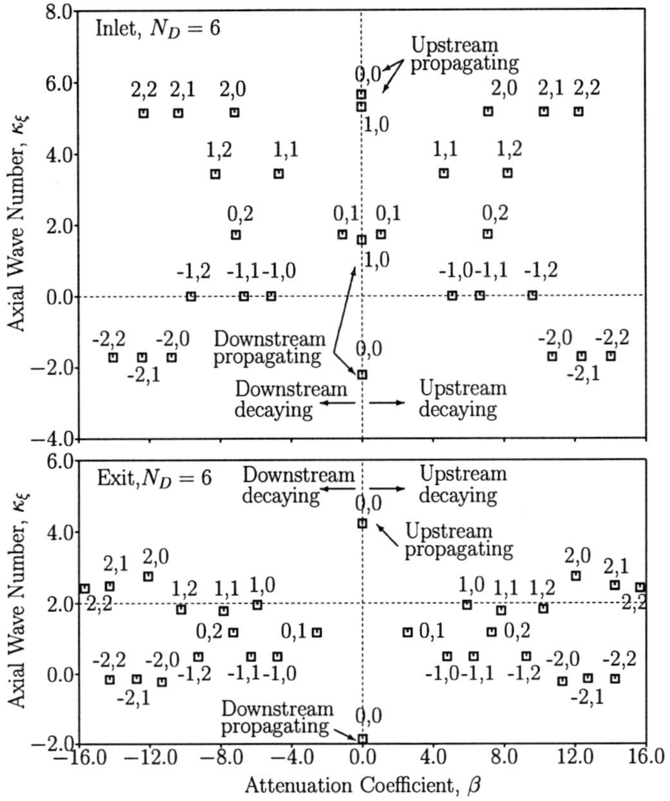

Figure 12. **Axial eigenvalues, $\chi = \beta + i\kappa_\xi$, for five circumferential ($m = -2, \ldots, 2$) and three radial ($\mu = 0, 1, 2$) modes of acoustic disturbances far upstream and far downstream of the 3D 10th Standard Cascade, for an unsteady flow at $\omega = 3.861$ and $N_D = 6$.**

5.2. ACOUSTIC EXCITATION

As our final example, we consider the response of the 3D Tenth Standard Cascade to an acoustic excitation from an adjacent upstream stator consisting of 18 vanes; i.e., an acoustic excitation from upstream at $\omega = -N_V\Omega = 3.861$ and $\sigma = 2\pi(N_B - N_V)/N_B = 90\,\text{deg}$ ($N_D = 6$). The far-field axial eigenvalues of the acoustic disturbances occurring in this unsteady flow are shown in Figure 12. These results indicate that there are propagating acoustic disturbances, at $(m, \mu) = (0, 0)$ and $(1, 0)$, far upstream, and at $(m, \mu) = (0, 0)$ far downstream. The remaining response modes attenuate. In the far-upstream region, the (0,0) and (1,0) downstream propagating (excitation) disturbances occur at axial wave numbers, κ_ξ, of -2.212 and 1.567, respectively, and the corresponding upstream propagating (response) disturbances occur at 5.647 and 5.308. In the far-downstream region, the (0,0) downstream-propagating (response) distur-

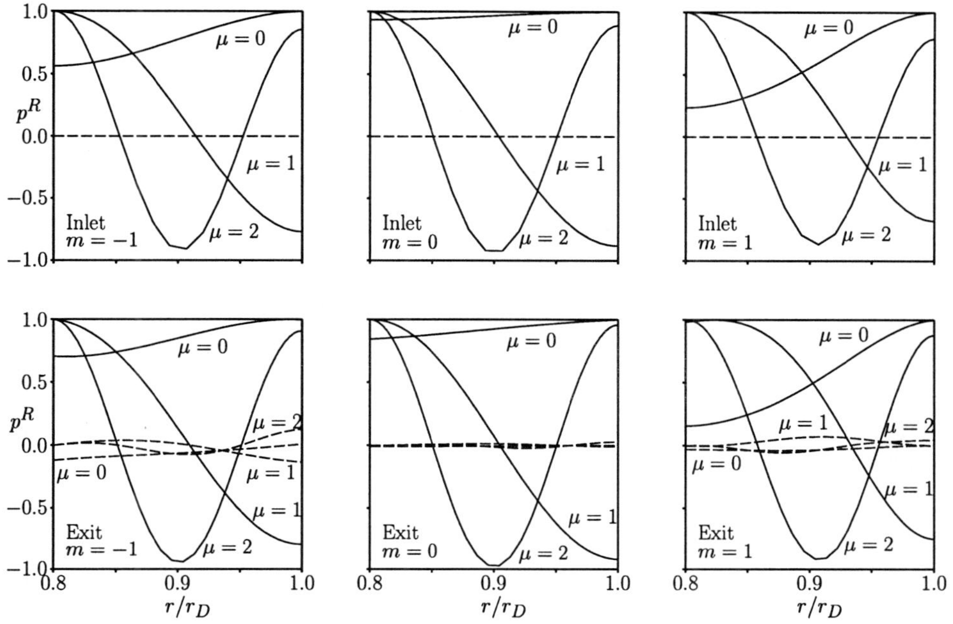

Figure 13. Radial pressure modes, $p^R_{m\mu}(r)$, $m = -1, 0, 1$, $(\bar{m} = -18, 6, 30)$, $\mu = 0, 1, 2$, of the acoustic response far upstream and downstream of the 3D 10th Standard Cascade at $\omega = 3.861$ and $N_D = 6$: (——) in-phase (real) component of $p^R_{m\mu}$, (– – –) out-of-phase (imaginary) component of $p^R_{m\mu}$.

bance occurs at $\kappa_\xi = -1.865$. Note that, in contrast to the far-upstream region, in the far-downstream region the axial wave numbers of attenuating disturbances in a given circumferential mode vary with radial mode number, μ, because the mean flow is nonuniform.

The pressure components of the radial eigenmodes associated with the acoustic responses far upstream and downstream are shown in Figure 13. Although the inlet and exit mean-flows differ, the upstream and downstream radial pressure modes are very similar. Note that the phase of a modal pressure disturbance is independent of radius upstream, where the absolute mean flow is uniform, but the phase varies with radius for the nonuniform mean flow downstream. Thus, the pressure modes, $p^R(r)$, are purely real upstream, but the downstream modes have some imaginary or out-of phase content. The downstream modes also show a somewhat greater radial variations than their upstream counterparts.

The unsteady surface pressure responses due to the (0,0) acoustic excitation from upstream are given in Figure 14. The 3D LINFLUX and 2D LINFLO results at midspan show substantial differences along the suction surface over the forward half of the blade, which we suspect are due to a lack of sufficient mesh resolution for the LINFLUX calculation. The com-

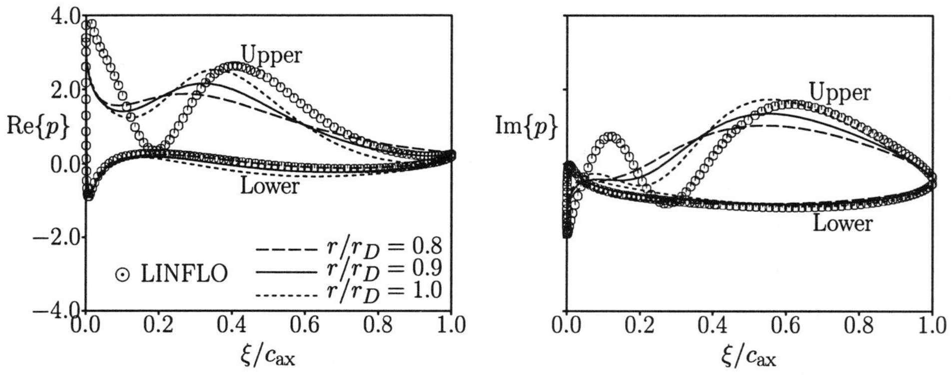

Figure 14. Unsteady surface pressure distributions due to the interaction of an acoustic excitation from upstream [$p_{A,\mathrm{Ref}}^{-} = (1,0)$, $\omega = 3.861$ and $\sigma = 90\,\mathrm{deg}$] with the 3D 10th Standard Cascade.

bination of nearly sonic flow in this region and a relatively high unsteady excitation frequency leads to local acoustic responses at high wave numbers, since the wave number scales with $\omega M/(1 - M^2)$. These high wave number, and thus short wavelength, acoustic responses are resolved on the LINFLO C-H grid, but are not resolved on the $141 \times 41 \times 11$ LINFLUX H-grid. The unsteady pressure distributions on the pressure surface, where steady Mach numbers are lower, are in good agreement.

The LINFLUX solution indicates that the unit-amplitude (0,0) acoustic excitation from upstream produces (0,0) and (1,0) acoustic response disturbances at amplitudes of 0.676 and 0.099 far upstream and a propagating (0,0) response at amplitude 0.768 far downstream. The corresponding LINFLO predictions are 0.667, 0.603 and 0.761, respectively. Thus, the LINFLUX and LINFLO predictions of the amplitude of the upstream-propagating acoustic response in the $m = 1$ circumferential mode differ. Both analyses indicate that the attenuating acoustic response waves are negligible at the computational inflow and outflow boundaries.

Contours of the in-phase component of the unsteady pressure, $Re\{p\}$, at midspan due to the acoustic excitation from upstream are shown in Figure 15. The LINFLUX and LINFLO results are in good qualitative agreement, but there are important quantitative differences along the suction surfaces of the blades, near the blade leading edges, and upstream of the blade row. These differences suggest the LINFLUX computational mesh should be refined near the forward portion of the suction surface.

We performed a brief study to test whether LINFLUX results approach those of LINFLO if a re-distributed and refined mesh is used for the LINFLUX calculations. For this purpose, we considered the unsteady flow through the 2D Tenth Standard Cascade excited by a unit-amplitude acous-

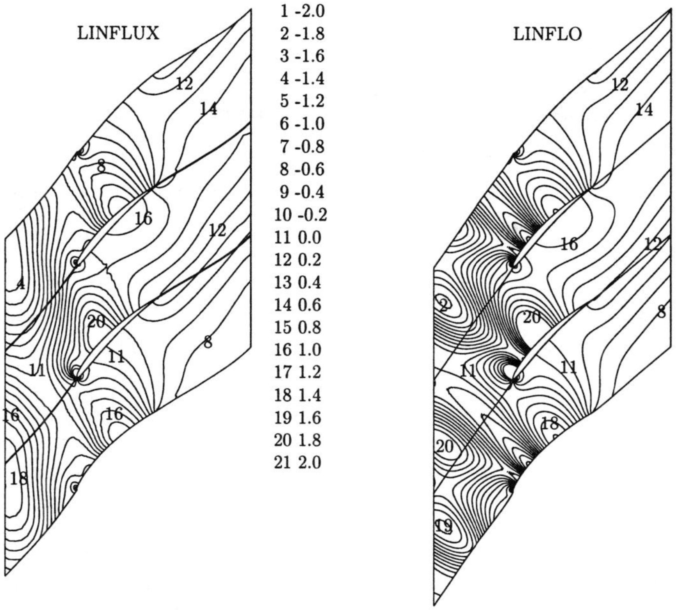

Figure 15. Contours of the in-phase component of the unsteady pressure at midspan due to the interaction of an acoustic excitation from upstream $[p^-_{A,\mathrm{Ref}} = (1,0)$, $\omega = 3.861$, $\sigma = 90\,\mathrm{deg}]$ with the 3D 10th Standard Cascade.

tic excitation from upstream at $\omega = 3.861$ and $\sigma = 90\,\mathrm{deg}$. The 2D LINFLUX results, determined on the 141 $\times$ 41 grid used at midspan in the 3D LINFLUX calculations, agree closely with the 3D LINFLUX predictions at midspan. The 2D LINFLUX calculations were then performed on a re-distributed 141 $\times$ 41 grid in which more axial lines were placed upstream and over the forward half of the blades. Following this calculation, 2D LINFLUX calculations were performed on refined meshes that were 1.5 and 2.0 times as dense as the re-distributed 141 $\times$ 41 mesh. The results are shown in Figure 16, where it can be seen that the re-distribution of the original 141 $\times$ 41 mesh leads to LINFLUX surface-pressure predictions that are in better agreement with LINFLO results. Moreover, as mesh density is increased, the LINFLUX predictions approach those of LINFLO, which are believed to be mesh converged.

6. Concluding Remarks

We have applied the 3D LINFLUX analysis to predict unsteady subsonic flows through a helical fan and a "real" blade row, i.e., a 3D version of the Tenth Standard Cascade Configuration. The flows considered are dominated by acoustic phenomena, with unsteady fluctuations due to prescribed blade vibrations or incoming acoustic disturbances. In each case, a nearly

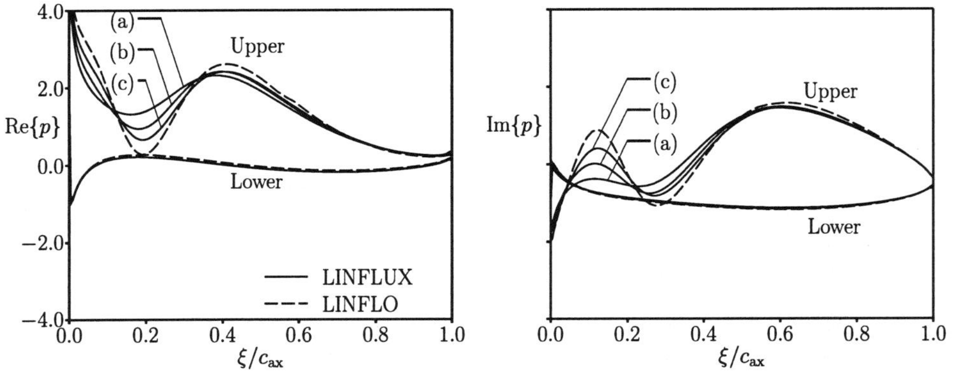

Figure 16. Unsteady surface pressure distributions, as predicted by the 2D LINFLUX and LINFLO analyses, due to the interaction of an acoustic excitation from upstream [$p_A^- = (1,0)$, $\omega = 3.861$, $\sigma = 90\,\text{deg}$] with the 2D 10th Standard Cascade. The LINFLUX solutions were determined on 141 × 41 (a), 211 × 61 (b), and 281 × 81 (c) H-meshes.

2D excitation has been imposed, so that LINFLUX could be validated against proven 2D analyses. The numerical results indicate that the 3D LINFLUX code gives accurate aerodynamic response information for acoustically dominated flows, provided that the grids employed are of sufficient overall density and grid lines are clustered in regions of high flow gradients. The mesh requirements for blade flutter applications, for which reduced frequencies are typically of order one, can usually be satisfied. However, requirements for forced vibration and aeroacoustic response studies, in which reduced frequencies on the order of 5 to 50 must be considered, will impose serious constraints on available computational resources.

For the helical fan, the LINFLUX results at midspan are in very good agreement with 2D predictions based on the Smith [4] and the LINFLO [5] analyses. The LINFLUX and LINFLO results at midspan for the 3D Tenth Standard Cascade are generally in good agreement, but in some cases significant quantitative differences occur along the suction surfaces of the blades, where steady Mach numbers are near sonic, and upstream of the blade row.

A number of computational strategies should be investigated to improve the LINFLUX analysis. In particular, improved surface boundary conditions should be installed to improve solution accuracy near blade surfaces. In addition, strategies for reducing the mesh densities required to achieve accurate field solutions, e.g., the use of more accurate field approximations and/or composite H–C meshes, should be investigated. Finally, because of the fine meshes required, a parallel version of LINFLUX should be considered for blade-row aeroacoustic response studies.

The far-field eigenanalyses have been shown to be successfully coupled to the near-field analysis. The former were used to determine the acoustic disturbances in the low-order radial modes. The higher-order acoustic disturbances, which are of negligible amplitude at the far-field boundaries, and the convected and nearly-convected components of the unsteady perturbation were convected out of the near-field domain through the computational outflow boundary. Thus, at this point, we have not applied the eigenanalyses to predict the axial eigenvalues and radial eigenmodes associated with nearly-convected disturbances. It may be necessary to provide accurate representations of such disturbances to predict the unsteady aerodynamic responses associated with vortical wake/blade-row interactions.

In the next LINFLUX development effort, we will focus on the unsteady flows associated with wake/blade-row interactions. For such applications, the wakes must, for swirling mean flows, be modeled as a superposition of convected and nearly-convected disturbances. Thus, the present far-field eigenanalysis must be extended to determine the nearly-convected modal disturbances. Following this, the LINFLUX predictions must be evaluated for unsteady flows associated with wake/blade-row interactions. Based on our earlier investigations with the 2D LINFLUX analysis [12, 13], we anticipate that higher-order surface boundary conditions will be needed to accurately model unsteady vortical behaviors near blade surfaces. To date, we have focused on demonstrating and validating the 3D LINFLUX code for unsteady subsonic flows. The 3D code can also be applied to unsteady transonic flows. But, again based on our earlier investigations with the 2D code, improvements in shock modeling will be required.

7. Acknowledgments

The research reported in [1] and in this paper was supported under NASA Contract NAS3–26618. The authors are indebted to D.L. Huff (NASA Lewis) and H.A. Chuang (UTRC) for providing helpful advice and assistance during the course of this effort.

References

1. M. D. Montgomery and J. M. Verdon. A 3D Linearized Euler Analysis for Blade Rows, Part 1: Aerodynamic and Numerical Formulations. *The 8th International Symposium on Unsteady Aerodynamics and Aeroelasticity of Turbomachines*, Stockholm, Sweden, September 14–18, 1997.
2. M. D. Montgomery and J. M. Verdon. A Three-Dimensional Linearized Unsteady Euler Analysis for Turbomachinery Blade Rows. NASA CR 4770, March 1997.
3. T. H. Fransson and J. M. Verdon. Standard Configurations for Unsteady Flow through Vibrating Axial-Flow Turbomachine Cascades. In *Unsteady Aerodynamics, Aeroacoustics and Aeroelasticity of Turbomachines and Propellers*, pages 859–889. Edited by H. M. Atassi, Springer-Verlag, New York, 1993.
4. S. N. Smith. Discrete Frequency Sound Generation in Axial Flow Turbomachines. R&M 3709, British Aeronautical Research Council, London, England, UK, March 1972.

5. J. M. Verdon. Unsteady Aerodynamic Methods for Turbomachinery Aeroelastic and Aeroacoustic Applications. *AIAA Journal*, 31(2):235–250, February 1993.
6. J. R. Caspar. Unconditionally Stable Calculation of Transonic Potential Flow through Cascades using an Adaptive Mesh for Shock Capture. *Transactions of the ASME: Journal of Engineering for Power*, 105(3):504–513, July 1983.
7. J. M. Janus, H. Z. Horstman, and D. L. Whitfield. Unsteady Flowfield Simulation of Ducted Prop-fan Configurations. Paper 92–0521, AIAA 30th Aerospace Sciences Meeting and Exhibit, Reno, Nevada, January 6-9 1992.
8. B. K. Soni and M. H. Shih. TIGER: Turbomachinery Interactive Grid GenERation. In *Proceedings of the Third International Conference on Numerical Grid Generation in CFD*, Barcelona, Spain, June 1991.
9. T. A. Beach and G. Hoffman. IGB Grid: User's Manual (A Turbomachinery Grid Generation Code). NASA CR 189104, January 1992.
10. J. M. Verdon. The Unsteady Aerodynamic Response to Arbitrary Modes of Blade Motion. *Journal of Fluids and Structures*, 3(3):255–274, May 1989.
11. T. C. Ayer and J. M. Verdon. Validation of a Nonlinear Unsteady Aerodynamic Simulator for Vibrating Blade Rows. ASME Paper 96–GT–340, International Gas Turbine and Aeroengine Congress & Exhibition, Birmingham, UK, June 10–13, 1996.
12. J. M. Verdon, M. D. Montgomery, and K. A. Kousen. Development of a Linearized Unsteady Euler Analysis for Turbomachinery Blade Rows. NASA CR 4677, June 1995.
13. M. D. Montgomery and J. M. Verdon. A Linearized Unsteady Euler Analysis for Turbomachinery Blade Rows Using an Implicit Wave-Split Scheme. In *Unsteady Aerodynamics and Aeroelasticity of Turbomachines*, pages 143–160. Edited by Y. Tanida and M. Namba, Elsevier, Amsterdam, 1995.

ANALYSIS OF NONLINEAR OSCILLATING CASCADE AERODYNAMICS INCLUDING SEPARATED FLOW

JAMES M. WOLFF
SANFORD FLEETER
Thermal Sciences and Propulsion Center
School of Mechanical Engineering
Purdue University
West Lafayette, Indiana 47907

Abstract. A mathematical model is developed to analyze the unsteady flow, including separated flow, through an harmonically oscillating cascade of airfoils. The computational model incorporates the time–marching Euler analysis, NPHASE, with an inverse integral boundary layer solution. Grid generation is simplified by utilizing an embedded composite grid formulation. Fourier series unsteady periodic boundary conditions are implemented, which reduce computational requirements. The integral turbulent boundary layer model is closed with steady correlations adopted in a quasi–steady manner. The viscous effect is modeled in the unsteady Euler solution in a quasi–steady manner by a transpiration boundary condition, thereby coupling the solution methods. An advanced design compressor geometry is utilized to study the effect of flow separation on the unsteady aerodynamics. An experimental unsteady aerodynamics data set for this configuration with separated meanflow executing torsional oscillations compared favorably with the analysis.

1. Introduction

Current unsteady aerodynamic cascade models are generally inviscid and time linearized. The validity of these linearized models is well established for subsonic flows for low aerodynamic loading (Whitehead, 1987). However these flow models are not valid at off-design flow conditions. Unfortunately, it is at these off-design conditions that serious aeroelastic problems are generally encountered in the development of new turbomachinery. To predict aeroelastic phenomena at off-design conditions, an unsteady cascade viscous analysis which properly models strong inviscid–viscous interaction effects such as boundary layer separation and shock/boundary layer interaction is needed.

Recently several time–marching solutions for the unsteady Reynolds–averaged Navier–Stokes equations have been developed to analyze blade row interactions (Giles and Haimes, 1991) and oscillating cascades (Huff, 1991). These Navier–Stokes solutions have demonstrated the feasibility of analyzing complex two and three dimensional turbomachinery flows. However, because of the large number of grid points and the requirement that the analysis be both accurate and stable, the maximum allowable time step in the calculations must be small, making these calculations prohibitively expensive for design use.

An alternate approach to using the time marching Reynolds–averaged Navier–Stokes equations for the computation of unsteady viscous cascade aerodynamics is a zonal solution method as opposed to the global solution method. One type of zonal method

T.H. Fransson (ed.), Unsteady Aerodynamics and Aeroelasticity of Turbomachines, 465-476.

used for inviscid–viscous interaction is an inviscid flow solution method for the flow away from the body matched to a viscous or boundary layer flow solution method for the flow near the body. For steady airfoil calculations, extensive inviscid–viscous interaction methods have been developed, for example Whitfield, et al. (1982). The boundary layer computational methods are either finite difference or integral methods. As well demonstrated for steady flow calculations, the integral form of the boundary layer equations leads to a more efficient way of including viscous effects in numerical solutions. Hence, an inviscid–integral viscous coupled approach seems particularly attractive for unsteady flow calculations.

Empirical correlations must be utilized in an integral boundary layer solution. Currently there are no unsteady flow correlations, only experiments on turbulent boundary layers subject to freestream oscillations which show that the steady correlations can be applied in a quasi–steady manner (Cousteix, et al., 1981, Karlsson, 1959). Very few attempts have been made to apply the viscous–inviscid interaction approach to unsteady turbomachinery flows. One attempt which shows encouraging results used a direct unsteady Euler and an entrainment integral boundary layer solution for unsteady flow around oscillating blades with a simultaneous coupling method (He and Denton, 1993).

In this paper, a computationally efficient mathematical model is developed to analyze the unsteady flow through an harmonically oscillating cascade of airfoils, including inviscid-viscous interactions. The model incorporates the time–marching Euler analysis NPHASE (Huff, 1991) with an inverse integral boundary layer solution. The code is then utilized to predict the unsteady aerodynamics of a cascade executing torsional oscillations including separated flow effects.

2. Inviscid Model

2.1 IMPLICIT FINITE VOLUME SCHEME

The inviscid flowfield computations are performed using an implicit finite volume Euler solver, NPHASE (Huff, 1991). The discretized integral form of the time-dependent curvilinear Euler equations is obtained by integrating the partial differential equations over a computational volume with the center denoted as (i,j) and changing the resulting volume integral to a surface integral using the divergence theorem

$$\frac{\partial Q}{\partial t} + \frac{\delta_i F}{\Delta \xi} + \frac{\delta_j G}{\Delta \eta} = 0 \qquad (1)$$

where the central difference operators $\delta_m(\bullet) = (\bullet)_{m+\frac{1}{2}} - (\bullet)_{m-\frac{1}{2}}$, m=i,j imply that the flux vectors are evaluated at the surfaces of a cell.

The dissipation aspect of this scheme is improved by a flux–difference split method for the residual based on the solution of approximate Riemann problems, with Roe averaging at the cell faces. With the determination of the eigensystem for the flux vectors using Roe averaging and the knowledge that the interface differential is proportional to the right eigenvectors, a first order flux is obtained. To obtain higher order spatial accuracy, a corrective flux is then added. Excellent results have been obtained by evaluating the residual term with the above flux–difference split method and the left hand side operator with the flux vector split scheme, which is approximately factored into the product of two operators (Huff, 1991).

2.2 EMBEDDED COMPOSITE COMPUTATIONAL GRIDS

The Euler solution is implemented for an airfoil cascade geometry by a computational C–grid embedded in a Cartesian H–grid. The embedded composite grid approach of Wolff and Fleeter (1994b) is utilized to extend the computational grid upstream and downstream of the cascade. The outer boundary condition for the inner C–grid is obtained by appropriate interpolation from the Cartesian H–grid, Figure 1, termed fringe points. The inner H–grid boundary is defined, thereby indicating the hole grid points that will not be solved by the flow solver, Figure 1. The grid points adjacent to the hole points are used as fringe points for the Cartesian H–grid, Figure 1. Bi–linear interpolation accounts for the transfer of information from one grid to another at the fringe points. The conservative variables ($\rho(\xi,\eta)$, $\rho u(\xi,\eta)$, etc.) are interpolated.

To ensure proper interpolation of information between the C–grid and the Cartesian H–grid, the outer portion of the C–grid remains fixed in space, with a deforming grid technique used to locate the position of the airfoil. Based on the grid point locations in the C–grid that are used for interpolating to the H-grid inner boundary, the user specifies the portion of the C–grid allowed to deform with the airfoil motion. This is accomplished by specifying a η grid line beyond which the C–grid will remain stationary, Figure 1. The grid lines connecting the inner (η=1) and outer (η=η_S) boundaries deform with the airfoil motion.

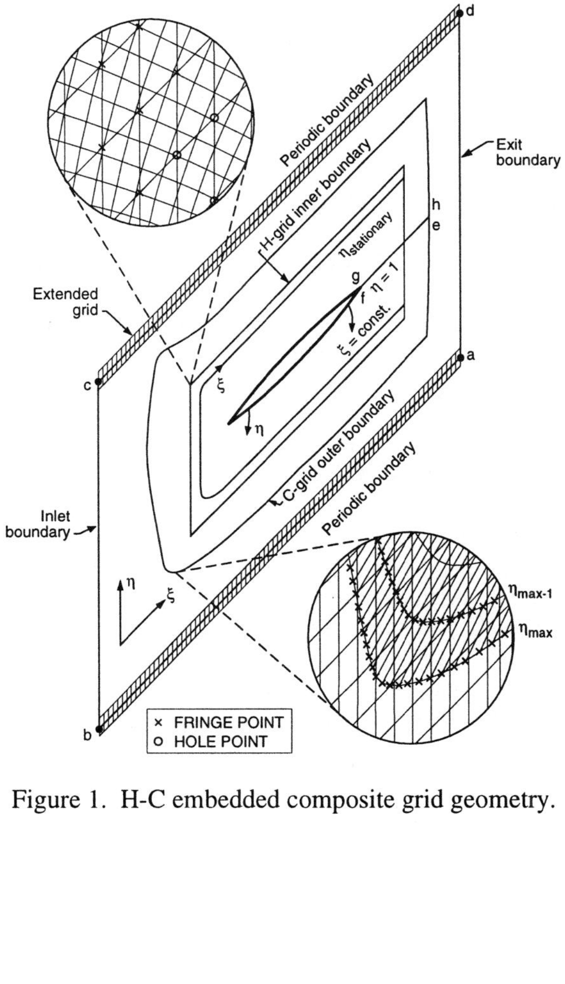

Figure 1. H-C embedded composite grid geometry.

2.3 BOUNDARY CONDITIONS

The solid surface boundary conditions, f–g in Figure 1, implement characteristic variable boundary conditions (CVBC). The CVBC's are consistent with the concept of upwinding in which the signs of the characteristic velocities determine the appropriate propagation directions. The flow variables are averaged across the interface aft of the airfoil solid surface to the exit, e–f and g–h.

CVBC are used for proper transmission of information into and out, b–c and a–d of the steady computational domain. The CVBC's are only valid for steady cascade flows because the exit boundary is assumed to have uniform static pressure.

Approximate non–reflecting unsteady inlet and exit boundary conditions are used by assuming that linear theory can be applied. First the steady flow is solved using the CVBC's. The linearized Euler equations are then solved at the inlet and exit boundary to determine the perturbation flow variables in terms of the characteristic variables. This allows time variations of static pressure at the exit and reduces reflections from the boundaries. For certain unsteady flows, reflections are still a problem. The composite grid method is used to extend the computational domain upstream and downstream to eliminate the reflections.

The flow variables at the unsteady periodic boundaries, a–b and c–d, are approximately expressed as a Fourier series for non-zero interblade phase angle (Wolff and Fleeter, 1994a). The Fourier series boundary conditions are applied by expanding the computational grid one node at the upper and lower periodic boundaries. At one node interior of the upper and lower periodic boundaries, the timewise integration for the Fourier coefficients are performed on the conservative variables. The solutions at the overlapped nodes for the lower and upper periodic boundaries are found by a timewise Fourier series.

3. Integral Boundary Layer Model

3.1 QUASI–STEADY ASSUMPTION

Steady predictions are assumed to apply in a quasi–steady manner to model the unsteady boundary layer behavior. This assumption has been shown to be valid by several researchers. Karlsson (1959) provided pioneering data on unsteady flat plate turbulent boundary layers with no pressure gradient obtained in a wind tunnel with velocity fluctuating sinusoidally in magnitude about a constant mean. The fluctuations varied from 8% to 34% of the free stream mean velocity, with $Re_x = 1x10^5$ and a reduced frequency range of 0.22 to 27.4. He found that the boundary layer momentum thickness did not vary from the quasi–steady thickness over the entire range of parameters.

Cousteix, et al. (1981) obtained experimental data on a similar configuration, but in an adverse pressure gradient. The fluctuations varied from 7% to 15% of the free stream mean velocity, with $Re_x = 1.5x10^6$ and a reduced frequency range of 0.76 to 5.0. These results support the validity of the quasi–steady boundary layer assumption for reduced frequency values of at least 5.0. Based on the above available experimental unsteady turbulent data, a quasi–steady model for unsteady boundary layer calculations is reasonable for unsteady turbomachine flows up to a reduced frequency of at least 5.0.

3.2 TURBULENT FLOW

The turbulent integral boundary–layer formulation is based on the mean flow kinetic energy. The formulation was developed originally by Whitfield (1978) and updated to include separated flow. The momentum integral equation and the mean–flow kinetic energy equation are written as:

$$\frac{1}{\rho_e u_e^2}\frac{\partial}{\partial x}\left(\rho_e u_e^2 \theta\right) + \frac{\delta^*}{u_e}\frac{\partial u_e}{\partial x} = \frac{c_f}{2} \tag{2}$$

$$\frac{1}{2\rho_e u_e^3}\frac{\partial}{\partial x}\left(\rho_e u_e^3 \theta^*\right) + \frac{\delta^{**}}{u_e}\frac{\partial u_e}{\partial x} = \frac{c_f D}{2} \tag{3}$$

The governing equations contain seven unknowns, u_e, δ^*, θ, c_f, δ^{**}, θ^* and D. An analytical expression for attached and separated turbulent boundary–layer velocity profiles permits the evaluation of certain auxiliary correlations necessary for the integral boundary–layer calculation method. A closure procedure is employed that expresses four of the unknowns, θ^*, δ^{**}, c_f and D, in terms of the remaining three, M_e, θ, and δ^*, by defining the following shape factors, H_{δ^*}, H_{θ^*}, $H_{\delta^{**}}$.

3.3 INVERSE SOLUTION

The integral boundary layer equations can be formulated to be solved by a direct method, where the pressure distribution is input to the solution. This method has a singularity associated with flow separation. This singularity is removed by specifying the displacement thickness distribution instead of the pressure distribution, i.e., an inverse method, and boundary-layer computations can proceed throughout separated regions, without the separation point being specified prior.

Equations 2 and 3 can be arranged to (Whitfield, et al., 1982).

$$\left(1 + H_{\delta^*}\right)\frac{d}{dx}\left[\ln(u_e)\right] - \frac{d}{dx}\left[\ln\left(H_{\delta^*}\right)\right] = \frac{c_f}{2\theta} - \frac{d}{dx}\left[\ln\left(\rho_e u_e \delta^*\right)\right] \tag{4}$$

$$\left(1 + 2\frac{H_{\delta^{**}}}{H_{\theta^*}} - H_{\delta^*}\right)\frac{d}{dx}\left[\ln(u_e)\right] - \frac{d}{dx}\left[\ln\left(H_{\theta^*}\right)\right] = \frac{c_f D}{\theta H_{\theta^*}} - \frac{c_f}{2\theta} \tag{5}$$

For a gas with constant specific heat ratio and constant total enthalpy at the boundary-layer edge, these equations can be written with H and M_e as dependent variables and solved by a Runge-Kutta scheme. The boundary layer calculations are performed separately on the upper and lower airfoil surfaces. The stagnation point is located. The turbulent boundary layer calculations are then started at 0.5% chord from the stagnation point on both surfaces. The calculations are continued past the airfoil surface to the computational domain exit using the same numerical scheme except assigning the skin friction, c_f, to 0.00001. For the steady calculations, a flat plate displacement thickness distribution is used to start the inverse boundary layer solution. In addition the starting shape factor, H, is specified as 1.2. For the unsteady calculations, the steady displacement thickness distribution is used along with the same starting shape factor.

3.4 INVISCID–VISCOUS COUPLED SOLUTION

A surface source model will be used as a coupling device between the steady viscous and inviscid computational regions. The surface source model imposes a mass flux on the solid wall and numerically simulates the displacement effect caused by the steady boundary layer. The surface model is derived by subtracting the two–dimensional continuity equation from the continuity equation evaluated at the edge of the boundary layer, with the resulting equation then integrated in the y–direction, knowing that ρv at the wall is zero. The result is

$$\frac{\partial}{\partial x}\left(\rho_e u_e \delta^*\right) = (\rho v)_n \qquad (6)$$

The steady surface source model is then applied in a quasi–steady manner.

The term $(\rho v)_n$ is updated as follows. The δ^* distribution at the previous time step is used in the inverse boundary layer code to solve for the distribution of $\rho_e u_e \delta^*$. This quantity is then used in Equation 6 to calculate $(\rho v)_n$ for the next inviscid calculation. The δ^* distribution for the next viscous calculation is obtained by the following expression

$$\delta^*_{new} = \delta^*_{old} + \omega \delta^*_{old}\left[\frac{u_{e,v}}{V_{w,i}} - 1\right] \qquad (7)$$

The inviscid solid wall boundary condition is implemented using a characteristic variable formulation. To transpire the proper amount of artificial mass to account for the viscous effect, Equation 6 is used in conjunction with the characteristic variables formulation.

For the steady–state analysis, the airfoil is held stationary as the flow solver is marched in time until steady conditions are reached. The viscous and inviscid analysis are calculated each time step. The relaxation parameter was set to 0.5 for the steady analysis and 1.0 for the unsteady analysis. The inverse integral boundary layer calculations are performed once per time step for the steady analysis, but for the unsteady analysis they are iterated until convergence. Convergence is achieved for each time step when the velocity from the viscous and inviscid solutions agree within 0.1%.

4. Results

In this paper, the two–dimensional unsteady flow through an harmonically oscillating cascade of airfoils is investigated using the time–marching inviscid–viscous interaction code. For comparison, solutions have been obtained for an isolated airfoil (RAE 2882) in a steady flow and for both flat plate and loaded (10th Standard Configuration) oscillating airfoil cascades (Wolff, 1995). Finally, an advanced design compressor geometry was utilized to investigate separated flow phenomena.

4.1 SEPARATED FLOW PREDICTIONS

Experiments have been performed in the NASA Lewis Transonic Oscillating Cascade facility on an airfoil cascade designed by Pratt & Whitney (Buffum, et al., 1996). The test configuration is a cascade of negatively cambered airfoils at 60 degrees stagger with a spacing to chord ratio of 0.657. A composite grid configuration, 217x39 C–grid, embedded in a 108x30 H–grid, with the inlet and exit plane locations extending 6 chords upstream and downstream is used for both the inviscid and inviscid-viscous interaction calculations.

Steady pressure distributions are available for a freestream Mach number of 0.5. Two different incidence angles, 0° and 10°, are considered. The corresponding pressure ratios, defined as the ratio of downstream static pressure to the upstream static pressure, are 0.93

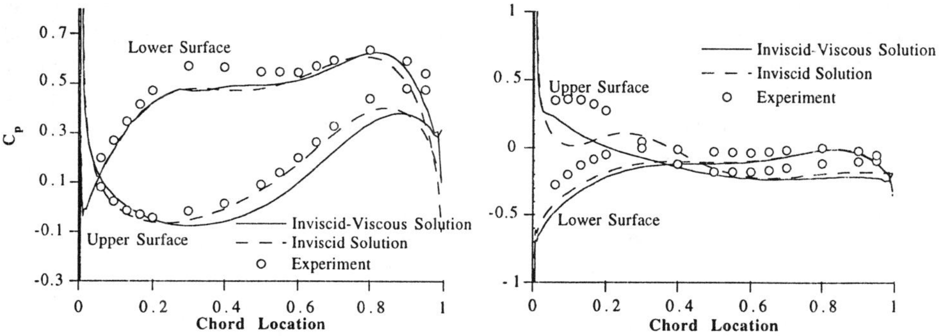

(a) $\alpha=+1°$, $p/P_0 = 0.784$ (b) $\alpha=+7°$, $p/P_0 = 0.868$

Figure 2. NASA Pratt & Whitney configuration steady pressure distribution: $M = 0.5$, $\zeta = 60°$, $S/C = 0.657$, $Re = 1x10^6$

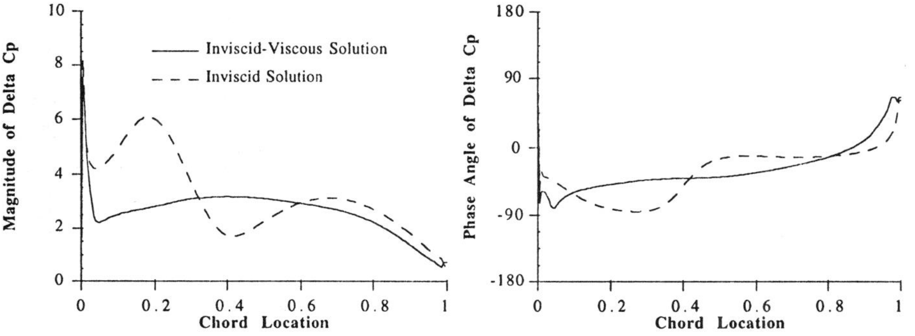

Figure 3. NASA Pratt & Whitney configuration unsteady pressure distribution: $M = 0.5$, $k=.4$, $\zeta = 60°$, $S/C = 0.657$, $\alpha=7°$, $p/P_0=.868$, $\alpha_1=1.2°$, $x_{ea}=.5$, $\sigma=0°$

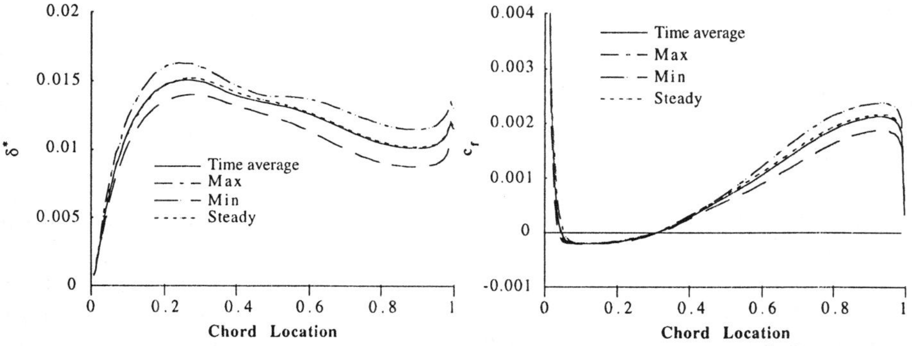

Figure 4. NASA Pratt & Whitney configuration unsteady boundary layer parameters: $M = 0.5$, $k=.4$, $\zeta = 60°$, $S/C = 0.657$, $\alpha=7°$, $p/P_0=.868$, $\alpha_1=1.2°$, $x_{ea}=.5$, $\sigma=0°$

and 1.03, respectively. The experimental incidence angle is set by positioning the upstream walls of the linear cascade to the specified inlet angle, with the actual incidence angle at the cascade inlet plane not measured. Thus, the incidence angle for the computations was determined by analyzing a limited range of incidence angles and selecting the one which best matched the experimental results. These tests were completed using the inviscid analysis.

Figure 2a shows the experimental steady pressure loading for the 0° incidence case together with the computational solutions obtained for a 1° incidence angle. The viscous effects are the strongest on the upper airfoil surface. The upper surface viscous effect moves the solution away from the data from 30% to 90% of the airfoil chord. Except for the upper surface viscous solution, the agreement between the predictions and the experimental data is generally good. Slight modifications to the airfoil shape could explain the differences at the trailing edge. To get a converged steady solution, the trailing edge had to be straighten to provide a better computational grid.

The experimental steady pressure loading for the 10° incidence angle case and the computational results obtained with a 7° incidence angle are shown in Figure 2b. The experimental data indicate flow separation on the upper surface from the 5% to 20% chord locations. The inviscid analysis on the upper surface under predicts the pressure through the separation zone and then over predicts the pressure until mid-chord. The upper surface viscous effect causes the pressure to flatten out as compared to the inviscid analysis. The pressure is still under predicted in the separated region, but the viscous solution tends toward the data. The viscous solution predicts boundary layer separation from 5% to 30% of the upper surface. The viscous effects on the lower surface are negligible. The lower surface pressure is under predicted over the first 80% of the chord. The agreement with the data shows that the general trendwise behavior of the flow is captured by the analysis. Transition was assumed at 0.5% chord, this limitation could be the cause of some of the error.

4.2 UNSTEADY TORSIONAL MOTION

The NASA Lewis Transonic Oscillating Cascade facility is designed to oscillate the cascade in a torsional motion about mid-chord with an amplitude of ±1.2 degrees. The torsional motion analysis at 7° of incidence with a reduced frequency of 0.4 and 0° interblade phase angle for the inviscid and inviscid-viscous, $Re=1x10^6$, interaction methods is shown in Figure 3. Because of periodicity problems (Buffum, 1996), experimental data is only available for σ=180°. The steady analysis computational grid was used. The steady results for this incidence angle showed the boundary layer to be separated from 5% to 30% chord location on the airfoil upper surface. The viscous effects on the unsteady loading, both the magnitude and phase of the unsteady pressure difference, are significant over the first 60% of the chord. There are large differences between the viscous and inviscid predictions of the unsteady pressure difference magnitude over the first 60% chord, with the viscous solution producing a greatly reduced magnitude that is flattened out through the separation zone. Compared to the inviscid analysis, the inviscid-viscous analysis predicts an increase in the unsteady magnitude from the 30% to 60% chord locations. The inviscid and viscous unsteady magnitude predictions agree over the last 40% chord. The unsteady phase pressure difference shows a sharp decrease at the beginning of the separated zone. The phase distribution then remains fairly constant until the trailing edge region where it increases.

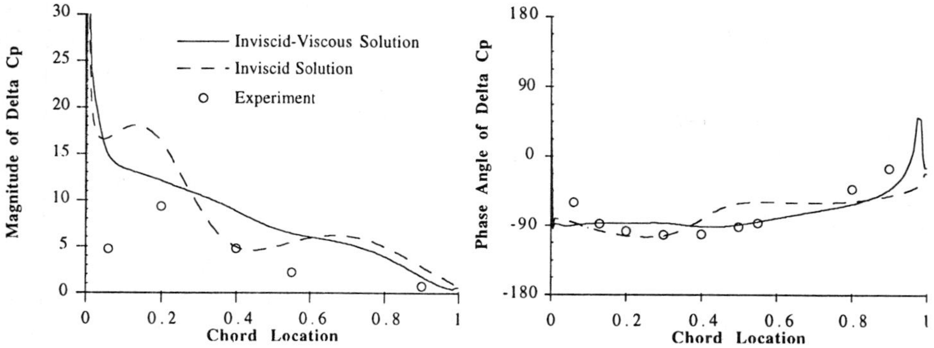

Figure 5. NASA Pratt & Whitney configuration unsteady pressure distribution: $M = 0.5$, $k=.4$, $\zeta = 60°$, $S/C = 0.657$, $\alpha=7°$, $p/P_0=.868$, $\alpha_1=1.2°$, $x_{ea}=.5$, $\sigma=180°$

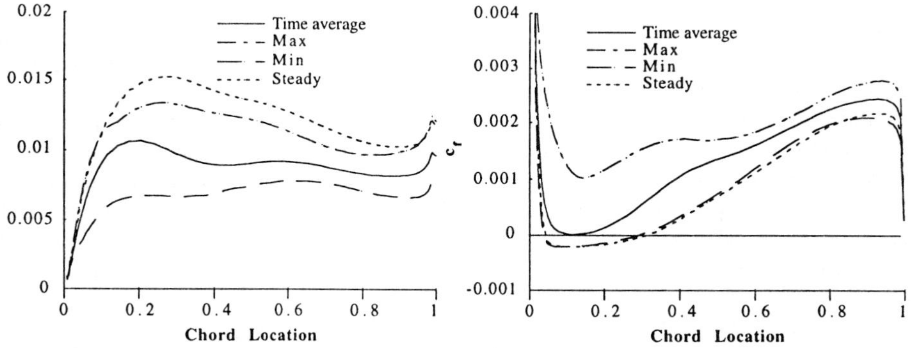

Figure 6. NASA Pratt & Whitney configuration unsteady boundary layer parameters: $M = 0.5$, $k=.4$, $\zeta = 60°$, $S/C = 0.657$, $\alpha=7°$, $p/P_0=.868$, $\alpha_1=1.2°$, $x_{ea}=.5$, $\sigma=180°$

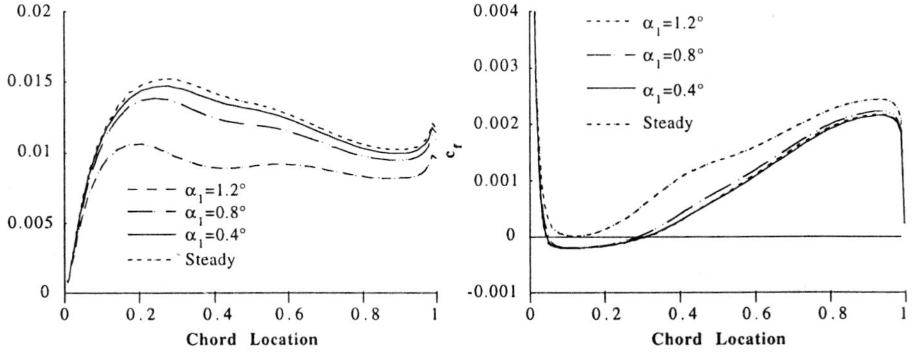

Figure 7. NASA Pratt & Whitney configuration time averaged boundary layer parameters: $M = 0.5$, $k=.4$, $\zeta = 60°$, $S/C = 0.657$, $\alpha=7°$, $p/P_0=.868$, $x_{ea}=.5$, $\sigma=180°$

Figure 4 shows the variation of the unsteady upper surface boundary layer parameters, δ^* and c_f, as a function of chord location. Surprisingly, the variation in the boundary layer parameters show no significant effects. The time average and steady values are the same. The flow remains separated over the complete oscillation cycle of the airfoil. The area of separation does change as the airfoil oscillates but this movement is not significant. The boundary layer parameters show the greatest unsteady variation near the trailing edge with a maximum unsteady envelope for δ^* of 27% and for c_f, 29% of their mean values. A negative c_f signifies separated flow, but its numerical size has no physical meaning. Therefore the maximum, minimum, and time average values of c_f in the separated region have the same negative value. The variation of δ^* shows that the unsteady boundary layer in the separated flow zone is changing as the airfoils oscillate. The conclusion drawn from these unsteady results is that mean separated flow conditions have significant effects on the unsteady pressure difference across the airfoil chord.

4.3 UNSTEADY NONLINEAR SEPARATED FLOW

Figure 5 shows the experimental data (Buffum, et al., 1996), inviscid, and inviscid-viscous, $Re=1x10^6$, interaction analysis prediction for the 7° incidence angle case but with a 180° interblade phase angle. Analogous to the previous 0° interblade phase angle separated unsteady results, there is a large difference between the inviscid and viscous predictions of the unsteady pressure difference magnitude in the region where the steady flow is separated, with the viscous prediction decreased. Both methods over predict the experimental magnitude, with the viscous solution providing a better trendwise agreement. The unsteady pressure difference magnitude predicted by the viscous analysis is increased as compared to the inviscid analysis from 30% to 60% of the airfoil chord, with minimal effect aft of 60% chord. The unsteady viscous phase distribution solution shows excellent agreement with the experimental results over the entire chord.

To study the unsteady effects on the boundary layer parameters, Figure 6 shows the upper surface boundary layer parameters, δ^* and c_f, as a function of chord location. Significant nonlinear effects can be seen in both boundary layer parameters. For the displacement thickness, the entire unsteady envelope is lower than the steady values. While for c_f, the minimum unsteady values are approximately the steady ones. Over part of the airfoil oscillation, the flow is completely attached on the upper airfoil surface, as shown by the c_f distribution. This causes significant differences between the steady and time average values of the skin friction.

At the 15% chord location, the time average value of δ^* is closer to the maximum and that of c_f nearer to the minimum. This means that the flow is separated over the majority of the oscillation cycle and is attached for only a short time. At the 30% chord location, the opposite is true. Therefore the flow is attached over the majority of the oscillation cycle and separated for a limited time. The unsteady boundary layer solution is significantly affected by the oscillation of the cascade as the steady δ^* values are never reached in the unsteady solution from 10% to 90% chord.

An explanation for this phenomena is obtained by considering the influence of the interblade phase angle on the flow physics. For the 0° interblade phase angle analysis, there was no difference between the time average and the steady boundary layer parameters. At 180°, the boundary layer is attached over part of the oscillation cycle. The main difference between a 0° and 180° interblade phase angle value is that at 180° the airfoils are out of phase with one other. When one airfoil is pitched maximum up, the adjacent airfoil is pitched maximum down. This causes a flow restriction between adjacent airfoils, with the adjacent pair of airfoils having an expanded flow area and more favorable flow conditions. Therefore, the flow will be attached on the upper surface of the airfoil that is experiencing the favorable flow conditions.

Assuming the nonlinear effects predicted with the inviscid-viscous interaction analysis at 180° interblade phase angle with ±1.2° torsional motion are real, then decreasing the amplitude of oscillation should eliminate the nonlinear effects. The time averaged separated unsteady boundary parameters for three different oscillation amplitudes, ±1.2°, ±0.8°, and ±0.4°, are shown in Figure 7, along with the steady solution. As the oscillation amplitude is decreased, the time average unsteady results approach the steady ones. The movement of the size of the separation region has significant effects on the downstream boundary layer values. As the separation zone gets smaller, the time average displacement thickness decreases.

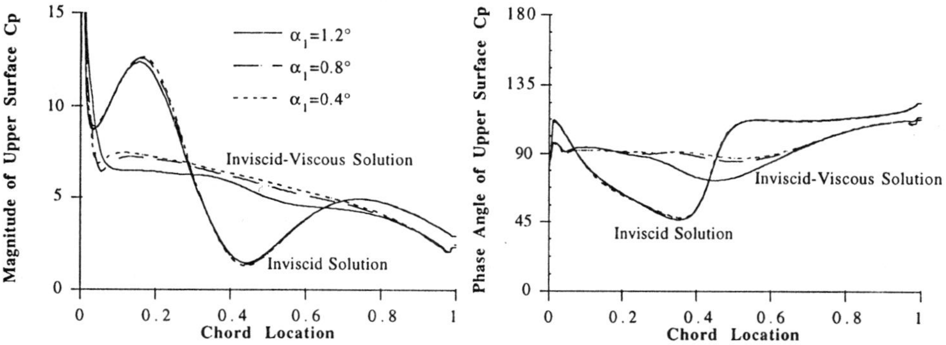

Figure 8. NASA Pratt & Whitney configuration unsteady upper surface pressure: $M = 0.5$, $k=.4$, $\zeta = 60°$, $S/C = 0.657$, $\alpha=7°$, $p/P_0=.868$, $x_{ea}=.5$, $\sigma=180°$

Finally, the effect of the nonlinear separated unsteady flow on the first harmonic upper surface pressure distribution is shown, Figure 8, for the three different oscillation amplitudes. The inviscid solutions show no nonlinear effects as the oscillation amplitude is increased from 0.4° to 1.2°. Significant nonlinearities in the inviscid–viscous solutions result from varying the oscillation amplitude. As the oscillation amplitude is increased, the magnitude of the unsteady upper surface pressure decreases from 5% to 80% of the airfoil surface. A decrease in the phase angle from 30% to 60% of the airfoil surface is caused by increasing the oscillation amplitude. This change in phase is caused by the significant movement of the boundary layer reattachment point for the 1.2° analysis.

4.4 COMPUTATIONAL RESOURCES

The computational time penalty required to do the inverse integral boundary layer calculations was approximately 25% of the time required for the inviscid analysis. The major computational savings over the Navier–Stokes methods are obtained by being able to use the same grid and time step for the inviscid-viscous interaction analysis as the inviscid analysis. Typically, the difference between inviscid and viscous computational times is 300% (Ayer and Verdon, 1995). The computational memory saving are significant for two reasons. First as the grid density is increased to model the viscous effects properly with the Navier–Stokes methods the memory requirements go up proportionally. Secondly, only a few boundary layer parameters need to be stored in one dimensional arrays compared to including all the viscous flux terms in the whole flow field.

5. Concluding Remarks

A computationally efficient mathematical model was developed to analyze the unsteady flow through an harmonically oscillating cascade of airfoils, including inviscid-viscous interactions. The flow solver which incorporates an inverse integral boundary layer solution with the time–marching Euler analysis NPHASE was demonstrated. The unsteady flow solver was compared to an unsteady oscillating cascade data set showing excellent agreement.

The conclusion of this study is that unsteady viscous effects increase the nonlinearity of the oscillating cascade flow. These effects are most significant on the boundary layer parameters, but nonlinear effects are apparent in the unsteady loading. Nonlinear viscous effects were largest for 180° interblade phase angle cascade flow conditions, with the viscous effects decreasing as the amplitude of oscillation decreased. In addition, an inverse integral boundary layer viscous/inviscid quasi-steady aerodynamic model has been shown to be a viable unsteady analysis method.

Acknowledgments: This research sponsored in part by the Army Research Office (ARO). This support is most gratefully acknowledged.

References

Ayer, T. C. and Verdon, J. M., 1995, "Numerical Unsteady Aerodynamic Simulator for Blade Forced Response Phenomena," *WL-TR-95-2011*.

Buffum, D. H., 1996, Private communication.

Buffum, D. H., Capece, V. R., King, A. J., and EL-Aini, Y. M., 1996, "Oscillating Cascade Aerodynamics at Large Mean Incidence," *ASME Paper 96–GT–339*.

Cousteix, J., Houdeville, R. and Javelle, J., 1981, "Response of a Turbulent Boundary Layer to a Pulsation of the External Flow With and Without Adverse Pressure Gradient," in "Unsteady Turbulent Shear Flows", R. Michel, J. Cousteix and R. Houdeville (ed.), Springer-Verlag.

Giles, M. B. and Haimes, R., 1991, "Validation of a Numerical Method for Unsteady Flow Calculations," *ASME Paper 91–GT–271*.

He, L. and Denton, J. D., 1993, "Inviscid–Viscous Coupled Solution for Unsteady Flows Through Vibrating Blades," *ASME Journal of Turbomachinery*, Vol. 115, pp. 101-109.

Huff, D. L., 1991, "Unsteady Flow Field Predictions for Oscillating Cascades," *Sixth International Symposium on Unsteady Aerodynamic, Aeroacoustics and Aeroelasticity of Turbomachines and Propellers*, Springer-Verlag, University of Notre Dame, U.S.A.

Karlsson, S. K. F., 1959, "An Unsteady Turbulent Boundary Layer," *Journal of Fluid Mechanics*, Vol. 5, pp. 622-636.

Swafford, T. W., 1980, "Analytical Approximation of Two-Dimensional Separated Turbulent Boundary-Layer Velocity Profiles," Arnold Engineering Development Center, Report # AEDC-TR-79-99.

Whitehead, D. S., 1987, "Classical Two - Dimensional Methods," AGARD Manual on Aeroelasticity in Axial-Flow Turbomachines, Vol. 1, Unsteady Turbomachinery Aerodynamics.

Whitfield, D. L., 1978, "Integral Solution of Compressible Turbulent Boundary Layers using Improved Velocity Profiles," Arnold Engineering Development Center, Report # AEDC-TR-78-42.

Whitfield, D. L., Swafford, T. W. and Donegan, T. L., 1982, "An Inverse Integral Computational Method for Compressible Turbulent Boundary Layers," in "Recent Contributions to Fluid Mechanics", W. Haase (ed.), Springer-Verlag.

Wolff, J. M. and Fleeter, S., 1994a, "Single Passage Euler Analysis of Oscillating Cascade Aerodynamics for Arbitrary Interblade Phase," *Journal of Propulsion and Power*, Vol. 10, pp. 690-697.

Wolff, J. M. and Fleeter, S., 1994b, "Euler Analysis of Oscillating Cascade Unsteady Aerodynamics Using Embedded Composite Grids," *AIAA Paper 94-0077*.

Wolff, J. M., 1995, "Nonlinear Separated Unsteady Flow Modeling," *Ph.D. Thesis*, Purdue University.

NUMERICAL SOLUTION OF THE NAVIER STOKES EQUATIONS FOR UNSTEADY UNSTALLED AND STALLED FLOW IN TURBOMACHINERY CASCADES WITH OSCILLATING BLADES

S. WEBER, H. E. GALLUS, AND D. PEITSCH
Institut für Strahlantriebe und Turboarbeitsmaschinen (IST)
Rheinisch-Westfälische Technische Hochschule Aachen
Templergraben 55, D-52064 Aachen, Germany

Abstract

The objective of this paper is the numerical analysis and prediction of unsteady quasi 3D flow through turbomachinery cascades with vibrating blades. The developed numerical algorithm SAFES1 solves the fully nonlinearized Navier Stokes equations on S_1-stream surfaces. It is able to cope with shock waves and areas of separation in laminar or turbulent flow.

To simulate turbulence, an algebraic model by Baldwin and Lomax is used. For computing the Euler fluxes, an approximate upwind flux difference splitting scheme by Roe is implemented which achieves third order spatial accuracy by applying the MUSCL technique together with a TVD scheme. Following a Newton subiteration for every time step, the implicitly formulated algorithm exactly describes the time dependent solution of the flow field with a temporal accuracy of second order. The chosen mode of harmonic blade oscillation can consist of bending or torsion or both simultaneously in a non-rotating or rotating frame of reference realized by a time dependent grid generation. To take into consideration a phaseshift between the motion of the neighbouring blades, which are discretized by C- and O-type grids, a direct store method is implemented. For analysis of the flow field, the time dependent axial and circumferential blade forces are determined as well as the resulting moment and damping coefficient.

In this work computations for flows through a turbine cascade and a compressor test facility are presented and discussed.

1. Nomenclature

B	Thickness of the S_1-stream surface	Re	Reynolds number
m, φ	Coordinates of the S_1-system	ξ, η	Contour conform coordinates
IBPA	Interblade phase angle	Ξ	Aerodynamic damping coefficient
r	Radius of the S_1-stream surface	τ	Time in transformed plane

T.H. Fransson (ed.), Unsteady Aerodynamics and Aeroelasticity of Turbomachines, 477-491.

SUBSCRIPTS

L, R Left- and rightsided value resp.
m, φ, ξ, η Partial derivative

SUPERSCRIPTS

ξ, η Direction of velocity

2. Introduction

Modern Turbomachinery building, either stationary gas turbines or jet engines, makes use of slender bladings for efficiency and flow reasons. Since the span is also increasing because of higher mass flows, flutter problems arise. The transonic flow regime leads to highly sensitive pressure distributions on the profiles' surfaces.

Current research work concerning flutter mainly concentrates on two areas: transonic/supersonic flutter and stall flutter. Both have an extremely negative impact on reliability and safetiness of the engine. Since experimental investigations are difficult and expensive, if they are performable at all, the main weight is put on numerical approaches.

From the point of view of code structure point there are two main areas of development. On the one hand linearized models have been established by various authors, e.g. Hall et al. (1993), Kahl and Klose (1993), Montgomery and Verdon (1995), and Hoff (1995), allowing for a fast simplified computation, and therefore useful for the basic design process of turbomachinery. On the other hand non-linearized methods are applied, if the oscillations or the flow behaviour diverts too much from the assumptions which have to be made for linearized flow. They are more complex, and require strong computer power. But since this requirement could be fulfilled more and more by computer technology during the last decade, time marching EULER codes have been developed, e.g. 2D solvers by Fransson (1986), Peitsch et al (1993), Huff et al. (1991), and Carstens (1991), and 3D codes e.g. by Gerolymos (1992), He and Denton (1993), Carstens (1995), and Peitsch et al. (1996). Most of these codes are built using upwind based methods with flux difference or flux vector splitting techniques.

Even more complex is the computation of unsteady viscous flows (Eguchi and Wiedermann, 1994, Grüber and Carstens, 1996), where one of the main working fields is the development of turbulence models for unsteady flow.

This paper shows new validation results obtained by application of the method presented by Weber et al. (1997) on the Standard-Configuration No. 4 (STC4) and on a compressor cascade by MTU. The quasi-3D algorithm used simulates sub- and transonic viscous flow on S_1-stream surfaces of revolution in single passages (Figure 1) The influence of the grid type used for STC4 on the damping coefficient is presented as well as a comparison of the calculated aerodynamic damping coefficient for several test cases of the STC4 with both measured data and results of Euler simulations. A parametric study has been carried out to investigate the influence of an increasing reduced frequency on a transonic compressor flow with a periodically separating boundary layer.

The computed results show the necessity of the inclusion of viscous effects into flutter analysis especially for transonic flows with shock boundary layer interaction.

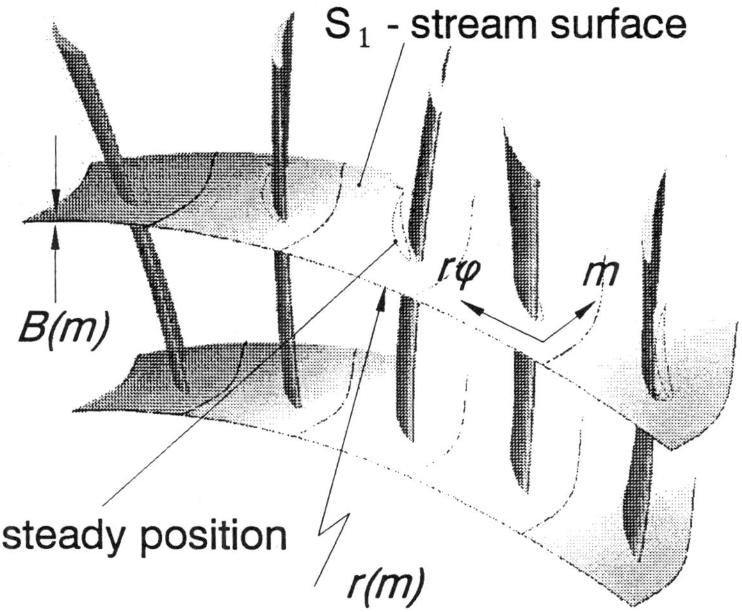

Figure 1. S_1-stream surface with distorted blades.

3. Mathematical Model

As shown by Benetschik (1991), the nondimensionalized Navier Stokes equations for unsteady compressible quasi 3D flow in a rotating frame of reference on S_1-stream surfaces of revolution can be derived as follows:

$$\frac{\partial \tilde{U}}{\partial \tau} + \frac{\partial}{\partial \xi}(\tilde{E} - \frac{1}{Re}\tilde{E}_v) + \frac{\partial}{\partial \eta}(\tilde{F} - \frac{1}{Re}\tilde{F}_v) = \tilde{Q}. \qquad (1)$$

They are transformed into a general curvilinear system of coordinates and Favre-averaged, in which $J = \partial(m,\varphi)/\partial(\xi,\eta)$ is the Jacobian of the transformation, with the state vector $\tilde{U}$ of the conservative flow quantities, and the Euler fluxes $\tilde{E}$, and $\tilde{F}$:

$$\tilde{U} = JrB\begin{pmatrix}\rho \\ \rho v_m \\ \rho r v_\varphi \\ e_{rot}\end{pmatrix}, \quad \tilde{E} = JB\begin{pmatrix}\rho W^\xi \\ \rho v_m W^\xi + rp\,\xi_m \\ r(\rho v_\varphi W^\xi + p\,\xi_\varphi) \\ e_{rot} W^\xi + p w^\xi\end{pmatrix}, \quad \tilde{F} = JB\begin{pmatrix}\rho W^\eta \\ \rho v_m W^\eta + rp\,\eta_m \\ r(\rho v_\varphi W^\eta + p\,\eta_\varphi) \\ e_{rot} W^\eta + p w^\eta\end{pmatrix}. \qquad (2)$$

the viscous fluxes $\tilde{E}_v$ and $\tilde{F}_v$:

$$\tilde{E}_v = JB \begin{pmatrix} o \\ r\tau_{mm}\xi_m + \tau_{m\varphi}\xi_\varphi \\ r(r\tau_{m\varphi}\xi_m + \tau_{\varphi\varphi}\xi_\varphi) \\ rE_{ev}\xi_m + F_{ev}\xi_\varphi \end{pmatrix}, \quad \tilde{F}_v = JB \begin{pmatrix} o \\ r\tau_{mm}\eta_m + \tau_{m\varphi}\eta_\varphi \\ r(r\tau_{m\varphi}\eta_m + \tau_{\varphi\varphi}\eta_\varphi) \\ rE_{ev}\eta_m + F_{ev}\eta_\varphi \end{pmatrix}, \tag{3}$$

$$E_{ev} = w_m\tau_{mm} + w_\varphi\tau_{m\varphi} + q_m,$$

$$F_{ev} = w_m\tau_{m\varphi} + w_\varphi\tau_{\varphi\varphi} + q_\varphi,$$

and the source vector $\tilde{Q}$:

$$\tilde{Q} = JrB \begin{pmatrix} 0 \\ \rho v_\varphi^2 \frac{1}{r}\frac{\partial r}{\partial m} + \frac{p}{rB}\frac{\partial rB}{\partial m} \\ 0 \\ 0 \end{pmatrix}, \tag{4}$$

with the transformed components of the stress tensor τ as well as the transformed heat fluxes q given in detail within Weber et al. (1997). The volume specific total energy e_{rot} and the rothalpy H_{rot} assuming a perfect gas are defined as:

$$e_{rot} = \frac{p}{\kappa - 1} + \rho\frac{v_m^2 + v_\varphi^2}{2} - \Omega r \rho v_\varphi, \quad H_{rot} = \frac{e_{rot} + p}{\rho}, \tag{5}$$

with κ as the ratio of specific heats, and Ω as the angular velocity. To simulate the turbulent stresses, an eddy viscosity is calculated in addition to a laminar viscosity following the mixing length turbulence model of Baldwin and Lomax (1978), whereas the laminar viscosity is prescribed by Sutherland's law. Furthermore, an effective thermal transport coefficient is introduced, using a laminar and a turbulent Prandtl number.

The equations are nondimensionalized referring to the inlet total density, the total velocity of sound, the chord length, and a total viscosity.

4. Numerical Method

The numerical solution of Eq. 1 is realized by employing a finite volume technique with a discretization in a node centered form. The computation of central and antisymmetric differences are used for the computation of the viscous fluxes following Benetschik (1991). By applying an upwind scheme based on a flux difference splitting method by Roe (1981) the Euler fluxes are calculated. This scheme approximatively describes a Riemann problem at each finite volume cell face, taking into consideration the physical propagation direction of disturbances. For example, the numerical flux function for the inviscid flow in ξ-direction across a cell face located at $(i+1/2,j)$ is defined as:

$$\tilde{E}(\tilde{U})_{i+\frac{1}{2},j} = \frac{1}{2}\left[\tilde{E}(\tilde{U}_L)_{i+\frac{1}{2},j} + \tilde{E}(\tilde{U}_R)_{i+\frac{1}{2},j} - \left(\tilde{R}^{\xi}|\tilde{\Lambda}^{\xi}|\tilde{L}^{\xi}(\tilde{U}_R - \tilde{U}_L)\right)_{i+\frac{1}{2},j}\right], \quad (6)$$

with the initial state vectors $\tilde{U}_L$ to the left and $\tilde{U}_R$ to the right of each cell face. $\tilde{L}^{\xi}$ and $\tilde{R}^{\xi}$ represent the matrices of left resp. right Eigenvectors, and $\tilde{\Lambda}^{\xi}$ the diagonal matrix of Eigenvalues with respect to the inviscid flux vector $\tilde{E}$. Each quantity of the matrices is obtained using the Roe average.

The characteristic wave decomposition realized by Eq. 6 results in two acoustic waves, one shear, and one entropy wave, where the Eigenvalues describe the characteristic speed and direction of the waves. Due to the particular Eigenvector formulation, the change of the characteristic variables across the cell face can be calculated in consistency with the characteristic compatibility relations.

A method by Harten (1983) is implemented to achieve a limit of artificial viscosity in any part of the flow field. Furthermore, use is made of corrected Eigenvalues by a proposal of Harten and Hyman (1983) to overcome non physical negative entropy changes across expansion fans. An extension to higher order spatial accuracy of the Euler fluxes by means of the MUSCL technique (van Leer, 1979) is applied. To avoid stability problems the TVD scheme by Harten (1983) with a limiter function following van Albada is incorporated.

The time marching flow field is computed by an implicit integration method proposed by Rai and Chakravarthy (1986) using a global time stepping. The method is extended to second order accuracy. Due to the nonlinearity of the Navier Stokes equations, a Newton iteration has to be undertaken for every single time step, while the system of equations itself is solved iteratively by a standard Gauss Seidel method.

4.1 BOUNDARY CONDITIONS

The numerical approach for the upstream and downstream boundary conditions follows a method of characteristics by Chakravarthy (1983), where the number of physical boundary conditions depends on the number of characteristics entering the computational domain. At the inlet boundary, initial values for total pressure, total temperature, and inflow angle are used in terms of the rotating frame of reference, while at the outlet boundary, only static pressure has to be imposed.

For viscous flow, assuming a vanishing normal pressure gradient and an adiabatic wall, Stokes' noslip condition is applied. All boundary conditions are linearized in time to be fully coupled into the implicit algorithm.

4.2 PERIODICITY CONDITION

The present method computes the unsteady flow field on a single passage which is a representative section of the complete annular cascade. To take into account harmonic blade oscillations for one and the same mode but with a constant interblade phase angle, a special periodic boundary treatment is introduced. Erdos et al. (1981) suggested the so called Direct Store method to account for the periodicity of the flow variables in time. By storing and updating the flow quantities at the periodic boundaries during every period, the time shifted fluxes can be computed. Convergence is reached if the difference between the flow quantities compared to the period before is below a certain limit. Large storage requirements are avoided by a technique by Peitsch et al. (1996).

4.3 UNSTEADY GRID GENERATION

After having computed the steady state solution on a Poisson generated C- or O-type grid, the unsteady flow simulation requires a blade movement corresponding to the chosen mode. Therefore, the entire grid is separated into three parts:

- a moved part including the blade and the fine grid for the boundary layer without deformation of the grid structure saving the orthogonality of the grid lines on the surface,
- a deformed part which is newly generated every time step by solving the Laplace equations,
- and an unmoved outer part

as given in Figure 2. The time dependent grid generation is realized by solving the Laplace equations.

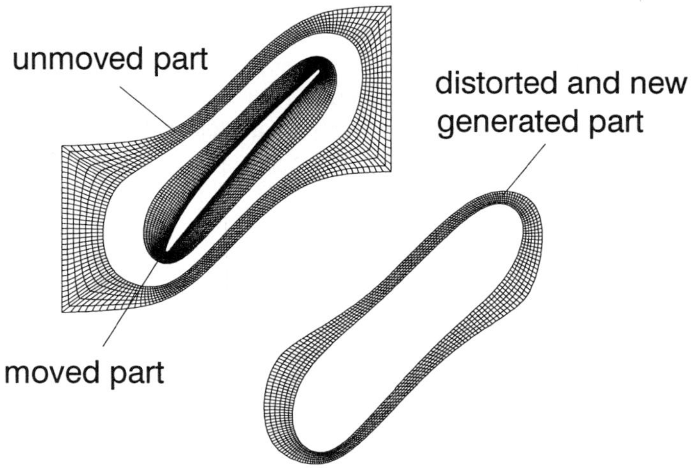

Figure 2. Separated O-type grid for the unsteady grid generation.

5. Results

To validate the present code SAFES1 and to prove the predictive capability for unsteady viscous flow through a turbine cascade, well known as the Standard-Configuration No. 4, and through a compressor cascade with an MTU designed transonic blade have been chosen.

5.1 FOURTH STANDARD-CONFIGURATION

This turbine cascade has been used for experimental flutter simulations at various flow conditions. Results of the measurements are reported in the aerolasticity workshop report by Bölcs and Fransson (1986). The selected investigations have been performed for the first bending mode and a constant IBPA of -90° with a variation of the exit Mach number (0.58, 0.76, 0.9, 1.02, and 1.19). Results are discussed for test case No. 2 and No. 3 corresponding to the exit Mach numbers of 0.76 and 0.9 resp. For the steady and unsteady predictions an O-type grid with 281x33 grid points with $y^+ < 1.5$ and a constant stream surface thickness B has been generated, taking care of a highly resolved leading and trailing edges. The 2D flow was assumed to be fully turbulent.

In Figure 3, the steady state pressure coefficient distributions versus the non-dimensional chord length for test case 2 (left) and 3 (right) are compared with measured data. PS denotes the pressure side and SS for the suction side.

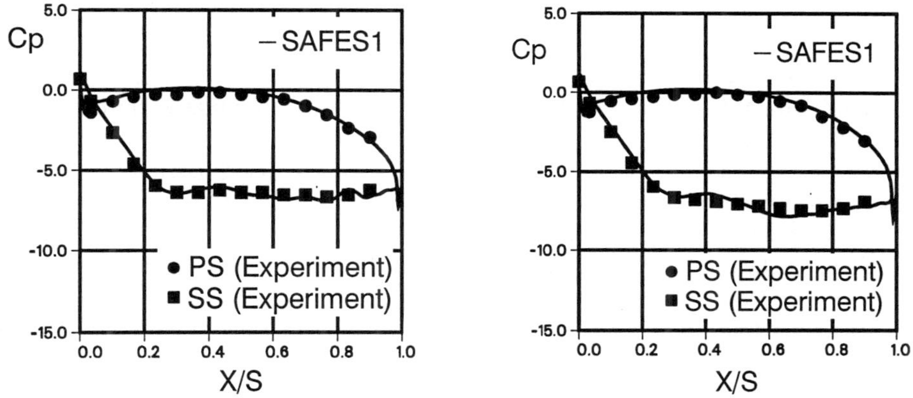

Figure 3. Comparison between measured and numerically predicted steady state pressure coefficient distribution for the STC4, at exit mach numbers of 0.76 (left) and 0.9 (right).

The results for test case No. 2 on both blade sides are in good agreement with the experimental data. For test case 3 the numerically predicted PS matches the measured data, while on the SS the pressure coefficient is slightly underpredicted at 40% chord length. Abhari and Giles (1995) found a small over prediction at this part, which has been computed as well by Peitsch et al. (1996). C-type grids were used here with nearly non orthogonal grid cells at the trailing edge for both viscous and inviscid flow. It should be stated that all calculations were performed with an inlet flow angle of 50°.

Once the steady state results have been achieved, the unsteady flow simulation can proceed. The blades were oscillating in the first bending mode with a frequency of 150 Hz and an IBPA of -90°. Convergence was obtained after about eight periods with 1500 time steps per period and three Newton steps per time step. Figure 4 contains unsteady results for the first harmonics of the unsteady pressure in amplitude and phase. For both test cases the prediction of the phase distribution on the SS as well as on the PS are in good agreement with the measurements.

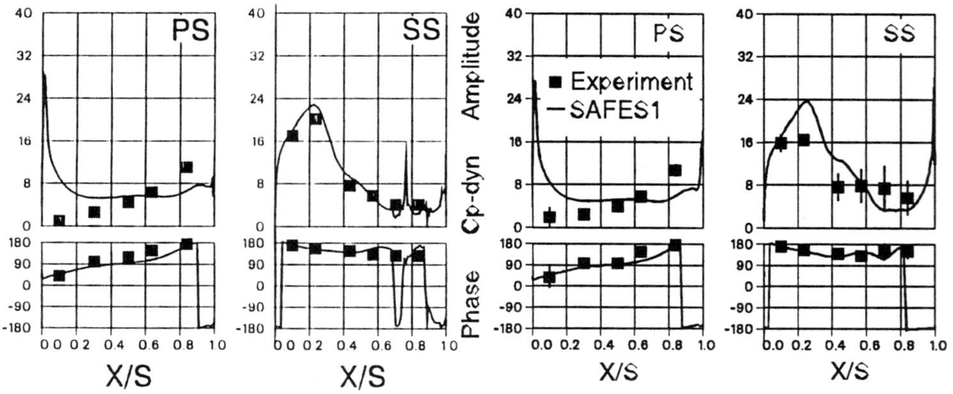

Figure 4. Comparison between measured and numerically predicted amplitude and phase distribution for the STC4, at exit Mach numbers of 0.76 (left) and 0.9 (right), and an IBPA of -90°.

Only for test case 2 (left) on the SS a jump in the phase angle distribution appeared at 75% chord length which has not been measured. This jump in phase is not due to the blade surface grid because the same grid has been applied to all test cases and only test case 1 (exit Mach number of 0.58) and 2 showed this particular behaviour. This jump can also be seen in the distribution of the amplitude on the SS. With exception of this peak and a small overprediction of the amplitude at 22% chord length on the SS the prediction matches the experimental data. On the PS the predicted amplitude of test case 2 shows larger discrepancies from leading edge to 25% chord length and from 60% up to the trailing edge. While at the trailing edge the correct tendency of an increasing amplitude can be observed, the amplitude is not reduced to nearly zero around the leading edge. But in the discussion of the data it has to be considered that the amplitudes are very small and difficult to measure and therefore, comprise a certain instability as well as the computed results. Looking at the distributions of the amplitude for test case 3 the same behaviour of the amplitude on the PS as for test case 2 is observed. On the suction side of test case 3 larger discrepancies have been found. The amplitude shows an overshoot at 25% X/S and is still too high at 42%. From 55% to the trailing edge it is nearly in the range of the measured distribution. In conclusion, it can be stated that all computations with an O-type grid lead to an improved prediction of the amplitude on the SS from 40% chord length to the trailing edge compared to the results for a C grid, where the amplitudes are underpredicted in this part of the blade.

The aim of flutter investigations is to predict whether the oscillation of the blade is stimulated or damped. This is resolved by calculation of the aerodynamic damping coefficient from the unsteady pressure distribution on the surface of the moving blades. As the result of this parametric study, the damping coefficients versus the isentropic exit Mach number Ma_{2is} are presented in Figure 5. This plot shows the measured data and the numerically simulated results for viscous (SAFES1) and inviscid (EULER: solving the Euler equations by SAFES1) flow. Compared to the experimental data, the viscous computations have predicted the measured tendency of the damping coefficient. Only for the subsonic test case 1 the oscillation is calculated to be stable, while it has been

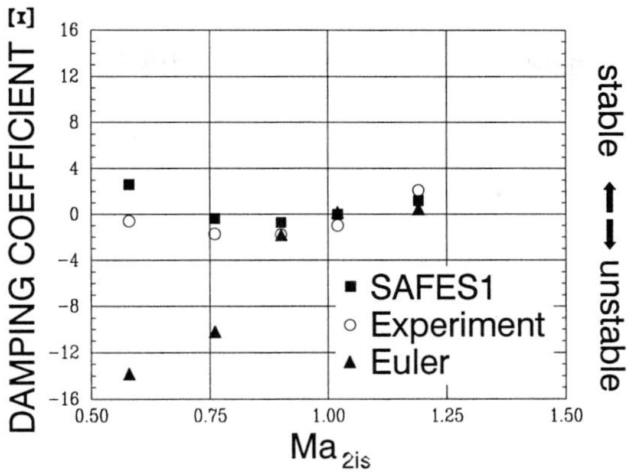

Figure 5. Comparison between measured and numerically predicted aerodynamic damping coefficient for the STC4, at exit Mach numbers of 0.58, 0.76, 0.9, 1.02, 1.19, and an IBPA of -90°.

stimulated in the experiment. An improvement compared to the EULER results could be achieved especially for the subsonic test cases 1 and 2, where the EULER results show a large discrepancy and not even correct trend of less stimulation at low speed conditions. In comparison with the inviscid results, the damping coefficient for transonic speed at highest exit Mach number could be slightly improved, whereas the damping coefficient of test case 3 for inviscid flow simulations is in better agreement with the experiment.

5.2 TRANSONIC COMPRESSOR BLADE DESIGNED BY MTU

Further computations have been performed for a transonic compressor blade used as a test case for experimental investigations of flutter behaviour in a 2D test facility at IST (Poensgen, 1994). The characteristic geometric values and the flow conditions for steady state are given in Table 1.

TABLE 1. Definition of the compressor blade configuration and the investigated test case.

chord length	41.2 mm	inlet total pressure	1.66 bar
tip to chord ratio	1.8	inlet total temperature	310.34 K
stagger angle	49.5°	inlet Mach number	0.88
		inflow angle	60°
		exit static pressure	1.22 bar

Although this test facility was designed for 2D flow investigations, the experiments showed strong corner stalls both at tip and hub. By means of an oil flow visualization technique the flow became apparent. In Figure 6, photographs of various flow conditions can be seen. For the numerical simulation of the flow it was hence assumed as fully turbulent. A variation of *B* from leading (LE) to the trailing edge (TE) up to the exit has been used to take into account the 3D influence on the midspan flow.

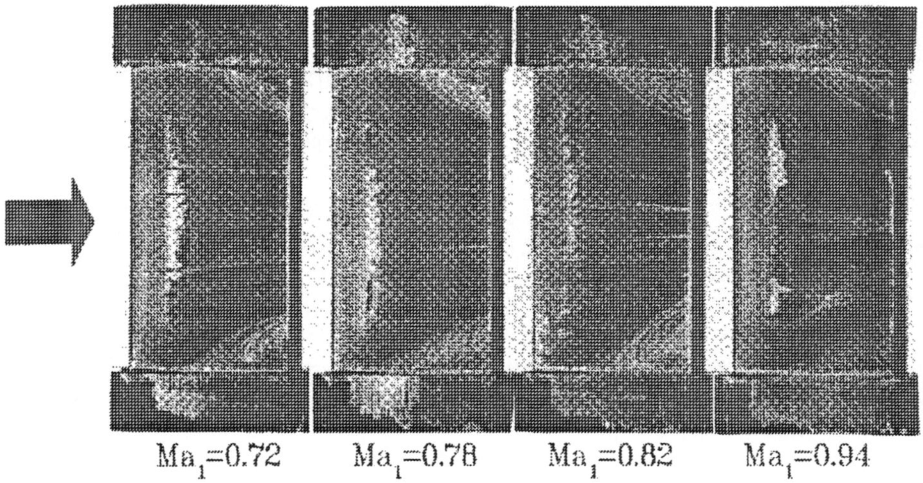

Figure 6. Photographs of oil flow visualization for various flow speed conditions (Poensgen, 1994).

The contour of the nondimensional ratio of B and B_{LE} versus the nondimensional m coordinate can be seen in the left part of Figure 7. Applying an O-type grid (241x41) with the given B leads to a steady solution shown in the right hand plot of Figure 7.

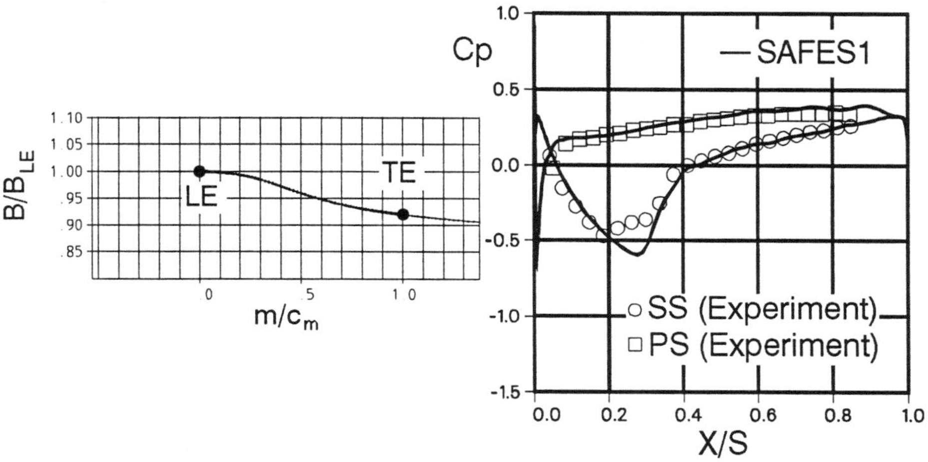

Figure 7. Contour of *B* (left) and comparison between measured and numerically predicted steady state pressure coefficient distribution of the compressor test case (right).

This plot contains the steady state pressure coefficient distribution ($y^+<1$) versus the chord length. On the pressure side the prediction is in good agreement with the measured data. The distribution of the suction side only differs from the experiment between 22% and 30% chord length. Here speed overshoot has been predicted because the code does not include a transition model and therefore, it is not able to predict the measured small laminar separation at this part of the SS, whereas the reduction of flow speed due to a strong shock nearly matches the measured data. Compared to the result, published by Weber et al. (1997), the steady state solution could be improved, especially the former overprediction of the pressure on the pressure side near to the leading edge vanished due to an improved *B* variation found by a 3D computation of the steady viscous flow.

The presented steady state solution has been used as a starting condition for an investigation of the impact of an increased reduced frequency on a shock influenced transonic viscous flow and its corresponding aerodynamic damping coefficient. Therefore, computations were performed at reduced frequencies of 0.15, 0.3, 0.6, 0.9, 1.2, and 1.5. For the first pitching mode an amplitude of 4° with an IBPA of 90° was applied. The center of rotation was on the chord at half chord length. The results of this parametric study are given in Figure 8. On the left hand side the distribution of amplitude and phase for a reduced frequency of 0.3, 0.9, and 1.5 is presented.

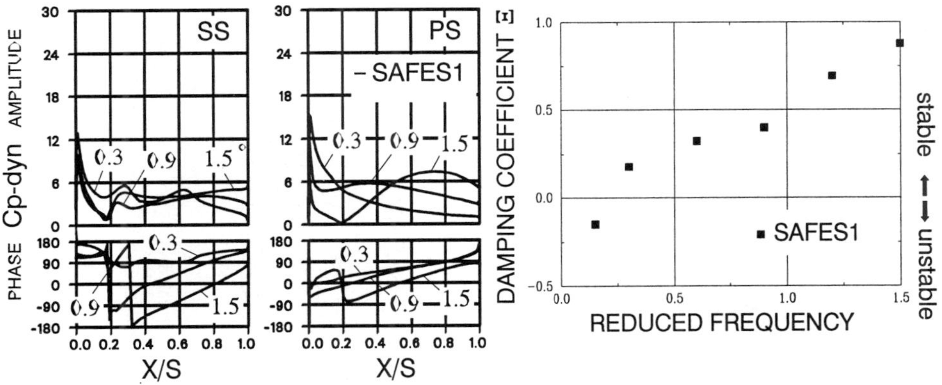

Figure 8. Distribution of amplitude and phase (left) and aerodynamic damping coefficient as a result of an increased reduced frequency with an pitching amplitude of 4° (right).

It can be seen that the amplitudes on the suction side are decreasing for a growing reduced frequency at 19% of X/S while they are increasing towards the trailing edge. The amplitude behaviour on the suction side still shows the influence of two strong shocks at 27% and 62% chord length for a reduced frequency of 0.3 which more and more vanished with the increased reduced frequency. On the pressure side the amplitude changed its behaviour significantly due to the reduced frequency. For the highest frequency the amplitude nearly vanished at 20% chord length and is growing up to a maximum of 7 at 70 % X/S. Comparing the distributions of the phase angles, it can be stated that the biggest jump in phase from 180° to -100° occurred on SS at 20% chord length for a reduced frequency of 0.9. A smaller jump in phase from 135° to 70° was found at 17% X/S for 1.5. While the distribution of the phase for a reduced frequency of 0.3 and 0.9 is similar without a jump in phase, the distribution for 1.5 turned out with a phase jump at 20% chord length. Out of phase and amplitude distributions the aerodynamic damping coefficients have been calculated and plotted in the diagram given on the right hand side of Figure 8. The distribution of the damping coefficient versus reduced frequency shows a nearly linear behaviour. As expected, the damping coefficients are increasing towards higher positive values with a higher reduced frequency. An unstable oscillation was predicted for a reduced frequency of 0.15. The critical frequency is around 0.25.

Figure 9 presents two snapshots of the unsteady Mach number distribution for oscillations with reduced frequencies of 0.15 (left) and 1.5 (right) at the same angle of distortion.

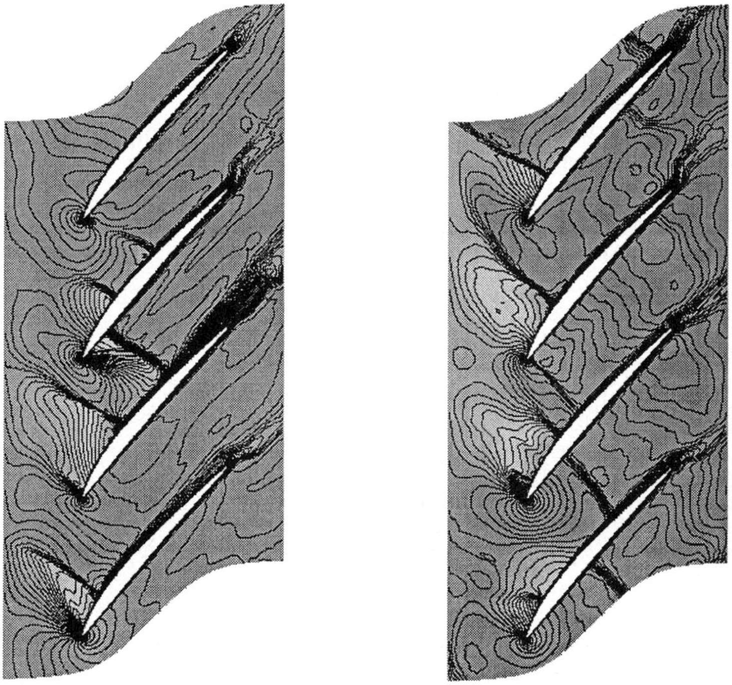

Figure 9. Snapshots of the unsteady Mach number distribution for reduced frequencies of 0.15 (left) and 1.5 (right) at the same position of the blades.

In the plot on the left hand side one shock free passage can be seen at the top while for all neighbouring passages moving shocks have been predicted. The strong passage shock inside the middle passage moves towards the leading edge during the oscillation. Furthermore, it can be seen that the passage shock forces the separated boundary layer to reattach at the shock position on the SS and to separate again behind the passage shock. Observing the unsteady flow, it can be seen that nearly 80% of the boundary layer on the suction side becomes separated. The separation did not occur with the same intensity at higher reduced frequencies and nearly vanished for 1.5, as can be seen in the plot on the right hand side. This is due to the high turning velocity of the blades which dramatically decreases the response time of the boundary layer.

6. Conclusions and Prospect

A CFD code for the prediction of the unsteady viscous quasi 3D flow around oscillating blades has been applied to various test cases. Due to the possibility of using C-type grids or O-type grids nearly orthogonal grid cells could be generated on the surface. Usage of an O-type grid rather than a previous used C-type grid for STC4 calculations improved the steady state results as well as the distribution of the amplitude on the suction side. Solving the Navier-Stokes instead of the EULER equations lead to a significantly better predicted aerodynamic damping coefficient for test case 1 and 2.

A numerical study to investigate the influence of an increased reduced frequency on the damping coefficient has been performed for a transonic flow with a strong shock on the suction side. The correct steady state position influenced by the corner stalls of the shock was predicted by means of a 3D Navier Stokes simulation to find the correct contour of B. Due to the pitching amplitude of 4° the boundary layer on the SS is separating periodically with an oscillating shock in front of the separation and a second shock is appearing, reaching from SS to PS.

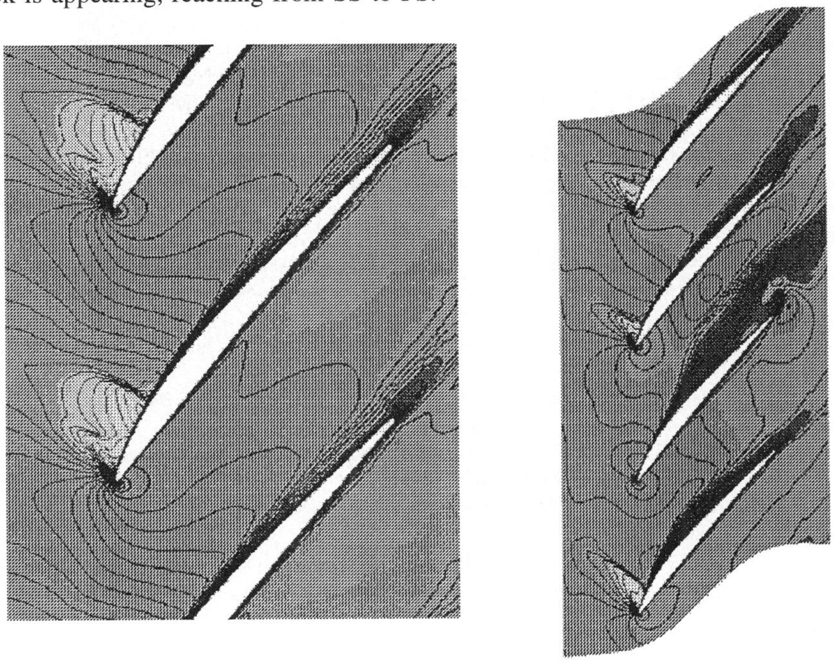

Figure 10. Distributions of the steady state (left) and unsteady (right) Mach number for an incidence angle of 4°, a reduced frequency of 0.15, an IBPA of 90°, and an amplitude of 2° (right).

While this strong separation occured for a reduced frequency of 0.15 resulting in a stimulated oscillation, it vanished at higher frequencies leading damped vibration of the blades.

The strong separation on the SS at 4° amplitude could not be found for a steady state solution of the flow with an incidence angle of 4° (left hand side of Figure 10). At 4° incidence angle the flow is still attached but separates periodically to a big stall for an oscillating blade with a pitching amplitude of 2° and a reduced frequency of 0.15 (right hand side of Figure 10). Further numerical investigations and measurements are necessary concerning stall onset and its influence on the damping behaviour.

7. References

Abhari, R.S. and, Giles, M. (1995) A Navier Stokes Analysis of Airfoils in Oscillating Transonic Cascades for the Prediction of Aerodynamic Damping, ASME-Paper 95-GT-182

Baldwin, B.S. and Lomax, H. (1978) Thin Layer Approximation and Algebraic Model for Separated Turbulent Flows, AIAA 16th Aerospace Science Meeting, Huntsville, Alabama

Benetschik, H. (1991) Numerische Berechnung der Trans- und Überschall-Strömung in Turbomaschinen mit Hilfe eines impliziten Relaxationsverfahrens (Numerical Computation of the transonic and supersonic flow in turbomachines by means of an implicit relaxation scheme), Doctorthesis, RWTH Aachen

Bölcs, A. and Fransson, T.H. (eds.) (1986) Aeroelasticity in Turbomachines - Comparison of Theoretical and Experimental Cascade Results, *Communication de Laboratoire du Thermique Appliquée et de Turbomachines*, No. 13, EPFL Lausanne, Switzerland

Carstens, V. (1991) Calculation of the Unsteady Transonic 2D Cascade Flow for Aeroelastic Applications, *International Forum on Aeroelasticity and Structural Dynamics*, Aachen, Germany

Carstens, V. (1995) Computation of Unsteady Transonic 3D Flow in Oscillating Turbomachinery Bladings by an Euler Algorithm with Deforming Grids", *Unsteady Aerodynamics and Aeroelasticity of Turbomachines*, Tanida, Y., and Namba, M. (eds.), Elsevier Science b.v., Amsterdam, The Netherlands

Chakravarthy, S.R. (1983) Euler Equations-Implicit Schemes and Boundary Conditions", *AIAA Journal*, Vol. 21, No. 5

Eguchi, T. and Wiedermann, A. (1994) Numerical Analysis of Unstalled and Stalled Flutter using a Navier-Stokes-Code with Deforming Meshes, *UAAT-Symposium*, Fukuoka, Japan

Erdos, J.I, Alzner, E., and McNally, W. (1981) Numerical Solution of Periodic Transonic Flow Through a Fan Stage, *AIAA-Journal*, Vol. 15

Fransson, T.H. (1986) Numerical Investigation of Unsteady Subsonic Compressible Flows Through Vibrating Cascades, Thesis, *Communication du Laboratoire de Thermique Appliquée et de Turbomachines*, 12, EPFL, Lausanne

Gerolymos, G.A. (1992) Advances in the Numerical Integration of the 3-D Euler Equations in Vibrating Cascades, ASME-Paper 92-GT-170

Grüber, B. and Carstens, V. (1996) Computation of the Unsteady Transonic Flow in Harmonically Oscillating Turbine Cascades taking into Account viscous Effects, ASME-Paper 96-GT-338

Hall, K.C., Clark, W.S., and Lorence, C.B. (1993) A Linearized Euler Analysis of Unsteady Transonic Flows in Turbomachinery, ASME-Paper 93-GT-94

Harten, A. (1983) High resolution Schemes for Hyperbolic Systems of Conservation Laws, *Journal of Computational Physics*, Vol. 49

Harten, A. and Hyman, J.M. (1983) A Self-Adjusting Grid for the Computation of Weak Solutions of Hyperbolic Conservation Laws, *Journal of Computational Physics*, Vol. 50

He, L., and Denton, J.D. (1993) Three Dimensional Time-Marching Inviscid and Viscous Solutions for Unsteady Flows around Vibrating Blades, ASME-Paper 93-GT-92

Hoff, K., (1995) Die Berechnung der quasidreidimensionalen, reibungsfreien, instationären, sub- und supersonischen Strömung durch schwingende Beschaufelungen mit einem linearisierten Euler-Verfahren (Computation of the quasi 3D, inviscid, unsteady, sub- and supersonic flow through oscillating cascades by means of a linearized Euler-method), Doctorthesis, RWTH Aachen

Huff, D.L., Swafford, T.W., and Reddy, T.S.R. (1991) Euler Flow Predictions for an Oscillating Cascade Using a High Resolution Wave Split Scheme, ASME-Paper 91-GT-198

Kahl, G., and Klose, A. (1993), Computation of Time Linearized Transonic Flow in Oscillating Cascades, ASME-Paper 93-GT-269

van Leer, B. (1979) Towards the Ultimative Conservative Difference Scheme V, A second order sequel to Godunov's Method, *Journal of Computational Physics*, Vol. 32

Montgomery, M.D., and Verdon, J.M. (1995) A Linearized Unsteady Euler Analysis for Turbomachinery Blade Rows Using an Implicit Wave-Split Scheme, *UAAT-Symposium*, Tanida, Y., and Namba, M. (eds.), Elsevier Science b.v., Amsterdam, The Netherlands

Peitsch, D., Gallus, H.E., and Kau, H.-P. (1993) Prediction of Unsteady 2D-Flow in Turbomachinery Bladings, *Unsteady Aerodynamics, Aeroacoustics, and Aeroelasticity of Turbomachines and Propellers*, Atassi, H.M. (ed.), Springer

Peitsch, D., Gallus, H.E., and Weber, S. (1996) Numerical Investigation of the Unsteady Transonic 3D-Flow in Stator and Rotor Cascades with Oscillating Blades, ASME-Paper 96-GT-307

Poensgen, C. A. (1994) Experimentelle Untersuchung an schwingenden Schaufeln (Experimental Investigations on Oscillating Blades), Final Report, TURBOTECH 1.1.3.1, Aachen

Rai, M.M., Chakravarthy, S.R. (1986) An Implicit Form for the Osher Upwind Scheme, AIAA Journal, Vol. 24, No. 5

Roe, P.L. (1981) Approximative Riemann Solvers, Parameter Vectors and Difference Schemes, *Journal of Computational Physics*, Vol. 34

Weber, S., Gallus, H.E., and Peitsch, D. (1997) A Numerical Approach to Unstalled and Stalled Flutter Phenomena in Turbomachinery, ASME-Paper 97-GT-102

9: Compressor Flow Instabilities: Theory

ACTIVE SUPPRESSION OF COMPRESSOR FLOW INSTABILITY

A. S. Abo El lail*, and A. S. Hassan*
*Mechanical Engineering Department, Faculty of Engineering,
Assiut University, Assiut, Egypt.

The present paper investigates the stabilization of compressor surge using an active method. An active element such as a damper is introduced to the compression system for increasing the surge margin. A movable plenum wall, driven by a signal proportional to the unsteady plenum pressure was used, to suppress the surge in the centrifugal compressors. Time variation of pressure and flow rate as well as so called surge cycle were mapped and discussed at different flow conditions and at different values of B-parameter. Effect of B-parameter on compression system stability is also discussed. The results of these analysis show that, using a damper as a controller with the plenum of the compression system (movable plenum wall) has reduced the surge flow coefficient of the compression system by about 12 % compared to the surge flow coefficient with the rigid plenum wall. The results of the present work showed a good agreement when compared with the theoretical and experimental work of Ffowcs and Huang [1].

1. Introduction

Centrifugal compressors and blowers are used in various engineering applications that cover, recycling the coolant gas in gas cooled reactors, power plant, supercharging reciprocating engines and different aspects in agricultural engineering. A few examples for using the centrifugal compressors and blowers in application of the agricultural engineering are ventilation of storages for seeds and grains and feeding placing for stocks and chickens as well as spraying systems.

The paper deals with the unsteady flow in centrifugal compressors. As the flow in a compressor is throttled from design points to the surge limit while the compressor speed is kept constant, the steady, axisymmetric, flow pattern that exists becomes unstable. A phenomena such as a surge may occur, which can results in periodically reversed flow and mechanical damage[2]. To avoid surge the compression system is generally operated away from the surge line which represents the boundary between stable and unstable operation on the pressure rise versus mass flow performance map. Operating the compressor on the negative sloped part of the compressor speedlines will ensure stable operation. However, this may result in a performance penalty since peak performance and efficiency often occur near surge line [3].

T.H. Fransson (ed.), Unsteady Aerodynamics and Aeroelasticity of Turbomachines, 495-509.

The scientific understanding of the surge phenomenon starts with the work of Emmons et al.[4], who showed that on a relatively low-speed machine, surge can be considered as a Helmholtz resonator type of oscillation in which the compressor forms the neck of the resonator and the plenum its volume. On the other hand, the experimental results on turbomachines showed that, the system is unstable when the slope of the compressor pressure rise-flow characteristic becomes unstable. This simple theory was expanded to a fully nonlinear version by Greitzer[5] and this can explain many observed features of compressor instabilities. In particular, the importance of the B-parameter (a ratio of the pressure forces to the inertia forces on the system) was highlighted. At small values of B-parameter, the dominant instability of the system is rotating stall and the dynamics of the system are dominated by the response of the compressor. At large values of B-parameter, the overall dynamics are governed by the emptying and filling time of the plenum and the mode of instability is a large amplitude relaxation oscillation. An entirely different approach to modeling high speed surge has been taken by Mazzawy [6], that noting, surge begins with a high wave traveling down the compressor. This wave is the cause of the large impulse loads that are produced during surge and are sufficient to bend blades and shafts.

There have been several investigations for surge suppression of the compression systems using the active and passive methods to increase surge margin. Various schemes have been put forward for the active control of the surge phenomenon, Epstein et al. [7]. It is argued that if the instabilities are described by linear dynamics, they are susceptible to feedback control. In this case, the stability of the system is measured by suitable sensors and the resulting signal is used to drive an actuator that stabilizes the system. The main point is that the feedback loop alters the dynamics of the system as a whole. Also experiments at MIT [8] and at Cambridge University [1, 9, 10, 11] have been most encouraging the use of active control methods. These experiments were on the type of low-speed machine with one dominant instability mode for which the prospect of success is greatest.

2. Experimental Facility

A centrifugal compressor of a turbocharger engine was tested at constant speed of 5000 rpm. The impeller had eighteen radial blades with hub to tip ratio of 0.53 and exit tip diameter of 0.32 m. At the exit of impeller, there was a diffuser of a thirteen straight blades. The compressor was delivering the mass flow to a plenum. The discharge is through a duct and throttled by a restricting valve. The compressor was tested without controller and with controller of a damper system. The damper has damping coefficient in a range of 1000 to 2000 Ns /m. The approach taken here, is to suppress surge by modifying the compression system dynamic behavior using structural feed back. One wall of a downstream volume, or plenum is constructed so as to move in response to small perturbations in pressure. The movable wall acts as a damper system driven by unsteady pressure perturbation in the plenum, and its motion is thus coupled to the compression system dynamics.

3. Derivation of Equation of Motion For the Compression system

Consider the compression system shown in Fig. 1(a,b) for the fixed plenum wall and movable plenum wall. The following assumptions are used: The flow in the inlet ducting is one-dimensional, incompressible, inviscid, and unsteady; the plenum pressure is spatially uniform and plenum processes are isentropic, the compressor follows a quasi-steady characteristic. The equations governing the dynamics of this compression system without controller are the particular cases of equations (11), (12) and (13) reported by Greitzer [12]. The detailed form has been found from experiment of Huang [9] to be:

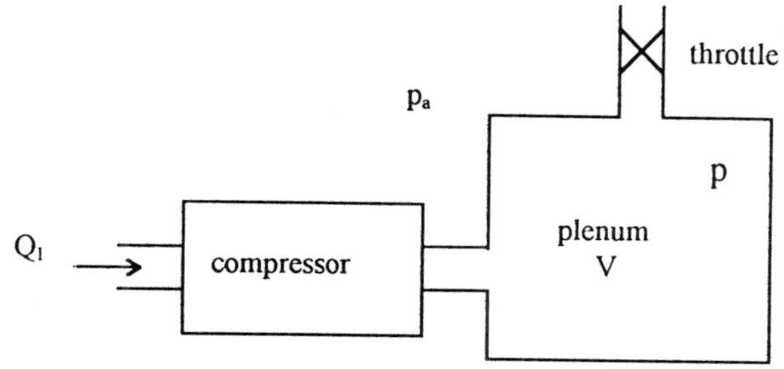

Fig. 1(a) Without controller

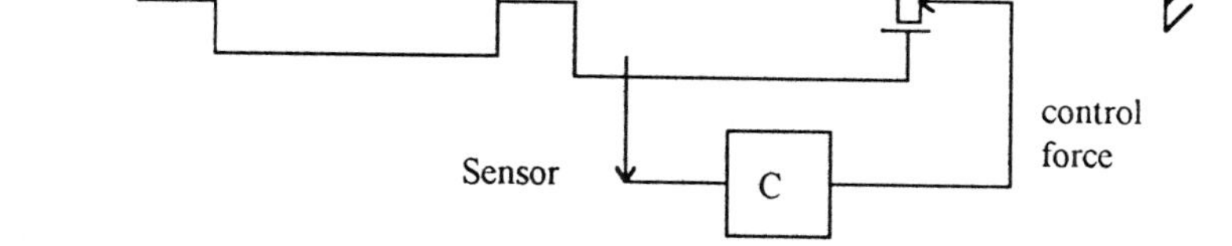

(b) With controller

Fig. 1 The compression system model (a) without controller, (b) with controller

$$\chi_c \frac{dQ_1}{dt} = f_1(Q_1) - (p - p_a) - \sigma_c(Q_1 - \overline{Q_1}), \tag{1a}$$

$$\frac{V}{a^2}\frac{dp}{dt} = Q_1 - Q_2 \ , \tag{1b}$$

$$\chi_t \frac{dQ_2}{dt} = (p - p_a) - f_2(Q_2), \tag{1c}$$

where, $\sigma_c(Q_1 - \overline{Q_1})$ is the flow resistance in the compressor duct, $x_c(dQ/dt)$ is the flow inertia in the compressor, x_c is ratio of compressor length to diameter, x_t is a similar parameter for the throttle duct, $f_1(\overline{Q_1})$ is the pressure rise, and p , p_a are the plenum pressure and atmospheric pressure respectively. The compressor mass flow rate in steady state condition is defined as $\overline{Q_1}$, and its pressure rise which is function of that mass flow is $f_1(\overline{Q_1})$. The compressor characteristic curve fit which has the approximate empirical equation of Ffowcs [1] is represented in Fig. 2. The mean pressure rise, $f_1(\overline{Q_1})$ is calculated as:

$$f_1(\overline{Q_1}) = \frac{1}{2}\rho U^2 \left\{ 0.045 \arctan[120(\overline{\phi}_1 - 0.052)] + 1.18 + 0.85\,\overline{\phi}_1 - 7.2\,\overline{\phi}_1^2 + 4.7\,\overline{\phi}_1^3 \right\} \tag{2}$$

where, $\overline{\phi}_1 = \overline{Q_1}/(\rho A_C U)$ is the nondimensional mass flow rate, U is the speed of impeller tip, ρ is the mean density, and A_c is the compressor duct cross sectional area at inlet.

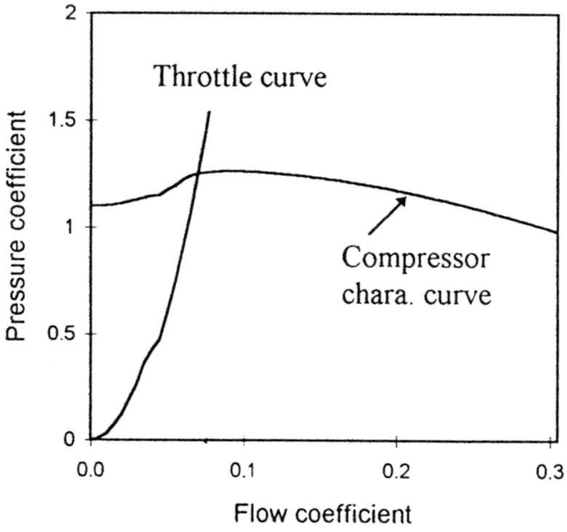

Fig.2 Compressor characteristic curve fit of empirical equation (2)

The throttle pressure drop $f_2(\overline{Q_2})$ is a function of the mass flow rate $\overline{Q_2}$ as follow

$$f_2(\overline{Q_2})=S\overline{Q_2}^2 \tag{3}$$

where, S is a coefficient depending on the throttle area A_t. If the mass flows are nondimensionalized by $\rho A_c U$, the pressure by $0.5\rho U^2$, and the time by the inverse Helmholtz frequency $1/\omega_H$, then the governing equations (1a to 1c) can be written in a nondimensional form as

$$\frac{d\phi_1}{d\tau}=B\,[\psi(\phi_1)-P]-\mu(\phi_1-\overline{\phi_1}),$$

$$\frac{dP}{d\tau}=\frac{\phi_1-\phi_2}{B}, \tag{4}$$

$$\frac{d\phi_2}{d\tau}=\lambda\,B[P-T(\phi_2)],$$

$$\phi_1=\frac{Q_1}{\rho A_c U},\quad \phi_2=\frac{Q_2}{\rho A_c U},\quad P=\frac{p-p_a}{\frac{1}{2}\rho U^2},\quad \psi=\frac{f_1(Q_1)}{\frac{1}{2}\rho U^2},$$

$$T=\frac{f_2(Q_2)}{\frac{1}{2}\rho U^2},\quad \tau=\omega_H t,\quad B=\frac{\frac{1}{2}U}{\omega_H A_c \chi_c},\quad \mu=\frac{\sigma_c}{\omega_H \chi_c},\quad \lambda=\frac{\chi_c}{\chi_t}.$$

4. Influence of the controller on the compression system stability

The movable wall surface area is defined as A_3 which is part of a damper system responding to the unsteady pressure fluctuation in the plenum and an externally induced control force; its displacement ξ, which is proportional to the driving force, produces a volumetric change $A_3\xi$ in the plenum, and the mass stored in the plenum is thus changed by $\rho A_3\xi$. This induced control force is generated by a feedback system which processes the signal detected by the pressure sensor which is located in the plenum. The controller was chosen as a variable area, C. In this case, equations of motion which describes the dynamic of the compression system are modified by the controller as follows:

$$\frac{V}{a^2}\frac{dp}{dt}=Q_1-Q_2-A_3\rho\frac{d\xi}{dt}, \tag{5a}$$

$$\chi_c\frac{dQ_1}{dt}=f_1(Q_1)-(p-p_a)-\sigma_c(Q_1-\overline{Q_1}), \tag{5b}$$

$$\chi_t\frac{dQ_2}{dt}=(p-p_a)-f_2(Q_2),\ . \tag{5c}$$

$$\frac{d\xi}{dt}=\frac{1}{D}(A_3+C)(p-p_a) \tag{5d}$$

Using the controller two new terms should be introduced, the term $(\eta\, P)$ is caused by the pressure induced motion of surface A_3 ; this reduces the resonance frequency of the system and is a destabilizing effect, and the control force term $(Z_\xi\, P)$ is aimed to increase the compression system surge limit. Adding these terms, equation (5a to 5c) can be written in non-dimensional form as:

$$\frac{d\phi_1}{d\tau} = B\,[\psi(\phi_1) - P] - \mu(\phi_1 - \overline{\phi_1}),$$
$$\frac{dP}{d\tau} = \frac{\phi_1 - \phi_2}{B} - (\eta + Z_\xi)P, \tag{6}$$
$$\frac{d\phi_2}{d\tau} = \lambda\, B[P - T(\phi_2)]$$

where, $\eta = \dfrac{\rho\, a^2 A_3^2}{V D\omega_H}$, $Z_\xi = \dfrac{\rho\, a^2 A_3}{V D\omega_H} C$

The operator Z_ξ in the present paper is equivalent to $Z_\xi\, K/(D\omega_H)$ as written by Ffowcs [1] (where K is the spring stiffness, and D is the viscous damping coefficient), $Z_\xi\, K/(D\omega_H\, M^2)$ as defined by Epstein et al.[7]; other parameters are the same as those in Eq.(4). The linear perturbation equations for the quantities $\delta\phi_1, \delta\,\phi_2, and\ \delta\, P$, from Eq.(6), will become

$$\frac{d\delta\phi_1}{d\tau} = B\,[\psi'(\delta\phi_1) - \delta P] - \mu\delta\phi_1\ ,$$
$$\frac{d\delta P}{d\tau} = \frac{\delta\phi_1 - \delta\phi_2}{B} - (\eta + Z_\xi)\delta P\ , \tag{7}$$
$$\frac{d\delta\phi_2}{d\tau} = \lambda\, B[\delta P - T'(\delta\phi_2)]$$

The linear analysis of the compression system stability which represented in equation (7) can be written in the matrix form

$$\begin{bmatrix} s + \mu - B\psi' & B & 0 \\ -\dfrac{1}{B} & s + \eta + Z_\xi & \dfrac{1}{B} \\ 0 & -\lambda B & s + \lambda B T' \end{bmatrix} \begin{bmatrix} \delta\phi_1 \\ \delta P \\ \delta\phi_2 \end{bmatrix} = \begin{bmatrix} 0 \\ 0 \\ 0 \end{bmatrix} \tag{8}$$

Any way, Equations (6) and (7) are basically identical to (22) and (23) in Epstien et al.[7], but in a form in which the flow acceleration in the throttle duct is considered, and they have solutions of the form $e^{s\tau}$. Equation (8) then reduces to the characteristic equation

$$s^3 + a_2 s^2 + a_1 s + a_0 = 0 \tag{9}$$

where

$$a_0 = \lambda\, BT'[(\eta + Z_\xi)(\mu - B\psi') + 1] + (\mu - B\psi')\lambda\,,$$

$$a_1 = (\eta + Z_\xi)(\mu - B\psi' + \lambda\, BT') + \lambda\, BT'(\mu - B\psi' + b_1)\,,$$

$$a_2 = \lambda\, BT' + (\mu - B\psi') + (\eta + Z_\xi)$$

$$\textit{and} \quad b_1 = 1 + \lambda$$

The Routh-Hurwitz stability gives necessary and sufficient conditions for the real parts of all roots of (9) to be negative

$$a_0 \succ 0,\ a_2 \succ 0,\ a_1 a_2 \succ a_0 \tag{10}$$

These constraints define the stable range of the compression system. It should be mentioned here that, the throttle characteristic was defined corresponding to that instability point as the surge boundary and this is mapped in Fig.3. By substituting the measured values of Ffowcs [1] and the functions into (10), the linear instability point at which the stability conditions (10) will just be broken and the compression system will surge can be predicted. The predicted surge point for the system without controller in the present work is satisfactorily close to experimental surge point of Ffowcs [1].

The compressor and throttle, which are two nonlinear components in the system, act mainly as energy source and sink. If the rotation of the compressor is stopped $\psi' = 0$ and close the throttle $T = 0,\ \lambda = 0$, then the characteristic equation (8) becomes

$$s^2 + (\mu + \eta + Z_\xi)\, s + \mu\,(\eta + Z_\xi) + 1 = 0 \tag{11}$$

With no gain ($z = \dfrac{Z_\xi}{1+\eta} = 0$), the solution of (9) denoted by

$$\begin{aligned} s_0 &= \alpha_0 \pm i\,\omega_0 \\ &= -\frac{1}{2}(\mu\varepsilon_0 + 1 - \varepsilon_0)/\varepsilon_0 \pm i\,\frac{1}{\varepsilon_0}\sqrt{\mu^2\varepsilon_0^2 - 2\,\mu\varepsilon_0 + 2\mu\varepsilon_0^2 - 3\varepsilon_0^2 + 1 - 2\varepsilon_0} \end{aligned} \tag{12}$$

where, $\varepsilon_0 = \dfrac{1}{1+\eta}$, and $z = \dfrac{Z_\xi}{1+\eta}$

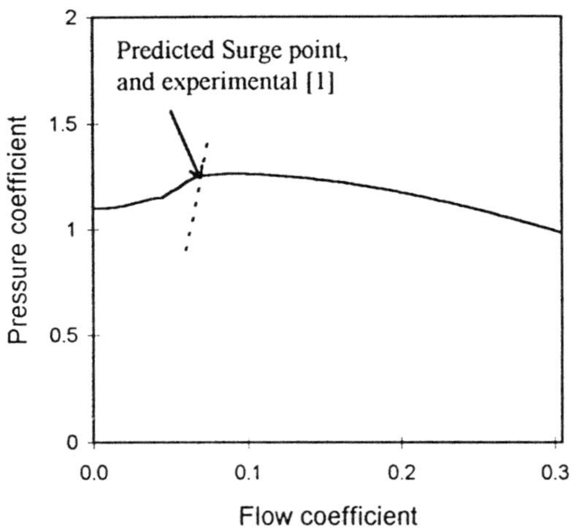

Fig.3 Limit of stability for the compression system without controller

The effect of the control parameter $\varepsilon_0 = \dfrac{1}{1+\eta}$ on the system natural frequency is represented in Fig. 4, where $\eta = \dfrac{\rho\, a^2\, A_3^2}{V\, D\, \omega_H}$. This figure clearly shows decrease in the system natural frequency with increasing either the damping coefficient D, or the plenum volume V. That is, with high value of the plenum volume the compression system will inter in surge at high flow rate than the system with low value of the system volume.

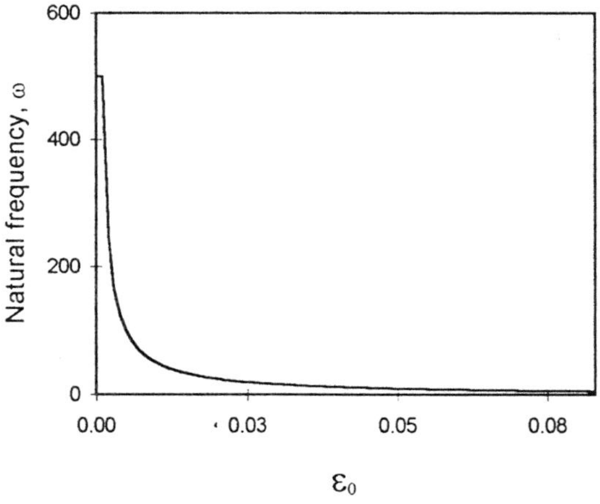

Fig.4 The effect of control parameter $\varepsilon_0 = \dfrac{1}{1+\eta}$ on the system natural frequency

The frequency is 31 Hz at value of the parameter-B ($\frac{U}{2\omega_H L_C} = \frac{U}{2a}\sqrt{\frac{V_P}{A_C L_C}}$)=0.05, compared to the predicted frequency of 7 Hz at value of the parameter-B= 0.1. That is the frequency decreases as B-parameter increased, That is the decrease in frequency is associated with the time needed for the plenum blow-down and pressurization. The time scale over which the surge suppression occurs is considerably longer than any time scale associated with the system oscillations because it is set by the filling time of the auxiliary plenum. The surge regime is one in which the oscillations greater than the two averaged mass flow. This indicated that the use of tailored structure can suppress surge even when the oscillations have large amplitude.

The present analysis using the Routh criterion show that the stability is dependent on the control parameter , Z_ξ. The calculated results, carried out around the natural surge point for special illustrative case where Z_ξ is arranged to be real and independent of the Laplace operator s. This shows that a real and negative Z_ξ will always improve the system stability in the range $-(1+\eta) \prec Z_\xi \prec 0$, while it will tend to destabilize the system when Z_ξ is a real and positive. In the absence of a control force, $Z_\xi = 0$, the system stability will be reduced in comparison with that of the hard walled plenum, because for a damper $\eta \succ 0$ and η has exactly the same effect on the compression system stability as does Z_ξ, a fact evident from equation (9). The controller can obviously have a wide range of influence because the control parameter Z_ξ can be arranged to have a variety of complex forms for various complex eigenfrequencies.

5. Nondimensional B-Parameter

The behavior of the compression system described by equation (6) has a complex parametric dependence involving nondimensional parameters. The B-parameter has a major influence on the surge dynamics of the compression system. The effect of B-parameter on surge cycle are displayed in Fig. 5, for both controlled and uncontrolled systems. This figure shows that the system without controller gives a large surge cycles at different values of B-parameter, but very small surge cycles are mapped with the movable plenum wall or controlled system. For the fixed wall compression system, the magnitude of the positive compressor characteristic slope required for disturbances to grow is set by the B-parameter, which can be viewed as providing a measure of coupling between mass flow oscillations through the compressor and through the throttle. The larger the B-parameter, the more isolated the throttle is from the compressor, and the less able to remove energy from flow disturbances. The surge occurs at a smaller positive compressor slope as B-parameter increases.

Equation (8) constitutes an eigenvalue problem for the complex growth rate, s, involving two coupled dynamic systems: the compression system, with variables, and the moving wall, with variable. Solving the eigenvalue problem as a function of marginal natural frequency, and the mean flow coefficient using the experimentally determined compressor characteristics of Ffowcs et al.[1] enables prediction of the instability onset condition for various system and control parameters. In addition, the control effectiveness decreases as either or both the B-parameter and the slope of the compressor characteristic increase. The physical reason for the decreased effectiveness with increase in B-parameter is connected to the overall decrease in stability that occurs as B increases. That is well known, and has been amply discussed [5]. Large B-parameter implies a more throttle is less coupled to the unsteady flow through the compressor. Control strategies using a downstream throttle would therefore be expected to lose effectiveness.

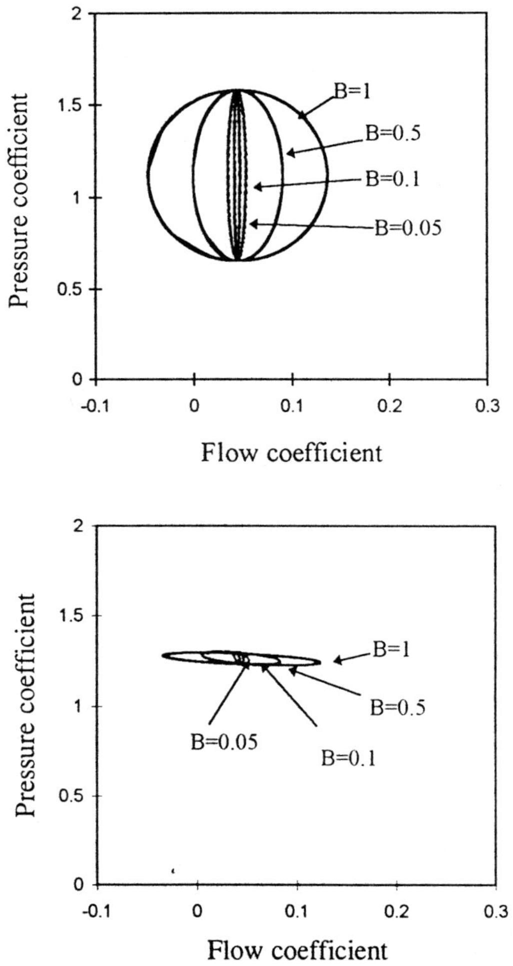

Fig. 5 Influence of B-parameter on surge cycles

The slope of the compressor characteristic is equal to 1/B at maximum gain as discussed with Pinsley, [10]. Using this result implies that the limit of stabilization for a given B, using the controller will thus occur at a point on the nondimensional compressor characteristic where the slope $(d\psi / d\bar{\phi}_1)$ is 1/B. This slope can be compared with that for the natural surge point, which occurs at a value of $d\psi / d\bar{\phi}_1$, given by

$$\frac{d\psi}{d\bar{\phi}_1} = \frac{1}{B^2(\partial\psi_T / \partial\bar{\phi}_1)_T}\text{, or for a parabolic throttle curve, } \frac{d\psi}{d\bar{\phi}_1} = \frac{\bar{\phi}_1}{2B^2\psi}$$

The maximum compressor characteristic slope for stable compressor operation versus B-parameter for the compression system without controller and with a controller of damper is represented in Fig. 6. However, there still remains the problem about the selections of the values for the parameters, the lumped parameter theory was able to be applied to the present study. Similar methods were successfully applied to the surge control of the centrifugal compressor with the high pressure ratio [10]; Gysling et al. [13], and the lumped parameter theory was also effective for the present results in the fundamental prediction.

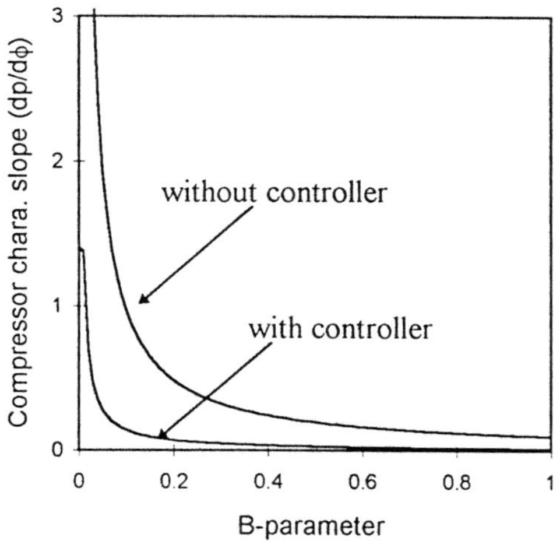

Fig. 6 Maximum compressor characteristic slope for stable compressor operation versus B-parameter

The system limit of stability or the surge point was determined from the Routh criterion and plotted in Fig. 7 with the compressor characteristic curve. This figure shows effect of damper as well as the spring as controllers for the system stability. The damper in fact is so efficient on the surge margin than the spring, and about 12% increase in stable flow range was predicted than with the system without controller. With the movable plenum wall the surge point is moved to the left in compressor performance curve. Moreover, the model of the present work has been used utilizing spring instead of the damper as a controller, where the surge point obtained from the model is almost identical to the experimental surge point presented in Ffowcs [1].

6. RMS of pressure fluctuation

A summarized view of the alteration in the unsteady behavior as a function flow rate is given in Fig. 8. This figure shows the root mean square (RMS) amplitude of pressure fluctuations for the compression system without controller as predicted in the present analysis and the experiment of Ffowcs [1], and with the controlled plenum using a spring as predicted in the present work. Also the predicted and experiment of Ffowcs [1], and predicted in the present work with damper as controlled plenum system are mapped in this figure. This figure clearly shows that the damper as a controller gives the best results for the surge suppression and limit of stability. The level of unsteadiness increases as the configurations approach their respective surge flow coefficient. The RMS functional dependence of the level of unsteadiness on flow coefficient in the three systems (without controller, spring as a controller, and damper as a controller). Any way, with damper system, from surge point (at mean flow coefficient = 0.035), the level of the RMS of pressure fluctuations is roughly 0.01, indicating that the magnitude of unsteadiness at these conditions is not greatly influence by aeromechanical feedback.

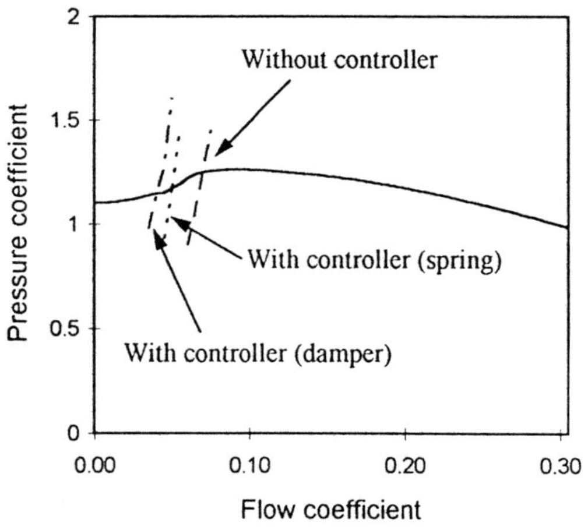

Fig. 7 Compressor characteristic curve and limit of stability for the system without controller and with controlled plenum

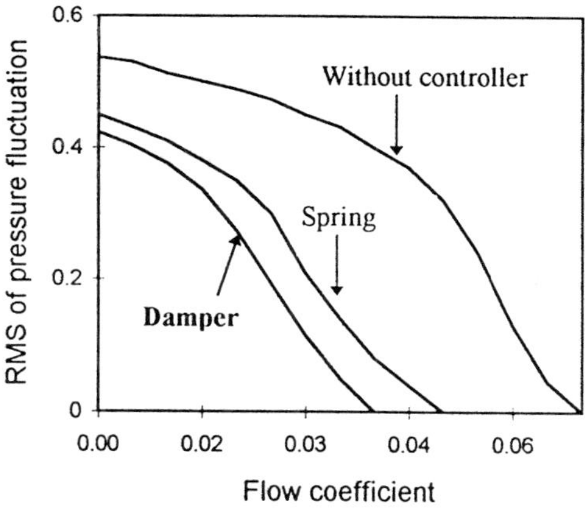

Fig. 8 RMS of pressure fluctuation for the compression system without controller predicted and experimental[1], and with controller (spring, predicted and experimental [1], and with damper)

7. Time variation analysis

The linear analysis yielded a set of optimized, nondimensional control parameters, which give large increases in the stable flow regime. Fig. 9 shows the time variation of the nondimensional pressure fluctuation, for both controlled and uncontrolled systems, at different flow coefficient, $\overline{\phi}_1 = 0.065, 0.055, 0.045, 0.035$ and 0.025. Pressure variation exhibit essentially damped harmonic motion. It is observed that the nonlinear computations showed that introducing wall motion into a fixed wall system undergoing deep surge cycles could suppress the surge. In the other words, even through the control scheme was designed bases on linear analysis, it was useful for oscillations that were strongly nonlinear. The magnitude of the flow perturbations increases as the flow coefficient is reduced.

8. Conclusions

In this paper a new method for dynamic control of compression system surge is introduced. An active element such as a damper is introduced to the compression system for increasing the surge margin. A movable plenum wall, driven by a signal proportional to the unsteady plenum pressure was used, to suppress the surge in the centrifugal compressors. Time variation of pressure and flow rate as well as so called surge cycle were mapped and discussed at different flow conditions and at different values of B-parameter. Effect of B-parameter on compression system stability is also discussed. As predicted by the model, a movable plenum wall (with damper) lowered the mass flow at

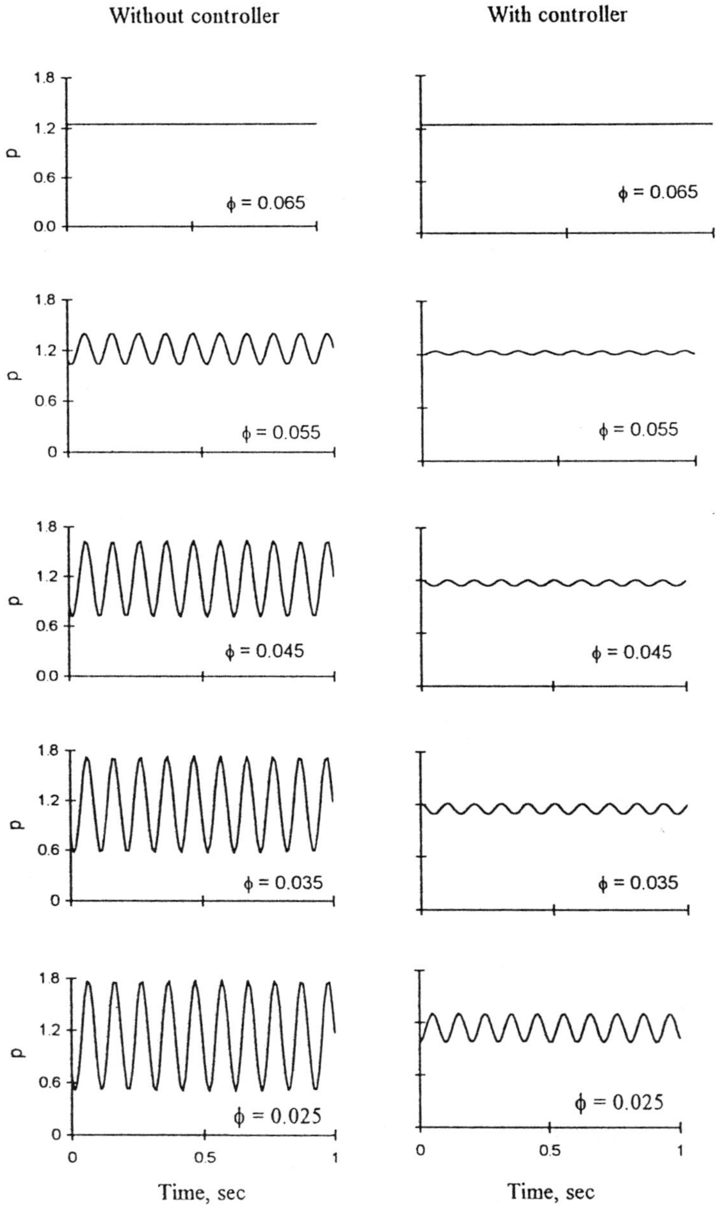

Fig.9 Fluctuation of system pressure rise with and without controlled plenum at different mean flow rates

which surge occurred by about 12 percent for a range of operating conditions. The approach taken is to suppress surge by modifying the compression system dynamic behavior using structural feed back. More specially, one wall of a downstream volume, or plenum is constructed so as to move in response to small perturbations in pressure. This structural motion provides a means for absorbing the stable operating range of the compression system, with no noticeable decrease in steady-state performance. Moreover, the model of the present work has been used utilizing spring instead of the damper as a controller, where the surge point obtained from the model is almost identical to the experimental surge point presented in Ffowcs [1].

Reference

1) Ffowcs Williams, J.E., and Huang, X.Y." Active Stabilization of Compressor Surge," J. of Fluid Mech., Vol.204, pp.245-262, 1995.
2) Pampreen, R.C., " Compressor Surge and Stall," Norwich, Vermont, USA 05055, 1995.
3) Dean, R.C., and Young, L.R., " The Time Domain of Centrifugal Compressor and Pump Stability and Surge ," Trans ASME, Journal of Fluids Engineering, Vol.99, pp. 53-64, 1977.
4) Emmons, H. W., Pearson, C. E., and Grant, H. P., " Compressor Surge and Stall Propagation," Trans ASME, Vol. 79, pp. 445-469, 1955.
5) Greitzer, E. M., " The Stability of Pumping Systems- The 1980- Freeman Scholar Lecture," Trans. ASME, Journal of Fluids Engineering, Vol. 103, pp. 193-242, 1981.
6) Mazzawy, R., " Surge-Induced Loads in Gas," Trans. ASME, Journal of Engineering for Power, Vol. 101, pp. xx-00, 1979.
7) Epstien, A. H., Ffowcs Willams, J. E., and Greitzer, E. M., "Active Suppression of Compressor Instabilities" AIAA 10th Aeroacoustics Conference Seattle, Paper 86-1994,1986.
8) Pinsely, J.E. ,"Active Control of Centrifugal Compressor Surge,"M.S. Thesis, Department of Aeronautics and Astronautics, MIT, Cambridge, MA, 1988.
9) Huang, X.Y.,"Active control of Aerodynamic Instabilities", Ph. D. Thesis, cambridge University, 1988.
10) Pinsely, J .E., Guenete, G.R., Epstien, A. H., and Greitzer, E. M.,"Active Stabilization of Centrifugal Compressor Surge," Trans. ASME, Journal of Turbomachinery, Vol.113, 1991.
11) Chen , G.T. ,"Active Control of Turbomachinery Instabilities-Initial Calculations and Results,"M.S.Thesis, Department of Aeronautics and Astronautics, MIT, Cambridge, MA, 1987.
12) Greitzer, E. M., "Surge and Rotating Stall in Axial Flow Compressors", ASME J. of Engg. for Power, Vol.98, pp190-217, 1976.
13) Gysling, D.L., Dugundji, Greitzer, E. M., Epstien, A. H.,"Dynamic Control of Centrifugal Compressor Surge Using Tailored Structures", Trans. ASME, Journal of Turbomachinery, Vol.113, 1991.

ON SUB-CELL STRUCTURE OF DEEP ROTATING STALL IN AN AXIAL COMPRESSOR

DAI KATO, EISUKE OUTA
Department of Mechanical Engineering, Waseda University
3-4-1 Okubo, Shinjuku-ku, Tokyo, 169, JAPAN
KAORU CHIBA
Research & Development Department, Aero-Engine & Space Operations, Ishikawajima-Harima Heavy Industries, Co., Ltd.
3-5-1 Mukoudai-cho, Tanashi-shi, Tokyo, 188, JAPAN

Abstract

Discussions are made on internal structure of a rotating stall cell in a single-stage compressor. Circumferentially extended part-span stall cells develop near the tip of the compressor. In the cells, tangential velocity exceeding even the blade speed is observed ahead of the rotor with flow reversal. Such measured features are studied by a 2-D Navier-Stokes numerical analysis. The computed stall cell is found to basically consist of a vortex ahead of the rotor and a counter-rotating vortex in the stator. This vortical feature is found to arise from a collision of the stage inlet flow with a reversed jet issuing from a high-pressure zone formed at the trailing side of the cell. The reversed jet assists to yield a significantly high tangential velocity. By computing with a large number of blade passages, the cell is found to extend circumferentially by splitting of the vortex into a train of multiple vortices ahead of the rotor. Comparing the flow velocity fluctuations with measured data, such a vortical flow pattern is thought to express the pattern which may actually exist near the tip. As a conclusion, it is suggested that a vortex forming from the tip inward to the hub is a sub-cell structure of the part-span cell, and that irregularly splitting and combining, multiple sub-cell vortices constitute the stall cell.

1. Introduction

Flow structure of rotating stall cells, full-span and part-span, have been studied experimentally, beginning with extensive measurements by Day and Cumpsty [1], and more recently by Poensgen and Gallus [2]. Puddu and Palomba [3] presented precise discus-

T.H. Fransson (ed.), Unsteady Aerodynamics and Aeroelasticity of Turbomachines, 511-524.

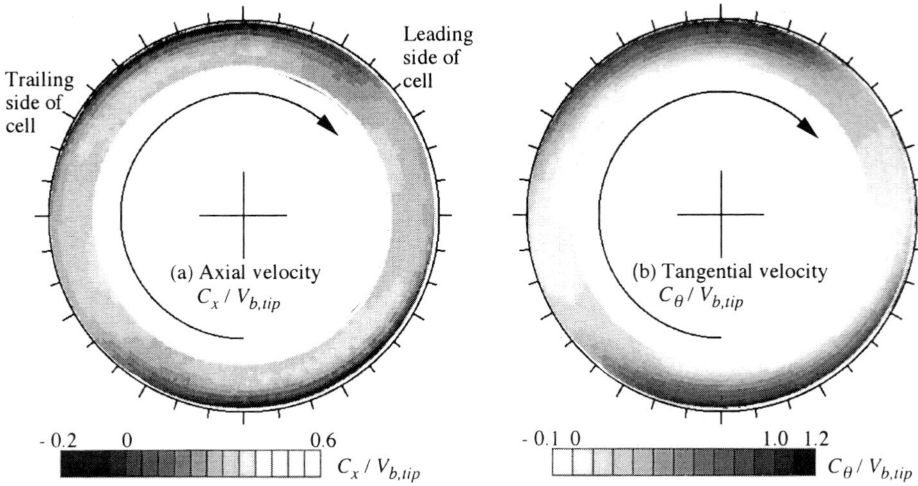

Fig.1 Phase-lock averaged velocity distribution ahead of rotor in fully-developed part-span rotating stall; flow coefficient ϕ_i = 0.22, $V_{b,tip}$ = 114 m/s. View from upstream, absolute frame.

sions on part-span cell structure. Despite these efforts, questions remain whether the cell is structured or unstructured. In this paper, part-span stall in a single-stage compressor with a high-staggered rotor is investigated. In the compressor, as shown in Fig.1, circumferentially extended regions of reversed flow are seen near the tip ahead of the rotor, and tangential velocity is seen to be very high exceeding the blade speed in these regions. How such a distorted cell pattern is brought about are the main interests of the study.

The flow in part-span stall is usually regarded as being more three-dimensional than that of full-span stall. Magnitude of radial velocity inside the cell of the present compressor is measured to be at most 10-15% of the absolute flow velocity of the compressor in the near-tip region. This is about a same level, for example, with 5-12% in full-span stall (Mathioudakis and Breugelmans [4]). Although secondary flow features affected by tip-clearance flow and by radial flow induced in the near hub region must be studied in three-dimensions, two-dimensional flow analyses may still be beneficial in forming a fluid mechanical insight about the inner-cell structure.

Based on such a consideration, the authors have developed 2-D Navier-Stokes analyses to study evolution process of stall cells [5,6] and some aspects of deep stall [7]. With increasing computer power, 2-D N-S analyses are utilized recently in a rotating stall/ stall flutter problem [8], and in the assessment of theoretical models of rotating stall [9].

The purpose of the present paper is to indicate, by studying the measured data based on the findings from a 2-D N-S numerical analysis, that a sub-cell structure exists in the part-span stall cell. In the N-S analysis, the computational region is expanded successively to allow the cell to grow without constraint. Vortical features of the computed

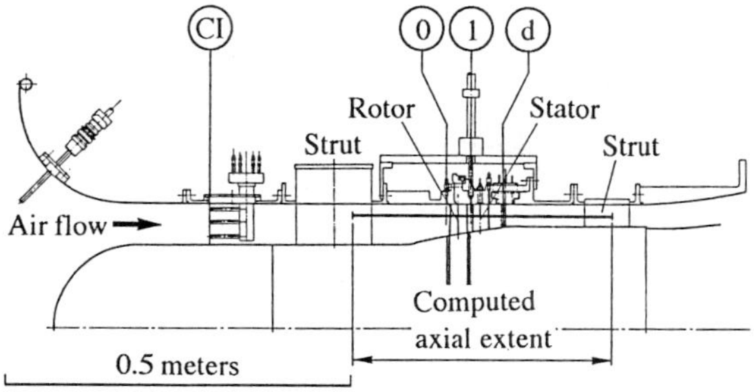

Fig.2 Cross-sectional view of single-stage compressor

Table 1 Compressor specifications

Design flow coefficient	0.45
Speed of rotation rpm	6000
Mass flow rate kg/s	2.5
Total pressure rise kPa	4.0
Tip diameter m	0.36
Blade Mach number	0.33 (tip)
Blade Reynolds number	3×10^5 (tip)
Hub-tip ratio	0.74(rotor inlet) 0.80(stator exit)
System surge/stall parameter B	0.7

deep stall cell are described in detail. From the comparison of the velocity fluctuations with the measured fluctuations, the N-S analysis is shown to be effective in expressing the cell features near the tip of the compressor. A specific configuration of the sub-cell structure is finally shown by referring to the measured and computed features.

2. Test Compressor and Flow Measurement

Fig.2 and Table 1 present a single-stage compressor in the present study. The hub radius increases axially to give passage contraction. In the computation, a stage section at 68% span from hub is used; Fig.3 and Table 2. It has high reaction with high-staggered rotor.

The compressor is run at 6,000 rpm, giving a blade-tip Mach number, $M_{b,tip}$, of 0.33. For the stage overall performance, time-averaged axial velocity and inlet total pressure at the computed radius are measured at location **0** by a yawmeter, and time- and space-averaged exit total pressure is measured at **d** by traversing a total pressure probe over two stator pitches. During stall, the inlet data are measured at **CI**, instead, by Pitot tubes.

Flow velocity in rotating stall of large angle change is measured by split-film probes

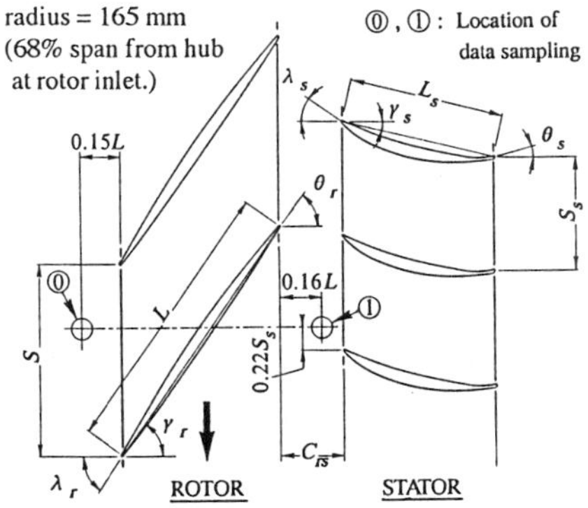

Fig. 3 Blade-to-blade geometry of rotor-stator cascades at radius of computed plane.

Table 2 Cascades geometry at computed plane

	Rotor	Stator
Profile	DCA	NACA 65
Number of blades	37	62
Chord length mm	L = 40.9	L_S= 23.0 (23.1)*
Solidity	L/S = 1.46	L_S/S_S = 1.37
Stagger angle deg	γ_r = 55.0	γ_s = 13.2
Inlet angle deg	λ_r = 56.8	λ_s = 37.2
Outlet angle deg	θ_r = 52.9	θ_s = -10.8
Axial Spacing	C_{rs} = 0.235 L	
Reaction	0.84	

* Modified for computation.

(DANTEC 55R57). Using developed calibration equations[10], angle variation of ±50 deg can be measured over 10-140 m/s with accuracies of ±2 deg and ±2 % of full speed.

3. Numerical Analysis of Unsteady Stage Flows

To gain insight into inner-cell features of rotating stall, a numerical simulation of multi-bladed cascades flows based on 2-D compressible N-S equations is conducted. Details of the developed numerical method are given in [5,6]. The governing equations are modified in the present study by spanwise averaging to include passage area contraction.

Boundary conditions are constructed so that the operating point can be determined and altered continuously by specifying only the rotor blade speed, V_b, and the load coefficient, ζ, of the exit throttle, just as in the actual compressor operation. Over the sliding boundary between the rotor and stator region, physical variables are transferred pointwise, so that flow disturbances, such as vortices and wakes shed from the rotor blades, and potential field of the blades/vanes pass smoothly from one region to the other.

A 3 rotor-blades/5 stator-vanes (3-R/5-S) region of almost the same pitch ratio as the actual ratio is chosen as a base unit. With increase in circumferential scale of stall cell, the region is expanded succesively,to 15-R/25-S, so as not to constrain the cell growth.

4. Comparison of the Computed Overall Characteristic with the Rig Test Data

Before discussing deep stall flow patterns, computed overall trends are compared with the measured trends. In Fig.4, the computed stage performance is presented in terms of

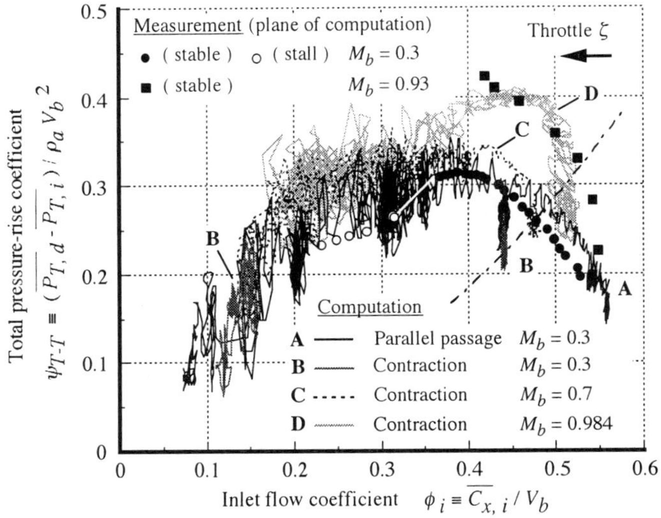

Fig.4 Computed stage total-pressure rise and comparison to rig test data.

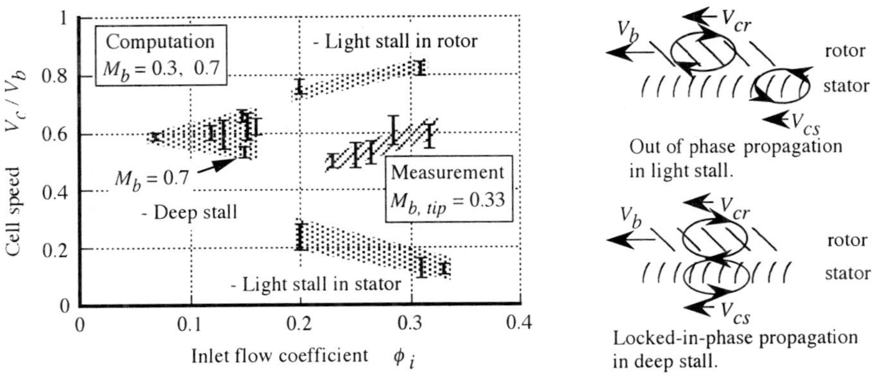

Fig.5 Comparison of computed cell speed with measured speed based on hot-film data.

inlet flow coefficient ϕ_i and total pressure rise coefficient $\psi_{T\text{-}T}$, which are defined as

$$\phi_i \equiv \overline{C_{x,i}} / V_b, \quad \psi_{T\text{-}T} \equiv \left(\overline{P_{T,d}} - \overline{P_{T,i}} \right) / \rho_a V_b^{\,2}. \tag{1}$$

Here, $\overline{C_{x,i}}$ and $\overline{P_{T,i}}$ are area-averaged axial velocity and inlet total pressure at axial section **0**, and $\overline{P_{T,d}}$ is area-averaged exit total pressure at axial section **d**. ρ_a is air density at compressor inlet. Loci **A** to **D** represent transition of the computed operating point as the flow is throttled under respective rotor speeds. High subsonic (M_b = 0.7) and transonic (M_b = 0.98) cases are indicated for reference. Irregular fluctuations in the loci correspond to the passings of vortices and reflected waves over the sampling planes. Compared to the measured data, overall trends of the stage performance are said to be properly simulated by the computation over a wide range of flow rate and also of blade Mach number.

As a characteristic quantity of rotating stall, propagation speeds of stall cells are shown in Fig.5 against flow coefficient. In the computation, a very short-scale stall cell covering at most one blade passage forms in the stator at ϕ_i = 0.33. With a slight load increase, a similar cell is formed in the rotor[5]. They propagate very slowly in the respective cascades (out-of-phase propagation). As the cells develop to deep stall by further throttle, the two cells eventually propagate together (locked-in-phase propagation).

In the measurement, stall cells are seen to rapidly develop after stall onset. This is presumably due to 3-D effects such as tip-leakage flows. Nevertheless, there still remains a possibility that this out-of-phase propagation of initial stall cells in each of the rotor and the stator at stall onset can be detected through careful measurements.

The computed cell speed in deep stage is found to agree well with the measured speed. Difference in the flow coefficient between the computation and the measurement is due to the part-span cell structure as will be shown later in Fig.15. In the measurement, large reversed flow is found to exist near the tip. However, near the hub, a high axial

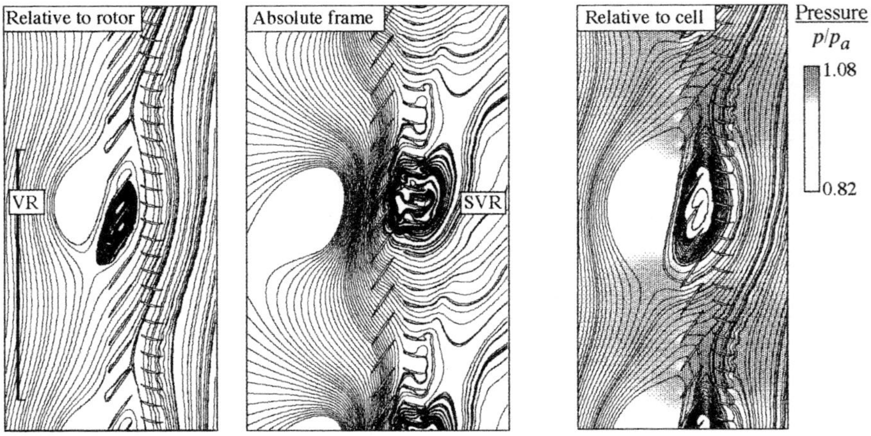

Fig. 6 Single vortex-pair structure of a deep stall cell indicated by computed streamlines and pressure field; M_b=0.3, 9-R/15-S, ϕ_i = 0.13. An instantaneous flow pattern is shown in three frames of reference.

velocity region still exists in the lowest flow coefficient. In the computation, 2-D nature misses this high axial velocity region. Then the flow coefficient is computed to be lower.

From the agreement in the cell speed, it may be said that the propagation of part-span stall is governed by the flow process in the tip similar to the computed flow process.

5. Two-Dimensional Flow Structure of Deep Rotating Stall

Computed flow patterns are studied in various frames to gain insight into the stall flow process near the tip of the compressor. Fig.6 shows a deep stall cell formed in 9-R/15-S system. The cell is found to consist of an axially-aligned pair of vortices, **VR** and **SVR**.

Relative to the rotor, a cell propagation mechanism in addition to the propagation of

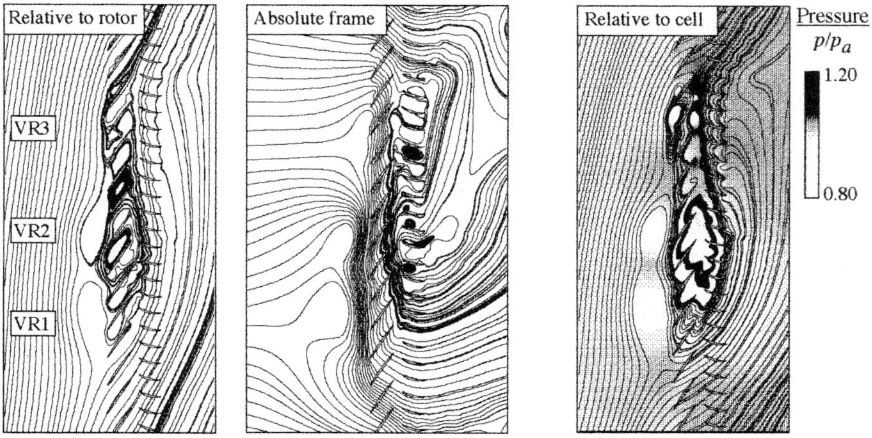

Fig. 7 Multiple-vortices structure of a deep stall cell indicated by computed streamlines and pressure field; M_b=0.3, 15-R/25-S, ϕ_i = 0.13. An instantaneous flow pattern is shown in three frames of reference.

blade leading-edge separation (Emmons et al [11]) is found. The stage inlet flow bifurcates upstream of the cell. A part of the flow that passes the rotor on the leading side of the cell is seen to be carried by the stator toward the trailing side of the cell, and to be reversed toward rotor upstream. The reversed flow generates a separation vortex at the trailing-edge of the rotor blade, which becomes a blockage. The reversed flow passage is forced by this blockage to propagate to the following blade passages.

Tangential velocity ahead of the rotor in the absolute frame is found to exceed the blade speed. If the vortical pattern is seen relative to the cell, it is clearly indicated that it takes a form of twin vortices. A high-pressure stagnant zone appears at the trailing side of the cell, partly by the action of the stator. Such a zone is also found in the measurement. A jet-like reversed flow sustained by the high-pressure zone collides with the stage inlet flow to form such twin vortices. The jet flow assists to accelerate tangential velocity of the cell flow to be higher than the blade speed. This will be shown in the measurement.

It is unlikely, however, that a single vortex covers a quarter of the compressor circumference. It is found, by computing with a larger number of blades, that a stall cell is composed not of a single vortex, but of a circumferentially-extended train of multiple vortices ahead of the rotor. Fig.7 shows a deep stall cell in 15-R/25-S, in which vortices **VR1**, **VR2**, and **VR3** form a single cell. Relative to the cell, three twin-vortices are found to align in the cell. Leading two twins appear distorted, since their propagation speeds are slightly lower than the last one. Vortices in the stator are seen to be closely packed.

These vortices ahead of the rotor appear as regions of high clockwise (CW) vorticity in the vorticity field shown in Fig.8. Trailing edge separation vortices in the rotor appear as regions of high counterclockwise (CCW) vorticity. It is seen that, at the trailing side of the stall cell, high CW vorticity of a blade leading-edge separation vortex is being ejected to rotor upstream by the reversed flow. Initial vortex **VR3** is found to be such an ejected leading-edge vortex. Since the blades are moving faster than the initial vortex, vorticity will be supplied

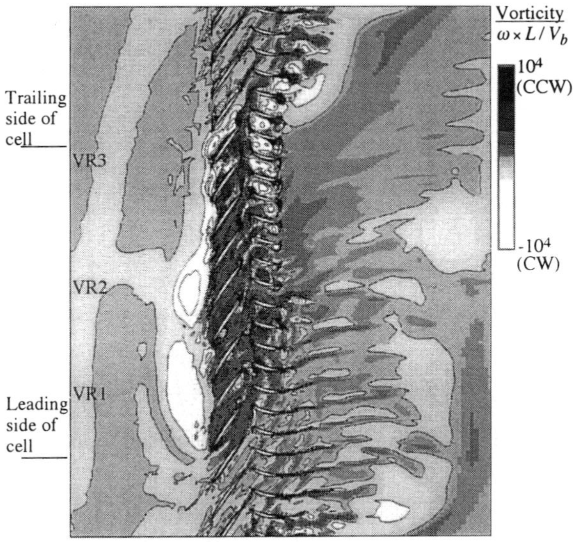

Fig.8 Ejection of vorticity from blade passage toward upstream of rotor at trailing side of stall cell; 15-R/25-S, $\phi_i = 0.13$.

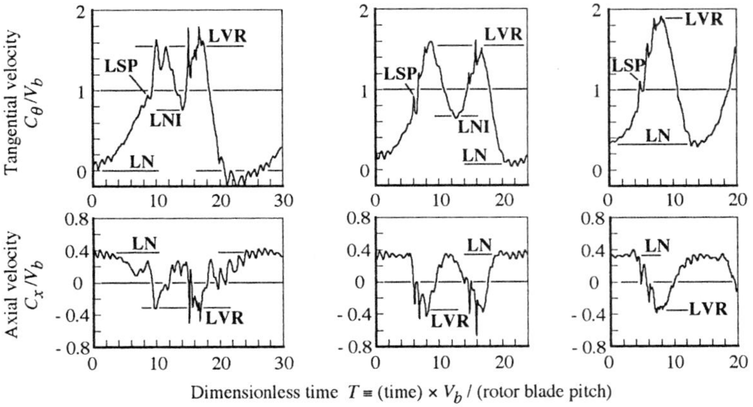

(a) a train of three vortices in a cell; 15-R/25-S, $\phi_i = 0.13$. (b) tandem-vortices in a cell; 12-R/20-S, $\phi_i = 0.12$. (c) single vortex-pair in a cell; 9-R/15-S, $\phi_i = 0.13$.

Fig.9 Computed time traces of flow velocity during deep stall; 15% rotor-chord ahead of rotor.

and accumulated into the vortex, which evolve into a developed vortex such as **VR2**.

Presence of a vortex at the leading side of a stall cell has also been suggested by Breugelmans, et al[12] by experiment. Development of such a vortex is thought to take a similar process as that described above based on the present computation.

6. Comparison of Computed Flow Velocity with Measured Velocity

In this section, the computed flow velocity in deep stall is compared with the flow velocity measured by split-film probes. Time traces of the computed velocity during one cell passing are shown in Figs.9(a)-(c), with different number of blades in the computational region. In all the cases, tangential velocity rises and axial velocity falls from the normal flow levels, **LN**, as the cell approaches the sampling point. Flow then reverses, and tangential velocity reaches a peak level, **LVR**, higher than the blade speed. In the15-R/25-S case, in which multiple vortices form the cell as seen in Fig.7, multiple peaks are seen inside the cell due to the passings of these vortices. Between the peaks, intermediate level **LNI** is seen due to the passings of unstalled flow paths between the vortices. The number of the peaks decreases and peak tangential velocity increases as the computational number of blades decreases. In the 9-R/15-S case, since the number of the blades is too limited, a single

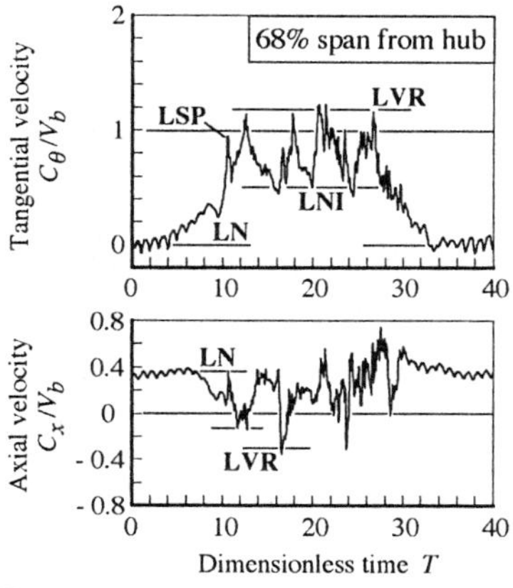

Fig.10 Measured flow velocity during deep stall, 15% rotor-chord ahead of rotor ; ϕ_i =0.23.

vortex develops excessively, and normal flow is seen not to return to axial direction.

Fig.10 shows the measured traces at 68% span from the hub. Multiple peaks (four in this case) in the velocity components are found inside the cell. Peak tangential velocity level **LVR** is found also to exceed the blade speed. Although the computed levels are still higher, measured levels at locations much closer to the tip are found to be as high as 160% of the blade speed, due to part-span cell configuration. Normal flow level **LN** as well as intermediate level **LNI** agree with the computed level. The time scale of each of the peaks are 4-6 blade passings, or 2-3 blade pitches in space. The measured traces are seen as a whole quite similar to the computed traces of the15-R/25-S case. From such a similarity, multiple sub-cell vortices are thought to exist in the actual compressor flow.

7. Temporal Behavior of Vortices and Sub-Cell Structure of Part-Span Stall Cell

Up to this point, a deep stall cell is shown to be composed of multiple sub-cell vortices. In this section, temporal behavior of these vortices is discussed. Finally, a sub-cell structure of the part-span stall cell is indicated from the computed and measured features.

Typical evolution and coalescence of vortices in the stall cell is shown in Fig.11. At $T = 35$, two vortices are seen in the leading portion of the cell. Vortex **VR2** is catching up with leading vortex **VR1**. At a later instant, $T = 39$, a third vortex, **VR3**, is seen to be generated behind **VR2** at the trailing side of the cell, by the vorticity ejection shown in Fig.8. As time elapses to $T = 43$, **VR3** evolves into a developed vortex, **VR2'**, while **VR1**

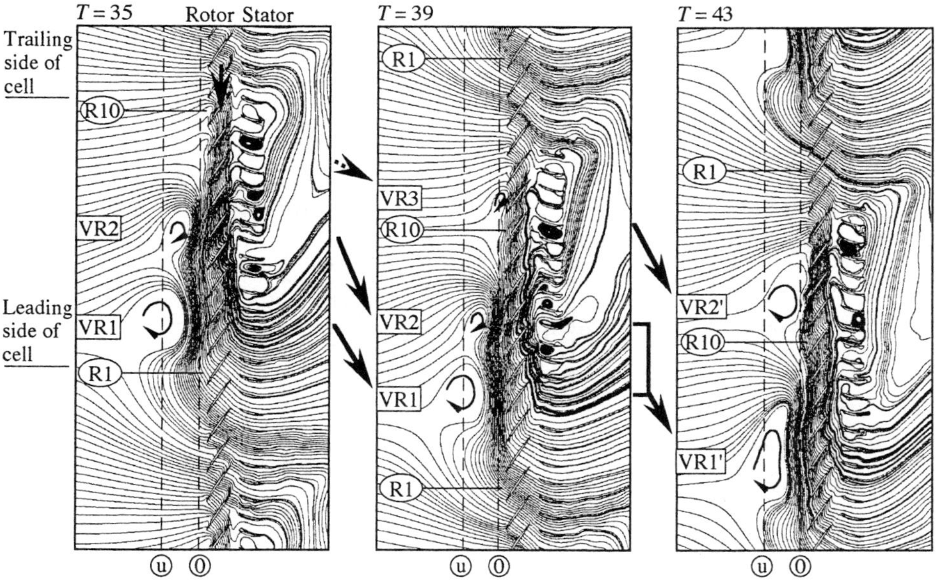

Fig.11 Computed instantaneous streamlines in absolute frame showing evolution of a vortex at the trailing side and coalescence of two vortices in a deep stall cell; 15-R/25-S, $\phi_i = 0.13$, $M_b = 0.3$.

and **VR2** coalesce into a new vortex, **VR1**'. Such inner-cell processes of the sub-cell vortices will naturally make the whole cell appearance irregular in time.

Behavior of sub-cell vortices affects forces on the rotor blades. Fig.12 shows time traces of the computed normal force coefficient, C_{FN}, defined as

$$C_{FN} \equiv F_N \,/\, (\rho_a V_b{}^2 L), \quad F_N = \boldsymbol{n}_c \cdot \left[\oint_B (-p\,\boldsymbol{n} + \boldsymbol{n} \cdot \boldsymbol{\tau})\, ds \right]. \tag{2}$$

Here, F_N is the force component in direction $\boldsymbol{n}_c$ normal to the chord. Blade force is obtained by integrating pressure p and viscous stress $\boldsymbol{\tau}$ on a blade surface of unit normal $\boldsymbol{n}$.

At T=35 (first picture in Fig.11), blade R10 entering the cell from the trailing side (T.S.) is found to receive the largest force among the blades, and blade R1 leaving the cell from the leading side (L.S.) is found to receive almost no force. Inside the cell, in which vortices **VR1** and **VR2** are arranged close to each other in the leading portion, blade R6 axially downstream of **VR2** is seen to receive the next largest force. At T=43, in contrast, vortices **VR1**' and **VR2**' develop relatively apart. In this mode, blade R11 traveling behind **VR2**' receive a large force exceeding that of R14 entering the cell. In this way, the blades receive significant force change while traveling through the cell.

In Fig.13, time traces of absolute flow velocity measured at locations **0** and **u**, shown in Fig.11, upstream of the rotor are presented for three spanwise positions. Six consecutive cell cycles are shown. Near the tip, vortices of similar size with the developed vortices in the computation are recognized at roughly one chord up-

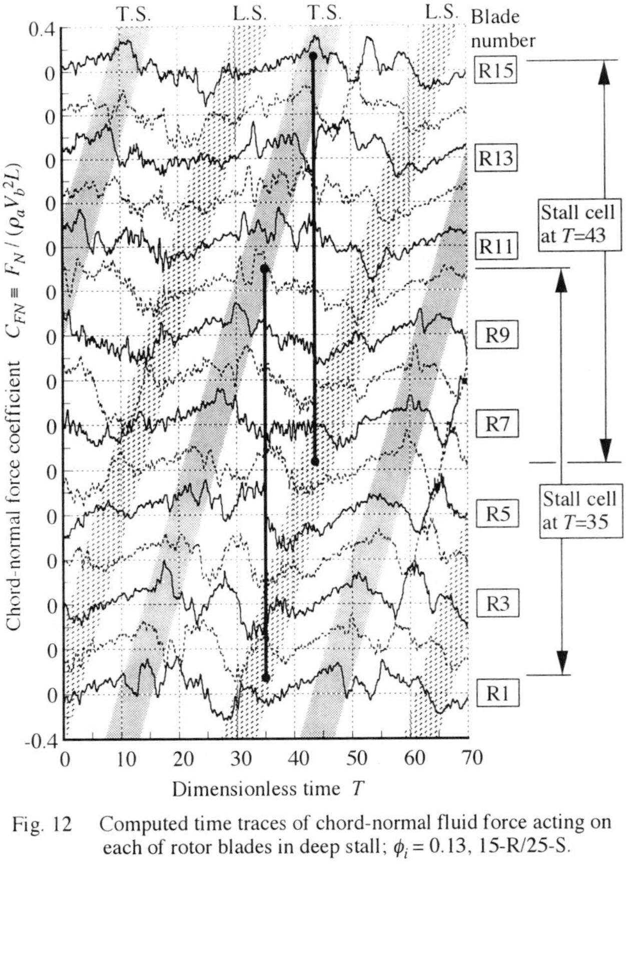

Fig. 12 Computed time traces of chord-normal fluid force acting on each of rotor blades in deep stall; $\phi_i = 0.13$, 15-R/25-S.

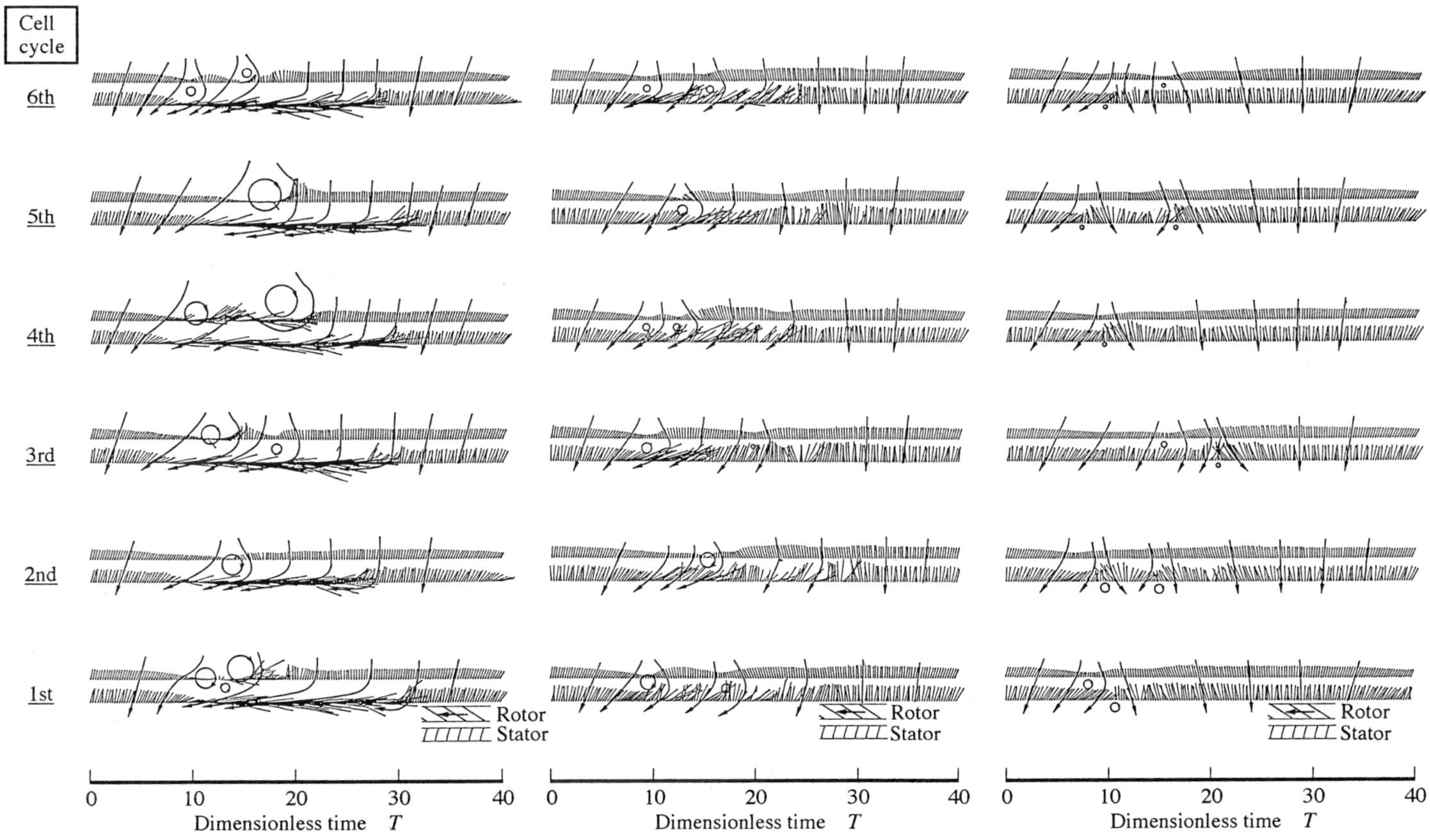

(a) Near tip (90% span from hub)

(b) Mid-span (50% span from hub)

(c) Near hub (20% span from hub)

Fig.13 Time histories of measured 2-D flow velocity and assumed sketches of vortices upstream of rotor showing cycle-by-cycle variation of vortical structure in deep stall operation; $\phi_i = 0.23$, $M_{b,tip} = 0.33$.

stream of the rotor. In the first cell cycle, two vortices are detected close to each other in the leading portion of the cell. In the second and third cycle, one vortex of smaller size is detected. Two large vortices are observed relatively apart in the fourth cycle. Such variations are thought to appear since the phases of evolution and combining process of the sub-cell vortices are different at each cycle. These vortices appear only from the leading to middle portion of the cell. Vortices are thought to be under development-phase in the trailing portion, similar to the computed development-phase described in Section 5.

At mid-span and at near the hub, vortices are found to be attenuated considerably in size, and their positions are seen to approach toward rotor inlet. Irregular vortices arrangement can also be observed at these spanwise positions.

Spanwise distribution of cell strength is shown in Fig.14. 2-D computed results are indicated for comparison. In this figure, circulation of the cell Γ_{cell} normalized by $\Gamma_{rot,\,des}$, circulation of a rotor blade design, is used to express the strength. Γ_{cell} is defined as the line integral of the 2-D flow velocity $\boldsymbol{C}$ along closed path ABCD surrounding the stall cell, as shown in an attached sketch of Fig.14:

$$\Gamma_{cell} = \oint \boldsymbol{C} \cdot d\boldsymbol{l}\,, \quad \text{along closed path ABCD in Fig.14} \tag{3}$$

In the computation, circulation $\Gamma_{cell,\,comp}$ is exactly calculated by Eq.(3). In the measurement, assuming identical velocity distribution along paths BC and DA, and negligibly small tangential velocity along path AB, circulation $\Gamma_{cell,\,meas}$ is approximated as:

$$\Gamma_{cell,\,meas} = \overline{C_{\theta 0,\,cell}} \cdot S_{cell}\,, \tag{4}$$

Here, $\overline{C_{\theta 0,\,cell}}$ is the mean tangential velocity along path CD (axial location **0**), and S_{cell} is

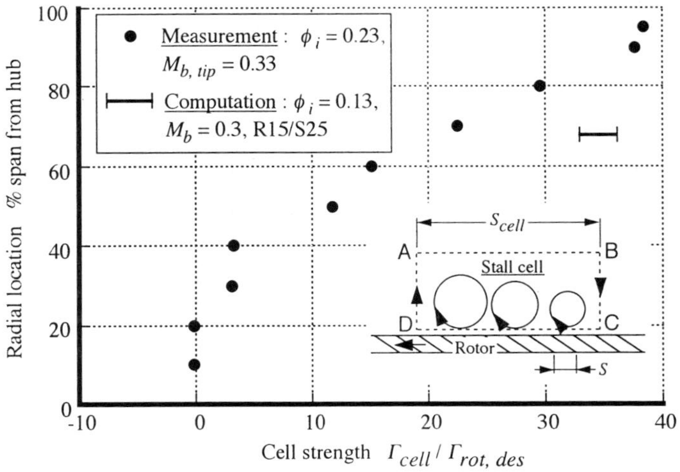

Fig.14 Radial distribution of cell strength ahead of rotor and comparison to computed cell strength.

the measured length of path CD. $\Gamma_{rot,\,des}$ is obtained by a usual expression in free-vortex design, using designed tangential velocity at rotor exit, $C_{\theta 1,\,des}$, and blade pitch S, as:

$$\Gamma_{rot,\,des} = C_{\theta 1,\,des} \cdot S \ . \tag{5}$$

The cell strength is seen to decrease from a maximum near the tip to zero at the hub, indicating that the cell is driven by strong vortices in the tip region. Computed cell strength is found to be of the same order of magnitude with the measured strength near the tip.

From the computed flow patterns and the measured flow velocity data, an assumed structure of the part-span stall cell is summarized in Fig.15. Multiple sub-cell vortices align circumferentially ahead of the rotor in the leading to the middle portion of the cell. These sub-cell vortices irregularly evolve and combine within the cell. Each of these sub-cell vortices forms from the tip inward to the hub, yielding a cone-shaped feature. Between the vortices, high axial velocity paths are formed in the near-hub region. The sub-cell vortex is generated and developed in the trailing portion of the cell near the tip, through the ejection of vorticity toward rotor upstream by the flow reversal. Many of the distorted features of the part-span cell in Fig.1 can be interpreted by such a structure.

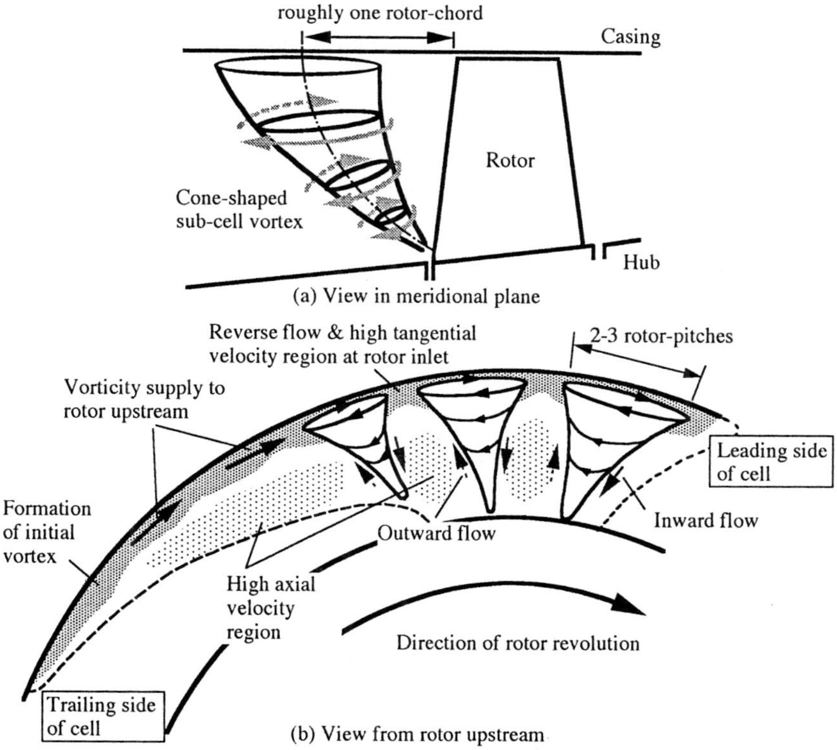

Fig. 15 Assumed sketch of multiple radial vortices and flow strucure upstream of the rotor inside part-span stall cell.

8. Conclusions

Vortical features of a deep stall cell are studied by an 2-D N-S analysis, and based on gained insights, measured data are studied to show a sub-cell structure of part-span stall in a single-stage compressor of high-staggered rotor. Following findings are obtained:

- Computed deep stall cell is found to basically consist of an axially-aligned pair of vortices, which arises from a collision of the stage inlet flow with a reversed jet sustained by a high-pressure zone formed at the trailing side of the cell.
- By computing in a region with a larger number of blades, the cell extends circumferentially by a splitting of the vortex ahead of the rotor into a train of multiple sub-cell vortices. These vortices are found to evolve by ejection of vorticity from the blade passages.
- By comparing with the measured data, the computed multiple vortices structure is indicated to express the actual sub-cell structure in the near-tip region of the compressor.
- The part-span stall cell is suggested to consist of irregularly evolving and combining, multiple sub-cell vortices. Each of the vortices form from the tip inward to the hub.

The study was supported by Grant in Aid for Scientific Research [No.05452157] of the Ministry of Education, and by Waseda University Grant for Special Research Projects [No.94B-7, No.96A-276, and No.97A-345]. The computation was also supported by Centre for Informatics of Waseda University.

References

1. Day, I. J., Cumpsty, N. A., 1978, Measurement and Interpretation of Flow within Rotating Stall Cells in Axial Compressors, Journal Mechanical Engineering Science, Vol.20, No.2, pp.101-114.
2. Poensgen, C. A., Gallus, H. E., 1996, Rotating Stall in a Single-Stage Axial Compressor, Journal of Turbomachinery, Trans. ASME, Vol.118, No.2, pp.189-196.
3. Puddu, P., Palomba, C., 1995, The Relative Flow Field Upstream and Downstream of an Axial-Flow Compressor Rotor during Rotating Stall, Proc. Yokohama International Gas Turbine Congress, pp.II-165-172.
4. Mathioudakis, K., Breugelmans, F. A. E., 1988, Three-Dimensional Flow in Deep Rotating Stall Cells of an Axial Compressor, Journal of Propulsion and Power, Vol.4, No.3, pp.263-269.
5. Outa, E., Kato, D., Chiba, K., 1994, An N-S Simulation of Stall Cell Behavior in a 2-D Compressor Rotor-Stator System at Various Loads, ASME Paper 94-GT-257.
6. Kato, D. Outa, E., Chiba, K., 1995, A Numerical Study of Propagating Stall in a 2-D Compressor Rotor/Stator Cascades System, Unsteady Aerodynamics & Aeroelasticity of Turbomachines, Elsevier, pp.609-624.
7. Kato, D. Hirose, E., Ichida, K., Outa, E., Chiba, K., 1995, Numerical and Experimental Study on Deep Stall Cell Behavior in an Axial Compressor, Proc. Yokohama International Gas Turbine Congress, pp.II-173-180.
8. He, L., 1995, Rotating-Stall/Stall-Flutter Prediction Using a Fluid/Structure Coupled Method, Unsteady Aerodynamics and Aeroelasticity of Turbomachines, Eds. Tanida, Y., Namba, M., Elsevier, pp.597-607.
9. Hoying, D. A., Greitzer, E. M., Tan, C. S., 1997, Evaluation of Compressor Stall Modeling Assumptions Utilizing Numerical Solutions of Rotating Stall, AIAA Paper 97-2771.
10. Chiba, K., Kobayashi, K., Nishiura, M., Outa, E., Kato, D., 1997, Investigation of Eddy Turbulent Diffusion and Radial Mixing in Axial-Flow Compressors, Proc.13th ISABE (to be published).
11. Emmons, H. W., Pearson, C. E., Grant, H. P., 1955, Compressor Surge and Stall Propagation, Trans. ASME, Vol.77, pp.455-469.
12. Breugelmans, F. A. E., Fabry, E., Mijs, J., 1993, The Leading Edge Vortex of a Rotating Stall Cell, Proc. 11th ISABE, pp.85-90.

STABILITY ANALYSIS OF SUPERSONIC CASCADE FLOW BY ACTUATOR DISK METHODS

KAZUHIKO YOKOTA, YUKIHARU IWAMOTO,
AND YOSHINOBU TSUJIMOTO
Division of Mechanical Science
School of Engineering Science
Osaka University
1-3, Machikaneyama, Toyonaka 560, Japan

1. Introduction

In recent years, the rotational speed of gas turbines is increasing more and more to increase the pressure ratio per one stage and hence to reduce the total number of stages. Accordingly, it is becoming more common that the relative inflow is supersonic near the tip of the fan and the first stage compressor cascades. At higher speeds, it is well-known that stable operating range becomes extremely narrow due to surge and choke. In this circumstances, it is important to understand the characteristics of the supersonic compressor in the off-design region as well as at the design point.

Freeman et al.[1] investigated the steady characteristics of the supersonic cascade by means of an analytical method using one dimensional flow model. It is indicated that the performance curve of the supersonic cascade has a positively sloped region, which is enlarged at higher speeds. The positive slope of the performance curve for the subsonic cascades causes flow instabilities, such as surge or rotating stall. Therefore, the results of Freeman et al. indicate the possibilities of the flow instabilities in supersonic cascades similar to those in subsonic ones. However, there exist no investigations of such instabilities. The present study investigates the flow stabilities in supersonic cascades with subsonic axial flow.

2. Nomenclature

Variables are denoted in the relative frame running with cascades if not stated otherwise.

a	: sound speed	k	: wave number in y-direction
i	: imaginary unit	M	: Mach number

T.H. Fransson (ed.), Unsteady Aerodynamics and Aeroelasticity of Turbomachines, 525-538.

p : static pressure
t : time
u, v : x-, y-direction velocities
x, y : Cartesian coordinate
κ : specific heat ratio
λ : wave number in x-direction
ρ : density
χ : stagger angle
ω : complex angular velocity

Subscripts
abs : in the absolute coordinates
D : in the downstream of cascade
im : imaginary part of complex
re : real part of complex
rel : in the relative frame
t : at the stagnation
U : in the upstream of cascade

Superscript
^ : disturbance component
 : disturbance amplitude

3. Actuator Disk Method

3.1. CASCADE MODEL

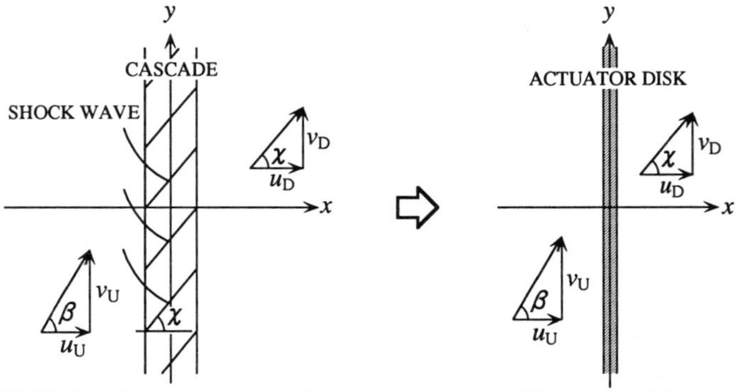

(a) Unstarted supersonic cascade (b) Actuator disk

Figure 1. Unstarted supersonic cascade and actuator disk

When the supersonic cascade with subsonic axial flow becomes unstarted, detached shock waves occur just upstream of the blade leading edges. Figure 1(a) shows the schematic of an unstarted supersonic flat plate cascade. In actuator disk methods, this cascade is modeled by an actuator disk with infinitesimal thickness as shown in Figure

1(b). In the present study, the y-axis, i.e. x=0, is considered to be an actuator disk as shown in Figure 1(b).

3.2. STEADY COMPONENTS

Here, we present the boundary conditions for the steady components. Generally, there exist the following relations between upstream and downstream flow variables. The continuity equation is given by

$$\rho_U u_U = \rho_D u_D , \tag{1}$$

the energy equation (the total enthalpy conservation equation) is given by

$$\frac{\kappa}{\kappa-1}\frac{p_U}{\rho_U}+\frac{1}{2}\left(u_U^2+v_U^2\right)=\frac{\kappa}{\kappa-1}\frac{p_D}{\rho_D}+\frac{1}{2}\left(u_D^2+v_D^2\right) , \tag{2}$$

and the Kutta's condition is given by

$$u_D \sin\chi - v_D \cos\chi = 0. \tag{3}$$

For compressible flows, another condition is needed to determine the relationship between upstream and downstream variables. We use the conservation equation of the momentum in the blade chord direction proposed by Freeman et al.[1] :

$$\begin{aligned} &p_U \cos\chi + \rho_U u_U \left(u_U \cos\chi + v_U \sin\chi\right) \\ &\quad = p_D \cos\chi + \rho_D u_D \left(u_D \cos\chi + v_D \sin\chi\right). \end{aligned} \tag{4}$$

From Eqs.(1) - (4), we obtain two Mach numbers. These correspond to two cases with or without shock waves. By selecting the lower Mach number solution, we obtain the solution that represents the flow field downstream of the supersonic cascade with shock waves. Freeman et al. note that this model gives a good prediction of the supersonic cascade characteristics.

3.3. DISTURBANCE COMPONENTS

If the amplitudes of the disturbance components are assumed to be infinitesimal, then the infinitesimal disturbance components satisfy the linearized continuity equation, the linearized two-dimensional Euler equations and the linearized energy equation given by

$$\frac{\partial\hat{\rho}}{\partial t}+u\frac{\partial\hat{\rho}}{\partial x}+v\frac{\partial\hat{\rho}}{\partial y}+\rho\left(\frac{\partial\hat{u}}{\partial x}+\frac{\partial\hat{v}}{\partial y}\right)=0, \tag{5}$$

$$\frac{\partial \hat{u}}{\partial t}+u\frac{\partial \hat{u}}{\partial x}+v\frac{\partial \hat{u}}{\partial y}=-\frac{1}{\rho}\frac{\partial \hat{p}}{\partial x}, \tag{6}$$

$$\frac{\partial \hat{v}}{\partial t}+u\frac{\partial \hat{v}}{\partial x}+v\frac{\partial \hat{v}}{\partial y}=-\frac{1}{\rho}\frac{\partial \hat{p}}{\partial y}, \tag{7}$$

$$\frac{1}{p}\left(\frac{\partial \hat{p}}{\partial t}+u\frac{\partial \hat{p}}{\partial x}+v\frac{\partial \hat{p}}{\partial y}\right)-\frac{\kappa}{\rho}\left(\frac{\partial \hat{\rho}}{\partial t}+u\frac{\partial \hat{\rho}}{\partial x}+v\frac{\partial \hat{\rho}}{\partial y}\right)=0. \tag{8}$$

Using the complex angular velocity ω of the disturbances, the wave numbers λ and k in the x- and y-directions, respectively, we assume the functional forms of the disturbance components as

$$\exp[i(\omega t-ky)+\lambda x]. \tag{9}$$

For example, using a infinitesimal density amplitude ρ', the disturbance component of density is given by

$$\hat{\rho} = \rho' \exp[i(\omega t-ky)+\lambda x]. \tag{10}$$

By putting subscripts 're' and 'im' on the real and the imaginary parts of wave numbers in Eq.(9), respectively, we can express Eq.(9) as

$$\exp[i(\omega_{re}t-k_{re}y+\lambda_{im}x)]\exp[-\omega_{im}t+\lambda_{re}x]. \tag{11}$$

Since the flow field is periodic in the y-direction, the y-direction wave number can be assumed to be real. Consequently, Eq.(11) does not include the imaginary part of k. The first exponential function in Eq.(11) shows that the angular velocity of the disturbance components is ω_{re}. The disturbances propagate with the wave numbers k_{re} and λ_{im} in the x- and y-directions, respectively. The second one indicates that the disturbances attenuate at the temporal rate ω_{im}, herein called the attenuation rate, and varies at the rate of λ_{re} in the x-direction.

By substituting the disturbance components in the forms of Eq.(9) into Eqs.(5) to (8), we obtain a homogeneous first-order simultaneous equations for the infinitesimal amplitudes. For the existence of non-trivial solutions, the determinant of the coefficient matrix must equal to zero. This leads to an algebraic fourth-order equation for λ. Solving the equation with respect to λ, we obtain the following expression of the wave number λ in the y-direction

$$\lambda_{1,2}=\frac{1}{a^2-u^2}\left\{i\,u(\omega-kv)+a\sqrt{(a^2-u^2)k^2-(\omega-kv)^2}\right\}, \tag{12}$$

$$\lambda_3 = -\frac{i}{u}(\omega - kv). \tag{13}$$

The wave numbers $\lambda_{1,2}$ have two values because the value in the radical sign in Eq.(12) is generally complex. One of them is represented by λ_1 and the other by λ_2. For the wave numbers λ_1 and λ_2, the disturbances appear only in pressure and not in vorticity. These wave numbers represent the pressure disturbance. Two wave numbers λ_1 and λ_2 represent the upstream and the downstream propagating pressure disturbances. In the present study, we denote λ_1 for the upstream x-direction wave number and λ_2 for the downstream one. On the other hand, for the wave number λ_3, the pressure disturbance becomes zero. This wave number λ_3 represents the disturbances of vorticity and entropy.

Since the sources of various disturbances are located on the cascade, we must appropriately choose the upstream and the downstream propagating pressure disturbances for the upstream and the downstream regions of the cascade, respectively. The method to assign λ_1 and λ_2 to upstream and downstream of the cascade, respectively, is as follows.

First, we consider the case in which the attenuation rate ω_{im} equals to zero, and a single disturbance source moves on the cascade in the y-direction at the velocity ω_{re}/k. The circular wave front of the disturbance generated from the source spreads at the sound speed a with being convected by the main flow velocity (u, v) as shown in Figure 2.

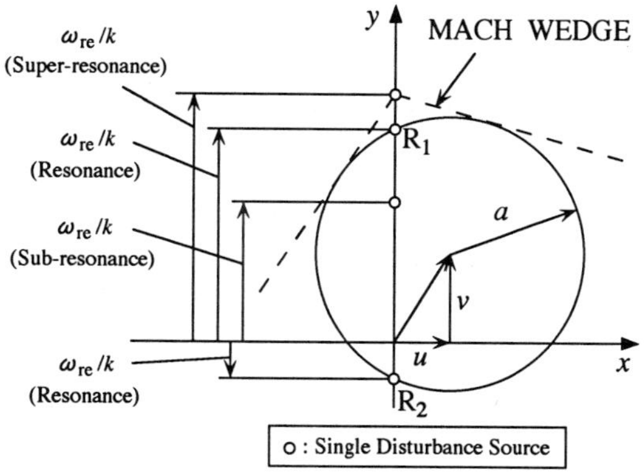

Figure 2. Relationship between circular wave front and moving single disturbance source

The total disturbances in the flow field is obtained by superposing all the circular waves generated from all the disturbance sources on the cascade. When the axial, i.e. x-directional, flow velocity is subsonic, a circular wave front have two intersections R_1 and R_2 with y-axis. The case in which the source on the cascade moves with the same velocity as these intersections is called "Resonance". At Resonance condition, the angular velocity ω_{re} gives zero to the radical sign in Eq.(12). The cases in which the source moves faster/slower than two intersections R_1 and R_2 are called Super-resonance/Sub-resonance, respectively. The disturbance patterns vary significantly depending on the value of ω_{re}, whether it is larger or smaller than the Resonance value. Therefore, we must choose the appropriate wave numbers λ_1 and λ_2 for Sub-resonance and Super-resonance cases.

For Super-resonance, a Mach wedge with its vertex at the disturbance source is generated as shown in Figure 2. For $\omega_{im}=0$, the slopes of the two sides of the Mach wedge equals to the ones of the isophase surfaces of Eq.(11), i.e. $\mathrm{Im}[\lambda_{1,2}/k]$, where Im[ ] represents the imaginary part. The two sides of the Mach wedge represent the two waves. One of the two waves extends upstream of the cascade, and the other downstream. Since the disturbance source is located on the cascade, we can choose the appropriate waves for the upstream and the downstream flow fields, based on the magnitudes of $\mathrm{Im}[\lambda_{1,2}/k]$.

For Sub-resonance, the Mach wedge is not generated. One of the real parts of λ_1 and λ_2 is positive and the other negative. We should assign λ_1 and λ_2 so that the disturbances remain finite at $x=\pm\infty$. This means that we should choose the wave numbers of which real parts are positive and negative real for the upstream and the downstream regions, respectively.

The above method for $\omega_{im}=0$ can be applied for $\omega_{im}\neq 0$. It should be noted that, although the angle formed by the two isophase surfaces becomes smaller than the Mach wedge vertex angle for $\omega_{im}\neq 0$ and Super-resonance, we can choose the appropriate wave numbers λ_1 and λ_2 based on the magnitudes of $\mathrm{Im}[\lambda_{1,2}/k]$ in the same manner as the $\omega_{im}=0$ case. With this method, the two wave numbers coincide the ones for the $\omega_{im}=0$ case when ω_{im} approaches zero.

Table 1 illustrates the behaviors of the disturbances with λ_1 and λ_2 chosen as the method mentioned above. Table 1 indicates that the amplitudes for λ_1 and λ_2 approach zero at $x=\pm\infty$ for Sub-resonance, but not always for Super-resonance. It is because all the disturbances from the sources on the cascade cancel out with one another for Sub-resonance, but the cancellation does not occur for Super-resonance.

Table 1 Amplification and attenuation of disturbance

	Amplifying ($\omega_{im}<0$)	Neutrally stable ($\omega_{im}=0$)	Attenuating ($\omega_{im}>0$)
Super-resonance	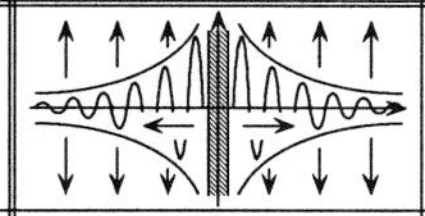	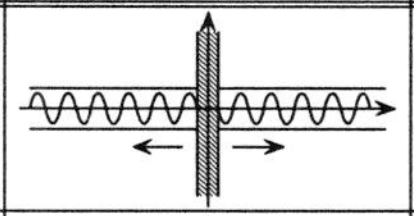	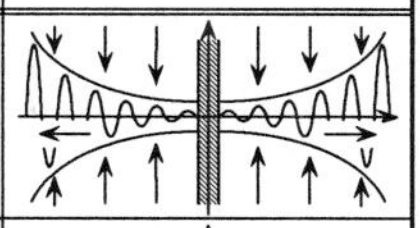
Sub-resonance	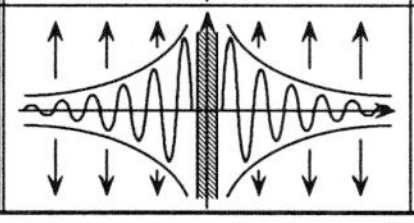	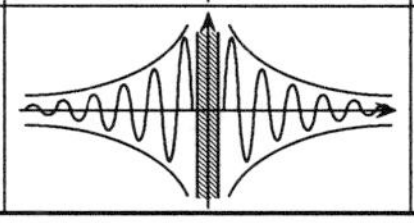	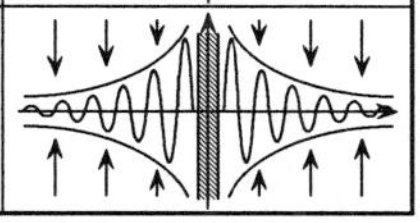

Note : Vertical arrows mean temporally amplifying or attenuating.
Horizontal arrows mean extending direction of isophase surface.

It is assumed that the upstream region is irrotational without incoming vortical disturbance but the downstream one is rotational and has the entropy disturbance with the vorticity shed from the cascade. Accordingly, there exist the disturbance with λ_{1U} in the upstream and the disturbances with λ_{2D} and λ_{3D} in the downstream of the cascade. For example, the density disturbances are given by $\hat{\rho}_U=\hat{\rho}_{1U}$, $\hat{\rho}_D=\hat{\rho}_{2D}+\hat{\rho}_{3D}$. The ratios of the disturbance components are determined as an eigenvector of the homogeneous simultaneous linear equations (5) to (8). Then we can choose $\hat{v}_{1U}$ with λ_{1U} in the upstream region, $\hat{v}_{2D}$ with λ_{2D}, $\hat{v}_{3D}$ and $\hat{\rho}_{3D}$ with λ_{3D} in the downstream region as unknown disturbances. These four unknowns are determined by the following four boundary conditions.

Here, we consider the cases in which the condition of Eqs.(1) to (4) can be applied quasi-statically. For the unsteady components, we obtain the following relations.

The linearized continuity equation for the infinitesimal disturbances is given by

$$\left(\hat{\rho}_U u_U+\rho_U \hat{u}_U\right)-\left(\hat{\rho}_D u_D+\rho_D \hat{u}_D\right)=0. \tag{14}$$

The linearized total enthalpy conservation equation for the infinitesimal disturbances is given by

$$\left\{\frac{\kappa}{\kappa-1}\frac{p_U}{\rho_U}\left(\frac{\hat{p}_U}{p_U}-\frac{\hat{\rho}_U}{\rho_U}\right)+u_U\,\hat{u}_U+v_U\,\hat{v}_U\right\} -\left\{\frac{\kappa}{\kappa-1}\frac{p_D}{\rho_D}\left(\frac{\hat{p}_D}{p_D}-\frac{\hat{\rho}_D}{\rho_D}\right)+u_D\,\hat{u}_D+v_D\,\hat{v}_D\right\}=0. \tag{15}$$

The linearized Kutta condition for the infinitesimal disturbances is given by

$$\hat{u}_D\cos\chi-\hat{v}_D\sin\chi=0. \tag{16}$$

The linearized momentum conservation equation in the blade chord direction for the infinitesimal disturbances is given by

$$\begin{aligned}&\left[\hat{p}_U+\left(\hat{\rho}_U u_U+\rho_U \hat{u}_U\right)\left(u_U+v_U\tan\chi\right)+\rho_U u_U\left(\hat{u}_U+\hat{v}_U\tan\chi\right)\right]\\&\quad-\left[\hat{p}_D+\left(\hat{\rho}_D u_D+\rho_D \hat{u}_D\right)\left(u_D+v_D\tan\chi\right)+\rho_D u_D\left(\hat{u}_D+\hat{v}_D\tan\chi\right)\right]=0\,,\end{aligned} \tag{17}$$

where $\hat{v}_{1U}$, $\hat{v}_{2D}$, $\hat{v}_{3D}$ and $\hat{\rho}_{3D}$ denotes the disturbance components. Then Eqs.(14) to (17) can be rewritten as a set of a homogeneous linear simultaneous equations for the disturbance amplitudes, v'_{1U}, v'_{2D}, v'_{3D} and ρ'_{3D}, given by

$$[\mathrm{M}]\begin{bmatrix}v'_{1U}\\ v'_{2D}\\ v'_{3D}\\ \rho'_{3D}\end{bmatrix}=[0]\ , \tag{18}$$

where $[\mathrm{M}]$ is the coefficient matrix. For the existence of non-trivial solutions, the determinant of the coefficient matrix should equal to zero, i.e.

$$\det[\mathrm{M}]=0. \tag{19}$$

This is an algebraic equation in terms of ω/k.

As mentioned above, for the complex ω and the positive real k, the real part ω_{re}/k of the complex ω/k represents the propagation velocity of the disturbances in the relative frame moving with the cascade. Since the steady components of the inflow is assumed to have no rotational, i.e. y-directional, velocity in the absolute frame in the present study, the rotational speed equals to the rotational relative flow velocity v_U. Therefore, the propagation velocity of the disturbances in the absolute frame is given by $v_U-\omega_{re}/k$. The propagation velocity ratio of the disturbances, i.e. the ratio of the propagation velocity to rotational speed, is given by $1-\omega_{re}/(k\,v_U)$. On the other hand, the imaginary part ω_{im}/k of ω/k represents the attenuation rate of the disturbances. That is, if $\omega_{im}/k>0$, the flow is stable because the disturbances attenuate temporally, but if $\omega_{im}/k<0$, the flow is unstable because the disturbances amplify temporally. To summarize, the propagation velocity ratio is given by $1-\omega_{re}/(k\,v_U)$ and the flow stability is judged from the sign of ω_{im}/k.

Equation (19) includes λ_{1U}/k and λ_{2D}/k with square root of the equation of ω/k as shown in Eq.(12). By transposing and multiplying by itself properly, we can eliminate the square root. As a result, we obtain a 12th order polynomial equation of ω/k. By solving this 12th order equation, for instance by using the DKA method, we obtain twelve solutions. These twelve solutions include solutions not satisfying Eq.(19), which appear from the process of transposing and multiplication by itself Eq.(19). True solutions should be identified by substituting the twelve solutions into Eq.(19). Based on this method, we can obtain the complete set of the solutions of Eq.(19).

4. Results and Discussions

4.1. STEADY PERFORMANCE CURVE

To verify the validity of our calculation code for the steady components, we calculated the steady performance map corresponding to the one given by Freeman et al. Since our results agree exactly with those by Freeman et al., we consider that the code validity is confirmed for the steady components.

Figure 3 shows the steady performance map for the flat plate rotor cascade, of which stagger χ is 60.0[deg.]. The horizontal and the vertical axes indicate the axial flow velocity ratio to the rotational speed and the pressure ratio of the downstream static pressure to the upstream total pressure, respectively. Since the steady components of the absolute inflow has no rotational velocity, the rotational speed is identical with v_U. The thin broken curve is connecting the maxima of the performance curves. On the thin dotted curve, the relative Mach number M_{Urel} of the inflow becomes unity.

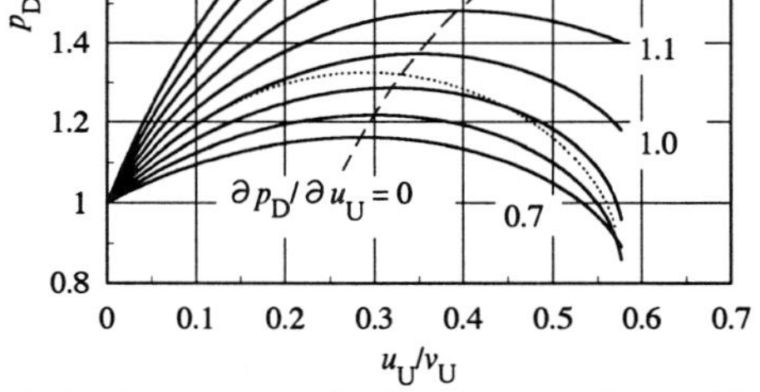

Figure 3. Performance map for flat plate cascade (χ= 60[deg.])

Figure 3 shows that even in the region where the rotational speed is relatively low, the performance curve is positively sloped. It is because the implicit incidence loss, which increases with the increase of incidence, is included in the present one-dimensional flow model with the momentum conservation equation in the blade code direction. Figure 3 also shows that with the increase of the rotational speed, the positively sloped region becomes larger and the slope increases. This is caused by the flow compressibility effect.

4.2. STABILITY ANALYSIS

We obtained eight kinds of solutions as the infinitesimal disturbances. Most of them are stable or neutrally stable in all operating conditions. It was found that two kinds of solutions may associate with the flow instability, i.e. become unstable.

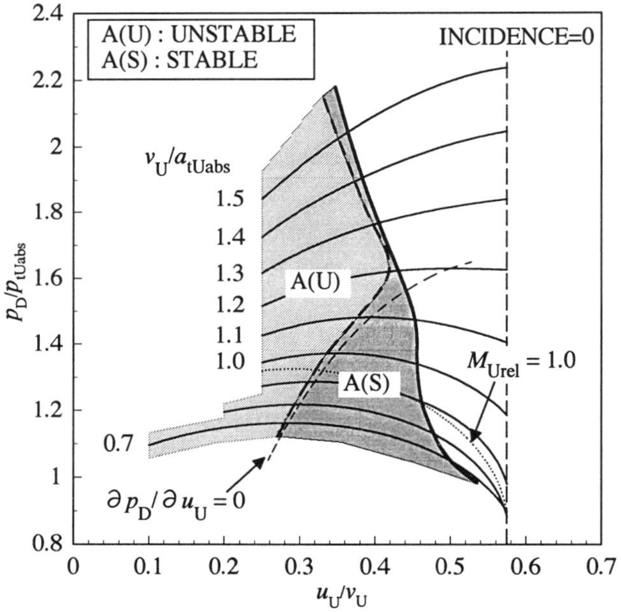

Figure 4. Stability map of solution "A"

Figure 4 shows the stability map of a solution named solution "A". The horizontal and the vertical axes are same as in Figure 3. The thin broken curve and the thin dotted curve are also same as in Figure 3.

Solution "A" exists in the left-side region of the thick solid curve. On the thick broken curve, solution "A" becomes neutrally stable. Solution "A" becomes stable in the region between the thick solid curve and the thick broken curve, which is indicated by A(S) and the darker gray region. Solution "A" becomes unstable in the left-side region of the thick broken curve, which is indicated by A(U) and the lighter gray region.

Figure 4 shows that solution "A" exists in the relatively low flow rate region and the performance curves are positively sloped in the region where solution "A" is unstable. It should be noted that solution "A" becomes unstable even in the region where there exists no shock wave ($M_{\mathrm{Urel}} < 1$) and the rotational speed is low, i.e. $v_{\mathrm{U}}/a_{\mathrm{tUabs}}$ is small.

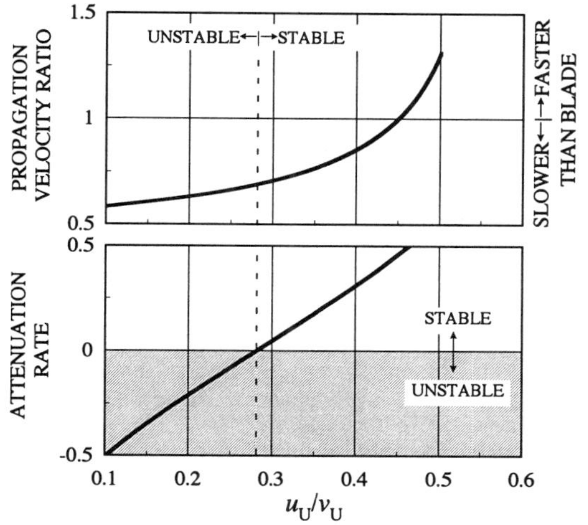

Figure 5. Propagation velocity ratio and attenuation rate of solution "A" for $v_{\mathrm{U}}/a_{\mathrm{tUabs}} = 0.7$

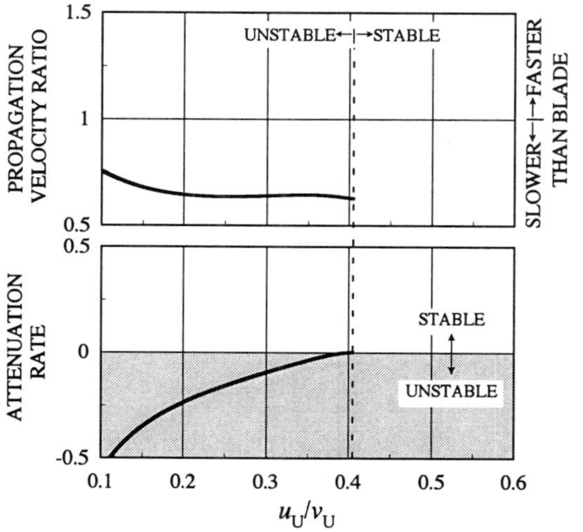

Figure 6. Propagation velocity ratio and attenuation rate of solution "A" for $v_{\mathrm{U}}/a_{\mathrm{tUabs}} = 1.3$

Figure 5 shows the propagation velocity ratio $1-\omega_{re}/(k\,v_U)$, and the attenuation rate ω_{im}/k, for v_U/a_{tUabs}=0.7. Figure 6 shows those for v_U/a_{tUabs}=1.3. The horizontal axis means the axial flow velocity ratio to the rotational speed.

These figures indicate that the propagation velocity ratio of solution "A" is smaller than unity when solution "A" is unstable. It is confirmed that the propagation velocity ratio of solution "A" is smaller than unity whenever solution "A" is unstable. On the other hand, the actuator disk analysis of the incompressible flow shows that the positive slope of the performance curve due to the flow loss produces the rotating stall, of which propagation velocity ratio is smaller than unity. Based on these results, solution "A" corresponds to the rotating stall solution also found in the incompressible flow.

Figure 7 shows the stability map of a solution named solution "B". The horizontal and the vertical axes are same as in Figure 3. The thin broken curve and the thin dotted curve are also same as in Figure 3.

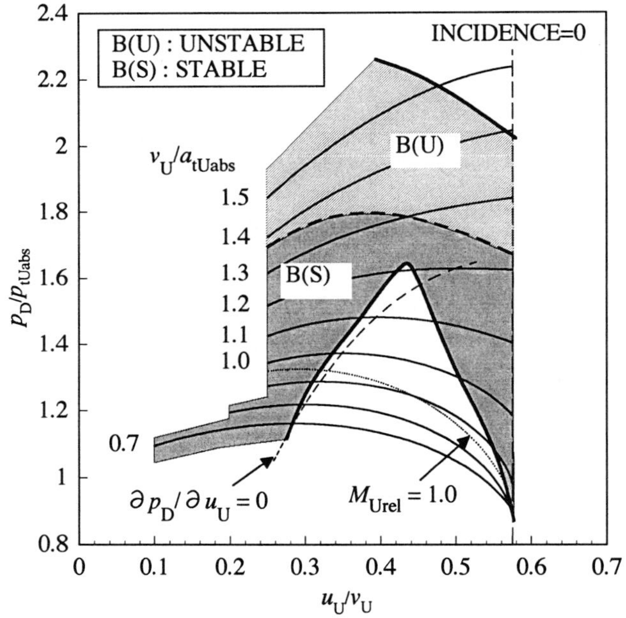

Figure 7. Stability map of solution "B"

Solution "B" exists between the two thick solid curves, i.e. in the gray sides of each thick solid curves. On the thick broken curve, solution "B" becomes neutrally stable. Solution "B" becomes stable in the region between the lower thick solid curve and the thick broken curve, which is indicated by B(S) and the darker gray region. Solution "B" becomes unstable in the region between the thick broken curve and the upper thick solid curve, which is indicated by B(U) and the lighter gray region.

Figure 7 shows that solution "B" occurs in the region where the rotational speed v_U is high, and the performance curves are positively sloped in the region where solution "B" is unstable. By comparing Figure 4 with Figure 7, we can see that there exists the region where solutions "A" and "B" occur. Therefore, we can say that they are mutually independent.

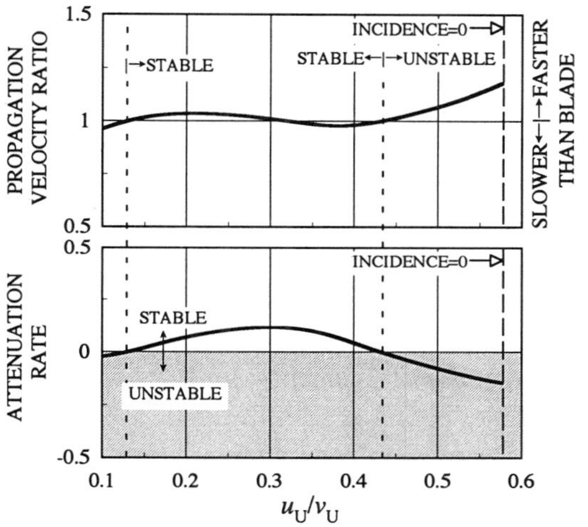

Figure 8. Propagation velocity ratio and attenuation rate of solution "B" for $v_U/a_{tUabs} = 1.3$

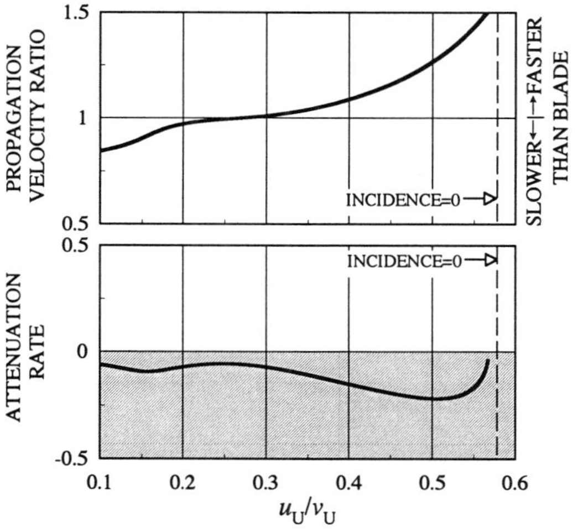

Figure 9. Propagation velocity ratio and attenuation rate of solution "B" for $v_U/a_{tUabs} = 1.4$

When the inflow is supersonic and incidence equals to zero, i.e. $u_U/v_U=\cot 60°$, the shock wave in the steady flow becomes normal. Even under this condition, there exists solution "B". Therefore, we can say that solution "B" is associated with the shock wave.

Figure 8 shows the propagation velocity ratio and the attenuation rate for $v_U/a_{tUabs}=1.3$. Figure 9 shows those for $v_U/a_{tUabs}=1.4$. The horizontal axis is same as in Figures 5 and 6.

These figures indicate that the propagation velocity ratio of solution "B" is mainly larger than unity where solution "B" is unstable, except in the low flow rate region.

5. Conclusions

From the present analysis, the following conclusions are obtained.

(1) The compressible flow also has the solution of the rotating stall corresponding to those found in the incompressible flow analysis. The slope of the performance curve is positive when this rotating stall occurs, and its propagation velocity ratio is smaller than unity.

(2) There exists the solution which indicates the flow instability associated with the shock wave. This instability occurs at higher speeds. The slope of the performance curve is positive in the onset region and its propagation velocity ratio is larger than unity except in the low flow rate region.

Reference

[1] Freeman, C. and Cumpsty, N. A., (1989), , ASME Paper No. 89-GT-326.

10: Aeroelastic Coupling - 2

SEMI-UNSTRUCTURED MESH GENERATOR FOR FLOW CALCULATIONS IN AXIAL TURBOMACHINERY BLADING

L. SBARDELLA, A. I. SAYMA AND M. IMREGUN
Imperial College of Science, Technology and Medicine
Mechanical Engineering Department
Exhibition Road, London SW7 2BX, UK

Abstract

This paper describes the development and application of a novel approach for discretizing turbomachinery blades by using a combination of structured and unstructured meshes, the former in the radial direction and the latter in the axial and tangential directions. The basic idea relies on the fact that blade-like structures are not strongly three dimensional since the radial variation is usually small. Compared with standard structured meshes for turbomachines, the proposed semi-unstructured grid demonstrates a significant improvement both in the smoothness of the grid-spacing and in resolving particular aspects of the passage geometry. Leading- and trailing-edge regions are discretized without generating superfluous points, while wake and shocks resolution can be obtained using a mesh refinement technique. The capability of the method is demonstrated using a semi-unstructured mesh for computing a steady-state viscous flow in a high pressure nozzle-guide-vane and in a transonic fan.

1. Introduction

It is well known that the grid structure must be selected carefully in order to achieve an accurate resolution of complex flow fields typical of axial-flow turbomachines. The minimization of skewness and the optimization of smoothness generally result in a faster convergence as well as less solution dependence on the grid density, therefore reducing computational cost both in terms of memory and CPU time. As a consequence, the grid generation procedure should be considered an integral part of the numerical method.

T.H. Fransson (ed.), Unsteady Aerodynamics and Aeroelasticity of Turbomachines, 541-554.

When performing a numerical simulation of turbulent-viscous flow in turbomachinery passage, the following aspects are of importance: 1) accurate leading- and trailing-edge flow descriptions, 2) wake resolution, 3) proper gridding in the throat area where most of the shock is expected to occur, and 4) imposition of periodicity.

Historically, mesh generation techniques for turbomachinery blades use structured hexahedral representations, the most commonly used ones being H-type, C-type, and O-type. These meshes are obtained either by using an algebraic approach or by solving a system of elliptic partial differential equations [1, 2, 3]. H-type meshes have been by far the most common choice in turbomachinery applications as they are very easy to generate, the imposition of periodicity is straightforward and the mesh density before, inside, and after the blade passage can be easily controlled. However, the leading- and trailing-edge descriptions are poor and a large amount of superfluous points is generated in the region between the inflow and the leading-edge. O-type grids are not very effective in capturing the wake and their quality outside the passage is very poor. This can smear the bow shock away from the leading edge of a transonic compressor or the outgoing shock of a transonic turbine blade. On the other hand, C-type meshes can capture the wake structure if they are carefully generated but their quality in the region between the inflow and the leading-edge is not suitable to resolve bow shock accurately.

A different approach is the use of unstructured triangular meshes for 2D turbomachinery calculations. Unstructured grids offer good flexibility and most of the flow features can be captured with good accuracy via mesh refinement. The use of unstructured meshes in 3D requires a large number of radial direction grid points so that the leading and trailing edges can be resolved. Moreover, they are best suited to inviscid flow computations.

The aim of this paper is to present a novel approach for discretizing turbomachinery blades by using a combination of structured and unstructured meshes, the former in the radial direction and the latter in the axial and tangential directions. The basic idea relies on the fact that blade-like structures are not strongly three dimensional since the radial variation is usually small. It is therefore possible to start with a structured and body-fitted two-dimensional O-grid around a given aerofoil section to resolve the boundary layer. This core mesh is then extended in an unstructured fashion up to the far-field boundaries, the triangulation being performed using an advancing front technique [4, 5]. Once this two-dimensional grid is generated, it is projected to the remaining radial sections via quasi-conformal mapping techniques. When all such radial sections are formed, a three-dimensional prismatic grid is obtained by simply connecting the corresponding points of different layers. In this way, hexahedral elements are generated in the viscous region and triangular prisms in the rest of the solution domain. In the work reported, these elements are split into tetrahedra for compatibility with the unstructured solver [6] used throughout this work.

2. Generation of Prismatic Grid

Since this three-dimensional mesh-generation procedure involves several distinct stages, like geometry modeling, two-dimensional hybrid mesh generation, mapping, etc., this section will describe the methodology first without undue details. The next section will focus on the proposed quasi-conformal mapping procedure which links the two-dimensional radial section meshes together. The method is based on five main stages:

- Mapping procedure to project all radial levels of the blade into two-dimensional planes, using local coordinates.
- Generation of a two-dimensional hybrid-mesh for a given radial section.
- Generation of a coarse body-fitted structured mesh for all radial sections.
- Inverse mapping of the unstructured mesh into a structured one at the same radial level.
- Direct mapping to obtain the final unstructured mesh at all radial levels.

Geometry modeling. Turbomachinery blade geometry is, usually, defined at a number of radial sections. Figure 1 shows four such sections for a typical nozzle guide vane (NGV). In the general case, these radial sections will lie on three-dimensional surfaces $\sigma_i(x, r, \theta)$, where i indicates the section index. Using parametric coordinates u and v, a typical surface can be defined as:

$$\begin{cases} x_\sigma &= x(u,v) \\ r_\sigma &= r(u,v) \\ \theta_\sigma &= \theta(u,v) \end{cases} \tag{1}$$

The starting point of the present method is the projection of the radial sections into parametric two-dimensional planes, using the local coordinate system u and v. In this way the mesh-generation procedure will deal with plane section only, thus the geometric dimension is reduced from three to two.

Unstructured mesh generation. Once all radial sections are mapped into two dimensional planes, a hybrid quadrilateral-triangular mesh is generated in a 2D plane which corresponds to a certain radial level (usually the middle one). The quadrilateral part of the mesh takes the form of a body-fitted

O-grid which is generated using a system of elliptic partial-differential equations. The orthogonality of this structured mesh is very important for an accurate resolution of the turbulent-boundary layer which originates from high Reynolds number flows.

The remaining part of the domain is discretized using an unstructured mesh generator which uses an advancing front algorithm [4, 5]. A distinctive feature of this method is that triangles and points are generated simultaneously. Such an approach enables the generation of elements with variable sizes and stretching and hence differs from Delaunay-type mesh generators [5, 7].

Figure 2 shows a two-dimensional viscous mesh for the middle section of the NGV shown in figure 1. Figures 3a and 3b show a zoom view of the mesh in the leading- and trailing-edge regions. The orthogonality and the smoothness of the mesh in these two regions are satisfactory for predicting the complex flow features associated with these regions. In particular, an accurate resolution of the trailing edge is essential in order to predict the wake angle and intensity correctly.

Quasi-conformal mapping. An important part of the mesh-generation strategy is the mapping procedure to project the unstructured mesh of figure 2 to all radial blade-definition levels. A necessary condition for a mapping function is that it must associate a given point of the first plane with one, and only one, point of the second plane. Moreover a mapping function should also guarantee that a given angle in one plane is mapped into a similar-valued angle in the target plane (quasi-conformal mapping). This last property is essential in order to minimize the skewness of the mesh, especially for highly-twisted fans. As will be illustrated in section 3, the quasi-conformal mapping procedure will make use of structured meshes.

Once the unstructured mesh has been mapped to all radial blade surfaces, a prismatic mesh is obtained by simply connecting the corresponding points at consecutive levels. Moreover, in order to enhance the quality of the three-dimensional mesh, a smoothing procedure is performed. This operation alters the positions of the interior nodes without changing the topology of the mesh. The element sides are considered as springs of stiffness proportional to the length of the side. The nodes are moved until the spring system is in equilibrium, the position of which is found by Jacobi iterations.

3. Quasi-Conformal Mapping

The starting point for the quasi-conformal mapping is the generation of coarse structured quadrilateral grids for all radial sections. Following the approach of Steger et al. [2], these meshes are obtained by solving a system of elliptic partial-differential equations. A standard O-type mesh, which is suitable for this purpose, is shown in figure 4. An essential requirement for such structured meshes is that they must be generated in exactly the same manner, i.e. with the same number of points and quadrilaterals. The mapping procedure is implemented as follows:

- **Geometry searching.** Each point J of the unstructured 2D mesh must be located on quadrilateral E of the structured mesh.
- **Inverse mapping.** The Cartesian coordinates $\vec{x}_{\rm J} = (x_{\rm J}, y_{\rm J})$ of point J, associated with the quadrilateral E are given by

$$\vec{x}_{\rm J} = \sum_{I=1}^{4} \vec{x}_{\rm I} N_{\rm I} \left(\xi_{\rm J}, \eta_{\rm J}\right) \tag{2}$$

  where $(\xi_{\rm J}, \eta_{\rm J})$ represent the local-coordinates, $\vec{x}_{\rm I}$ represents the Cartesian coordinates of nodes $I = 1, \ldots, 4$ of the quadrilateral E and $N_{\rm I}$ is the standard finite-element bilinear shape-functions which take the form [8]

$$N_{\rm I} = \frac{1}{4} \left\{ \begin{array}{l} (1-\xi)\,(1-\eta) \\ (1+\xi)\,(1-\eta) \\ (1+\xi)\,(1+\eta) \\ (1-\xi)\,(1+\eta) \end{array} \right. \tag{3}$$

  A Newton-Raphson method is used in order to obtain the values of $\xi_{\rm J}$ and $\eta_{\rm J}$.
- **Direct mapping.** Once all points J of the unstructured mesh are associated with quadrilateral E and the local-coordinates $(\xi_{\rm J}, \eta_{\rm J})$ are determined, the coordinates $\vec{x}_{\rm J}$ of the points on the remaining radial sections are obtained directly using equation (2).

The above-described steps are shown in figure 5. The mapping method becomes fully conformal if a given element of the 2D hybrid mesh lies within a single quadrilateral of the structured grid. In addition the angles of this quadrilateral must remain the same for all radial sections [8]. If the above two conditions are not satisfied, the conformal property is not guaranteed and for this reason the procedure has been labelled quasi-conformal here. Finally, it is worth noting that the algorithm is not CPU-intensive, since the geometry searching and the inverse-mapping are performed once only.

4. Applications

The proposed semi-unstructured mesh generator has been applied to two different turbomachinery blades, namely the NGV of figure 1 and a transonic fan. The prismatic elements of these meshes have been split in to tetrahedra in order to compute a solution with the solver presented in [6].

Figures 6a and 6b show two different views of a three-dimensional viscous mesh for the NGV of figure 1. Figure 7 shows the same mesh at the hub and tip sections as well as the steady-state iso-Mach contours produced by the solver. The wake behind the trailing edge is smeared because of the mesh coarsering in that region. Figure 8a shows a zoom view, at the hub section, of a semi-unstructured mesh generated by the proposed methodology in the case of a transonic fan. Figure 8b shows the iso-pressure contours in the suction surface of the blade at 66% of the design speed. In this configuration the inlet flow becomes supersonic towards the tip section generating a shock wave very close to the leading-edge. Figures 8c, 8d and 9 compare the mesh and the solution at tip and hub sections. It is worth to note that even for such heavily-twisted fan, the quality of the grid both in the tip and hub section is satisfactory.

Semi-unstructured meshes generated by the present method have been successfully used for stator-rotor interaction as well as flutter analysis of turbine and fan blades as shown in reference [9].

5. Conclusion and Further Work

A method to generate semi-unstructured prismatic meshes for turbomachinery blades has been presented. The unstructured mesh in the axial and tangential directions offers more flexibility than standard structured O-type and C-type meshes, both in terms of skewness minimization and smoothness optimization.

At the moment prismatic grids are split into tetrahedra because the solver can only deal with such elements. This represents a serious limitation because of the inefficiency of tetrahedral grids for computations of the high-Reynolds number viscous flows. One of the reasons of this inefficiency is explained by counting the number of flux computations required by edge-based schemes running on tetrahedral grids. An hexahedral mesh of N points contains 3N sides if the boundary effects are neglected. If the same mesh is transformed by dividing each hexahedron into 6 tetrahedra, the number of unknown remains the same but the number of sides becomes 7N. If the total computational time is proportional to the number of flux computations and the flux is computed at each side, then a numerical scheme on the tetrahedral mesh is more than two times expensive in CPU times to that on the corresponding structured grid for the same number of points.

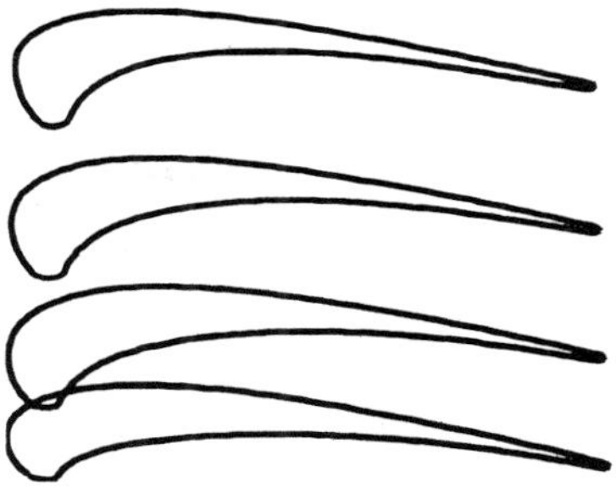

Figure 1. Blade profiles at different radial levels

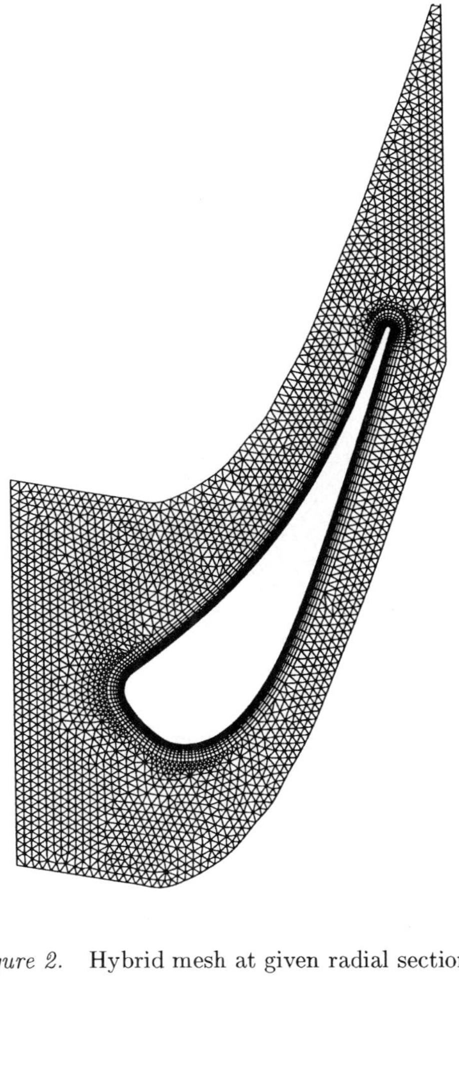

Figure 2. Hybrid mesh at given radial section

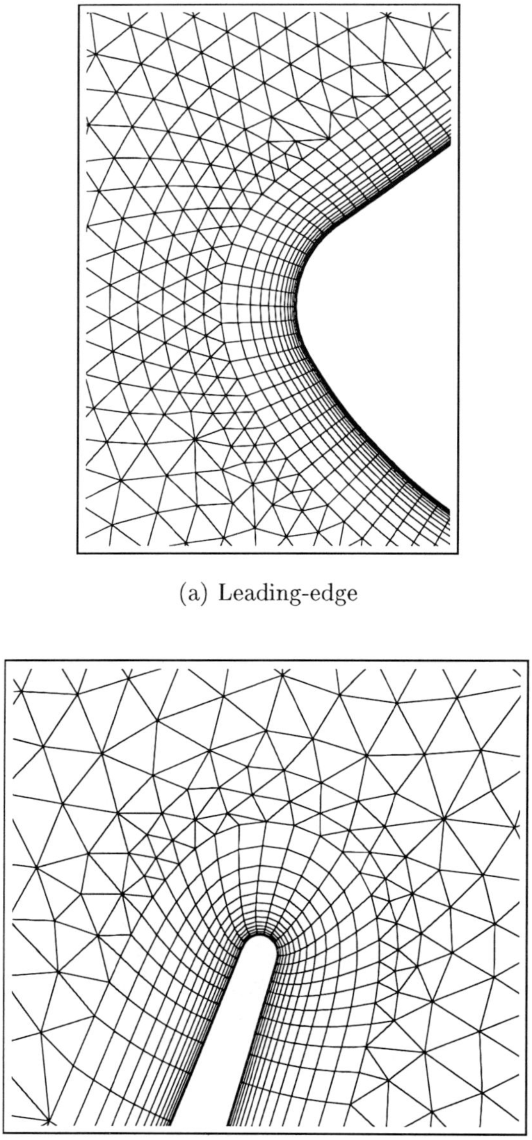

(a) Leading-edge

(b) Trailing-edge

Figure 3. Leading- and trailing-edge regions of the mesh

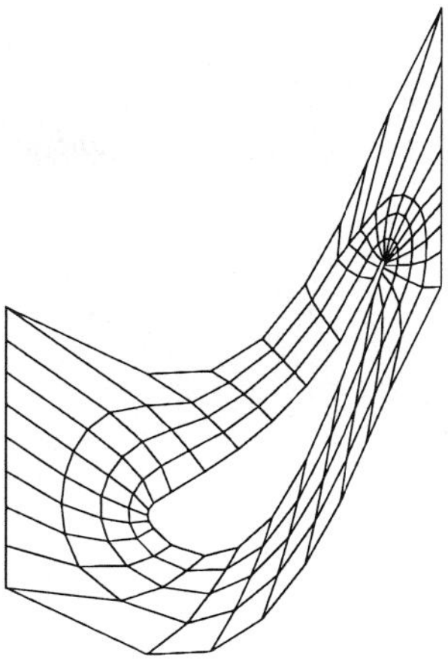

Figure 4. Structured mesh for mapping

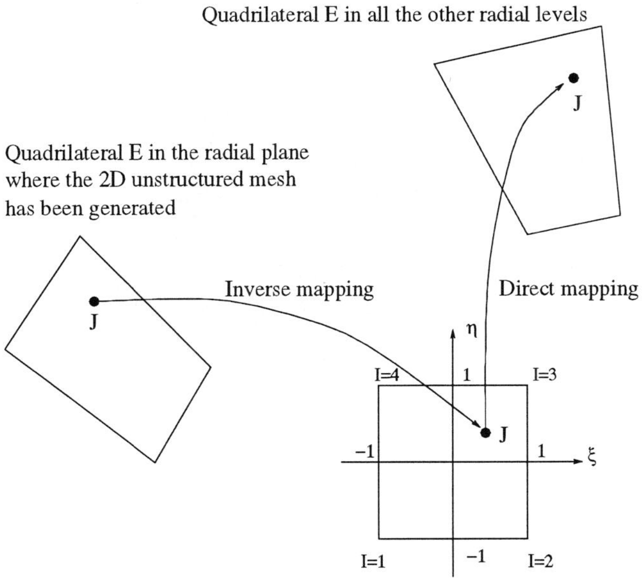

Figure 5. Mapping procedure

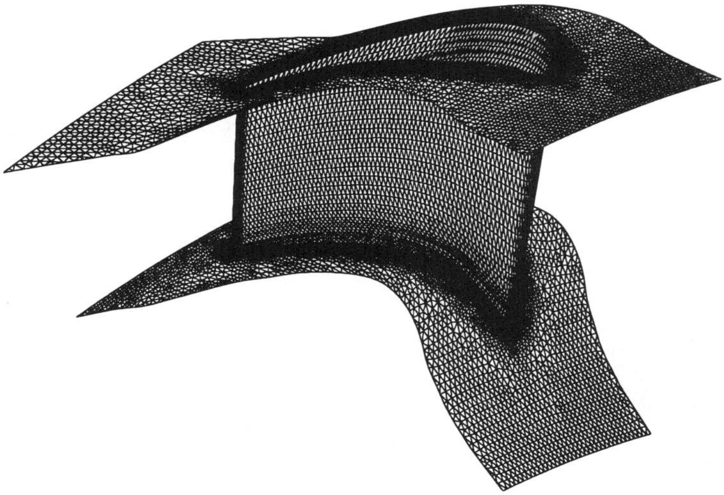

(a) Pressure surface view

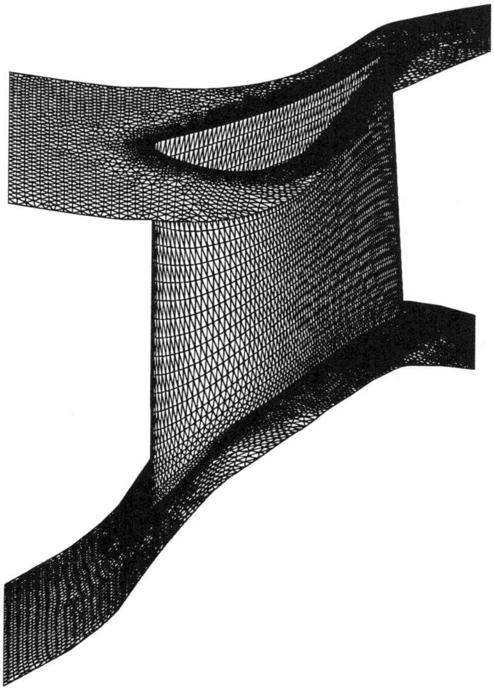

(b) Suction surface view

Figure 6. 3D mesh for high-pressure NGV

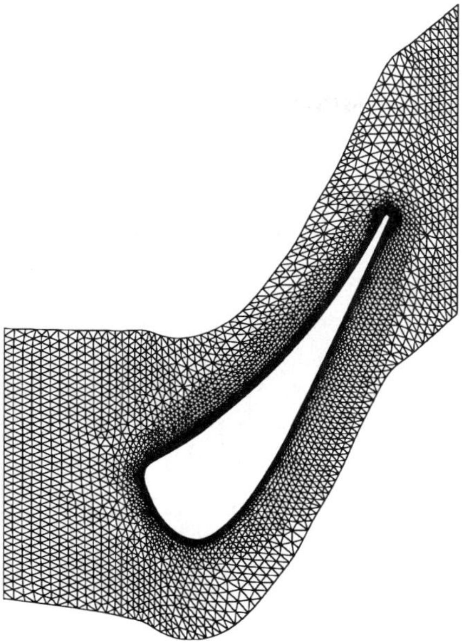

(a) Hub section: mesh

(b) Tip section: mesh

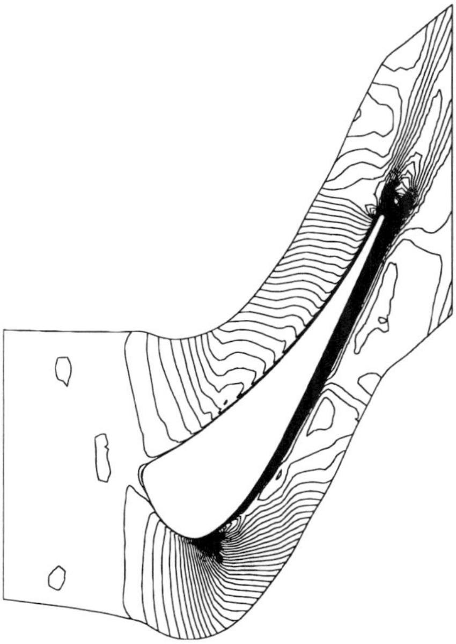

(c) Hub section: Mach iso-contours

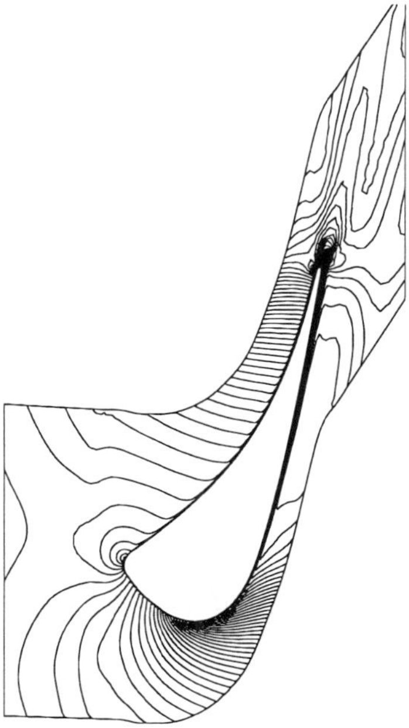

(d) Tip section: mach iso-contours

Figure 7. NGV: mesh and iso-Mach contours of two plane section

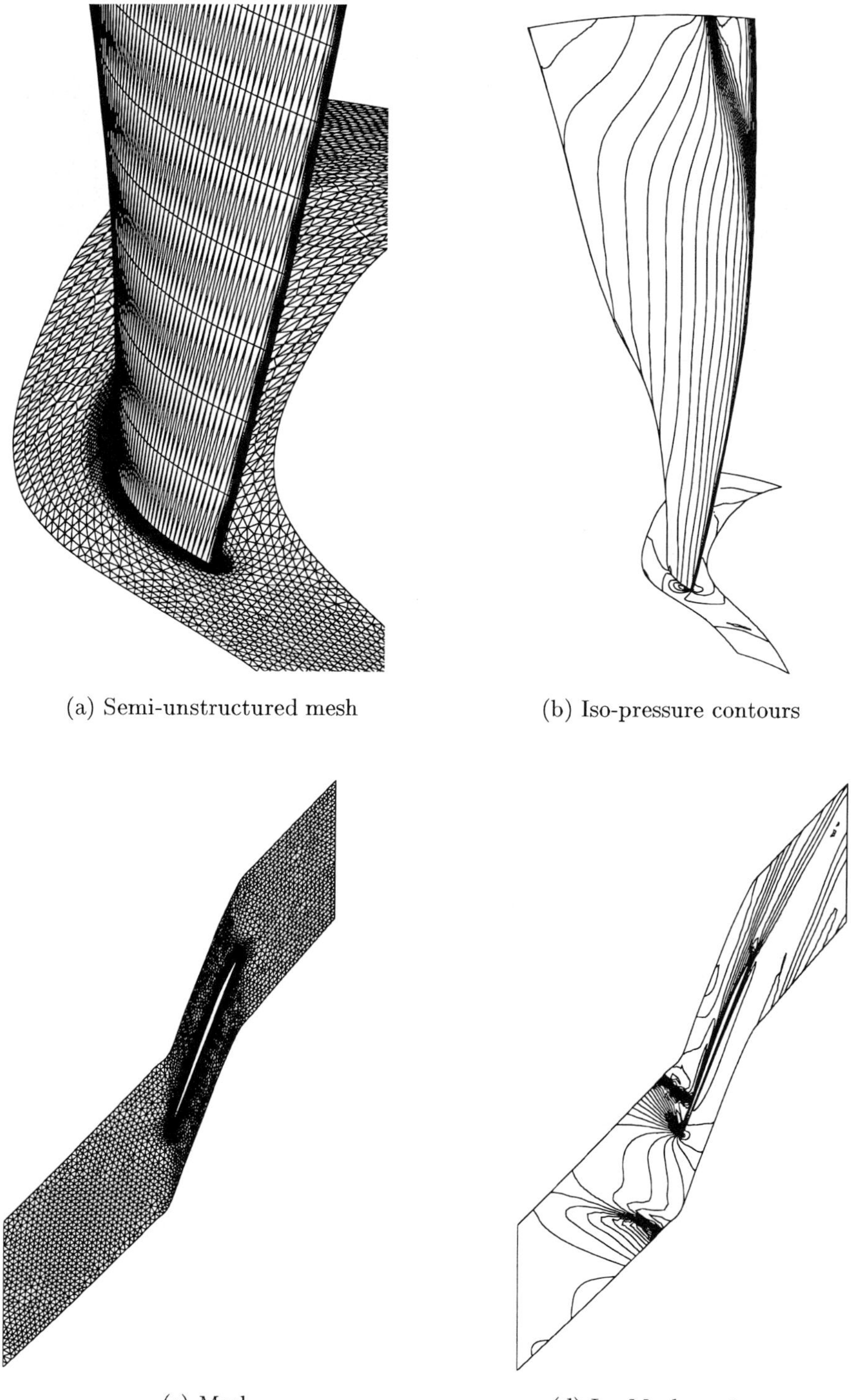

(a) Semi-unstructured mesh (b) Iso-pressure contours

(c) Mesh (d) Iso-Mach contours

Figure 8. Transonic fan: mesh and solution on the blade and at tip section

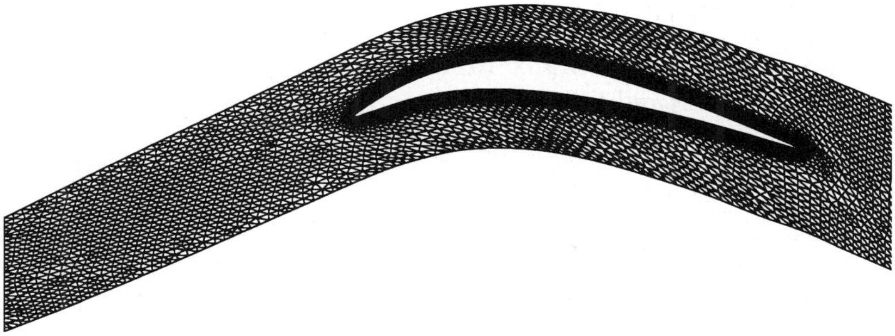

(a) Mesh

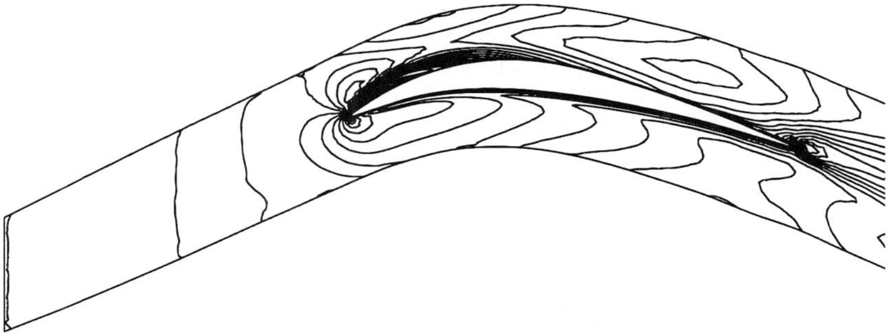

(b) Mach iso-contours

Figure 9. Transonic fan: mesh and solution at hub section

Acknowledgments

The authors would like to thank Rolls-Royce plc for both sponsoring the work and allowing its publication.

References

1. Thompson, J. F., Thames, F. C., and Mastin, C. W. (1974) 'Automatic Numerical Generation of Body-Fitted Curvilinear Coordinates System for Field Containing any Number of Arbitrary Two-Dimensional Bodies', *Journal of Computational Physics*, **15**, 299-319.
2. Steger, J.L. and Sorenson, R.L. (1979) 'Automatic Mesh-Point Clustering Near a Boundary in Grid Generation with Elliptic Partial Differential Equations', *Journal of Computational Physics*, **33**, 405-410.
3. Brackbill, J. U., and Saltzman, J. S. (1982) 'Adaptative Zoning for Singular Problems in two Dimensions', *Journal of Computational Physics*, **46**, 342-368.
4. Morgan, K., Peraire, J. and Peiró, J. (1992) 'Unstructured Grid Methods for Compressible Flows', In AGARD Report 787, Special Course on Unstructured Grid Methods for Advection Dominated Flows, 5.1 - 5.39.
5. Mavriplis, D. J. (1995) 'Unstructured Mesh Generation and Adaptivity', ICASE Report No. 95-26.
6. Vahdati, M. and Imregun, M. (1995) 'Non-linear aeroelasticity analysis using unstructured dynamics meshes', In Unsteady Aerodynamics and Aeroelasticity of Turbomachines, Tandia, T. and Namba, M. (editors), 177 - 196.
7. Baker, T. J. (1990) 'Unstructured Mesh Generation by a Generalized Delaunay Algorithm', In Applications of Mesh Generation to Complex 3-D Configurations, AGARD Conference Proceedings No. 464, 20.1 - 20.10.
8. Zienkiewicz, O.C. and Morgan, K. (1983) 'Finite Elements and Approximation', John & Sons Ltd.
9. Sayma, A. I., Vahdati, M., Green J. and Imregun, M. (1997) 'Whole-assembly Flutter analysis of a Low Pressure Turbine Blade', In Proceeding of 8th International Symposium on Unsteady Aerodynamics and Aeroelasticity of Turbomachines, Stockholm.
10. Kallinderis, Y. and Ward, S. (1992) 'Prismatic Grid Generation with an Efficient Algebraic Method for Aircraft Configurations', AIAA Paper 92-2721.

AEROELASTIC TAILORING FOR ROTOR BLADES OF TRANSONIC SWEPT FANS

ZHOU SHENG
Chinese National Aircraft Engine Lab.
Beijing University of Aeronautics and
Aeronautics, Beijing, China 100083

ABSTRACT

For further raising fan stage's Pressure ratio, there are two main restricting factors. First, increasing rotor's tip wheel speed will strengthen the shock waves in rotor, and it is possible to decrease the stage efficiency. Secondly, as the result of increasing rotor's tip wheel speed, strengthening the shock waves in rotor will induce the onset of stall flutter. For the turbofans with small bypass ratio, the phenomenon of stall flutter is the main type of aeroelastic instabilities of transonic fans.

Both the above-mentioned two restricting factors for further raising the stage pressure ratio connect closely with 3D shock wave's structure and the large-scale separating zone in the rotor, which is induced by 3D interaction of shock wave with boundary layer mainly. However, the 3D shock wave's structure in rotor connects with the leading-edge spatial curve of the rotor blade closely. So, it becomes an important frontier problem among the turbomachinery aerodynamics —— how to design the blade's leading-edge spatial curve carefully. During the designing process of some swept fans, a kind of composite aerodynamic/aeroelastic tailoring program has been used. The aeroelastic tailoring, one of main parts of the composite tailoring has been presented, and an example has been shown in this paper too.

* Supported by National Natural Science Foundation of China.

1. Introduction

It has important meaning for raising the thrust/weight ratio of turbofan to increase fan stage pressure ratio.

For further raising the fan stage's pressure ratio, there are two main restricting factors. First, increasing rotor's tip wheel speed will strengthen the shock wave's in rotor, and it is possible to

T.H. Fransson (ed.), Unsteady Aerodynamics and Aeroelasticity of Turbomachines, 555-564.

decrease the stage efficiency. Secondly, as the result of increasing rotor's tip wheel speed, strengthen the shock waves in rotor could in duce the onset of stall flutter. For the turbofans with small bypass ratio. the phenomenon of stall flutter is the main type of aeroelastic instabilities of transonic fans[1].

Both the above-mentioned two restricting factors for raising the fan stage's pressure ratio connect closely with 3D shock wave's structure and the large-scale separating zone, which is induced by 3D interaction of shock wave with boundary layer mainly. For the onset of stall flutter of rotor, the existence and intensity of shock waves in rotor has clear effect to the unsteady pressure distribution on blade surfaces. Therefore, the connection between shock wave and aeroclastic stability is definite[2]. Besides, the interaction of shock wave with boundary layer could induce the unsteady large-scale separating zone on the suction surfaces of oscillating blades. The response of flow field to the small amplitude oscillations of blades inside the separating zones is different to the response beyond the separating zones. This feature will influence the onset of stall flutter[3,4].

In the face of shock wave's structure in rotor, there are two kinds of design philosophies.The first one is to seek for accomplishing the shock free flow in rotor[5,6]. Another kind is quite different from the first one. It means that the shock waves in rotor play important beneficial role in doing work to the flow by blades, then, it isn't an optimum selection seeking the shock free flow in rotor. Of course, in the meantime of playing the beneficial role, the opening separating flow zone, which is induced by the interaction of shock wave with boundary layer, will bring evident unfavorable effect to the rotor's efficiency and aeroelastic stability. So, from the viewpoint of synthesizing both beneficial and unfavorable effects to the rotors, Some engineering criteria for controlling the intensity of shock waves in the rotor must be used during design process[7,8].Perhaps the second kind of design philosophies is more reality.

In fact the incoming relative Mach number at front of shock waves isn't the unique parameter influencing the intensity of shock waves in rotor. The project of incoming relative velocity at front of shock wave to the normal direction of shock wave's surface will determine the shock wave's intensity immediately. According to the mathematic model suggested by Ref.[7,8,9],the 3D shape of shock wave surface connects with the spatial curve of rotor's leading edge. However, the shape of spatial curve of rotor's leading edge hasn't been considered in detail during the classical design process.

After a lot of research works, the importance of the spatial curve of blade's leading edge to performance has been understood. At present, it has become one of the fronts of turbomachinery aerodynamics —— how to design the shape of rotor's L.E. spatial curve for controlling the 3D shock wave's structure [10].

Concrete speaking, after picking out a model describing the 3D shock wave's structure[7,8,9], We could give a mathematical array, which is a group of main geometric parameters, to represent the 3D blade's shape. With the specified aerodynamic parameters at inlet and exit boundaries, the mathematical array of main geometric parameters will correspond a definite 3D shock wave's structure inside the rotor uniquely. Then, during the design process, we could control the shock wave's intensity and improve the aerodynamic performance by means of adjusting the mathematical array. We named the adjusting process as the aerodynamic tailoring. Of course, the fan's aeroelastic stability must be guaranteed. Then, the corresponding tailoring process could be named as the composite aerodynamic and aeroelastic tailoring. The aeroelastic tailoring, one of the two main parts of the composite tailoring, has been presented in this paper.

2. Influence of Swept Aerodynamic Design upon Rotor's Structural Dynamics

As above-mentioned, the essence of fan's swept aerodynamics means how to control the shock wave's structure in rotor mainly. According to the model suggested by Ref[7-10].it means using a mathematical array of key geometric parameters to represent the complicated 3D shape of rotor's blade. Comparing with the rotor's swept aerodynamic design, there are no enough immediate connections of rotor blade's geometric shape with the shock wave's structure for classical transonic fan's aerodynamic design procedures. Precisely because of this, it becomes one of the main developing direction for modern fans how to connect the shock wave's structure with the blade's shape?

From the viewpoint of structure dynamics, the recent swept aerodynamic design procedure bringes some problems. The first one is concerning the static stress near blade's root. In order to control the intensity of 3D shock waves, the shape of blade L.E. spatial curve must be designed carefully. Then, we must increase design freedoms for changing the shape of blade L.E. spatial curve during the tailoring process. In the classical procedures of stacking blade elements, the curve of stacking gravity center of every blade element along span usually are the radial line basically. In order to match the stacking with swept aerodynamic design, the curve of stacking blade elements along span must be changed as an arbitrary spatial curve. It means that the curve of stacking the gravity centers of blade element along span could be backward or forward swept in project plane r-z (Fig.1(a)), and backward or forward bend in project plane r- θ (Fig.1(b)). It could increase the design freedoms, using the arbitrary spatial curve as stacking curve. Then, we could control the shock wave's intensity and improve the rotor's aerodynamic performance. But the improvement of aerodynamic design could weaken the static stress situation of rotor blade.

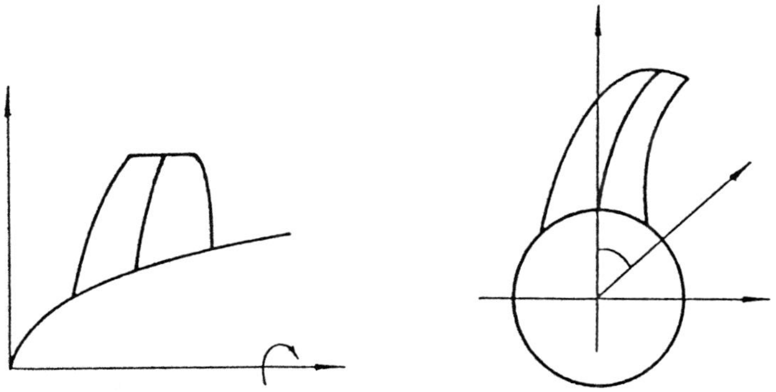

(a) r-z plane (b) r- θ plane

Fig.1 Swept and bend blade of rotor

The second problem is concerning the aeroelastic stability. The recent swept aerodynamic design and corresponding stacking procedure could benefit the aeroelastic stability, because the shock wave's intensity will be controlled by the blade's L.E. spatial curve. However, the analyze of stall flutter connects with both structure dynamics and unsteady aerodynamics. The swept aerodynamic design procedure bringes more complicated 3D shape of blades, and increases the complex of analyzing the aeroelastic stability from the side of structure dynamics. In brief, the onset of stall flutter has brought on so much troubles for classical transonic fans, which stage pressratio are about 1.60 level. For the modern fan stages, the typical stage pressure ratio has increased to more than 2.0, and the rotor's tip wheel speed is about more than 500m/sec. Therefore, the dangers of stall flutter couldn't be neglected.

3. Aeroelastic Tailoring of Rotor Blades

For modern transonic swept fans, the main object of composite aerodynamic & aeroelastic tailoring of rotor blades is to accomplish higher stage pressure ratio with higher stagy efficiency, mean while, there is main constrained condition —— there is no onset of stall flutter. In order to accomplish the composite tailoring smoothly, following procedure has been used. A mathematical array of main geometric parameters, which represents a design project, is picked out. According to the model of 3D shock waves in rotor, we could calculate the intensity and loss of 3D shock waves corresponding to the selected project. While the design project is changed, the shock wave's surface is changed too. So, the through-flow aerodynamic design and stacking procedure could combine with the 3D shock wave's intensity immediately. It means that the jump of doing work value and the jump of total pressure ratio across the 3D shock wave surface could be calculated for every selected project. Besides, the process of repeated tailoring will change the aeroelastic stability, of course, the constraint for acroelastic stability must be satisfied. Then ,the prediction of rotor's stall flutter must be included in the composite tailoring process.

Most of engineering predicting procedures for the onset of rotor's stall flutter belong to empirical methods[11]. However, the experimental data for modern swept fans with stage pressure ratio more than 2.0 are lacked for us yet. Using CFD for prediction of stall flutter is the future direction. But we can't to predict the boundary of stall flutter on the stage performance map satisfactorily at present, although we try to do that for many years.

Therefore, the 3D deforming actuator-disk model[12]-[16]is picked up for predicting the onset of stall flutter. It is well known that the deforming actuator disk models belong to a kind of semi-empirical methods. It is convenient for aeroelastic tailoring, because its CPU time is so short for every design project, then we can know the situation of aeroelastic stability for every selected project rapidly. We use the deforming actuator-disk models for predicting the onset of stall flutter since 1980. However, the former engineering backgrounds just corresponded classical transonic fan stages, which pressure ratio near 1.6 ~ 1.7 and the rotor tip wheel speed near 400m/sec. At present, we face the modern swept fans, which stage pressure ratio more than 2.0, and the tip wheel speed about 500m/sec. Therefore, the deforming actuator-disk model isn't the same as that in Ref. [12],there are a lot of mends from the original model in Ref. [12].

Since the branch "Aeroelasticity in Turbomachinery" was founded, some basic assumptions have been used, one of them is that the inter-blade phase angle distribution along the blade row is constant. However, that basic assumption has been proved to be invalid by the fact that neither dynamic stress nor inter-blade phase angles along the blade row is constant, when the stall flutter occurs. With that assumption abandoned, a new model and corresponding code have been developed [13.14]. Therefore, we could calculate the number of blades, which have negative aerodynamic damping, N_f, $N_f < N$, where N represents the number of blades in the blade row. So, the indices for assessing the aeroelastic stability could be two parameters, one is the minimum negative aerodynamic damping value, and the another is N_f, the number of blades having negative aerodynamic damping. Especially for the process of composite aerodynamic and aeroelastic tailoring, N_f is a very useful index, because the values of minimum negative aerodynamic damping are near zero usually.

Secondly, the new deforming actuator-disk model is a 3D one [15. 16]. As above-mentioned, the essence of aerodynamic & aeroelastic tailoring means controlling the shape and intensity of 3D shock wave in rotor by means of designing the shape of rotor L.E. spatial curve. For the process of aeroelastic tailoring, it means the variation of 3D shock wave's intensity with the shape of rotor L.E. spatial curve mainly. Besides, the derivation angles along span, which reflects the flow separation, has been included in the new model too. The model of 3D shock wave's structure in rotor has been shown in Fig.2[8]. The variation of minimum aerodynamic damping along a blade row with the shock wave

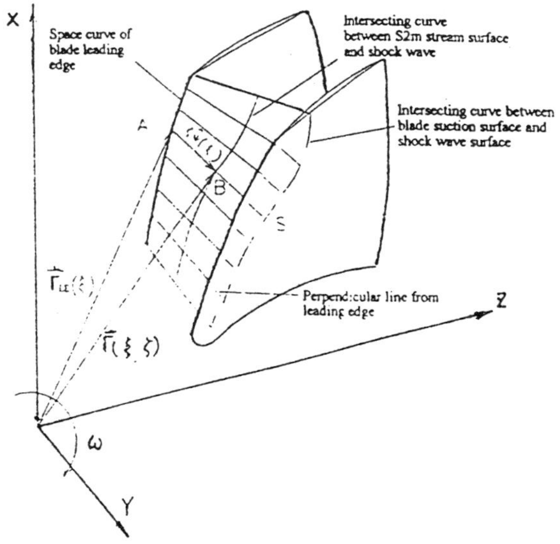

Fig.2 The model of 3Dshock wave's structure in rotor

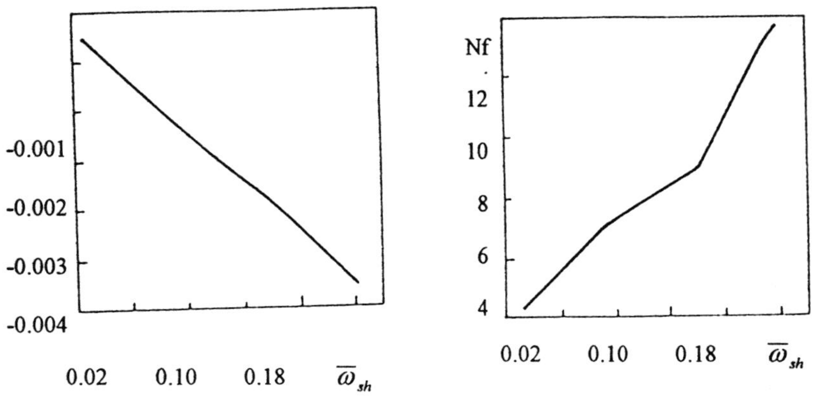

(a) $\delta_{min}=f(\overline{\omega}_{sh})$ **(b) $N_f=f(\overline{\omega}_{sh})$**

Fig. 3 Varation of δ_{min} and N_f with shock wave loss coefficient $\overline{\omega}_{sh}$

loss coefficient at a characteristic blade element has been shown at Fing.3(a). The variation of the number of blades, which have negative aerodynamic damping, with the shock wave loss coefficient at the same characteristic blade element has been shown at Fig.3(b). Above calculation corresponds the HTFC rotor, which number of rotor blades is 20[17].

4. The Process of Aeroelastic Tailoring for a Single stage Fan Rotor

There is a reduced-scale model of transonic fan stage in our Lab. The main parameters are as follows: tip wheel speed 495m/sec: outer diameter 440mm; inlet hub/tip ratio 0.4; number of rotor blades 14; number of stator vanes 34; air flow rate 25 kg/sec. The main design philosophy is considering the favorable role of shock waves to rotor's pressure radio.

In order to control the harmful effect of large-scale separation induced by the shock waves mainly, an engineering criteria, which takes the shock wave's intensity into account, has been used[7,8].Finally, a transonic stage has been obtained, which rotor has small aspect ratio and suitable swept and bending degrees. The Mach number contour patterns, obtained from a 3D N-S code, have been shown at Fig. 4(a) and (b).

When we form the blade's shape, the arbitrary spacial curve will be selected as the curve for staking gravity centers of blade elements. In order to control the 3D shock wave's structure in rotor, the shape of blade's L.E. spatial curve has keen designed carefully.

During the process of accomplishing the aeroelastic tailoring, We use two indices for reflecting the onset situation of flutter: N_f the number of blades having negative aerodynamic damping, and δ_{min} —— minimum negative aerodynamic damping's value among the rotor blades. The predicted variations of N_f and δ_{min} with reduced velocity and incidence for this model rotor have been shown at Table 1 and Table 2.

The aerodynamic damping's value of every blade around the model rotor have been presented at Fig.5, which correspondes the working point at the performance map of the model rotor. From the Fig.5,we could say that there is no onset of stall flutter in that working point.

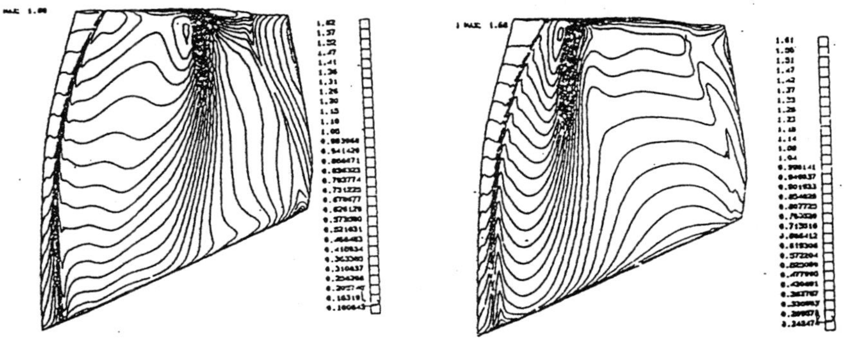

(a) near suction surface (b) near pressure surface

Fig.4 Mach number contour

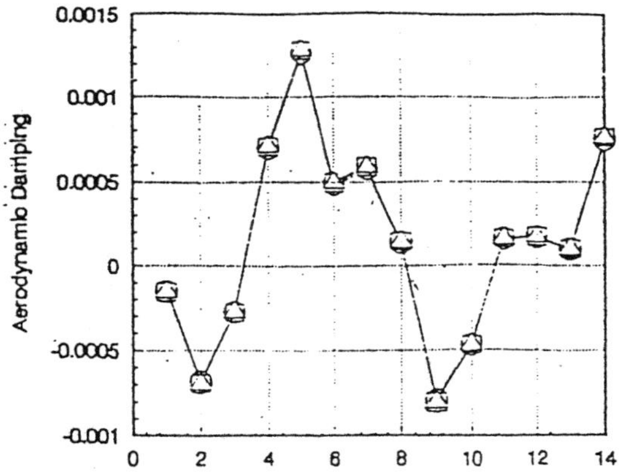

Fig.5 Aerodynamic damping of every blade around rotor

5. Concluding Remarks

1) Many works have shown that the 3D shock wave's structure in rotor of a fan stage will play important roles to the pressure ratio and doing work to the air. Meanwhile, the opened large-scale

Table 1. N_f with reduced velocity and incidence

$\bar{\upsilon}$ \ i	2.6 °	3.0 °	3.5 °	4.5 °	5.0 °	6.1 °	7.4 °
1.65	9	9	9	9	9	9	9
2.15	10	10	10	8	8	8	8
2.30	0	2	2	5	5	6	9
2.48	0	0	0	3	5	9	11
2.65	0	2	4	6	8	10	11
2.82	10	10	10	8	8	8	8
3.15	5	5	5	5	5	7	7

Table 2. δ_{min} with reduced velocity and incidence

$\bar{\upsilon}$ \ i	2.6 °	3.0 °	3.5 °	4.5 °	5.0 °	6.1 °	7.4 °
1.65	-9.10	-9.33	-9.76	-10.60	-11.10	-12.02	-13.21
2.15	-5.20	-6.13	-7.50	-10.43	-11.36	-13.32	-16.09
2.30	>0	-0.465	-3.24	-7.06	-9.37	-12.83	-17.71
2.48	>0	>0	>0	-4.24	-3.53	-17.32	-26.30
2.65	>0	-2.03	-2.17	-23.71	-25.30	-39.40	-76.51
2.82	-19.31	-21.62	-24.22	-29.13	-36.11	-45.52	-76.23
3.15	-10.71	-11.92	-18.54	-12.91	-18.92	-24.81	-48.62

* $\delta_{min} \times 10^{-4}$ *in table*

separating zone, which is induced by the interaction between shock wave and boundary layer, will effect efficiency and aeroelastic stability of the rotor. According to the model presented at ref.[4,7,8,9,10],the shock wave's structure in rotor connects with the shape of L.E. spatial curve of rotor blade immediately. Therefore, from the viewpoint of design, the shape of L.E. spatial curve of rotor blade will be an important means for controlling the intensity of 3D shock wave in rotor. However, it is arbitrary that forming the L.E. spatial curve of rotor blade in classical stacking blade elements procedures, and it doesn't connect with the shock wave's structure. So, it becomes a main task for aerodynamic & aeroelastic tailoring process —— how to design the L.E. spatial curve of rotor blade for optimum controlling the shock wave's structure in rotor.

2) In order to increase the freedoms of aerodynamic design and to form the 3D blade's geometric shape, the staking line of blade element's gravity centers along span will be a selected spatial curve, and the aeroelastic tailoring is necessary for guaranting non-onset of stall flutter at the anticipated region of the fan's map.

3) During the process of aeroelastic tailoring, the 3D deformed actuator-disk model with variable interblade phase angle distribution along blade row has been used as the main tool for predicting the onset of stall flutter. The value of minimum negative aerodynamic damping δ_{min} and the number of blades having negative aerodynamic damping N_f have been used as the two indices for assessing the situation of aeroelastic stability.

4) A reduced-scale model of transonic fan stage has been picked out as the calculating example for composite aerodynamic & aeroelastic tailoring. According to the final result of aeroelastic tailoring. there is non-onset of stall flutter at the anticipated region on the fan stage's map.

References

[1] Low, C.H. and Wadia, A.R., (1992). " Low Aspect Ratio Transonic Rotors: Part 1 — Baseline Design Performance " ,ASME92-GT-185,.

[2] Shen, M.Y., Zhou, S., Lin, B.Z. (1988). " Transonic Flow in Turbomachinery " , Academic Press, Beijing, 338-341,.

[3] He L. and Zhou. S. (1985). " Onset Mechanism of Balde Stall Flutter " .ACTA AERONAUTICA ET ASTRONAUTICA SINICA, Vol.6. No.5, 411-416, .

[4] Zhang, Y. J., (1994). " Some Discussions on The Models of Aerodynamics & Aeroelasticity of Transonic Fan " , Dissertation of BUAA for Ph.D.

[5] Neubert, R. J., Hobbs, D.E., Weingold, H.D. (1990). " Application of Sweep to Improve the Efficiency of a Transonie Fan, Part I — Design " AIAA 90-1915,.

[6] Moore, R.D., Tweedt, D.L., Chima R.V., (1992). " NASA Supersonic Throgh-flow Fan Program " , NASA TM-103248,.

[7] Zhou.S. (1996). " Some Progress on Aerodynamic Design of Modern Fans " , Proc. of 8th Symposium of Turbomachinery, 1 ~ 9. Zhu Hi, china, NoR.

[8] Zhou.S., (1995). " From Flow Fields in Turbomachinery to Their Mathematical Models " , Invited

Lecture in 2nd International Conference on Pumps & Fans. Beijing, Oct.
[9] Gua, X.M., (1993). " The Research on the Model of Controlable Shock Wave used in the Design
of Transonic Axial Fan/Compressor Stage " Dissertation of BUAA for ph.D.
[10] Shan. P., (1996). " Kinematic Problems of L.E. Supersonic Sweep-Bend in Supersonic & Transonic Axial Compressors " , Dissertation of BUAA for Ph.D.
[11] Casey, J.K., (1988). " Empirical Flutter Prediction Method " , AD/A 195699,
[12] Zhou. S., et al, (1989). " Introduction of Aeroelasticity in Turbomachinery " , China Defence Industry Press. Beijing,
[13] Yang, X.D., Tao, D.P., Zhou. S., (1991). " Concerning a Basic Assumption for Aeroelasticity in Turbomachinery " , 746 ~ 754, Vol. 34, No.6, Science in china (Ser.A.), .
[14] Zhou, S., (1993). " Some Improvements on the Models regarding Unsteady Flow in Turbomachinery " , Invited Report, International Conf. on Aero-Hydro Elasticity. Beijing. oct.
[15] Yang, X.D., (1988). " the Research for Basic Model of Aeroelasticity in Turbomachinery " , Dissertation of BUAA for ph.D.
[16] Yang, X.D. and Feng, Y.C., (1989). " The Interaction between Distortion of Inlet Flow and Blade Stall Flutter in Axial Flow Compressor ", 167 ~ 177, in Proc. of 5th International Symposium for Unsteady Aerodynamics & Aeroelasticity of Turbomachines & Propellers, Beijing, Sept.
[17] Wennerstrom. A.J., (1984). " Experimental Study of a High Through-flow Transonic Axial Compressor Stage " , " J. of Eng. for Gas Turbine & Power " , 552 ~ 560,Vol. 106. No.3.

METHODS FOR AEROELASTICITY IN PROPULSION

HANS MÅRTENSSON AND PIETER GROTH
Volvo Aero Corporation
S-461 85 TROLLHTTAN
SWEDEN

Abstract. Methods for aeroelasticity based on CFD are described as used in component analysis and design for forced response and flutter blade systems and side loads on rockets. Attention is given to the use of advanced methods from the perspective of their use in design and for particular components. Comments are made as to some issues that need to be addressed. The strength of advanced methods is that more can be analysed and more of the physics can be brought into the analysis of a component, the major weaknesses are speed in use, and insight into the coupling from engine to the simplifications still embedded in a very advanced analysis method.

1. Introduction

Interaction of fluid and structure in propulsion has profound effects on component functionality, life and safety. For aeronautical propulsion systems two lines of development combine in increasing the need to address aeroelasticity issues. As specific thrust is a major characteristic that is pushed higher and higher the material in general becomes more highly loaded and less rigid, which shifts the resonant frequency in the structure. This tends to increase the susceptibility to fatigue as high cycle fatigue driven crack growth is overlaid on top of low cycle fatigue initiation. Drastically decreased allowable run times may result from this combination.

The aeroelastic phenomena of most concern to aeroengine designers are forced response and flutter. Forced response are vibrations induced by inhomogeneinites in the flow field, e.g. gusts wakes from upstream blade rows. Flutter is a self-excited vibration with an amplitude which increases exponentially in time. It is caused by unsteady forces induced by the motion of the structure. Data being available from field operation of an engine are

T.H. Fransson (ed.), Unsteady Aerodynamics and Aeroelasticity of Turbomachines, 565-582.

crude and it can not always be identified immediately which class of problems that is the source. Design engineers, in particular those dealing with this type of problems has many constraints to work with and often a number of possible origins of a problem, not exclusively aeroelastic. Solutions chosen are therefore most often derived from a trade-off among several goals. Non-linear interaction with aerodynamics, and structural contacts may transfer energy among modes, thereby making the prediction more difficult although this may in practice work to limit stress levels.

Indications of aeroelastic problems in engines are often vibration levels, accelerated crack growth or even blade out. These indications could result from various sources which means that some further identification may have to be done as to the source of the problem. Identifying aeroelasticity as prime source may be done by fitting strain gauges on the rotor in engine tests or by vibration measurements on engine shell exterior and correlating measured frequencies with known sources and resonance frequencies in the engine.

If flight safety is affected different measures are taken to solve the problem temporarily and permanently:

- Restrictions of flight envelope in order to avoid sensitive points
- Changes in control system
- Redesign of aerodynamical or structural nature
- Run-time limitation on component

2. Methods used

For some of the 2D flutter calculations in-house codes VOLFAP and LINFAP are used. These codes are based on unstructured 2D finite element methods as described in (Groth, 1994) and (Siden, 1991). For all other calculations here general purpose in-house codes are used, referred to as VOLSOL. These are based on 3rd order upwind biased finite volumes for multiblock grids as described in (Eriksson, 1990),(Mårtensson, 1994),(Groth *et al.*, 1996),(Mårtensson, 1995).

VOLFAP solves the compressible Navier-Stokes equations in the temporal domain, along with an algebraic turbulence model for the boundary layer. The computational grid is moving with the blade surface. Uneven inter blade phase angles are accounted for by the use of a procedure usually referred to as "direct store". Absorbing boundary conditions are used for inflow and outflow boundaries. A hybrid grid arrangement is used where a structured grid embed the blade for resolution of the boundary layer. On the structured grid an implicit factorisation scheme is used for increasing the CFL numbers near the wall.

LINFAP solves the linearised harmonic Euler equations on an unstructured grid that deforms to conform with the blade motion. Exact 2D non-reflecting boundary conditions are imposed at the inflow and outflow boundaries, which makes the solution independent of the computational domain. At the boundaries between adjacent blade passages phase-lagged periodic boundary conditions are imposed to handle arbitrary inter blade phase angles. The code is based on the pseudo-time technique by (Ni and Sisto, 1976). By marching the governing equations in pseudo-time to convergence the code yields the field of complex perturbation amplitudes, i.e. a frequency domain solution is obtained. One of the most attractive features of a linear method is that the principle of superposition may be applied to a set of solutions for different blade motions to obtain the solution for any blade motion which is a linear combination of those computed.

VOLSOL is a general purpose finite volume based package that may be used with general geometries and a broad range of boundary conditions and physical models for turbulence, species and reactions. For the applications shown here moving grids are used for flutter calculations. The technique used for the moving grid extends the finite volume to from 3 spatial dimension to include the time axis direction. Absorbing boundary conditions on inflow and outflow are use. Forced response calculations are made by modelling stators and rotor on fixed grids in their respective frame of reference, and coupling the 2 over the sliding interface. The sliding interface coupling is such that local conservation of mass momentum and energy is satisfied equally to the interior of the computational domain. The spatial discretisation is a 3rd order upwind biased reconstruction technique for finite volumes. For shock capturing and discontinuities 2nd difference non-linear switches or TVD procedures may be used. Time marching may be performed by either explicit Runge-Kutta methods or an implicit multigrid procedure depending on the problem at hand. Parallelisation is used over clusters of workstations in order speed up computational times.

3. Turbine forced response

Forced response analyses as used here are aimed at computing excitation frequencies and amplitudes. For turbines all variations in the tangential direction will be produce a periodical disturbance on the rotor and thus drive a structural vibration. Typical sources for tangential variations are the inhomogeneous temperature field from the combustor and wakes, shocks and interaction of potential fields from blade rows upstream and downstream, in particular if throat area variations in stators occur.

For detuning the cyclic loading caused by upstream blade rows a stator and a rotor will have odd combinations of number of blades. The straight forward way of handling this and model all blades would lead to enormous models as typical blade counts are between 30 and 60 for one blade row. Methods as in (Saxer, 1992),(Giles, 1988) are therefore introduced for imposing the phase lagged periodicity. Several proposals to this end have been devised. The problem is theoretically difficult as stability and accuracy may easily be affected.

Shifting the eigenfrequencies by design and adding damper elements in the structural design is done. Increasing damping action is typically done in the structural design in order to bring down vibrational stress levels. One way to accomplish this is to increase the flexibility of the shanks often by making these longer and thereby transfer more energy through the damper element. The modelling and design of dampers is a research area in itself. Aerodynamically the excitation may be lowered by considering such design features as leaning stators and uneven pitch of stators.

3.1. DESIGN CALCULATIONS

The first example of forced response analysis is a two-stage supersonic turbine. The computational model used consists of 25 blades in two-stages. The structure in the analysis remains stiff and it has been presumed that the induced vibrations will not affect the aerodynamics. Unsteady compressible 2D Euler analysis is used for coupled rotor and stator analyses. Figure 1 shows the non-stationary field in 2-stage supersonic turbine designed for the Ariane launcher. Computational pressure time histories can in this case be compared to measured pressures from a single sensor located at mid-span on the pressure side of the blade, Figure 2. In this case fair enough agreement is achieved for the lower frequencies which are of highest interest for the structural assessment. The test made in the example is running in air while the machine in operation is running on exhaust gases from gas generator.

3.2. SINGLE STAGE OFF-DESIGN EFFECTS

In the single stage analysis shown in figs 3a,b a comparison can be made of the non-stationary effects at design and when running the stage severely off design. Shock patterns and shock induced separation are outstanding features of the off-design case. These will inevitably give rise to broad spectrum excitation of the rotor, whereas in the design point lower speed case a more narrow band of excitation may be expected. It should be noted that this is not a supersonic design to start with as may be concluded from the rather thick leading edges.

3.3. DISCUSSION OF ANALYSIS METHODS

The modal analysis framework is an incredibly strong and successful foundation for the structural dynamics. Most of the structural behaviour is likely to be well described and understood by using a limited number of eigenmodes with the addition of non-linear mechanisms. For aerodynamically coupled 3D effects more exploration is needed for understanding of energy transfer mechanisms between flow field and structure. Phase lagged periodicity, in particular for multiple sources of non- periodicity will need some attention, we may have for example 9 combustors 31 stator blades 41 rotor blades. In general the subharmonics is not fully clear, in compressors a well known important phenomenon is rotating stall. Also important is turbulence modelling for sensitive cases such as wakes, boundary layers and shock/boundary layer interaction. Numerics using fast implicit methods and rational descriptions of circumferential variations are needed for design calculations.

4. Flutter

Flutter may be characterised as a stability problem where, for a given movement of the structure, energy may be transferred from the airstream to the structure. Given that this happens at an eigenfrequency of the structure this leads to an exponential divergence of the blade. The problem arise for new designs, but flutter also notably arise with increasing operating hours in an engine, and may differ engine to engine. Blades are worn down, which leads to changes in the aerodynamics of the blade, also indirectly due re-matching as performance changes. Shrouds, inserts and balancing all can change with time. For a production series individual performance differences arise as manufacturing tools are worn and from changes in manufacturing procedures. Based on the information from analyses the structure may be altered, modification of shrouds, dampers, or the blade loading by aerodynamic redesign. Being a very elusive problem, cut and try has been, and in part still is, used to alleviate problems in engines. Given an obviously sensitive problem and plenty of uncertainty we may expect that the full problem can not be solved for a long time to come. Robustness in analyses is needed so that improvements in calculated stability limits can be used with higher degree of confidence.

4.1. 2D DESIGN CALCULATIONS

As an example a DCA compressor blade is analysed with LINFAP under realistic conditions for a mode shape that in the section is turning the profile around mid chord. The mean flow analysis show a typical compressor flow pattern with a shock through the passage, Figure 5. From a linearised Euler solution Figure 6 shows maps of calculated amplification factors, where negative values of torsion load points to amplification. The maps of axial and tangential loads show that combinations of mode shapes may be destabilised or stabilised by a basic motion, i.e. combined modes may be unstable even if all 3 generic modes are stable. Figure 4a-d shows the stationary and Figure 5 the non-stationary pressures over the blade surface. Notably the shock is an outstanding feature of also the non- stationary pressure profiles. For a case where 5 span wise sections are analysed in this manner a total of 1500 linear 2D solutions and 5 non-linear 2D mean field solutions or 1 non-linear 3D solution is needed for one point in the compressor map. 100 CPU hours on a modern fast work station can be expected with a week of work. This type of analysis is informative, and gives some objective basis for changes in design, but is not accurate to the point where calculated results can be guaranteed to come through in the engine. The advantage of linear inviscid calculations made this way is of course speed. For flutter near stall where the boundary layer behaviour is critical of course this must at some point fail. It is to be expected that a linear inviscid method is not able to capture the physics embodied in this very complex physics. It would seem necessary to move to using the Navier-Stokes equations in this case. The N-S calculation is done in the time domain and takes a factor of 100 times longer time than the linear analysis described before, which limits the analysis to a select number of modes and inter-blade phase angles. Turbulence modelling is critical, and difficult enough, for the stationary case and is a source of uncertainty also for this application. Validation in this area therefore becomes a very difficult task with increasing uncertainty in the physical modelling. Such methods are anyway in use for the analysis of blade flutter.

4.2. 3D METHODS DEVELOPMENT

For presenting 3D compressible Euler/Navier-Stokes methods single delta wings at transonic speeds will be used. The rigid wing profile is NACA64A006. Its span, root chord, and sweep angle are 134 mm, 230 mm, and 55°, respectively. In the wind tunnel the wing is set in motion through a hydraulic actuator system attached to a pivot axis at the wing root (Johansson and Winzell, 1995). The rigid motion here is pitch oscillation. Good agreement of predicted and measured non-stationary pressures is obtained in the cases of:

Subsonic:	$M = 0.7$	$\alpha = 0°$	amplitude 0.278°	f=190 Hz
Transonic:	$M = 0.95$	$\alpha = 0°$	amplitude 0.53°	f=95 Hz

The flexible wing is a flat plate of thickness 1 mm. Its span, root chord, and sweep angle are 147 mm, 176 mm, and 45°, respectively. The motion used in test is the first torsion-bending mode, self excited. The corresponding tests have not yet been reported or fully evaluated so only computations are reported. In 3D the introduction of a complex mode shape becomes necessary. Figure 8. shows the Mach number distribution around the wing and an exaggerated view of the actual mode. The use of linearised methods is attractive also in 3D for the same reasons of speed as in 2D. For the delta wing validation is therefore made as to the amplitude dependence of the solutions. At $M = 0.82, \alpha = 0°$, 3 amplitudes with maximal displacements of $5.4, 10.8, 21.6$ mm respectively were calculated with the full 3D compressible Euler equations. The resulting non-stationary pressure profiles shows that a gradual deviation from small to large amplitudes indicating a smooth limiting behaviour. Linearisation should therefore be a viable route also in 3D towards faster methods.

4.3. DISCUSSION ON ANALYSIS METHODS

Today the typical calculation of flutter is done by forcing the blade and calculating the net energy transfer to or from the structure in order to conclude whether flutter will occur, certainly in both 2D and 3D such calculation have a reasonable accuracy. Candidates for differences to engine tests are influences from upstream and downstream blade rows. The practice of using non-reflecting inflow and outflow conditions are fine for consistency and for validation of codes against wind tunnel data, but it is hardly representative of engine conditions. Very good throughflow data are needed as sensitivity is high to blade loading. The effects of propagation in 3D with rotation and the influence of a 3D structural motion added will need exploration, as analyses in 2D sections shows the different section being stable and unstable. The coupled system may not be well described by a single mode being forced, interacting modes may well exhibit other stability scenarios. Certainly for the purpose of eventually calculating amplitude limits this will be the case. A general way of giving the impact of a design change a quantitative meaning, based on rational arguments, would improve understanding.

5. Rocket nozzle side load

A rocket nozzle for satellite launchers is designed for optimal performance over a flight envelope that extends outside the atmosphere. Expansion ratios for a fixed nozzle will by necessity be high in order to perform, and subsequently at sea level the nozzle is over expanded resulting in flow separation during the start transient. A side load may occur during this process that coupled with a flexible nozzle will jeopardise the installation. For the nozzle structure critical design loads are the buckle load and side load, designs with reinforcement rings are used to prevent buckling. This in turn drives the necessity too control thermal stress fatigue incurred by a cold and stiff reinforcement ring connected to a hot nozzle wall. Better understanding and predictive ability of the side load has a weight reducing potential for design with nozzle expansion ratios currently used. For future designs the potential is to improve performance by allowing the use of a higher expansion ratio. In general much higher expansion ratios would be beneficial for improving pay load cost. The use of CFD for this purpose is relatively new and makes use of technology derived for turbomachinery applications. An account for the problem and for a simpler method for estimating side loads is given in (Pekkari, 1993),(Pekkari, 1994). Figure 9 shows diagrams of measured side load in tests of the Vulcain nozzle.

5.1. NON-STATIONARY SIDE LOAD ON A MOVING NOZZLE

In Figure 10 to 12 the results of an analysis of an oscillating exhaust nozzle are presented. The nozzle is overexpanded and shock system is formed in the nozzle. There exists a natural flow instability that is not aeroelastically driven which may be correlated by Strouhal number of around 0.2, based on the Mach disc diameter and the velocity upstream of the shock.. This instability does not couple with the structure and is at an order of magnitude higher frequency than the frequency giving response in tests. Figure 10 shows a series of Mach number plots at different times from the results of an unsteady Navier-Stokes calculation on a nozzle moving in a pendulum mode. In the time history, and Fourier analysis, of the integrated side load, the natural instability shows up as overlaid on the forced motion, Figure 12. In this case the response phase is such that it tends to stabilise the mode, i.e. the motion is aeroelastically stable. The modal form in the analysis is a simplified form of the real mode. The mode analysed turns out to be stable also according to a simplified analysis as in (Pekkari, 1994).

5.2. DISCUSSION OF ANALYSIS METHODS

For this application exploration is needed to assess uncertainties and understand which phenomena are most important. The simplified analysis show very clearly the sensitivity to modal shape so great care will have to be take towards the introduction of modes, and integration with structure.

6. Conclusions

Many propulsion components operate in a very complex environment. Engine conditions are always vital, and the robustness of methods for use in engine design. Building confidence in new methods is therefore a long and difficult task. Learning how to utilise new methods does have a learning curve, but still the use of very complex tools has developed and this will certainly continue. What seems clear is that although advanced techniques are becoming accessible to aid designers is that a lot remains to be done. Apart from straight forward correlations 2 levels of computational modelling will be needed, a complex modelling where exploration can be done into physical phenomena in addition to verification of a design, and well founded design modelling where speed and consistency are the priorities. For these areas there exist a wealth of physical and mathematical research opportunities both on a theoretical and an applied level. The general impression in our perspective is that the development of more capable computational tools have advanced faster than the ability to assimilate and make use of the information generated. Not the least to have technical assessment capabilities that keep up with increasingly powerful computational wizardry. There is a need for explanatory models to help interpret very complex computational results, and also most likely an extended terminology to give a language for new understanding as new phenomena may be studied.

Acknowledgements.The authors wish to thank Mr Lennart Kjellén for sharing with us his vast experience in Gas Turbine Engineering.

References

Eriksson, L.-E. (1990) A 3rd order upwind scheme for unsteady compressible flow, Technical report,VAC **9970-154**

Giles, M. (1988) UNSFLO: A numerical method for unsteady inviscid flow in turbomachinery. GTL report **195**, MIT

Groth, P. (1994) A linearised Euler method for unsteady flow in axial-flow turbomachines, PhD Thesis, Chalmers University of Technology, Dept. of theoretical and applied mechanics, Göteborg, Sweden

Groth, P., Mårtensson,H., Eriksson, L.-E. (1996) Validation of a 4D finite volume method for blade flutter. ASME-paper 96-GT-429

Johansson, U., Winzell, B. (1995) Steady and unsteady measurements on a 55° delta wing and comparison with computation, Royal Aeronautical Society: Int. forum on aeroelasticity and structural dynamics, Manchester, U.K.

Mårtensson, H. (1994) Rotor-stator interaction - method and 2D tests, Technical report,VAC **9970-1003**

Mårtensson, H.(1995) 4D finite volume metrics formulation for moving grids, Technical report,VAC **9970-1193**

Ni, R.H., Sisto, F. (1976) Numerical computation of non-statinaryaerodynamics of a flat plate cascade in compressible flow, *ASME Journal of Eng. for Power*, **Vol. 98**, pp. 165-170

Pekkari ,L.-O. (1993) Aeroelastic stability of supersonic nozzles with separated flow, AIAA-93-2588, 29th Joint propulsion conference and exhibit, June 28-30, Monterey, CA

Pekkari ,L.-O. (1994) Aeroelastic analysis of supersonic nozzles with separated flow, AIAA-94-3377, 30th Joint propulsion conference, June 27-29,Indianapolis, IN

Saxer, A. (1992) A numerical analysis of 3D inviscid stator/rotor interactions using non-reflecting boundary conditions, GTL report **209**, MIT

Siden, L.G.D. (1991) Numerical simulation of viscous compressible flows applied to turbomachinery blade flutter, PhD Thesis, Chalmers University of Technology, Dept. of thermo and fluid mechanics , Göteborg, Sweden

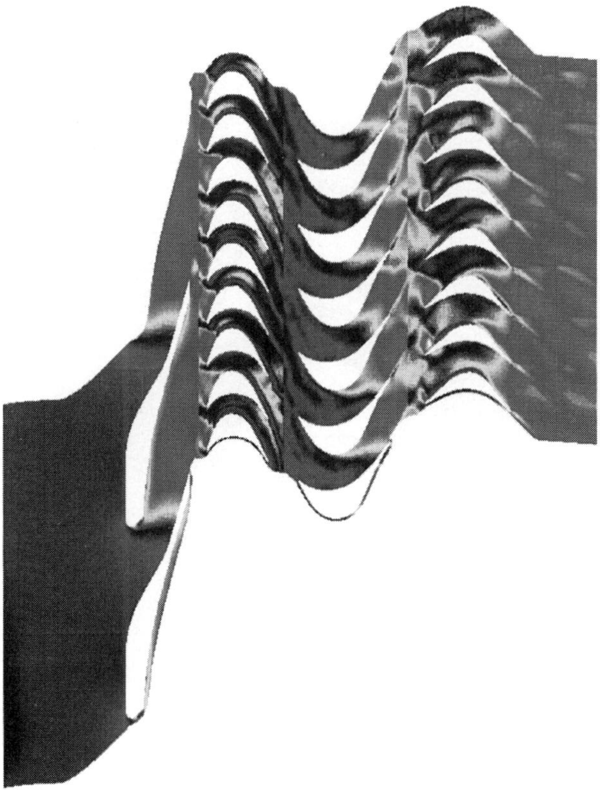

Figure 1. Snapshot at one time point of unsteady Mach numbers in relative frames of reference

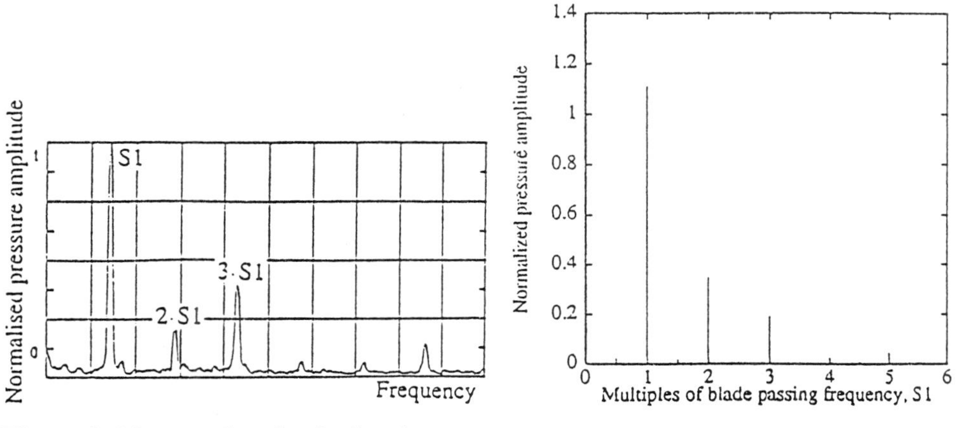

Figure 2. Measured and calculated pressures

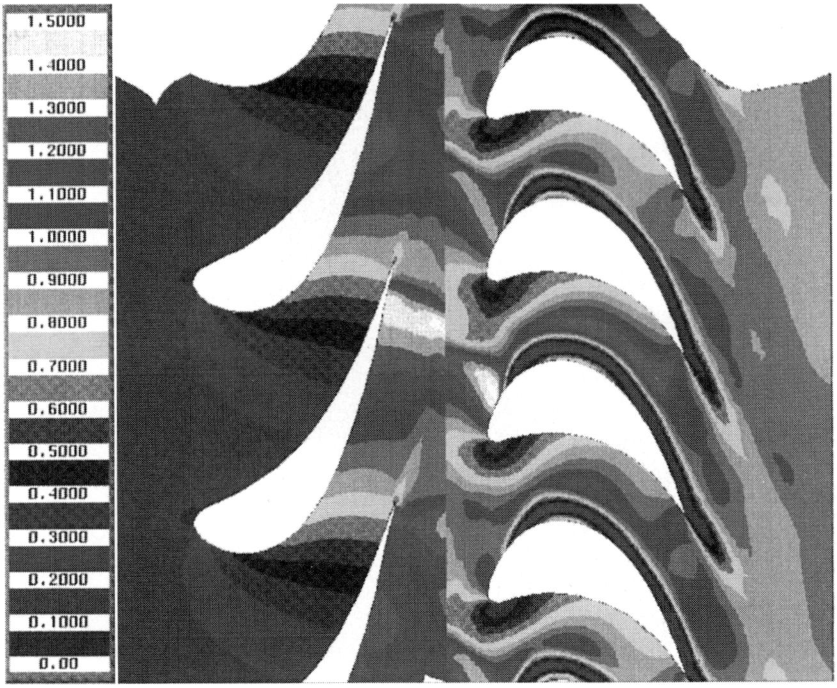

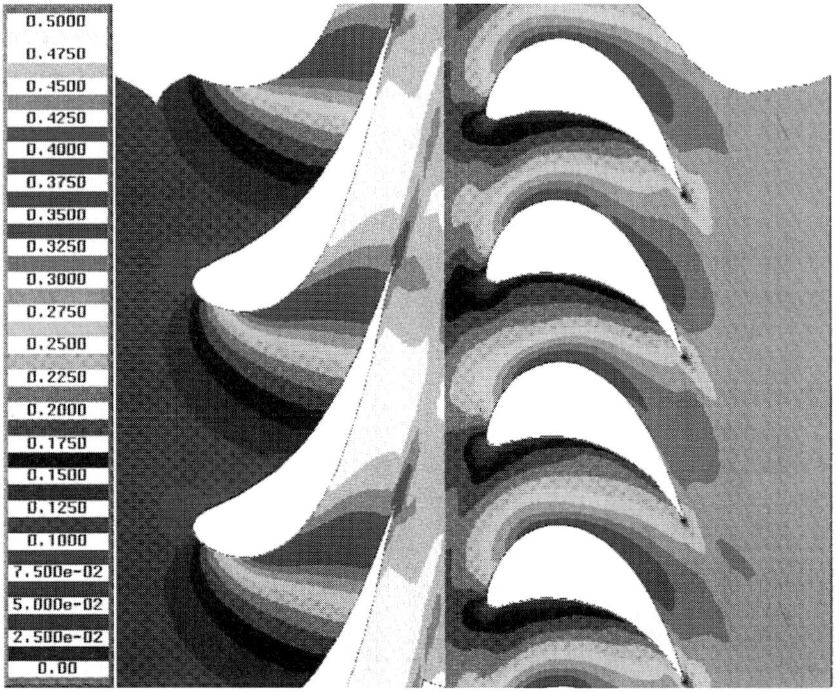

Fig 3b Snapshot Transient Calculation - Mass flow 3320 kg/s m
- Rotor speed 100 m/s

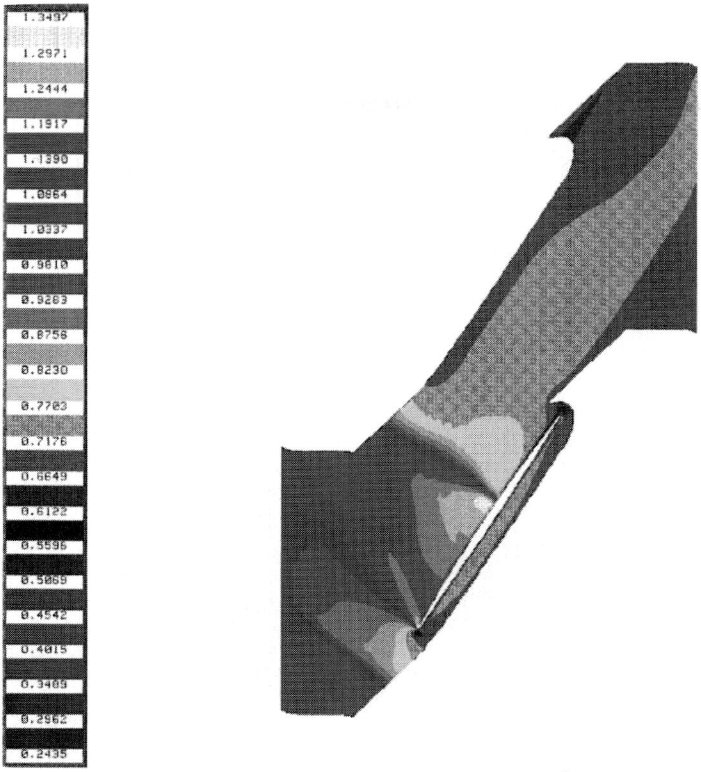

Figure 4a Mach numbers of the mean field around a DCA compressor blade

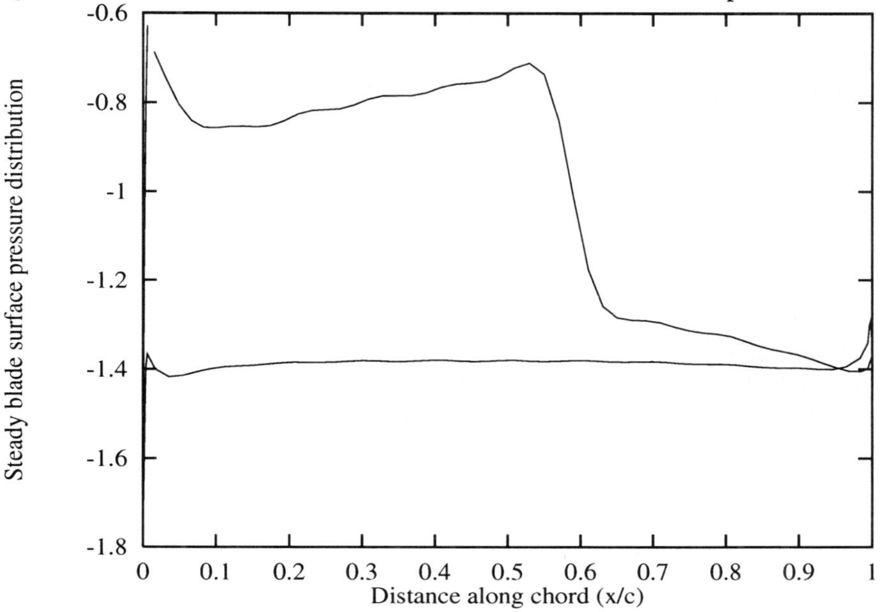

Figure 4b Mean normalised pressure (Cp) distribution around the DCA compressor blade.

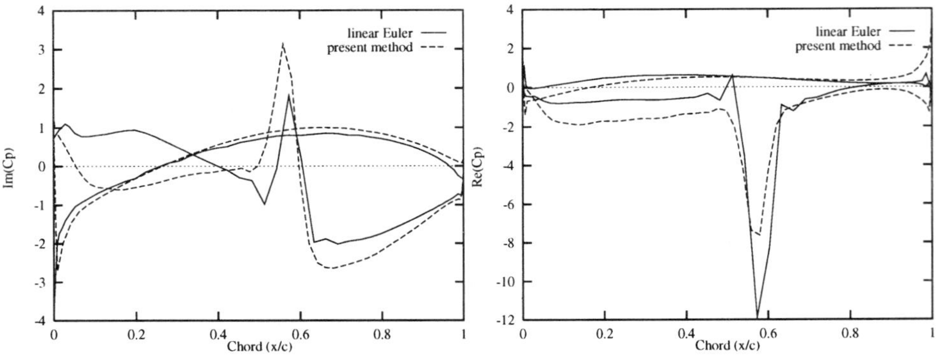

Figure 4c Real and Imaginary parts of unsteady pressures (Cp) around the DCA blade

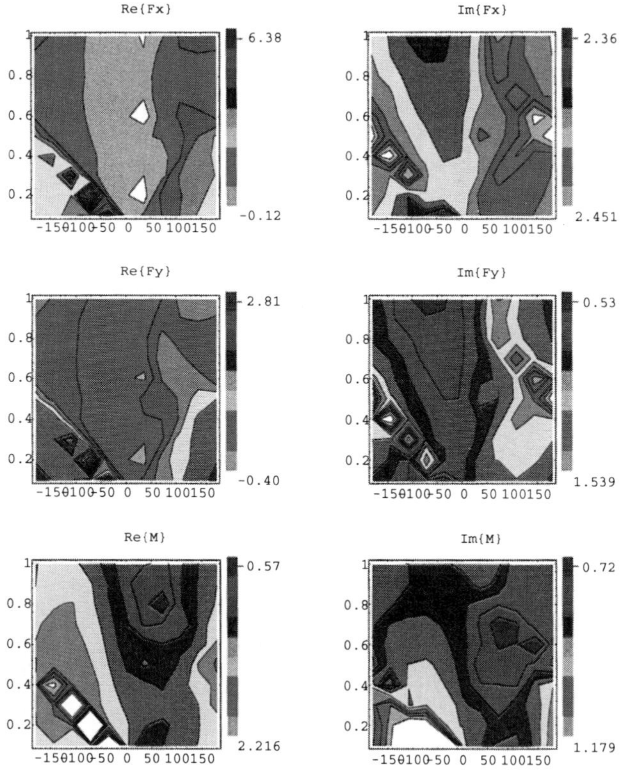

Figure 5 Map of unsteady load on DCA blade over inter blade phase angles and reduced frequencies.

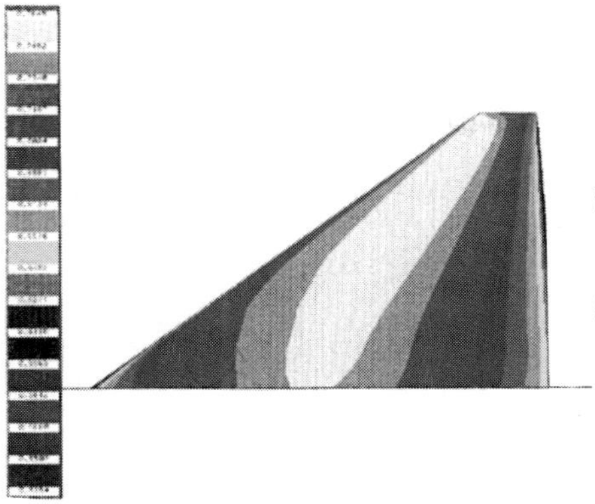

Figure 6a Mach numbers over wing ,subsonic

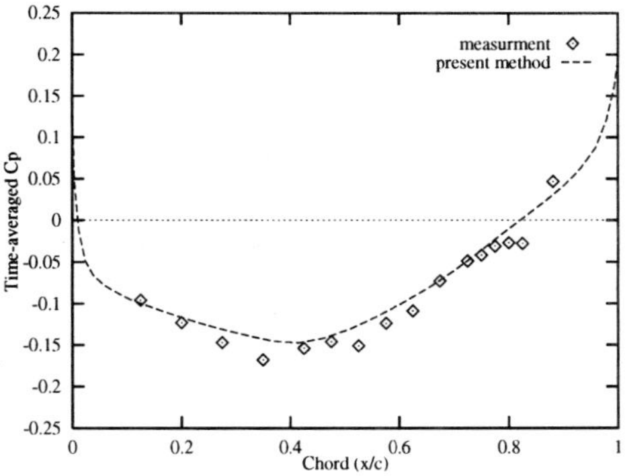

Figure 6b Mean Cp distribution,

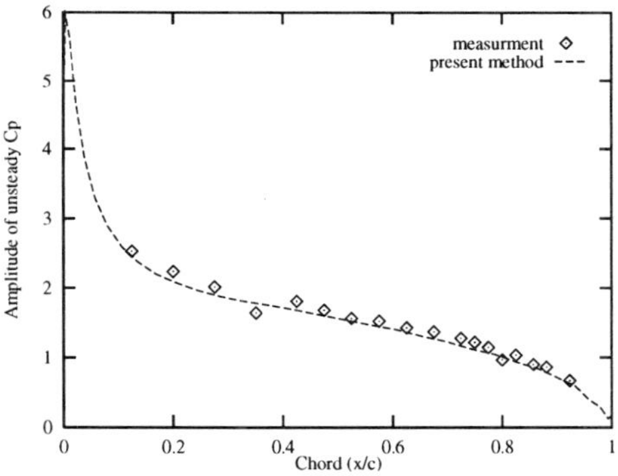

Figure 6c Non-stationary Cp distribution amplitude, subsonic

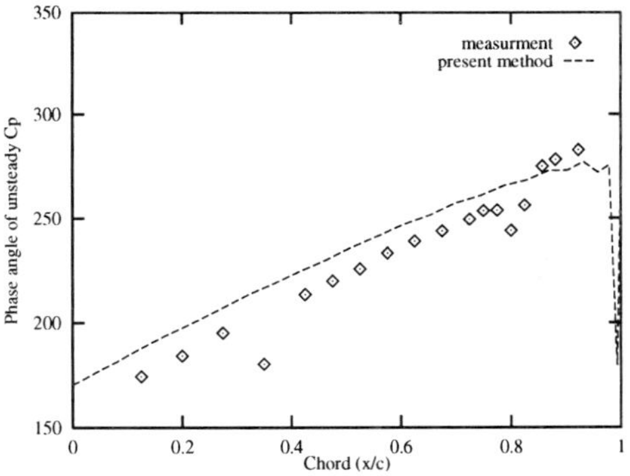

Figure 6d , Non-stationary Cp phase, subsonic

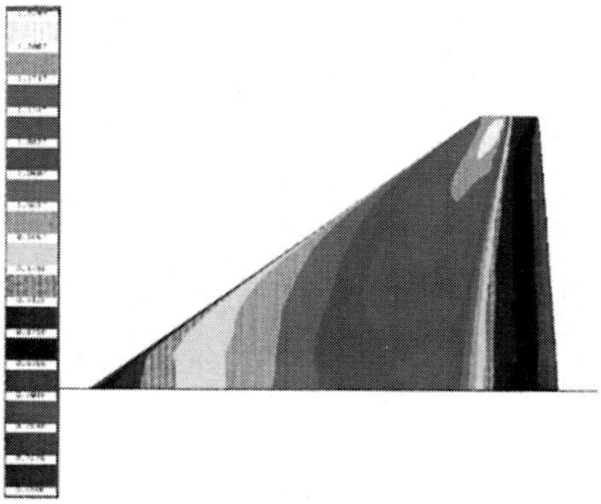

Figure 7a Mach numbers over wing transonic

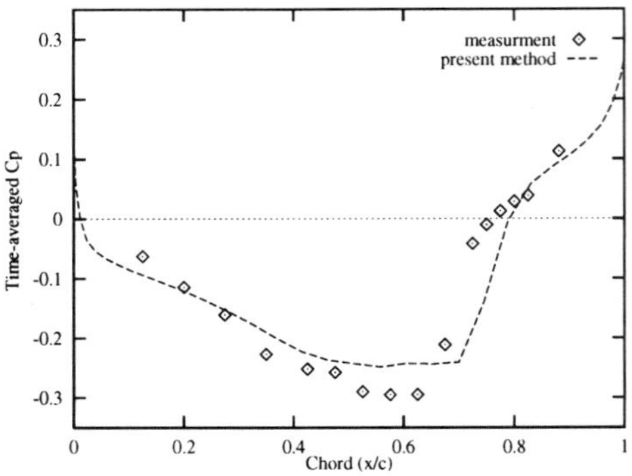

Figure 7b Mean Cp distribution,

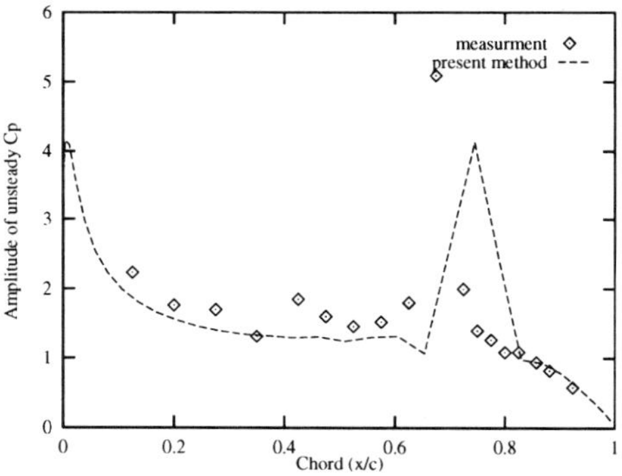

Figure 7c Non-stationary Cp, amplitude transonic

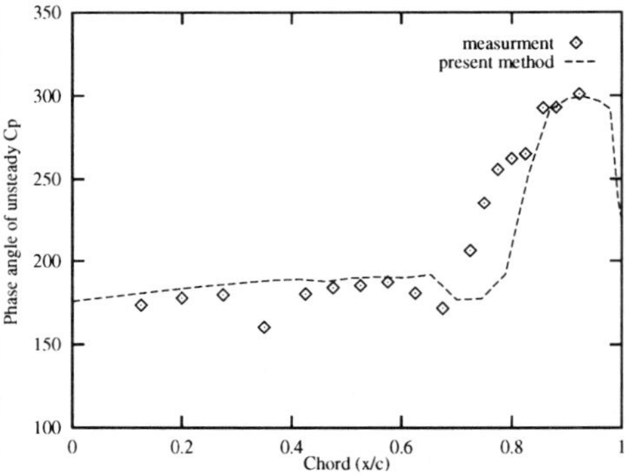

Figure 7d , Non-stationary Cp phase, transonic

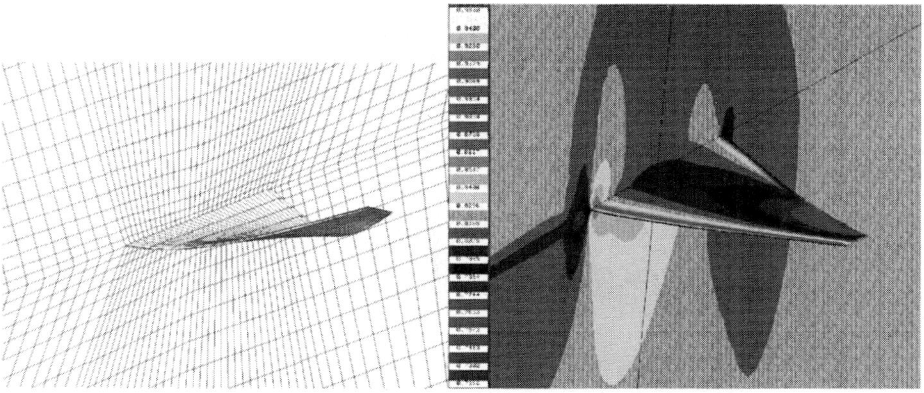

Figure 8. Snapshot Mesh and Mach numbers over deltawing at one point in time.

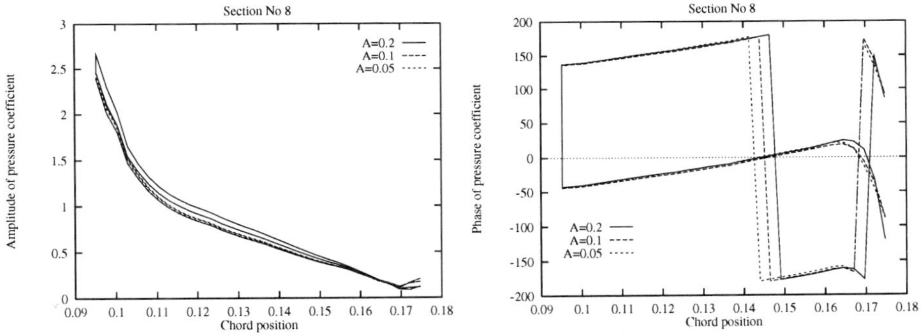

Figure 9 Cp distribution calculated at different amplitudes

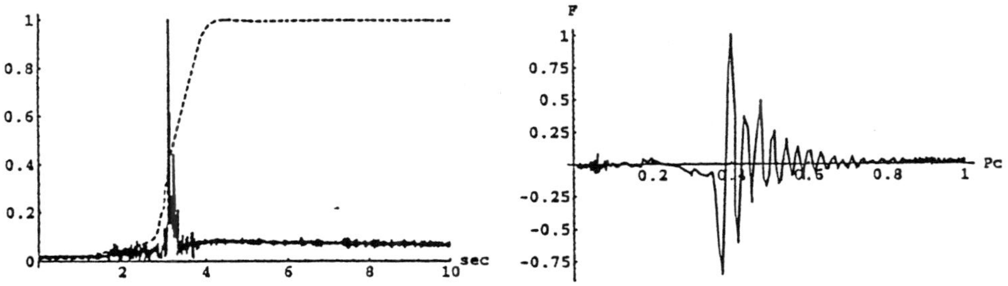

Side load (—) & chamber pressure (- -) during the start transient with the Vulcain engine

Side load versus chamber pressure during the start transient with the Vulcain engine

Figure 10 Side load from test & by simplified model

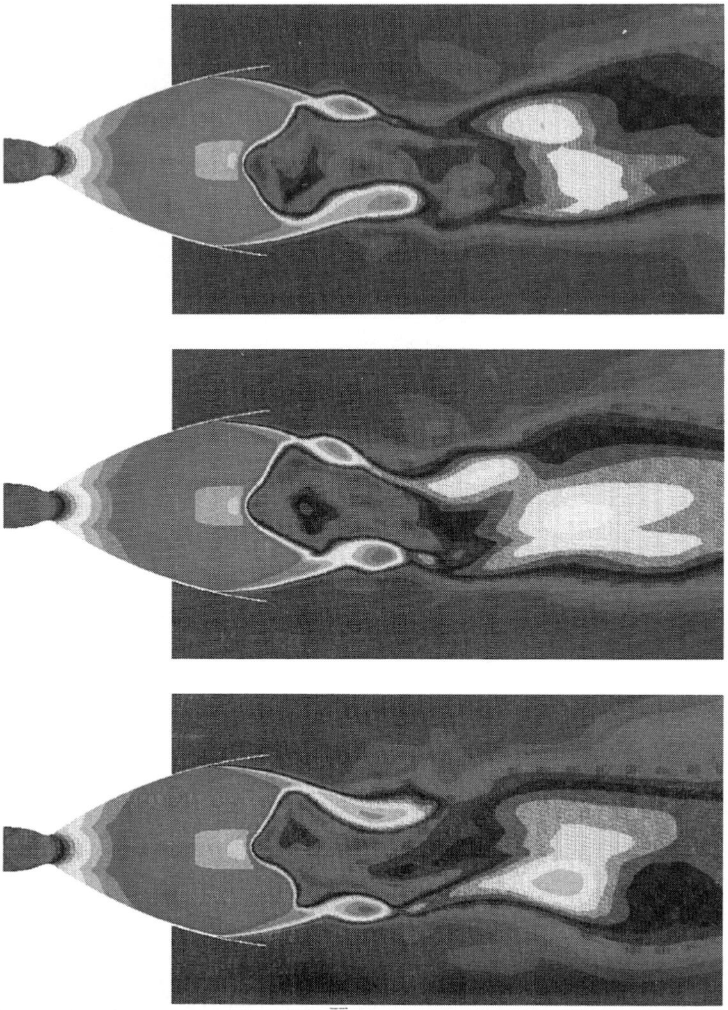

Figure 11. Snapshot of Mach numbers in over expanded rocket nozzle at 3 time instances.

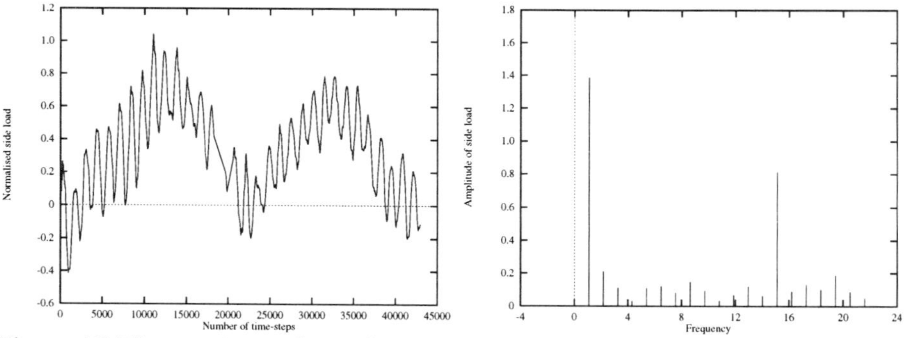

Figure 12 Time series and amplitude spectrum of nozzle side load

11: Theoretical Blade Row Interactions - 2

PRELIMINARY RESULTS OF A UNSTEADY THROUGHFLOW CODE TO SIMULATE THE START-UP OF A MULTISTAGED TURBOPUMP

M.DEBARD, P.FERRAND, F.LEBOEUF
Laboratoire de Mécanique des Fluides et d'Acoustique de l'Ecole Centrale de Lyon.
email: ferrand@mecaflu.ec-lyon.fr

1 Introduction

The start-up of a turbomachine or, more generally, the large and quick variation of its rotational speed is rarely studied. Knapp[1] has shown that in the case of a slow variation, the behavior of a turbomachine is quasi-steady. If the variation is quick, this is not true. This fact was confirmed by authors who have studied experimentally quick start-up of centrifugal hydraulic pumps. Tsukamoto and Ohashi[2] have explained the basic mechanism of deviation from quasi-steadiness as a consequence of impulsive pressure and lag in formation of circulation around the blades. They have proposed a non-dimensional parameter of quasi-steadiness. Barrand and al[3] have studied in depth the complex interaction between the pump and the components (ie pipelines, tank, valves..) of the hydraulic circuit. They have also shown the effects of unsteady cavitation. Nguyen and al[4] have analysed the behaviour of liquid hydrogen pumps during the start-up of space-rocket engines: they have concluded that the quasi-steady assumption is valid for these machines, a coherent result regarding the Tsukamoto-Ohashi parameter.

There are practically no calculations of fast start-up flows. Tsukamoto and Ohashi have designed a 2D, blade-to-blade code using singularities method which gives good qualitative but poor quantitative results and does not provide opportunities of taking into account the whole circuit. Computing flows in such conditions is certainly a difficult task. This is precisely the scope of our work at LMFA.

T.H. Fransson (ed.), Unsteady Aerodynamics and Aeroelasticity of Turbomachines, 585-598.

The method we have designed is a throughflow calculation because we are mainly interested in the unsteady characteristics curve of the pump. Througflow calculations are also handy tools in multistaged turbomachines. In this context, 3D calculations are too big and 1D (transfer-matrix representation) simulations precisely cannot compute the unsteady characteristics. Our method uses a tangential mass-average formulation which gives an unsteady axisymetric representation of the flow. Two source terms appear in the averaged equations representing the pressure difference on the blade surfaces and the effect of flow non-axisymetry upon its axisymetric component. We call them subsequently blade forces -Fp- and 3D forces -F_{3D}. The first term is the most important as it represents the work exchanged. The later is often neglected but some authors[5] have shown that its contribution may be important.

Our unsteady method needs unsteady source terms. In most throughflow methods, source terms are usually given by means of a steady blade-to-blade calculation. Here the throughflow calculation is coupled with a blade pressure calculation using the Boundary Element Method (BEM) and yielding unsteady blade forces at each time step. We have chosen BEM because we need to know the pressure field on blades only and because it fully takes into account the geometry of the turbomachine. The pressure calculation requires some 3D informations: they are provided by a third calculation which also computes F_{3D}. The whole calculation takes the form of a three boxes coupling depicted on fig.1.

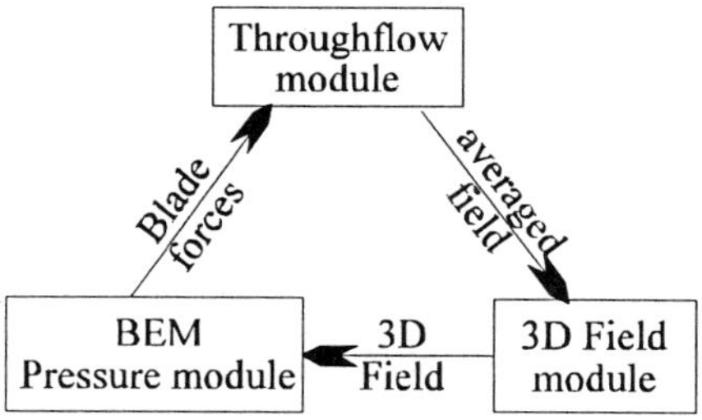

fig.1: structure of the method

Steady results on various geometries have been successfully compared with a steady 3D Euler code using Clebsch formulation[6]. The results we present in this paper primarily show the ability of our method to handle unsteady flows in turbomachines. Viscous effects and cavitation are neglected.

2 Numerical Method

2.1 Throughflow calculation

We start with the Euler equations expressed in cylindric coordinates (z,r,θ) in a relative frame of reference animated by a rotational motion $\Omega\vec{z}$:

$$\frac{\partial q}{\partial t}+\frac{\partial E}{\partial z}+\frac{\partial F}{\partial r}+\frac{1}{r}\frac{\partial G}{\partial \theta}+H=0,\ \text{with}\ q=\left(\rho,\rho u,\rho v,\rho w,\rho e^{*}\right) \tag{1}$$

Our method uses the following averaging operator :

$$\tilde{A}(z,r)=\frac{\overline{\rho A}}{\overline{\rho}}=\frac{1}{\overline{\rho}}\frac{1}{b}\int_{\theta_{ps}}^{\theta_{ss}}\rho A(z,r,\theta)d\theta\ \text{with}\ b=\theta_{ss}-\theta_{ps} \tag{2}$$

Its main mathematical properties are:

$$A(z,r,\theta)=\tilde{A}(z,r)+A'(z,r,\theta) \tag{3}$$

$$\overline{\rho AB}=\overline{\rho}\tilde{A}\tilde{B}+\overline{\rho A'B'} \tag{4}$$

$$\overline{\frac{\partial A}{\partial \alpha}}=\frac{1}{b}\frac{\partial\left(b\tilde{A}\right)}{\partial \alpha}-\frac{1}{b}\left[A\frac{\partial \theta}{\partial \alpha}\right]_{\theta_{ps}}^{\theta_{ss}}\ \text{for}\ \alpha=z,r \tag{5}$$

and

$$\overline{\frac{\partial A}{\partial \theta}}=\frac{1}{b}\left[A\right]_{\theta_{ps}}^{\theta_{ss}} \tag{6}$$

The averaged Euler equation system is expressed by:

$$\frac{\partial \overline{q}}{\partial t}+\frac{\partial \overline{E}}{\partial z}+\frac{\partial \overline{F}}{\partial r}+H=F_P+F_{3D},\ \text{with}\ \overline{q}=\left(\overline{\rho},\overline{\rho u},\overline{\rho v},\overline{\rho w},\overline{\rho e^{*}}\right) \tag{7}$$

- blade forces: $$F_P=\frac{1}{\theta_{ss}-\theta_{ps}}\left[p.\vec{n}\right]_{ps}^{ss},\ \text{with}\ \vec{n}=\left(\frac{\partial \theta}{\partial z},\frac{\partial \theta}{\partial r},\frac{1}{r}\right) \tag{8}$$

- fluctuation forces: $$F_{3D}=f\left(\frac{\partial \overline{\rho a'b'}}{\partial \alpha}\right),\ \text{with}\ a',b'=u',v',w',e^{*\prime} \tag{9}$$

The reader may find a full account of the equations in Debard[7].

Equation (7) is solved by the throughflow "box", F_p and F_{3D} being known at each time step. This "box" is a 2D, inviscid, compressible, unsteady solver using finite volumes. Its spatial scheme is a "mixed scheme"[8] combining a Van-Leer flux-splitting scheme and a centered scheme. The centered scheme being not dissipative, it requires an artificial fourth order dissipative operator. The temporal scheme is a four step Runge-Kutta with fourth order precision in time. The time step is subject to a CFL condition with typical values of CFL number between 0.5 and 1.5.

Boundary conditions are:

- upstream: stagnation pressure and temperature, two velocity angles and a fifth condition in the case of an upstream supersonic flow.
- downstream: static pressure or nothing if the downstream flow is supersonic.

The domain boundaries are treated with either compatibility relations or non-reflexive conditions. Because the former induce undesirable reflexions and the later do not enforce boundary conditions, we have also used on the downstream boundary a linear combination of both treatments such as the one proposed by Jameson and al[9]. This treatment induces a relaxation of the boundary condition which can be viewed as a pipe inertia effect.

2.2 blade pressure calculation

The formulation and discretization of the pressure calculation with the Boundary Element Method have been established by Larrey and al[10] and we give here a brief description of them.

2.2.1 Boundary Element Method

We consider the divergence of the momentum conservation equation:

$$\vec{\nabla}.(\rho\frac{\partial\vec{W}}{\partial t}+\rho\vec{W}\circ\vec{\nabla}(\vec{W})+\rho\frac{\partial\vec{\Omega}}{\partial t}\wedge\vec{R}+2\rho\vec{\Omega}\wedge\vec{W}-\rho\vec{\nabla}(R^2\Omega^2)) = -\Delta P \qquad (10)$$

Taking into account the peculiar form of the blade forces source term in equation (7), we only need to know pressure on blade surfaces. This is the reason why we use BEM to solve (10). The problem to be solved is basically a Poisson problem.

Equation (10) is multiplied by the Green function u^*_M ,integrated on Ω and then transformed using the Green integration theorem :

$$\int_\Omega \Delta u^*_M P d\omega - \int_{\Gamma 2} P\frac{\partial u^*_M}{\partial n}d\gamma + \int_{\Gamma 1} u^*_M \frac{\partial P}{\partial n}d\gamma = \int_\Omega \Delta u^*_M s d\omega + \int_{\Gamma 1} f_1 \frac{\partial u^*_M}{\partial n}d\gamma - \int_{\Gamma 2} f_2 u^*_M d\gamma \quad (11)$$

M being a given point located on the surface boundary.

The problem is discretized by meshing the boundary Γ in n surface elements γ_i, each element being defined by m nodes, and the the integration volume Ω in p elements ωj, each element being defined by q nodes. At each mesh point are defined the values of pressure P and normal gradient pressure $\partial P / \partial n$. Their evolution on a surface element are given by interpolation of their nodal values. Taking into account the interpolation functions ϕ_k, the discretized equation (11) is rewritten as (12):

$$\int_\Omega \Delta u^*_M P d\omega - \sum_{i=1}^{i=n}\sum_{k=1}^{k=m} P_{ik}\int_{\gamma_i}\phi_k\frac{\partial u^*_M}{\partial n}d\gamma_i + \sum_{i=1}^{i=n}\sum_{k=1}^{k=m}\frac{\partial P}{\partial n}_{ik}\int_{\Gamma_i} u^*_M\frac{\partial \phi_k}{\partial n}d\gamma_i = \sum_{j=1}^{j=p}\int_{\omega_j}\Delta u^*_M s d\omega_j$$

making appear the Dirichlet and Neumann influence coefficients:

$$H_{Mik} = \int_{\gamma_i}\phi_k\frac{\partial u^*_M}{\partial n}d\gamma_i \ (13) \text{ and } G_{Mik} = \int_{\gamma_i} u^*_M\frac{\partial \phi_k}{\partial n}d\gamma_i, \ (14)$$

They are functions of geometry and interpolation.

The different integrals are computed by the Gauss method. The number of gaussian points and their respective weights are functions of the integration point M and the surface (volumic) element considered.

2.2.2 Boundary conditions

On solid surfaces, ie blades, hub and casing, we impose the normal pressure gradients:

$$-\frac{\partial P}{\partial n} = \rho\frac{W^2}{R_C} + (\rho\frac{\partial\vec{\Omega}}{\partial t}\wedge\vec{x}_M + 2\rho\vec{\Omega}\wedge\vec{W} - \rho\,\mathrm{grad}\left(R^2\Omega^2\right))\bullet\vec{n} \qquad (15)$$

where Rc is the curvature radius, computed from the surface geometry. On the upstream boundary, we impose the normal pressure gradient. On the downstream boundary, we impose the static pressure. On the lateral (non-solid) boundaries, we use periodicity conditions (fig.3).

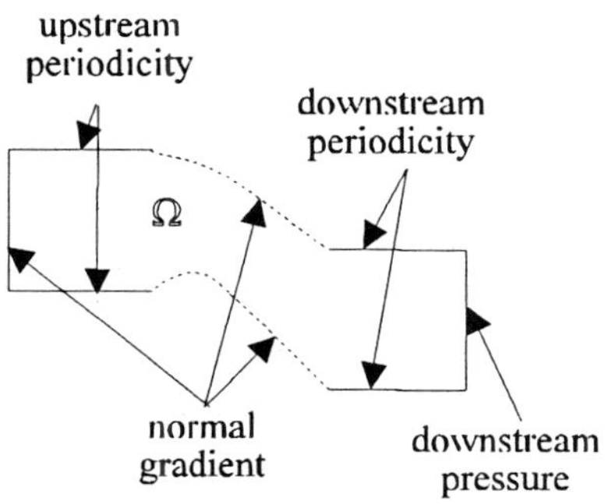

fig.2: BEM boundary conditions

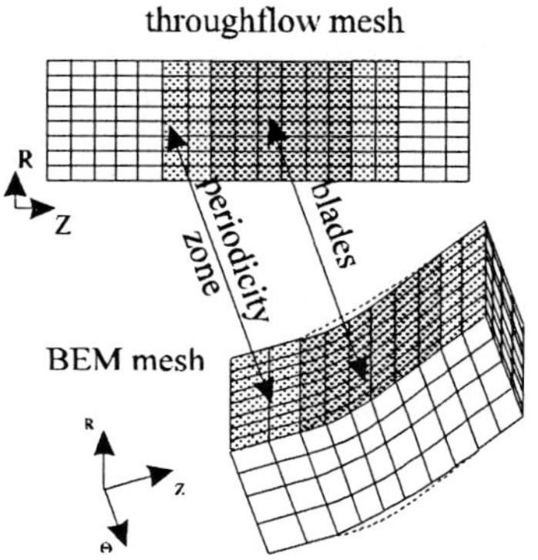

fig.3: throughflow and BEM meshes

We finally obtained the following linear system:

$$[[H][G]]_u \circ \begin{pmatrix} P_u \\ \partial P_u \end{pmatrix} = [S]\circ(s) - [[H][G]]_b \circ \begin{pmatrix} P_b \\ \partial P_b \end{pmatrix} \qquad (16)$$

where the indices u and b stand for unknow and boundary (known) variables.

The construction of the matrices in (16) are all executed as preliminary computations. The left-hand matrix of (16) is also inversed using LU factorization. So in the context of the whole method, BEM is reduced to a matrix product giving the unknown blade pressure performed at each time-step of the throughflow module.

2.3 3D forces module

The reader may have noticed in the precedent pages that the BEM module needs some 3D (here taken in the sense of blade-to-blade) informations. This is the task of the 3D forces (or fluctuation) module. At its core is a polynomial construction of the blade-to-blade evolution of the velocity field. Velocity components are written as:

$$A(z,r,\theta) = A(z,r,x) = a_0(z,r) + a_1(z,r)x + a_2(z,r)x^2 \qquad (17)$$

with A=u,v,w and $x = \dfrac{2\theta - \theta_e - \theta_i}{\theta_e - \theta_i} \in [-1,+1]$.

The computation of the 9 unknown polynomial coefficients $a_i(z,r)$ is done by resolution of 9 constraints equations:

- 2 equations for the no-sleep condition on blades,
- 3 equations for the respect of the azimuthal average, meaning that the average of a reconstructed component must be equal to its throughflow value,
- 2 equations of velocity gradients on the pressure side and the suction side,
- 2 closure equations , stating that the orthomeridian velocity is axisymetric.

This system is resolved with an iterative Gauss-Newton method. In a periodicity zone (upstream and downstream of the blades), we use another model. The polynoms are replaced by a sinus and a cosinus function to enforce periodicity of the velocity field. The no-sleep and gradient equations are replaced by hypothesis of axisymetry of the averaged velocity angle.

With the 3D reconstructed velocity field, we are now able to compute not only F_{3D} in (7) but also the volumic source term in (10) and the pressure gradients in (15). In this last equation, we have stated that the curvature radius is computed from the geometry: this is not the case at leading edge where we must take into account incidence effects. This is done by a model computing an equivalent curvature radius. These models (polynoms, incidence) have given excellent results on sample 2D geometries (see fig.4).

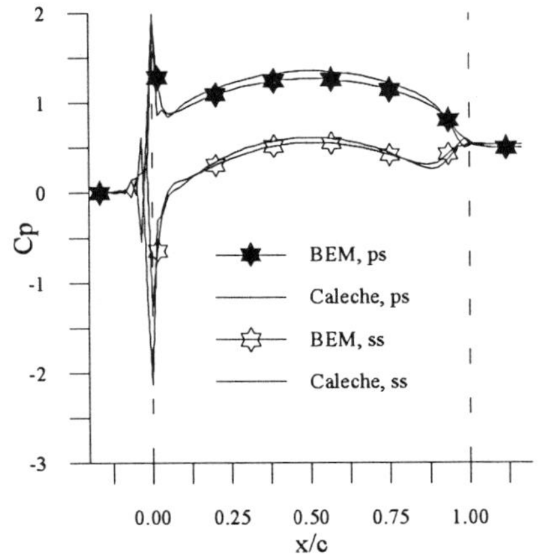

fig.4: pressure coefficient on a cambered profile

3 Unsteady Calculations on a 3D inducer

The geometry studied (fig.5) is the 3D inducer of a liquid hydrogen pump fitting the Vulcain rocket-engine on which the code has been validated for steady flows. The fluid is air with a perfect gas law.

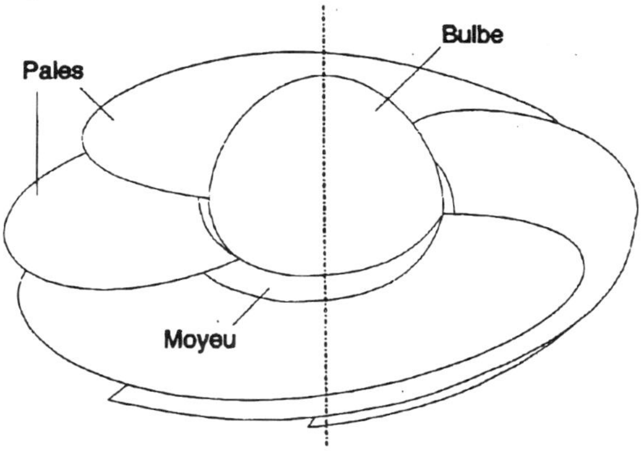

fig.5: 3D inducer

3.1 Variation of downstream pressure

We have simulated the effect of a quick variation of downstream pressure. This is a common problem in a rocket-engine with many possible sources such as opening/closing valves, ignition of the combustion chamber,....The following "triangle" evolution of pressure is imposed, T being the pressure variation time:

$$\text{for } 0 < t \leq \frac{T}{2}, Cp = Cp_0 + \Delta Cp \frac{2t}{T} \text{ and for } \frac{T}{2} \leq t < T, Cp = Cp_{max} - \Delta Cp \frac{2\left(t - \frac{T}{2}\right)}{T} \quad (18)$$

Steady conditions prevail before and after this evolution. Upstream conditions and rotational speed stay constant in time. We have investigated the effects of T, ΔCp and rotational speed.

We can draw two conclusions from fig.6:

- the flow-rate response to amplitude ΔCp is linear.
- for the largest value of T (0.1s), the flow-rate response is similar to the pressure signal, thus indicating a quasi-steady behaviour. For a small value of T (0.01s) flow-rate lags. The oscillations on fig.6 have the same period as pressure-wave reflection on the upstream and downstream boundaries.

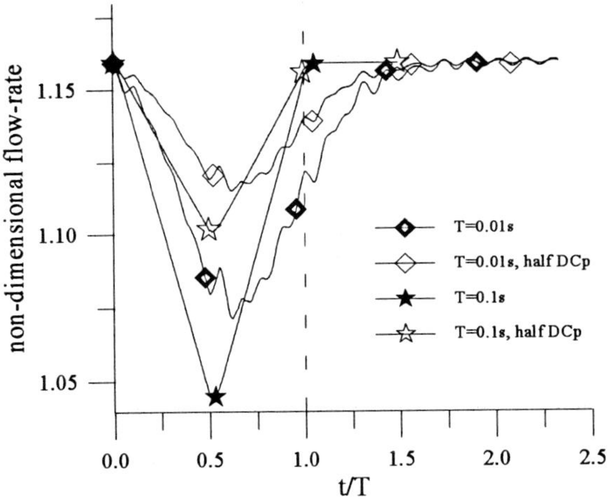

fig.6: non-dimensional flow-rate vs non-dimensional time, 8000 rpm, effect of amplitude and period

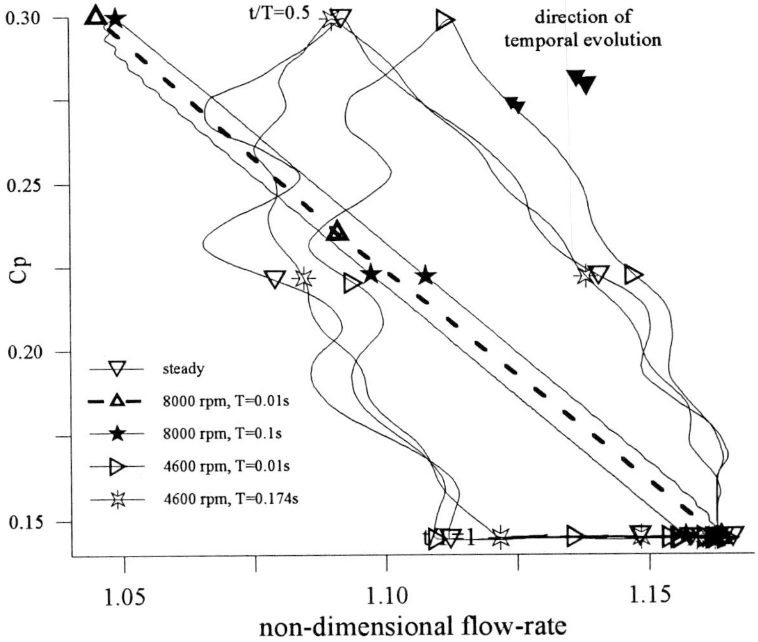

fig.7: characteristics curve, effect of period and rotational speed, two consecutive symbols on a given curve are equally spaced in time

We have represented on fig.7 characteristics curves to show the effects of rotational speed and T. The response of the inducer to a pressure fluctuation appears to follow a similarity law in ΩT. We must indeed take into account the inherent time response or inertia of the inducer:

$$T_r = \frac{\text{inducer length}}{\text{flow speed}} = \frac{c_m}{\phi_m R_t \Omega} \tag{19}$$

where c_m is the meridian chord, ϕ_m the meridian flow coefficient and R_t the tip radius. The response of the inducer is then a function of the ratio of T to T_r:

$$\frac{T}{T_r} = \frac{\phi_m R_t}{c_m} \Omega T \tag{20}$$

which is precisely the inverse of the reduced frequency of the problem. On fig.7, this ratio equals 1 for the most unsteady case (the rightmost curve), 1.7 for the intermediate cases and 17 the quasi-steady case. Our method thus seems to yield an adequate description of this problem.

3.2 Variation of rotational speed

We present here results of a simulation of the variation of rotational speed. We have investigated the influence of 2 operation parameters: the transient time T and the final value Ω_f of rotational speed. This is not a simulation of start-up as it begins at a non-zero value of initial rotational speed Ω_0 (2300 rpm). This restriction is a consequence of the complexity of the geometry, the inherent limitation of the numerical scheme at very low Mach numbers and the problems posed by the fluctuation module at high incidence angles (low flow-rates). It may actually turn to an advantage as it enable us to study the validity of the quasi-steady assumption.

We consider the following evolution of rotational speed:

$$0 \le t \le T,\ \Omega(t) = \Omega_0 + 3(\Omega_f - \Omega_0)\frac{t^2}{T^2} - 2(\Omega_f - \Omega_0)\frac{t^3}{T^3}, \tag{21}$$

with T=0.2s or 0.02s and Ω_f=4600, 9200 or 13800 rpm. Typical transient parameters of this inducer are T=1s and Ω_f=30000 rpm.

Upstream boundary conditions are constant in time (stagnation pressure and temperature, velocity angles). On the downstream boundary we impose a constant pressure coefficient and consequently an unsteady static pressure:

$$P_{S,down}(t) = P_{0,up}(t) + \tfrac{1}{2}\rho_{up} U_{tip}^2(t) \cdot Cp \tag{22}$$

With a constant pressure coefficient and if the inducer behaves quasi-steadily, then all non-dimensional parameters of the problem must be constant in time.

Fig.8 shows the evolution of pressure coefficients for Ω_f=13800 rpm and T=0.02s. Cp decreases sharply at leading edge and increases in the same manner at trailing edge. This consequence of inertia has been observed experimentally and may have the undesirable effect of triggering cavitation at leading edge as shown by Barrand and al[3].

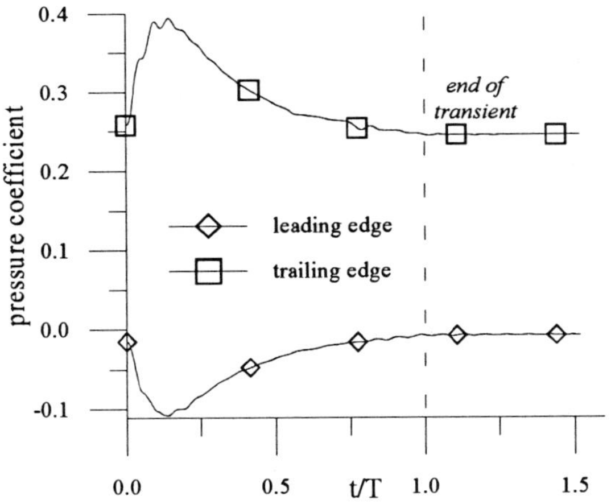

fig.8: pressure coefficient vs non-dimensional time, sharpest acceleration

The unsteady evolution of the non-dimensional mean blade forces (a mean value of Fp over the blade span) on fig.9 shows that the quasi-steady assumption is nearly correct for T=0.2s but wrong for T=0.02s. Non-dimensional blade forces evolve like the ratio of acceleration to the square of rotational speed. They are a direct result of the pressure calculation using BEM.

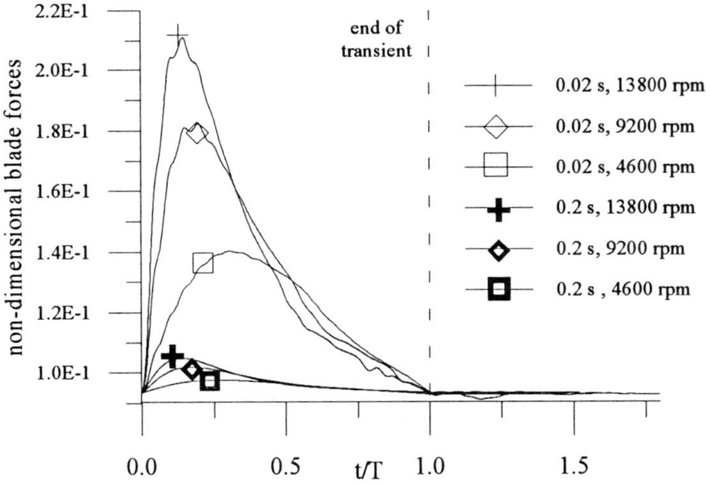

fig.9: non-dimensional mean blade forces vs non-dimensional time

The unsteady behaviour of the inducer is summarized on fig.10 and 11 by a representation of its characteristics curves (mean blade forces versus non-dimensional flow-rate). The steady characteristics curve is also represented, having been previously computed. Deviation from quasi-steadiness is obvious on fig.10. The interesting thing here is that the non-dimensional time required to come back to a quasi-steady state (ie a point located on the steady characteristics curve) increases as Ω_f decreases. This result fully agrees with what has been observed by Ohashi and Tsukamoto[2].

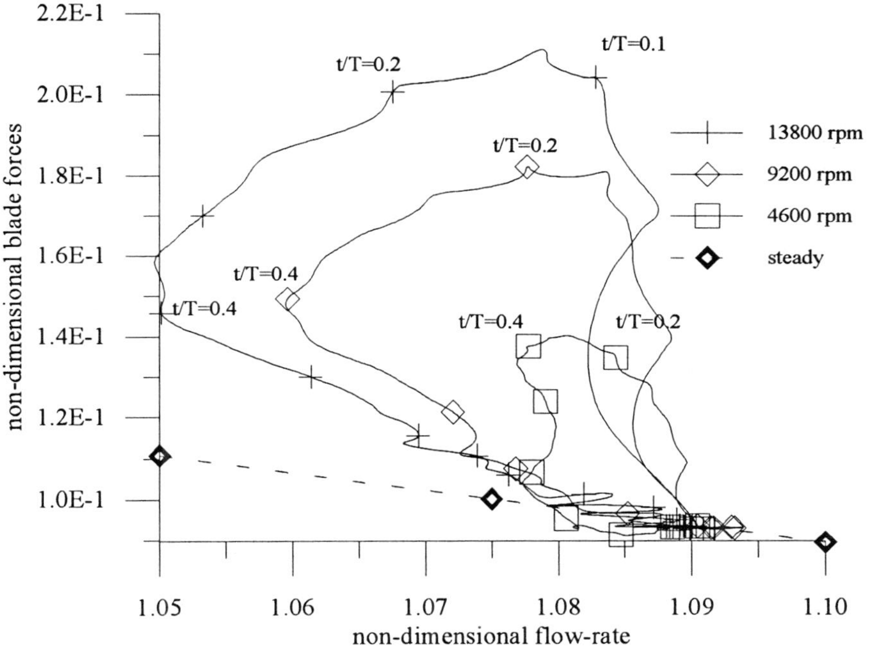

fig.10: characteristics curves for T=0.02s, consecutive symbols on a given curve are equally spaced in time

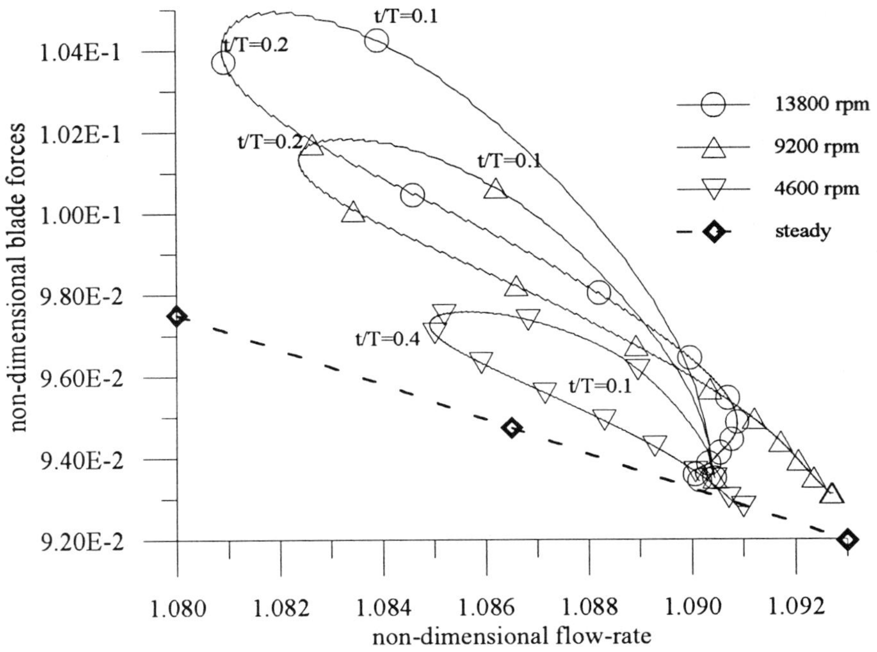

fig.11: characteristics curves for T=0.2s, consecutive symbols on a given curve are equally spaced in time

Because the quasi-steady assumption is very practical but clearly not valid for quick transients, it would be useful to have a criterium of its validity. Ohashi and Tsukamoto have proposed the parameter $\Omega_f T$ interpreting it as the number of rotations needed to reach Ω_f. Based on our work, we propose an extension of this parameter, in the form of:

$$\frac{\phi_m R_t}{c_m} \frac{T}{\Delta\Omega} \Omega_f^2 \tag{23}$$

where ϕ_m is the flow coefficient, c_m the meridian chord and R_t the tip radius. It may of course be interpretated as the ratio of steady forces to inertia, but the similarity between (20) and (23) suggests that the convergence to a quasi-steady state is connected here to a reduced frequency, ie the ratio of transient time to convection time.

Conclusion

We have designed a new unsteady throughflow method in turbomachinery and have computed unsteady flows in a 3D inducer during a quick variation of rotational speed. This method has some limitations at the present time but offers opportunities of improvement, especially the fluctuation module. Taking into account the rarity of such unsteady computations and their difficulty, we are rather satisfied of the results obtained as they have enabled us to analyse this problem. An unsteady validation is still to be performed on a centrifugal pump, but the combination of a throughflow code and an unsteady pressure calculation can nonetheless be credited for this achievement as it stands at the core of the ability of the method to handle unsteady flows in transient operating turbomachines.

Acknowledgements

The authors would like to thank SEP and CNES for the financial support.

References

1. Knapp R.T., Complete Characteristics of Centrifugal Pumps and their Use in the Prediction of Transient Behaviour, *ASME Transactions*, Vol.59, pp.683-689

2. Tsukamoto H., Ohashi H., Transient Characteristics of a Centrifugal Pump During Starting Period, *Transactions of the ASME*, Vol.104, march 1982, pp.6-14

3. Barrand J.P., Ghelici N., Caignaert, G., Unsteady Flow During the Fast Start-Up of a Centrifugal Pump, *FED-Vol.154, Pumping Machinery*, ASME 1993

4. Nguyen Duc J.M., Von Kaenel A., Danguy F., Transient Behaviour of Liquid Hydrogen Pumps During Start-up and Shutdown of Rocket Engines, *FED-Vol.154, Pumping Machinery*, ASME 1993

5. Perrin G., Leboeuf F., Investigation of through-flow hypothesis in a turbine cascade using a 3D Navier-Stokes computation, *ASME Paper n°93-GT-21*, 1993

6. El Ghazanni E.M., Sannino J.M., Maitre T., Développement du code de calcul Caleche (c) et ses applications au cas des turbopompes spatiales, *Colloque Ecoulements propulsifs dans les systèmes de transport spatial*, Bordeaux 1995, vol.2, pp. 680-713

7. Debard M., Simulation de l'écoulement lors du démarrage d'une turbopompe spatiale: contribution au développement d'un outil numérique, *Thèse de Doctorat de l'Ecole Centrale de Lyon*, 1997
8. Ferrand P., Aubert S., A New Mixed Van-Leer Flux-Splitting for Transonic Viscous Flows, *AIAA Journal*, Vol.34, n°1, pp.193-197, January 1996
9. Jameson A., Schmidt W., Turkel E., Numerical Solution of the Euler Equations by Finite Volume Methods Using Runge-Kutta Time Stepping Schemes, *AIAA 5th Computational Fluid Dynamics Conference*, AIAA Paper 81-1259
10. Larrey E., Ferrand P., Leboeuf F., Resolution of the pressure equation by the boundary element method in multistaged turbomachines, *Boundary Elements XV*, Vol.1, EM, CMP/Elsevier, pp.147-161, 1993

Glossary

Z, R, Θ :	axial, radial and azimuthal coordinates	(m)
c :	chord	(m)
b :	blockage factor	(rad)
Ω :	rotational speed	(rad.s^{-1})
U :	entrainment velocity	(m.s^{-1})
W :	relative velocity	(m.s^{-1})
e^* :	relative stagnation energy	(J.m^{-3})
P, P_s :	static pressure	(Pa)
P_0 :	relative stagnation pressure	(Pa)
Cp :	pressure coefficient	
n :	normal vector	
ps, ss (subscripts):	pressure side, suction side	
m (subscript):	meridian	
t (subscript):	tip	
E,F,G :	convective fluxes	
H :	curvature and mobile frame of reference terms	
[H], [G], [S]:	BEM influence coefficients	

INVESTIGATION OF STATOR-ROTOR INTERACTION PHENOMENA IN AN CENTRIFUGAL PUMP

YVES P. MARX
Institut de Machines Hydrauliques et de Mécanique des Fluides
Ecole Polytechnique Fédérale de Lausanne
CH-1015 Lausanne, Switzerland.

1. Introduction

In a cooperative effort with the Sulzer company, a research project was set up to investigate the stator-rotor mechanisms on the middle stage of a pump ($n_q = 33$). Detailed flow measurements using Laser Particle Tracking Velocimetry and Laser Doppler Anemometry were performed at Sulzer Innotec [1] to improve the understanding of the interaction phenomena and to build a data base for validating the numerical flow simulations. Steady numerical simulations were carried out on the different elements of the pump [2] and unsteady simulations were performed in the vane diffuser, using the measured unsteady profiles as inlet boundary conditions [3]. In the following, coupled unsteady simulations of the impeller are presented.

The flow leaving a centrifugal impeller presents non-uniformity between the hub and the shroud as well as in the circumferential direction. Under the action of the centrifugal and Coriolis forces [4], the low-momentum fluid accumulates in the shroud-suction side corner. This migration led to the so-called "jet-wake" model [5]. To reduce the complexity of the flow analysis, only 2D simulations were performed, even if it appears that hub-shroud variations are more important than circumferential non-uniformity, [6, 7]. The objective of this work was to simulate stator-rotor interaction flows and to analyze this flow, since the "numerical pump" can be exhaustively instrumented.

The flow solution of the 2D pump exhibits two peculiarities, fig.(1); most of the fluid in the impeller is flowing in a jet and a massive separation is present in the diffuser. The jet structure in the impeller is ideal since stator-rotor interactions are investigated, therefore it is preferable to have a flow

T.H. Fransson (ed.), Unsteady Aerodynamics and Aeroelasticity of Turbomachines, 599-612.

leaving the impeller with significant non-uniformity in the circumferential direction. Furthermore, this jet flow can be perceived as a model of a real 3D flow leaving a centrifugal impeller with the "jet-wake" structure. The separation in the diffuser can be also representative of a 3D flow, since the thick boundary layers growing on the walls can produce a mismatch in the vane incidence flow angle and induce a separation, even at nominal flow rate [7]. Additionally, it may be interesting to study the influence of the unsteady forcing generated by the rotation of the impeller on the stability of separated flow in the diffuser.

Several simulations were performed to investigate the flow phenomena generated by stator-rotor interactions. An unsteady simulation on a complete pump (7 impeller blades and 14 diffuser vanes) was made to verify that the flow in each impeller channel and in every two diffuser channels are the same. Since the computed flow presents effectively a $2\pi/7$ circumferential periodicity, the remaining sliding simulations were conducted on only one impeller and two diffuser channels. For these simulations 100 time steps were used for representing the sliding of an impeller blade across two diffuser channels ($\Delta\theta = 2\pi/7$).

To estimate the influence of the diffuser on the pump flow, steady simulation on a pump with a vaneless diffuser was also calculated and compared with the time-averaged unsteady sliding simulation.

Steady mixing plane solution on an impeller and two diffuser channels was calculated to assess the importance of the unsteady effects. An unsteady mixing plane solution on the complete pump was calculated also, for a reason described in §3.1.

To quantify the importance of potential interaction effects in the impeller, an Euler simulation on a segment of disk ($r \in [r_1, r_2]$ and $\theta \in [0, 2\pi/7]$ with r_1 and r_2 respectively the impeller inlet and outlet radii) was made with the following boundary conditions. At r_1 an uniform inlet flow was prescribed with the same mass flow as for the pump. At r_2 the time-averaged solution was imposed and periodic conditions were specified on the lateral boundaries.

Finally, unsteady simulations on the diffuser and on the impeller alone were made using unsteady boundary conditions.

2. Numerical Procedure

The stator-rotor simulations were performed using a computer program developed within the institute. This program solves the unsteady Reynolds-averaged Navier-Stokes equations for an incompressible fluid on 3D block-structured meshes. The equations are discretized using a cell-centered finite-volume method. The artificial compressibility technique is used for coupling the incompressible velocity field to the pressure field. A high-order upwind spatial discretization scheme based on the approximate Riemann solver of Roe is employed for approximating the advection terms. The diffusion terms are discretized using a centered approximation. A dual-time step procedure with the implicit two-stage Runge-Kutta scheme recommended in [8] is used for integrating in time the unsteady Navier-Stokes equations. At each time step, a non-linear system resembling the equations for stationary flow is solved using an ADI method.

Steady simulations were conducted using a mixing plane interface condition at the block boundary and unsteady simulations were made using a sliding block technique. The flow equations are solved in a fixed frame in the diffuser and in a rotating frame in the impeller. The sliding technique consists to move the impeller mesh at each time step and to calculate the cell overlapping (patches) between the impeller outlet and the diffuser inlet face meshes. The flux at the block interface is then obtained by adding the flux contribution of each "patch". This flux is calculated using techniques similar to the one employed for a cell face inside a block mesh. The mixing plane technique was derived such that (i) it has a minimal effect on an usual flux evaluation and (ii) any dependence on the relative position between the impeller and the diffuser is suppressed. This was achieved by averaging circumferentially any impeller quantity required in the flux evaluation, at the block boundary, for updating the flow in the diffuser. Similarly, diffuser quantities are circumferentially averaged when the impeller flow is updated.

For all the simulations, the Baldwin-Lomax turbulence model with wall functions was used to estimate the Reynolds stresses.

3. Simulations and discussions

3.1. REMARKS ON THE FLOW SOLUTIONS

Stability of the flow From the various simulations, it is observed that a solution in which both diffuser channels are performing identically, is not stable when circumferentially uniform flow is imposed at the inlet of the diffuser. The stable solution is such that the flow in both diffuser are different; in one diffuser channel the flow is almost blocked, and in the other channel it is fully attached. This "non-periodic" solution was found with steady mixing plane simulation of the pump (1 impeller and 2 diffuser channels), with steady simulation of the diffuser alone and with unsteady mixing plane solution of the full pump. In this latter case, a visualization of the unsteady flow shows clearly how the "periodic" solution destabilizes slowly and how the "non-periodic" solution forms. On the other hand, when sliding mesh simulation and when diffuser simulation with unsteady inlet flow are performed, the time-averaged solution was always found to be identical in both diffuser channels.

In the remaining, to produce a fair comparison with the time-averaged solution, the steady simulations were carried out by imposing a $2\pi/14$ periodicity at the lateral boundaries of the diffuser.

Computation of the average Since the computations are performed in a rotating frame in the impeller and in a fixed frame in the diffuser, it is natural (and easier) to calculate the time-averaged solution in these frames. With this procedure the time-averaged solution will be discontinuous across the impeller-diffuser boundary. Such a discontinuity will show up also when the evolution of the unsteady part (difference between the instantaneous signal and the time-averaged solution) is displayed. In order to calculate averaged solutions that are continuous across the sliding boundary, it is necessary to perform two averages; one in the rotating frame and one in the fixed frame. The averages are calculated from

$$\bar{q} = \int_0^T q(X,t)H(X,t)\,dt \Big/ \int_0^T H(X,t)\,dt$$

where

- X is the location of the point (in the fixed or rotating frame),
- $H(X,t)$ is a function having two values; $H(X,t) = 1$ when the point is in the fluid and $H(X,t) = 0$ when it is in the blade.

These averages have the advantage to enable the comparison between the steady solution in the vaneless diffuser and the time-averaged solution in the vane diffuser §3.2. The averages allow also to quantify the importance of potential interaction effects in the impeller §3.3.

3.2. EVOLUTION OF THE FLOW IN THE DIFFUSER

The development of the flow after it leaves the impeller is given in figures (2-3). The vaneless and vane diffuser cases are compared. The vane diffuser solution is the time-averaged solution performed in the rotating frame. To allow a good comparison, the mean values were subtracted out of the distributions. In the diffuser channel, fig.(3 $R = 1.1$) indicates that the pressure oscillations are larger in the vane diffuser than in the vaneless diffuser case. It is observed that the extrema are located at the same positions. From figure (3) it is seen that the wake in the vaneless gap is only slightly damped by the diffuser vanes. Again, the extrema observed in both cases are well correlated. Inside the vane diffuser (R=1.1) it is found that the vanes modify completely the velocity distribution. Since u_r is almost constant at R=1.1, the velocities in the vane diffuser are steady and do not depend on the position of the impeller blades.

The pressure and velocity distributions displayed in figures (4,5) were obtained when the time averaging is calculated in the fixed frame. The "diffuser" distributions were obtained from steady simulations on the diffuser using the time-averaged profiles as inlet boundary condition. In the vaneless gap, while the velocities are in good agreement the pressure distributions show noticeable differences, the larger deviation being observed with the mixing plane solution. In the diffuser channel all solutions are similar.

3.3. EVOLUTION OF THE FLOW IN THE IMPELLER

In figure (6) pressure distributions in the impeller are given for three flow simulations. Apart from the mean pressure differences which are due to the reference pressure that was set at the pump exit, the pressure distributions and the velocity distributions are almost not affected by the presence of the diffuser.

Figure (7) illustrates clearly that unsteady fluctuations in the impeller result mostly from potential effects. In these figures the variations of the time-averaged velocity distributions (the averaging being performed in the fixed frame) are compared with the variation of velocities obtained from the calculation of an inviscid flow on a portion of disk with boundary conditions described in section §1. The agreement is very good for the radial velocity, fig.(7), but also for the circumferential velocity (not shown). It should be stressed that such a good agreement is not observed on the pressure distribution and is found only when the time-averaged solution is used at the outlet boundary and when the incoming Riemann invariant are employed as outflow boundary condition. With other treatment, e.g. imposing the pressure at the outlet, the agreement is poor.

3.4. UNSTEADY SIMULATIONS

Animations of the impeller-diffuser interaction were made. Some of the animations can be viewed on W^3 at the address

`http://imhefwww.epfl.ch/lmf/animation/pump`

These include the evolution of the pressure field, the evolution of the absolute and relative velocity norms and the variation of unsteady velocity (difference between the instantaneous velocity and the time-averaged velocity). The pressure animation shows clearly the potential effect when the impeller trailing edge passes in front of the diffuser vane leading edge. Pressure perturbations due to the vortex shedding behind the impeller blades can be observed as well. The velocity animations (viewed by an observer attached to the impeller or to the diffuser) indicate that whereas the separated zone in the diffuser seems to be unaffected by the impeller motion, the jet is periodically cut by the diffuser vanes. When examining carefully behind the trailing edge of the impeller blade, a small vortex shedding is noticed. Animations of the unsteady velocity seen by observers attached to the diffuser or to the impeller are very different. Viewed from a fixed frame, it appears that most of the perturbations are cycling, following the impeller blades, around the pump and that layers of low and high velocities are piling in the diffuser. On the other hand, an observer attached to the impeller sees spots of high and low velocities moving with the diffuser vane. These spots are generated by the lateral motion of the jet when it passes in front of the vanes. In this latter animation, an excellent correlation between the spots and the pressure contours is found in the interaction zone.

The time evolution of the pressure and velocity norm were recorded at several locations in the pump, fig.(8, 9), (intersection of mean flow streamlines with constant radius lines). The numerical probe points are labeled $P_1, P_2, \ldots$. On a radius line, the labels are increasing in the counterclockwise direction. Only the unsteady part are displayed. These signals were shifted vertically to reduce their overlapping. The signals displayed are non-dimensionalized by the blade speed and by the pressure difference in the pump. The first feature to notice is that all the signals are exactly periodic. The periodicity is $2\pi/7$ in the diffuser and $2\pi/14$ in the impeller.

On the pressure signals in the diffuser, fig.(10-12), five vertical lines were traced. These corresponds to specific instants identified in the animations. At time $t \simeq 2.75$ the impeller trailing edge is passing in front of the diffuser vane 1 leading edge . Time $t \simeq 3.2$ is the equivalent of time $t \simeq 2.75$ for the middle vane 2. At time $t \simeq 2.8$ the impeller is passing in front of the low pressure zone on the suction side of the vane 1. Finally, times $t \simeq 2.86$ and $t \simeq 3.31$ correspond to the instants when the jet leaving the impeller is cut by the diffuser leading edge. At $t \simeq 2.86$ the spitting is done by the vane 3 and at $t \simeq 3.31$ by the middle vane 2). Other instants can be identified —for instance, the local peak around $t = 2.97$ observed on the signal measured at point P_1 on the diffuser radius D_1, fig.(12), is related to the vortex shedding. The lines marked lines were chosen because they influence all the points in the diffuser. From figure (10), it appears that the signals at the different points are synchronized and the unsteadiness is larger on the suction side. Upstream in the diffuser, fig.(11), the signals are also synchronized. The signal measured at point P_4 seems shifted, but it is because the point lies in another channel than points $P_1 - P_3$. The comparison of the signal issued from point P_4 with the other points indicates that the sectioning of the jet leaving the impeller influences both channels. On the other hand, when the impeller passes in front of the low pressure zone (time $t \simeq 2.8$) it has a large effect on the diffuser channel between vanes 1 and 2, but its influence on the neighboring channel can hardly be seen. Further upstream in the diffuser, fig.(12), the pressure signals are far more complex. The synchronization is not as obvious as before because the position of the blade introduces locally large perturbations. The symbols marked in figure (12) indicate the time when the blade is passing in front of the numerical probe. It is noticed that the perturbation introduced by the blade increases as the blade is approaching the diffuser leading edge.

In the impeller, the influence of the vane when the blade is passing in front of the low pressure zone of the diffuser suction side was not marked since it affects only the pressure signals near the outlet of the impeller. Conversely, a peak is visible at $t \simeq 2.96$, fig. (13), but its origin is unclear. Figure (13) indicates that the pressure unsteadiness underwent a minimum about the middle of the impeller which has some resemblance with the minimum observed in [9].

The velocity signals recorded are much smoother than the pressure signals and they decay more rapidly. For instance, the jet strength induces a variation of 10% of the velocity norm at radius D_1, figure (14), but only a variation of around 1% at radius D_4. In the impeller, the velocity signals present also some unsteadiness. These unsteadiness are increasing toward the pressure side. The animation of the unsteady component of the velocity (viewed in the impeller frame) indicates that spots of unsteadiness form behind the impeller blade. They follow the diffuser motion and simultaneously grow. The signals recorded at the different probes indicate that the velocity extrema are neither synchronized nor linked directly to the diffuser motion, except just behind the impeller.

Comparisons between the signals recorded during the sliding block simulation of the pump and those obtained in the diffuser, showed that the sliding block computation gives larger pressure but smaller velocity unsteadiness. The pressure signals observed in the unsteady diffuser simulation are also much simpler. The variations generated by the unsteady inlet conditions are roughly independent of the angular position of the probe which contrasts with the increase in amplitude, observed in sliding block computations near the diffuser leading edge. The amplitude of pressure unsteadiness in the diffuser simulation are between 1/3 and 1/2 less than those of the pump simulation. On the other hand, the unsteady velocity displays more extrema, with larger amplitude in the unsteady diffuser simulation, fig.(15). However, the velocity unsteadiness are reduced more quickly in the diffuser simulation than in the coupled simulation.

4. Conclusions

Unsteady, sliding block computations in a centrifugal pump were performed and compared with simplified simulations. From the different simulations conducted the following was found.

- The flow in the diffuser obtained from a sliding block simulation is the same for each channel. In a mixing plane computation it tends to be different.
- When periodicity in the diffuser is controlled, the mixing plane and time-averaging of the unsteady simulation differ by few percent.
- The presence of the diffuser enhance only slightly the mixing of the wake leaving the diffuser.
- The unsteadiness generated in the impeller are due to potential effects.
- Pressure unsteadiness can be as high as 20% of the pump head. The velocity unsteadiness are about 10% of the impeller trailing edge velocity.
- Unsteady signals in the diffuser can be reproduced only roughly using unsteady inlet conditions.

Acknowledgments — This work received financial support from the Commission pour la Technologie et l'Innovation and Sulzer Brothers S.A. under Contract Number 3062.1

References

1. M.V. Casey, M. Eisele, Z. Zhang, Gülich J., and Schachenmann A. Flow analysis in a pump diffuser, part 1 : LDA and PTV measurements of the unsteady flow. *ASME Paper*, FED Summer Meeting, Symposium on Laser Anemometry, 1995.
2. M.V. Casey, M. Eisele, Muggli F.A., Gülich J., and Schachenmann A. Flow analysis in a pump diffuser, part 2 : Validation of a CFD code for steady flow. *ASME Paper*, FED Summer Meeting, Symposium on Laser Anemometry, 1995.
3. F. A. Muggli, D. Wiss, K. Eisele, Z. Zhang, M. A. Casey, and P. Galpin. Unsteady flow in the vaned diffuser of a medium specific speed pump. *ASME Paper*, 96-GT-157, 1996.
4. M. W. Johnson. Secondary flows in rotating bends. *ASME J. of Eng. for Power*, 100:553–560, 1978.
5. R.C. Dean and Y. Senoo. Rotating wakes in vaneless diffusers. *ASME J. of Basic Engineering*, 82:563–574, 1960.
6. S. Baghdadi. The effect of rotor blade wakes on centrifugal compressor diffuser performance — a comparative experiment. *ASME J. of Fluids Engineering*, 99:45–52, 1977.
7. W.N. Dawes. A simulation of the unsteady interaction of a centrifugal impeller with its vaned diffuser: flows analysis. *ASME J. of Turbomachinery*, 117:213–222, 1995.
8. Y.P. Marx. Time integration schemes for the unsteady incompressible Navier-Stokes equations. *J. of Comput. Physics*, 112(1):182–209, 1994.
9. N. Arndt, A. J. Acosta, C. E. Brennen, and T. K. Caughey. Experimental investigation of rotor-stator interaction in a centrifugal pump with several vaned diffusers. *ASME J. of Turbomachinery*, 112:99–108, 1990.

Figure 1. Mean flow in the 2D pump

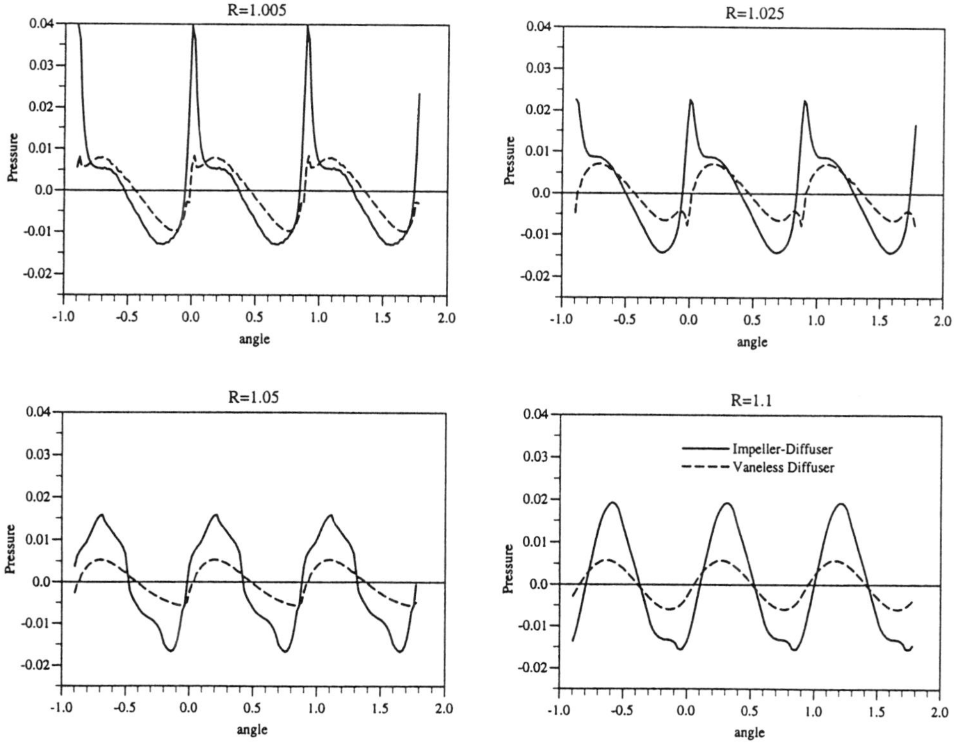

Figure 2. Mean pressure in the diffuser (rotating frame)

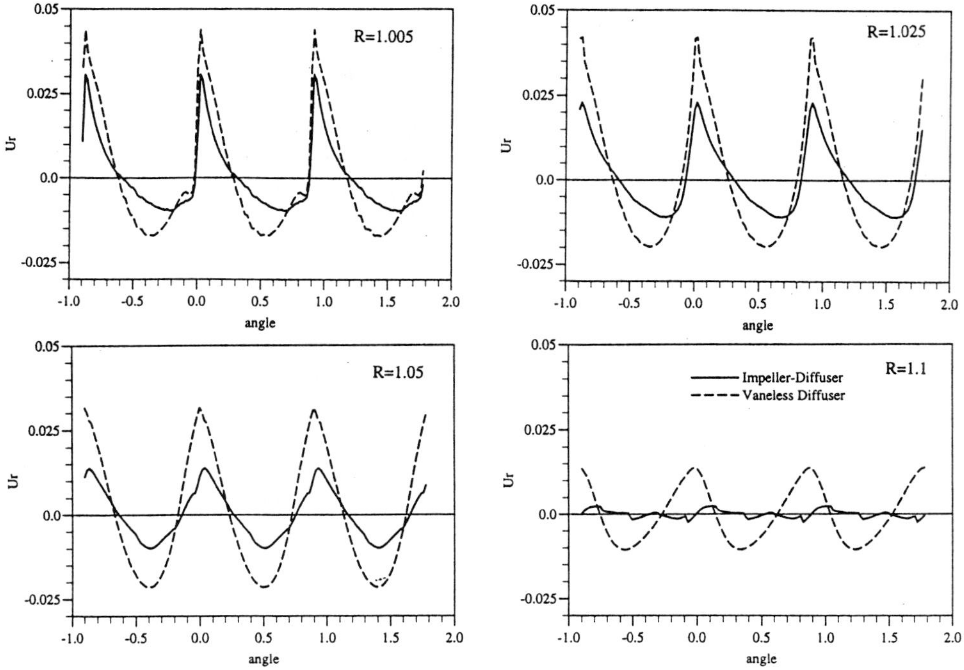

Figure 3. Mean velocity in the diffuser (rotating frame)

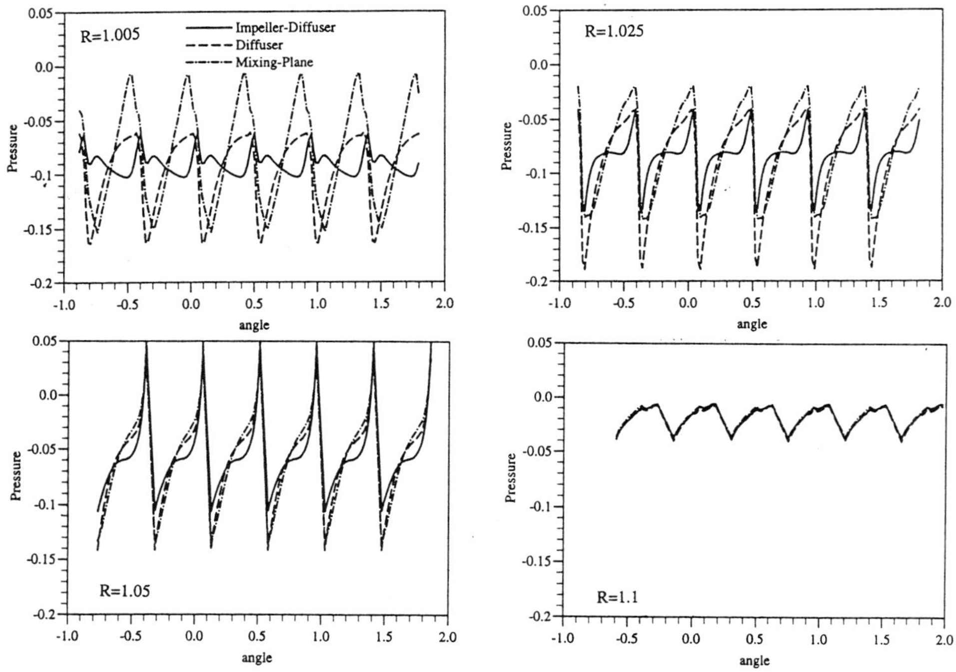

Figure 4. Mean pressure in the diffuser (fixed frame)

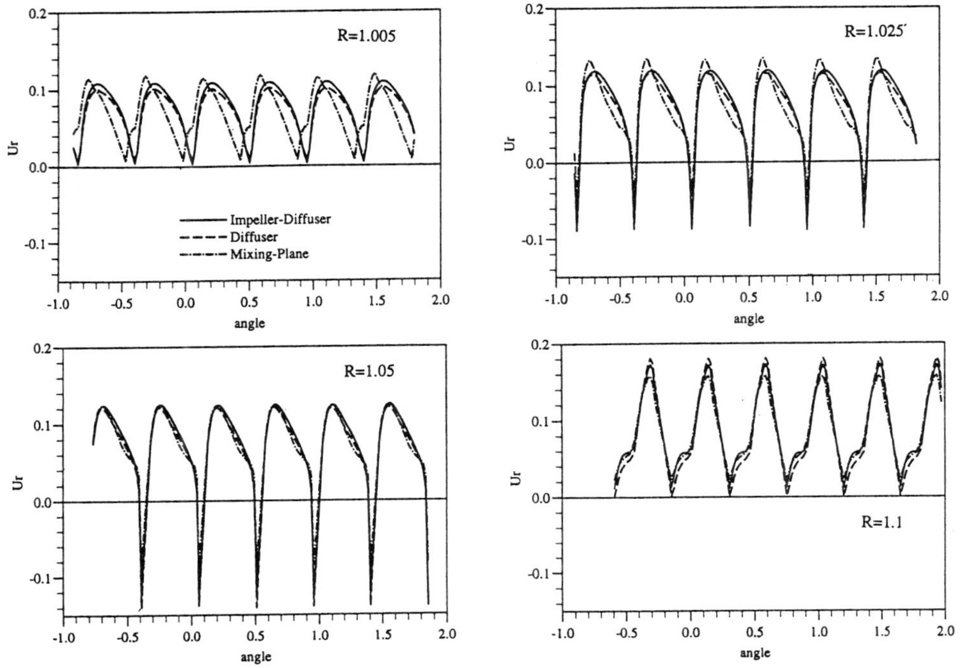

Figure 5. Mean velocity in the diffuser (fixed frame)

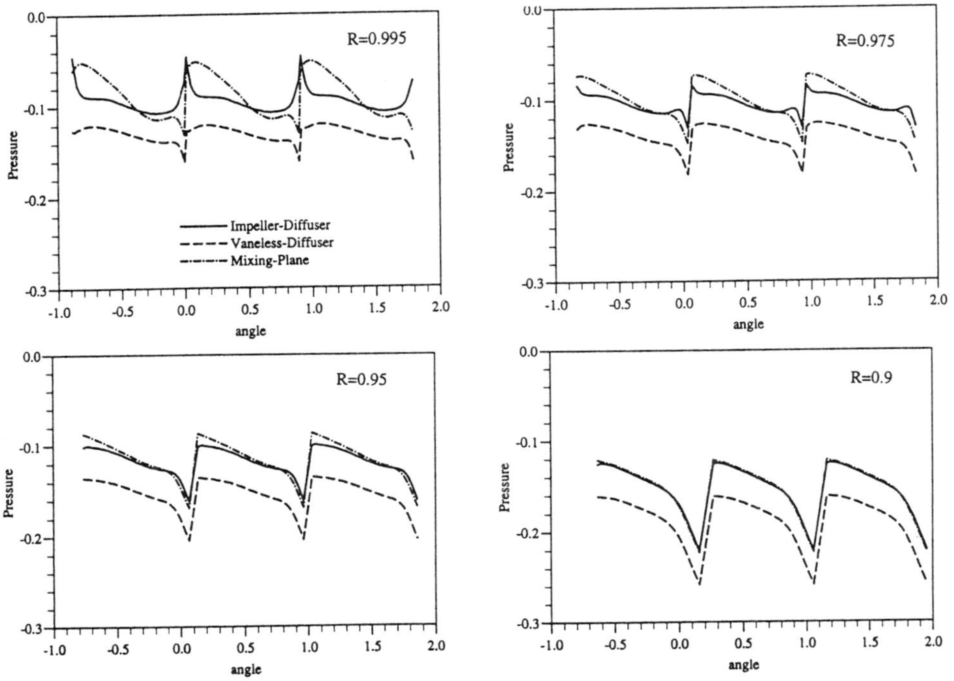

Figure 6. Mean pressure in the impeller (rotating frame)

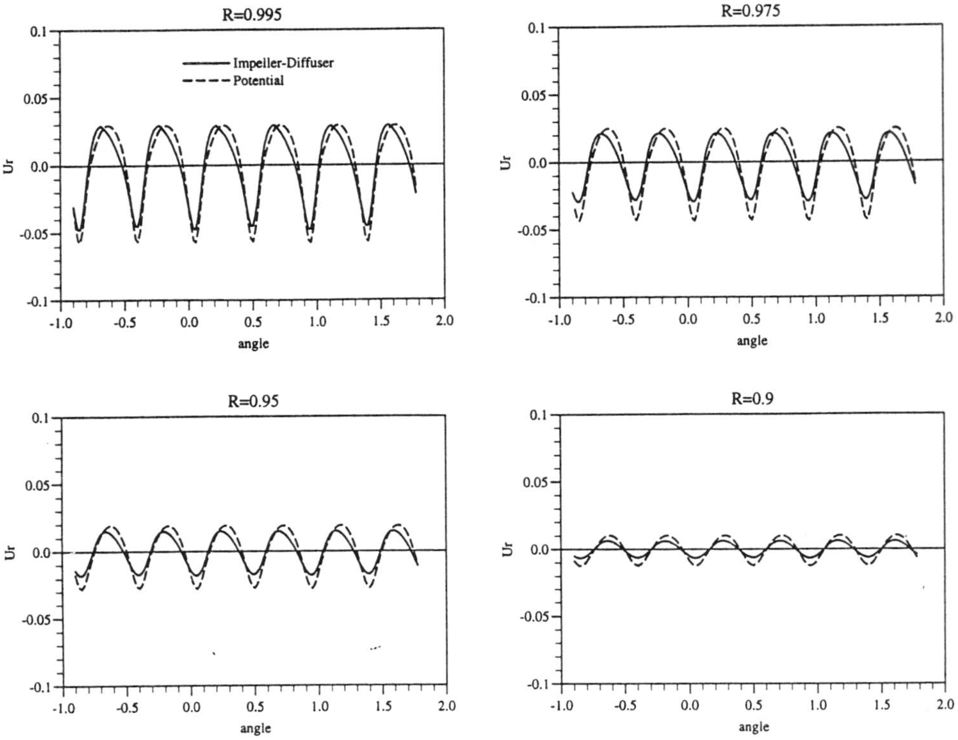

Figure 7. Velocity variations in the impeller (fixed frame)

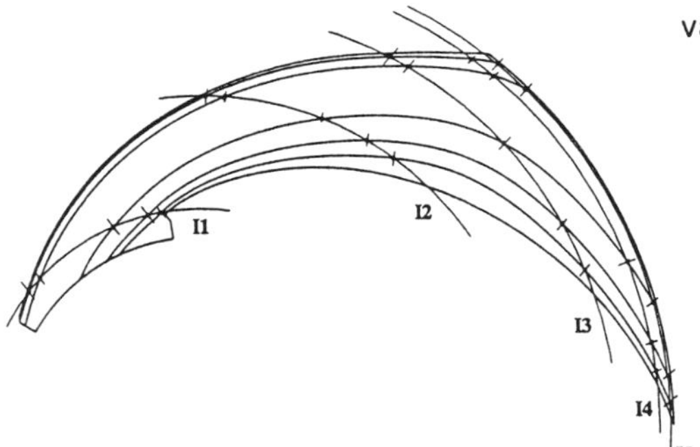

Figure 8. Probe locations in the impeller

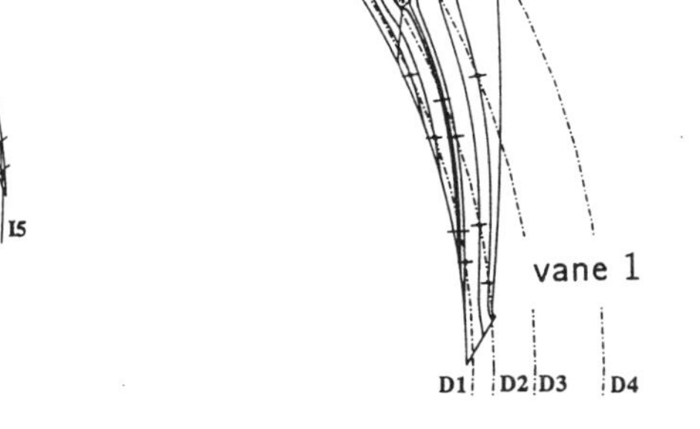

Figure 9. Probe locations in the diffuser

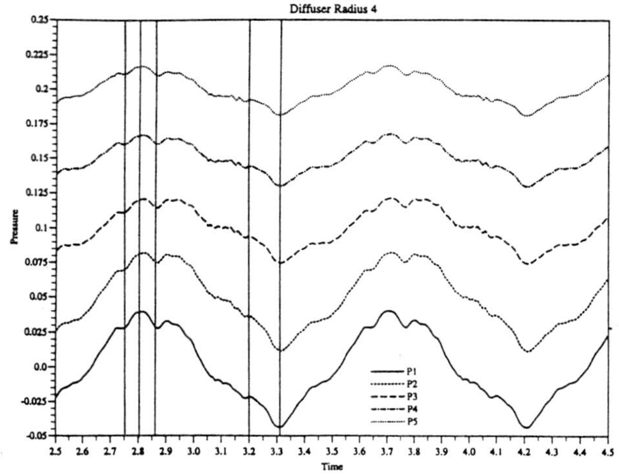

Figure 10.

Unsteady pressure signals at radius D_4 in the diffuser

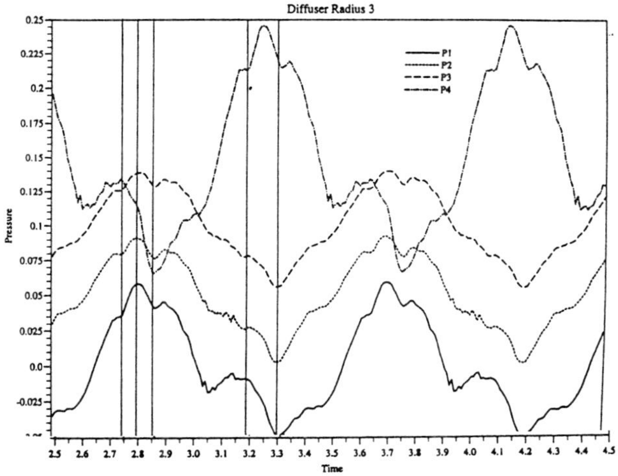

Figure 11.

Unsteady pressure signals at radius D_3 in the diffuser

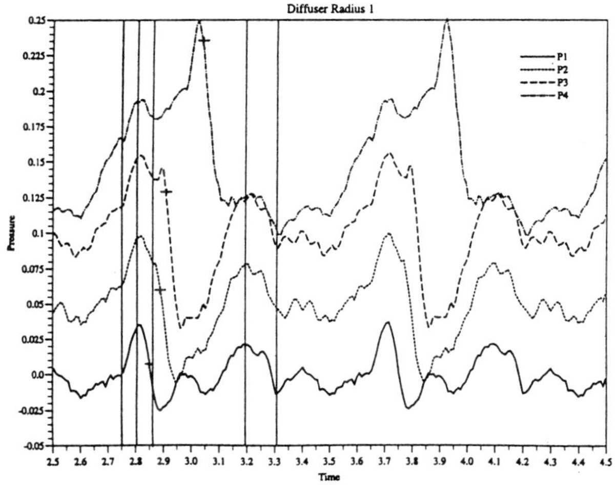

Figure 12.

Unsteady pressure signals at radius D_1 in the diffuser

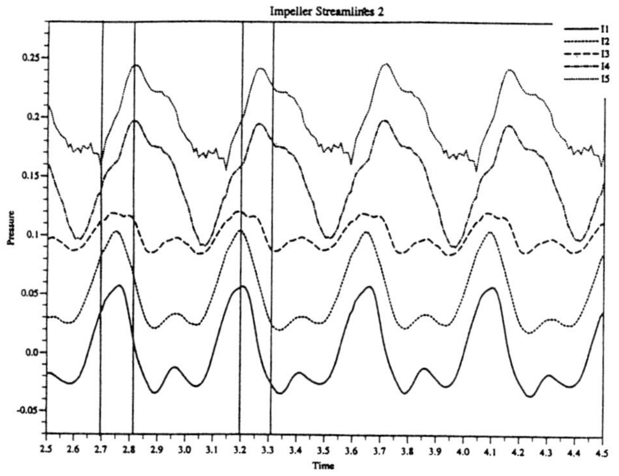

Figure 13.

Unsteady pressure signals along streamlines 2 in the impeller

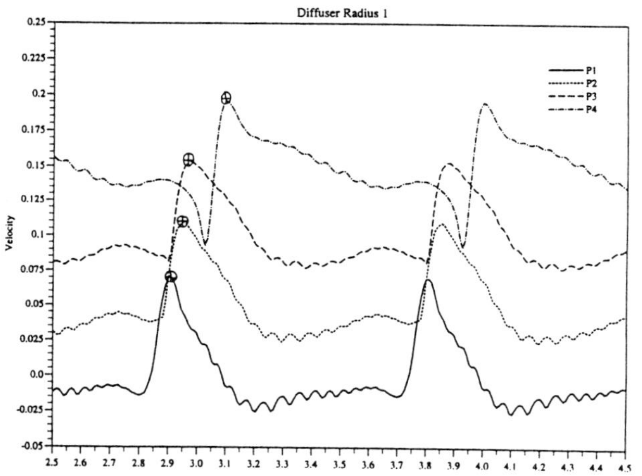

Figure 14.

Unsteady velocity signals at radius D_1 in the diffuser

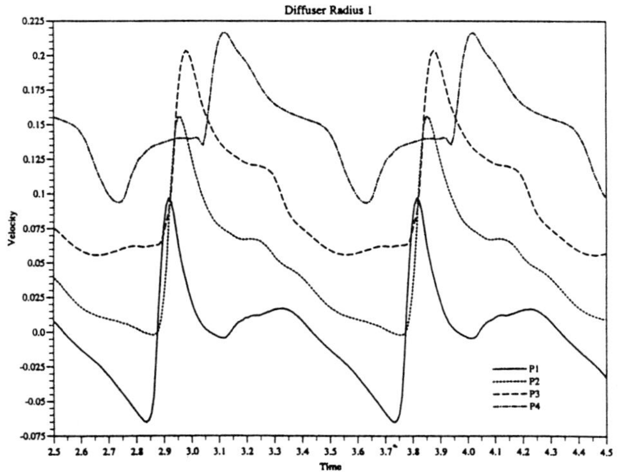

Figure 15.

Unsteady velocity signals at radius D_1 in the diffuser, Unsteady diffuser simulation

The 3D Unsteady Aerodynamic Forces: The Forced Vibration of Bladed Discs

V. GNESIN
Department of Hydromechanics of Power Machines,
Institute for Problems in Machinery
Ukrainian National Academy of Sciences
Pozarsky St. 2/10 Kharkov, 310046, Ukraine

R. RZĄDKOWSKI
Institute of Fluid-Flow Machines, Polish Academy of Sciences,
Gdańsk 80 – 952, *ul.Fiszera 14, Poland.*

1.Introduction

The main problem for a bladed disc assembly under running conditions is the response levels when the excitation frequencies coincide natural frequencies of the structure during operation. Generally in the bladed analysis the excitation force are assumed in advance, although there are papers where these forces were calculated from the very simple flow model (see Irretier and Rzadkowski [3], Rzadkowski [8]).

In this paper the forced vibration analysis of the bladed disc will be considered for the nozzle and low frequency excitations. The unsteady forces will be calculated from the 3D flow model.

T.H. Fransson (ed.), Unsteady Aerodynamics and Aeroelasticity of Turbomachines, 613-632.

2. Structure model

The blade model applied here is a one-dimensional beam described by an extended beam-theory including all important effects on a rotating blade. The model was described in detail in reference (see Rzadkowski [6]). The disc is modelled by a moderately thick plate theory. Assuming rigidly fixed blades on the disk rim, the displacements of any particle of the bladed disc are found.

$$\mathbf{u} = \mathbf{U}\mathbf{q}, \tag{1}$$

where

$$\mathbf{u} = col(\mathbf{u}_1, ..., \mathbf{u}_N, \mathbf{u}_d),$$

$$\mathbf{q} = col(\mathbf{q}_1, ..., \mathbf{q}_N, \mathbf{q}_d),$$

$$\mathbf{U} = \begin{bmatrix} \mathbf{U}_1 & \mathbf{0} & \dots & \mathbf{0} & \mathbf{U}_{sp1} \\ \mathbf{0} & \mathbf{U}_2 & \dots & \mathbf{0} & \mathbf{U}_{sp2} \\ \cdot & \cdot & \cdot & \cdot & \cdot \\ \mathbf{0} & \mathbf{0} & \dots & \mathbf{U}_N & \mathbf{U}_{spN} \\ \mathbf{U}_{sp1}^T & \mathbf{U}_{sp2}^T & \dots & \mathbf{U}_{spN}^T & \mathbf{U}_d \end{bmatrix},$$

$\mathbf{u}_i$ is the vector of generalized displacements of each point of the ith blade, $\mathbf{u}_d$ is the vector of general displacements of each point of the disc, $\mathbf{q}_i$ are the unknown functions of the ith blade and the $\mathbf{q}_d$ are the unknown functions of the disc, both describing the displacement field. The matrices $\mathbf{u}, \mathbf{q}, \mathbf{U}_i, \mathbf{U}_d, \mathbf{U}_{spi}$, i=1, ... , N are shown in (Rzadkowski [6]).

The blades and disc are made of a Hooke's material.

In order to find an approximate solution of the forced vibrations of the mistuned bladed disc, the Hamilton principle can be used:

$$\delta \int_{t_1}^{t_2} (E - K)\,\mathrm{dt} = \int_{t_1}^{t_2} \delta(W - D)\,\mathrm{dt}, \tag{2}$$

where E and K are elastic and kinetic energies, respectively, given by

$$E = 0.5 \int_B \mathbf{C} \cdot (\mathbf{E} \otimes \mathbf{E})\mathrm{dV}, \quad K = 0.5 \int_B \rho \dot{\mathbf{u}}\dot{\mathbf{u}}\mathrm{dV}, \tag{3}$$

where $\mathbf{C}$ and $\mathbf{E}$ are the tensors of elastic coefficients and the Green deformation, respectively, and δW and δD are the variations of external work and the internal dissipative forces, respectively.

Following the above equations, the equation of motion of the bladed disc for the forced vibration is represented by the equation

$$\mathbf{M\ddot{u}} + \mathbf{C\dot{u}} + \mathbf{Ku} = \mathbf{F}, \tag{4}$$

where **M**, **C**, **K** are the mass, damping and stiffness matrix of the bladed disc, respectively, and **F** is the vector of external forces.

It is assumed that the damping is proportional i.e.

$$\mathbf{P}_i^T\mathbf{C}\mathbf{P}_j = 2\omega_i\xi_i\delta_{ij}, \tag{5}$$

where $\mathbf{P}_i$ are the eigenvectors of the undamped system, ω_i are the natural frequencies of the system, ξ_i are the modal damping ratios and δ_{ij} is the Kronecker delta. A proportional damping matrix **C** is of the form:

$$\mathbf{C} = \alpha\mathbf{M} + \beta\mathbf{K}, \tag{6}$$

where α and β are constants related to the modal damping ratio by:

$$2\xi_i = \frac{\alpha}{\omega_i} + \beta\omega_i. \tag{7}$$

Taking two values of $\xi_1 = 0.001$ and $\xi_2 = 0.01$ and the two corresponding natural frequencies, the values for α and β were determined.

The unsteady forces are periodic functions in time and can be expressed by the Fourier-series

$$\mathbf{F}(t) = \sum_{k=0}^{K}(\mathbf{F}_{1k} + i\mathbf{F}_{2k})e^{i\nu_k t}, \tag{8}$$

where

$$\nu_k = kN_1\Omega, \tag{9}$$

N_1 is the number of nozzles, Ω is the angular velocity of the bladed disc, and $\mathbf{F}_{1k}$ and $\mathbf{F}_{2k}$ the real and imaginary parts of the amplitudes. Generally, the force vectors $\mathbf{F}_{1k}$ and $\mathbf{F}_{2k}$, in the structure model consists of

$$\mathbf{F}_{1k} = col(\mathbf{P}_{1kx_{11}}, \mathbf{P}_{1kx_{21}}, \mathbf{P}_{1kx_{31}}, \mathbf{M}_{1kx_{11}}, \mathbf{M}_{1kx_{21}},$$

$$\ldots, \mathbf{P}_{1kx_{1N}}, \mathbf{P}_{1kx_{2N}}, \mathbf{P}_{1kx_{3N}},$$

$$\mathbf{M}_{1kx_{1N}},, \mathbf{M}_{1kx_{2N}}, \mathbf{M}_{1kx_{3N}})_k \tag{10}$$

where $\mathbf{P}_{1kx_{ij}}, \mathbf{M}_{1kx_{ij}}$ are the forces and moments acting on the jth blade with respect to the directions $x_{1_{ij}}$ for kth harmonic.

The displacement vector $\mathbf{u}$ is of the form (see Bathe et al. [1]):

$$\mathbf{u}(t) = \sum_{k=1}^{K} \mathbf{P}_k x_k(t). \tag{11}$$

So, the first n frequencies and modes of the undamped system are found, and the modal transformation equation (see equation (11)) and equation (8) are applied to the system (4), which yields

$$\ddot{x}_k + 2\omega_k \xi_k \dot{x}_k + \omega_k^2 x = r_k. \tag{12}$$

It is assumed that x_k is in the form

$$x_k = \sum_{j=0}^{K} (a_{kj} + ib_{kj}) e^{i\nu_j t}, \tag{13}$$

so, one can find

$$\begin{aligned} a_{kj} &= (\mathbf{P}_k^T \mathbf{F}_{2j} + \tfrac{\omega_k^2 - \nu_j^2}{2\omega_k \nu_j \xi_k} \mathbf{P}_k^T \mathbf{F}_{1j}) / (2\omega_k \nu_j \xi_k + \tfrac{(\omega_k^2 - \nu_j^2)^2}{2\omega_k \nu_j \xi_k}), \\ b_{kj} &= ((\omega_k^2 - \nu_j^2) a_{kj} - \mathbf{P}_k^T \mathbf{F}_{1j}) / (2\omega_k \nu_j \xi_k). \end{aligned} \tag{14}$$

3. Flow model

The mathematical model (see Gnesin [4]- [5]) assumes inviscid flow and uses the fully three-dimensional Euler equations to analyze unsteady transonic flow characteristics in a stage with arbitrary pitch ratios, in which the unsteadiness is due to the relative blade rows motions. This model is used as a basis for the prediction of both the unsteady aerodynamic forces acting on turbine blades, and the time-averaged flow field in a highly loaded stage, in which the 3D effects are essential. Unsteady Euler equations are integrated using an explicit monotonous second-order accuracy Godunov's type difference scheme.

4. Numerical results

In this paper the unsteady forces will be calculated by using a 3D Flow model of a complete 200 MW steam turbine stage (see Figures 1, 2, 3). In this stage the number of stator blades is equal to 54 and the number of rotor blades is equal to 94. The geometrical parameters of the blade and the

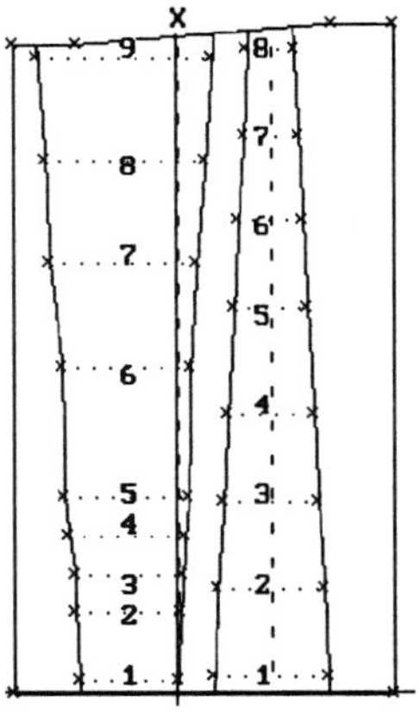

Figure 1: The Stator-Rotor.

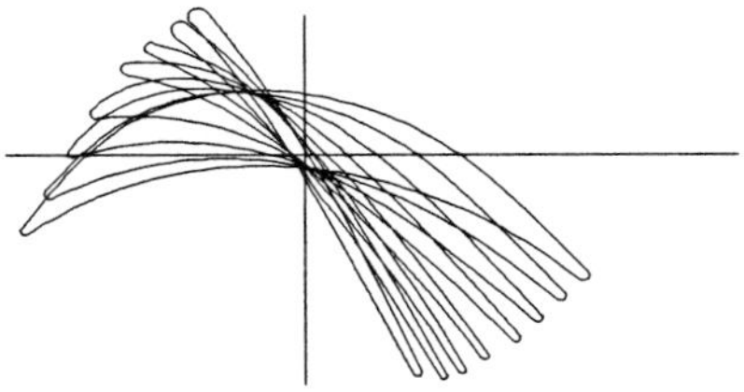

Figure 2: The rotor blade.

disc are presented in Rzadkowski [6]. As a first step, the steady forces were calculated using the design program for the areo-thermodynamic computation of a flow system of known geometry, developed by Blazko and Lidke [2]. By using this program the parameters of an impulse or reaction stage, according to one dimensional flow models can be calculated. In numerical calculations the rotor blade was divided into 11 segments. The basic results of the considered stator-rotor stage, in the middle of each segment, as velocities (see Figure 4) $u_1, u_2, w_1, w_2, c_{2s}$ (inlet stator velocity) c_1, c_2, Mach number M_1, relative Mach number $M_{1w} = w_1/a, M_{2w} = w_{2w}/a$, the angles $\alpha_1, \beta_1, \alpha_2, \beta_2$ and forces in axial and tangential directions P_a, P_u are presented in Table 1. The mass flow ratio G=65.19 kg/s, the pressure and temperature in front of the stage p_1=12600 Pa, t_1 =50.84°C, the pressure and temperature behind the stage p_2=2250 Pa, t_1 =19.74° C.

Figure 5 presents the forces P_a, P_u along the blade length.

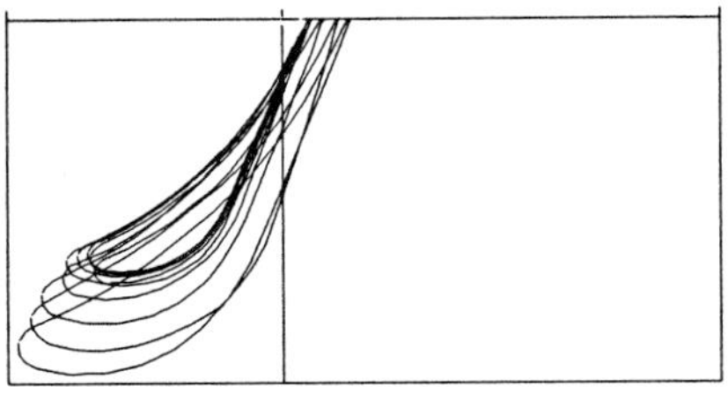

Figure 3: The stator blade.

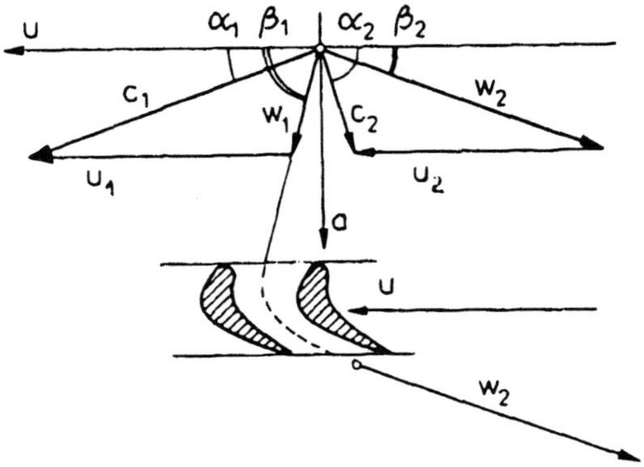

Figure 4: The velocity triangles of rotor blades.

Next, for the same flow parameters the unsteady forces in a complete stage of an axial turbine are calculated using a model developed by Gnesin [4]- [5]. The mathematical model assumes inviscid flow and uses the fully three-dimensional Euler equations to analyze unsteady transonic flow characteristics in a stage with arbitrary pitch ratios, in which the unsteadiness is due to the relative blade rows motions. This model is used as a basis for the prediction of both the unsteady aerodynamic forces acting on turbine blades, and the time-averaged flow field in a highly loaded stage, in which the 3D effects are essential. Unsteady Euler equations are integrated using an explicit monotonous second-order accuracy Godunov's type difference scheme.

The forces (the nozzle excitation) acting on the rotor blades , in the axial, tangential and radial directions $P_a, P_u, P_r,$ in form of, $P_i = \sum_{k=0}^{K} P_{ki} sin(kN_1 \Omega t + \varphi_{ki})$ (where i=a, u, r) are presented in Table 2. The forces were calculated at different cross-sections along the blade length. In column 0, the steady part of the forces is presented, in columns 1 and 2 the unsteady forces P_{i1}/P_{io}, P_{i2}/P_{io} (where i=a, u, r) were shown.

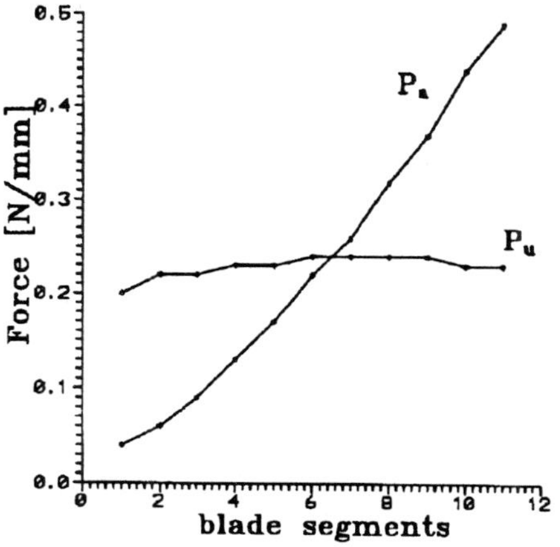

Figure 5: The forces P_a, P_u along the blade length.

The input data for the program consists of: the geometry of the stator and rotor blade, the angles of incidence of the stator δ_s, the gas constant R=287 J/(kg K), the adiabatic coefficient κ=1.44 and p_o, t_o, p_1 and t_1. The values of the steady forces, calculated by these two programs (see Table 2 and 1), are similar. There are greater differences between the velocities and Mach numbers.

From Table 2 and Figures 6 and 7 it is seen that the steady force P_a are changing from 0.049 to 0.42 N/mm and the unsteady first harmonic from 0.006 (0.6 %) to 0.23 (23 %) of the steady part. The steady force P_u are changing from 0.16 to 0.186 N/mm and the unsteady first harmonic from 0.002 (0.2 %) to 0.15 (15 %) of the steady part. The steady force P_r are changing from 0.006 to 0.018 N/mm and the unsteady first harmonic from 0.004 (0.4 %) to 0.268 (26,8 %) of the steady part. The values of the unsteady forces in the first cross-section of the blade are relatively high in comparison to the other values. The reason of that is that in this cross-section the large separation (stall) occur.

In Table 3 the unsteady forces are presented for the same stage and the different pressure p_2=6000 Pa. Generally the zero's harmonic decrease although the 1 and 2 harmonics slightly increase in comparison to the unsteady forces calculated for p_2=2250 Pa.

Table 1: The data and results of the areo-thermodynamic calculations.

	u_1 (m/s)	u_2 (m/s)	w_1 (m/s)	$w_2(m/s)$	c_{2s} (m/s)
1	221.68	220.63	409.91	481.43	193.47
2	243.130	242.13	348.04	490.53	193.72
3	264.58	264.32	290.40	500.42	195.88
4	286.03	286.17	239.61	509.64	195.38
5	307.48	308.02	196.05	522.16	193.03
6	328.92	329.87	160.78	535.32	190.62
7	350.37	351.72	136.13	549.42	187.71
8	371.82	373.56	123.86	558.70	185.19
9	393.27	395.41	126.12	569.41	182.02
10	414.72	417.26	139.51	580.56	178.19
11	436.17	439.11	160.49	586.05	173.00

c_1 (m/s)	c_2 (m/s)	M_1	M_{1w}	M_2
595.23	354.61	1.54	1.04	1.37
557.38	359.05.	1.46	0.88	1.39
520.48	367.11	1.36	0.74	1.41
487.59	371.77	1.27	0.61	1.43
458.58	375.48	1.19	0.5	1.45
433.03	378.48	1.13	0.42	1.48
410.97	378.48	1.07	0.36	1.50
391.15	373.31	1.02	0.34	1.53
372.96	372.37	0.97	0.35	1.56
356.46	358.03	0.93	0.38	1.59
341.53	351.61	0.89	0.44	1.62

$\alpha_1(°)$	$\beta_1(°)$	$\alpha_2(°)$	$\beta_2(°)$	P_a (N/mm)	P_u (N/mm)
26.89	41.04	68.57	43.31	0.04	0.20
23.57	39.78	72.34	44.23	0.06	0.22
21.55	41.10	76.36	45.47	0.09	0.22
20.29	44.74	79.27	45.71	0.13	0.23
19.50	51.08	80.86	45.24	0.17	0.23
19.03	60.87	82.10	44.47	0.22	0.24
18.77	74.70	82.57	43.14	0.26	0.24
18.63	92.16	83.29	41.63	0.32	0.24
18.70	110.55	84.50	40.69	0.37	0.24
18.72	126.11	83.57	37.88	0.44	0.23
18.78	137.75	85.30	36.83	0.49	0.23

Table 2: The results of the 3-D Flow calculation of 94 rotor blades and p2=2250 Pa.

P_a			P_u			P_r		
0	1	2	0	1	2	0	1	2
N/mm	–	–	N/mm	–	–	N/mm	–	–
0.049	0.23	0.037	0.16	0.15	0.008	0.006	0.268	0.019
0.083	0.05	0.038	0.18	0.07	0.01	0.007	0.096	0.007
0.117	0.02	0.006	0.19	0.04	0.02	0.007	0.041	0.029
0.152	0.01	0.014	0.186	0.028	0.028	0.006	0.045	0.054
0.186	0.006	0.014	0.182	0.011	0.023	0.007	0.029	0.048
0.22	0.01	0.01	0.181	0.003	0.012	0.009	0.013	0.021
0.27	0.009	0.005	0.18	0.002	0.006	0.013	0.006	0.008
0.315	0.008	0.001	0.179	0.011	0.007	0.017	0.007	0.002
0.364	0.008	0.001	0.178	0.015	0.001	0.0177	0.004	0.002
0.42	0.012	0.003	0.176	0.02	0.003	0.018	0.004	0.004

φ_{ka}			φ_{ku}			φ_{kr}		
0	1	2	0	1	2	0	1	2
	[rad]	[rad]		[rad]	[rad]		[rad]	[rad]
	(-0.042)	(1.282)		(0.302)	(2.324)		(3.593)	(-1.986)
	(0.676)	(3.597)		(0.584)	(-1.231)		(3.819)	(1.746)
	(0.305)	(3.564)		(0.315)	(-0.514)		(3.771)	(2.616)
	(0.725)	(3.319)		(0.190)	(-0.377)		(3.857)	(2.868)
	(1.183)	(3.301)		(0.638)	(-0.194)		(-1.826)	(2.953)
	(0.204)	(3.872)		(3.470)	(0.695)		(-0.858)	(3.447)
	(0.128)	(-2.103)		(2.174)	(1.077)		(0.161)	(3.825)
	(-0.722)	(-1.796)		(1.529)	(1.889)		(-0.464)	(3.843)
	(-1.721)	(2.404)		(1.187)	(-1.460)		(-0.872)	(2.786)
	(-2.215)	(3.169)		(0.914)	(0.097)		(-1.618)	(2.753)

The stall flow in the cross-section means that this cross-section was not properly design. In order to improve the flow conditions, the numerical simulation were made to change slightly geometry of the cross-sections. The results of the unsteady forcers and their harmonics of a modernize stage were presented in Tables 4 and 5 and Figures 6 and 7. From Table 4 it is seen that the first harmonics of the unsteady force P_a are up to 5.7 % and P_u are up to 4.5 %, which is acceptable. In case of pressure p_2=6000 Pa the first harmonics of P_a are up 8.9 % and P_u are up to 13.8%, but the steady parts (0 harmonics) of the unsteady forces are smaller in comparison to the case of p_2=2250 Pa.

By using of this 3D Flow model the unsteady forces were also calculated for a stage with the same stator and rotor blade geometry (number of rotor blades and stator blade equal to 36).

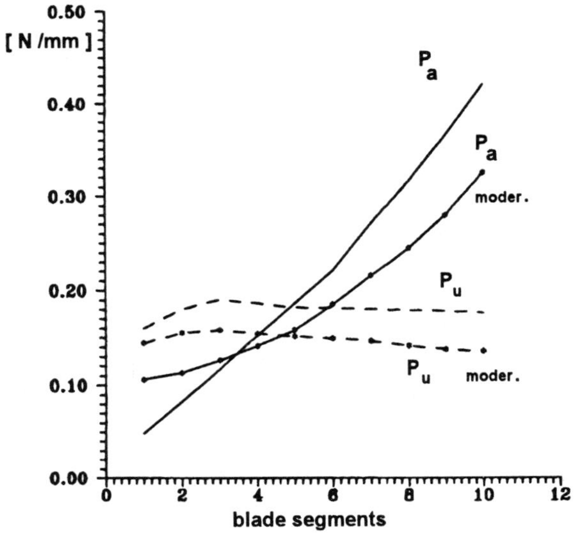

Figure 6: The forces P_a, P_u along the blade length, the high frequency excitation, 94 blades .

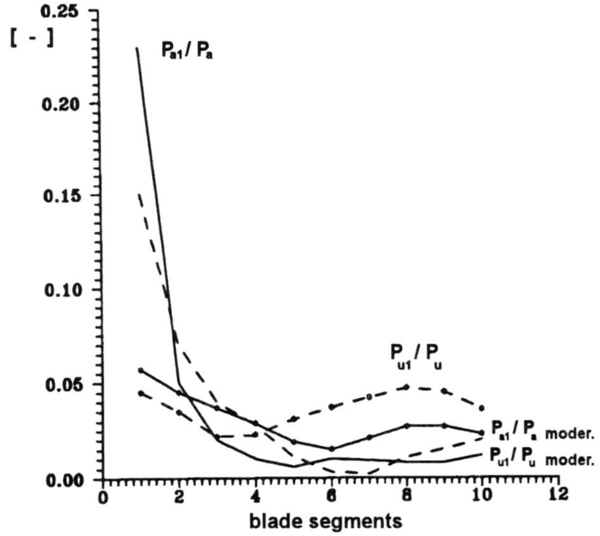

Figure 7: The forces $P_{a1}/P_a, P_{u1}/P_u$ along the blade length, the high frequency excitation, 94 blades.

Table 3: The results of the 3-D Flow calculation of 94 rotor blades and p2=6000 Pa.

P_a			P_u		
0	1	2	0	1	2
N/mm	–	–	N/mm	–	–
0.0	0.0	0.0	0.069	0.11	0.017
0.027	0.08	0.038	0.078	0.05	0.013
0.053	0.044	0.017	0.079	0.05	0.036
0.072	0.013	0.016	0.0789	0.04	0.041
0.090	0.042	0.026	0.0782	0.095	0.051
0.113	0.052	0.026	0.080	0.079	0.045
0.136	0.026	0.014	0.080	0.025	0.026
0.155	0.008	0.004	0.075	0.006	0.014
0.168	0.011	0.002	0.067	0.011	0.008
0.181	0.013	0.002	0.058	0.012	0.0089

φ_{ka}			φ_{ku}		
0	1 [rad]	2 [rad]	0	1 [rad]	2 [rad]
	(3.197)	(1.540)		(0.833)	(0.093)
	(2.851)	(1.032)		(0.656)	(0.818)
	(2.879)	(0.329)		(0.642)	(1.215)
	(3.889)	(-1.095)		(1.927)	(1.233)
	(-0.404)	(-1.294)		(2.908)	(1.513)
	(0.04)	(-1.276)		(3.309)	(1.732)
	(0.211)	(-1.278)		(3.337)	(1.668)
	(0.534)	(-1.699)		(2.363)	(1.206)
	(1.008)	(-1.885)		(3.440)	(1.381)
	(0.916)	(-2.211)		(3.529)	(1.398)

Tables 6 and 7 and Figures 8 and 9 present the unsteady forces for different values of p_2.

From Tables 6 it is seen that the unsteady 0-harmonics of the force P_a is changing from 0.14 to 0.526 N/mm and the unsteady 1-harmonics is changing up to 12.8%.

In this paper the unsteady 1-harmonic amplitude force $P_{o_{x_2}}$ (see Figure 1 in Rzadkowski [6]) was assumed to be equal to 0.3/2 =0.15 N/mm, and for a 36 mistuned bladed disc the maximal stress of the blade, for a run-up was about 230 N/mm^2. From the above calculation (see Table 6) it is seen that the unsteady 1-harmonic amplitude $P_{o_{x_2}}$ is in the range from 0.049 to 0.071 N/mm along the blade length, so the maximal stress of the blades during the transient state are lower. The stresses of the blades around the disc during the steady force vibration, were calculated for above presented unsteady forces.

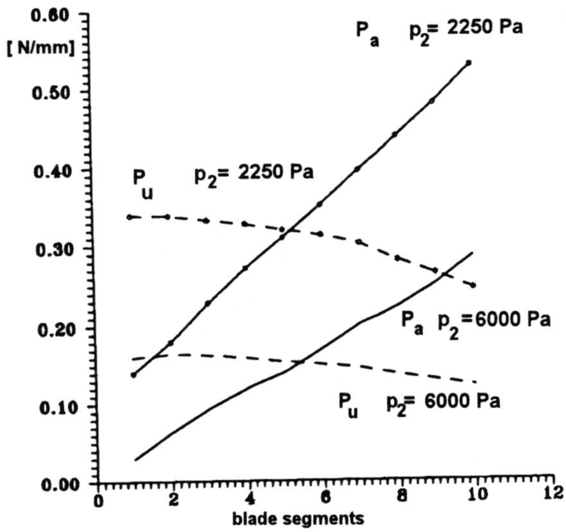

Figure 8: The forces P_a, P_u along the blade length, the high frequency excitation, 36 blades .

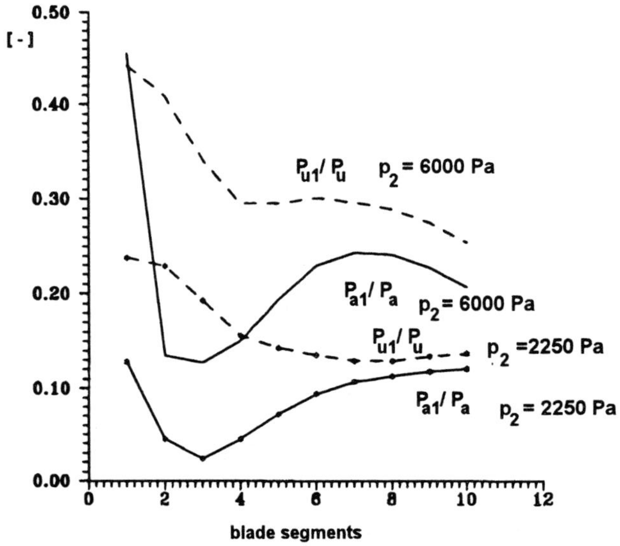

Figure 9: The forces $P_{a1}/P_a, P_{u1}/P_u$ along the blade length, the high frequency excitation, 36 blades.

Table 4: The results of the 3-D Flow calculation of 94 rotor modernize blades and p2=2250 Pa.

P_a			P_u		
0	1	2	0	1	2
N/mm	–	–	N/mm	–	–
0.106	0.057	0.005	0.144	0.045	0.006
0.113	0.045	0.002	0.155	0.035	0.005
0.126	0.037	0.002	0.158	0.022	0.004
0.141	0.029	0.003	0.154	0.023	0.004
0.158	0.019	0.003	0.151	0.031	0.004
0.185	0.015	0.003	0.149	0.037	0.005
0.215	0.021	0.004	0.146	0.042	0.006
0.244	0.027	0.004	0.141	0.047	0.006
0.279	0.027	0.003	0.137	0.045	0.005
0.324	0.023	0.002	0.135	0.036	0.003

φ_{ka}			φ_{ku}		
0	1	2	0	1	2
	[rad]	[rad]		[rad]	[rad]
	(1.563)	(1.334)		(0.392)	(2.252)
	(1.705)	(1.177)		(0.271)	(1.708)
	(1.749)	(1.470)		(0.414)	(1.222)
	(1.846)	(1.726)		(0.989)	(0.891)
	(2.194)	(2.027)		(1.214)	(0.664)
	(2.979)	(2.466)		(1.279)	(0.567)
	(3.646)	(2.923)		(1.319)	(0.639)
	(-2.229)	(3.335)		(1.391)	(0.829)
	(-1.793)	(3.786)		(1.569)	(1.083)
	(-1.39)	(-2.10)		(1.813)	(1.312)

It was found that the unsteady forces for an assumed number of stator blades caused negligible stresses in comparison to the centrifugal stresses (root stress 228.67 N/mm^2).

This type of excitation is called the nozzle excitation or high frequency excitation. During a steady forced vibration, the nozzle excitation frequencies were different from the natural frequency of the system, so the stress caused by unsteady forces was very small.

The second type of excitation is the low frequency excitation arise mainly from low frequency harmonic distortions of the flow field. Harmonic flow distortions may arise from the inlet flow guide passage spacing, e.g., from guide vane pitching and, or gauging errors during manufacture or assembly of the guide rows.

Table 5: The results of the 3-D Flow calculation of 94 rotor modernize blades and p2=6000 Pa.

P_a			P_u		
0	1	2	0	1	2
N/mm	–	–	N/mm	–	–
0.049	0.045	0.012	0.074	0.072	0.005
0.055	0.037	0.008	0.073	0.029	0.005
0.063	0.033	0.005	0.071	0.026	0.005
0.069	0.035	0.003	0.064	0.034	0.005
0.076	0.042	0.003	0.059	0.051	0.006
0.089	0.052	0.003	0.058	0.073	0.006
0.103	0.069	0.002	0.056	0.103	0.007
0.115	0.089	0.003	0.052	0.136	0.008
0.130	0.082	0.005	0.05	0.138	0.009
0.155	0.062	0.005	0.05	0.108	0.009

φ_{ka}			φ_{ku}		
0	1	2	0	1	2
	$[rad]$	$[rad]$		$[rad]$	$[rad]$
	(3.075)	(-0.336)		(0.861)	(-2.347)
	(3.261)	(-0.505)		(1.010)	(3.050)
	(3.251)	(-0.443)		(1.207)	(2.281)
	(3.264)	(-0.041)		(1.307)	(2.027)
	(3.376)	(0.399)		(1.254)	(2.062)
	(3.578)	(0.401)		(1.153)	(2.113)
	(3.767)	(0.052)		(1.083)	(2.091)
	(-2.353)	(-0.489)		(1.079)	(2.152)
	(-2.146)	(-0.550)		(1.162)	(2.347)
	(-1.981)	(-0.515)		(1.238)	(2.512)

These excitations also arise from construction features such as diaphragm joints, and from radial or angular eccentricity of the moving row and the nonsymmetrical pressure distribution behind the exhaust hood in circumferential direction. This type of excitation was simulated for a 36 rotating ($\Omega/(2\pi)$=50 Hz) bladed disc with 36 nozzles. It was assumed that the pressure behind the exhaust hood is changing in the circumferential direction (measured by the angle α_o around the axis X_1, see Figure 1 in Rzadkowski [6]) in the following way. For angle α_o=0 °, p_2=9000 Pa, α_o= 90 °, p_2=7500 Pa, α_o= 180 °, p_2=6000 Pa, α_o= 270 °, p_2=7500 Pa. For a such distribution of p_2 and a rotational frequency 50 Hz, the unsteady forces acting on the rotor blades, in the axial, tangential and radial directions P_a, P_u, P_r, in form of, $P_i = \sum_{k=0}^{K} P_{ki} sin(k\Omega t + \varphi_{ki})$ (where i=a, u, r) and moment (of the blade shear centre, see [6]) were presented in Figures 10 and 11 and Table 8.

Table 6: The results of the 3-D Flow calculation of 36 rotor blades and p2=2250 Pa.

P_a			P_u		
0	1	2	0	1	2
N/mm	–	–	N/mm	–	–
0.140	0.128	0.072	0.339	0.238	0.008
0.180	0.045	0.043	0.338	0.229	0.020
0.228	0.024	0.019	0.332	0.193	0.022
0.271	0.045	0.006	0.327	0.156	0.018
0.309	0.072	0.004	0.319	0.143	0.016
0.350	0.094	0.007	0.312	0.135	0.013
0.394	0.107	0.008	0.301	0.129	0.012
0.437	0.113	0.009	0.280	0.129	0.012
0.479	0.118	0.010	0.263	0.134	0.013
0.526	0.121	0.011	0.244	0.137	0.014

φ_{ka}			φ_{ku}		
0	1	2	0	1	2
	[rad]	[rad]		[rad]	[rad]
	(0.496)	(1.649)		(1.500)	(3.443)
	(0.376)	(1.991)		(1.509)	(3.212)
	(-0.697)	(2.133)		(1.598)	(3.164)
	(-0.872)	(1.761)		(1.777)	(3.031)
	(-0.857)	(-0.024)		(1.929)	(2.777)
	(-0.897)	(-0.593)		(2.023)	(2.521)
	(-0.907)	(-0.883)		(2.116)	(2.318)
	(-0.856)	(-1.022)		(2.222)	(2.261)
	(-0.769)	(-0.995)		(2.314)	(2.285)
	(-0.715)	(-0.921)		(2.356)	(2.289)

In the first column the steady part was presented. In the second P_{i1}/P_{io} (where i= a, r) for frequency $\Omega/(2\pi)$=50 Hz, in the third P_{i2}/P_{io}(where i= a, r) for frequency $50 \cdot 36$ Hz. The first natural frequency of the rotating bladed disc is equal to 108 Hz, so this type of excitation for the excitation frequency equal to 50 Hz cause negligible blades stress.

Next, the pressure behind the exhaust hood was assumed in order to obtained the unsteady forces for an excitation frequency equal to 108 Hz. So, the pressure behind the exhaust hood was changing in the circumferential direction in the following way. For angle α_o=0 °, p_2=6000 Pa, α_o= 60 °, p_2=7500 Pa, α_o= 120 °, p_2=7500 Pa, α_o= 180 °, p_2=7500 Pa, α_o= 240 °, p_2=9000 Pa and α_o= 300 °, p_2=7500 Pa. For such distribution of p_2 and the frequency of excitation 108 Hz the unsteady forces and moment were calculated (see Figures 10 and 11 and Table 9).

Table 7: The results of the 3-D Flow calculation of 36 rotor blades and p2=6000 Pa.

P_a			P_u		
0	1	2	0	1	2
N/mm	–	–	N/mm	–	–
0.029	0.455	0.069	0.160	0.442	0.008
0.063	0.135	0.021	0.165	0.409	0.013
0.094	0.127	0.013	0.163	0.341	0.012
0.120	0.151	0.009	0.159	0.295	0.011
0.140	0.194	0.006	0.154	0.295	0.012
0.170	0.229	0.004	0.150	0.301	0.011
0.201	0.243	0.005	0.145	0.296	0.009
0.222	0.241	0.007	0.136	0.289	0.010
0.250	0.227	0.009	0.129	0.275	0.014
0.284	0.207	0.012	0.122	0.254	0.017

φ_{ka}			φ_{ku}		
0	1	2	0	1	2
	$[rad]$	$[rad]$		$[rad]$	$[rad]$
	(1.010)	(1.726)		(2.231)	(-1.343)
	(0.202)	(2.108)		(2.146)	(-1.448)
	(-0.512)	(2.187)		(2.115)	(-1.765)
	(-0.608)	(1.947)		(2.188)	(-2.231)
	(-0.637)	(1.408)		(2.259)	(3.569)
	(-0.696)	(0.395)		(2.299)	(3.263)
	(-0.727)	(-0.505)		(2.335)	(2.865)
	(-0.721)	(-0.9923)		(2.375)	(2.413)
	(-0.677)	(-1.051)		(2.424)	(2.225)
	(-0.654)	(-1.020)		(2.438)	(2.137)

For these forces and moments (see Table 9) the stresses in blades were equal to 407 N/mm^2, although the centrifugal stress was equal to 225 N/mm^2. This value is dangerous from the stress point of view, because for the steel X 10 Cr 13, the yield stress for an operating temperature of 300 °C is equal to 495 N/mm^2 and the fatigue strengths is equal to 275 N/mm^2.

All numerical calculations were made in the Academic Computer Center TASK, in Gdańsk.

Table 8: The results of the 3-D Flow calculation of 36 rotor blades, the low frequency excitation, f=50 Hz.

P_a			P_u			M_3		
0	1	36	0	1	36	0	1	36
N/mm	–	–	N/mm	–	–	N	–	–
0.032	0.087	0.193	0.089	0.509	0.385	0.81	0.418	0.356
0.054	0.082	0.056	0.087	0.566	0.374	1.22	0.186	0.093
0.072	0.179	0.067	0.082	0.632	0.346	1.49	0.106	0.0356
0.084	0.266	0.107	0.075	0.697	0.334	1.50	0.062	0.128
0.091	0.355	0.164	0.068	0.776	0.372	1.57	0.023	0.194
0.100	0.432	0.222	0.065	0.837	0.414	1.86	0.008	0.219
0.111	0.491	0.262	0.060	0.896	0.441	2.34	0.006	0.222
0.121	0.545	0.283	0.054	0.996	0.472	2.81	0.003	0.224
0.132	0.589	0.280	0.049	1.072	0.473	3.15	0.008	0.213
0.15	0.596	0.254	0.046	1.089	0.438	3.47	0.139	0.176

φ_{ka}			φ_{ka}			φ_{km}		
0	1	36	0	1	36	0	1	36
	[rad]	[rad]		[rad]	[rad]		[rad]	[rad]
	(4.592)	(1.694)		(4.634)	(-0.675)		(4.622)	(2.1385)
	(1.508)	(0.878)		(4.633)	(-0.659)		(4.620)	(2.088)
	(1.494)	(-0.219)		(4.6314)	(-0.692)		(4.616)	(-1.139)
	(1.492)	(-0.428)		(4.6309)	(-0.708)		(4.610)	(-1.23)
	(1.491)	(-0.543)		(4.6309)	(-0.714)		(4.578)	(-1.30)
	(1.490)	(-0.603)		(4.6307)	(-0.795)		(4.513)	(-1.27)
	(1.489)	(-0.603)		(4.6304)	(-0.651)		(4.505)	(-1.18)
	(1.489)	(-0.565)		(4.6302)	(-0.531)		(4.604)	(-1.09)
	(1.489)	(-0.485)		(4.6302)	(-0.598)		(4.622)	(-1.04)
	(1.489)	(-0.441)		(4.63)	(-0.487)		(4.625)	(-1.05)

5. Conclusions

The forced vibration analysis of the bladed disc will be considered for the nozzle and low frequency excitations. The unsteady forces will be calculated from the 3D flow model. During a steady forced vibration, the nozzle excitation frequencies were different from the natural frequency of the system, so the stress caused by unsteady forces was very small. There is the pressure distribution behind the exhaust hood, for which the low frequency excitation can cause the dangerous stress in the blades placed on the disc.

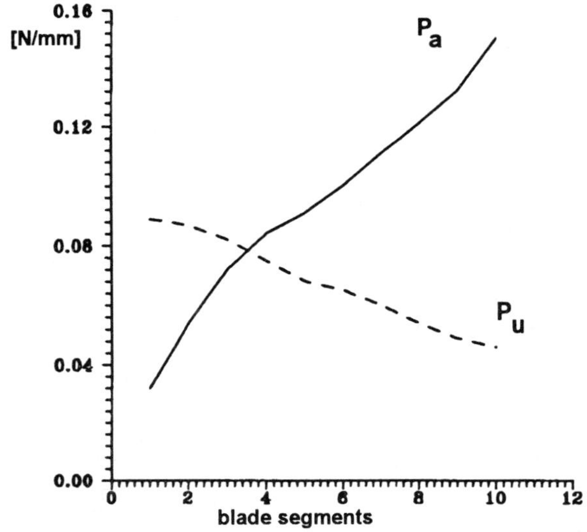

Figure 10: The forces P_a, P_u along the blade length, the low frequency excitation, f=108 Hz and f=50 Hz.

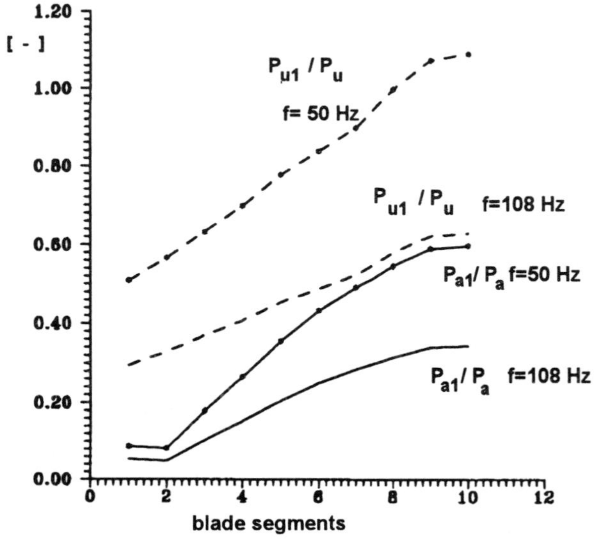

Figure 11: The forces $P_{a1}/P_a, P_{u1}/P_u$ along the blade length, the low frequency excitation.

Table 9: The results of the 3-D Flow calculation of 36 rotor blades, the low frequency excitation, f=108 Hz.

P_a			P_u			M_3		
0	1	2	0	1	2	0	1	2
N/mm	–	–	N/mm	–	–	N	–	–
0.032	0.053	0.065	0.089	0.296	0.393	0.80	0.243	0.317
0.054	0.048	0.064	0.087	0.329	0.437	1.29	0.108	0.142
0.072	0.104	0.138	0.081	0.369	0.489	1.49	0.062	0.082
0.084	0.154	0.205	0.075	0.407	0.540	1.51	0.037	0.050
0.090	0.206	0.274	0.068	0.453	0.602	1.58	0.018	0.026
0.099	0.251	0.333	0.065	0.488	0.648	1.87	0.012	0.018
0.111	0.285	0.378	0.060	0.522	0.693	2.35	0.008	0.013
0.120	0.316	0.420	0.054	0.579	0.769	2.82	0.016	0.021
0.132	0.341	0.453	0.049	0.621	0.825	3.15	0.050	0.067
0.15	0.344	0.457	0.047	0.628	0.834	3.45	0.082	0.109

φ_{ka}		φ_{ka}		φ_{km}	
0 1	26	0 1	2	0 1	2
[rad]	[rad]	[rad]	[rad]	[rad]	[rad]
(3.93)	(-1.146)	(4.15)	(-1.242)	(4.110)	(-1.24)
(1.122)	(1.83)	(4.148)	(-1.244)	(4.141)	(-1.28)
(1.03)	(1.88)	(4.152)	(-1.25)	(4.239)	(-1.387)
(1.014)	(1.89)	(4.154)	(-1.25)	(4.427)	(4.696)
(1.006)	(1.896)	(4.152)	(-1.25)	(-1.390)	(4.21)
(0.999)	(1.902)	(4.146)	(-1.25)	(-1.028)	(3.876)
(0.992)	(1.905)	(4.139)	(-1.25)	(-1.066)	(3.91)
(0.988)	(1.908)	(4.131)	(-1.235)	(4.260)	(-1.44)
(0.981)	(1.915)	(4.117)	(-1.222)	(3.950)	(-1.08)
(0.969)	(1.927)	(4.104)	(-1.209)	(3.929)	(-1.05)

References

[1] Bathe K., Wilson E.: *Numerical Methods in Finite Element Analysis* , Prentice-Hall, Inc., Englewood Cliffs, New Jersey, 1976.

[2] Blażko E., M :A Program of Thermodynamic Computations of a Flow System of Known Geometry, *Report Polish Academy of Sciences, Institute of Fluid-Flow Machinery*, 79/87, Gdansk 1987.

[3] Irretier H., Rzadkowski R.: Forced Vibration of Mistuned Bladed Discs in Stationary Operations, *Proceedings IFToMM Fourth International Conference on Rotor Dynamics*, Chicago, September 7-9, pp.53-61, 1994.

[4] Gnesin V.I.: Numerical Study of Low-Frequency Flow Instability in an Axial Transonic Compressor Stage, *Proceedings of Fifth International Conference of Fluid Mechanics*, Cairo, pp.257-268, 1995.

[5] Gnesin V.I: 3D-Unsteady Stator-Rotor Interaction in a Turbine Stage, *VDI Berichte 1186, Turbomachinnery -Fluid Dynamic and Thermodynamic Aspects*, pp. 202-215, 1995.

[6] Rzadkowski R.: The General Model of Free Vibrations of Mistuned Bladed Discs. Part I. Theoretical Model. Part II. Numerical Results, *Journal of Sound and Vibration*, 173(3), pp.395-413, 1994.

[7] Rzadkowski R.: Transient Nozzle Excitation of Mistuned Bladed Discs, *Journal of Sound and Vibration*, 190(4), pp. 629-643, 1996.

[8] Rzadkowski R.: Resonance Vibration of Mistuned Bladed Discs in Stationary Operations, *ASME Turbo Expo 97*, June 2-5, *ASME paper 97-GT-499*, 1997.

PARALLEL COMPUTATION OF ROTOR-STATOR INTERACTION

PAUL CIZMAS
Westinghouse Electric Corporation
Science & Technology Center
Pittsburgh, Pennsylvania 15235-5098

AND

RAVISHANKAR SUBRAMANYA
Pittsburgh Supercomputing Center
Pittsburgh, Pennsylvania 15213-2683

Abstract.
Numerical simulation of rotor-stator interaction is crucial for turbomachinery improvement since it allows the optimization of blade path design. In addition, the numerical simulation of rotor-stator interaction makes it possible to predict the location and variation of hot spots on blades, which is extremely useful in the turbine thermal design. However, simulation of rotor-stator interaction is notorious for being computationally expensive. To reduce the turnaround time and cost/MFLOP, a parallel code based on the *message-passing interface* libraries was developed. This code runs on symmetric multi-processors (Silicon Graphics Challenge) and massively parallel processors (Cray T3E). The parallel code was used to compute the rotor-stator interaction in two turbine configurations. A super-linear speedup of 15.6 was obtained on a 12 CPU Silicon Graphics Challenge for a geometry distributed over ten processors.

1. Introduction

The need for increased performance and improved reliability in turbomachinery has motivated designers to identify and evaluate unsteadiness effects. An important part of the unsteady effects in turbomachinery is the rotor-stator interaction. The main sources of unsteadiness present in the rotor-stator interaction are potential flow interaction and wake interaction.

T.H. Fransson (ed.), Unsteady Aerodynamics and Aeroelasticity of Turbomachines, 633-645.

Additional sources of unsteadiness are hot streak interaction, vortex shedding, shock/boundary layer interaction and flutter.

Potential flow interaction is a purely inviscid interaction due to the pressure variation caused by the relative movement of the blades and vanes. The effect of potential flow interaction increases if the rotor-stator gap decreases and if the flow is supersonic. The unsteady forces developed on the blades due to the potential flow interaction can be as large as 20% of the steady forces [14].

Wake interaction is the unsteadiness generated by the vortical and entropic wakes shed by one or more upstream rows. These wakes interact with the downstream airfoils which are moving relative to the wakes. Wake interaction is the primary contributor to unsteady forces on the blade for a large rotor-stator gap. In this case, the unsteady forces on the blade can be as large as 5% of the steady forces. Wake interaction induces intermittent boundary layer transition which causes a variation of the heat transfer between a low, laminar value to a high, turbulent value [2]. Consequently, the mean heat transfer and the mean overall loss of the blade increase.

Experimental and numerical investigations of rotor-stator interaction were developed to understand and control the governing physical mechanisms. Dring and his colleagues have studied experimentally the aerodynamic interaction between the blades and vanes of a large scale axial turbine. The turbine facility was assembled in both single stage [5] and $1\frac{1}{2}$-stage [6, 7, 8, 9] configurations with various axial spaces between adjacent rows. These experimental results show that reducing the axial gap from 65% of the blade axial chord to 15% has a strong effect on the unsteady pressure field.

The effects of axial spacing on the aerodynamic characteristics of a turbine were also experimentally investigated and numerically simulated by Dunn and his colleagues [10]. The results of this investigation were in agreement with Dring's conclusions, namely, increasing the axial gap between vane and blade reduces the unsteadiness in the pressure fields of both rows.

Huber *et al.* experimentally investigated vane indexing of the space shuttle main engine turbine [20]. The experimental results showed a 0.8% efficiency variation for the clocking positions investigated. The highest efficiency was obtained when the wake of the first vane was aligned with the second vane's leading edge. A two-dimensional numerical analysis [15] of the flow at the turbine midspan was also performed using STATOR2 code [16]. The numerical simulation predicted correctly the maximum efficiency indexing positions of the first and second stage vanes. However, the efficiency yielded by the numerical simulation was higher than the efficiency obtained from experiment. In addition, the efficiency variation as a function of clocking position was smaller in the case of the numerical simulation than in the case of experiment.

In spite of the differences which still exist between experimental data and numerical results, the numerical simulation of stator-rotor interaction is a useful tool for turbomachinery improvement. The numerical simulation of unsteady flows in turbomachinery started with the work of Erdos *et al.* [11] who solved the inviscid, compressible, two-dimensional, unsteady flow on a blade-to-blade stream surface through a compressor stage. The first three-dimensional model of the stator-rotor interaction was solved by Koya and Kotake [24]. Similarly to Erdos code, Koya's code used phase-lagged periodic boundaries. The first spatially periodic boundary conditions, as well as the first model using Navier-Stokes equations to model the stator-rotor interaction, were introduced by Rai [28] in 1985. Since then, numerous computer codes for stator-rotor interaction simulation have been developed, especially in the last two years [1, 12, 22, 26, 27, 31].

Some of the unsteady multi-row codes solve the two-dimensional flow [12, 13, 21, 25, 27, 28] and some solve the three-dimensional flow [1, 22, 26, 30, 31]. The computation time for solving the three-dimensional stator-rotor interaction problem is very large, especially if airfoil clocking is the final goal. In the case of a $1\frac{1}{2}$-stage turbine, with the simplest vane/blade count of 1:1:1, the computation time reported by Gundy-Burlet and Dorney [19] is about 21 days on the Cray YMP C90. Obviously, the computation time becomes prohibitive for a realistic vane/blade count.

To reduce the high turnaround time and the associated cost of such a simulation, a parallel code was developed for the more cost effective massively parallel processing platforms. This code is based on the STATOR2 code developed at NASA Ames [16]. Similarly to Eulitz's code [12], the code presented in this paper uses message-passing interface (MPI) libraries.

The next section briefly presents the numerical method inherited by the parallel code from the sequential code. Details about the parallel computation are given in section three. Section four presents some rotor-stator interaction results for two turbine configurations. Conclusions and future work suggestions are covered in the last section.

2. Numerical Model

The two-dimensional, unsteady, compressible flow through a multistage axial turbomachine with arbitrary blade counts is modeled by using the Navier-Stokes/Euler equations. The computational domain associated with each airfoil is divided into an inner region, near the airfoil, and an outer region, away from the airfoil. The thin-layer Navier-Stokes equations are solved in the regions near the airfoil, where viscous effects are strong. Euler equations are solved in the outer region, where the viscous effects are weak. Some viscous effects are present in the outer region due to the wakes coming from the upstream rows. However, the numerical dissipation in the outer region is much larger than the viscous effects, even if the Euler equations are replaced by fully viscous Navier-Stokes equations [15].

The integration method of the Euler/Navier-Stokes equations written in strong conservation form is a third-order-accurate, iterative, implicit, upwind scheme. The method is presented in detail in the paper by Rai and Chakravarthy [29]. The nonlinear finite-difference approximation is solved iteratively at each time level using an approximate factorization method. Three Newton iterations are used at each time step to reduce the associated linearization and factorization error.

2.1. GRID GENERATION

Two types of patched and overlaid grids are used to discretize the flow field surrounding the rotating and stationary grids. An inner "O" grid is used to resolve the viscous effects near the airfoil. The "O" grid which discretizes the Navier-Stokes equations is generated using an elliptical method. An outer "H" grid is used to resolve the Euler equations away from the airfoil. The "H" grid is algebraically generated. The "O" and "H" grids are overlaid. The flow variables are communicated between the "O" and "H" grids through bilinear interpolation. The outer grids corresponding to consecutive blades and vanes slip past each other to model the relative motion.

2.2. TURBULENCE MODEL

The flow is assumed to be fully turbulent. The eddy viscosity is computed using Baldwin-Lomax model and the kinematic viscosity is computed using Sutherland's law. The Baldwin-Lomax turbulence model is written in a stationary reference frame for the vanes and in a rotating reference frame for the blades [16].

2.3. BOUNDARY CONDITIONS

Since multiple grids are used to discretize the Euler/Navier-Stokes equations, two classes of boundary conditions must be enforced on the grid boundaries: natural boundary conditions and zonal boundary conditions. The natural boundaries include inlet, outlet, periodic and the airfoil surfaces. The zonal boundaries include the patched and overlaid boundaries.

Boundary conditions enforced at the airfoil surface are: the "no slip" condition, the adiabatic wall condition, and the zero normal pressure gradient condition. The adiabatic wall condition and the pressure gradient condition are implemented in an implicit manner. For the blade, the "no slip" boundary condition enforces that the fluid velocity at the blade surface is equal to the rotor speed [28].

The inlet boundary conditions specify flow angle, average total pressure and upstream Riemann invariant. The downstream Riemann invariant is extrapolated from the interior of the domain. At outlet, the average static pressure is specified, while the upstream Riemann invariant, circumferential velocity, and entropy are extrapolated from the interior of the domain. Periodicity is enforced by matching boundary conditions between the lower surface of the lowest "H" grid of a row and the upper surface of the top most "H" grid of the same row.

For the zonal boundary conditions of the overlaid boundaries, data is transferred from the "H" grid to the "O" grid along the "O" grid's outermost grid line. Data is then transferred back to the "H" grid along its inner boundary. At the end of each iteration an explicit, corrective, interpolation procedure is performed. The outer boundaries of the "O" grids are interpolated from the interior grid points of the "H" grids. The inner boundaries of the "H" grids are interpolated from the interior grid points of the "O" grids. This information update is performed using a nonconservative, linear interpolation technique. Stability improves by increasing the overlap area of the "O" and "H" grids [16]. The patch boundaries are treated in a nonconservative manner as well, using linear interpolation to update data between the adjoint grids [28].

3. Parallel Computation

The high turnaround time and the associated cost of running a sequential code to simulate the rotor-stator interaction is unacceptable for the turbomachinery design process. To reduce the turnaround time and cost/MFLOP, a parallel code was developed based on the sequential code STATOR2 [16]. The parallel code uses message-passing interface (MPI) libraries and runs on symmetric multi-processors (Silicon Graphics Challenge) and massively parallel processors (Cray T3E). The parallel code should run with minor modifications on all platforms that support MPI.

The development of the quasi three-dimensional parallel code was done such that a three-dimensional parallel version can be an easy extension. As a consequence of this requirement, one processor was allocated for each airfoil in the two-dimensional simulation. Consequently, the number of processors necessary for a typical three-dimensional turbomachinery configuration will not exceed the number of processors available on today's computers.

The processors allocation is presented in Figures 1 and 5. One processor is allocated for each inlet and outlet "H" grid. One processor is allocated

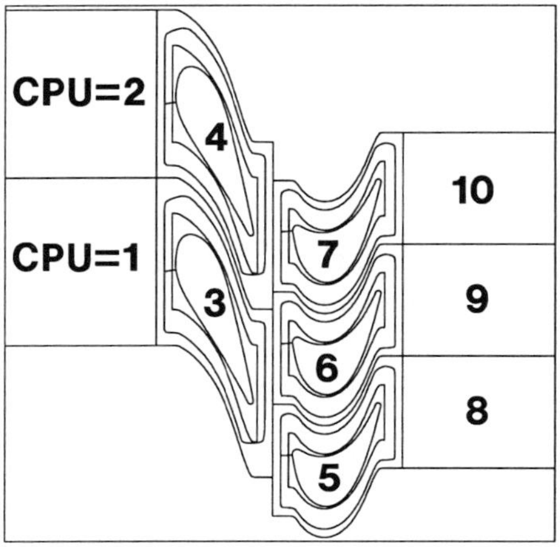

Figure 1. Processors allocation for single stage geometry.

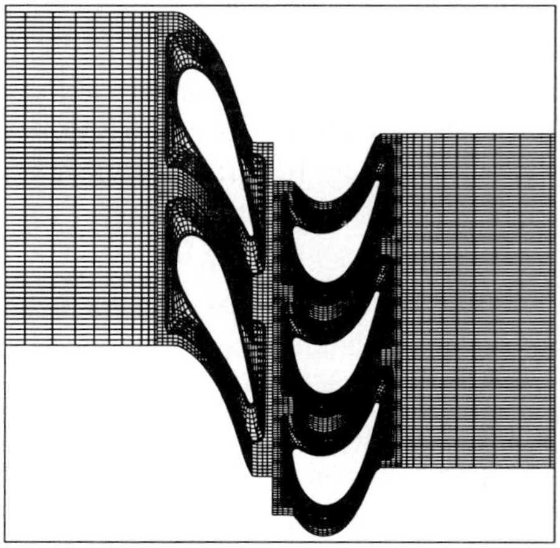

Figure 2. Grid of single stage geometry (every other point in each direction shown).

for the "O" and "H" grids corresponding to each airfoil. In the case of the single stage turbine, the airfoil count allows that the computational domain be reduced to two passages for the vane row and three passages for the blade row, such that the number of allocated processors is ten.

Interprocessor communication is used to match boundary conditions between grids. Periodic boundary conditions are imposed by cyclic communication patterns within rows. Slip boundary conditions are imposed by *gather-send_receive-broadcast* communication patterns between adjacent rows. Load imbalance issues need to be considered at grid generation time to reduce synchronization overhead.

4. Results

In this section, results of using the parallel code are presented for two turbine configurations. In the first case, single stage turbine results are presented and the CPU time compared to that of the sequential code. The second geometry investigated is a $1\frac{1}{2}$-stage turbine.

4.1. SINGLE STAGE TURBINE

The turbine airfoil count allows that the computation domain be reduced to two vane and three blade passages, as shown in Figure 2. The number of grid points is presented in Table 1. For the given "O" grid point distributions and flow conditions, the average value of the y^+ number is 1.

The sequential code STATOR2, which was used to develop the paral-

	Grid points
H inlet	2x(30x65)
H vane/ O vane	2x(61x95) / 2x(151x30)
H blade / O blade	3x(53x95) / 3x(151x45)
H outlet	3x(30x65)
Total grid points	65890

TABLE 1. Grid points for single stage turbine.

Serial	SGI Octane	397.2x10-6 second/(iteration * grid point)
Serial	Cray YMP C90	33.5x10-6 second/(iteration * grid point)
Parallel	SGI Challenge	25.5x10-6 second/(iteration * grid point)
Parallel	Cray T3E	25.0x10-6 second/(iteration * grid point)

TABLE 2. Wall-clock running time.

lel code presented in this paper, had been extensively calibrated against experimental data [15, 16, 17, 18]. As a result, the focus will be here on comparing the performance of the parallel code to that of the sequential code. The parallel code was implemented on a Silicon Graphics Challenge computer and on a Cray T3E. The sequential code was implemented on a Silicon Graphics Octane and a Cray YMP C90. A comparison of the wall-clock time is presented in Table 2. Parallelization yields better turnaround times on low end symmetric multi-processor (SMP) machines such as Silicon Graphics Challenge as compared to a serial vectorized code on a Cray YMP C90. The economic benefits of this are readily apparent. Not only is the turnaround time shorter for the parallel code, but also the computing cost is smaller, since the ratio of the aggregated CPU time on the parallel code by the CPU time on the sequential code is exceeded by the ratio of the sequential CPU time cost by the parallel CPU time cost.

Parallelization of the code resulted in super-linear speedups on the Silicon Graphics Challenge of 15.57 as opposed to the expected speedup of 10, due to the 10 processors used. The reason for this greater than expected speedup is the availability of 10 times the cache of a single processor on the SMP. Increased cache availability leads to faster memory accesses for the CPU and consequent higher MFLOPS.

The Cray T3E is a distributed memory machine and should have greater overhead for communication than does the Silicon Graphics Challenge which is a shared memory machine. The faster turnaround time is due to the slight edge in performance of the 21164 Alpha Microprocessor (450 MHz) as compared to the R10000 MIPS processor (200 MHz). Clock speeds do not

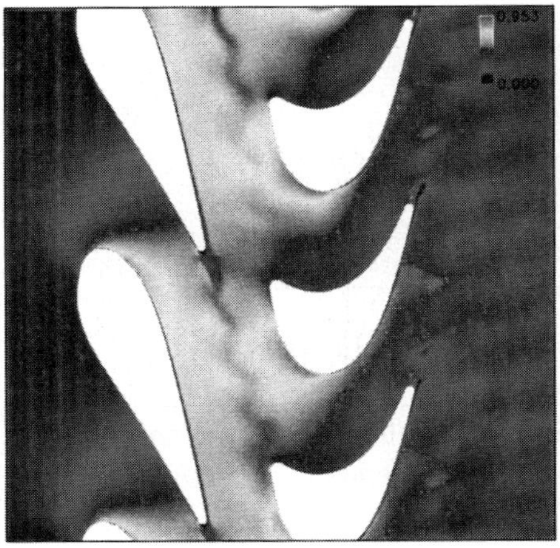

Figure 3. Instantaneous Mach contours. *Figure 4.* Instantaneous entropy contours.

directly translate into sustained MFLOPS due to chip architecture issues, such as functional unit design and on-chip cache sizes.

Instantaneous Mach contours are presented in Figure 3 in order to visualize the velocity distribution. Instantaneous entropy contours, shown in Figure 4, were generated to visualize the airfoil wakes. The wakes of the first vanes are chopped by the blades and then interact with the wakes of the blades, increasing blades loses.

4.2. $1\frac{1}{2}$-STAGE TURBINE

In this section the flow characteristics of a $1\frac{1}{2}$-stage turbine are presented. The specifications of the test turbine are similar to those of the single stage case except for the pressure ratio. The first two rows of this turbine are identical to those presented in the previous section. The number of grid points of the third row are 97x33 for the "O" grid and 61x65 for the "H" grid. The total number of grid points is 78272. The number of allocated processors is 11, as shown in Figure 5.

Pressure coefficient, C_p, is defined as $C_p = (p - p^*_{inlet})/\frac{1}{2}\rho_{inlet}U_m^2$, where U_m is the wheel speed at mid-span. Pressure coefficient variation on the airfoils of each row is presented in Figures 7, 8 and 9. The largest variation between the maximum and minimum values of pressure coefficient occurs on the suction side of the blade, where the variation is $\Delta C_p/C_p = 24.5\%$.

At a rotational speed of 3600 rpm and the given blade count, the time while a composite pitch (two vane passages or three blade passages) is swept is T=1.04 ms. During one period T, three blades perturb the flow in a passage of the first row of vanes. Figure 10 shows the pressure variation at 50% axial chord on the suction side of the first vane. As expected, three

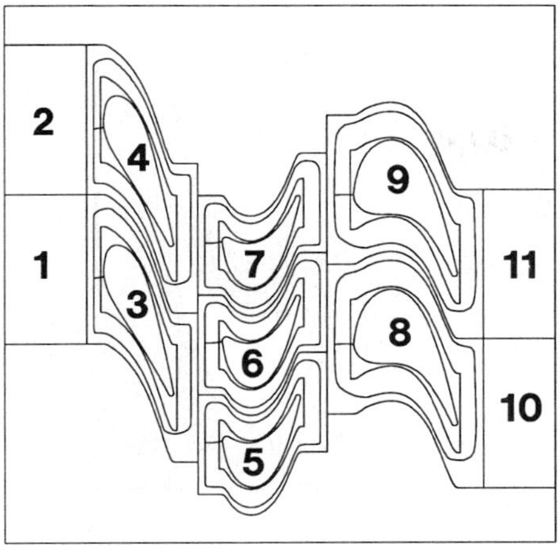

Figure 5. Processors allocation for $1\frac{1}{2}$-stage geometry.

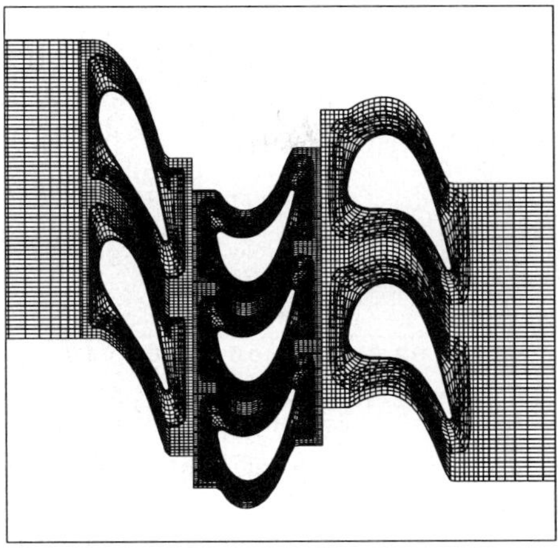

Figure 6. Grid of $1\frac{1}{2}$-stage geometry (every other point in each direction shown).

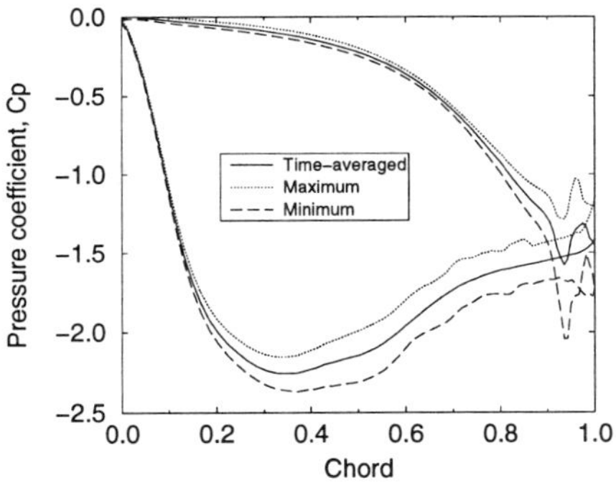

Figure 7. Pressure coefficient variation on the first vane.

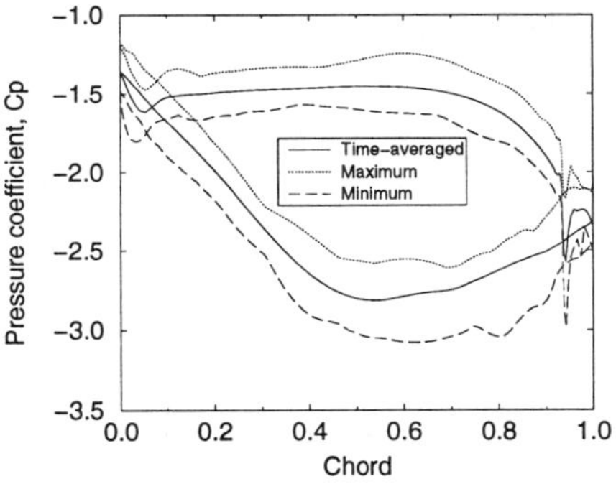

Figure 8. Pressure coefficient variation on the first blade.

perturbations can be observed during one period T, due to the potential flow interaction with the blades.

The pressure variation at 25% axial chord on the suction side of the first blade is presented in Figure 11. During one period T one observes two spikes due to the interaction with two vanes from the adjacent row. The pressure variation on the blade is more jagged than the pressure variation on the first vane. This can be caused by the fact that while the first vane is exposed only to potential flow interaction, the blade is perturbed by additional sources of unsteadiness such as wakes and vortices shed from the first row of vanes. Also, the potential flow interaction of the blade is more complex than the potential flow interaction of the first row of vanes, since the blade

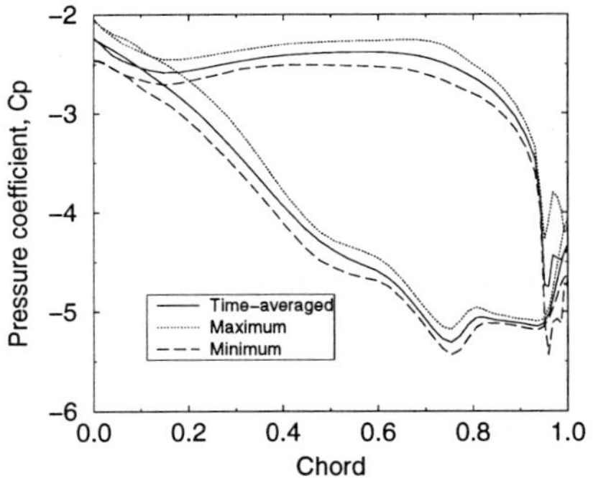

Figure 9. Pressure coefficient variation on the second vane.

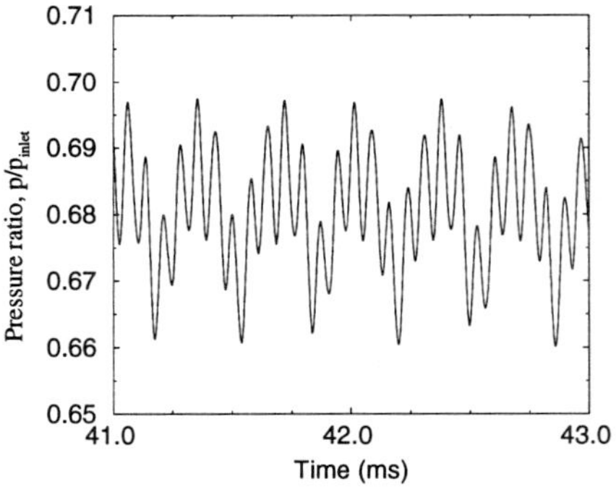

Figure 10. Pressure variation on the suction side of the first vane, at 50% axial chord.

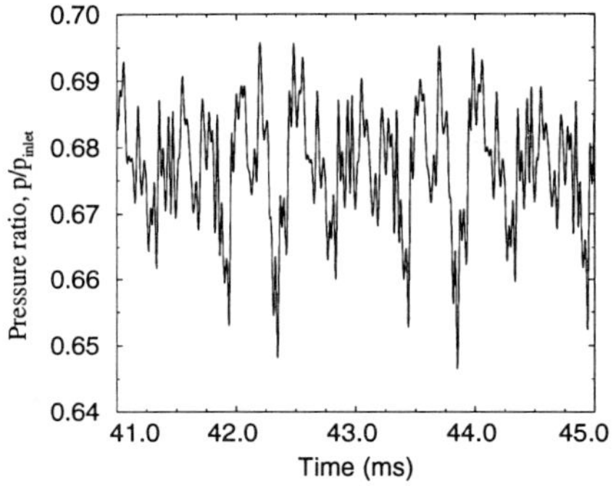

Figure 11. Pressure variation on the suction side of the first blade, at 25% axial chord.

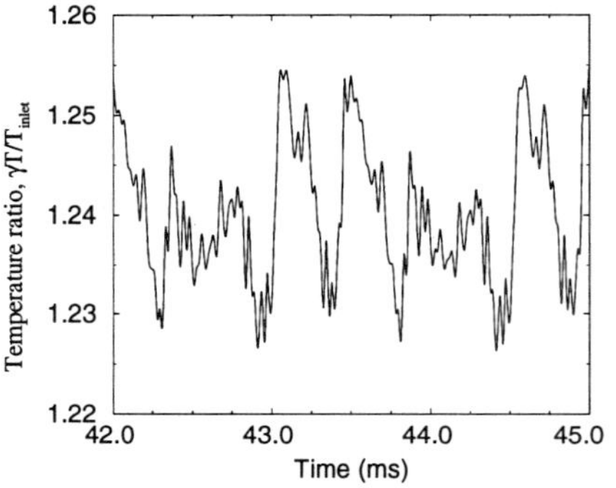

Figure 12. Temperature variation on the suction side of the first blade, at 25% axial chord.

interacts with two adjacent rows of vanes. In Figure 11 it is shown that the pressure pattern repeats after 1.5 ms. Maximum pressure variation at 25% axial chord on the suction side is 7.4%. The temperature variation at the same point on the blade is shown in Figure 12. The maximum temperature variation at this point is 2.3%.

5. Conclusions

This paper presents the development of a two-dimensional, time-accurate, Navier-Stokes equations parallel code from a sequential multiple-grid code. The primary objective of this investigation was to compare the performance of the parallel code to that of the sequential code. The secondary objective was to use the parallel code for the numerical simulation of flow in a single stage and $1\frac{1}{2}$-stage turbine.

Parallelization of the code resulted in super-linear speedups of more than 50% for a simulation running on 10 processors. The reason for this greater than expected speedup is the availability of more cache on the SMP. Increased cache availability leads to faster memory accesses for the CPU and consequent faster MFLOPS.

6. Acknowledgments

The authors wish to thank the Westinghouse Power Generation and Westinghouse Science & Technology Center for supporting this work. The authors are thankful to Pittsburgh Supercomputing Center for making the computing resources available. The authors are especially grateful to Dr. Karen Gundy-Burlet of NASA/Ames Research Center for her help in using the STATOR2 code and to Ms. Anjana Kar of Pittsburgh Supercomputing Center for her help in visualizing the numerical results.

References

1. Davis, R. L., Shang, T., Buteau, J., Ni, R. H. (1996) Prediction of 3-D Unsteady Flow in Multi-Stage Turbomachinery Using an Implicit Dual Time-Step Approach, AIAA Paper 96-2565, Lake Buena Vista, Florida.
2. Doorly, D. J. (1987) Modeling the Unsteady Flow in a Turbine Rotor Passage, Gas Turbine Conference and Exhibition, Anaheim, California, 87-GT-197.
3. Dorney, D. and Gundy-Burlet, K. (1995) Hot Streak Clocking Effects in a 1-$\frac{1}{2}$ Stage Turbine, ASME Paper 95-GT-202, Houston, Texas.
4. Dorney, D. and Sharma, O. P. (1996) A Study of Turbine Performance Increases Through Airfoil Clocking, AIAA Paper 99-2816, Lake Buena Vista, Florida.
5. Dring, R. P., Joslyn, H. D., Hardin, L. W., and Wagner, J. H. (1982) Turbine Rotor-Stator Interaction, *ASME Journal of Engineering for Power* **104**, pp. 729-742.
6. Dring, R. P., Blair, M. F., Joslyn, H. D., Power, G. D., and Verdon, J. M. (1986) The Effects of Inlet Turbulence and Rotor/Stator Interactions on the Aerodynamics and Heat Transfer of a Large-Scale Rotating Turbine Model: I - Final Report, NASA Contractor Report 4079.
7. Dring, R. P., Blair, M. F., Joslyn, H. D., Power, G. D., and Verdon, J. M. (1986) The Effects of Inlet Turbulence and Rotor/Stator Interactions on the Aerodynamics and Heat Transfer of a Large-Scale Rotating Turbine Model: II - Heat Transfer Data Tabulation 15% Axial Spacing, NASA Contractor Report 4079.
8. Dring, R. P., Blair, M. F., Joslyn, H. D., Power, G. D., and Verdon, J. M. (1986) The Effects of Inlet Turbulence and Rotor/Stator Interactions on the Aerodynamics and Heat Transfer of a Large-Scale Rotating Turbine Model: III - Heat Transfer Data Tabulation 65% Axial Spacing, NASA Contractor Report 4079.
9. Dring, R. P., Blair, M. F., Joslyn, H. D., Power, G. D., and Verdon, J. M. (1986) The Effects of Inlet Turbulence and Rotor/Stator Interactions on the Aerodynamics and Heat Transfer of a Large-Scale Rotating Turbine Model: IV - Aerodynamic Data Tabulation 65% Axial Spacing, NASA Contractor Report 4079.
10. Dunn, M. G., Bennett, W. A., Delaney, R. A., and K. V. Rao (1992) Investigation of Unsteady Flow Through a Transonic Turbine Stage: Data/Prediction Comparison for Time-Averaged and Phase-Resolved Pressure Data, *ASME Journal of Turbomachinery* **114**, pp. 91-99.
11. Erdos, J. I., Alzner, E., and McNally, W. (1977) Numerical Solution of Periodic Transonic Flow through a Fan Stage, *AIAA Journal* **15**, pp. 1559-1568.
12. Eulitz, F., Engel, K., Gebbing, H. (1996) Numerical Investigation of the Clocking Effects in a Multistage Turbine, ASME Paper 96-GT-26, Birmingham, UK.
13. Fourmaux, A. (1986) Unsteady Flow Calculation in Cascades, ASME Paper 86-GT-178, Dusseldorf, West Germany.
14. Giles, M (1989) Numerical Methods for Unsteady Turbomachinery Flow, CFDL-TR-89-3, M.I.T.
15. Griffin, L. W., Huber, F. W., Sharma, O. P. (1996) Performance Improvement Through Indexing of Turbine Airfoils: Part2 - Numerical Simulation, *ASME Journal of Turbomachinery* **118**, pp. 636-642.
16. Gundy-Burlet, K. L., Rai, M. M., and Dring, R. P. (1989) Two-Dimensional Computations of Multi-stage Compressor Flows Using a Zonal Approach, AIAA Paper 89-2452, Monterey, California.
17. Gundy-Burlet, K. L., Rai, M. M., Stauter, R. C., and Dring, R. P. (1992) Temporally and Spatially Resolved Flow in a Two-Stage Axial Compressor: Part 2 - Computational Assessment, *Journal of Turbomachinery* **113**, pp. 227-232.
18. Gundy-Burlet, K. L. (1992) Unsteady Two-and Three-Dimensional Navier-Stokes Simulations of Multistage Turbomachinery Flows, *Computing Systems in Engineering* **3**, pp. 231-240.

19. Gundy-Burlet, K. L., Dorney, D. J. (1996) Three-Dimensional Simulations of Hot Streak Clocking in a 1-$\frac{1}{2}$ Stage Turbine, AIAA Paper 96-2791, Lake Buena Vista, Florida.
20. Huber, F. W., Johnson, P. D., Sharma, O. P., Staubach, J. B., and Gaddis, S. W. (1996) Performance Improvement Through Indexing of Turbine Airfoils: Part1 - Experimental Investigation, *ASME Journal of Turbomachinery* **118**, pp. 630-635.
21. Jorgenson, P. C. E. and Chima, R. V. (1988) An Explicit Runge-Kutta Method For Unsteady Rotor/Stator Interaction, AIAA Paper 88-0049, Reno, Nevada.
22. Jung, A. R., Mayer, J. F., Stetter, H. (1996) Simulation of 3D-Unsteady Stator/Rotor Interaction in Turbomachinery Stages of Arbitrary Pitch Ratio, ASME Paper 96-GT-69, Birmingham, UK.
23. Kelecy F. J., Griffin, J. W., and Delaney, R. A. (1995) The Effect of Vane-Blade Spacing on Transonic Turbine Stage Performance, 85th Propulsion and Energetics Panel Symposium on Loss Mechanisms and Unsteady Flows in Turbomachines, Derby, UK.
24. Koya, M., Kotake, S. (1985) Numerical Analysis of Fully Three-Dimensional Periodic Flows Through a Turbine Stage, ASME Paper 85-GT-57, Houston, Texas.
25. Lewis, J. P., Delaney, R. A., and Hall, E. J. (1989) Numerical Prediction of Turbine Vane-Blade Aerodynamic Interaction, *ASME Journal of Turbomachinery* **111**, pp. 387-393.
26. Liamis, N., Bacha J. L., and Burgaud, F. (1995) Numerical Simulation of Stator-Rotor Interaction on Compressor Blade Rows, 85th Propulsion and Energetics Panel Symposium on Loss Mechanisms and Unsteady Flows in Turbomachines, Derby, UK.
27. Michelassi, V., Adami, P., and Martelli F. (1996) An Implicit Algorithm For Stator-Rotor Interaction Analysis, ASME Paper 96-GT-68, Birmingham, UK.
28. Rai, M. M. (1985) Navier-Stokes Simulation of Rotor-Stator Interaction Using Patched and Overlaid Grids, AIAA Paper 85-1519, Cincinnati, Ohio.
29. Rai, M. M. and Chakravarthy, S. (1986) An Implicit Form for the Osher Upwind Scheme, *AIAA Journal* **24**, pp. 735-743.
30. Rai, M. M. (1985) Unsteady Three-Dimensional Navier-Stokes Simulation of Turbine Rotor-Stator Interaction, AIAA Paper 87-2058, San Diego, California.
31. Tsung, F. L., Loellbach, J., and Hah, C. (1996) Development of an Unsteady Unstructured Navier-Stokes Solver for Stator-Rotor Interaction, AIAA Paper 96-2668, Lake Buena Vista, Florida.

12: Experimental Unsteady Flow Around Oscillating Blades - 2

CORRELATION OF PRESSURE AND HOT FILM DATA ON AN OSCILLATING 2D NACA63A006 SINGLE AIRFOIL AT HIGH INCIDENCE

SVENSDOTTER, S.
Chair of Heat and Power Technology
Royal Institute of Technology
Stockholm
Sweden

JOHANSSON, U.
Volvo Aero Corporation
Trollhättan
Sweden

KEMPPAINEN, A.[*]
Energy Systems Division
Ericsson Component AB
Kista
Sweden

FRANSSON, T.
Chair of Heat and Power Technology
Royal Institute of Technology
Stockholm
Sweden

1. Abstract

Turbomachine design, even though as today often performed with an increasing degree of sophistication as regards to the use of numerical models, still is in need of high quality data in order to reduce the machine life cycle cost are various flow conditions, for example high angle of attack and stall under transonic flow conditions at high oscillating frequencies. The absence of reliable experimental data under these flow conditions provides problems for the validation of numerical models. Therefore the presented measurements were performed, with the objective to obtain detailed data for the pressure field and the boundary layer state at high angle of attack and high Mach number, with special emphasise to establish how the flow field on the surface side changes from attached to separated. The frequency and angle of attack range for the tests were chosen to address dynamic stall conditions for compressor blades.

In the presented paper investigations on an oscillation blade at an inlet Mach number of 0.5 has been performed with oscillation frequencies from 60 to 410Hz, mean incidence angles from 0 to 10° and amplitudes up to 3°. The steady state results show that the transition point is moving towards the leading edge for increasing angle of attack. For angles of attack above 9°, the suction surface seems to be entirely separated. The hot film and pressure data show the moving separation point at the same part of the oscillation cycle, while the reattachment point is shown earlier in the oscillation cycle by the hot film than the pressure data, probably due to the time for

---

[*] Formerly Chair of Heat and Power Technology, Royal Institute of Technology, Stockholm, Sweden

T.H. Fransson (ed.), Unsteady Aerodynamics and Aeroelasticity of Turbomachines, 649-663.

the newly established attached flow to build up the potential field again. The separation point is dependent on oscillation amplitude and somewhat on the oscillation frequency.

2. Nomenclature and Abbreviations

c	Chord length	[m]
Cp	Pressure coefficient, $(P - P_\infty)/q$	[-]
f	Oscillation frequency	[Hz]
e	Hot film output voltage	[V]
k	Reduced frequency, $\pi fc/U_\infty$	[-]
M	Mach number	[-]
P_∞	Tunnel free-stream static pressure	[Pa]
P	Local mean wall pressure on model	[Pa]
q	Dynamic pressure, $½*\rho*U_\infty^2$	[Pa]
U_∞	Free-stream velocity	[m/s]
α	Incidence angle	[deg]
$\tilde{\alpha}$	Top amplitude, $\Delta\alpha*\sqrt{2}$	[deg]
α_{tot}	Total incidence $(\alpha+\tilde{\alpha})$	[deg]
$\Delta\alpha$	Oscillation amplitude (rms)	[deg]
ρ	Density	[kg/m^3]

Suffix

c	chord-related
mean	Arithmetic mean value
rms	Root-mean-square value

Abbreviations

LE	Leading edge
LP	Lower blade position
UP	Upper blade position

3. Introduction

The interaction between unsteady flow and vibrating blades in turbomachinery can sometimes reach dangerous levels as regards to high material stresses, which has a serious effect on the safe operation of the engine.

In order to achieve knowledge of the flow conditions leading to vibrations, three ways are open: theoretical analysis of the flow conditions [Lin et al. 1997], numerical calculations [Ekaterinaris, 1995, Wernert et al., 1996] and experimental investigations [Lorber, 1988, He and Denton, 1991, van Dyken et al., 1996, Wernert et al., 1996].

Even though the role of CFD calculations during the design process has become strong the last decennial, the development of computational prediction models is still an important field for research. One factor is that it is still not possible to accurately predict the flow conditions in boundary layers and separated flow areas, and to understand some of the most dangerous flow conditions which includes high angle of attack, close or above the stall angle under transonic flow conditions. Newly developed computer models have to be compared with experimental data or with other validated CFD prediction codes.

Experimental data describing the appearance and development of boundary layer separation during high frequency vibrations is necessary in order to bring the development of numerical prediction models further, formulate theoretical models and to get insight and understanding of the physics of the unsteady flow. Carr [1986] mentions that more investigations have to be performed concerning the effects of Reynolds numbers and compressibility on dynamic stall.

Most studies on dynamic stall, reported so far, have investigated oscillating blades with oscillation frequencies not higher than 20Hz and at low air speeds ($M \leq 0.3$) [He and Denton, 1991, van Dyken et al., 1996, Wernert et al., 1996]. Chandrasekhara and Brydges [1990] have performed studies at Mach numbers up to 0.45 and reduced frequencies up to 0.1.

4. Objectives

The objective of the investigations discussed in this paper was to achieve experimental data at relatively high oscillation frequencies and high inlet Mach numbers in order to extend existing data bases in this field and to correlate the boundary layer state, measured with hot film technique, with unsteady pressure measurements on the blade surface.

5. Experimental Equipment

5.1. THE WIND TUNNEL

The wind tunnel used for the experiments was wind tunnel No 5 at VAC which is connected to a large air reservoir providing the tunnel with air of constant inlet conditions for continuous testing. The test section has the span width of 150mm and the height of 180mm. The top and bottom ceilings are slotted in the stream-wise direction in order to prevent reflecting shock waves and to bleed off the boundary layer which will build up at the tunnel walls. The cross section of the stagnation chamber is $0.35m^2$ and of the measurement section $0.027m^2$. This gives a contraction ratio of 12.96. The inlet Mach number can be varied continuously from 0.3 to 1.6 and the inlet stagnation temperature is about 280K [Johansson, 1992]. The free stream static pressure can be varied between slightly below atmospheric pressure and up to 400kPa at Mach 1. The free-stream turbulence intensity is below 3%.

5.2. THE PROFILE

Two symmetrical 2D airfoil NACA 63A006 with a chord length of 80mm (figure 1) and a span of 150mm were used for the tests. The first model was instrumented with 67 hot film gauges along the mid span chord around the profile surface. The total thickness of the hot film sheet and the adhesive is 0.16mm, which adds a total 0.32mm to the profile thickness (30% of maximum blade thickness). The leading edge radius increases from 0.2mm to 0.36mm.

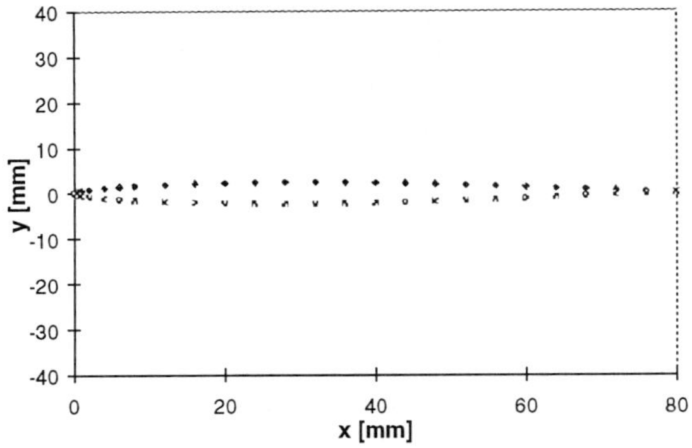

Figure 1. The profile geometry

The second profile is equipped with 13 pressure transducers on the suction surface of the blade. Pressure gauge 1-11 are mounted in cavities connected through drilled holes to the surface (∅ 0.6mm). Gauge 12 and 13 are flush mounted with the surface.

Both airfoils are mounted with plain bearings in the side walls with two shafts. The pitching axis is at 43% of the chord and the torque was transmitted to the airfoil by the starboard side shaft.

5.3. THE EXCITATION SYSTEM

Accurate time history traces of the motion is needed for the performance of the experiments, and to achieve this a known excitation of the blade is required. The oscillations are performed with an hydraulic actuator, of ONERA type [Cristoux and Hutin, 1991], run with oil of 200bars. The oscillation frequencies and maximum oscillating angles can be seen in table 1. Details of the excitation system can be found in Sjunnesson [1992]. The actuator motion can be detected by a high precision potentiometer on the actuator shaft.

5.4. THE MEASUREMENT EQUIPMENT

The equipment used for the hot wire measurements was a 16 channel constant temperature anemometer system from TSI (IFA300).

A Senflex hot film sheet with in total 67 sensors was mounted on the blade. 14 sensors at the time were multiplexed to the anemometer. The 15:th channel was used to record the oscillating angle. The last channel was not used in these tests. Due to sensor failures, the most forward position measured was at 15% chord.

The hot film signals were sampled with 50kHz, low-passed filtered at 20kHz and thereafter recorded on the PC hard disk. The signals were normalised with the mean voltage, not as common with the mean voltage at zero flow conditions. This was due to the fact that no zero flow measurements were obtained.

The unsteady pressure data were recorded on a 28 channel Honeywell 101 tape recorder with 15kHz bandwidth. The signals were low pass filtered with 10kHz as the upper limit. Thereafter, the data was digitised with a sampling frequency of 4kHz and analysed on a HP microcomputer based system called TRAS-A.

The tunnel state, as well as the rms levels of all pressure gauges, was recorded with a PSI 8400 pressure measurement system.

5.5. ACCURACY

The pressure measurement system used for controlling and recording the tunnel state has an accuracy of ±500Pa (±0.15%F.S), which gives about ±0.5% accuracy at the pressure level (100kPa) used at the measurements.

The AD card for the hot film system has a resolution of 12 bits and a range of ±5V. This gives a digitisation error of ±0.002V. The error due to temperature drift is ±0.3μV/°C and the noise give an error of 1.7nV/√Hz. The bridge accuracy is ±0.1%±0.01. The total accuracy is then ±0.003V, considering an error of 10°C in temperature drift and temperature compensation.

The unsteady pressure measurement system has an accuracy of ±300Pa (±1.5% in the calibrated range ±20kPa) and the tape recorder gives an inaccuracy of within 1dB.

6. Method of Attack

TABLE 1. Oscillating frequencies for M=0.5, Re_c=850 000

[Hz]	$\Delta\alpha_{rms}$ [°]				
α [°]	7	3	1.9	0.6	0.2
0	5	60	60, 110	60, 110	60, 110, 410
1		60	60, 110	60, 110	60, 110, 410
3		60	60, 110	60, 110	60, 110, 410
6		60	60, 110	60, 110	60, 110, 410
7		60	60	60	60, 410
9		60	60	60	60, 410
10		60	60	60	60, 410

To achieve the objectives outlined in the previous section, the following experimental approach was selected. The investigated flow conditions were specified to consist of high oscillating frequencies (up to 410Hz), from attached to near or at stall conditions. The inlet Mach number was 0.5 and the chord based Reynolds number was 850 000. The details of the oscillation data can be found in table 1. Special focus on the development of the boundary layer on the suction side is important since the prediction of boundary layer transition is more complicated on the suction side than on the pressure side. Another important phenomenon is that the boundary layer separates on the suction side at high angles of attack. Hot films and pressure measurements were performed at conditions given in table 1, which gives reduced frequencies between 0.1 and 0.6.

7. Steady State Flow Case

Figure 2 gives information concerning the boundary layer state for the steady state flow case for Mach number 0.5, measured by hot film. For incidence angles 0 a clear transition zone can be seen from the peak of the rms values of the hot film signals, starting close to the leading edge and stretching until 60% chord. For incidence angle 3° the suction side has laminar flow at the leading edge, turning turbulent towards the trailing edge, but without a significant transition zone. This can bee seen by the fact that there is no peak in the rms level, but towards the trailing edge of the profile, the level is the same as for the zero incidence angle and it can thus be concluded that the flow is turbulent here. For higher incidence angles the suction side undergoes a rapid bypass transition to turbulent flow already at the leading edge. The steady state separation can be suspected when the rms level is increasing, but it can not clearly be stated from the hot film signals exactly at which angle it occurs.

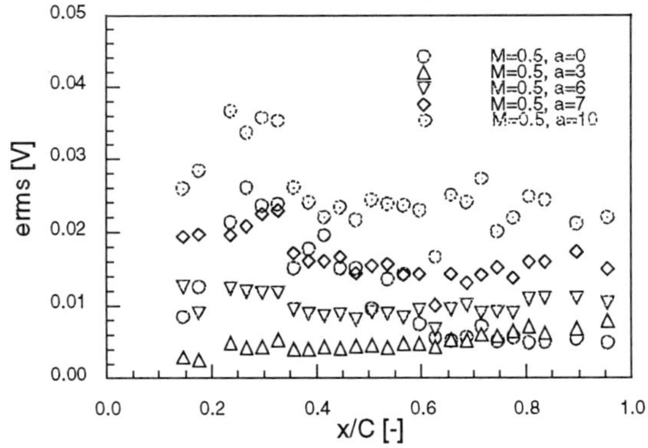

Figure 2. Hot film rms output for M=0.5

Pressure data (figure 3) show that there is a local separation until about 15% chord at 7°, with the decrease in Cp, and at 9° incidence the flow on the suction side is entirely separated. Since, for this profile, the steady state stall angle is between 6° and 8°, the mean angle of 6 ° has been chosen for further evaluation of the unsteady state.

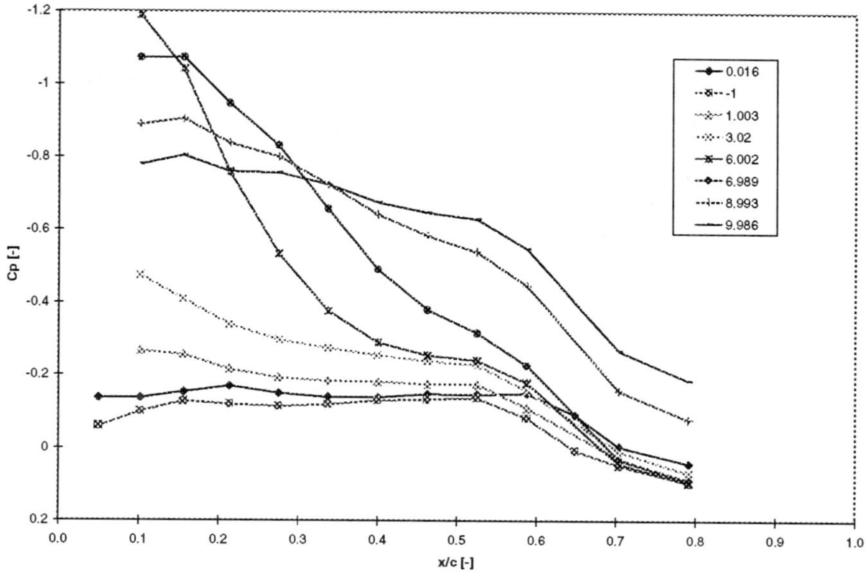

Figure 3. Cp versus x/c at α=-1° to 10° for M=0.5

8. Time-Dependent Flow Cases

In figures 4 and 5 is illustrated how the separation and reattachment is determined from the pressure and hot film signals. For the pressure signal, the interpretation of a separation start is when the pressure sensors shows a rapid increase in pressure, less lifting force and disorder in the periodic reading. The reattachment is interpreted when there is a re-established order in the periodic pressure time-readings and as the point where the profile surface pressure starts to decrease. During the separated part of the oscillating cycle, there is a large scatter of the signals over several cycles, while the attached part of the cycle shows a very repetitive behaviour. Figure 4 shows as an example the signal from the pressure sensor at 15% chord for a mean incidence angle of 6° an oscillation frequency of 60Hz and a rms amplitude of 1.9°. The separation is detected at about 2° amplitude, relative to the mean incidence angle, on the blade down-stroke and the reattachment is determined at about -1° amplitude on the blade up-stroke.

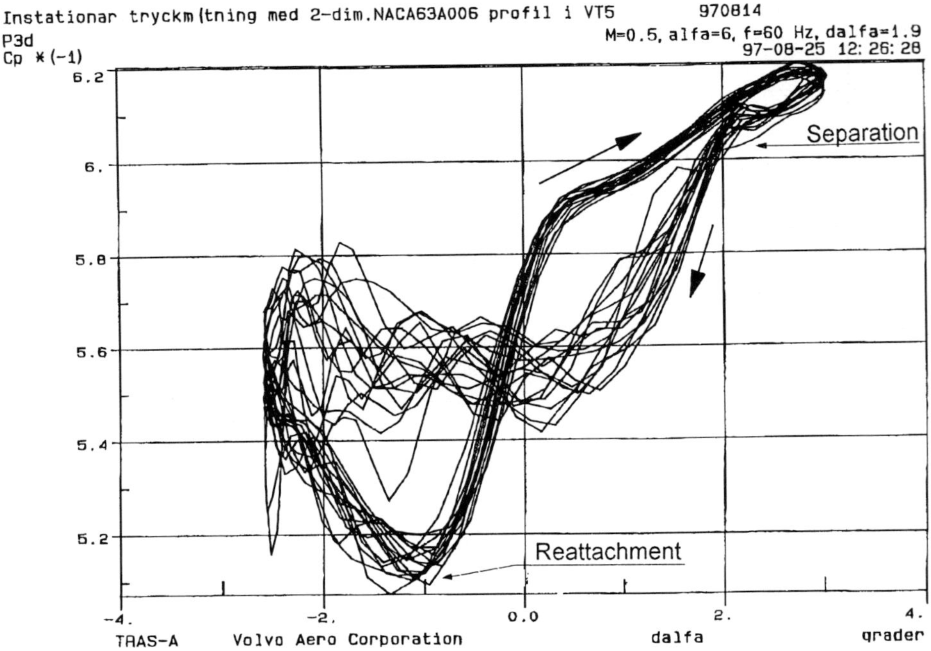

Figure 4. Example of pressure time trace at 15% chord, M=0.5, f=60Hz, α=6°, Δα=1.9°

Figure 5 shows the hot film trace for the same flow case as above, also for the sensor positioned at 15% chord. Separation gives a lower heat transfer between the blade (i.e. the hot film sensor) and the fluid, which is shown as a rapid decrease of the level of the hot film signal. The end of separation is seen as a recovery of heat transfer. Increase in heat loss is occurring during transition while a decrease is typical for separation. For figure 5 the separation is determined at about 2° amplitude on the blade up-stroke and the reattachment is seen at about zero amplitude on the blade down-stroke.

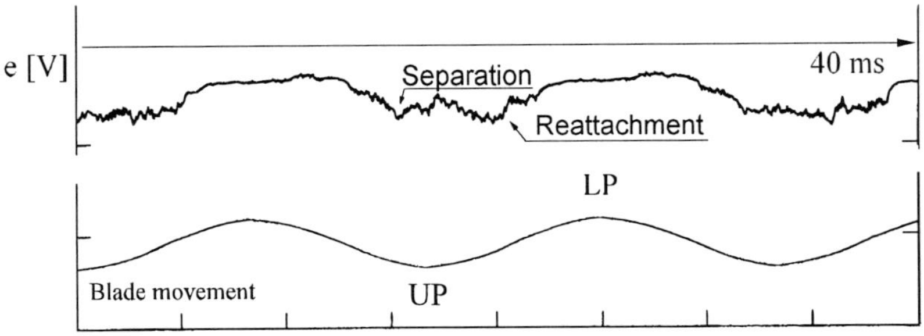

Figure 5. Example of hot film time trace at 15% chord, M=0.5, f=60Hz, α=6°, Δα=1.9°

It can be seen, especially from the pressure signals, that the reattachment process is slower than the initiation of the separation process. The part of the oscillation cycle that experiences separated flow increases and the region becomes wider (in downstream direction) for increasing α. The difference in the response of the pressure and hot film signals for the separation and reattachment points is discussed in the following chapter.

8.1 INCIDENCE ANGLE OF 6°

Figure 6 shows the separation and reattachment positions over the blade chord for 60Hz oscillation frequency. It can be seen that the highest amplitude gives the highest total angle for the separation point and that the separated part of the oscillation cycle is longer at the aft part of the blade. The 0.2° amplitude does not show any difference from the steady state flow case and has been excluded from the figure. The reattachment point seems to be fairly constant around 6° total incidence, independent of amplitude.

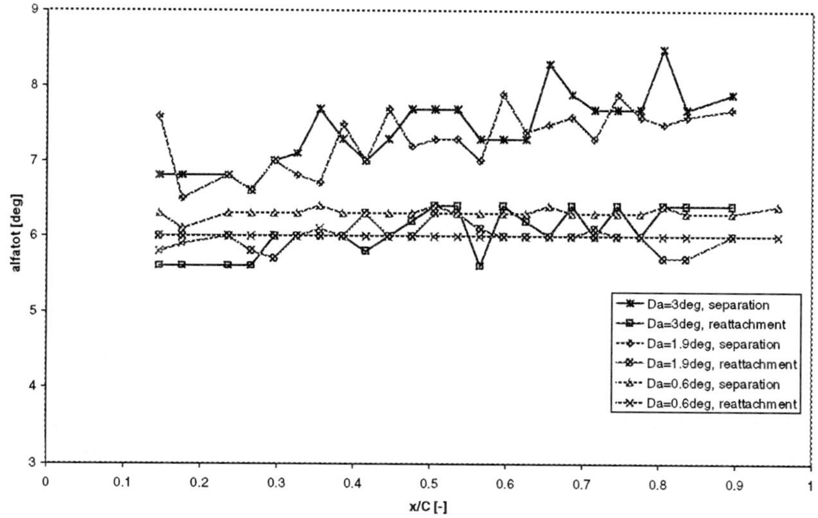

Figure 6. Separation region versus x/c, α=6°, f=60Hz, hot film

The dynamic pressure measurements (figure 7) at 60Hz and $\Delta\alpha$=1.9° suggests that the separation starts from the aft part of the blade at the upper turning point (8.4° total incidence angle). This separation starting at the aft part, propagates upstream along the chord until the whole suction side flow is separated on the down-stroke at about 7.9° total incidence angle. The hot film and the pressure sensors give similar results for the separation start. Both pressure and hot film data show the same trends, the separated part of the oscillation cycle increases in extension towards the trailing edge. As was mentioned in the previous chapter, there is however a discrepancy for the reattachment between the hot film and the pressure signals. The hot film shows a reattachment at about the centre position (α_{tot}=6°) on the down-stroke, while the pressure signals show a reattachment at angles below the centre position, on the up-stroke. It is possible that the hot film readings which are sensitive to the boundary layer cooling effect, detect a reattached flow at the instant point where it occurs and that the pressure on the surface, which is created by the potential field around the whole blade, is delayed because it takes a while until the pressure is built up by the newly established flow along the blade surface. At 60Hz and 1.9° amplitude, 1° delay gives the same period of time as the time it takes for the air to travel along the blade chord. This can be a good measure of what kind of time period it takes to establish the potential flow again after complete stall. A similar reasoning can be used to explain the short delay of the separation point detected from the pressure sensors.

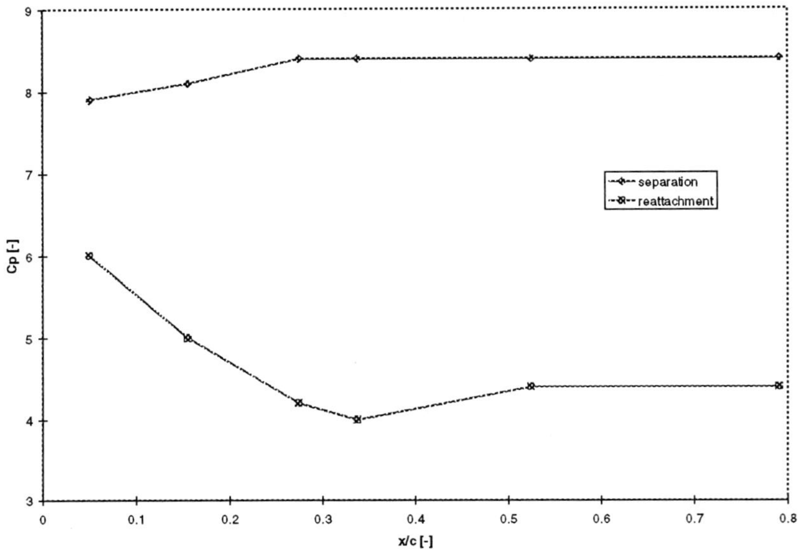

Figure 7. Separation region versus x/c, α=6°, f=60Hz, pressure

The lowest oscillation amplitude (Δα=0.2°) gives the same response as the steady state case, also for the pressure measurements. One reason for this could be that the changes in the boundary layer might be too small for the sensors to detect. Another explanation is that for an amplitude of 0.2°, the movement of the blade is of the same scale as the thickness of the boundary layer. This could mean that the flow does not have the time to react to the blade movement. For this reason no effect can be noted from the 410Hz case, since this could not be run at a higher amplitude

One exception is the oscillation frequency 110Hz (figure 8) and incidence angle 6°, which shows a very small effect from the oscillation at Δα=0.2°. This is not seen in any other test case and the reason for this is unknown. Figure 8 shows that a higher oscillation frequency increases the total incidence angle for stall, while the reattachment position does not seem to be affected.

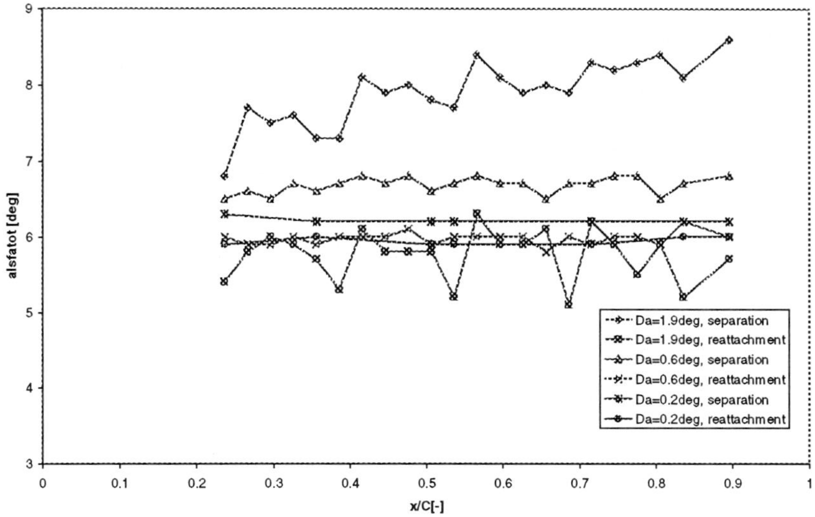

Figure 8. Separation region versus x/c, α=6°, f=110Hz, hot film

When studying the separation start and the flow reattachment as functions of amplitude and frequency (figures 6 to 8) it can be noted that the separation start is at a higher angle for higher oscillation amplitude, probably due to the higher speed at which the blade is moving on the up and down stroke part of the period at higher amplitude. This has also been mentioned by Lorber and Carta [1988]. Amplitude and frequency dependence has also been discussed by Chandrasekhara and Brydges [1990] for an oscillating NACA 0012 blade with inlet Mach numbers between 0.15 and 0.45 and reduced frequencies below 0.1.

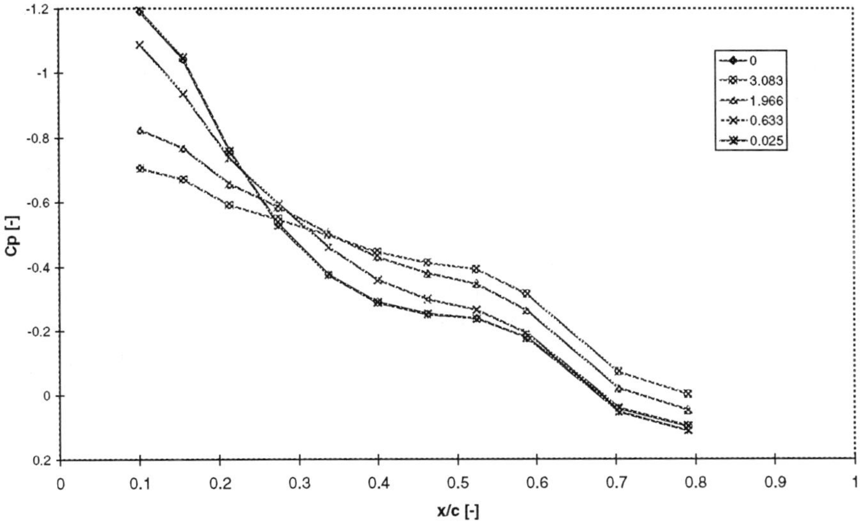

Figure 9. Mean Cp versus x/c at α=6° and f=60Hz as a function of Δα

Figure 9 shows the mean Cp for the suction side at varying amplitudes for 60Hz oscillation. It can be seen that the suction peak mean level is decreasing with increasing amplitude. This is a consequence of the fact that the mean incidence angle is close to the stall angle and when the oscillation amplitude is added, the total incidence angle reaches values above the stall angle. A higher amplitude will result in a longer time period spent with stalled flow conditions.

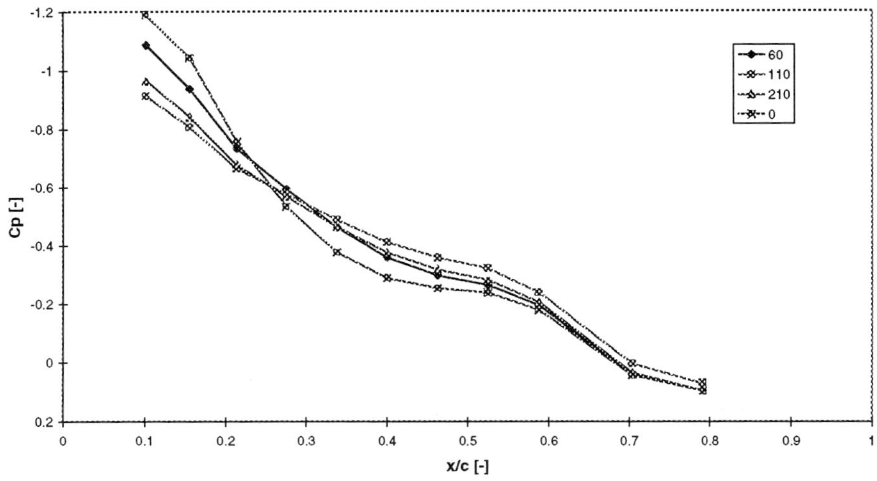

Figure 10. Mean Cp versus x/c at α=6° and Δα=0.6° as functions of frequency

From figure 10, showing Cp as a function of frequency for a constant amplitude at 0.6°, can be seen that the suction peak level on the forward part of the profile is decreasing with increasing frequency. This can be explained by looking back at figure 4. The reattachment delay time is longer than the separation time delay after respective blade turning point, which means that the separated part of the oscillation cycle is larger then the attached flow part. If the frequency is increased the total time of the cycle is decreased. Assuming that the delay time is dependent on the flow speed, which is probable, and that the separation respectively the reattachment process is triggered by the turning of the blade motion direction and the lower motion speed part of the cycle, then the delay times are constant while the total cycle time is decreasing with increasing frequency. This will lead to a larger portion of separated flow over the cycle, which results in a lower mean level.

The mean level of Cp for an oscillation of 110Hz is seen in figure 11. As for the 60Hz oscillation frequency, the suction peak mean level is decreasing with increasing amplitude. Also here is shown an effect of the amplitude of 0.2°, as did the hot film data (figure 8).

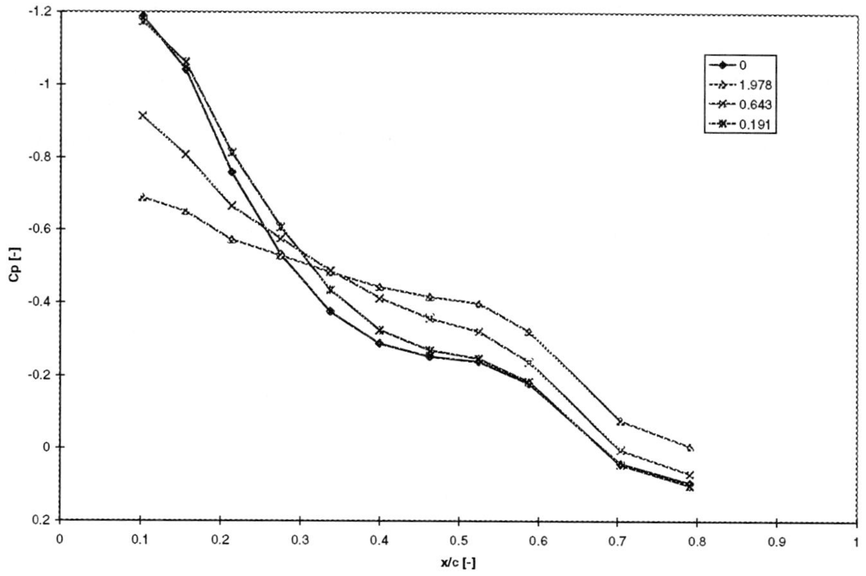

Figure 11. Mean Cp versus x/c at α=6° and f=110Hz, as a function of Δα

The Cp plots for α=6° (figures 9 to 11) show that the part of the cycle which has a separated suction side flow increases with increasing amplitude as well as frequency. Both the 60Hz (figure 9) and the 110Hz (figure 11) case show the same behaviour, with the 110Hz oscillation giving a slightly higher suction peak than 60Hz when looking at the same amplitude. The pressure distribution is stronger dependent on amplitude than the frequency as far as these results show. However that can be a result of the fact that comparisons between frequencies only can be done for the lower amplitudes.

8.2 OTHER TEST CASES

The other test cases have so far only been evaluated from the hot film signals. It can briefly be mentioned that low mean incidence angles show traces of both one leading edge separation and one trailing edge separation, while at increasing angles, the trailing edge separation seems to take over. If the mean incidence angle is above the static stall angle, decreasing amplitude gives separation over a larger part of the oscillation cycle, while the opposite is valid for mean incidence angles below the static stall angle. This will be reported in an internal report at HPT, which is under progress (Svensdotter and Johansson, 1997).

9. Conclusions

Pressure and hot film data for Mach number 0.5 has been presented for oscillation frequencies of 5, 60, 110 and 410Hz (and 210Hz for pressure data).

The steady state flow case shows the steady state stall angle to be at an incidence angle between 6° and 9°.

The hot film and pressure data gives the separation point at about the same total incidence angle but differs considerably concerning the reattachment position.

Separation occurs at a higher angle of attack for both increasing amplitude and frequency, but the separation and reattachment points are mostly dependent of the oscillation amplitude and only some of the oscillation frequency.

For an amplitude of 0.2°, no effect on the flow can be seen neither by the hot film nor the pressure sensors, regardless of oscillation frequency. Either the sensors can not resolve the flow, or the blade movements are too small for the flow to react on.

10. Future Work

The above presented data was only a small part of the measured test cases. The evaluation will continue, aiming on stability influence of the dynamic stall.

It would also be of interest to study the effect of the extra thickness from the hot film sheet on the measured result, as well as the difference in surface roughness, with for example flow visualisation and numerical calculations.

11. Acknowledgements

This study was financed by Försvarets Materialverk (FMV), through Volvo Aero Corporation, and the Chair of Heat and Power Technology. Support was also obtained from Rolls-Royce, Derby, and ABB, Baden.

12. References

Chandrasekhara, M.; Brydges, B.; 1990
"Amplitude Effects on Dynamic Stall of an Oscillating Airfoil"
28:th Aerospace Sciences Meeting, January 8-11, 1990, AIAA-90-0575

Cristoux, C.; Hutin, P.; 1991
"Hydraulic servoactuator controler, Notice"
ONERA, Rapport Technique No 4/8443 RY 010 R, Octobre 1991

Ekaterinaris, J.; 1995
"Numerical Investigation of Dynamic Stall of an Oscillating Wing"
AIAA Journal, Vol. 33, No. 10, October 1995, pp. 1803-1808

He, L.: Denton, J.; 1991
"An experiment on Unsteady Flow Over an Oscillating Airfoil"
ASME 91-GT-181

Johansson, U.; 1992
"Instationära luftkrafter, 2D profil NACA 63A006. Etapp 1, Stationära mätningar."
Internal report VAC 9370-532

Lin, H.; Vezza, M.; McD. Galbraith, R.; 1997
"Discrete Vortex Method for Simulating Unsteady Flow Around Pitching Airfoils"
AIAA Journal, Vol. 35, No. 3, March 1997, pp 494-499

Lorber, P.; Carta, F.; 1988
"Airfoil Dynamic Stall at Constant Pitch Rate and High Reynolds Number"
Journal of Aircraft, Vol. 25, No. 6, June 1988, pp 548-556
Sjunnesson, I.; 1992
"Experiments on an oscillating 2D aerofoil, NACA 63A006. Test of oscillating system. Preliminary results"*Internal report VAC 6825-229*
Svensdotter, S.; Johansson, U.; 1997
"Experiments on an oscillating 2D NACA 63A006 isolated airfoil.
Internal report HPT/VAC, UNDER PROGRESS
Van Dyken, R.; Ekaterinaris, J.; Chandrasekhara, M.; Platzer, M.; 1996
"Analysis of Compressible Light Dynamic Stall Flow at Transitional Reynolds Numbers"
AIAA Journal, Vol. 34, No. 7, July 1996, pp 1420-1427
Wernert, P.; Geissler, W.; Raffel, M.; Kompenhaus, J.; 1996
"Experimental and Numerical Investigations of Dynamic Stall on a Pitching Airfoil"
AIAA Journal, Vol. 34, No. 5, May 1996, pp 982-989

A BASIC FORCED RESPONSE EXPERIMENT

M. C. NOWINSKI, P. OTT
Ecole Polytechnique Fédérale de Lausanne
EPFL
DGM-LTT
CH-1015 Lausanne

Abstract

A new test facility has been constructed to investigate the unsteady aerodynamics around a single compressor blade. A series of preliminary experiments was conducted to quantify the unsteady blade loading produced by a downstream aerodynamic gust, a sinusoidal forced plunging motion of the blade, and a controlled combination of the two. These measurements were made in an effort to improve basic understanding of the forced response problem and in particular the interaction between the unsteady flow fields induced by the flow perturbation and the blade vibration. Evidence is presented which indicates that for reduced frequencies between 0 and 0.2, the unsteady loading resulting from the combined excitation of the blade can be accurately predicted by a linear superposition of the individual gust and blade vibration-induced flow fields. These results are presented for both subsonic and transonic flow conditions. For the transonic case, some discrepancies exist in the region near the suction-side shock, but it is likely that these differences arise from a small shift in the mean shock location on the blade between the two individual superimposed cases.

Nomenclature

c	chord length [m]
$\tilde{c}_P$	unsteady pressure coefficient, $\tilde{p}/(p_{T1} - p_1)*100$
$\bar{c}_P$	steady-state pressure coefficient, $p/(p_{T1} - p_1)$
f	excitation frequency [Hz]
i	flow incidence angle [°]
k	reduced frequency, $f\pi\, c/U_1$
$\tilde{p}$	unsteady pressure [Pa]
p_{T1}	upstream total pressure [Pa]
p_1	upstream static pressure [Pa]
PS	pressure side of the blade
SS	suction side of the blade
x	axial direction (origin at blade leading edge, positive in direction of flow) [m]
U_1	upstream velocity [m/s]
y	vertical direction (origin at blade leading edge, positive from PS to SS) [m]
α	angle of attack [°]
θ	phase angle between excitation sources [°]

T.H. Fransson (ed.), Unsteady Aerodynamics and Aeroelasticity of Turbomachines, 665-678.

1. Introduction

Forced response phenomena in turbomachinery remain a primary concern in gas turbine engine design and have important implications for engine life and reliability. The forced response problem encompasses potential flow interactions between fixed and rotating blade rows, as well as the periodic impingement of upstream-generated wakes or shocks on downstream blading. In cases where the frequency of the aerodynamic forcing function matches the eigenfrequency of the blade structure, significant vibrations can occur leading to structural fatigue and eventual failure.

To date, a significant research effort has been devoted to the quantification of the unsteady blade loading induced by such aerodynamic disturbances. Combined with knowledge of the structural characteristics of the blade assembly, this information can be used to predict blade vibration amplitudes and ultimately to determine the susceptibility of the blade to forced response phenomena. This complex problem has commonly been divided into two parts: (1) aerodynamic gust investigations, such as those conducted by Fleeter, et al. (1978), Capece and Fleeter (1988), and Manwaring and Wisler (1992), which measure the unsteady loading induced on a fixed (non-vibrating) blade by an imposed periodic flow disturbance; and (2) aerodynamic damping investigations, such as those conducted by Bölcs, et al. (1989), Buffum and Fleeter (1990), and Körbächer and Bölcs (1994), which measure the unsteady loading on a blade forced to vibrate in a uniform flow. Other work, such as that conducted by Manwaring et al. (1996) and Hennings (1997), more closely simulates the real turbomachinery environment by measuring the unsteady loading induced on elastic blade structures. These types of experiments, however, typically require relatively expensive and complicated test setups. Therefore, a need still exists today to understand how the results from (1) and (2) above can be combined to accurately predict the overall forced response of the blade. Poengsen and Gallus (1991) presented some experimental evidence which indicates that it may be possible to consider the complete forced response problem as a linear superposition of the gust- and blade vibration-induced flow fields. However, additional information over a wide range of flow conditions and reduced frequencies is required to validate this approach.

In order to provide an improved understanding of this unsteady interaction and of forced response phenomena in general, a new test facility was constructed and a basic experiment was performed to approximate the unsteady interaction between a rotor and a downstream vane row. In this experiment, a single compressor blade is subjected to a downstream aerodynamic disturbance, a forced bending motion, and a combination of the two. In the latter case, the forced excitation and the flow perturbation share a common frequency and the phase angle between them can be varied. In this way, an approximation of the structural response of the blade to the imposed flow disturbance is realized. This paper describes the new test facility in detail and presents some preliminary steady and unsteady experimental results.

2. Experiments

2.1 TEST FACILITY

The experimental measurements were conducted in the newly constructed Unsteady Transonic Wind Tunnel (refer to Figure 1) at the Ecole Polytechnique Fédérale de Lausanne. Air is supplied to this open cycle facility on a continuous basis by a four-stage centrifugal compressor with a pressure ratio of 3.5 and maximum flow rate of 10 kg/s. The air enters the tunnel through an inlet settling chamber and is accelerated by a converging nozzle into a rectangular test section measuring 200 mm (height) by 40 mm (width). The air exiting the test section is vented through a downstream valve to the atmosphere.

A single compressor blade with a NACA 3506 profile and a chord length of 78 mm is centrally mounted in the test section. A vibration system driven by a hydraulic motor is externally attached to the blade (refer to Figure 2). The rotational motion of the motor is translated to a plunging blade vibration perpendicular to the chord via an eccentric shaft and connecting arm. The vibration frequency of the system can be varied between 0 and 100 Hz, and the average vibration amplitude is approximately 0.4 mm. The entire blade-motor assembly is mounted to the tunnel side-wall via a circular disc to allow easy adjustment of the inlet flow incidence angle. Within this disk, the blade entry passage is equipped with a specially designed labyrinth sealing system (refer to Figure 2) which is described in further detail by Bölcs and Norryd (1994). This system significantly reduces the effects of flow leakage on the steady and unsteady measurements, while still ensuring an unrestricted motion of the blade.

Approximately 375 mm downstream of the blade mid-chord, a small flat plate (15 mm high, 39 mm wide, and 2 mm thick) is mounted within the rotating shaft of a second hydraulic motor (refer to Figure 3). The rotational frequency of this assembly can be varied between 0 and 100 Hz, producing upstream-running waves at the doubled frequency (i.e. 0 to 200 Hz).

The above excitation systems can be precisely synchronized and the phase lag between them can be freely varied. This allows for unsteady measurements to be conducted in the presence of only the downstream flow perturbation, only the blade vibration, or a combination of the two for different phase angles.

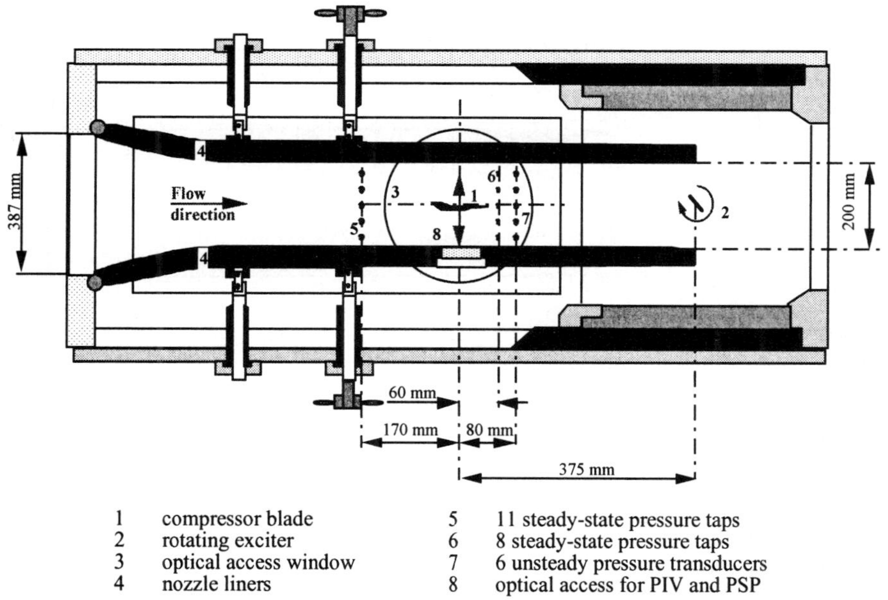

1	compressor blade	5	11 steady-state pressure taps
2	rotating exciter	6	8 steady-state pressure taps
3	optical access window	7	6 unsteady pressure transducers
4	nozzle liners	8	optical access for PIV and PSP

Figure 1. Unsteady Transonic Wind Tunnel

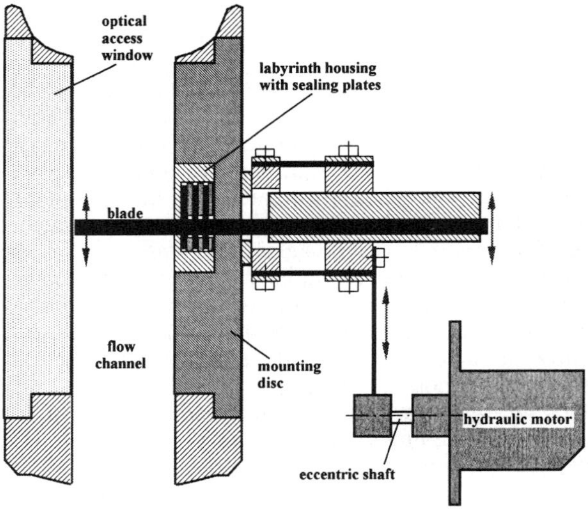

Figure 2. Blade Oscillation System

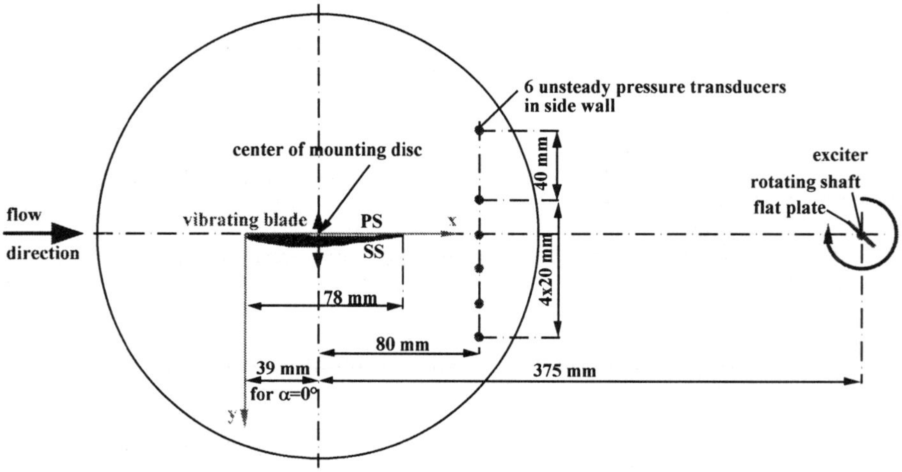

Figure 3. Unsteady Measurement Specifications

2.2 DATA ACQUISITION AND PROCESSING

The steady-state flow conditions are determined from the inlet plenum total pressure and temperature and static pressures measured along the test section side wall at axial stations up- and downstream of the compressor blade (refer to Figure 1). The steady-state pressure distribution along the blade surface at mid-span is obtained using a series of 8 pressure taps along the pressure side and 10 pressure taps along the suction side. The exact pressure tap locations along the blade are given in Appendix 1.

For the unsteady measurements, a second blade is equipped with high-response, piezoresistive pressure transducers. A total of 8 transducers are installed in the blade at mid-span, with 3 installed along the pressure side and 5 installed along the suction side. Additional specifications are given in Appendix 2. As shown in Figure 4, the pressure transducers are mounted within a cavity fabricated near the blade surface on the opposite side of the desired measuring location. This cavity is then connected to the measuring location by a small hole. Remaining cavities or gaps are sealed with epoxy. This installation method ensures a smooth measuring surface and protects the transducers from the outside environment.

A second set of unsteady measurements is obtained along the test section side-wall. A vertical line of unsteady pressure transducers are mounted within the optical access window at a distance of 80 mm downstream of the blade mid-chord (refer to Figure 3). Information from these transducers is used to quantify the periodic, upstream-running aerodynamic gust.

The voltage signal from each transducer is passed through a DC-filter, amplifier, and low-pass filter. Using a PC-based data acquisition system (Keithley Metrabyte DAS-1800), the analog data signals are digitized at a constant sampling rate yielding at least fifteen samples per period and are then stored directly to hard disk. The calibration of the transducers is performed dynamically using a periodically oscillating reference pressure over the desired range of frequencies. For the cases in which the blade was vibrating, a correction is also added to compensate for the false content present in the pressure transducer signals which arises from the blade acceleration. These corrections are typically small and obtained under a no-flow condition.

Approximately 25,000 data points are obtained for each channel during a typical unsteady measurement. This data is then sub-divided into overlapping blocks of 1024 points and each data block is Fourier decomposed. To minimize the effects of spectral leakage, a Hanning window is applied to each data block prior to the decomposition. The spectral components corresponding to the downstream disturbance and/or blade vibration frequency are extracted and averaged over the individual blocks. The pressure transducer sensitivity values, low-pass filter corrections, and blade acceleration corrections (if applicable) are applied to these average values. Statistical quantities are calculated using the average values and the individual data block values. For the unsteady pressure measurements presented here, the corresponding 95% confidence intervals were calculated as approximately ±3% for the amplitude and ±3° for the phase lag.

The control and synchronization of the dual-motor excitation system is performed using a PC-based system (with a National Instruments AT-MIO-16E I/O card). The control algorithm employs a basic PDI approach coded in Labview as described by Vogel (1997). The reference shaft position signals are derived from optoelectronic detectors. With this system, the motor frequencies and phase lag can typically be maintained within errors of approximately ±0.3 Hz and ±5°, respectively.

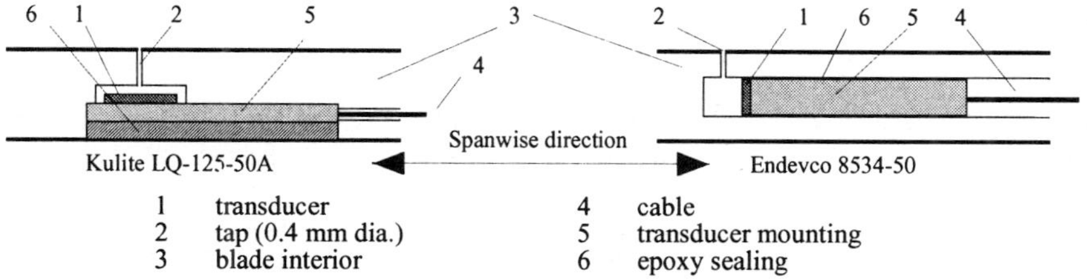

Figure 4. Unsteady Pressure Transducer Mounting Scheme

3. Results

3.1 TEST CASES

In this section, the results from a series of three-part measurements performed under common steady-state flow conditions are presented. These measurements were conducted for a:

(1) fixed blade subjected to a downstream flow disturbance (aerodynamic gust response)
(2) vibrating blade in uniform flow (aerodynamic damping)
(3) vibrating blade subjected to a downstream flow disturbance for a common frequency and specific phase angle (combined excitation).

Testing was conducted for a single flow incidence angle of 0° and under both subsonic and transonic flow conditions. Measurements were obtained for (1) and (2) for reduced frequencies, k, between 0.02 and 0.18. Due to some start-up difficulties with the motor synchronization, experimental data for (3) are only available for selected phase angles and for $f = 60$ Hz. The objective of these tests is to investigate the interaction of the gust- and blade vibration-induced unsteady flow fields, and particularly to verify whether a linear superposition of (1) and (2) is a valid method to obtain the combined flow field (3).

3.2 STEADY-STATE RESULTS

Cases 1-3 and 4-6 shown in Table 1 were chosen to represent typical measurements obtained in the subsonic and transonic regimes, respectively. The steady-state blade pressure distributions for these two sets of tests are shown in Figure 5. For Cases 4-6, a shock is present on the blade suction side at approximately $x/c = 0.1$.

It should be noted that for all test cases involving the downstream exciter, the steady-state conditions were measured with the flat plate in rotation. The rotating exciter appears in the steady state as a fixed reduction to the downstream exit area. This results in a shift in the measured flow conditions on and around the blade. However, it is this flow condition which represents the true stationary flow component onto which the perturbation is imposed.

In addition, in order to be able to use the results obtained in (1) and (2) above to compare with results from the combined case (3), it is critical to closely match the respective steady-state flow conditions. This is particularly true for the transonic flow cases, in which the position of the shock is very sensitive to changes in the flow conditions. This matching process is somewhat complicated by the fact that, as described above, the steady flow conditions for the downstream gust measurements are obtained with the flat plate in rotation. However, it was determined that it is possible to obtain essentially identical flow conditions by adjusting the stationary position of the flat plate during the aerodynamic damping measurements (2). The general level of steady-state matching obtained is demonstrated in Figure 5.

TABLE 1. Test Cases

Case	M_1	M_2	*DS Exciter*	*Blade*	f [Hz]	k	θ [°]
1	.41	.42	Rotating	Fixed	20-100	.04-.18	N/A
2	.41	.42	Fixed	Vibrating	20-100	.04-.18	N/A
3	.41	.42	Rotating	Vibrating	60	.11	-25
4	.61	.69	Rotating	Fixed	20-100	.02-.12	N/A
5	.61	.69	Fixed	Vibrating	20-100	.02-.12	N/A
6	.61	.69	Rotating	Vibrating	60	.07	0

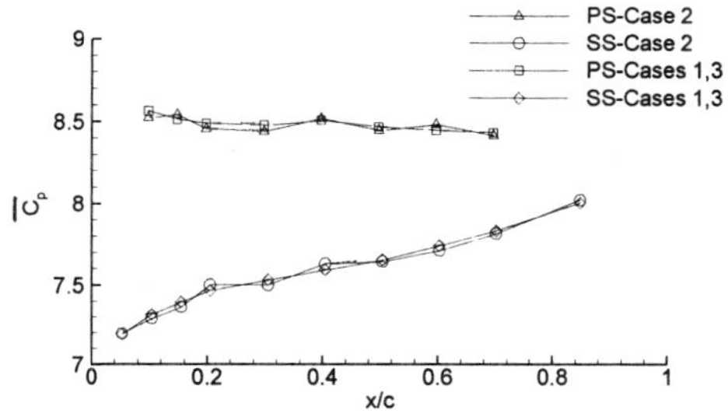

(a)

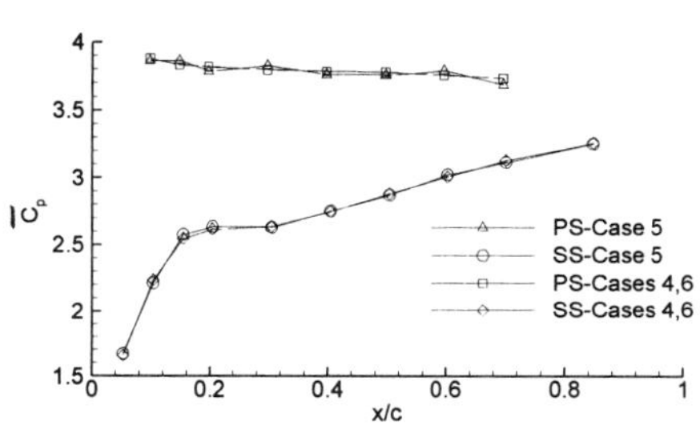

(b)

Figure 5. Steady-State Blade $\bar{c}_p$ Distribution: (a) Cases 1-3, (b) Cases 4-6

3.3 AERODYNAMIC GUST RESPONSE RESULTS

Clearly, to produce an aerodynamic forcing function in this test facility which accurately simulates the interaction between a rotor and downstream vane is a very difficult task. Hence, the downstream exciter described in Section 2.1 was constructed to instead provide a simple and easily quantified approximation to these potential interactions. For each unsteady test, the characteristics of the downstream disturbance are quantified using the side-wall unsteady pressure transducers described in Section 2.2. For Cases 1 and 4, unsteady tests were performed on a fixed blade for flow perturbation frequencies between 20 Hz and 100 Hz (k = 0.02 and 0.18) in steps of 20 Hz. For all measured cases, it was observed that the imposed flow disturbance is a vertically planar, upstream-running wave. An example of a typical unsteady side-wall measurement is given in Figure 6.

Representative measurements for Cases 1 and 4 are given in Figures 7 and 8 for k = 0.11 and 0.07, respectively. Figure 7 demonstrates that in the subsonic Case 1 the unsteady pressure coefficient amplitude generally follows the trend of the steady-state Mach number (i.e. rises on the suction side toward the leading edge, and diminishes on the pressure side). The phase angle is only slightly changed across the blade chord. For the transonic Case 4, Figure 8 presents the effects of the suction-side shock on the upstream running wave. Following the acoustic blockage theory proposed by Atassi, et al. (1994), the peak unsteady pressure amplitude at x/c = 0.1 can be considered indicative of the approximate shock location. This can be further verified by studying the raw data signal obtained by the pressure transducer at this location, as shown in Figure 9a. The square nature of this signal indicates that the shock position is oscillating back and forth directly over this measuring position, applying alternately the high pressure- and the low pressure-sides of the shock at the measuring tap. Furthermore, the symmetric nature of the signal over a given cycle implies that the mean position of the shock is coincident with this location. Experimental and computation results described by Ott, et al. (1993) provide evidence that the shock position is indeed closely synchronized with the incoming downstream disturbance. In addition, only a negligible phase difference between the blade surface unsteady pressures measured in the vicinity of the shock and the oscillation of the shock position was observed. Hence, the unsteady signal presented in Figure 9a can be considered as a time-accurate representation of the instantaneous shock position. Without additional information in this region, however, the extent of the shock displacement cannot be determined.

Another interesting unsteady feature for this case is the rather abrupt drop in unsteady pressure amplitude and shift in phase angle at x/c = 0.2 on the suction side, which is just upstream of the shock position. This phenomenon has been observed in earlier tests using flat plates (Körbächer, 1997), but is difficult to explain without more detailed steady-state and unsteady measurements in the region around the shock. One possible source for this effect is a shock-induced boundary layer separation downstream of the shock position.

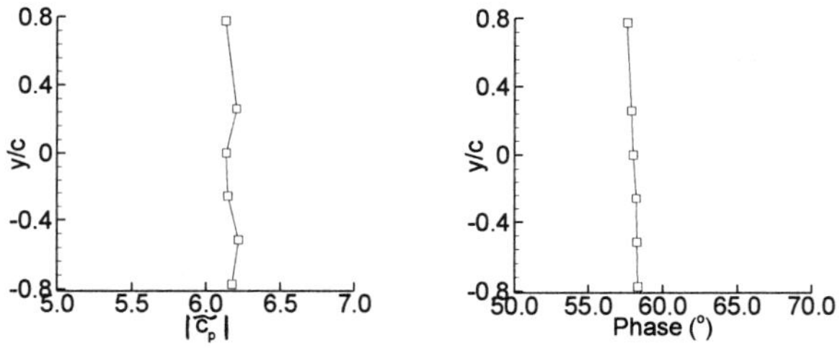

Figure 6. Side-Wall Unsteady Pressure Distribution: Case 1, $k = 0.11$

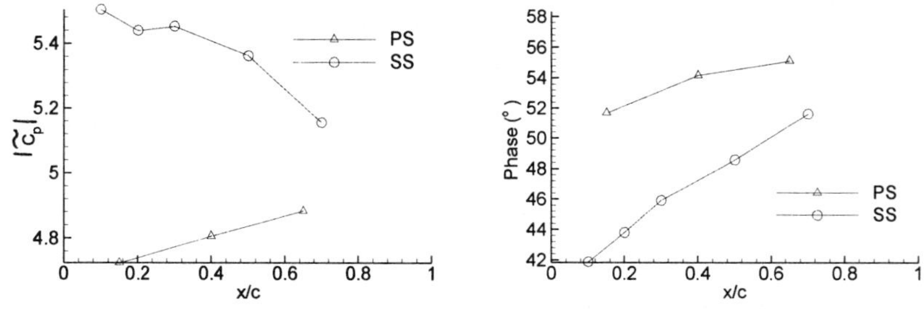

Figure 7. Unsteady Blade Pressure Distribution: Case 1, $k = 0.11$

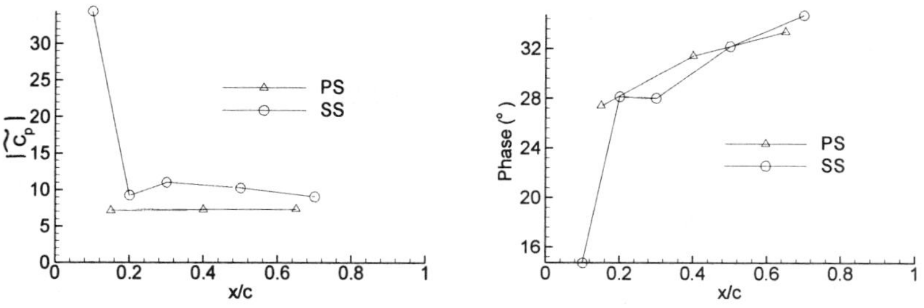

Figure 8. Unsteady Blade Pressure Distribution: Case 4, $k = 0.07$

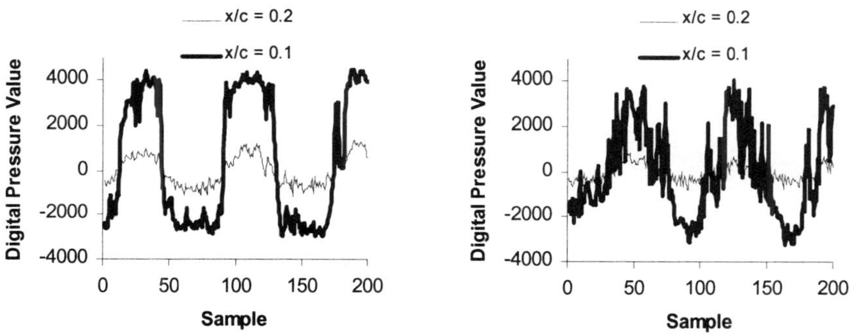

(a) (b)

Figure 9. Raw Unsteady Pressure Data, $k = 0.07$: (a) Case 4, (b) Case 5

3.4 AERODYNAMIC DAMPING RESULTS

Unsteady measurements on a vibrating blade in uniform flow were conducted for steady-state flow conditions and reduced frequencies matching those of the aerodynamic damping tests described in the previous section. The blade surface unsteady pressure distribution for two representative cases, Cases 2 and 5, are shown in Figures 10 and 11 for $k = 0.11$ and 0.07, respectively. For the subsonic Case 2, the unsteady pressure amplitudes steadily increase toward the leading edge. The phase angles on the two surfaces remain relatively constant across the blade chord, implying a mostly quasi-steady behavior. In addition, for this case, as well as for the other measured reduced frequencies, the phase angles on the opposing sides of the blade are approximately 180° out of phase. Hence, the unsteady forces on the two sides of the blade generally act in a coincident direction during the vibration cycle.

For the transonic Case 5, the predominant feature is again the suction-side shock. At $x/c = 0.1$, close to the shock mean position, the unsteady pressure amplitude rises abruptly. However, the overall magnitudes measured at $x/c = 0.1$ for Case 5 are generally less than for Case 4. Analyzing the raw unsteady signal measured here reveals a different behavior than that described in Section 3.3. As can be seen in Figure 9b, the unsteady pressure measured at $x/c = 0.1$ is primarily sinusoidal in nature. This is a likely indication that, despite the steady-state matching detailed in Section 3.2, the mean shock position for this case in slightly different from Case 4. This discrepancy has some important effects for the combined excitation results, as described in next section.

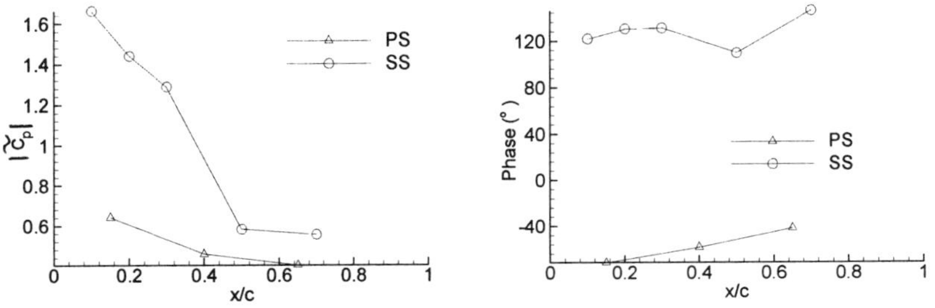

Figure 10. Unsteady Blade Pressure Distribution: Case 2, $k = 0.11$

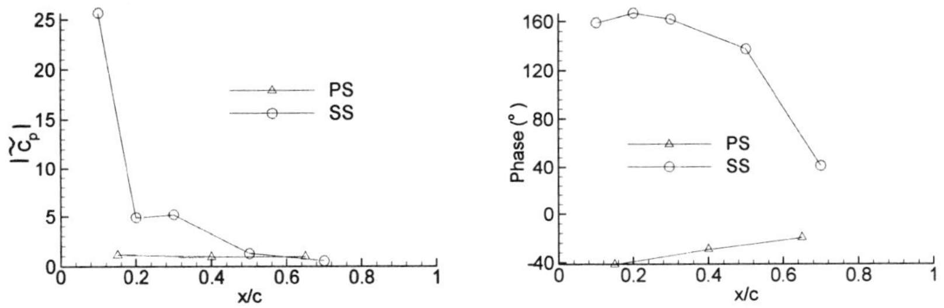

Figure 11. Unsteady Blade Pressure Distribution: Case 5, $k = 0.07$

3.5 COMBINED EXCITATION RESULTS

In this section, measurements obtained on a vibrating blade in the presence of an imposed downstream gust are compared with calculated values obtained by linearly superimposing the results described in Sections 3.3 and 3.4. The objective of this comparison is to provide some quantitative information concerning how the gust- and blade vibration-induced flow fields interact in the forced response problem. These preliminary results are currently only available for a single excitation frequency, 60 Hz (corresponding to $k = 0.11$ for Cases 1-3, and $k = 0.07$ for Cases 4-6), and for selected excitation phase angles between the two sources of excitation. The excitation phase angle, θ, represents the time lag between the occurrence of the vibrating blade mean position and the downstream rotating exciter vertical position. It is defined as positive when the vibrating blade leads the rotating exciter.

A typical subsonic test, Case 3, is shown in Figure 12 for $\theta = -25°$. The same figure includes predicted values for this case based on a linear superposition of the flow fields presented in Figures 7 and 10 for Cases 1 and 2, respectively. The agreement for both unsteady pressure amplitude and phase angle is excellent along the pressure and suction sides. The maximum error is on the order of 5% for the amplitude and 3° for the phase angle. Similar close comparisons were obtained for other excitation phase angles. These results indicate that for this subsonic case and reduced frequency, the combined excitation flow field is an almost exact linear superposition of the two single excitation flow fields.

The results for a typical transonic test, Case 6, are given in Figure 13 for $\theta = 0°$. The same figure includes predicted values for this case based on a linear superposition of the flow fields presented in Figures 8 and 11 for Cases 4 and 5, respectively. Two distinct levels of comparison can be observed. For regions away from the shock (all of the pressure side, and downstream of approximately $x/c = 0.3$ on the suction side), the agreement between the measured and predicted flow fields is excellent with a maximum amplitude error of less than 1%, and maximum phase angle error of 2°. However, near the shock at $x/c = 0.1$ and 0.2, some relatively significant discrepancies are apparent. The maximum unsteady pressure amplitude error is approximately 25% and is located at $x/c = 0.1$, closest to the shock position. An error of 10° for the phase angle can also be observed at this position. One possible explanation of these discrepancies is that the interaction of the two unsteady flow fields in the region near the shock has some non-linear qualities, and hence cannot be accurately predicted by a linear superposition. Another possibility, however, which was alluded to in Section 3.4, is that because the shock is located essentially coincident with the pressure transducer measuring tap at $x/c = 0.1$ for Case 4, relatively small shifts in its mean position for Case 5 can still lead to quite significant changes in the measured unsteady pressures. Therefore, the source of these discrepancies could be associated with a slight mismatch in the Case 4 and Case 5 steady-state conditions. These observations bring to light the inherent difficulties in obtaining sufficiently matched mean transonic conditions for these types of tests.

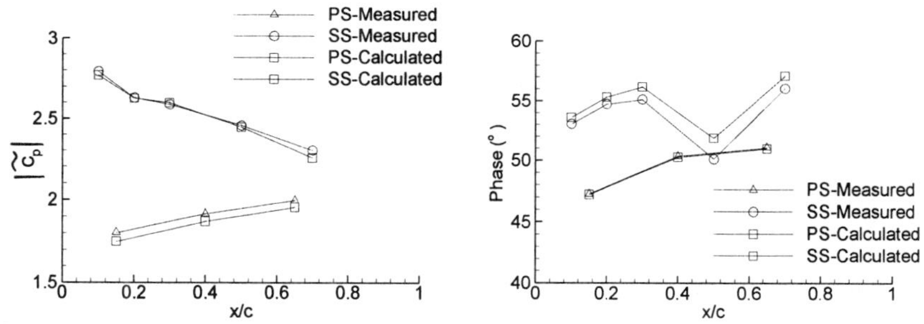

Figure 12. Unsteady Blade Pressure Distribution: Case 3, $k = 0.11$, $\theta = -25°$

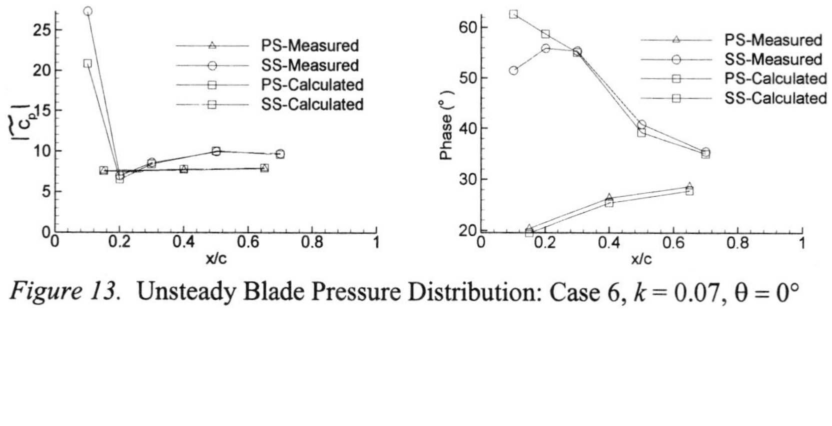

Figure 13. Unsteady Blade Pressure Distribution: Case 6, $k = 0.07$, $\theta = 0°$

4. Summary and Conclusions

A new test facility focused on the experimental investigation of unsteady flow phenomena around a single compressor blade has been introduced and described in detail. A series of preliminary experiments were conducted in this facility to measure the unsteady pressures produced on the blade in response to a downstream aerodynamic gust, a forced vibration, and a controlled combination of the two. The objective of these tests was to provide some fundamental knowledge with regards to the forced response problem, and in particular to give insight into the interaction of the gust- and blade vibration-induced unsteady flow fields. Tests were conducted for reduced frequencies between 0 and 0.2, and two representative cases, one with subsonic steady-state flow conditions and the other with transonic conditions, were presented. From this work, the primary observations and conclusions are as follows:

- for all transonic cases, the suction side shock is associated with an abrupt increase in the blade surface unsteady pressure amplitude. For the case involving measurements on a fixed blade subjected to a downstream gust, the oscillation of the shock location on the blade surface is clearly tracked by the unsteady pressure signal measured at $x/c = 0.1$ and is closely coupled with the incoming disturbance

- outside of the shock region, the unsteady pressures measured on the blade for the combined excitation cases can, in general, be accurately predicted by a linear superposition of the individual gust- and blade vibration-induced flow fields for reduced frequencies between 0 and 0.2

- some discrepancies with regards to this linear superposition were evident for suction side locations near the shock. It is likely, though, that rather than being due to some type of non-linear interaction, these differences are instead derived from small shifts in the mean shock location between the two individual superimposed cases

The results presented in this paper represent only preliminary measurements. Continuing work will concentrate on investigating the linear superposition principle for forced response over a much wider range of reduced frequencies and steady-state flow conditions, including off-design conditions involving separated flow. In addition, a corresponding computational effort is being undertaken at the DLR, Institut für Aeroelastik.

Acknowledgements

This work is supported by the Ecole Polytechnique Fédérale de Lausanne. The authors wish to express their appreciation for permission to publish these results.

References

Atassi, H.M., Fang, J., and Ferrand, P., 1994, “A Study of the Unsteady Pressure of a Cascade Near Transonic Flow Condition,” ASME Paper 94-GT-476.

Bölcs, A., Fransson, T.H., and Schläfli, D., 1989, "Aerodynamic Superposition Principle in Vibrating Turbine Cascades," AGARD 74th Specialists' Meeting of the Propulsion and Energetics Panel on Unsteady Aerodynamic Phenomena in Turbomachines, Luxembourg, August 28-September 1.

Bölcs, A. and Norryd, M., 1994, "Development of a Sealing Construction to Prevent Leakage Flow During Flutter Tests in a Linear Cascade," 12th Symposium on Measuring Techniques for Transonic and Supersonic Flow in Cascades and Turbomachines, Prague, 12-13 September 1994.

Buffum, D.H., and Fleeter, S., 1990, "Oscillating Cascade Aerodynamics by an Experimental Influence Coefficient Technique," Journal of Propulsion, Vol. 6, No. 5, Sept.-Oct. 1990, pp. 612-620.

Capece, V.R., and Fleeter, S., 1988, "Experimental Investigation of Multistage Interaction Gust Aerodynamics," ASME Paper 88-GT-56.

Fleeter, S., Jay, R.L., and Bennet, W.A., 1978, "Rotor Wake Generated Unsteady Aerodynamic Response of a Compressor Stator", ASME Journal of Engineering for Power, Vol. 100, pp. 664-675.

Hennings, H, 1997, "Windkanal-Flatteruntersuchungen an einem endlichen, geraden 2d-Verdichtergitter in inkompressibler Strömung," DLR-FB 97-20, Deutsche Forschungsanstalt für Luft- und Raumfahrt, Göttingen.

Körbächer, H., 1997, Private communication.

Körbächer, H., and Bölcs, A., 1994, "Experimental Investigation of the Unsteady Behavior of a Compressor Cascade in an Annular Ring Channel," Paper presented at the 7th ISUAAT Conference in Fukuoka, Japan, September 25-29.

Manwaring, S.R., Rabe, D.C., Lorence, C.B., and Wadia, A.R., 1996, "Inlet Distortion Generated Forced Response of a Low Aspect Ratio Transonic Fan," ASME Paper 96-GT-376.

Manwaring, S.R., and Wisler, D.C., 1992, "Unsteady Aerodynamics and Gust Response in Compressors and Turbines," ASME Paper 92-GT-422.

Ott, P, Bölcs, A., and Fransson, T.H., 1993, "Experimental and Numerical Study of the Time-Dependent Pressure Response of a Shock Wave Oscillating in a Nozzle," ASME Paper 93-GT-139.

Poengsen, C.A., and Gallus, H.E., 1991, "Experimental Investigation on the Unsteady Pressure Field on a Vibrating Cascade in Steady and Unsteady Flow," Unsteady Aerodynamics of Turbomachines and Propellers, University of Notre Dame, September 15-19, pp. 583-601.

Vogel, G. (1997) Equipement du stand Laval II en système de vibration et de perturbation déphasé (Simulation sur des moteurs électriques), Internal Report LTT-97-09, EPF-Lausanne.

APPENDIX 1: Steady-State Blade Surface Pressure Tap Locations (*x* is measured along the chord line)

Suction Side *x/c* [-]
0.05
0.10
0.15
0.20
0.30
0.40
0.50
0.60
0.70
0.85

Pressure Side *x/c* [-]
0.10
0.15
0.20
0.30
0.40
0.50
0.60
0.70

APPENDIX 2: Unsteady Pressure Transducer Locations (*x* is measured along the chord line)

Name	Manufacturer, Type	Location *x/c* [-]	Sensitivity [mV/kPa]
SS 1	Endevco 8534A-50	0.10	0.7672
SS 2	Endevco 8534A-50	0.20	0.7672
SS 3	Kulite LQ-125-50A	0.30	0.3049
SS 4	Kulite LQ-125-50A	0.50	0.2994
SS 5	Endevco 8534A-50	0.70	0.7462
PS 1	Endevco 8534A-50	0.15	0.7401
PS 2	Kulite LQ-125-50A	0.40	0.2858
PS 3	Endevco 8534A-50	0.65	0.7425

VIBRATION CHARACTERISTICS OF A TRANSONIC TURBINE CASCADE

T. WATANABE, M. AOTSUKA, and Y. MACHIDA
Department of Aeronautics and Astronautics, University of Tokyo
7-3-1 Hongo, Bunkyo-ku, Tokyo 113, Japan

Abstract

The vibration characteristics of a transonic steam turbine cascade with a thin airfoil were experimentally studied by the influence coefficient method in a linear cascade tunnel. Two types of flow field were adopted for the experiment: one in near-design condition and the other in off-design condition of low pressure ratio. In the results for the near-design case, a large unstable region of the vibration was observed in the positive side of the interblade phase angle, which corresponded to flutter instability. In the off-design case, however, the instability did not appear for almost all the interblade phase angle. A drastic change was found in the phase angle of the unsteady aerodynamic force between the two cases, which change was a dominant factor for the unstable oscillation. The local fluctuating pressure data indicated that the unsteady pressure on the shock foot point had no significant effect in the present vibration instability. A CFD simulation was also performed, and the results showed similar instability of the blade oscillation.

1. Introduction

In advanced types of jet engines, gas turbines and steam turbines, the turbine blades frequently meet air flows in transonic regime for their high performance. These transonic turbine blades accompany various shock waves around them, and some of them have very thin tip cross sections. They are accordingly susceptible to such blade vibrations as cascade flutter or forced vibration. The fundamental knowledge of unsteady aerodynamic and aeroelastic characteristics is thus required for the transonic turbine blades.

Around the tip section of the last stage of steam turbine blades, some types of forced vibration and/or cascade flutter have been experienced so far in a stage of their development. Such turbines sometimes operate in quite complicated, hence,

T.H. Fransson (ed.), Unsteady Aerodynamics and Aeroelasticity of Turbomachines, 679-692.

unstable flow field conditions with reversed flow regions or severely three-dimensional flows completely different from the designed condition. Since the studies on the vibration characteristics of such transonic steam turbine blades are rather limited up to now [1][2], the fundamental research and data accumulation are essential for the advanced design of high-performance turbine cascade.

In the present study, an experimental investigation was performed on the oscillating turbine cascade in transonic flows. The purpose was to reveal the unsteady aerodynamic characteristics of the tip section of the last stage cascade of a transonic steam turbine. The effect of oscillation of the shock waves generated in the turbine flow passages was an important factor throughout the experiment. The unsteady aerodynamic force acting on the blades was measured by the influence coefficient method in a linear transonic cascade tunnel. A CFD simulation was performed as well for understanding the physical mechanism of the experimentally observed unsteady phenomena.

2. Experimental Method

The unsteady aerodynamic force acting on turbine blades were measured with strain gauges mounted on the root of the blades. The unsteady surface pressure was additionally detected on some points of the blade surfaces so that the contribution of local pressure fluctuation could be clarified. The standard steady pressure measurement was also made on the blade surfaces and the wall surfaces of the test section. A flow visualization with schlieren technique was carried out for grasping the detailed view of the shock wave behavior.

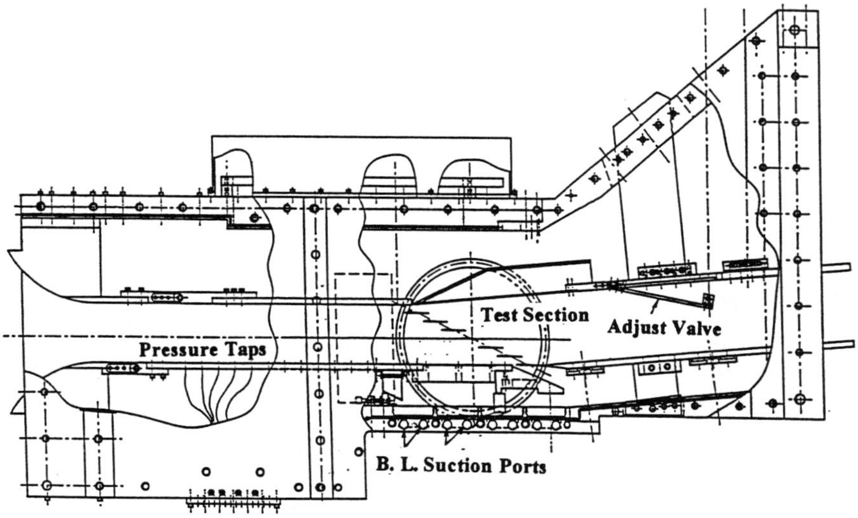

Figure 1. Schematic View of the Transonic Cascade Tunnel

2.1. TEST FACILITY

A blow-down type, transonic linear cascade tunnel was used in the present experiment, the schematic view of which is shown in Fig. 1. The test section had a rectangular cross section 50 mm wide (in the blade span direction) and 102 mm high. The pressure ratio was changed by controlling the total pressure in the upstream settling chamber. The flow exit was open to the atmosphere through a silencer. The boundary layer suction was made on all the walls of the upstream flow channel.

The test cascade had a typical configuration of the transonic steam turbine tip section (VKI steam turbine tip section 3 [3]). The configuration is schematically shown in Fig. 2, and its specification is summarized in Table 1. At the design point, the ratio of the inflow total pressure to downstream static pressure is 4.936, and the isentropic outlet Mach number is 1.7. The cascade was composed of seven blades which were numbered as shown in Fig. 2. The central blade (No.0) could be oscillated by a mechanical vibration system with a piston-crank mechanism. The oscillation was in a translational mode, and its amplitude was 2 mm. The frequency was changed as 10 Hz, 43 Hz, and 82 Hz. The reduced frequency k based on the blade half-chord length ranged between 2.5×10^{-3} and 2×10^{-2}.

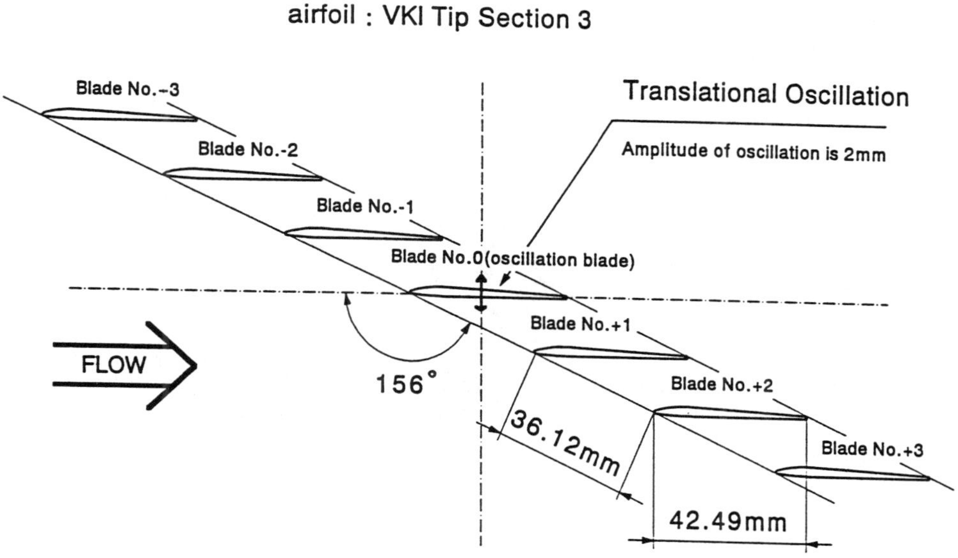

Figure 2. Cascade Configuration

2.2. MEASUREMENT OF UNSTEADY AERODYNAMIC FORCE

The unsteady aerodynamic force was measured on the central blade (No.0) and the two neighboring blades (No.+1 and -1). Superposition of the data on these three blades approximately provides the unsteady aerodynamic force when all blades are oscillating with an arbitrary interblade phase angle through the influence coefficient method [4]. On each blade, the aerodynamic force was measured with strain gauges mounted on the support of the blade. The setup of the strain gauge is shown in Fig. 3.

2.3. MEASUREMENT OF UNSTEADY SURFACE PRESSURE

To make out how the local pressure fluctuation contributed to the unsteady aerodynamic force acting on the blade, we carried out the unsteady surface pressure measurement on some points of suction and pressure surfaces of No.0

TABLE 1. Specification of test cascade

Geometry		VKI tip section 3	
Inlet flow angle	β_1	156.0	deg
Outlet flow angle	β_2	5.0	deg
Stagger angle	γ	156.0	deg
Pitch	s	36.12	mm
Chord length	c	42.49	mm
Solidity	s/c	0.85	
Span length	b	50.0	mm
Aspect ratio	b/c	1.18	

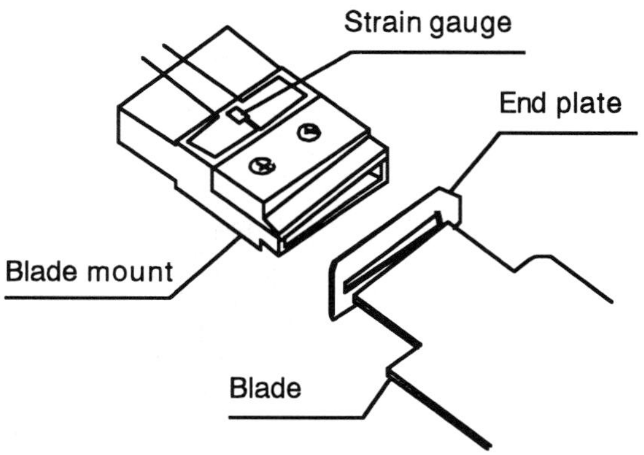

Figure 3. Strain Gauge Equipment

TABLE 2. Position of unsteady pressure sensors

Suction Surface			
Sensor Number	S1	S2	S3
Position (% chord)	23.7	50.0	76.5
Pressure Surface			
Sensor Number	P1	P2	
Position (% chord)	23.4	76.4	

and No. ±1 blades. The positions of pressure sensors are indicated in Table 2. Unsteady pressure sensors 1.0 mm wide, 3.0 mm long and 0.2 mm thick (made by TOYODA) were mounted inside the pressure holes on the surfaces.

2.4. FLOW VISUALIZATION

The flow field around the cascade was visualized with schlieren technique. The clear images of the behavior of shock waves were obtained for studying the role of shock oscillation in the unsteady aerodynamic characteristics of the oscillating blades.

3. Results and Discussions

3.1. STEADY FLOW FIELDS

In the design condition of the present cascade, a shock wave is generated at the trailing edge of each blade, which impinges on the trailing edge of the pressure-side adjacent blade [3]. In the experiment, two types of the steady flow field were adopted, one of which, called case 1, was a flow field near the design point, the other of which, called case 2, was an off-design flow field where the back pressure was set higher than in the design point. The features of these flow fields are summarized in Table 3, and the schlieren photos of them are shown in Fig. 4 together with the sketches made from the photos.

(a) Schlieren Photo: Case 1

(b) Schlieren Photo: Case 2

Figure 4. Visualized Flow Field (continued)

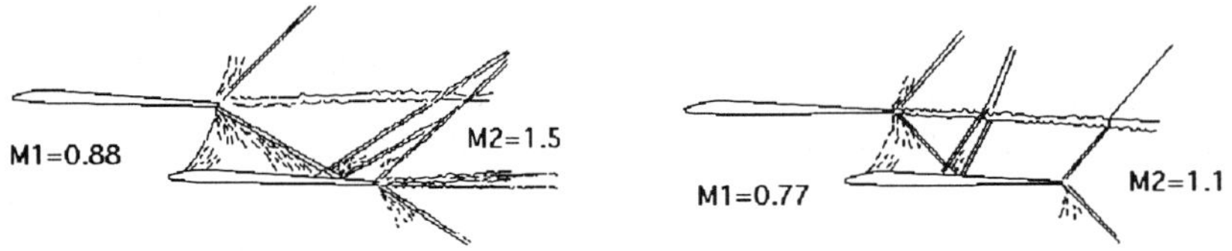

(c) Sketch: Case 1 (d) Sketch: Case 2

Figure 4. Visualized Flow Field (concluded)

TABLE 3. Flow field description

Case 1 (Fig. 4 (a) and (c)): Near-design condition
Static pressure ratio 2.2 Outlet Mach number 1.5
A shock wave from a blade trailing edge impinges on a point at around 80% chord length position on the suction surface of the adjacent blade, and reflects there. The shock wave is stable.
Case 2 (Fig. 4 (b) and (d)): Off-design condition, High back-pressure
Static pressure ratio 1.4 Outlet Mach number 1.1
A shock wave from a blade trailing edge impinges on a point around the half-chord length position on the suction surface of the adjacent blade, and reflects there. The shock wave shows intense oscillation even in the steady flow condition.

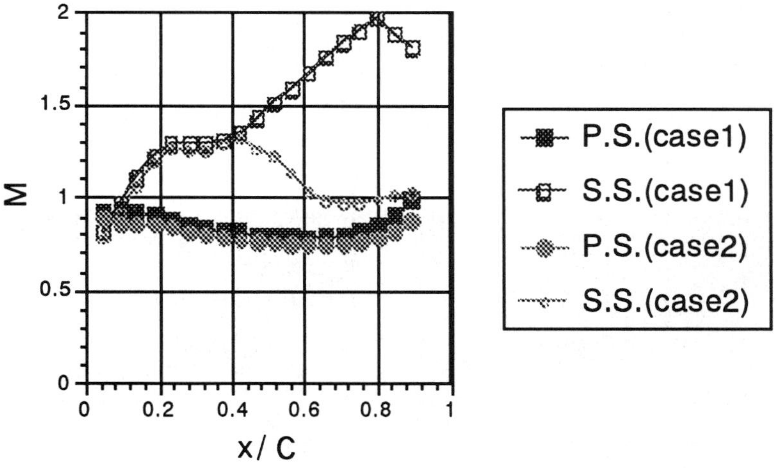

Figure 5. Surface Mach Number Distribution

Figure 5 shows Mach number distributions along the blade surface obtained by surface pressure data. The Mach number distribution along the pressure surface is identical for the two cases. The distributions also agree well on the suction surface before the 40% chord length position where the shock wave from the adjacent blade reflects in the case 2. In this case, the Mach number decreases after the shock reflection, while the Mach number in the case 1 further increases until the maximum value of about 2.0 at the 80% chord length position on which the shock impinges in the case 1.

3.2 UNSTEADY BLADE FORCE

The central blade (No.0) was oscillated in the two flow conditions mentioned above. The unsteady aerodynamic forces measured on the blades No.0, No.+1 and No.-1 are shown in Fig. 6 against the reduced frequency. Figure 6 (a) indicates the data in the case 1, while Fig. 6 (b) shows the data in the case 2. The unsteady aerodynamic force coefficient C_F is expressed by its amplitude $|C_F|$ and the phase angle ϕ, which are plotted against the reduced frequency. C_F is defined as follows:

$$C_F = \frac{F}{\frac{1}{2}\rho U^2 ch}, \tag{1}$$

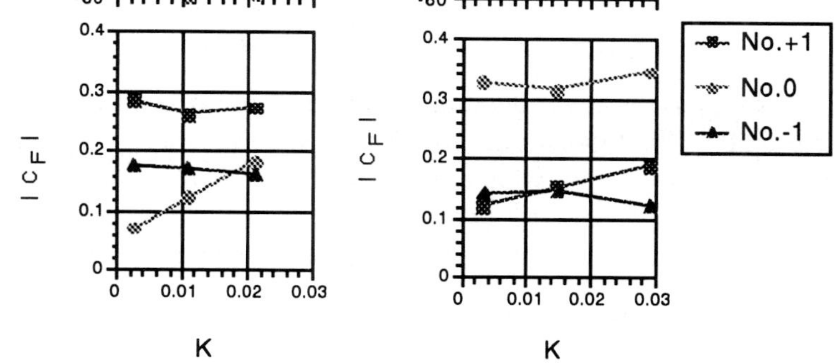

(a) Case 1 (b) Case 2

Figure 6. Unsteady Aerodynamic Force

where F is the measured unsteady force, ρ the density of air, U the upstream uniform velocity, c the blade chord length, and h the oscillation amplitude. The phase angle is measured as its difference from the blade displacement, the lag being positive.

In the unsteady aerodynamic force on the blade No.0 shown in Fig. 6, a remarkable phase change of 180° is observed between the case 1 and the case 2. The phase angle in the case 1 is in advance, while it shows a lag in the case 2. The amplitude of the unsteady force is larger in the case 2 (with lower pressure ratio) than in the case 1 (with higher pressure ratio). The amplitude in the case 1 increases with increasing reduced frequency, while it is almost constant in the case 2.

A phase change similar to the data for the blade No.0 can be seen in the unsteady force on the blade No.+1, though the phase angle has a 180° lag in the case 1 on the contrary to that on the blade No.0. In the case 2, the phase lag slightly increases with increase in reduced frequency. The amplitude is larger in the case 1 than that in the case 2 in contrast to the results for the blade No.0. As for the unsteady aerodynamic force on the blade No.-1, neither the amplitude nor the phase angle differ so much between in the case 1 and in the case 2.

From the above results, it is noticeable that the unsteady flow field in the blade pressure side is not affected by the pressure ratio, while the flow field in the suction side is greatly changed by the pressure ratio.

3.3. STABILITY ANALYSIS

The exciting energy input can be calculated by the influence coefficient method [4]. Let $C_{F,m}$ be the unsteady force coefficient of the blade No.m obtained from the measurement. The overall unsteady aerodynamic force coefficient $C_{F,\text{all}}$, that is, the coefficient in the case when all blades are oscillating in an interblade phase angle σ, is calculated as follows:

$$C_{F,\text{all}} = \sum_m C_{F,m} \cdot \sin(\phi_m - m \cdot \sigma). \tag{2}$$

In the present experiment, the unsteady aerodynamic force on the blades No.0 and ±1 were used for the calculation. The nondimensional energy input during one period of the blade oscillation is obtained from the following equation:

$$E = \frac{\int \frac{\partial h}{\partial t} C_{F,\text{all}} dt}{\pi h}. \tag{3}$$

The calculated energy is plotted against the interblade phase angle in Fig. 7. If the energy input E is positive, the blade oscillation is unstable. In Fig. 7 (a) for the case 1 near the design condition, it can be seen that the unstable oscillation occurs in the positive interblade phase angle region. This instability corresponds to the cascade flutter in which the suction-side blade advances in the phase angle. On the other hand, the unstable region almost completely disappears in Fig. 7 (b) for the case 2 in the off-design condition where the shock wave impinges on around the mid-chord position of the suction surface. The oscillation is gradually damped with the increase in reduced frequency in both cases.

Figure 8 shows the energy coefficient of each blade E_m which contributes to the overall energy input shown in Fig. 7. E_m is evaluated with the same equation as Eq. (3) by the replacement of $C_{F,\text{all}}$ with $C_{F,m}$. In Fig. 8, the results for the vibration frequency of 10 Hz are indicated. In Fig. 8 (b) for the case 2, we can realize that the energy inputs from the blade No.+1 and No.-1 cancel out each other for whole region of the interblade phase angle. The blade oscillation consequently becomes stable in this case. In Fig. 8 (a) for the case 1, on the contrary, the energy inputs from the blades ±1 are added each other, and, hence, the blade oscillation becomes unstable in the positive region of the interblade phase angle.

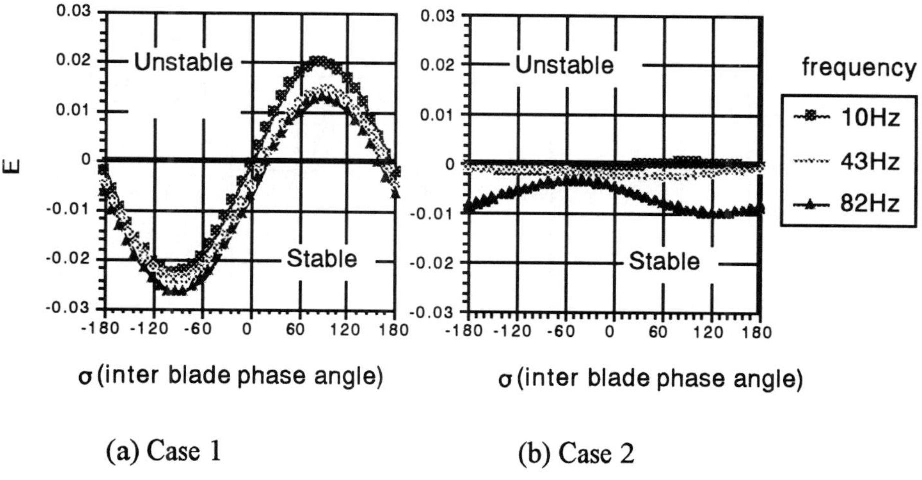

Figure 7. Exciting Energy

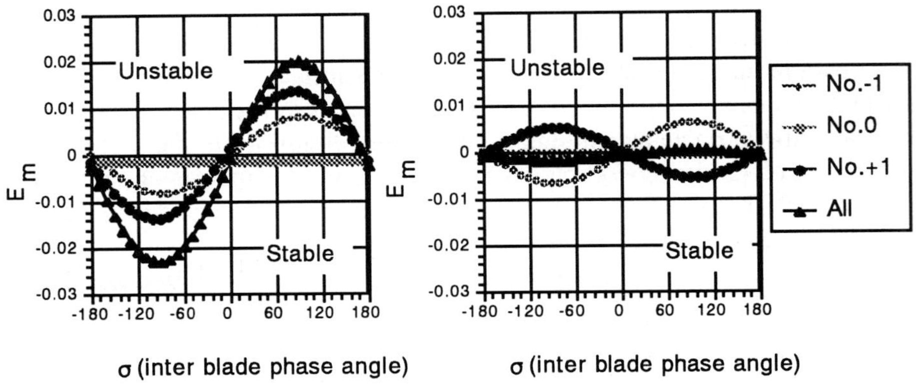

(a) Case 1 (b) Case 2

Figure 8. Energy Input from Each Blade

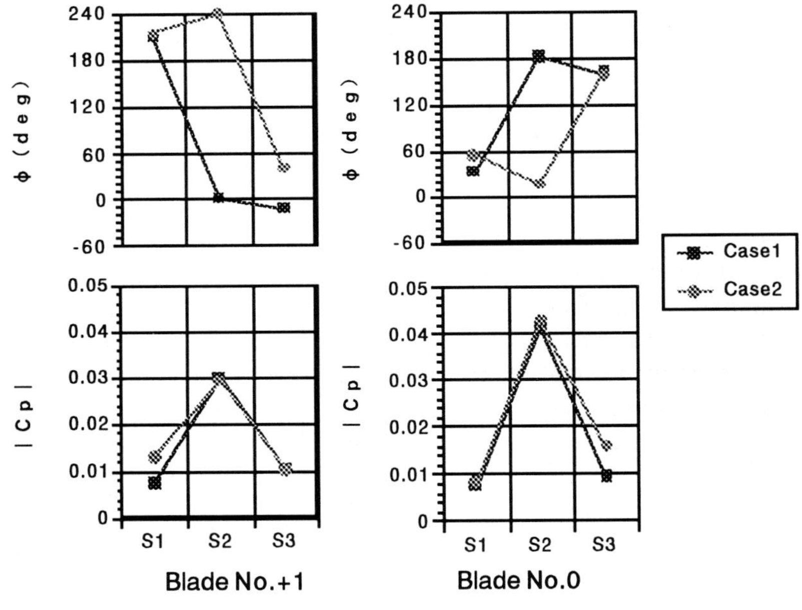

Figure 9. Unsteady Surface Pressure (Suction Surface)

The unsteady surface pressure was measured at three points on the blade suction surface and at two points on the pressure surface. The results are shown in Fig. 9. and Fig. 10 as the normalized pressure coefficient C_p whose amplitude is $|C_p|$ and phase angle is ϕ. In Fig. 9, C_p on the suction surface are plotted for the blades No.0 and No.+1. The data both in the case 1 and in the case 2 are indicated

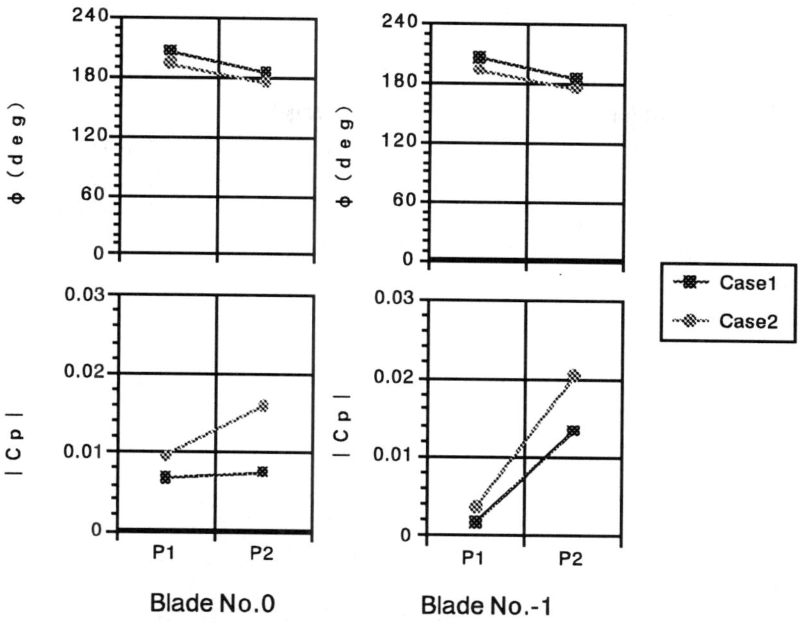

Figure 10. Unsteady Surface Pressure (Pressure Surface)

together. The pressure amplitude on the blade No.-1 is so small that the data are not shown. It is remarkable that the amplitude of S2 is quite large compared with that of the other positions for all the data. Besides this, the phase angle on the S2 position shows a drastic change between the case 1 and the case 2. Concerning the unsteady pressure on the blade pressure surface, which is shown in Fig. 10, the results are almost identical for the case 1 and the case 2. It is thus realized that the energy input previously shown are mainly dominated by the unsteady pressure around mid-chord position on the suction surface, and that the change in phase angle should be crucial for the vibration instability. The amplitude of the unsteady pressure always has the maximum at the position of S2 , though the shock wave impinges on the blade surface at the position of S3 in the case 1. Hence the unsteady pressure generated at the foot of the shock wave is judged not to be critical in the present case. Instead, the change in the unsteady flow characteristics due to the change in the pressure ratio would be more important here. The cause of the phase change should be further investigated.

3.4. CFD SIMULATION

The unsteady flow field in the experiment was simulated with an Euler code developed in the laboratory. In the numerical code, the two-dimensional Euler equation was discretized by an upwind TVD scheme in the space domain. The time integration was performed implicitly by the LU-ADI scheme. Figure 11 shows the H-grid used for the computation. The flow field composed of six channels were simulated for the case in which the central blade was oscillating in a translational mode, that was correspondent to the experiment.

Figure 12 shows the steady Mach contour diagrams obtained from the computation. The flow condition for Fig. 12 (a) is similar to the case 1 condition, and that for Fig. 12 (b) is similar to the case 2 condition. In both figures, we can observe the almost identical shock wave configurations to those observed in the experiment shown in Fig. 4.

The computed energy input of the blade vibration is shown in Fig. 13, which is compared with Fig. 7 for the experiment. The vibration frequency of the numerical simulation was changed as 82 Hz, 200 Hz and 500 Hz. The results of 82 Hz oscillation is comparable with the experimental results. In Fig. 13 (a) for the

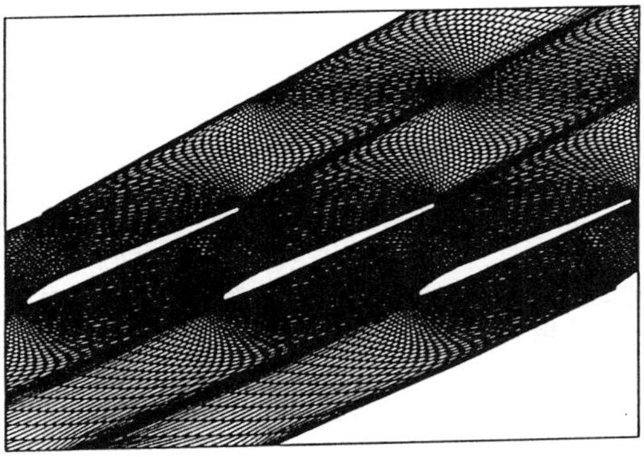

Figure 11. Computational Grid (H-type, 140×30)

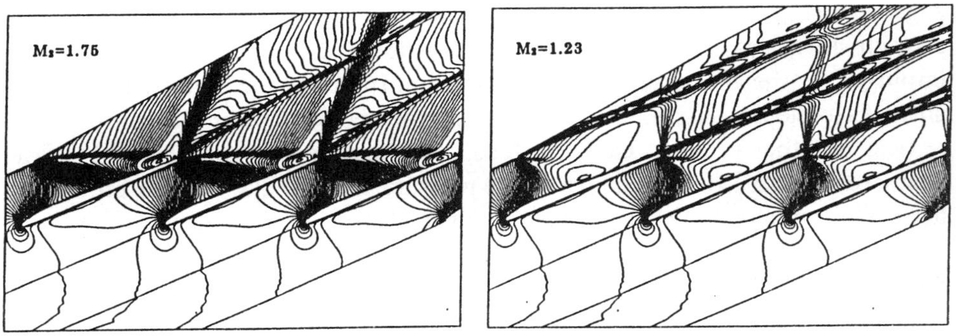

(a) correspondent to case 1 (b) correspondent to case 2

Figure 12. Computed Mach Contour Diagram

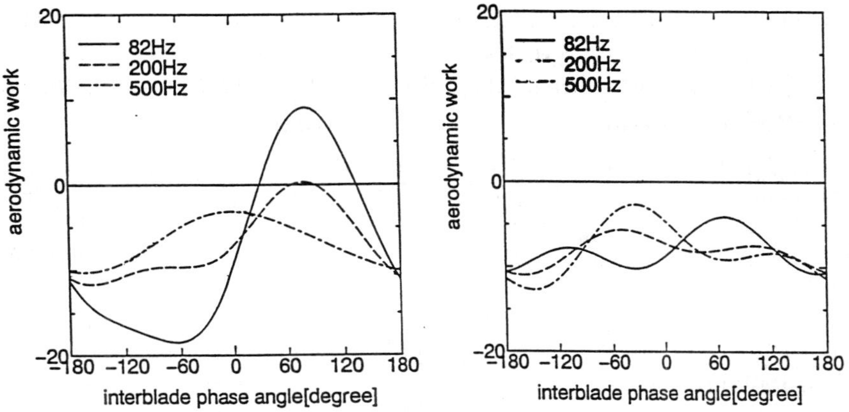

(a) correspondent to case 1 (b) correspondent to case 2

Figure 13. Computed Aerodynamic Work

condition correspondent to that of the case 1, it is clearly seen that there is a large unstable region of the blade oscillation in the positive range of the interblade phase angle. Figure 13 (b) correspondent to the case 2 shows no instability, on the other hand. These results agree very well with the experimental ones in a qualitative sense, and thereby support the experimental findings in the previous sections.

4. Conclusions

The vibration characteristics of a transonic steam turbine blades with a thin airfoil were experimentally studied by the influence coefficient method. Two typical flow conditions, one of which corresponded to the design point, the other of which corresponded to an off-design condition of a low pressure ratio, were adopted, and the results were compared each other. The conclusions are summarized as follows.

(1) The stability of the blade vibration considerably differs between the tested two flow conditions. The vibration is much more unstable in the case of near-design point condition than in the case of an off-design condition. This difference comes mainly from the change in phase angle of the unsteady aerodynamic force.

(2) In the off-design case when the shock wave impinges on around the mid-chord position, the oscillation energy input from the adjacent two blades in the suction

side and the pressure side of the oscillating blade are canceled out each other, which situation thus leads to the less unstable vibration. On the contrary, the energy input are added each other in the case near the design point, which results in the unstable oscillation.

(3)The local unsteady force generated at the impingement point of the shock wave has no significant role in the present vibration instability of the blades.

(4) The results of a CFD simulation qualitatively support the experimental findings.

Acknowledgment

This research was supported under the Kurata Research Grant. The authors gratefully acknowledge the financial support from the Kurata Foundation. They also thank to Mr. Nakata, a former graduate student of University of Tokyo, who developed the CFD code used in the present study.

References

1. Bölcs, A., Cargill, A., Fransson, T.H., Suddhoo, A., and Vogeler, K., "Time-Dependent Predictions and Analysis of Turbine Cascade Data in the Transonic Flow Region", Proceedings of the International Symposium on Aeroelasticity in Turbomachines and Propellers, Notre Dame, 1991, pp.289-308.

2. Bölcs, A., Fransson, T.H., and Körbächer, H., "Time-Dependent Pressure Fluctuations on an Oscillating Turbine Cascade at Transonic Off-Design Flow Conditions", Proceedings of the International Symposium on Aeroelasticity in Turbomachines and Propellers, Notre Dame, 1991, pp.547-565.

3. Sieverding, C., Decuypere, R., Hautot, G., "Investigation of Transonic Steam Turbine Tip Sections with Various Suction Side Blade Curvatures", I Mech E C195/79, 1979.

4. Hanamura, Y., Tanaka, H. and Yamaguchi, K., "A Simplified Method to Measure Unsteady Forces Acting on the Vibrating Blades in Cascade", Bulletin of JSME, Vol.23, No.180, 1980, pp.880-887.

PERIODIC FLUCTUATION OF SHOCK WAVES IN TRANSONIC CASCADE FLOWS

T. SHIRATORI
Tokyo Metropolitan Institute of Technology
Hino, Tokyo, JAPAN

M. MATSUSHITA
Tokyo Metropolitan Institute of Technology
Hino, Tokyo, JAPAN

Y. NOGUCHI
University of Salford
Salford, England, UK

Abstract

An experimental study has been carried out to investigate the behavior of passage shock waves with the shock induced boundary layer separation in two-dimensional transonic cascade flows. The experiments are performed for an unstaggered case and a staggered case with 30 degrees. The time dependent streamwise shock wave positions are measured by a specially designed measuring system, of shock wave position. The results show that the self-excited shock wave oscillations occur in a range of back pressures in unstaggered and staggered cascades. The fundamental frequencies of the shock oscillation are around 400 Hz ~ 900 Hz in the tested conditions. An approximately linear relation between the frequency and the mean shock wave position is observed in the unstaggered case, but is not clearly seen in the staggered case.

1. Introduction

Aeroelasticity problems for fan and compressor blades operating in the transonic or supersonic region with shocks necessitate information of the shock wave behavior and the relation to other parameters. It may be easily understood that the movement of the shock wave caused by airfoil deformation, small disturbances in the flow field and the shock-boundary layer interaction, results in large changes in airfoil surface pressure which in turn lead to changes in the lift and the pitching moment. In case of a single airfoil oscillating in pitch in transonic flow, Shiratori and Tanida[1] showed the role of

T.H. Fransson (ed.), Unsteady Aerodynamics and Aeroelasticity of Turbomachines, 693-704.

shock movement on the unsteady aerodynamic force by measuring shock motion and pitching moment simultaneously. In oscillating cascade flows, the influence of shock motion on flutter of transonic cascades were discussed by calculations (Bendiksen[2], Watanabe *et al.* [3], Shiratori and Noguchi[4]) and by experiments (Shibata et al.[5], Kobayashi *et al.*[6]).

However, even without blade oscillations, spontaneous shock wave oscillation can occur due to shock-boundary layer interaction, etc., causing unsteady aerodynamic forces on the blade. The self excited oscillation of normal shocks were studied in a transonic diffuser (Bogar[7]) and in an isolated airfoil (McDevitt[8]). Similar phenomenon on a single airfoil in two-dimensional channel flow was reported by Shiratori and Tanida[9] and then studied by Yamamoto and Tanida in details[10]. Geometrically, this case is regarded as a part of the unstaggered linear cascade. Their studies gave the frequency range, the mechanism of the self-excited shock wave oscillation and the strong coupling of the shock behavior at the upper and lower passages in the channel flow.

From the above points of view, it may be suggested that the generation of self excited oscillation of passage shock waves may be one of the causes for the onset of the shock associated flutter, or at least, for the change of the flutter generation region in transonic cascade flows. Moreover, the experimental data of the shock wave behavior become increasingly important to evaluate the results of the computational estimations in transonic cascade flows.

In the present study, the behavior of passage shock waves in a two-dimensional transonic cascade flows is investigated experimentally. The shock wave positions along streamwise direction are measured by a specially designed system of measuring shock wave position which consists of a set of schlieren optical equipment, a line scan camera and an associated data processing computer[10]. Here, the shock wave behavior must be investigated at higher than hundreds of Hz. This is because the flow field being affected by the shock wave fluctuation would be relatively small and therefore the frequency is high. The experiments are performed for a unstaggered case and a staggered case with 30 degrees. The results of the unsteady shock position which include the periodic high frequency fluctuation of shock waves, are presented.

2. Experimental Apparatus

2.1 TESTED CASCADE

Experiments are performed using the blow-down type high-subsonic wind tunnel at Tokyo Metropolitan Institute of Technology. The test section is 100 mm wide, 150 mm high and 400 mm long. The cascade consists of three symmetrical circular-arc airfoils of 10 % thickness, 75 mm chord length and are cantilevered on a side wall of the test section with a 46 mm blade spacing (pitch-chord ratio, s/c=0.61). Figure 1 shows the configuration of the tested cascade and the dimensions. The present cascade consisting of three airfoils is considered to enable to take into account the flow interaction with the

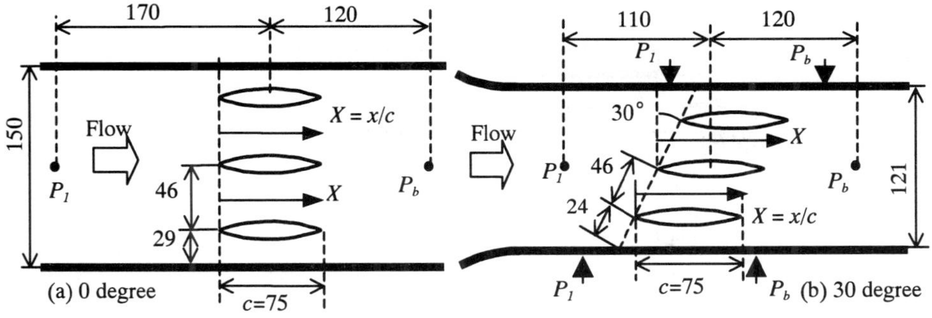

Figure 1. Arrangement of airfoils.

neighboring blade or passage. However it should be noted that more airfoils may be necessary to gain more periodic cascade flow, and the mirror image relation at the walls does not satisfy the linear cascade conditions in the staggered case. Because the flow is choked, experimental flow conditions are adjusted by the upstream total pressure and the replaceable rear throat which also suppresses the propagation of the flow disturbances from the downstream side into the test section.

Experiments are performed with the flow conditions of freestream Mach number of M=0.61, back pressure ratios of $P_b \fallingdotseq 0.63$~0.79 for the unstaggered case, and of M=0.58, $P_b \fallingdotseq 0.63$~0.76 for the staggered case, where P_b are back pressures (downstream static pressures normalized by an upstream stagnation pressure). At these flow conditions, normal shock waves appear in the passage between the blades.

2.2 MEASURMENT OF SHOCK WAVE POSITION

The shock wave positions, measuring system comprises a schlieren optical system arranged by reflection, a line scan camera using a linear image sensor (DALSA, CL-C9-512M), and a data acquisition and processing computer as shown in Figure 2. The sensing photoelements have an array of 512 charge coupled photodiodes with 14 x 10^{-3} mm spacing and driven by a main clock frequency of 16 MHz. The output video signal of the line scan camera gives light intensity distributions along a straight line where the image of shock wave is observed as a negative peak. Number of lines per second, i.e. the line scan frequency, is 25kHz. The system has the capability of detecting streamwise positions of shock waves with an optical space resolution of 0.094mm, at the fluctuation frequency and moving velocity limit of approximately 2 kHz & 50 m/s respectively. The linear image sensor is located at the center of the lower or upper flow passages to measure the shock positions along the streamwise direction (see Figure 1). The thickness of the schlieren image of shock wave is generally influenced by the three-dimensional flow effects near side walls, and the adjustment of knife edge. In the present study, these influences are presumed insignificant in respect of the determination of the shock wave position. The knife edge is set normal to the main flow and adjusted so that the shock wave image is clearly visualized in the schlieren image in later experiments.

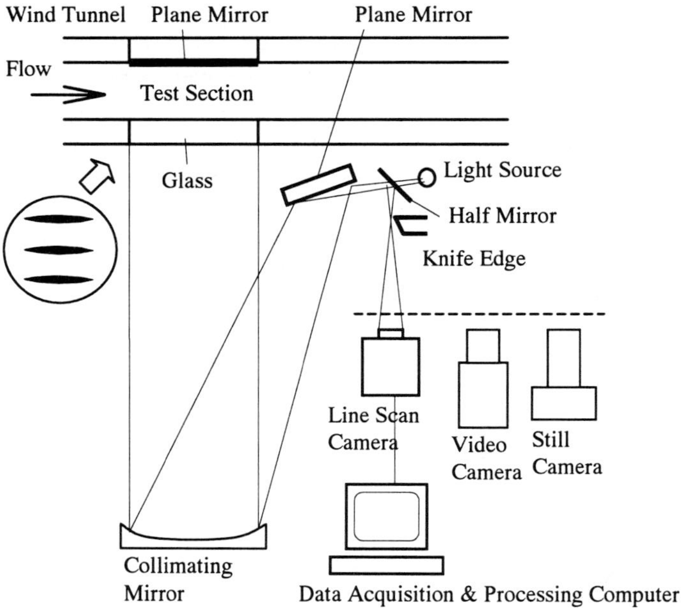

Figure 2. System of measuring shock wave position.

3. Results

3.1 FLOWFIELD AROUND CASCADE

Flow fields around the cascade exhibiting the oscillation of shock waves are visualized by the schlieren system and recorded by a 35 mm still camera with a spark light and a

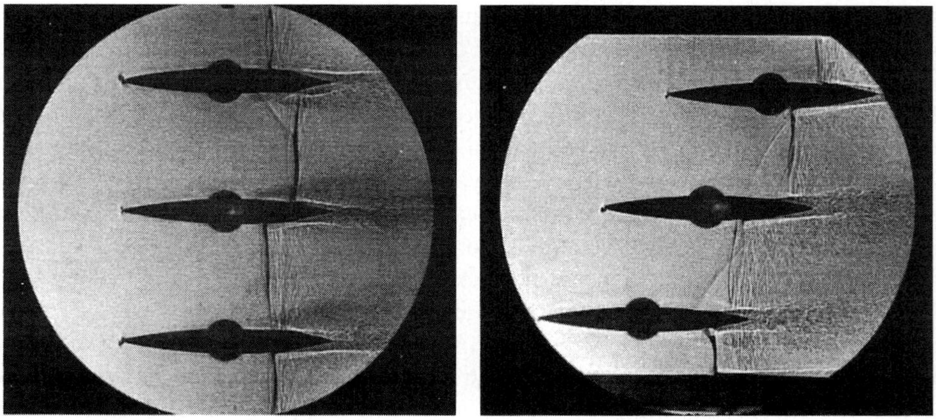

(a) 0 degree, (M=0.613, P_b=0.724) (b) 30 degree, (M=0.579, P_b=0.724)

Figure 3. Spark schlieren photographs of tested cascade flows.

standard video camera (30 flames/second with the exposure time of 1/10000 second) with a continuous light. Figure 3 shows the examples of the spark schlieren photographs of transonic flow around (a) the unstaggered and (b) staggered cascades. The near normal shocks are interacting with the boundary layers and the flow separations are observed in these photographs. In the unstaggered case, the upper passage shock wave position is different from the lower passage ones. The observations by the standard video camera show that the shock fluctuation is occurred in a range of back pressures and the upper and lower passage shock waves move in almost antiphase to each other. This observation is almost the same results of transonic flows around a single airfoil in two-dimensional channel (Ref.[9], [10]). In the staggered case, the position difference between the upper passage shock and the lower passage shock is not clearly seen in this photograph, but the pictures of video camera show that the position of shock waves are not stable.

3.2 INSTANTANEOUS SHOCK WAVE POSITION

The instantaneous positions of shock wave are obtained from the instantaneous schlieren photographs and video camera pictures. Figure 4 shows the variation of the instantaneous shock positions against the back pressure. In this figure, the average positions obtained by the line scan camera and the results of Ref.[9] and [10] are also included. The shock waves move rearward with the decreasing in back pressures. These positions do not remain in a fixed point and widely dispersed against the given value of back pressure. This indicates the occurrence of the self-excited shock oscillation and the irregular fluctuation which are shown in details in the next section. In the staggered case, The qualitative relation between the shock wave positions and the back pressures is the same as that of the unstaggered case. However, the mean shock

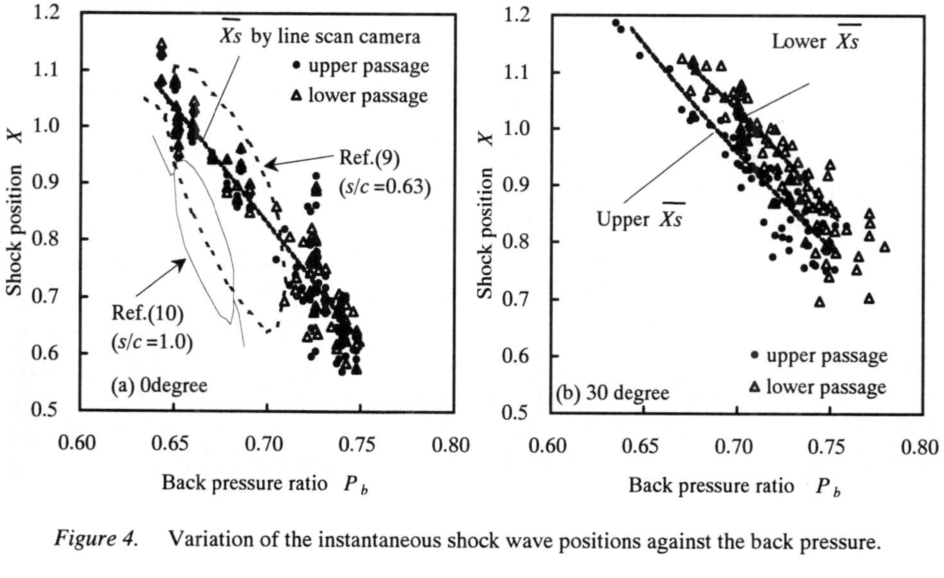

Figure 4. Variation of the instantaneous shock wave positions against the back pressure.

positions at the lower passage is located downstream as compared with those at the upper passage. These differences imply that the periodicity of the time averaged flow of the present cascade is slightly insufficient to assume as a linear cascade.

3.3 PHENOMENA OF SHOCK WAVE OSCILLATION

Time histories of the linear image of the shock wave motions along the center line of upper and lower passages are recorded by using the line scan camera to observe the quantitative behavior of the fluctuating shock wave in details. Figure 5a (0 degree) and Figure 5b (30 degree) show the time series of the linear images, i.e. the movements of the shock wave for various back pressures. Appearances of shock wave oscillation varies with the back pressure and the stagger angle. In the unstaggered case, weak shock wave appears, moves forward and disappears at first (P_b=0.742). This phenomenon is not a continuous oscillation but almost an irregular fluctuation. As back pressure decreases, the average shock wave position moves downstream, becomes a continuous and stable oscillation (P_b=0.723, 0.712) with increase in frequency (P_b=0.699). Finally, the shock wave fluctuates irregularly with small amplitude and high frequency oscillation (P_b=0.670). The shock behavior of upper and lower passages are almost the same. In the staggered case (see Figure 5b), the general trend is similar to the unstaggered case. However the amplitude of oscillation is smaller and the oscillation itself is less continuous as compared with the unstaggered case. This is due to more moderate change in the flow passage area. This causes the occurrence of the weaker shock waves and less strong coupling between the upper and lower passage flow phenomena (or the neighbor blades) than those of the unstaggered case. In addition, the shock wave oscillations between the upper and lower passages are different at the similar back pressures as shown in figure 5b. Although, the behavior of the shock wave and boundary layer at the upper passage may be seemingly predominant, it may be the influence of the difference between the mean shock wave positions at the upper passage and the lower passage.

3.4 WAKE FLUCTUATION AND SHOCK WAVE OSCILLATION

To study the relation between the wake behavior near the trailing edge and the shock wave oscillation around the center airfoil, the measurements of the wake position and the shock wave position at the 10 % chord distance away from the airfoil chord are carried out. The measurements are made to the cases that the self-excited shock wave oscillation are clearly observed (0 degree :$P_b \doteqdot 0.71$, 30 degree :$P_b \doteqdot 0.73$). In the wake measurements, knife edge is set to horizontal in order to obtain sharp images of the wake. The results are shown in Figure 6a (0 degree) and 6b (30 degree). While the shock waves near the both surfaces are oscillating, the wake also fluctuates up and down to the chord line in the unstaggered case (Figure 6a). The frequencies of the wake are approximately 910 Hz at just behind the trailing edge and 960 Hz at the 20% chord length downstream of the trailing edge which are similar to the range of shock wave oscillation frequencies (820 Hz at the upper surface and 860 Hz at the lower surface).

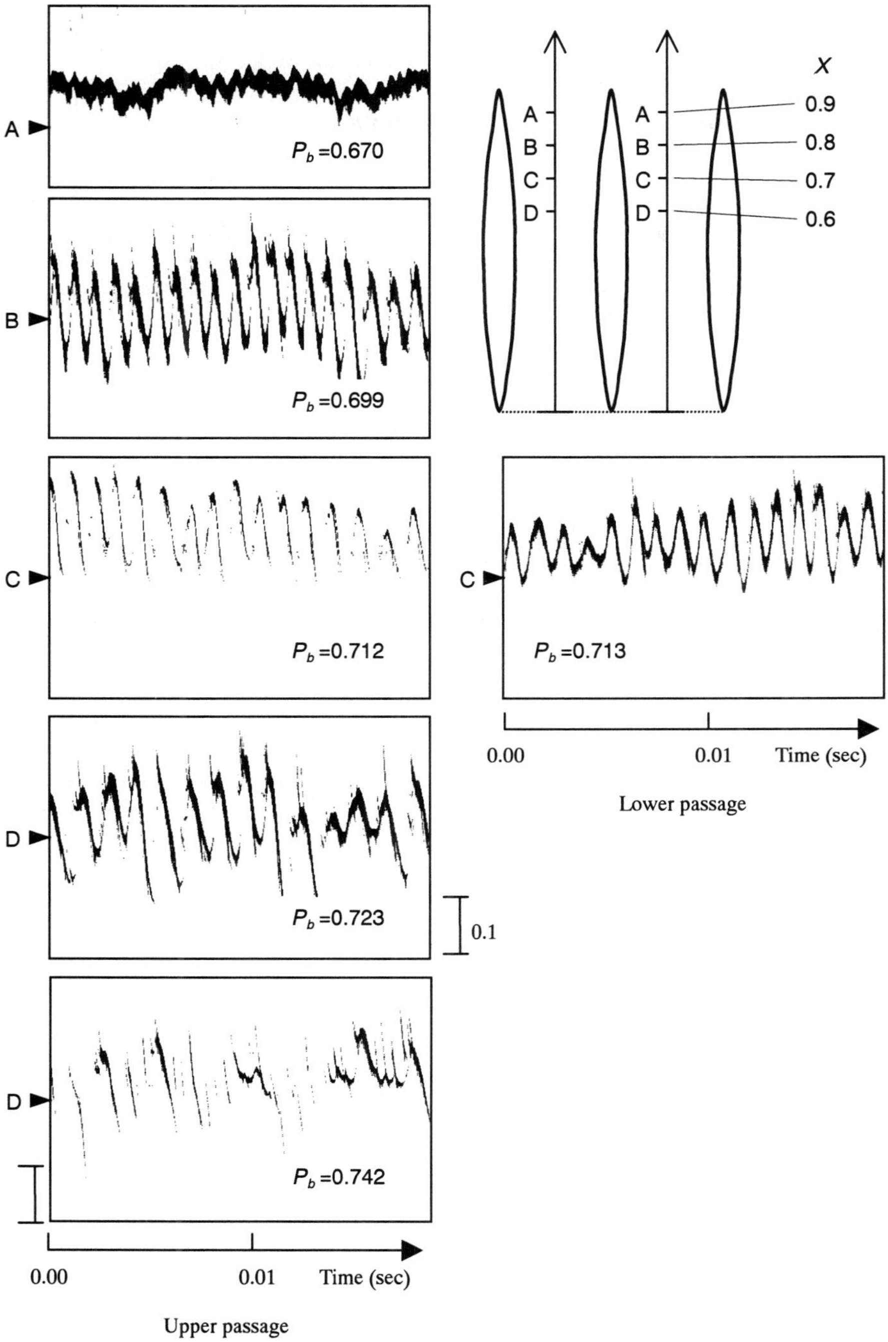

Figure 5a. Movement of shock wave along a center line of flow passage (Stagger angle = 0 degree).

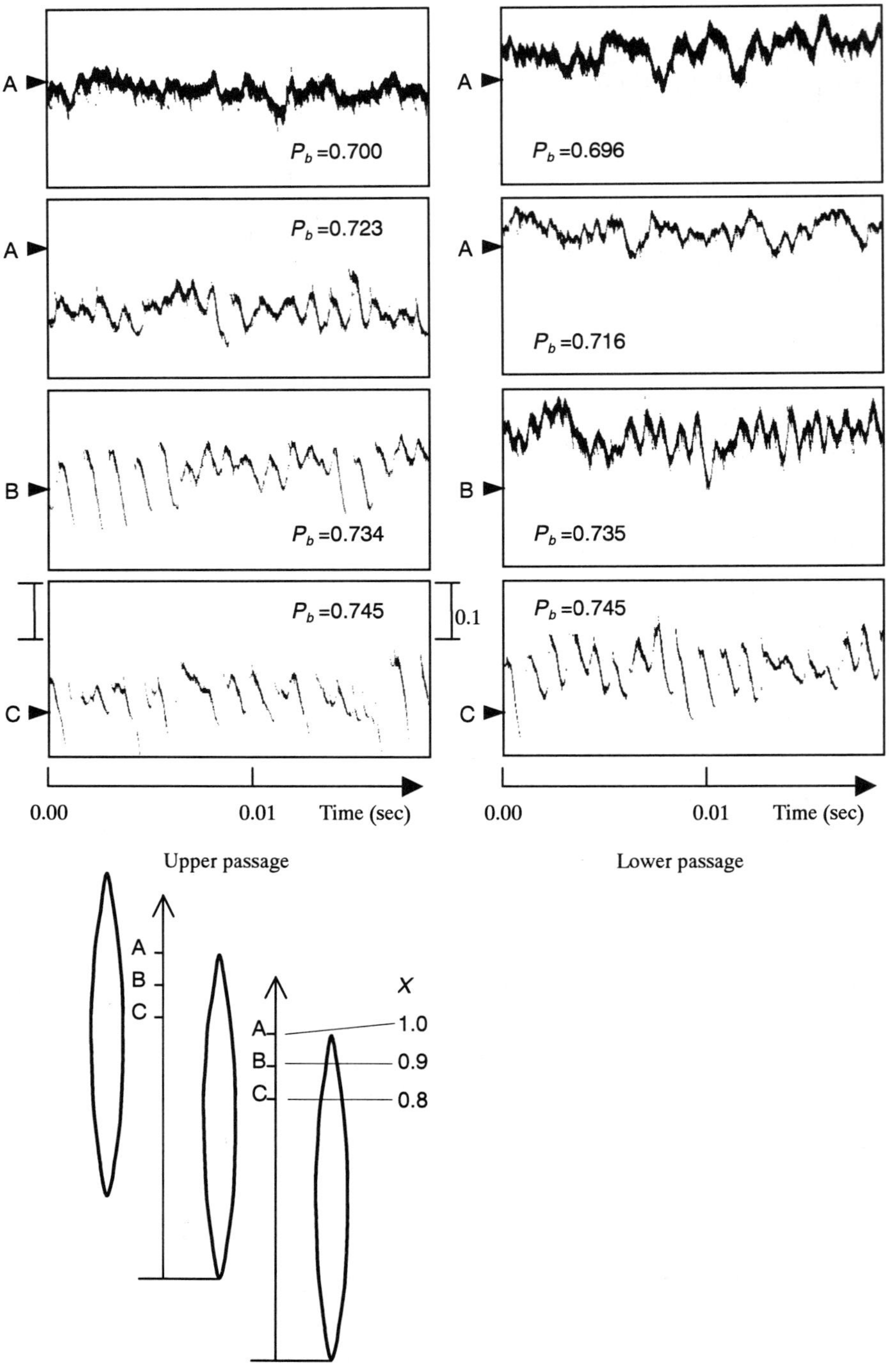

Figure 5b. Movement of shock wave along a center line of flow passage (Stagger angle = 30 degree).

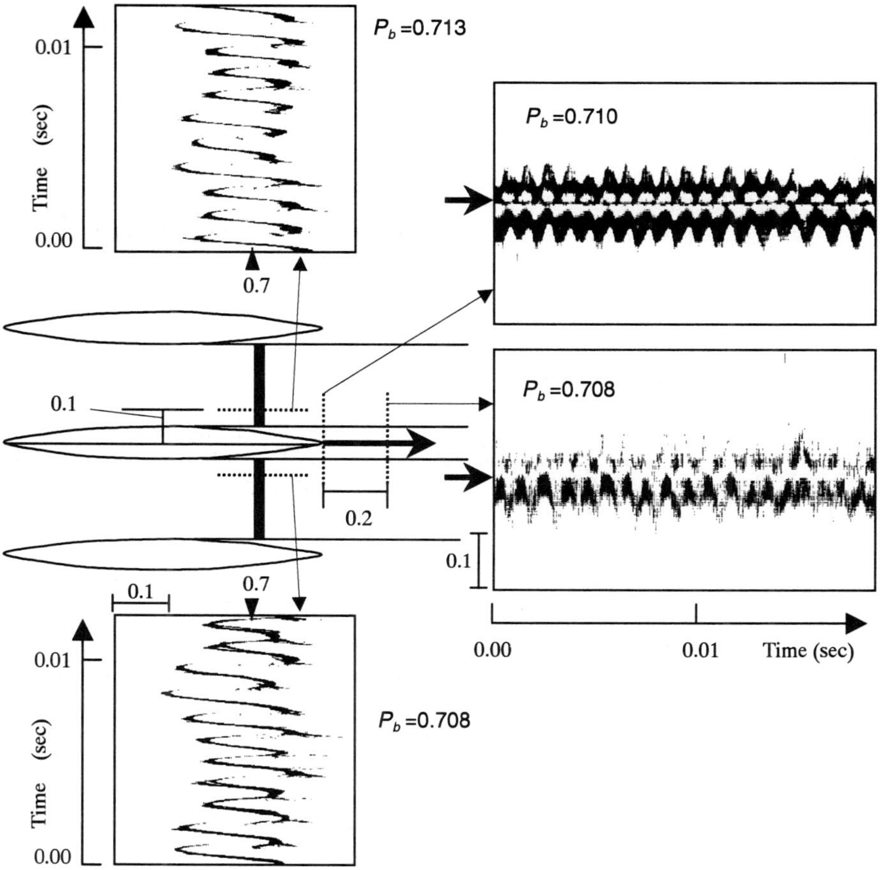

Figure 6a. Shock wave oscillation and movement of wake position (Stagger angle = 0 degree).

These results support that the mechanism of the shock wave oscillation of the present unstaggered case is almost equal to that of the channel flow. On the other hand, the staggered case gives a slight different feature. The shock wave oscillation have the frequency of 690 Hz at the upper surface and higher frequency (1500 Hz) with moderate fluctuation (230 Hz) at the lower surface. The frequency of wake fluctuation are 760 Hz and 780 Hz at just behind the trailing edge and at the 20% chord length behind trailing edge respectively which are similar to that of the shock wave oscillation at the upper surface.

3.5 FREQUENCY OF SHOCK WAVE OSCILLATION

The time series data of the shock wave position giving the negative peak are extracted from the light intensity distributions of the line scan camera signals. Subsequently, a spectral analysis of the time series data of the position is carried out to get the

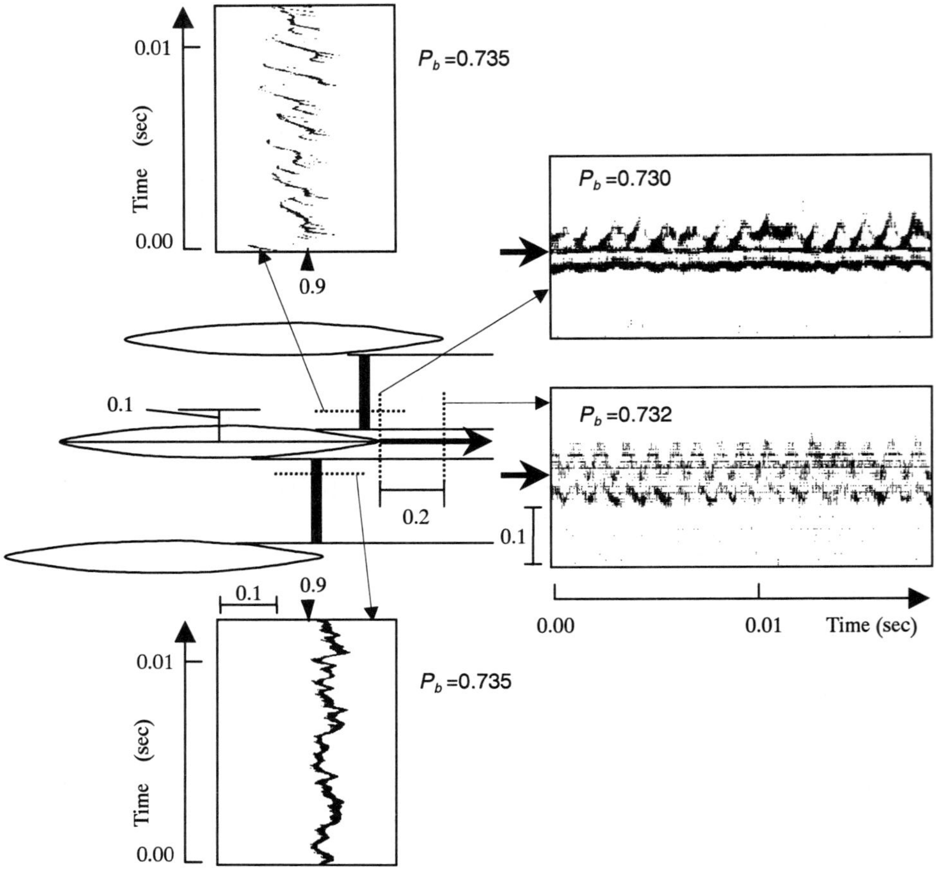

Figure 6b. Shock wave oscillation and movement of wake position (Stagger angle = 30 degree).

fundamental frequency of the shock oscillation which is valuable to the shock induced flutter analysis. The frequencies of shock wave oscillation are normalized by using the half chord length and the speed of sound at the inlet. Figure 7 shows these frequencies against the mean shock wave positions, $\overline{X_S}$. The Figure also includes the results of the channel flow. The unstaggered case (pitch-chord ratio, s/c=0.61) shows that the frequency of the shock wave oscillation increases linearly with the rearward movement of the mean shock wave position and is also almost equal to the results of Ref.[9] (equivalent pitch-chord ratio, s/c=0.63). These results are a conclusive evidence that the mechanism of the shock wave oscillation of the present unstaggered case is almost equal to that of the channel flow. The another result of the channel flow (Ref.[10], equivalent pitch-chord ratio, s/c=1.0) gives slightly smaller values than that of the present study and Ref.[9]. These differences can be explained as follows. The change of the flow cross-sectional area along the streamwise direction becomes gentler when the

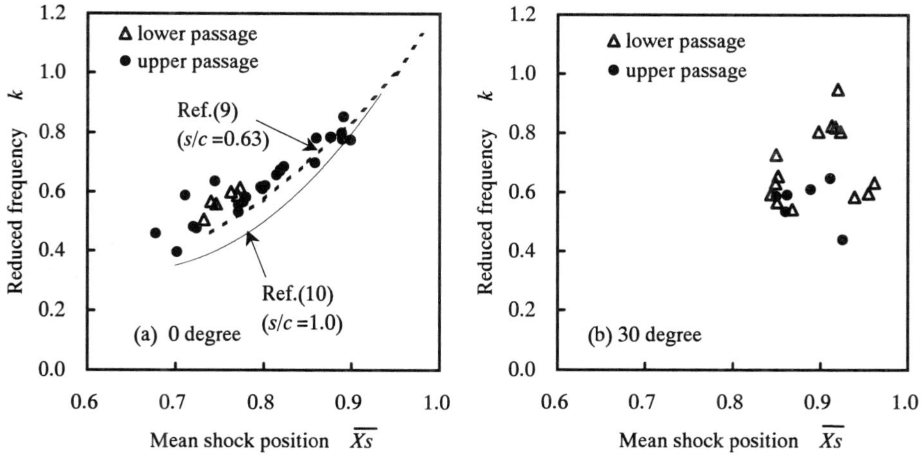

Figure 7. Frequency of shock wave oscillation against the mean shock wave position.

pitch-chord ratio becomes larger. This leads to weakening of the shock wave, then the rise of Mach number at the flow region behind the shock wave which makes the propagation time from the rear to the shock wave longer. As a results, the frequency of the shock wave oscillation becomes lower. In the staggered case, although the range of the frequencies is approximately the same as the unstaggered case, the linear relation between the frequency and the mean shock wave position is not clearly seen.

4. Concluding Remarks

To investigate the behavior of the passage shock wave in a two-dimensional transonic cascade, experimental study has been performed for a unstaggered case and a staggered case with 30 degrees. The fluctuating shock wave positions in the streamwise direction in the blade passages and also the wake fluctuation are measured by the specially designed system. The occurrence of the phenomenon of the self-excited shock wave oscillation is confirmed clearly in the unstaggered case. The results show that the shock wave positions and the frequencies in the unstaggered case are almost equal to the values of the equivalent channel flow. It is therefore concluded that the mechanism of the shock wave oscillation of the present case is almost equal to that of the channel flow. The self-excited shock wave oscillation was also observed in the staggered case. However, the oscillation itself is less continuous as compared with the unstaggered case. The approximately linear relation between the frequency and the mean shock wave position which is obtained in the unstaggered case, is not clearly seen in the staggered case. This implies that although the mechanism of staggered case is similar to that of the unstaggered case, other factors such as non-uniformity of propagation time in space behind shock waves and the difference of shock-boundary layer interactions at the upper and lower airfoil surfaces may influence the flow field. In the present study, the coupling of phenomena and phase difference between the upper and lower passages (or surfaces) could not be clearly explained. From this point, it is necessary to carry out a study to investigate these points through the simultaneous measurements of shock waves of neighboring passages and wakes.

Acknowledgments

The authors wish to express our thanks to Mr. C. Sakurai and Mr. Y. Iwasaki for their numerous assistance during the preparation and experiments. The work was supported in part by the Ministry of Education and Science in Japan under the Grant-in-Aid of Scientific Research and by the Tokyo Metropolitan Government under the Grant-in-Aid of Specific Research.

References

1. Shiratori, T. and Tanida, Y., Aerodynamic Characteristics of An Airfoil Oscillating in Transonic Flow Between Parallel Walls, *Proceedings of the Tokyo International Gas Turbine Congress*, 1983, 659-666.
2 Bendiksen, O.O., Role of Shocks in Transonic/Supersonic Compressor Rotor Flutter, *AIAA J.*, Vol.24, July, 1986, 1179-1186.
3. Watanabe, T., Doi, H. and Kaji, S., Numerical Analysis of the Unstarted Flutter in a Compressor Cascade, *Unsteady Aerodynamics and Aeroelasticity of Turbomachines* (The 7th ISUAAT), 1995, 207-222.
4. Shiratori, T. and Noguchi, Y., Unsteady Aerodynamic Characteristics of a Pitch Oscillating Transonic Cascade and the Role of Shock Wave Behavior, *Unsteady Aerodynamics and Aeroelasticity of Turbomachines* (The 7th ISUAAT), 1995, 197-206.
5. Shibata, T. and Kaji, S., Experimental Study of the Unstarted Flutter in a Compressor Cascade, *Unsteady Aerodynamics and Aeroelasticity of Turbomachines* (The 7th ISUAAT), 1995, 365-381.
6. Kobayashi, H., Oinuma, H. and Araki, T., Shock Wave Behaviour of Annular Blade Row Oscillating in Torsional Mode with Interblade Phase Angle, *Unsteady Aerodynamics and Aeroelasticity of Turbomachines* (The 7th ISUAAT), 1995, 347-364.
7. Bogar, T.J., Structure of Self-Excited Oscillations in Transonic Diffuser Flows, *AIAA J.*, vol. 24, 1986, 54-61.
8. McDevitt, J.B., Levy Jr., L.L. and Deiwert, G.S, Transonic Flow about a Thick Circular-Arc Airfoil, *AIAA Journal*, Vol.14, No.5, 1976, 606-613.
9. Shiratori, T. and Tanida, Y., Self-Excited Oscillations of Shock Wave on an Airfoil in Transonic Internal Flow, *Memoirs of Tokyo Metropolitan Institute of Technology*, No.1, 1987, 31-36 (in Japanese).
10. Yamamoto, K. and Tanida, Y., Self-Excited Oscillation of Shock Waves on an Airfoil in Two-dimensional Transonic Channel Flow, *Proceedings of 4th Symposium on Unsteady Aerodynamics and Aeroelasticity of Turbomachines and Propellers*, 1987, 769-783.
11. Shiratori, T. and Matsushita, M., A Measurement of Moving Shock Wave Positions By A linear Image Sensor and Schlieren System, *Proceedings of The 1st Pacific Symposium on Flow Visualization and Image Processing*, 1997, 412-417.

13: Theoretical Studies of Unsteady Flow Around Oscillating Blades - 4

PART-SPEED FLUTTER ANALYSIS OF A WIDE-CHORD FAN BLADE

J. W. CHEW, J. G. MARSHALL
Technology Department
Rolls-Royce plc.
P.O. Box 31, Moor Lane
Derby DE24 8BJ, UK

M. VAHDATI & M. IMREGUN
Mechanical Engineering Department
Imperial College
Exhibition Road
London SW7 2BX, UK

Abstract

Experimental and theoretical results are reported in this paper for a flutter research rig with 26 wide-chord fan blades. The structural model was based on a 3D finite element eigensolution while two different 3D unsteady aerodynamic codes were used for the fluid. The first one is a thin-layer Reynolds-averaged Navier-Stokes solver using structured meshes and it was used for specified blade motion and interblade phase angle. The second code is an implicit upwind solver that uses unstructured meshes of tetrahedra elements. An inviscid flow representation, together with a viscous loss model, was used for whole annulus calculations with flexible blades. The second approach is an integrated aeroelasticity analysis in the sense that the structural and fluid equations are solved in a coupled fashion by exchanging data at each step of the time integration. Flutter predictions were attempted at 75%, 80% and 85% speeds and the least stable vibration modes were found to lie within the 0-3 nodal diameter range, in agreement with the experimental observations. The predicted variation of flutter margins with respect to speed was found to exhibit fair correlation with available experimental data. A detailed investigation of the blade's surface pressure fluctuations revealed a linear perturbation trend from the mean flow solution and highlighted the importance of getting the correct shock position in the steady flow calculations. Also, the results provided support for an inviscid flutter mechanism which is discussed in some detail.

1. Introduction

Aeroelasticity phenomena are characterised by the interaction of fluid and structural domains, whose describing equations are non-linear in the general case. In turbomachinery environments, the problem is further compounded by the fact that blades vibrate with a relative phase with respect to each other, the value of which is not necessarily known. Most prediction methods tend to treat the two domains separately and they usually assume some critical interblade phase angle for which the flutter

T.H. Fransson (ed.), Unsteady Aerodynamics and Aeroelasticity of Turbomachines, 707-724.

analysis is carried out for a single passage. More recent approaches, the so-called integrated or coupled methods, link the structural and fluid domains via a set of boundary conditions that must be satisfied simultaneously throughout the solution phase. Multi-stage, multi-passage models, including whole annulus representations, are also starting to emerge as computing power becomes more readily available. A literature review is beyond the scope of this paper but a survey of aeroelasticity methods with emphasis on turbomachinery applications can be found in Marshall and Imregun (1996).

Of particular interest here are the integrated methods but such approaches are not widespread in the literature. Bendiksen (1991) investigated a 2D turbomachinery geometry while a 2D cascade was considered by Abdel-Rahim et al. (1991). Vahdati & Imregun (1994) have applied a fully integrated method to a 3D blade, further results being reported by Marshall & Imregun (1995).

This paper will focus on a particular aeroelastic phenomenon, fan blade flutter. Flutter is a particularly difficult problem in a turbomachinery environment since there are many complex features with not fully-understood consequences: flow distortions due to up and downstream blade-rows, effects of the intake duct, coupling of structural modes, and loss of spatial periodicity due to aerodynamic effects and/or blade-to-blade manufacturing differences - known as mistuning. As is well known in the turbomachinery community (see, for example, Cumpsty 1989), flutter can occur in various regions of the fan operating range. Here attention is focused on part-speed assembly mode flutter that may be observed as the mass flow is reduced on a constant speed characteristic. This particular type has sometimes been called supersonic stall flutter. This work has been carried out with the support of Rolls-Royce plc. and the Department of Trade and Industry.

2 Formulation

2.1 STRUCTURAL MODEL

The structural equation of motion can be written as:

$$[\mathbf{M}]\{\ddot{\mathbf{q}}\}+[\mathbf{C}]\{\dot{\mathbf{q}}\}+[\mathbf{K}]\{\mathbf{q}\}=\{\mathbf{p}(t)\} \tag{1}$$

where [**M**], [**C**] and [**K**] denote the structural mass, viscous damping and stiffness matrices, {**q**} is a vector of the vibration shape and {**p**} is a time-dependent vector of the unsteady aerodynamic forces. In aeroelasticity analyses below, structural damping is set to zero as its effects can be added later in a linear fashion.

The undamped eigensolution for [**C**]=[0] and {**p(t)**}={0}yields the natural frequencies, ω_r and the associated mass-normalised modeshapes [Φ]. The principal co-ordinates {η} can then be defined as:

$$\{\eta\} = [\Phi]^T\{\mathbf{q}\} \qquad \text{or: } \{\mathbf{q}\} = [\Phi]\{\eta\} \tag{2}$$

Inserting (2) into (1) and pre-multiplying by $[\Phi]^T$ yields:

$$\{\ddot{\eta}\} + [diag(2\zeta_r\omega_r)]\{\dot{\eta}\} + [diag(\omega_r^2)]\{\eta\} = [\Phi]^T\{\mathbf{p}(t)\} \tag{3}$$

where ζ_r is the damping ratio for mode r.

In this form, the equations of motion can be solved by using any standard integration method and a Newmark-β scheme was used here.

2.2 FLUID MODEL

The integral form of Reynolds-averaged mass, momentum and energy conservation equations with an eddy viscosity turbulence model and a moving cylindrical co-ordinate system can be written in the following vector notation:

$$\frac{\partial}{\partial t}\iiint_{\Delta V}\mathbf{U}dV + \oiint_{dA}\{(\mathbf{F}-\mathbf{U}u_g)n_x + (\mathbf{G}-\mathbf{U}v_g)n_\theta + (\mathbf{H}-\mathbf{U}w_g)n_r\}dA = \iiint_{\Delta V}\{\mathbf{S}_i + \mathbf{S}_v\}dV \tag{4}$$

where (u_g, v_g, w_g) are the components of the grid velocities relative to the rotating frame, t is time, dV is an elemental volume, $\mathbf{n}$ is the area normal vector, and dA is the area elemental volume. The state vector **U**, flux vectors **F**, **G**, & **H**, inviscid source term $\mathbf{S}_i$ are given by:

$$\mathbf{U} = \begin{bmatrix}\rho \\ \rho u \\ \rho vr \\ \rho w \\ \rho E\end{bmatrix}, \mathbf{F} = \begin{bmatrix}\rho u \\ \rho uu + p \\ \rho uvr \\ \rho uw \\ (\rho E + p)u\end{bmatrix}, \mathbf{G} = \begin{bmatrix}\rho v \\ \rho vu \\ (\rho vv + p)r \\ \rho vw \\ (\rho E + p)v\end{bmatrix}, \mathbf{H} = \begin{bmatrix}\rho w \\ \rho wu \\ \rho wvr \\ \rho ww + p \\ (\rho E + p)w\end{bmatrix}, \mathbf{S}_i = \begin{bmatrix}0 \\ 0 \\ 0 \\ \rho v^2 / r \\ 0\end{bmatrix} \tag{5}$$

where ρ is density, (u, v, w) is the fluid velocity, p is the pressure, E is the total internal energy, and r is the radius. The viscous source term $\mathbf{S}_v$ is obtained by a standard eddy viscosity turbulence model. Two different CFD codes were used in this study: AV3D and AU3D. The AV3D code is based on the work of He (1993a) and uses a two-grid acceleration method with a Runge-Kutta time-stepping scheme; the turbulence model is based on a mixing length scheme. AU3D employs an implicit time-stepping algorithm and unstructured meshes of tetrahedral elements (Vahdati & Imregun 1994), and uses a Baldwin-Barth turbulence model. Both codes allow mesh movement to accommodate the blade motion and use the Newmark-β scheme to advance coupled blade motion-fluid calculations.

For unsteady flow calculations, the aeroelastic mesh is moved at each time step to ensure correct modelling of the fluid-structure interface. The motion of the blade is obtained from the available FE eigensolution and the motion of the aerodynamic grid points are calculated by linear interpolation and spring analogy, their amplitude matching the actual blade displacements at the solid boundaries and reducing to zero at the far field. The aerodynamic boundary conditions are described in Section 3.2.

2.3 AEROELASTICITY MODEL

The fluid and the structural equations are solved in a closely coupled manner as shown in **Fig. 1**.

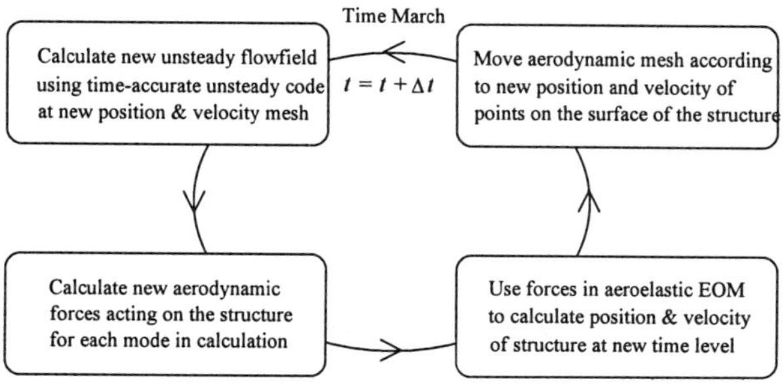

Fig. 1 Solution of the aeroelastic equations of motion

The timestep used is the same for both fluid and structural calculations, and is fixed throughout the time integration. In flutter calculations, interest is confined to the first 10 modes or so and hence the fluid solver has a much more restrictive timestep limit. However a minimum number of 50 timesteps per period must be used for the structural modes in order to ensure accuracy of the Newmark method, which begins to deteriorate at 30 timesteps per period.

Single passage flutter calculations were performed for specified blade vibration in first flap mode, with a maximum blade displacement of about 1 mm. For full assembly calculations, the structure was given an initial velocity in all first flap nodal diameter modes. In principle, such a solution should yield the same response irrespective of starting conditions. However, the solutions could only be run for a limited time period, and the stability of the various modes had to be gauged from their response over this limited period.

2.4 PRESENTATION OF FLUTTER RESULTS

A convenient and traditional parameter for characterising flutter stability is the logarithmic decrement, which is frequently abbreviated to logdec, δ, defined as:

$$\delta = \frac{1}{n}\log\left(\frac{x_0}{x_n}\right) \tag{6}$$

where n is the number of cycles of an oscillation required for the amplitude to reach x_n from an initial value of x_0. Thus for decaying or growing oscillations logdec is positive or negative, respectively. Such an approach is useful in interpreting the results but should be treated with some caution as the logdec is a linear tool and, in the case of non-linear time histories, it can only be applied to a given time interval and, as such, it cannot properly represent the non-linear interactions that may be significant in the calculation.

Estimates of logdec may also be derived from uncoupled analyses for which the blade motion must be specified. Following the analysis of Giles (private communication), the equation of motion for a blade vibrating in a single mode can be written as:

$$\frac{d^2\eta}{dt^2} + \omega^2\eta = P \tag{7}$$

where η, t, ω and P denote modal displacement, time, blade vibration frequency and modal force respectively. A solution of the following form is then assumed:

$$\eta = r(t)\cos\left[(\omega + \delta\omega)t\right] \tag{8}$$

where $\delta\omega \ll \omega$ and r varies slowly with time so that terms involving d^2r/dt^2 and higher derivatives of r are negligible. Substituting (8) into (7) gives:

$$-2\omega\left\{\delta\omega r\cos(\omega t) + \frac{dr}{dt}\sin(\omega t)\right\} \approx P \tag{9}$$

Taking Fourier components gives:

Amplification factor: $$\frac{dr}{dt} = -\int_0^{2\pi/\omega} \frac{P}{2\pi}\sin(\omega t)dt \tag{10}$$

Frequency shift: $$\delta\omega = -\int_0^{2\pi/\omega} \frac{P}{2r\pi}\cos(\omega t)dt \tag{11}$$

For small values of logdec, δ, it can be shown that:

$$\delta = \frac{-W}{2KE} \tag{12}$$

where W is the work done on the blade by the air in one period and KE is the maximum blade kinetic energy during the oscillation.

$$W = \int_0^{2\pi/\omega} P\frac{d\eta}{dt}dt \tag{13}$$

$$KE = \omega^2 r^2/2 \tag{14}$$

3. Cases Studies

The case study is for a 26-bladed rig fan for which some measured flutter data are available. The method used here is to approach the flutter boundary progressively by changing the downstream aerodynamic boundary conditions, keeping rotational speed constant. The success of the analysis can then be assessed by how accurately one can define the flutter boundary by considering several speed characteristics. The aerodynamic conditions at which flutter predictions were made are shown in **Fig. 2** together with the experimentally-determined flutter boundary. Results from two codes, namely AV3D and AU3D are also shown and will be discussed in Section 3.3.

3. 1 STRUCTURAL MODELLING

The blade-alone modes, listed in **Table 1** for various fan speeds, were used for all cases reported. The rigid disk assumption was justified through numerical investigations of the effects of disk flexibility. A representative disc sector was added to the FE model and a fixed condition was applied at the inner circumference of the disc. Cyclic symmetry conditions were then applied for predicting the nodal diameter assembly modes of interest. It was found that there was little natural frequency variation in each of the nodal circle families, the predicted assembly frequencies remaining close to the grounded blade natural frequencies.

TABLE 1. Predicted single blade natural frequencies (Hz) mode shapes
F: flap; E: edgewise, T: torsion; S: stripe

Mode shape	75% speed	80% speed	85% speed
1F	178.5	186.0	193.5
1E	376.8	390.9	405.4
1T	704.0	704.4	704.9
2F	793.6	811.7	830.4
2T	1170.7	1171.7	1172.6
2E	1386.0	1399.5	1413.3
1S	1496.9	1500.9	1505.7
3F	1856.1	1864.2	1873.0

3. 2 AERODYNAMIC MODELLING

A simple structured H mesh with 81 axial, 25 circumferential and 25 radial stations was used for viscous, single-passage AV3D calculations. The same grid was also used in viscous single-passage AU3D calculations with each box element being divided into 6 tetrahedral cells. For multi-passage calculations, the AU3D mesh was reduced to 65x11x15 and the analysis was conducted inviscidly, the viscous effects being approximated by a loss model. The full assembly inviscid mesh and steady state blade surface pressures contours are shown in **Fig. 3**.

The codes were run with specified values for exit static pressure, inlet total pressure, inlet total temperature and inlet swirl velocity. Although unsteady, non-reflective inlet and exit boundary conditions are available in AV3D difficulties had been encountered with their use in earlier studies and they have not been used here. In AU3D modelling, it was attempted to avoid boundary condition reflections by adopting the method of Carstens et al (1993), though complete success is not guaranteed. At the solid boundaries, both codes use wall functions which are implemented via a slip condition at the wall. For steady flows or unsteady flows with zero inter-blade phase angle, periodic boundary conditions are required for single-passage analyses. For non-zero inter-blade phase angles, a single passage analysis with prescribed blade motion through phase-shifted periodic boundary conditions is used in AV3D (He 1993b). A more straightforward multi-passage analysis is conducted in AU3D. Neither approach models over-tip leakage, the meshes used do not resolve the end-wall boundary layers and the relative rotation of the annulus wall is not accounted for.

3.3 STEADY-STATE SOLUTION

As mentioned before, the aerodynamic conditions at which the flutter calculations were made are shown in **Fig. 2.** The steepness of the AU3D constant speed lines (relative to the test) is due to the choice of exit boundary conditions which do not quite match the test conditions. This point is receiving attention in subsequent studies. Generally speaking, AU3D results were found to be more reliable and less prone to numerical instability as the stall boundary was approached. The AU3D-predicted mass flow and pressure ratios are listed in **Table 2** in normalized form. The compressor characteristics are plotted in Fig. 2 along with the AV3D predicted values and the test data. The maximum difference in the predicted efficiencies from AU3D to the test data were of the order of 5%. Given the modelling approximations made and the relative coarseness of the mesh[1], this level of agreement between AU3D and the test data is perhaps as good as could be expected.

TABLE 2. AU3D predictions vs measured data for fan efficiency
(Normalized to arbitrary value)

Rig speed (%)	Calculation point in Fig. 2	Normalized Total Pressure Ratio (AU3D)	Normalized Mass Flow Rate (AU3D)
75	A	0.857	0.738
80	B	0.891	0.830
85	C	0.960	0.868

[1] For aerodynamic design calculations, it is not unusual to use about 300,000 grid points in a single passage.

The effectiveness of the AU3D loss model is illustrated in **Fig. 4** where two steady state solutions are compared for **Case B** of **Table 2**. The first one is a viscous solution on a fine mesh and the second one is an inviscid solution on the standard (relatively coarse) mesh together with the loss model. With this particular implementation of the loss model, source terms are defined on the coarse mesh such that the coarse mesh steady state solution must agree with an interpolation of a fully viscous solution on the finer mesh. As expected, the isentropic surface Mach numbers for both approaches are in good agreement in **Fig. 4**.

A number of numerical difficulties were encountered with the AV3D solution at 75% and 85% speeds. A high degree of solution sensitivity to the specified exit pressure, to the type of exit boundary conditions and to the choice of numerical parameters (such as time-step) was encountered. In some cases solutions would fail to converge with the calculated mass flow reducing to very low values. Further investigation of such behaviour was considered to be beyond the scope of the present study and so only the 80 % speed AV3D results will be included here. The steady state results from the AU3D code are illustrated in **Fig. 5**. At these conditions, the shock is expelled from the passage and intersects the blade suction surface at a forward position.

The relative success of AU3D may be due to the Baldwin-Barth turbulence model which is claimed to behave relatively well in separated flows (Baldwin &Barth, 1991). The implicit solution algorithm used might also have helped stability.

4. Flutter Predictions at 80% Speed

Minimum flutter margin is observed for the 80% speed condition which will be studied in most detail. AV3D results were obtained from uncoupled, single passage analyses while AU3D results were obtained from coupled full-assembly calculations. The AV3D model included the first flap vibration mode only and the initial condition was specified as a modal velocity of 0.5 m/s√kg, corresponding to tip displacements of order 1mm. A typical AV3D solution took around 25 periods to reach a periodic state. The AU3D model included all nodal diameter assembly modes corresponding to the first 5 blade modes. Previous experience indicated that the inclusion of the higher modes had little effect. The initial blade motion in the AU3D calculations was generated by imparting modal velocities of 0.2, 0.05, 0.02, 0.01 and 0.005 m/s√kg in the assembly modes associated with the first to fifth blade modes, respectively.

In AU3D calculations, the modal time histories were computed for about 0.03 seconds and the stability was inferred from examination of the modal responses, some examples being given in **Fig. 6** for the first flap mode[2]. It is seen that the fan appears to be marginally unstable to flutter in the 1 nodal diameter (ND) mode. It can be seen that the

---

[2] A typical unsteady AU3D analysis requires about 8 days CPU time on an IBM RS/6000 model 390 workstation and a corresponding AV3D analysis takes about 16 hours on the same machine.

decay (or growth) rates for the oscillations may vary in a non-linear fashion with time, leading to some uncertainty in estimating stability.

For AV3D, aero damping (logdec) values were obtained using the procedure of **Section 2.4**. The results, presented in the form of logdec vs. nodal diameter plots, are summarised in **Fig. 7** for calculation Point B. There is good overall agreement between the coupled full-assembly (AU3D) and uncoupled, single-passage (AV3D) analyses, both methods indicating that flutter is most likely to occur with the lower interblade phase angles corresponding to 0 to 2 nodal diameter modes. It should be noted that the whole-assembly analysis of AU3D yields all nodal diameter modes in one calculation while the single passage analysis of AV3D requires several runs to obtain the same information. Therefore, only a limited number of modes were considered in the AV3D calculations. This result also agrees reasonably well with experimental observations which are summarised in **Table 3.**

TABLE 3. Summary of rig test results

% Speed	Max response frequency (Hz)	Phase angle (blades 15 to 26)	Interblade phase angle (degrees)	Corresponding nodal diameter number. (IBPA)
75	169	110	10.0	3 (41.5deg)
80	174	-66	-6.0	2 (27.7deg)
85	162	162	14.7	1 (13.8 deg)

Fig. 8 shows plots of the first flap modal forces calculated by AV3D. These are separated into components for the suction and pressure surfaces on neighbouring blades and are referenced to steady state values. The modal velocities for the two surfaces are also shown. As indicated by equations (12) and (13), stability is governed by the product of velocity and force perturbations from their time-averaged values. When the time-average of this product is negative, the blade motion is being damped by the fluid. However, when it is positive, fluid energy is being transferred to the blade, causing increase in the vibration amplitude. The results show that considerably larger modal forces arise when the motion of adjacent blades is out of phase. In these calculations the phase shift of the modal forces is such that the net effect of the forces on the suction and pressure surfaces are destabilising for the 0 ND mode and stabilising for the 2ND mode. For the 6 ND mode, both pressure and suction surface pressures have a stabilising effect.

Further insight into the flutter mechanism can be observed from **Figs 9** to **12** which show the AV3D results for the 0, 2 and 6 ND modes, all calculations being at aerodynamic point B. The mean surface pressures of **Fig. 9** were obtained for the 0 ND case by averaging the (unsteady) pressures over four equally-spaced time points during the last vibration cycle of the computation. The pressure perturbations from these mean values are plotted in **Fig. 10** at the same four time points. Similar pressure fluctuation plots are given in **Figs. 11** and **12** for the 2 and 6 ND cases for which the mean surface pressures are indistinguishable from those in **Fig. 9**. Also shown in **Figs 10** to **12** are the blade positions and the direction of the blade motion at the instants considered. A striking feature of the pressure perturbation plots is the antisymmetry of the results at time intervals of π/ω. The instantaneous position and velocity of each blade are reversed at these time intervals. The fact that force perturbations are also approximately reversed suggests an essentially linear perturbation from the mean flow solution, at least for the particular flow condition and perturbation amplitude considered here. Another feature is that, for 0 and 2 nodal diameter nodes, the sign of the pressure perturbation on the suction surface is reversed on passing through the shock. However, this is not the case when the velocity perturbations of adjacent blades are of opposite sign, as for the 6 ND case where there are particularly high pressure fluctuations in the region of the shock.

Referring to **Figs 10** and **11**, which illustrate cases closest to the a flutter condition, it appears that the largest pressure perturbations occur when the blade velocities and positions combine to give the maximum change in flow incidence to the blade. Towards the leading edge, these perturbations are stabilising, tending to oppose the blade motion. On the other hand, the pressure perturbations change sign on the suction surface and become destabilising as the shock is traversed. A change of sign also occurs on the pressure surface, although no obvious explanation of this feature can be given at this stage. This predicted behaviour is consistent with experimental observations that flutter of wide-chord fans occurs when the shock is in a forward position and that there is no direct correlation between stall and flutter onset. The present results seem therefore to indicate that flutter may well have an inviscid mechanism.

5. Effect Of Speed On Flutter Predictions

The effect of speed on flutter calculations was investigated using AU3D and it was found that the 75 and 85% speed cases showed similar trends to those for the 80% speed case which was discussed in detail in **Section 4**. Again the lower nodal diameter modes were found to be those most susceptible to flutter, an observation which is again in line with the experimental findings.

Assuming that the flutter onset occurs at a condition for which the aerodynamic damping is zero, a flutter boundary was estimated via linear interpolation from the various speed runs and it is plotted in **Fig. 13** together with its experimentally-derived counterpart. Although there are some differences between the two boundaries, the agreement is very encouraging in terms of predicting the correct trends with speed. Also, the calculations assume zero structural damping, the inclusion of which would make the two curves closer.

6. Concluding Remarks

(i) Despite the preliminary nature of some of the results presented here, important conclusions can be drawn from this work. Not the least of these is that part-speed assembly mode flutter is susceptible to CFD analysis, and the physical mechanism for this phenomenon appears essentially inviscid. The term "supersonic stall flutter" for this particular flutter regime is misleading as the results indicate that it can occur without stall actually taking place.

(ii) The main modelling difficulties encountered were associated with the calculation of the steady-state aerodynamics rather than that of the unsteady aeroelastic time histories. Premature "numerical stall" was encountered at near-stall, part-speed running conditions. This is believed to be sensitive to turbulence modelling and constitutes a major challenge to CFD researchers.

(iii) The visualisation of the predicted blade surface pressure perturbations has shown the importance of predicting the steady-state shock position correctly. For a perfectly tuned assembly vibrating in its least stable mode, such as those investigated here, suction surface pressure perturbations at the shock position are expected to be small. Pressure perturbations upstream of the shock position tend to stabilize the blade motion while those downstream are destabilizing.

(iv) There is broad agreement between the two non-linear models in spite of the marked difference in representing the structural motion. Both approaches predict instability in the 0 to 3 forward travelling nodal diameter mode and this finding is consistent with the test results. An examination of the solution indicates that a linearised unsteady calculation would also have the same level of success in this particular case.

(v) Further efforts are being directed towards the development and evaluation of models for this class of flutter. These include improvement of the steady-state flow modelling, and increasing the computational efficiency of the unsteady calculations.

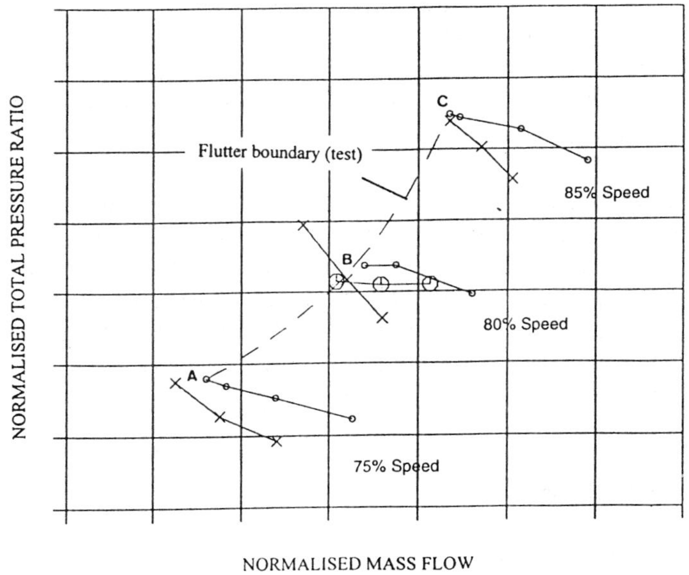

Fig. 2 Compressor map showing the aerodynamic conditions for flutter calculations
o: Test, x: AU3D, O: AV3D

Fig. 3 Inviscid mesh for 26-bladed assembly and steady-state surface pressure contours

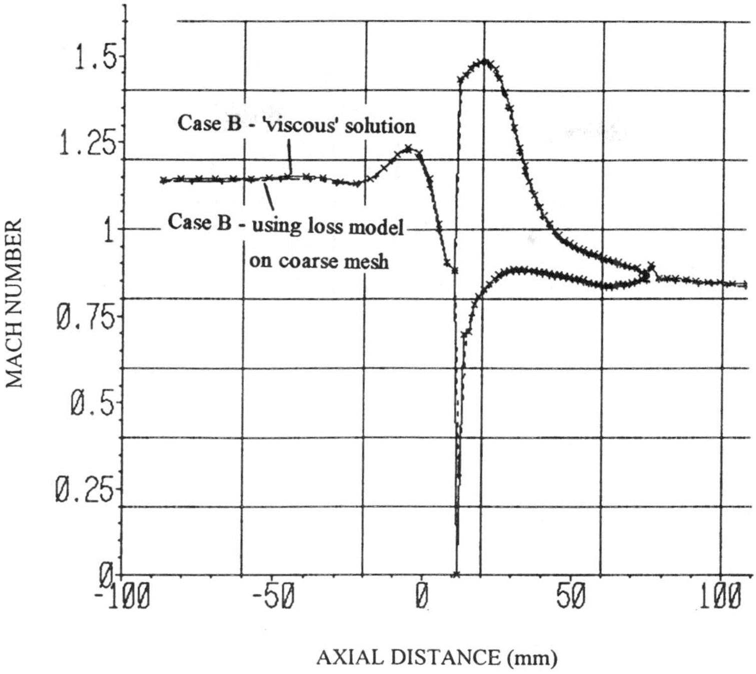

Fig. 4. Fine mesh viscous solution vs coarse mesh inviscid solution with loss model - AU3D, Case B.

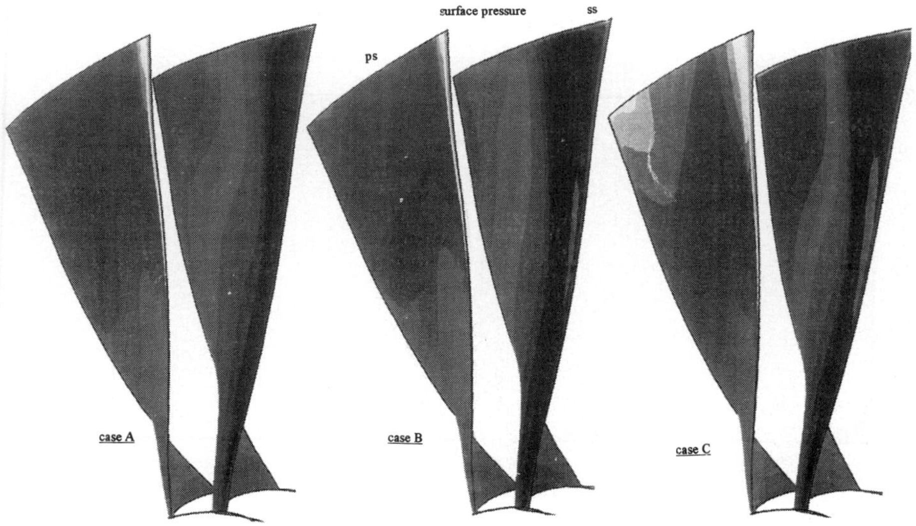

Fig. 5 Steady-state blade surface pressure contours - AU3D, Cases A to C.

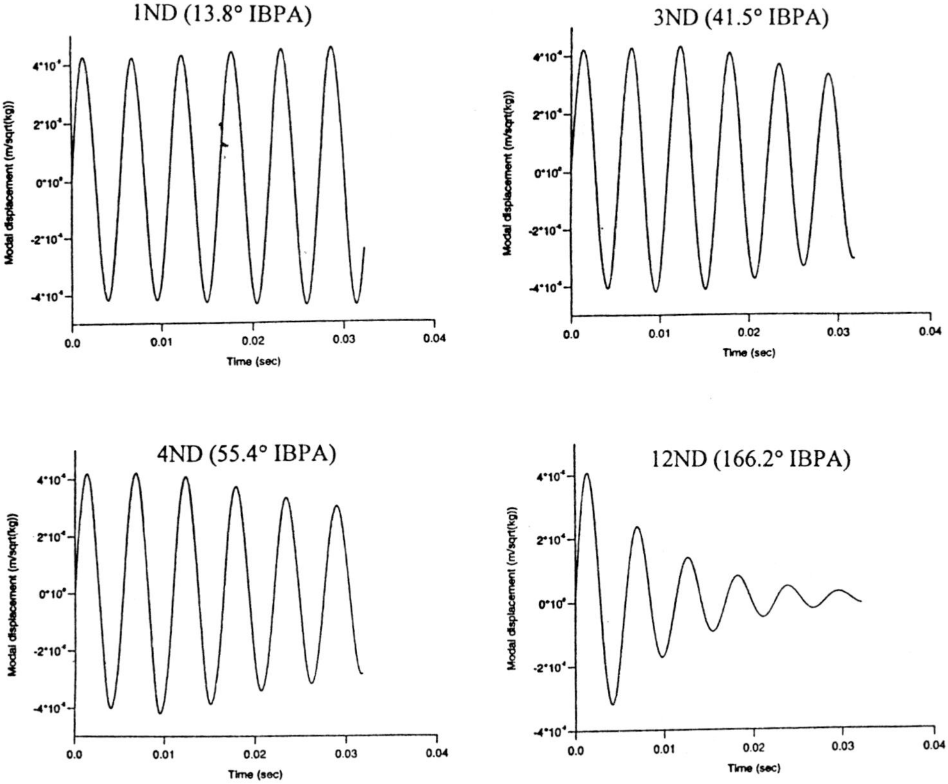

Fig. 6 First flap modal displacements - AV3D, Case B

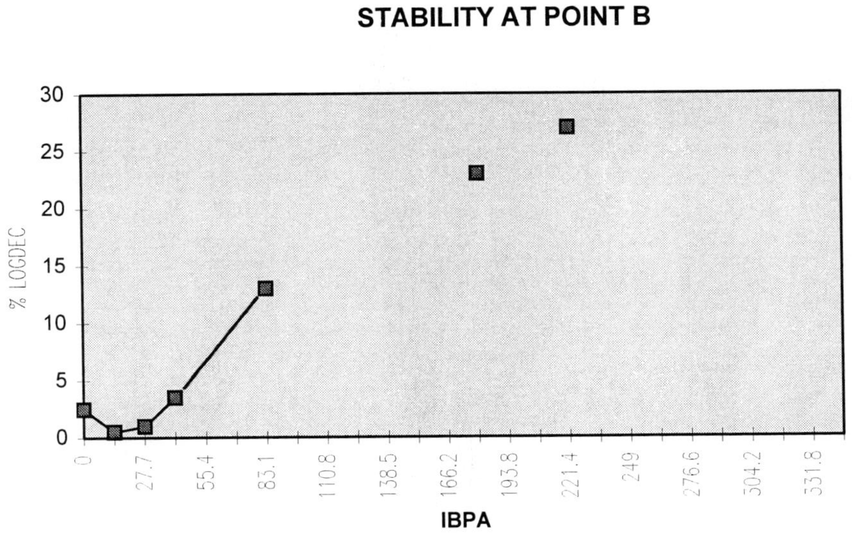

Fig. 7. Stability at point B - AV3D, Case B

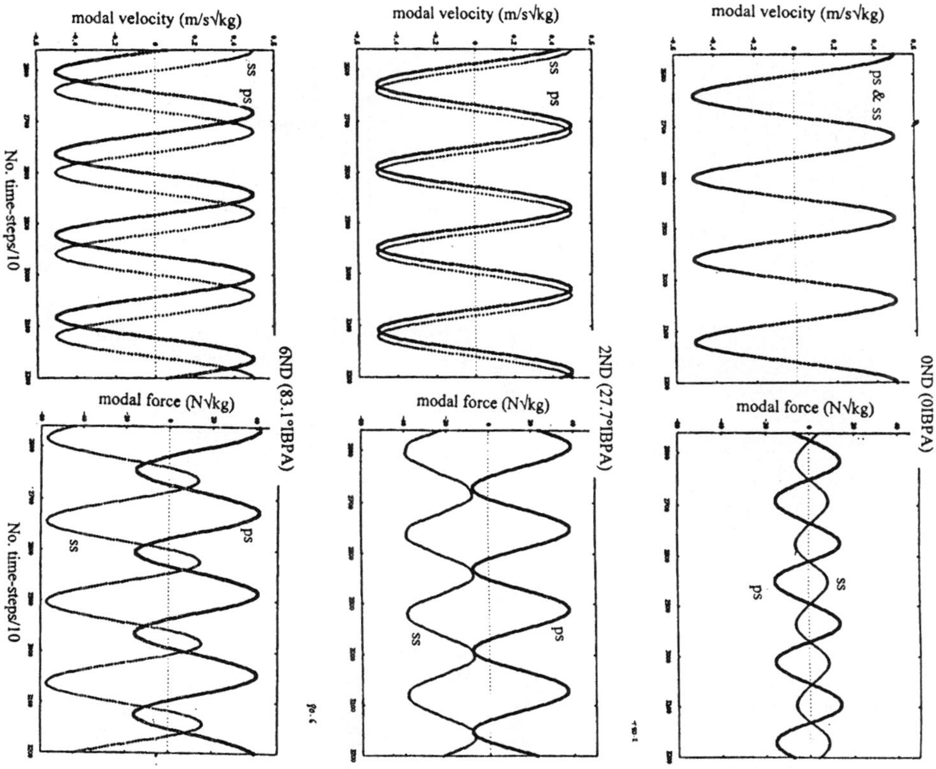

Fig. 8 First flap modal forces and velocities - AV3D, Case B

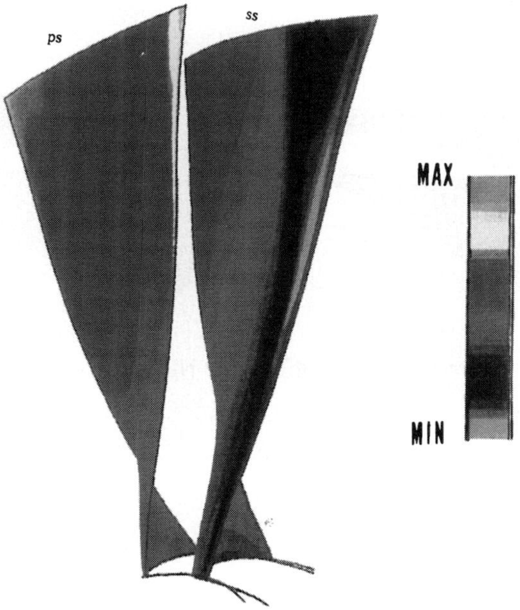

Fig. 9 Time-averaged surface pressures - AV3D, Case B

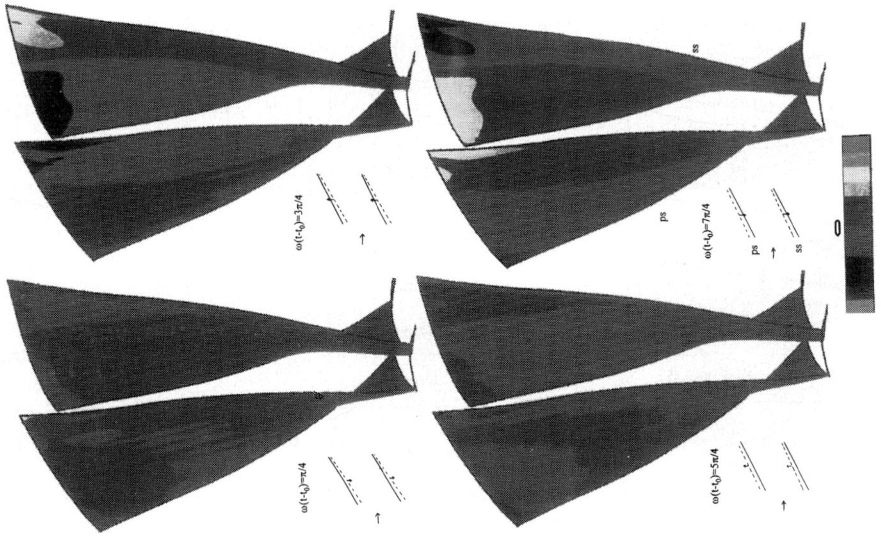

Fig. 10 Pressure fluctuations in 0 ND mode - AV3D, Case B

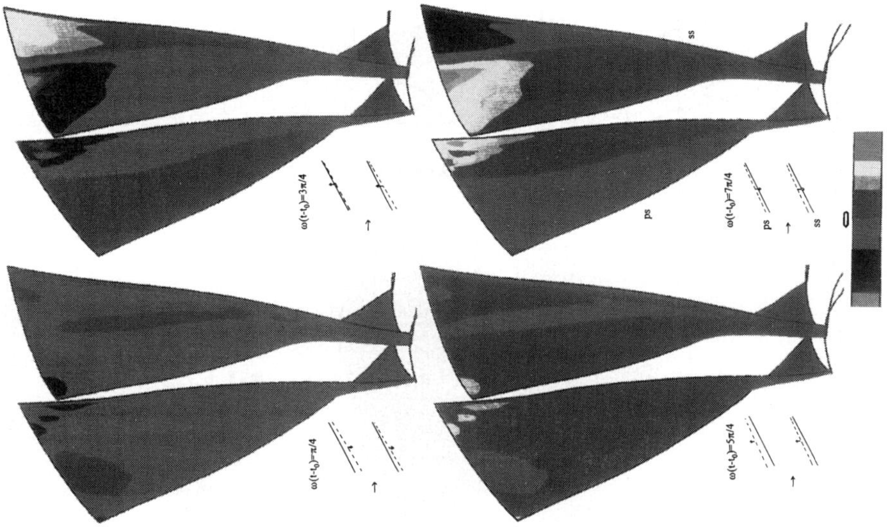

Fig. 11 Pressure fluctuations in 2 ND mode - AV3D, Case B

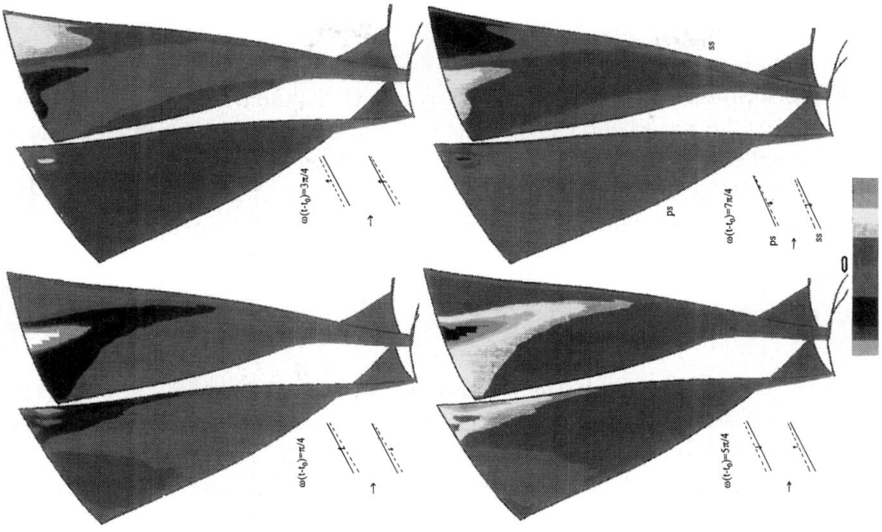

Fig. 12 Pressure fluctuations in 6 ND mode - AV3D, Case B

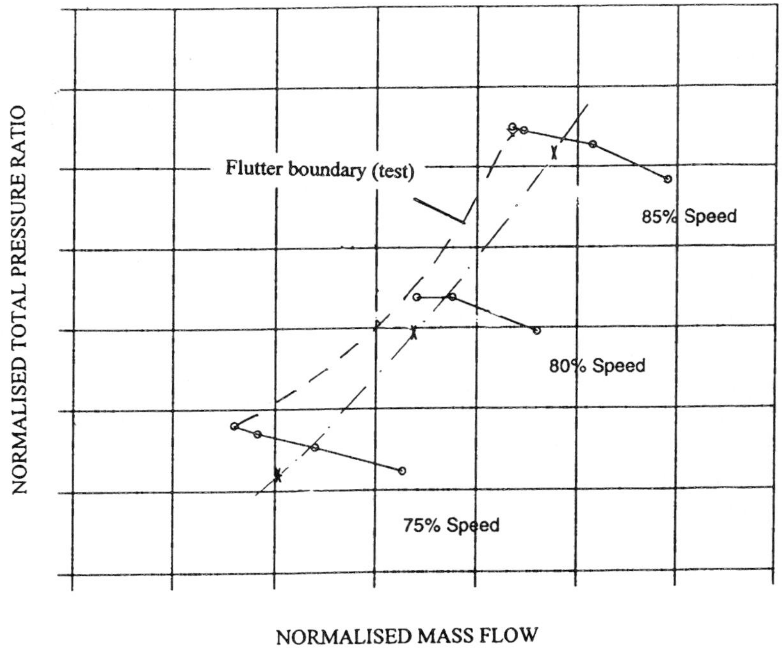

Fig. 13 Comparison of predicted (_ . _ .) and measured (– – –) flutter boundaries
x: AU3D calculation point, o:Test

7. Acknowledgements

The authors would like to thank Rolls-Royce plc. for their support and allowing the publication of this paper, and Dr. L He for supplying the aerodynamic part of the AV3D code, and colleagues in Compressor Engineering at Rolls-Royce plc. for supplying the experimental data. The authors also wish to thank the Ministry of Defence (MOD), DERA, and the Department of Trade and Industry (DTI) for their support.

8. References

Abdel-Rahim, A., Sisto, F., & Thangam, S., 1991. Computational Study of Stall Flutter in Linear Cascades, ASME Paper 91-GT-5, IGTI Gas Turbine Conference, Orlando FL, June 1991

Bendiksen, O.O., 1991. A New Approach to Computational Aeroelasticity, AIAA Paper 91-0939, 1991

Baldwin, B.S. and Barth, T. J., 1991. A One Equation Turbulence Transport Model for High Reynolds Number Wall-bounded Flows. AIAA paper 91-0610.

Carstens, V., Bolcs, A. and Korbacher, H., 1993. Comparison of Experimental and Theoretical Results for unsteady Transonic Cascade Flow at Design and Off-design Conditions. ASME paper 93-GT-100.

Cumpsty, N., 1989, Compressor aerodynamics, Longman Scientific & Technical, New York.

Giles, M.B., 1990. Non-reflecting Boundary Conditions for Euler Equation Calculations, AIAA Journal, Vol. 28, No. 12 p. 2050, November 1990

He, L., 1993a. New Two-Grid Acceleration Method for Unsteady Navier-Stokes Calculations, AIAA Journal of Propulsion and Power, Vol. 9, No. 2, p. 272, March-April 1993

He, L., 1993b. Three-Dimensional Time-Marching Inviscid and Viscous Solutions for Unsteady Flows around Vibrating Blades, ASME Paper 93-GT-92, IGTI Gas Turbine Conference, Cincinnati OH

Marshall, J. G. & Imregun, M., 1996. A Review of Aeroelasticity Methods with Emphasis on Turbomachinery Applications, Journal of Fluids and Structures, Vol. 10, pp 237-267, 1996

Marshall, J.G. & Imregun, M., 1995. A 3D Time-Domain Flutter Prediction Method for Turbomachinery Blades. Proc. Int. Symp. Aeroelasticity and Structural Dynamics, Manchester, Royal Aeronautical Soc., 42.1-42.14.

Vahdati, M. & Imregun, M., 1994. Non-Linear Aeroelastic Analyses Using Unstructured Dynamic Meshes, Proc of the 7th Int Symp. on Unsteady Aerodynamics and Aeroelasticity of Turbomachines, Fukuoka, Japan, September 1994

The Effect of Blade Vibration Mode on a Flutter in a Transonic Fan

Kousuke ISOMURA
IHI Co. Ltd. Aero-Engine & Space Operations
229 Tonogaya Mizuho-machi Tokyo 190-12, JAPAN

1.Introduction

After more than 30 years of research, flutter is still a major headache when developing jet engines. In modern transonic fans, requirements of high efficiency at high speed results in flat and thin blades which are especially vulnerable to a new type of flutter. The mechanism of this flutter was revealed by Isomura [1] and Isomura and Giles [2], that the shock instability at the point of unstarting is responsible. In their paper, numerical studies were performed assuming the blades are oscillating in pure bending mode. As an approximation to study the mechanism, this assumption is acceptable because the torsional component in the first flex mode, which is the typical oscillation mode in transonic fans, is small. But in reality, fans will seldom oscillate in pure bending mode in two dimensional section aligned to the flow direction, instead the blade oscillation mode usually contains torsional components. For designers of transonic fans, information on the effect of the torsional component on this shock-induced flutter will be important to push their designs to the limit of high efficiency. Flutter is often very costly to fix and can result in fatal accident when it occurs.

In this paper, the effect of the torsional component on the shock-induced flutter is studied by numerical simulation.

2.Numerical Scheme

The numerical code used for current study is based on quasi three dimensional this shear layer Navier-Stokes equations. The code uses upwinding alternating direction implicit (ADI) scheme formulated by Giles [3] with Roe's third order flux differencing for inviscid flux [4], and second order central differencing for viscous flux. Johnson and King's turbulence model [5] with later modification by Johnson and Coakley [6] is used for the turbulence closure. Transition points are modeled by the e^n method, but with the envelope of Orr-Sommerfeld spatial amplification curve used instead of the amplification factor, n, to simplify the calculation. The downstream motion of the transition points are limited by one-half of the boundary layer edge velocity to suppress an unphysical jump. One-dimensional non-reflecting boundary conditions formulated by Giles [7] were

T.H. Fransson (ed.), Unsteady Aerodynamics and Aeroelasticity of Turbomachines, 725-732.

used for both inflow and outflow. The code is second order time accurate. The detail of the scheme is described in ref.[1].

C-type grids are used for the calculations. The grid size for one blade-to-blade passage contains 223 nodes in ζ-direction aligned to the blade surface, and 70 nodes in η-direction approximately normal to ζ-direction. Out of 70 η nodes, approximately 30 nodes are put in the boundary layer grid, and approximately 7 nodes are put in the laminar sub-layer. Passages of C-type grids are connected to form a multi-passage calculation domain. The calculation grid is shown in Fig.1.

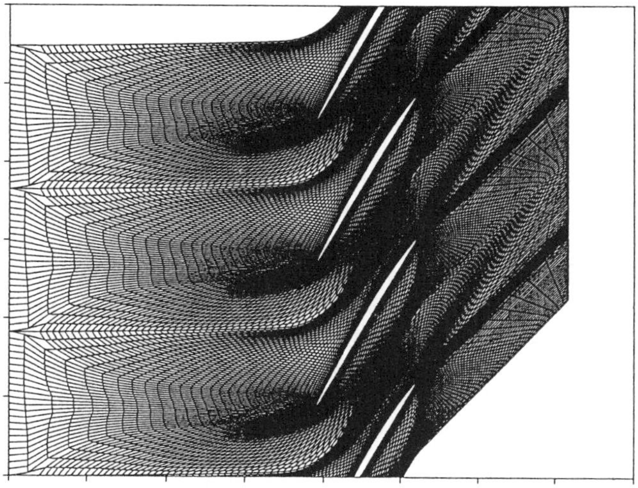

Figure 1. C-grid for the unsteady transonic fan calculations

3.Numerical Simulations

The numerical simulations were performed at a near flutter condition measured in an internal research fan rig test of Ishikawajima-Harima Heavy Industries (IHI) Co. Ltd. The fan has 22 blades of multiple circular arc (MCA) type blade section with very small positive camber angle at the front section, which is a typical of modern transonic fan. The input data for the numerical simulations were generated by through-flow calculations using the rig-measured data. The input data is tabulated in Table-1. The inter-blade phase angle for the numerical simulations were -32.7 degrees, which is a counter rotating mode at nodal diameter 4. This was the mode observed in the fan rig test, and was also an inter-blade phase angle to give a least stable aerodynamic work in the previous calculations reported in ref.[1].

86% span position was chosen for the numerical simulation with quasi three dimensional code. This span-wise position was that the relative Mach number just exceeded unity, and the shock wave is unstable. The steady Mach contour of the flow field is shown in Fig.2.

Table.1 Input parameters for calculations

Inflow Mach number	1.01		
Static pressure ratio	1.32		
Inflow angle	70.1		[degree]
Reynolds number		1.77×10^6	
Reduced frequency		0.44	

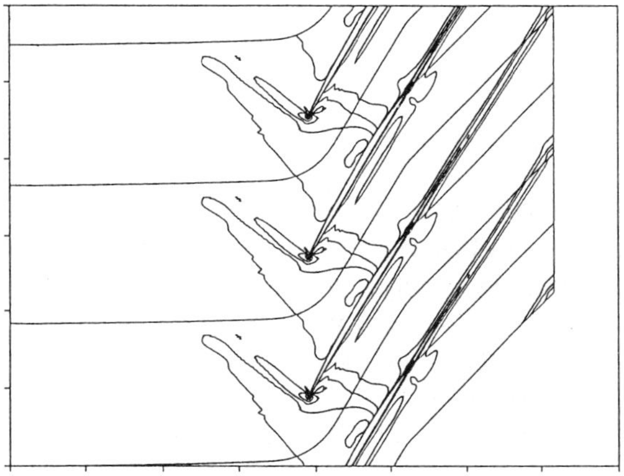

Figure 2. Mach contour of the steady flow field

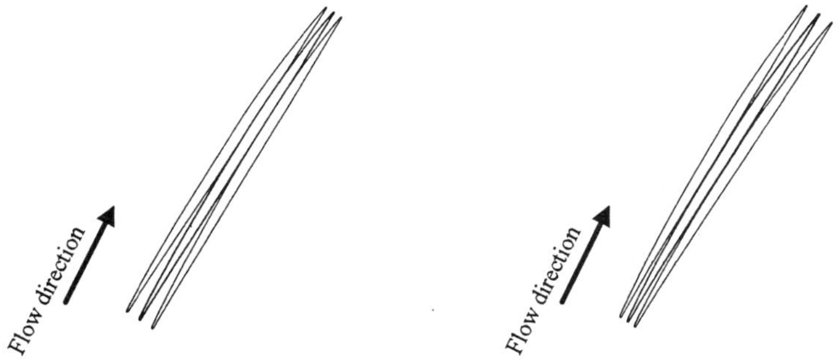

(a) with +1.0 degree torsion (b) with -1.0 degree torsion

Figure 3. Blade oscillation mode of 4.5% pitch bending with torsion coupled

Numerical simulations were performed at a fixed bending oscillation amplitude of 4.5% of the pitch (3.3% of the chord), which corresponds to the blade oscillation limit during the rig test. 5 different tortional oscillation modes were combined to this bending oscillation; $\pm$1.0 degree, $\pm$0.5 degrees, and 0 degrees. These angle correspond to $\pm$26.2%, $\pm$13.1%, 0% of the amplitude of the bending oscillation at the leading edge. The positive angle indicates the torsional oscillation in phase with the bending oscillation, which the blade "heads up" when the mid-chord moves upward. The negative angle indicates the torsional oscillation in counter phase with the bending oscillation, which the blade "heads down" when the mid-chord moves upward. The blade vibration mode of 4.5% pitch bending with ±1.0 degree of torsional components are shown in Fig.3.

4. Calculation Results

The variation of the blade section's total aerodynamic work is shown in Fig.4. The ordinate of the figure is the normalized aerodynamic work integrated over the chord. The normalized aerodynamic work over a cycle of oscillation is defined by $\frac{1}{2\pi}\oint pu_n dt = \frac{1}{2}p_a x_a \sin\theta$, where p is the blade surface pressure, u_n is the blade surface normal velocity, p_a and x_a are the amplitudes of the first harmonics of the unsteady pressure and blade oscillations respectively, and θ is the phase difference between the two. The value is normalized by the inflow density and the cube of the inflow sound velocity. Positive work corresponds to blade excitation, while negative work corresponds to blade damping. Fig.4 shows that the in-phase torsional component damps the blade and the counter-phase torsional component excites the blade, relative to the pure bending oscillation.

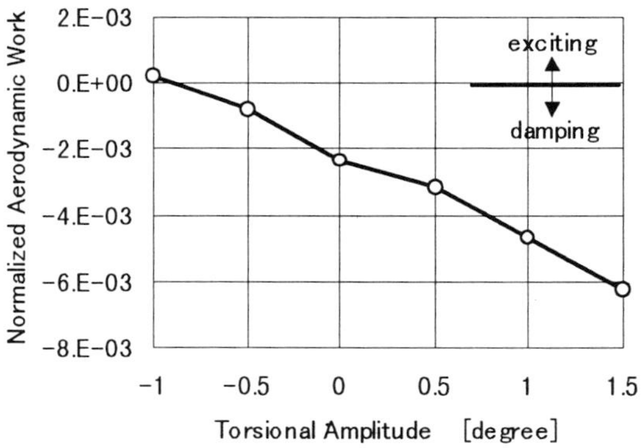

Figure 4. Variation of the blade exciting work due to the blade exciting mode.

The blade surface distributions of the normalized aerodynamic work of these five cases are shown in Fig.5. In pure bending oscillation, the shock foot on the suction surface damps the blade, while that on the pressure surface excites. In the bending with counter-phase torsional oscillation, this tendency is the same and enhanced. As the in-phase torsional component increases, this trend is reversed. The aerodynamic work at the shock foot on the pressure surface decreases, and changes to large damping work, while that on the suction surface shifts to the blade exciting work. The blade exciting work on the suction surface suggests that the flutter will transfer to classical stall flutter by further increasing the positive torsional amplitude.

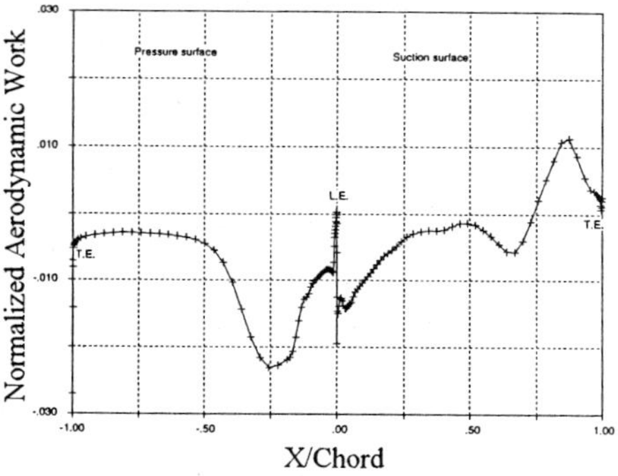

(a) with +1.0 degree torsion

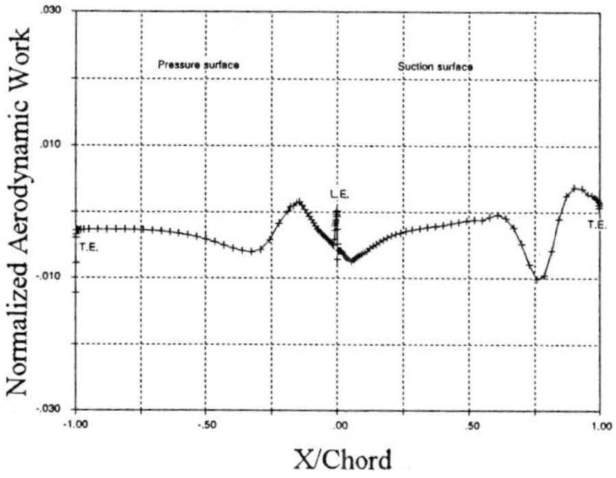

(b) with +0.5 degrees torsion

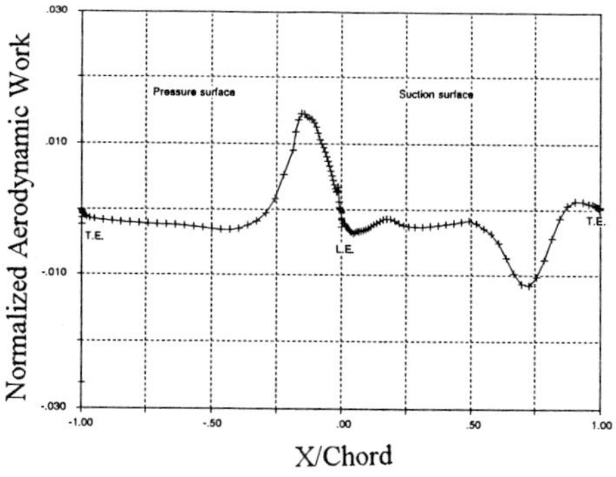

(c) pure bending

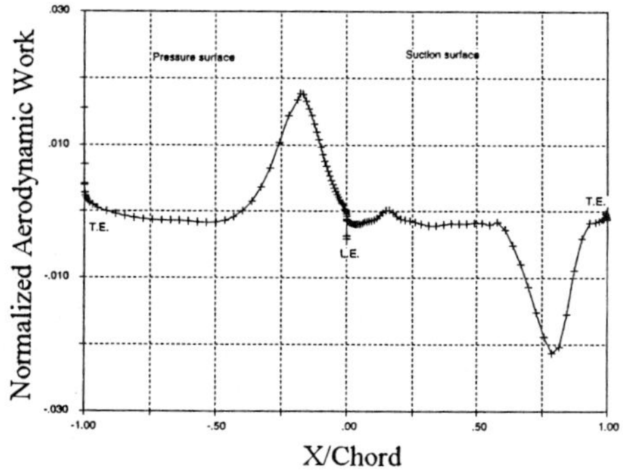

(d) with -0.5 degrees torsion

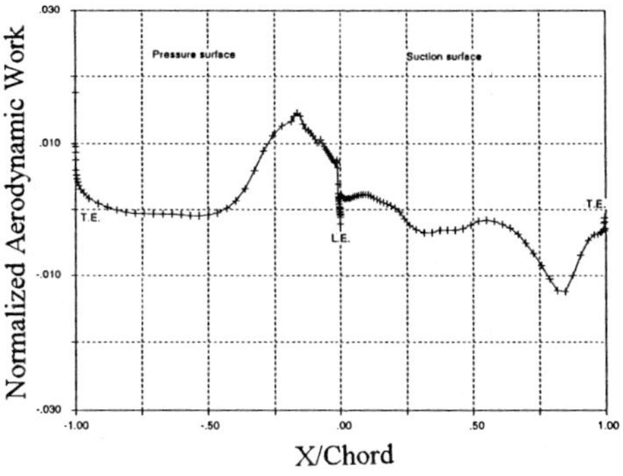

(e) with -1.0 degree torsion

Figure 5. Blade surface normalized aerodynamic work distributions.

Amplitude and phase distributions of the case of bending with ± 0.5 degrees of torsion are shown in Fig.6. The results show that the phase angles are increased about 60 degrees when the torsional component switched from counter phase to in-phase to the bending oscillations. No significant change is seen in the amplitudes distribution.

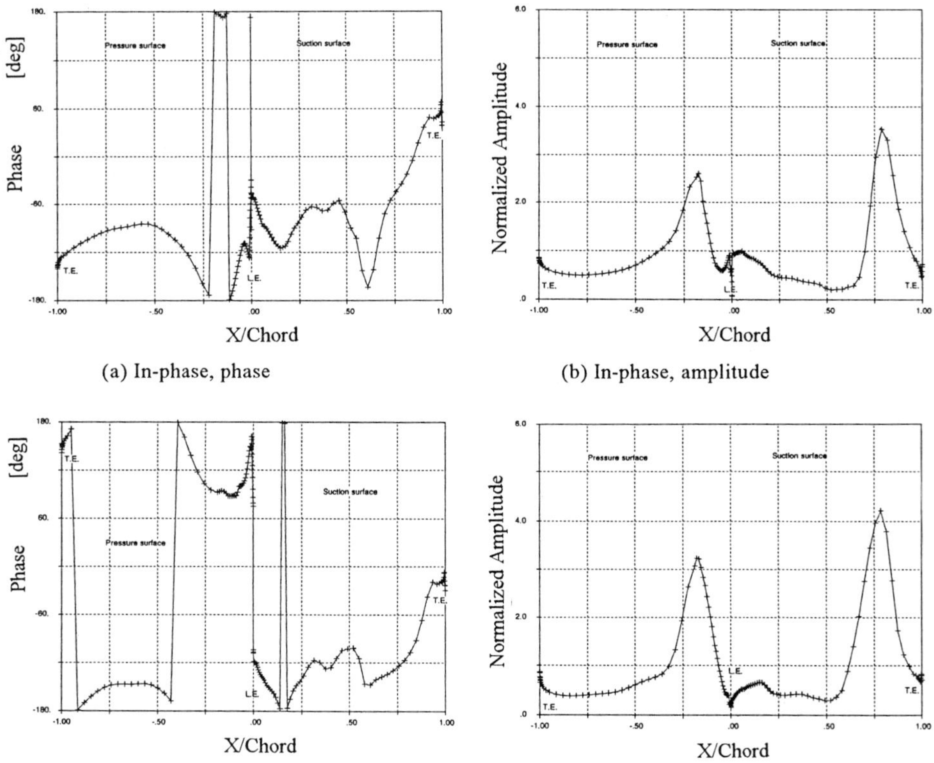

(a) In-phase, phase (b) In-phase, amplitude

(c) Counter-phase, phase (d) Counter-phase, amplitude

Figure 6. Change of phase and amplitude distribution due to the blade oscillation mode.

The blade surface distribution of the normalized aerodynamic work of the blade in pure torsional oscillation is shown in Fig.7. The figure shows that the shock feet on both the suction surface and the pressure surface damp the torsional oscillation. This implies that the shock motion induced by the torsional component excites the bending motion on the suction surface and damps it on the pressure surface when the both oscillations are in phase. Since the amplitude of the unsteady pressure induced by the torsional oscillation is larger on the pressure surface, the torsional component damps the blade when the torsional oscillation is in-phase to the bending, and excites when it is counter-phase.

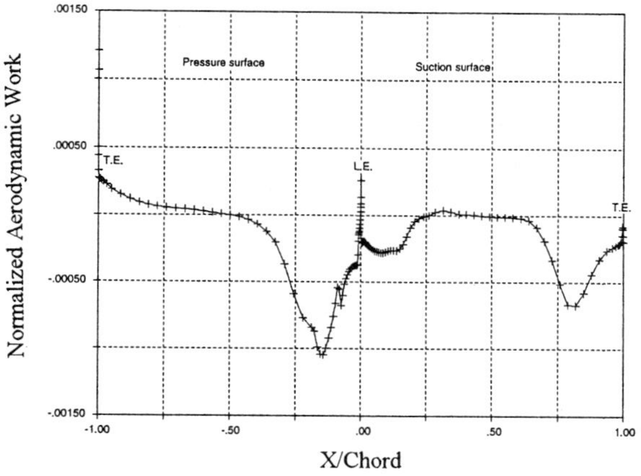

Figure 7. Surface work distribution of the blade in a pure torsional oscillation. (±0.5 degrees)

Thus the blade exciting work reduced by adding the torsional component at in-phase motion at constant inter-blade phase angle of -32.7 degrees. Next we have to check if the blade exciting work shows larger value at other inter-blade phase angles. The effect of the inter-blade phase angle at plus 0.5 degrees torsional mode is shown in Fig.8. The result shows larger blade exciting work at inter-blade phase angle 0 degrees. This suggests that the fan is more vulnerable to flutter when in-phase 0.5 degrees torsional mode is added, than pure bending mode. However, at the inter-blade phase angle 0 or -16.4 degrees, the nodal diameters are 1 and 2, respectively. Such flutter will require coupling with the shaft vibration to occur, and can easily be damped by altering the vibrational character of the shaft. Therefore adding the torsional component at in-phase to bending is still effective to push the flutter boundary up in fan characteristic maps.

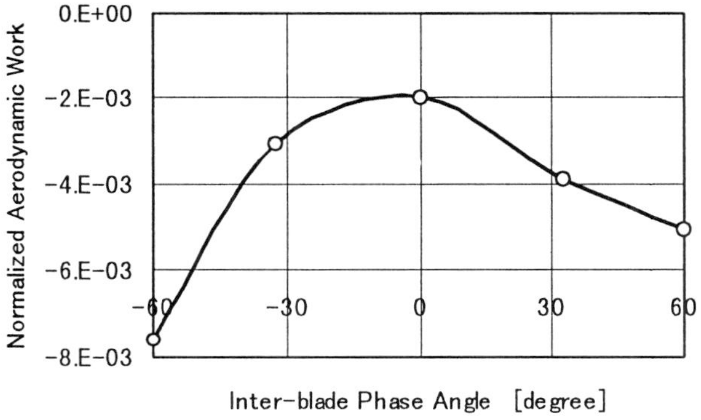

Figure 8. The effect of the inter-blade phase angle.(with +0.5 degrees torsion)

5. Conclusions

(1) The torsional component in the first flex mode has been shown to damp the blade oscillation when it is in-phase to the bending oscillation, and excites the blade when it is counter-phase to the bending oscillation.

(2) The mechanism is that, at the pressure surface shock foot, the aerodynamic work due to the torsional component of the blade motion damps the blade when it is in-phase, and excites when it is counter-phase, to the bending component.

(3) When the torsional component in-phase to the bending oscillation is added, the inter-blade phase angle which gives the largest blade exciting work shifts to 0 degrees. However, a coupling with the shaft vibration is needed for the flutter at such inter-blade phase angle to occur. Therefore adding torsional component in-phase to the bending oscillation will be effective to suppress the flutter.

(4) When the torsional amplitude increases, the region on the suction surface behind the shock foot has shown to be a blade exciting aerodynamic work, which suggests that the flutter proceeds to a classical stall flutter.

References

[1] K. Isomura, A Numerical Investigation of a Flutter in a Transonic Fan. M.I.T. GTL Report #223, October 1996.

[2] K. Isomura, M. B. Giles, A Numerical Studyof a Flutter in a Transonic Fan. ASME 97-GT-235, June 1997.

[3] M. B. Giles, Numerical Methods for Unsteady Turbomachinery Flow. M.I.T. CFDL-TR-89-3, April 1989.

[4] P. L. Roe, Characteristic-Based Schemes for the Euler Equations, vol18, pages 337-365. ARF, 1986.

[5] D. A. Johnson, L. S. King, A Mathematically Simple Turbulence Closure Model for Attached and Separated Turbulence Boundary Layers. AIAA-84-0175, January 1984.

[6] D. A. Johnson, T. J. Coakley, Improvements to a Nonequilibrium Algebraic Turbulence Model. AIAA Journal, 28(11):2000-2003, November 1990.

[7] M. B. Giles, Non-reflecting Boundary Conditions for Euler Equation Calculations. AIAA Journal, 28(12), December 1990.

ROLE OF SHOCK STRUCTURES IN TRANSONIC FAN ROTOR FLUTTER

TAKANORI SHIBATA[1] AND SHOJIRO KAJI
Department of Aeronautics and Astronautics
University of Tokyo
Bunkyo-ku, Tokyo 113, Japan

Abstract. A linearized Euler analysis for two-dimensional unsteady flows through harmonically oscillating cascades is presented. A nonlinear Euler analysis, which is based on the implicit time-marching procedure, is also developed and used to validate the shock capturing capabilities of the linearized Euler analysis. Extensive studies on the flutter problem of transonic fans indicate that chordwise motion of the blade is very important to predict the correct flutter boundary at highly loaded conditions. It is also shown that shock structures play a dominant role in determining the aeroelastic response of the cascade and the unsteady load due to shock motion is the main destabilizing source of bending oscillations.

1. Introduction

Some experimental and computational studies have shown that the unsteady aerodynamic forces due to shock motions have a significant effect on aeroelastic stabilities of a cascade. However, because results are limited, overall characteristics of transonic rotor flutter, especially the couplings between shock structures and the oscillating cascade blades, remain unknown. Although modern fans and compressors operate under several kinds of shock systems, the flutter in bending motion is observed over only a limited range of their operating conditions. This suggests that shock structures may have a pronounced influence on the occurrence of cascade flutter in this mode.

Bending mode supersonic flutters at highly loaded conditions are generally classified into two regimes on the compressor performance map [1] : (i) the portion at mid-speed and close to the surge line of a stage; and (ii) the portion at high- or over-speed and high back pressure conditions. The

[1]Presently at Hitachi Ltd., Power & Industrial Systems R & D Division

T.H. Fransson (ed.), Unsteady Aerodynamics and Aeroelasticity of Turbomachines, 733-747.

flutter in region (i) is, in particular, frequently encountered in the development of large fan and compressor rotors. Although it is presumed that the passage flow is unstarted near the flutter boundary of the region (i), the relation between the unstarted flow and the flutter mechanism has not been clarified yet.

The objective of the present paper is to determine the importance of in-passage shock structures in transonic rotor flutter problems.

2. Numerical Method

2.1. GOVERNING EQUATIONS AND NUMERICAL SCHEME

We consider inviscid, adiabatic, supersonic flow which contains strong (in the sense that the flow goes from supersonic to subsonic) nearly normal shocks in the blade passage. The Euler equations are adopted because flow is rotational due to the existence of strong shock waves. For the two-dimensional curvilinear coordinate system (ξ, η), the Euler equations in the strong conservation-law form are given by

$$\frac{\partial \widehat{\boldsymbol{U}}}{\partial \tau} + \frac{\partial \widehat{\boldsymbol{F}}}{\partial \xi} + \frac{\partial \widehat{\boldsymbol{G}}}{\partial \eta} = 0 \tag{1}$$

where

$$\widehat{\boldsymbol{U}} = \frac{\boldsymbol{U}}{J}, \quad \widehat{\boldsymbol{F}} = \frac{\xi_t \boldsymbol{U} + \xi_x \boldsymbol{F} + \xi_y \boldsymbol{G}}{J}, \quad \widehat{\boldsymbol{G}} = \frac{\eta_t \boldsymbol{U} + \eta_x \boldsymbol{F} + \eta_y \boldsymbol{G}}{J} \tag{2}$$

The vector of conservation variables $\boldsymbol{U}$ and flux vectors $\boldsymbol{F}$, $\boldsymbol{G}$ are defined by

$$\boldsymbol{U} = \begin{bmatrix} \rho \\ \rho u \\ \rho v \\ e \end{bmatrix}, \boldsymbol{F} = \begin{bmatrix} \rho u \\ \rho u^2 + p \\ \rho u v \\ (e+p)u \end{bmatrix}, \boldsymbol{G} = \begin{bmatrix} \rho v \\ \rho v u \\ \rho v^2 + p \\ (e+p)v \end{bmatrix} \tag{3}$$

where ρ is the density, p is the pressure, u, v are the axial and circumferential components of velocity, and $e = p/(\gamma - 1) + \rho(u^2 + v^2)/2$. The Jacobean of the transformation and the other metric quantities are given by

$$J = \xi_x \eta_y - \xi_y \eta_x = 1/(x_\xi y_\eta - x_\eta y_\xi) \tag{4}$$

$$\xi_x = J y_\eta, \quad \xi_y = -J x_\eta, \quad \xi_t = -x_\tau \xi_x - y_\tau \xi_y \tag{5}$$

$$\eta_x = -J y_\xi, \quad \eta_y = J x_\xi, \quad \eta_t = -x_\tau \eta_x - y_\tau \eta_y \tag{6}$$

Equation (1) can be solved on deforming grid using a time-marching, implicit algorithm. To obtain second-order temporal accuracy, time discretization is based on the three-point, backward implicit scheme.

$$\frac{1.5 \Delta \widehat{\boldsymbol{U}}^n - 0.5 \Delta \widehat{\boldsymbol{U}}^{n-1}}{\Delta t} + \frac{\partial \widehat{\boldsymbol{F}}^{n+1}}{\partial \xi} + \frac{\partial \widehat{\boldsymbol{G}}^{n+1}}{\partial \eta} = 0 \tag{7}$$

To reduce factorization and linearization errors, the equation (7) is solved at each time step using a Newton iterative method:

$$\left[\boldsymbol{I}+\frac{2\Delta t\theta}{3}\frac{\partial\widehat{\boldsymbol{A}}^{p}}{\partial\xi}+\frac{2\Delta t\theta}{3}\frac{\partial\widehat{\boldsymbol{B}}^{p}}{\partial\eta}\right]\Delta\widehat{\boldsymbol{U}}^{p}=$$

$$-\left(\widehat{\boldsymbol{U}}^{p}-\widehat{\boldsymbol{U}}^{n}\right)-\frac{2\Delta t}{3}\left[\frac{\partial\widehat{\boldsymbol{F}}^{p}}{\partial\xi}+\frac{\partial\widehat{\boldsymbol{G}}^{p}}{\partial\eta}\right]+\frac{1}{3}\Delta\widehat{\boldsymbol{U}}^{n-1}\qquad(8)$$

where $\widehat{\boldsymbol{A}}=\frac{\partial\widehat{\boldsymbol{F}}}{\partial\widehat{\boldsymbol{U}}}$, $\widehat{\boldsymbol{B}}=\frac{\partial\widehat{\boldsymbol{G}}}{\partial\widehat{\boldsymbol{U}}}$, $\Delta\widehat{\boldsymbol{U}}^{p}=\widehat{\boldsymbol{U}}^{p+1}-\widehat{\boldsymbol{U}}^{p}$, and p=1,2,... is the sub-iteration index at each time step. When the iteration converges, the second-order temporal accuracy is maintained. The left hand side of eq.(8) is factored using LU-ADI [2] which is computationally stable and efficient. The right hand side of eq.(8) is evaluated using the upwind total variation diminishing (TVD) scheme proposed by Yee et al. [3, 4]. Spatial accuracy is second-order. A minmod function is used as a limiter function to preserve the monotonicity property in the numerical solutions.

Because a nonlinear Euler analysis is too expensive to use to evaluate the effects of many parameters associated with blade vibration, a linearized Euler analysis has been developed. We restrict our consideration to the small-amplitude blade motions which have a harmonic time-dependence. Hence, the time-dependent inviscid flow can be regarded as a small perturbation of an underlying nonlinear mean or steady background flow. Thus, the nonlinear conservative variables can be expressed as follows:

$$\boldsymbol{U}(\xi,\eta,\tau)=\overline{\boldsymbol{U}}(\xi,\eta)+\widetilde{\boldsymbol{u}}(\xi,\eta)e^{i\omega\tau}\qquad(\widetilde{\boldsymbol{u}}<<\overline{\boldsymbol{U}})\qquad(9)$$

where $\overline{\boldsymbol{U}}$ is the vector of conservative variables representing the zeroth order or mean flow field, and $\widetilde{\boldsymbol{u}}$ is the vector of the small complex amplitudes of conservative variables representing the first order or unsteady flow, i.e. $\widetilde{\boldsymbol{u}}=[\widetilde{\rho},\ \widetilde{\rho u},\ \widetilde{\rho u},\ \widetilde{e}]^{T}$. ω is the angular frequency of blade vibration. In the following formulation, the superscript symbols $^{-}$ and $^{\sim}$ are used to denote time averaged and unsteady small-perturbation variables, respectively.

The motion of the grid is defined by

$$x=\overline{x}+\widetilde{x}e^{i\omega t}\qquad y=\overline{y}+\widetilde{y}e^{i\omega t}\qquad(10)$$

Substitution of eqs. (8) and (9) into eq. (1) and collection of the terms that are first order in the perturbations, results in the linearized Euler equations,

$$\frac{\partial}{\partial\tau}\overline{\left(\frac{1}{J}\right)}\widetilde{\boldsymbol{u}}+i\omega\overline{\left(\frac{1}{J}\right)}\widetilde{\boldsymbol{u}}+\frac{\partial\widetilde{\boldsymbol{f}}}{\partial\xi}+\frac{\partial\widetilde{\boldsymbol{g}}}{\partial\eta}+\boldsymbol{S}_{M}=0\qquad(11)$$

where $\widetilde{\boldsymbol{f}} = \left\{\overline{\left(\frac{\xi_x}{J}\right)}\,\overline{\frac{\partial \boldsymbol{F}}{\partial \boldsymbol{U}}} + \overline{\left(\frac{\xi_y}{J}\right)}\,\overline{\frac{\partial \boldsymbol{G}}{\partial \boldsymbol{U}}}\right\}\widetilde{\boldsymbol{u}}$, $\widetilde{\boldsymbol{g}} = \left\{\overline{\left(\frac{\eta_x}{J}\right)}\,\overline{\frac{\partial \boldsymbol{F}}{\partial \boldsymbol{U}}} + \overline{\left(\frac{\eta_y}{J}\right)}\,\overline{\frac{\partial \boldsymbol{G}}{\partial \boldsymbol{U}}}\right\}\widetilde{\boldsymbol{u}}$ and $\boldsymbol{S}_M$ is a source term that depends on the background flow and the prescribed grid motion:

$$\boldsymbol{S}_M = \widetilde{\left(\frac{\xi_t}{J}\right)}\frac{\partial \overline{\boldsymbol{U}}}{\partial \xi} + \widetilde{\left(\frac{\xi_x}{J}\right)}\frac{\partial \overline{\boldsymbol{F}}}{\partial \xi} + \widetilde{\left(\frac{\xi_y}{J}\right)}\frac{\partial \overline{\boldsymbol{G}}}{\partial \xi} + \widetilde{\left(\frac{\eta_t}{J}\right)}\frac{\partial \overline{\boldsymbol{U}}}{\partial \eta} + \widetilde{\left(\frac{\eta_x}{J}\right)}\frac{\partial \overline{\boldsymbol{F}}}{\partial \eta} + \widetilde{\left(\frac{\eta_y}{J}\right)}\frac{\partial \overline{\boldsymbol{G}}}{\partial \eta}.$$

The linearized Euler equations are discretized on a deforming grid by the same TVD scheme as the nonlinear analysis, and then solved by shock capturing method using explicit pseudo-time marching procedures. A local time stepping technique is adopted to accelerate the convergence rate of the frequency domain solution. The validity of this linearized analysis is checked against the nonlinear Euler analysis.

2.2. BOUNDARY CONDITIONS

In both nonlinear and linearized Euler analyses, the same boundary treatments are imposed (fig. 1). At the inlet and outlet computational boundaries, nonreflecting boundary conditions are applied to prevent spurious reflections of unsteady disturbances [5, 6]. Flow tangency conditions are applied at solid surfaces, that is, the relative velocity perpendicular to the blade surfaces is forced to zero. As the side boundary condition, phase-lagged, blade-to-blade periodicity is assumed.

3. Results and Discussion

3.1. 10TH STANDARD CONFIGURATION (NACA5506)

In order to validate our numerical code, we calculated unsteady aerodynamic responses for the 10th Standard Configuration [7]. The cascade consists of NACA5506 airfoils, and has a stagger angle of 45 deg and a solidity of unity. The blade oscillates normal to its chord at an amplitude, h, of one percent of the chord length. The computational grid used for the calculation consists of a 171×21 node O-type grid around the airfoil and two 41 × 25 node H-type grids as shown in fig. 2.

Steady Mach number distributions for subsonic and transonic operating conditions are shown in fig. 3. Inlet Mach number, M, and flow angle, Ω, are 0.7 and 55 deg, respectively, for the subsonic case, and 0.8 and 58 deg for the transonic case. For the subsonic operating condition, the maximum Mach number on the blade surface is 0.92, which corresponds to that of Clark and Hall [8]. For the transonic operating condition, a normal shock appears on the blade suction surface. Although the present solver captures the shock over only three grid points, shock resolution is not so good because of the coarse grid in the vicinity of the shock wave.

Next, we computed unsteady flow fields around the airfoils vibrating in these steady flows using the linearized Euler analysis. Unsteady pressure distributions (nondimensinalized by $\rho(u^2+v^2)h$) are shown in fig. 4 for the subsonic cascade vibrating in a bending mode at a reduced frequency ω (based on the upstream velocity and blade chord) of 1.29 and an interblade phase angle, σ, of -90 deg. The results of the present method agree well with those of Clark and Hall [8]. This shows that when the flow is totally subsonic, the present method can correctly predict the unsteady flow field around the oscillating airfoils.

For the linearized Euler analysis with shock capturing, it is important to have the capability to predict the correct moving shock effect. Hall et al. [9] have shown that the shock impulse can be properly captured using the linearized Euler equations, provided that the discretization scheme is stable, consistent, and conservative. Lindquist and Giles [10, 11], on the other hand, have proved that the artificial viscosity added to smear the shock effect does not affect the global lift perturbation to leading order, and that a relatively large amount of viscosity can be added to capture the shock effect.

In the limit as the width of the shock, or the effect of numerical viscosity, goes to zero, the shape of the shock impulse approaches a delta function. Such a delta function should be smeared into an impulse with finite width, because it is impossible to capture a delta function accurately with a few grid points. In most of our numerical experiments, shocks were moderately smeared by the inherent properties of the numerical scheme, and no special viscosity was needed to capture the shock impulses accurately. If grid points, however, are highly clustered in the near vicinity of shocks, smoothing

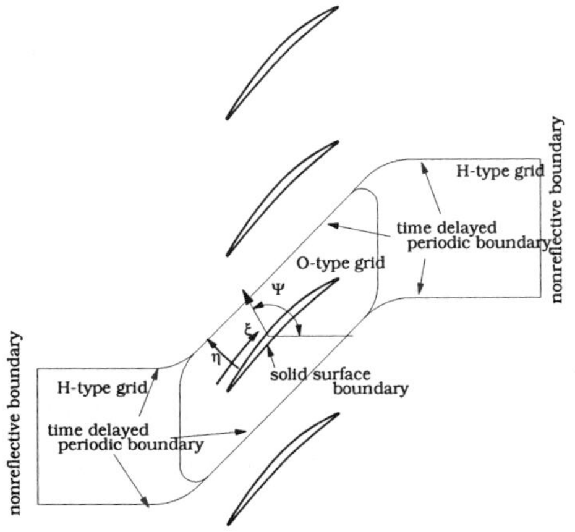

Figure 1. Boundary conditions.

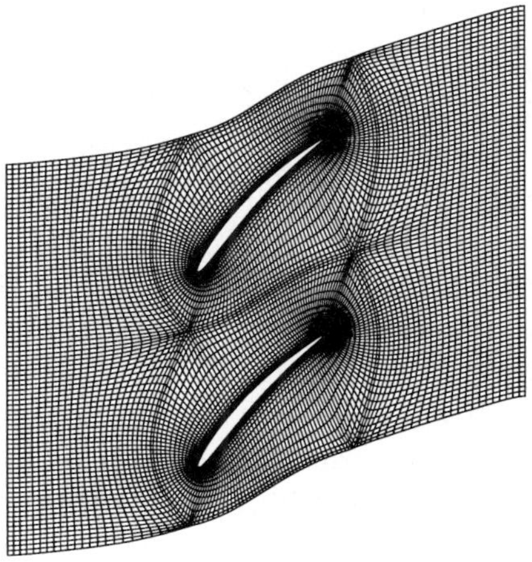

Figure 2. Computational Grid.

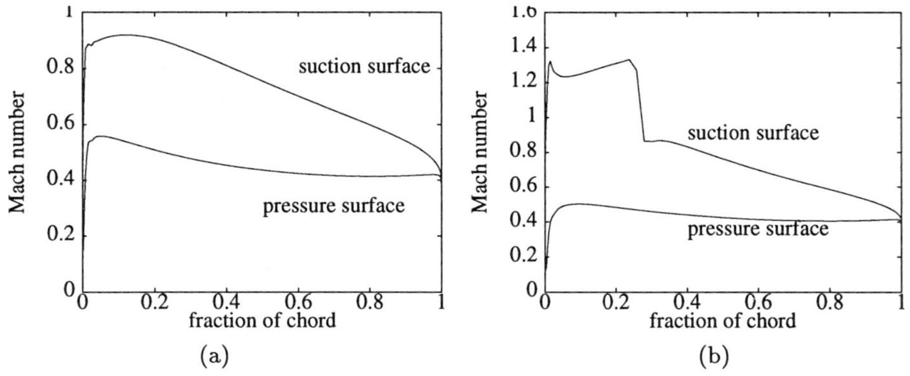

Figure 3. Steady Mach number distributions for 10th standard configuration. (a): subsonic (M = 0.7, Ω = 55 deg), (b): transonic (M = 0.8, Ω = 58 deg).

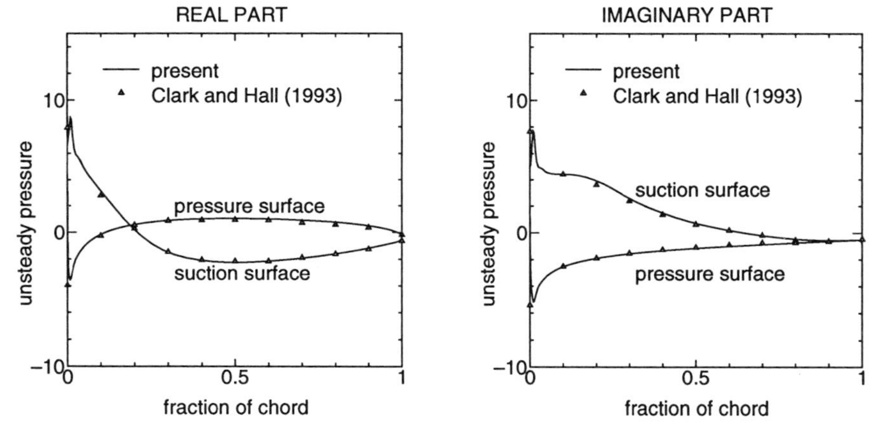

Figure 4. Unsteady pressure distributions for 10th Standard Cascade vibrating in bending mode. M = 0.7, Ω = 55 deg, ω = 1.29, σ = -90 deg).

should be explicitly added to smear the shock impulse properly. It should be noted that the shock impulse captured by a nonlinear Euler analysis will also be smeared, not because of the viscosity, but because of the unsteady shock displacement, which is the actual phenomenon of unsteady flows.

To demonstrate the shock capturing capabilities of our linearized Euler code, we compared the results of the linearized Euler analysis with those of time accurate nonlinear Euler analysis for the transonic operating conditions.

Unsteady aerodynamic work distributions (nondimensionalized by $\rho(u^2+v^2)h^2$) for the transonic flow are shown in fig. 5. For the linearized Euler analysis with shock capturing, the shape of the shock impulse depends

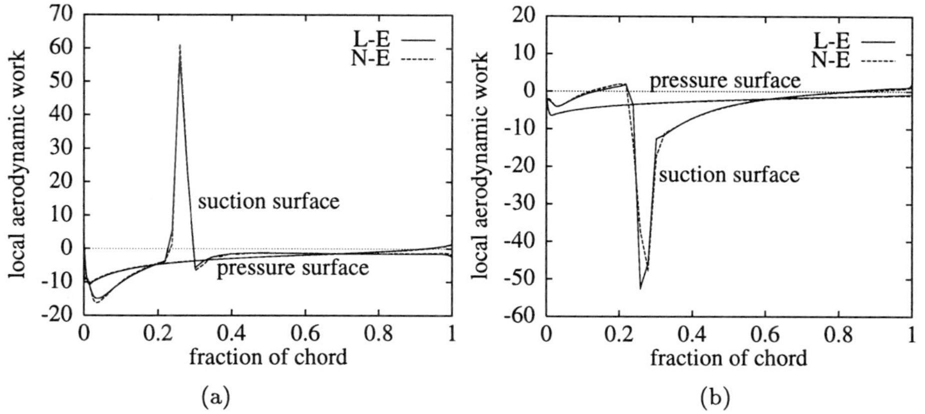

Figure 5. Unsteady aerodynamic work distributions for 10th Standard Cascade vibrating in bending mode. M = 0.8, Ω = 58 deg, ω = 1.00. (a): σ = 90 deg, (b): σ = -90 deg.

on the amount of the numerical viscosity, but the area under the impulse is independent of it. In other words, the numerical viscosity, which may influence *local* aerodynamic work distributions around shock waves, does not influence the *global* aerodynamic work acting on the oscillating blade. Although there are some differences in the local work distribution in the vicinity of the shock, global unsteady aerodynamic works of the linearized Euler analysis agree well with those of the nonlinear Euler analysis (L-E:-4.57 , N-E:-4.61 for σ = 90 deg; L-E:-7.04, N-E:-7.18 for σ = -90 deg). Hence, the unsteady shock impulse is correctly captured by this linearized Euler analysis.

The same numerical experiment was performed for the other cascade geometries and shock structures. As for the essentially nonlinear flow, which includes a large shock movement or a disappearing process of shock waves, the linearized code does not converge well and shows poor agreement with the nonlinear one. However, for most transonic flows containing shocks, the linearized Euler analysis is found to be in good agreement with the nonlinear one. That was true for unstarted flows mentioned below. Hence, we discuss only results of the linearized Euler analysis hereafter.

3.2. TRANSONIC ROTOR TIP SECTION

3.2.1. *Steady and unsteady characteristics of a high speed cascade*

Typical shock structures for a high speed transonic rotor tip section are depicted in fig. 6. The in-passage shock structure is strongly affected by the back pressure and the rotor speed. As shown in fig. 6(a), in-passage shock waves emanate from the trailing edge at the lower pressure ratio. When the back pressure is raised from 1.70 to 1.90 at constant rotor speed, the trailing edge shock goes upstream, and then the shock impinges on both sides of the blade (fig. 6(b)). The important difference in these shock patterns is whether the strong shock impinges on the suction surface, or originates at the trailing edge. We note that though at the condition of fig.6(a) the back pressure is relatively low and the in-passage shock exists relatively downstream, this is a highly loaded condition because the rotational speed is higher than that of the design operating condition.

If the back pressure is raised enough with this inlet Mach number to make the passage flow unstarted, the detached bow shock goes upstream unceasingly, and the cascade stalls due to the strong detached bow shock. When the inlet Mach number is too high, the detached bow shock becomes too strong, and no stable unstarted flow can exist. In the case of this cascade geometry, unstarted flow only exists up to the inlet Mach number of approximately 1.15. When the wheel speed is lowered and the back pressure is controlled, the passage flow can be unstarted. As shown in fig. 6(c), at the unstarted condition, the pressure surface of the airfoil is totally subsonic, and the detached bow shock is located upstream of each airfoil.

Several researchers have already shown the overall unsteady characteristics of flat plates or lightly loaded cascades, but those of the high speed cascade which has a real blade geometry has not been clarified yet. Kaji [12] showed the importance of the direction of bending oscillation on the flutter boundary by the semi-actuator disk method. Bendiksen [13] also pointed out that chordwise motion of the blades should be included when modeling the aeroelastic effects of oblique shock wave motion. Since his results, however, were limited to flows with only weak oblique shocks, we investigated the effect of chordwise motion using the flow field with strong shocks in blade passages.

Chordwise motion is given by the oscillation angle, Ψ, from the axial direction of the cascade as shown in fig. 1. Since airfoils are usually twisted and the node of a bending oscillation is located around the blade hub, the direction of oscillation is generally not normal to the chord line of the blade tip section. A real rotor blade which has no part span shroud, in general, oscillates in bending at a smaller angle than that of pure bending.

Figure 7(a) shows the effect of oscillation angle on global aerodynamic work. Chordwise motion of the blade has a pronounced influence on the aeroelastic stabilities of the cascade. The peak of aerodynamic work per cycle increases and the unstable range of interblade phase angle spreads out gradually, as the angle of oscillation decreases from that of pure bending i.e. $\Psi = 143.42$ deg.

Figure 7(b) shows the local aerodynamic work distributions along blade chord at an interblade phase angle of 90 deg, which corresponds to the

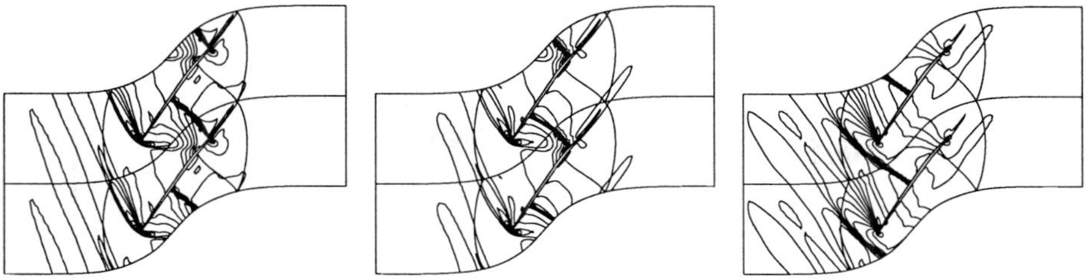

(a) (b) (c)

Figure 6. Mach number contours. (a) M = 1.25 PR = 1.70 (b) M = 1.25 PR = 1.90 (c) M = 1.09, PR = 1.61, ΔM = 0.05.

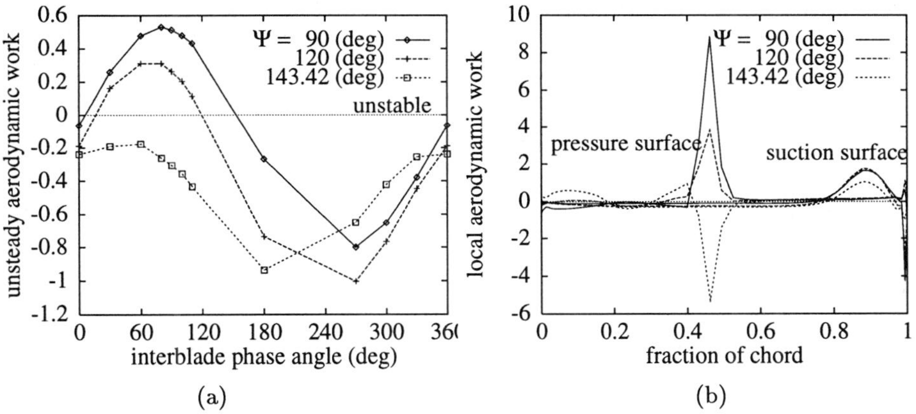

Figure 7. Effect of oscillation angle on aerodynamic work, FAN B tip, M = 1.25, PR = 1.70, $\omega = 0.084$. (a) Work per cycle vs. interblade phase angle, (b) local work distributions at σ = 90 deg.

most unstable interblade phase angle. It is clearly shown that the local aerodynamic works in the vicinity of the normal shock are strongly affected by the direction of the blade translational motion. The shock load acting on the pressure surface changes from stabilizing to destabilizing, as the angle of oscillation is decreased. The circumferential direction of oscillation, that is, Ψ = 90 deg, is the most dangerous because the shock has a large destabilizing effect in the forward-traveling-wave oscillation. These results show that chordwise motion of the blade must be included to predict the correct flutter boundary for a transonic rotor with strong shocks, since the relative motion of shocks to the blade is fairly dependent on the angles of the blade oscillation.

Chordwise motion of the blade is also important for the other shock geometries. Then, as the angle of oscillation is decreased from that of pure bending to the circumferential direction, the cascade is more susceptible to bending flutters.

3.2.2. *Importance of shock structures in flutter problems*

The difference among the shock structures of fig. 6 may have a dominant influence on the aeroelastic stability of the cascade. Figure 8 shows that the increase in back pressure from 1.70 to 1.90, that is, the change in shock structure from fig. 6(a) to fig. 6(b), has a stabilizing effect on the plunging oscillations. The reason for this drastic change of aeroelastic stability is easily understood by checking the local work distributions along the blade surfaces. In these cases, unsteady aerodynamic loads on the airfoil are dominated by the shock impulses as can be seen from fig. 7(b) and fig. 9. At the lower back pressure, since one end of the shock wave is fixed at the trailing edge, no large shock impulse may appear on the suction surface, and the shock load acting on the pressure surface is sufficiently destabilizing to cause a bending flutter in the forward traveling wave. At the higher back pressure, on the other hand, the shock load on the suction surface has a stabilizing effect large enough to cancel out the destabilizing shock load on the pressure surface(fig. 9). Consequently, no bending flutter occurs in this higher back pressure case. These results show that the shock structures are very important in determining the aeroelastic response of oscillating cascades.

Mach number effect on the flutter boundary was evaluated for the flow field which contains a strong shock at the blade trailing edge. If the inlet Mach number is only varied with a constant pressure ratio, the shock pattern will also change, and thus the aeroelastic response of the cascade will become completely different. Hence, both the back pressure and the inlet Mach number are varied so as not to change the shock structure. Figure 10

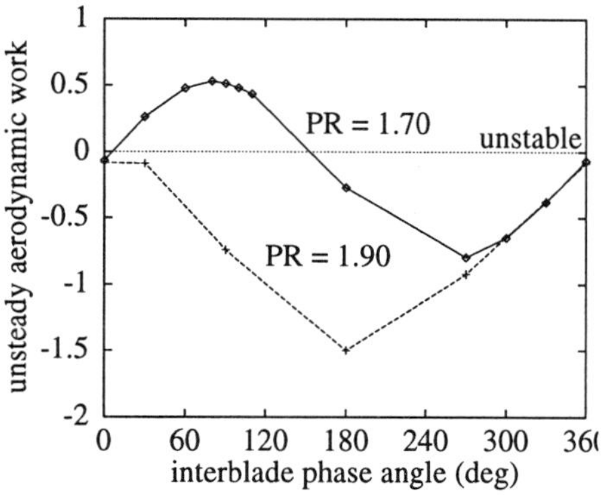

Figure 8. Work per cycle vs. interblade phase angle, M =1.25, ω = 0.1.

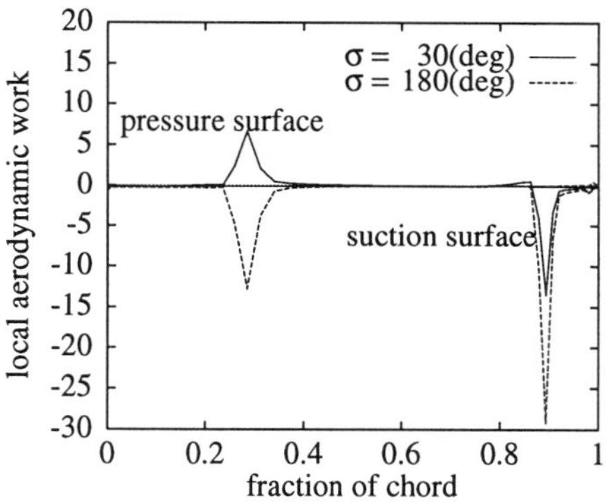

Figure 9. Local work distributions, M = 1.25, PR = 1.90, ω = 0.1.

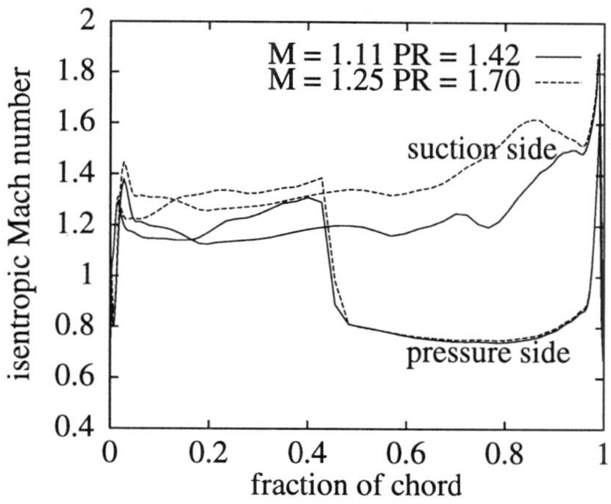

Figure 10. Variations of steady Mach number distributions with the inlet Mach number.

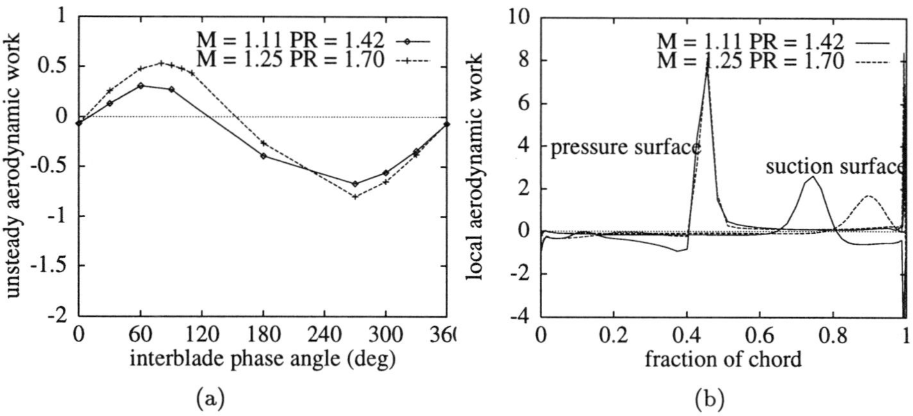

Figure 11. Effect of the inlet Mach number on aerodynamic work, $\omega \approx 0.1$. (a) Work per cycle vs. interblade phase angle, (b) local work distributions at $\sigma = 60$ deg.

shows the variation of steady Mach number distributions. Shock strength increases with increasing inlet Mach number, but the shock impinging point is not altered by the change of inlet Mach number. The oblique shock from the adjacent blade is reflected at 70 % and 90 % of the chord, where a slight decrease in Mach number can be seen. When the inlet Mach number is increased, the normal shock strength increases and the oblique shock reflection point moves aft.

Variations of aerodynamic work per cycle with the inlet Mach number are shown in fig. 11(a). This figure indicates that increasing the inlet Mach number is destabilizing on the blade oscillation. As shown in fig. 11(b), the unsteady shock load on the pressure surface is not varied so much, whereas the other part of the blade surface is influenced by the inlet Mach

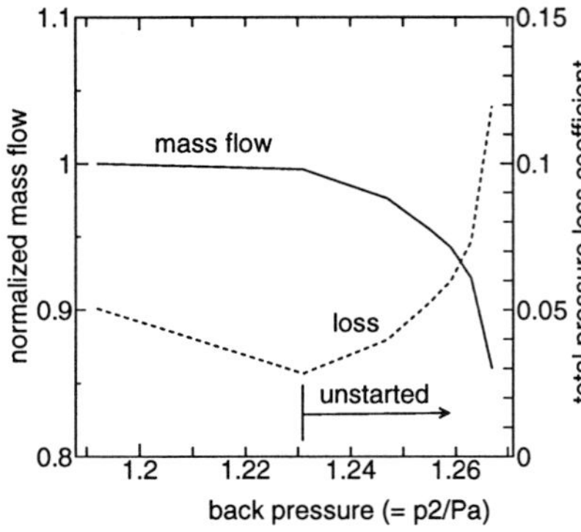

(a)

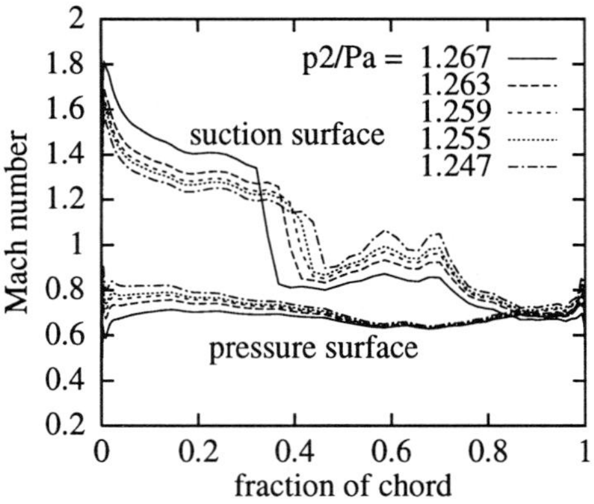

(b)

Figure 12. Effect of the back pressure on steady state parameters at unstarted conditions. (a) loss coefficient and normalized mass flow rate, (b) Mach number distributions. (M $\approx$ 1.05 $\sim$ 1.12, Pa : upstream total pressure in non-rotating frame)

number. When the inlet Mach number is increased, aerodynamic damping away from the in-passage shock decreases, contributing to the reduction in aeroelastic stabilities of the cascade. Hence, this type of the flutter which is induced by the trailing-edge shock may be observed at high inlet Mach number and high pressure ratio, i.e. the over-speed highly loaded condition. If the back pressure is high enough to make in-passage strong shocks move upstream and detach from the trailing edges, the cascade instability in bending oscillation will rapidly disappear.

3.2.3. *Bending flutter in unstarted flow*

Though the unstarted flow has the possibility of playing an important role in flutter problems, even its steady state characteristics are not understood well, because of the complexity and difficulty of modeling the detached bow shock system. The bending flutter at mid-speed and near-surge line operating conditions is frequently encountered in turbomachinery rotors which have an S-shape, i.e. negative camber followed by some positive camber, blade tip sections. This blade geometry may have a considerable influence on the detached bow shock structure at unstarted condition, and hence the pre-compression type blade geometry was used here.

In fig. 12(a), variations of two typical steady state parameters, a loss coefficient, $\overline{\omega}$, and a mass flow rate, are shown. The loss coefficient is given by $\overline{\omega} = (Pt1 - \overline{Pt2})/(Pt1 - p1)$ (Pt, p : total and static pressure; $\bar{\ }$: mass averaged property; 1, 2: upstream and downstream) and the mass flow rate is normalized by the mass flow rate at the choke or the unique incidence condition. When the back pressure is increased at constant wheel speed, the loss coefficient increases, and the mass flow rate decreases drastically. As

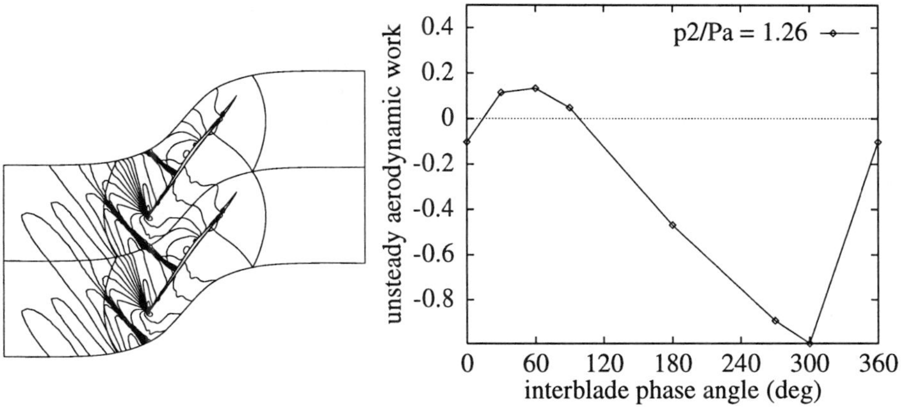

(a) (b)

Figure 13. Characteristics of unstarted flow for a pre-compression blade. (a) Steady state Mach contours, M = 1.07, p2/Pa = 1.26, (b) work per cycle vs. interblade phase angle, bending:$\omega = 0.1$, $\Psi = 90$ deg.

the back pressure is increased, the mass flow rate, or the axial flow velocity, decreases while keeping its circumferential velocity constant, and consequently the incident angle of the flow increases with respect to the back pressure. Figure 12(b) shows the steady Mach number distributions along the blade surfaces. As can be seen from this figure, detached bow shock strength and its stand-off distance increase with increasing back pressure. Because of this increase in shock strength, large shock loss is generated by the detached bow shock at the high back pressure condition. Though total pressure loss due to viscous effects should be included in the loss coefficient of actual turbomachinery, an essential part of the total pressure loss of transonic cascades is due to shock waves arising in the entrance region and the passage of the blades [14]. Furthermore, shock loss, rather than viscous loss, is more dependent on the inlet Mach number or the pressure ratio. This change of shock strength with the back pressure may contribute to an abrupt increase in the loss coefficient experimentally observed near the flutter boundary in rotating rig tests [15].

As shown in fig. 13(b), the least stable interblade phase angle is approximately 60 deg for a typical unstarted configuration (fig. 13(a)). This interblade phase angle is in agreement with experimental observations cited in Adamcyck et al. [16], though the present value for reduced frequency is lower than theirs. Figure 14(a) illustrates that the shock-induced aerodynamic loading on the suction surface is the largest destabilizing source for the unstarted flutter. The effect of back pressure on the global aerodynamic work per cycle is shown in fig. 14(b). It is found that the aerodynamic work does not increase monotonically with respect to the back pressure, in other

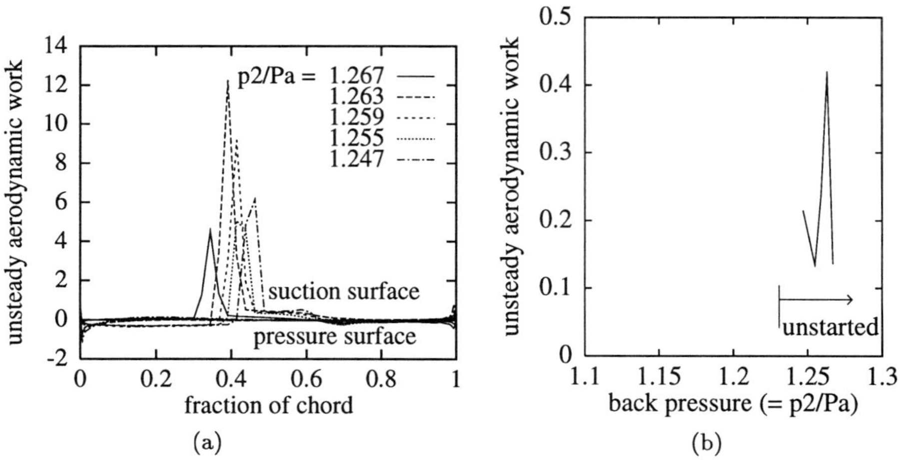

Figure 14. Effect of back pressure on the aerodynamic work at unstarted conditions. Bending: $\omega = 0.1$, $\sigma = 60$ deg, $\Psi = 90$ deg. (a) Local aerodynamic work distributions, (b) global work per cycle vs. back pressure.

words, the cascade reaches peak instabilities at a certain back pressure. These results also show that the flutter boundary for this unstarted background flow is very sensitive to the level of back pressure, and this sensitivity may contribute to the steep flutter boundary observed experimentally.

4. Conclusions

Nonlinear and linearized Euler analyses were presented and verified against two-dimensional, subsonic and transonic unsteady flows through 10th Standard Configuration. The linearized Euler analysis was found to give results in good agreement with the nonlinear Euler analysis results even for the transonic conditions.

Extensive studies on flutter problems of transonic fans indicated that chordwise motion of the blade was very important to predict the correct flutter boundary, especially when strong shocks existed in blade passages. Shock structures were found to play a dominant role in determining the aeroelastic response of the cascade. Two distinct types of shock structures were responsible for the occurrence of bending flutter; one was the strong shock emanating from the trailing edge, and the other was the detached bow shock at the entrance of the blade passage. At these shock structures, bending flutter in the forward traveling wave became unstable, and the unsteady load due to the shock motion was found to be the main destabilizing source of blade oscillations.

References

1. Sisto, F., "Introduction and Overview," AGARD Manual on Aeroelasity in Axial-Flow Turbomachines, vol. 1, AGARD-AG-298, March 1987, Chap. 1.
2. Obayasi, S., Matsushima, K., Fujii, K., and Kuwahara, K., "Improvements in Efficiency and Reliability for Navier-Stokes Computations using the LU-ADI Factorization Algorithm, " AIAA Paper 86-338, Jan. 1986.
3. Yee, H. C., Warming, R. F. and Harten, A., "Implicit Total Variation Diminishing (TVD) Schemes for Steady-State Calculations," Journal of Comput. Phys., vol. 57, 1985, pp.327-360.
4. Yee, H. C., Klopeer, G. H. and Montagne J. L. , "High-resolution Shock Capturing Schemes for Inviscid and Viscous Hypersonic Flows," Journal of Comput. Phys., vol. 88, 1990, pp.31-61.
5. Hall K. C. and Crawley, E. F., "Calculation of Unsteady Flows in Turbomachinery Using the Linearized Euler equations," AIAA Journal, Vol. 27, No. 6, 1989, pp.777-787.
6. Verdon, J. M. and Caspar, J. R., "Subsonic Flow Past an Oscillating Cascade with Finite Mean Flow Deflection," AIAA Journal, Vol. 18, 1980, pp.540-548.
7. Fransson, T. H., and Verdon, J. M., "Panel Discussion on Standard Configurations for the Determination of Unsteady Flow Through Vibration Axial-Flow Turbomachine Cascades," Proc. of the Sixth International Symposium: Unsteady Aerodynamics of Turbomachines and Propellers, Sept. 1991, pp.859-889.
8. Clark, W. S., and Hall, K. C., "Prediction of Unsteady Flows in Turbomachinery Using the Linearized Euler Equations on Deforming Grids," NASA-CR-192919, April 1993.
9. Hall, K. C., Clark W. S., and Lorence C. B., "A Linearized Euler Analysis of Unsteady Transonic Flows in Turbomachinery," Trans. of the ASME: Journal of Turbomachinery, Vol. 116, July 1994, pp.477-488.
10. Lindquist, D. R., and Giles, M. B., "On the Validity of Linearized Unsteady Euler Equations with Shock Capturing," AIAA Paper 91-1598, June 1991.
11. Lindquist, D. R, and Giles, M. B., "Validity of Linearized Unsteady Euler Equations With Shock Capturing," AIAA Journal, Vol. 32, No. 1, 1994, pp.46-53.
12. Kaji, S., "Stall Flutter of Linear Cascade in Compressible Flow," IMechE, C282/80, 1980.
13. Bendiksen, O. O., "Role of Shocks in Transonic/Supersonic Compressor Rotor Flutter," AIAA Journal, 24-7 (1986), p1179
14. Schreiber, H. A. and Starken, H., "Experimental Cascade Analysis of a Transonic Compressor Rotor Blade Section," Trans. of the ASME: Journal of Engineering for Gas Turbines and Power, vol. 106, April 1984, pp.288-294.
15. Isomura, K., "A Numerical Investigation of a Flutter in a Transonic Fan," Ph. D. Dissertation, Dept. of Aeronautics and Astronautics, Massachusetts Inst. of Technology, Cambridge, MA, June 1996.
16. Adamczyk, J. J., Stevans, W., and Jutras, R., "Supersonic Stall Flutter of High-Speed Fans," Trans. of the ASME: Journal of Engineering for Power, vol. 104, July 1982, pp .675-682.

COMPARISON OF NUMERICAL SCHEMES TO INVESTIGATE BLADE FLUTTER

L. SMATI, S. AUBERT, P. FERRAND AND F. MASSÃO
Ecole Centrale de Lyon / Université Claude Bernard,
Lab. Mécanique des Fluides et Acoustique, UMR CNRS 5509,
ECL, BP 163, 69131 Ecully Cedex, FRANCE

1. Introduction

For flutter investigations on oscillating transonic turbomachine blade rows, the knowledge of unsteady air loads is fundamental. For this reason flow solvers have been developed to predict unsteady pressure, lift and moment coefficients. Thus linearized aerodynamic methods, nonlinear Euler and Reynolds averaged Navier-Stokes solvers have been available in Computational Fluid Dynamics for application to turbomachinary for some years. The early work in this field established the feasibility of providing reasonable predictions. Nowadays the aims are more ambitious and include the detailed analysis of the flow. The need thus arises for numerical schemes of higher precision in both space and time. Despite their high accuracy, such schemes still generate non-physical numerical disturbances in the predicted flowfield.

This work analyses the influence of various numerical treatments on flutter prediction. Using an isolated airfoil configuration as test case, Euler results are compared to viscous simulations to estimate how phenomena such as flow separation and shock-boundary layer interaction affect unsteady pressure and damping coefficients. From a numerical point of view the influence of the convective fluxes discretization, which fixes the level of artificial dissipation, is evaluated. Eventually, a compressor cascade configuration is simulated and results discussed.

2. Numerical method

The analysis presented is conducted using PROUST, a CFD tool developed by the authors [1]. It has previously been shown to successfully simu-

T.H. Fransson (ed.), Unsteady Aerodynamics and Aeroelasticity of Turbomachines, 749-763.

late steady and unsteady, inviscid and viscous flows. The equations solved are the fully 3D unsteady, compressible, Reynolds-averaged, Navier-Stokes equations cast in the absolute frame where the laminar viscosity is assumed constant or calculated by the Sutherland's law.

The space discretization is based on a MUSCL finite volume formulation with moving structured meshes, which utilizes vertex variable storage. The convective fluxes are evaluated using an upwind scheme. Three different schemes were implemented: Van Leer's Flux Vector Splitting with the Hanel correction [2], Roe's Approximate Riemann Solver [3] and Liou's Advection Upwind Splitting Method [4]. These are subsequently referenced as *Hanel scheme*, *Roe scheme* and *Liou scheme*. An hybrid method combining the advantages of the central scheme in subsonic regions with the properties of the upwind scheme through discontinuities has been introduced to reduce the numerical losses in very low Mach number regions [5]. Schemes based on this approach will be referenced as *mixed schemes*, whereas the published formulations of the upwind schemes will be referenced as *pure schemes*. The viscous terms are computed by a second order centered scheme. Turbulence effects are described either by basic algebraic eddy viscosity models (Michel [6] or Baldwin-Lomax [7]) or by two-equations models (k-ε [8] or k-ω [9]). Compatibility relations are used to take into account physical boundary conditions [10]. Ghost cells for which the equations are not solved are built around the domain to simulate geometrical boundary conditions, like periodicity and symmetry.

The resulting semi-discrete scheme is integrated in time using two procedures. The first one is an explicit one or five steps Runge-Kutta time marching algorithm [11]. The second one is fully implicit and it is based on a first order Taylor expansion of the forward Euler scheme. The key point of this approach is to use the numerical explicit part to accurately compute the Jacobian matrix composing the implicit terms [12]. BI-CGSTAB algorithm with an implicit preconditionning is then used to iteratively solve the achevied linear system [13].

At the trailing edge node, pressure side and suction side subsonic flow fields should be identical. This condition is not easily verified by the numerical procedure, and it is quite difficult to ensure the flow field continuity. To solve this problem, two solutions are computed at the trailing edge point, one on the pressure side, the other on the suction side. These two sets of conservative variables are then averaged to define a unique solution. This is the simplest treatment leading to a stable result. Nevertheless, it generates localized errors for non-conservative variables (total pressure, Mach number, ...).

The presented results were achieved using various convective schemes, pure or mixed, with Mulder limiter [14] to treat discontinuities. Turbulence effects were described by $k-\omega$ model with constant turbulent Prandtl number set to 0.9. The laminar viscosity was assumed constant. Convergence toward steady state was accelerated using local time stepping and residual smoothing [15]. The unsteady integration in time used the backward Euler scheme with CFL number around 0.5 and global time stepping with no residual smoothing.

3. Isolated 2D airfoil configuration

The first computed configuration is based on NASA Ames Research Center studies on oscillating profiles. It is referenced as case CT6 (Reference Index 55) and it was extensively described by Davis [16].

3.1. GEOMETRY

The geometry consists of a NACA64A010 isolated profile. The airfoil leading edge (LE) is round, whereas the trailing edge (TE) is sharp. The profile is slightly non symmetric. The mesh used is a structured 2D C-grid extending 4 chords away from the airfoil. It is refined close to the wall, around the leading and trailing edges and in the shock region. It comprises 299×35 nodes for inviscid simulations and 299×50 nodes for turbulent simulations. Both node distributions along the profile are identical, with 120 nodes per side. There are around 25 nodes inside the boundary layer.

3.2. FLOW CONDITIONS

The fluid is modeled as a perfect gas. The specific heat ratio is $\gamma = 1.4$, and the perfect gas constant equals $r = 287\ J.kg^{-1}.K^{-1}$. The laminar dynamic viscosity and the thermal conductibility are assumed constant, $\mu = 1.81\ 10^{-5}\ kg.m^{-1}.s^{-1}$ and $\kappa = 2.54\ 10^{-2}\ m.kg.K^{-1}.s^{-3}$.

The steady far field conditions are such that the total pressure $P_{t\infty}$ equals 203 321 Pa, the total temperature $T_{t\infty}$ is 339.7 K, the static $P_{s\infty}$ is 133 913 Pa and the incidence angle α_∞ equals 0 deg. The Mach number M_∞ is thus around 0.8 and the velocity U_∞ is close to 275 $m.s^{-1}$. The turbulent quantities specified upstream are the kinetic energy k_∞ equals $103.63\ m^{-2}.s^{-2}$ and the specific dissipation rate ω_∞ is $8.86\ 10^5\ s^{-1}$. These values correspond to a turbulence intensity of 3% and to a turbulent to laminar viscosity ratio of 10. The Reynolds number based on chord ($C = 0.5\ m$) is equal to $11.8\ 10^6$.

The steady flow is transonic and almost symmetric. The iso-Mach number lines are shown in figure 1. The maximum Mach number achieved, assuming a non viscous fluid, is 1.24, the shock being situated at 0.54 C (see figure 1 - up). Assuming a turbulent flow, the maximum Mach number is 1.12 and the shock is situated at $0.48\,C$ (see figure 1 - down). The shock is too weak to induce the separation of the boundary layer, which just thickens quickly because of the adverse pressure gradient.

The unsteady flow field is due to the prescribed harmonic oscillating motion of the airfoil. The rotation center is positioned at $0.25C$ from leading edge and the oscillation is set to 1.01 *deg*. The frequency equals 34.4 Hz, corresponding to a reduced frequency based on half chord, $k = 2\pi f C/2U_\infty$, of 0.197. The shock motion covers about 8% of chord.

3.3. STEADY INVISCID RESULTS

Firstly, six steady inviscid solutions using three different upwind schemes, pure or mixed, were computed. The mixed results were achieved using the same hybridization Mach number set to 0.8, i.e. convective fluxes where the contravariant Mach number was above 0.8 were calculated with the unmodified upwind scheme. The artificial dissipation level was fixed to 0.01.

The static pressure coefficient is given in figure 2. Negative values of x/C correspond to suction side (upper surface), positive values to pressure side (lower surface); x/C equals zero is the leading edge position. Experimental data are time averaged values, whereas numerical data are steady results. This could explain why the predicted inviscid pressure jump is sharper compared to the experimental one, which is also smeared by the shock-boundary layer interaction not simulated in this case. The computed levels are slightly too low, what could be related to the upstream Mach number being too high. It is nevertheless consistant for inviscid simulations that the pressure jump is larger and that the shock position is closer to trailing edge. It will be noticed that between leading edge and mid chord, the predicted pressure profile is wiggly. Other published results exhibit the same behaviour [17]. This is due to the provided airfoil definition which generates uneven variations of wall curvature radius. Comparing the six numerical solutions, it is striking that these cannot be differentiated, all the C_p curves being superimposed. The convergence histories, not shown here, are different, but starting from the same initial guess, stability and convergence speed are comparable between all the schemes.

To analyse more precisely the influence of the convective terms evaluation, total pressure and total temperature profiles along the airfoil surface on the pressure (PS) and suction (SS) sides are presented in figures 3 and 4. The total pressure variation ΔP_t, usually generated by viscous effects or shocks, is defined as $\Delta P_t = (P_t - P_{t\infty})/P_{t\infty}$. The total temperature variation ΔT_t, usually associated with non adiabatic flows, is defined as $\Delta T_t = (T_t - T_{t\infty})/T_{t\infty}$. These quantities are constant and uniform except across the shock since the flow simulated is steady, adiabatic without source terms, and the boundary conditions are uniform. ΔP_t and ΔT_t profiles are plotted as function of the streamwise index, what emphasises the leading edge and the shock regions where the grid is very fine. For all six schemes, small oscillations are located around the leading edge where the flow is subject to large gradients. Plateaux are clearly predicted before and after the shocks, which are captured over two cells (three nodes). Overshoots and undershoots are predicted through the shocks because of the unphysical mixing of supersonic and subsonic information across the discontinuity [18]. There are peaks also at the trailing edge but as mentioned previously, these are due to the averaging procedure used at this particular node. Results achieved using Roe and Liou schemes are very similar, over/undershoots being however lower with Liou scheme. Hanel scheme is too dissipative in very low Mach number regions. Mixed schemes treat the same way the stagnation point region, and the profiles are almost identical.

3.4. STEADY TURBULENT RESULTS

Next, four steady turbulent solutions using the two best upwind schemes, Roe and Liou, are compared. Only mixed Liou scheme was tested, with hybridization Mach number set to 0.8, and with two different artificial dissipation levels, 0.01 and 0.08.

C_p distributions along the profile, obtained for steady inviscid and turbulent computations with pure Roe scheme are compared with measured values in figure 5. The turbulent curve is very similar to the experimental one. The shock position and the pressure jump are better predicted taking into account the turbulence effects.

Figure 6 is an enlarged view of the previous graph. C_p along the pressure surface, around the shock region, is plotted for the four convective schemes. The two pure schemes exhibit the same behaviour, and the two mixed schemes an other one. In spite of the low artificial dissipation level, mixed schemes are more diffusive than pure schemes. This is obvious in the wake region, one chord length downstream the trailing edge (see figure 7). Mixed schemes generate a total pressure deficit of 12%, compared to 11% for pure schemes.

Static pressure transverse distributions at $x/C = 0.58$ are shown in figure 8. Pure Liou scheme behaviour is strongly different from the other schemes: oscillations originating from the boundary layer region are clearly visible. These are probably due to the center approximation of the sound velocity used in the fluxes splitting. Pure Liou scheme is then not dissipative enough, and odd-even modes appear. Mixed Liou scheme with the lowest artificial dissipation level exhibits also some small oscillations in the same region, where the mixed scheme is close to the pure center scheme because the contravariant Mach number is low.

3.5. UNSTEADY INVISCID RESULTS

Finally, unsteady inviscid solutions achieved with pure Roe and pure Liou schemes are presented. Time averaged static pressure coefficient along the airfoil is in fairly good agreement with the experimental data. Both schemes predict again almost the same profiles, with the shocks being slightly too downstream (see figure 9). It is quite different for the moment coefficient evolution (see figure 10). The Roe scheme curve is not too far from the experimental points, whereas the discrepancies are significant using Liou scheme. These differences are due to a 20 *deg* shift of the phase lag between the profile incidence and the shock position predicted by the two numerical approaches.

Figure 11 presents the real component of the first harmonic of C_p Fourier transform. Roe computation shows good agreement with the experiment. Liou computation, even though it follows the experimental trends, underpredicts C_p fluctuation upstream of the shock, and overpredicts it downstream. On the other hand, the imaginary component is coincidentally close to the measured values, except in the shock region (see figure 12). The unsteady C_p evolution predicted using Liou scheme is then slightly different from the Roe results: the phase but also the magnitude are not the same.

4. Compressor cascade configuration

The last test case studied is a compressor cascade in supersonic flow. It is known as the Seventh Standard configuration [19].

4.1. GEOMETRY

The configuration is representative for the tip sections of turboreactor fan stages (multiple circular arc transonic profiles). Each blade has a chord length and a span of $0.0762\,m$, the stagger angle is $61.55\,deg$. The mesh used is a structured 2D H-grid, comprising 130×63 nodes. There are 12 nodes inside the boundary layer near the leading edge, and 20 at mid chord. There are 67 nodes along the blade surface. The mesh extends from $X = -0.0885\,m$ upstream to $X = 0.2097\,m$. The axial position of the leading edge is $X = 0.0575\,m$, and $X = 0.1335\,m$ for the trailing edge. The upstream boundary is far enough to avoid its interaction with the leading edge weak shock.

4.2. FLOW CONDITIONS

The fluid properties are identical to previous test case. The boundary conditions correspond to case No. 9 among the 12 referenced aeroelastic cases. The inlet conditions are such that the total pressure $P_{t-\infty}$ equals 99501 Pa, the total temperature $T_{t-\infty}$ is 297.3 K, and the imposed transverse velocity component is calculated from the experimental freestream Mach number (1.315) and the incidence angle (2.45 *deg*). The static pressure ratio was adjusted to 1.53, instead of 1.45, to ensure the same steady flow structure compared to measurements (see figure 13): a normal shock from leading edge to suction side, the aft 40% of chord being separated [20].

The inlet turbulent conditions are the turbulent kinetic energy $k_{-\infty}$ equals 415.6 $m^2.s^{-2}$, which corresponds to a turbulence intensity of 4.2%, and the specific dissipation rate $\omega_{-\infty}$ is 1.148 10^6 s^{-1}. The dynamic turbulent to laminar viscosity ratio is then around 11.

4.3. STEADY TURBULENT RESULTS

The solutions presented were achieved using pure Roe scheme with $k-\omega$ turbulence model. The inlet conditions being supersonic, the blade geometry and the freestream Mach number set the unique incidence angle upstream the leading edge weak shock. If the initial conditions are incompatible with this angle, the flow adjusts itself to the right incidence. Using the published blade definition [19], a Mach number equal to 1.22 and an incidence angle of 2.6 *deg* were predicted. When the incidence angle decreases, the Mach number decreases too, so the computated results cannot be improved to fit the experimental conditions. The authors suspected then an error in the blade geometry definition. Indeed, the sketch of the blade in [20] is not in agreement with the table in [19] describing the coordinates of the profile. The sketch shows the same blade, but with pressure and suction surfaces reversed. Solution computed with the new geometry, i.e swaping upper and lower surfaces in table [19], is in better agreement with inlet experimental data. The predicted Mach number is 1.331 and the incidence angle equals 2.43 *deg*. In 1991, questions about the blades geometry were already raised. According to the experimenter's answer, the blade definition is correct, and the sketch is wrong. However, the differences between experimentai data and all the published computated results may be due to the inadequate experimental conditions: the number of blades may be too low to reproduce the periodicity of the whole row.

In supersonic flows, the position and the intensity of shock waves are of the highest importance. So it is possible to obtain good results with a slightly wrong geometry, but with the correct shock waves. The solution achieved by the authors with the inversed blade geometry is better than previous computations. Figure 14 shows the steady static pressure distribution along the blade on pressure (PS) and suction (SS) sides. The numerical solution is in good agreement with the experiment.

4.4. UNSTEADY TURBULENT RESULTS

The blades oscillate in pitching mode around a pivot axis at mid chord, with a frequency of 724 Hz corresponding to a reduced frequency based on half chord, $k = 2\pi f C/2U_\infty$, of 0.44 ($U_\infty = 395.05\, m.s^{-1}$). The amplitude is 0.00202 rad (0.116 deg); the interblade phase angle is 0 deg. The blades motion is simulated with transpiration velocity approximation. The mesh is not moved; only the velocity level specified as boundary condition along the adiabatic walls is modified in time to reproduce the solide surface motion.

Only the unsteady blade pressure coefficient is presented. Figure 15 shows computed and measured time averaged C_p levels. The numerical results fit the experimental data quite well. Moreover, the magnitude of the first harmonic of C_p Fourier transform is in reasonable agreement with the experiment (see figure 16 - left), expect in the shock regions, near the leading edge on pressure side and at 75% chord on suction side. The phase on the pressure side is in qualitative agreement with the experimental data (see figure 16 - right), even if it is observed that the phase angle lagges by approximately 90 deg. On the suction side the numerical results are rather poor. These discrepancies may be due to the wrong definition of the blade geometry. Another explanation could be in the difficulty encountered in experiment to keep constant the amplitude and interblade phase angle over the five blades in the cascade, because of the high vibration frequency.

5. Conclusion

The analysis of the results achieved on the isolated airfoil configuration showed the treatment of the stagnation point to be crucial. With a fine enough mesh, Roe and Liou schemes lead to nearly identical steady solutions, whereas Hanel scheme is more dissipative. Mixed schemes are as good as, or better in Hanel case, pure schemes. For unsteady simulations with moving mesh, Roe scheme was found superior to Liou scheme in predicting the phase lag. Results achieved with Roe scheme in the compressor cascade configuration confirmed the abality of this scheme to predict flutter in turbomachinery.

Acknowledgments

The authors would like to thank F. Pleynet who processed the unsteady results presented in this paper.

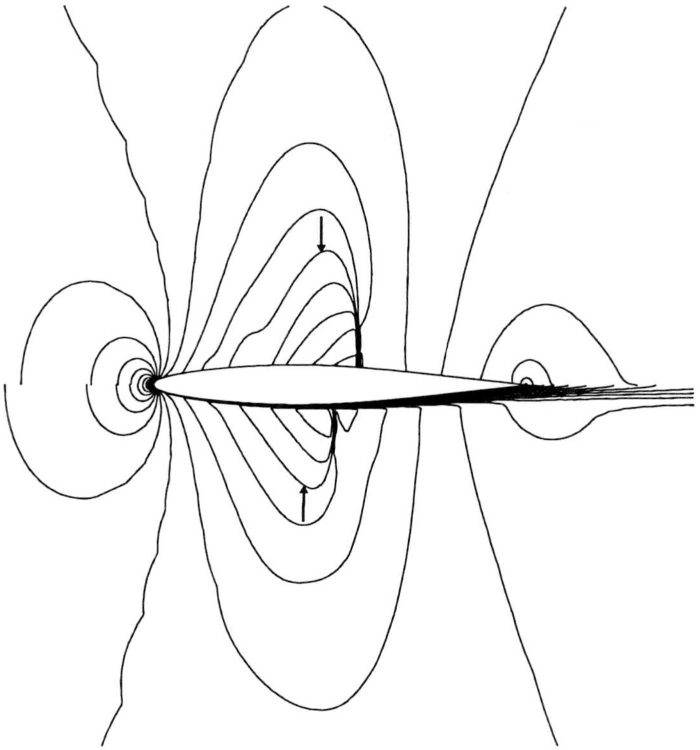

Figure 1. NACA 64A010 - Mach number iso-lines ($\Delta M = 0.05$). Arrows point at the sonic line ($M = 1$). *up:* Inviscid simulation. *down:* Turbulent simulation.

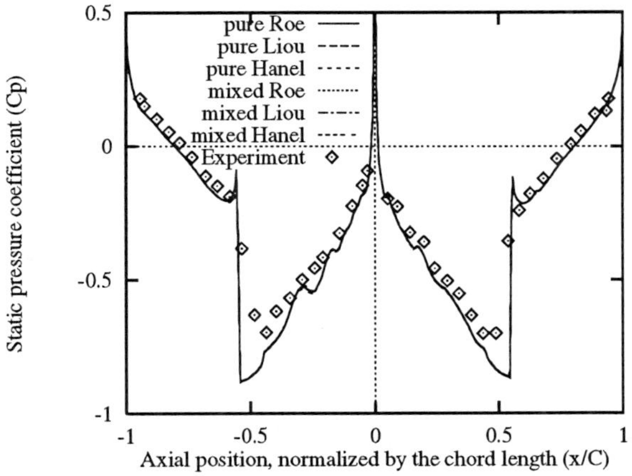

Figure 2. NACA 64A010 - Steady static pressure coefficient versus axial position, using different inviscid schemes.

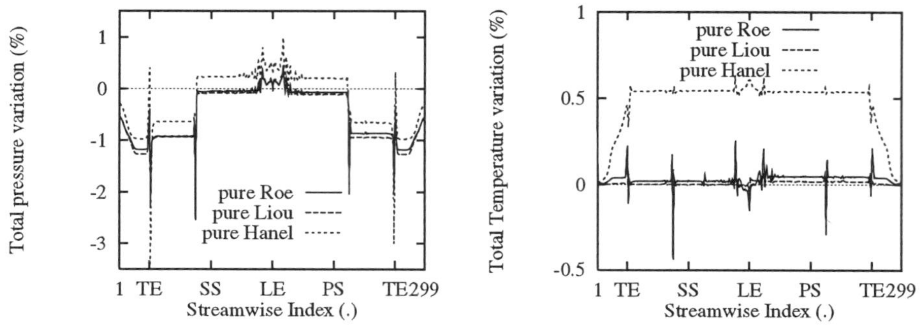

Figure 3. NACA 64A010 - Total pressure and total temperature variations versus streamwise index along the blade surface, using different inviscid pure schemes.

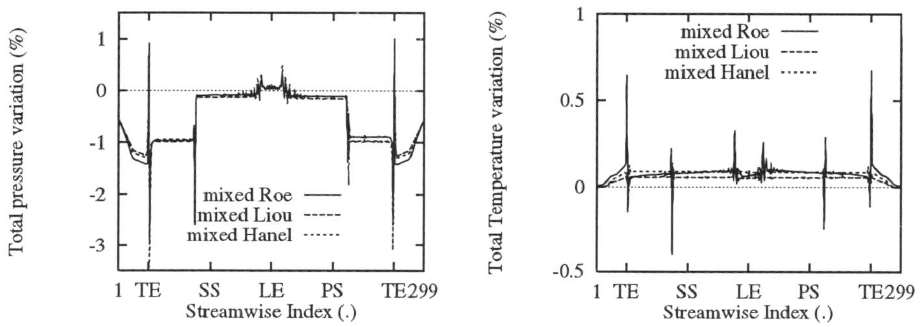

Figure 4. NACA 64A010 - Total pressure and total temperature variations versus streamwise index along the blade surface, using different inviscid mixed schemes.

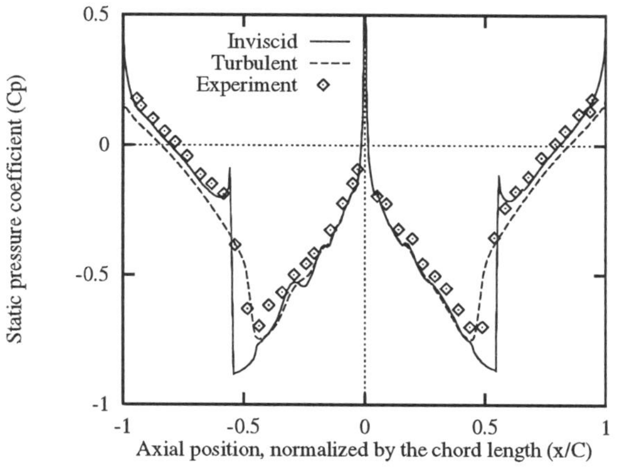

Figure 5. NACA 64A010 - Steady static pressure coefficient versus axial position, using inviscid and turbulent pure Roe scheme.

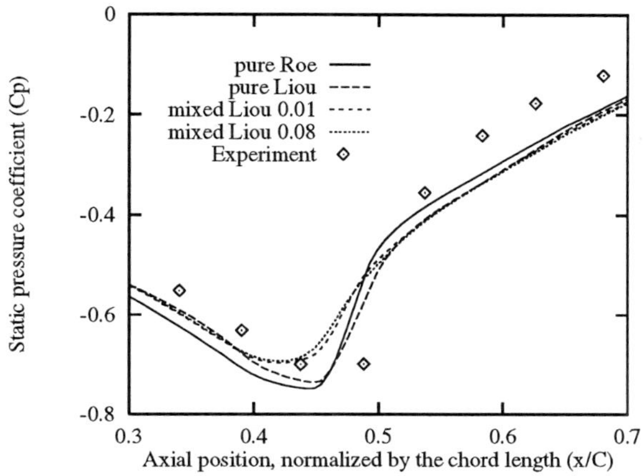

Figure 6. NACA 64A010 - Steady static pressure coefficient versus axial position, using different turbulent schemes.

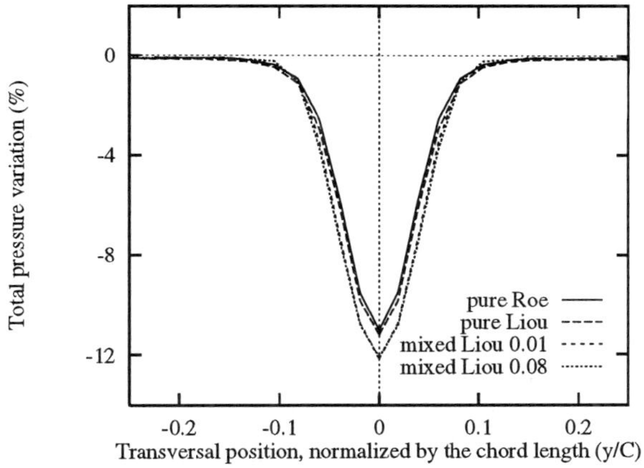

Figure 7. NACA 64A010 - Total pressure variation in the wake one chord downstream the trailing edge, using different turbulent schemes.

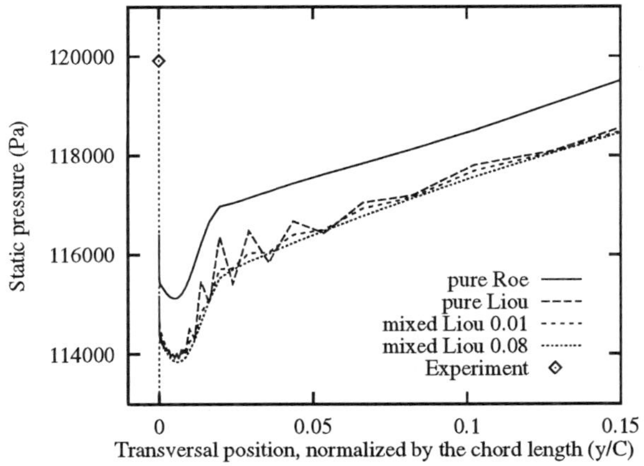

Figure 8. NACA 64A010 - Transverse static pressure distribution downstream of the shock on pressure side, $x/C = 0.58$, using different turbulent schemes.

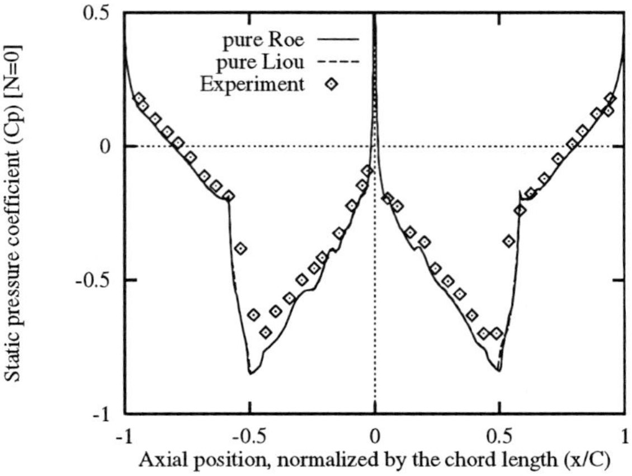

Figure 9. NACA 64A010 - Time averaged static pressure coefficient versus axial position, using different inviscid pure schemes.

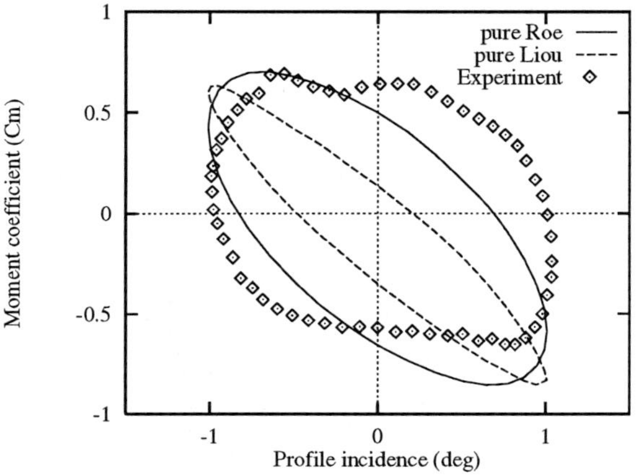

Figure 10. NACA 64A010 - Moment coefficient versus profile incidence, using different inviscid pure schemes.

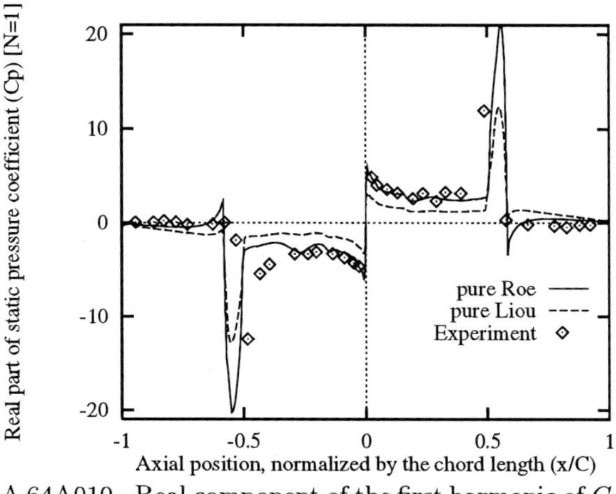

Figure 11. NACA 64A010 - Real component of the first harmonic of C_p Fourier transform versus axial position, using different inviscid pure schemes.

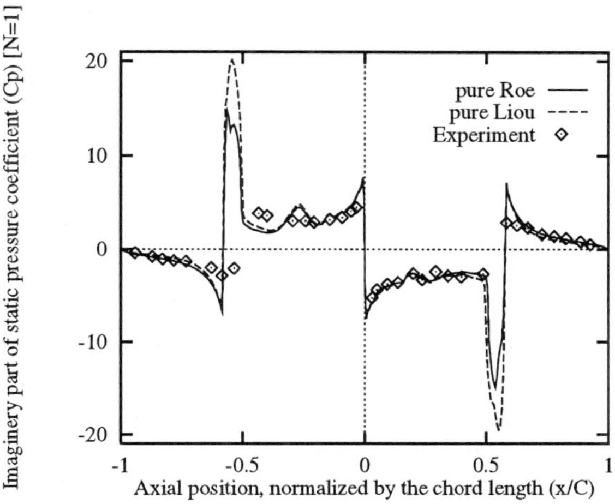

Figure 12. NACA 64A010 - Imaginary component of the first harmonic of C_p Fourier transform versus axial position, using different inviscid pure schemes.

Figure 13. 7th standard configuration - Mach number iso-lines ($\Delta M = 0.05$). Arrows point at the sonic line ($M = 1$).

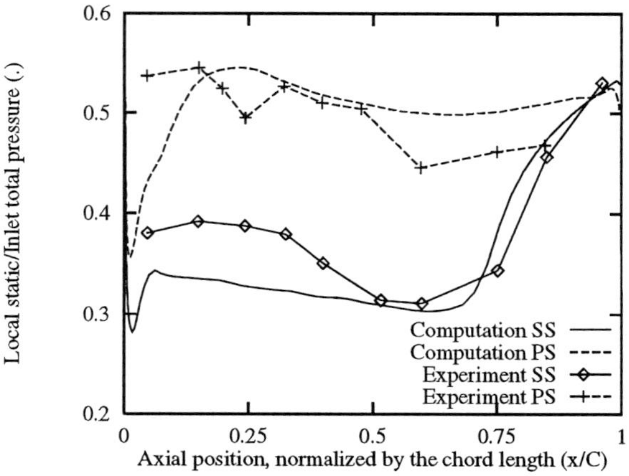

Figure 14. 7th standard configuration - Steady static to total pressure ratio versus axial position.

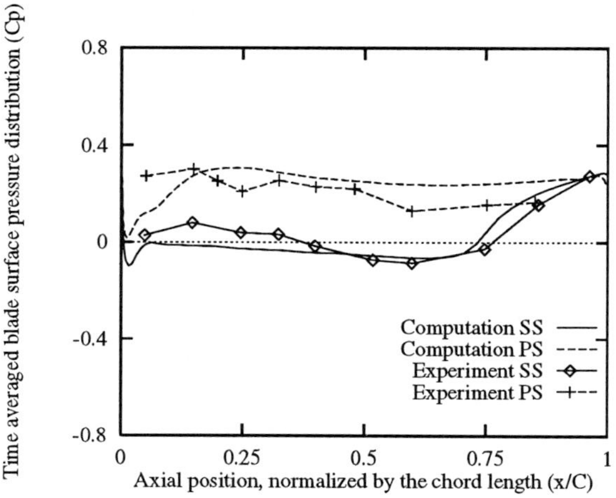

Figure 15. 7th standard configuration - Time averaged static pressure coefficient versus axial position.

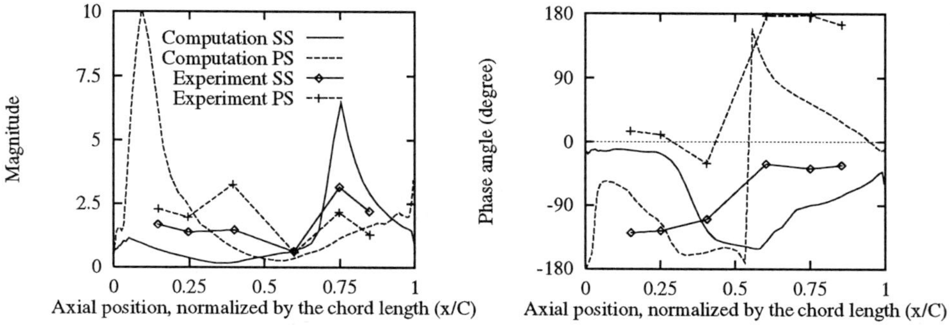

Figure 16. 7th standard configuration - Magnitude and phase of C_p first harmonic versus axial position.

References

1. Aubert, S., and Ferrand, P., "Study of unsteady transonic flows with a new mixed Van Leer flux splitting method," *7th International Symposium on Unsteady Aerodynamics and Aeroelasticity of Turbomachines*, Tanida and Namba Editors, Elsevier, September 1994, pp. 23-38.
2. Hanel, D., Schwane, R., and Seider, G., "On the accuracy of upwind schemes for the solution of the Navier-Stokes equations," AIAA Paper 87-1105, 1987.
3. Roe, P.L., "Approximate Riemann solvers, parameters vectors, and difference schemes," *Journal of Computational Physics*, Vol. 43, No. 2, 1981, pp. 357-371.
4. Liou, M.S., "A continuing search for a near perfect numerical scheme. Part 1: AUSM+," NASA TM 106524, 1994.
5. Ferrand, P., and Aubert, S., "A new mixed Van Leer flux splitting for transonic viscous flow," *AIAA Journal*, Vol. 34, No. 1, January 1996, pp. 190-193.
6. Michel, R., Quémard, C., and Durant, R., "Application d'un schéma de longueur de mélange à l'étude des couches limites turbulentes d'équilibre," ONERA NT No. 154, 1969.
7. Baldwin, B.S., and Lomax, H., "Thin layer approximation and algebraic model for separated turbulent flows," AIAA Paper 78-257, January 1978.
8. Launder, B.E., and Jones, W.P., "Sink flow turbulent boundary layers," *Journal of Computational Physics*, Vol. 38, 1969, p. 817.
9. Wilcox, D.C., "Reassessment of the scale-determining equation for advanced turbulence models," *AIAA Journal*, Vol. 26, No. 11, November 1988, pp. 1299-1310.
10. Chakravarthy, S.R., "Euler equations-Implicit schemes and boundary conditions," *AIAA Journal*, Vol. 21, No. 5, May 1983, pp. 699-706.
11. Jameson, A., "Artificial diffusion, upwind biasing, limiters and their effect on accuracy and multigrid convergence in transonic and hypersonic flows," AIAA Paper 93-3359, July 1993.
12. Issman, E., Degrez, G., and Deconinck, H., "Implicit iterative methods for a multidimensional upwind Euler/Navier-Stokes solver on unstructured meshes," AIAA Paper 95-1653-CP, 1995.
13. van der Vorst, H.A., "BI-CGSTAB: a fast and smoothly converging variant of BI-CG for the solution of nonsymmetric linar systems," *SIAM J. Sci. Stat. Comput.*, Vol. 13, 1992, pp. 531-544.
14. Mulder, W.A., and Van Leer, B., "Implicit upwind methods for the Euler equations," AIAA Paper 83-1930, July 1983.
15. Lerat, A. ,"Implicit methods of second order accuracy for the euler equations," *AIAA Journal*, Vol. 23, No. 1, January 1985, pp. 33–40.
16. Davis, S., "NACA 64A010 (NASA Ames Model) oscillatory pitching," AGARD R-702, "Compendium of Unsteady Aerodynamic Measurements," August 1982, pp. 2.1–2.21.
17. Jouet, C., and Cheret, V., "Méthode numérique appliquée à des profils oscillants," *31ème Colloque d'Aérodynamique Appliquée*, Association Aéronautique et Astronautique de France, Ministère de l'Enseignement Supérieur et de la Recherche, March 1995.
18. Aubert, S., Hallo, L., Ferrand, P., and Buffat, M., "Numerical simulations of a three-dimensional nozzle, from inviscid to turbulent flows," *Int. J. Num. Meth. Heat Fluid Flow*, Vol. 5, No. 10, December 1995, pp. 889–905.
19. Bölcs, A., and Fransson, T.H., "Aerolasticity in turbomachines, comparison of theorical and experimental cascade results," *Communication du Laboratoire de Thermique Appliquée et de Turbomachines*, École Polytechnique Fédérale de Lausanne, 1986, pp. 153–166.
20. Riffel, R.E, and Rothrock, M.D., "Experimental determination of unsteady blade element aerodynamics in cascades." *Torsion mode final report, Volume I*, Detroit Diesel Allison, 1980.

14: Theoretical Studies of Unsteady Flow Around Oscillating Blades - 5

EIGENMODE ANALYSIS OF UNSTEADY VISCOUS FLOWS IN TURBOMACHINERY CASCADES

R. FLOREA, K. C. HALL
Duke University
Department of Mechanical Engineering and Materials Science
Durham, NC, U.S.A. 27708-0300

AND

P. G. A. CIZMAS
Westinghouse Electric Corporation
Science & Technology Center
Pittsburgh, PA, U.S.A. 15235-5098

1. Introduction

Stall flutter and rotating stall in turbomachinery cascades are recurring problems in the development and operation of gas turbine engines [see e.g. Sisto (1987)]. In this paper, we present a novel technique for analyzing unsteady flows called "reduced order modelling." The method, which is based on the eigenmodes (natural modes) of unsteady fluid motion through a cascade, is useful for predicting unsteady flows in turbomachinery over a range of reduced frequencies. Furthermore, the resulting model is low-order, and provides physical insight into certain aeroelastic and/or fluid dynamic instabilities, e.g. rotating stall.

Over the past several years, significant advances have been made in the development of reduced order models of unsteady aerodynamic flows. Mahajan *et al.* (1991) computed the eigenvalues of unsteady flow over isolated airfoils. The unsteady flow was modelled with a nonlinear Navier-Stokes solver. Later, Mahajan *et al.* (1993) computed the eigenvalues of a coupled fluid/structure system for the case of compressible potential flow over an elastic airfoil. Hall (1994) computed the eigenvalues and eigenmodes of an unsteady incompressible vortex lattice model of flows about two-dimensional airfoils, cascades, and three-dimensional wings, and then used this information to construct reduced order models. Similarly, Hall *et al*

T.H. Fransson (ed.), Unsteady Aerodynamics and Aeroelasticity of Turbomachines, 767-782.

al. (1994) and Florea & Hall (1994) constructed eigenmode based reduced order models of unsteady compressible potential flows around cascades and isolated airfoil. These later studies were important in that they showed the necessity of using so-called static correction techniques to reduce the number of eigenmodes required for accurate reduced order models. Later, Florea *et al.* (1996) applied similar techniques to create reduced order models of viscous flows in cascades. Around the same time, Romanowski & Dowell (1994) and computed the eigen-information of a time-domain Euler solver and constructed reduced order models of flows about isolated airfoils. More recently, Romanowski (1996) has used Karhunen-Loeve eigenmodes as basis functions. For a recent review of eigenmode based reduced order models, see Dowell *et al.* (1997).

In this paper, the previously developed reduced order modelling techniques are applied to viscous flows in a low-speed compressor cascade. We model the unsteady flow using a frequency-domain viscous-inviscid (VII) technique. The inner viscous and outer inviscid regions are simultaneously coupled [see Cizmas & Hall (1995)] so that unsteady flows with moderately large flow separations can be analyzed. Because we are interested in the onset of instabilities, both fluid dynamic (rotating stall) and aeroelastic (flutter), we assume that the unsteadiness in the flow is small compared to the mean flow. Therefore, the unsteady flow is decomposed into a nonlinear mean flow plus a linear unsteady perturbation flow which is assumed to vary harmonically in time. So-called time-linearized (frequency-domain) equations are obtained that describe the unsteady small-disturbance viscous flow about the cascade.

With a time-linearized model in hand, we then construct "reduced order models" of the unsteady flow, using the eigenmodes of the time-linearized model as basis vectors. We compute the lowest-frequency eigenvalues and corresponding eigenvectors using a nonsymmetric Lanczos algorithm developed for this purpose. Finally, we use the mode summation method, together with a static correction technique, to reconstruct the unsteady flow solution at arbitrary frequencies.

2. Conventional Time-Linearized Unsteady Aerodynamic Model

2.1. AERODYNAMIC THEORY

In this paper, we model the two-dimensional unsteady viscous flow through a single blade row using a viscous-inviscid interaction (VII) technique. To begin, consider the inviscid outer flow. For subsonic flutter problems, the flow will be irrotational. Thus, the velocity may be represented by the gradient of a homentropic scalar velocity potential ϕ. The inviscid flow is governed by the well-known full potential equation, i.e.

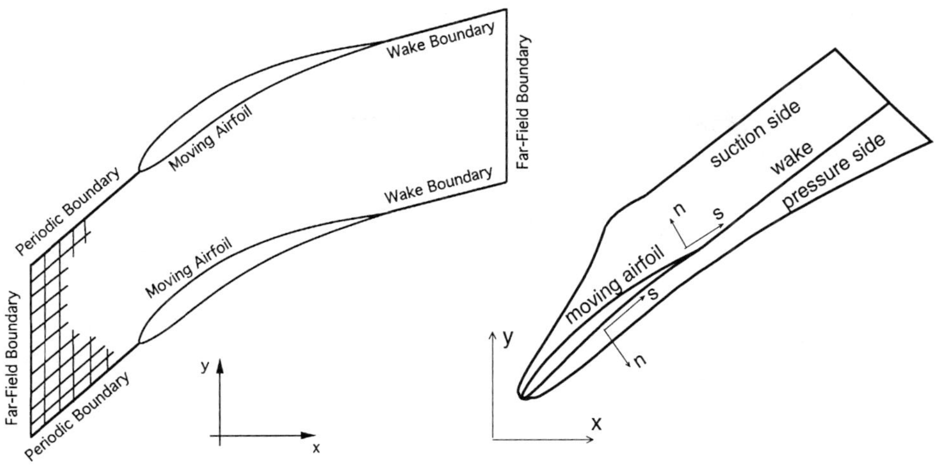

Figure 1. Computational domains used to compute flow through cascade. Left: Inviscid outer flow domain. Right: Inner viscous flow domain around airfoil and wake.

$$\nabla^2\phi = \frac{1}{a^2}\left[\frac{\partial^2\phi}{\partial t^2} + 2\nabla\phi\cdot\nabla\frac{\partial\phi}{\partial t} + \frac{1}{2}\nabla\phi\cdot\nabla\left(\nabla\phi\right)^2\right] \tag{1}$$

where a is the speed of sound equal to $\sqrt{\gamma p/\rho}$, and where ρ and p are the static density and pressure, respectively. Note that the full potential equation is nonlinear for compressible flows.

Next, we consider the viscous inner flow near airfoil surfaces and in the wakes trailing behind airfoils. In the present analysis, we are interested in low-speed flows. However, the Reynolds number based on the aerodynamic chord Re_c is assumed to be large, and boundary layers and flow separations are assumed to be thin. Under these assumptions, the viscous inner flow is governed by the incompressible boundary layer equations. Thus, the conservation of mass is

$$\frac{\partial u}{\partial s} + \frac{\partial v}{\partial n} = 0 \tag{2}$$

where s and n are orthogonal curvilinear coordinates in the directions tangent to and normal to the airfoil and wake centerline, as shown in Figure 1, and u and v are the corresponding velocity components. Note that s and n are noninertial coordinates since they are "attached" to the vibrating airfoil.

Similarly, the conservation of streamwise momentum is given by

$$\rho_0\left(\frac{\partial u}{\partial t} + u\frac{\partial u}{\partial s} + v\frac{\partial u}{\partial n}\right) = \frac{\partial}{\partial n}\left(\mu\frac{\partial u}{\partial n} - \rho_0\overline{u'v'}\right) - \frac{\partial p_e}{\partial s} - \rho_0\frac{\partial^2 q_a}{\partial t^2} \tag{3}$$

where q_a is the displacement of the airfoil surface in the direction tangent to itself. The last term in Eq. (3) is a D'Alembert force arising from the use of noninertial coordinates (s,n). Also, since the flow is assumed to be incompressible, the density ρ_0 is constant, taken here to be ρ_T. The term $-\rho_0\overline{u'v'}$ is the so-called Reynolds stress arising from turbulent velocity fluctuations u' and v'.

In Eq. (3) the usual boundary layer approximation that the pressure is constant through the boundary layer is used. The pressure at any point in the boundary is then equal to the pressure at the edge of the boundary layer (denoted by the subscript "e"), given by

$$\frac{\partial p_e}{\partial s} = -\rho_e\left[\frac{\partial u_e}{\partial t} + u_e\frac{\partial u_e}{\partial s} - \frac{\partial}{\partial s}\left(u_e\frac{\partial q_a}{\partial t}\right)\right] \tag{4}$$

where here u_e is the velocity measured in the *inertial* frame of reference.

By introducing the stream function ψ, Eqs. (2) and (3) can be reduced to a single third-order scalar equation for the unknown stream function. To reduce the size of the computational grid required to obtain accurate solutions, we also use a coordinate rescaling in the normal direction, i.e.

$$s = s, \qquad \zeta = \frac{n}{l^*} \tag{5}$$

where l^* is a characteristic boundary layer thickness which may vary with s. For details, see Florea *et al.* (1996).

The Reynolds stress in Eq. (4) is computed using a slightly modified form of the Spalart-Allmaras (1994) turbulence model. The original Spalart-Allmaras model is a one-equation partial differential equation for the eddy viscosity derived using a combination of empiricism and physical modelling. Here we make the usual boundary layer assumptions to convert the turbulence model to a boundary layer form. We place turbulent "trips" just aft of the stagnation point on both the pressure and suction surfaces.

2.2. LINEARIZATION OF THE GOVERNING EQUATIONS

For many flows of practical interest – for example the onset of stall flutter and rotating stall – the unsteadiness in the flow is small compared to the mean flow, and is harmonic in time. The governing equations, therefore, can be linearized to obtain linear equations describing the unsteady perturbation flow field. For example, the outer inviscid potential flow can be represented by

$$\phi(x,y,t) = \Phi(x,y) + \phi'(x,y)\exp(j\omega t) \tag{6}$$

where Φ is the nonlinear mean flow potential, and ϕ' is the complex amplitude of the unsteady perturbation potential.

Substitution of Eq. (6) into the full potential equation and collection of zeroth-order (steady) terms gives

$$\nabla^2\Phi = \frac{1}{A^2}\left[\frac{1}{2}\nabla\Phi\cdot\nabla\left(\nabla\Phi\right)^2\right] \tag{7}$$

where A is the mean flow speed of sound. Similarly, the unsteady potential ϕ' is described by the first-order perturbation equation

$$\nabla\cdot R\nabla\phi' - \nabla\cdot\left[\frac{R}{A^2}\left(\nabla\Phi\cdot\nabla\phi' + j\omega\phi'\right)\nabla\Phi\right] - \frac{R}{A^2}\left(j\omega\nabla\Phi\cdot\nabla\phi' - \omega^2\phi'\right) = 0 \tag{8}$$

where R is the mean flow density. To improve the accuracy of the perturbation potential flow solver, a harmonically deforming computational grid which conforms to the motion of the airfoil is used, resulting in an additional inhomogeneous term on the right-hand side of Eq. (8) (Hall, 1993). A similar linearization is performed for both the boundary layer and turbulence model equations. The resulting set of differential equations describing the unsteady flow through the cascade is quadratic in the excitation frequency ω.

2.3. BOUNDARY CONDITIONS

Boundary conditions are required to complete the problem specification. On airfoil surfaces, the no-slip condition applies. Similarly, the turbulent viscosity is prescribed to be zero on the airfoil surfaces. Also, we require that the velocities and turbulent viscosity be continuous across the centerline of the wake, i.e. at $\zeta = 0$, where the upper and lower wake grids are connected.

At the edge of the boundary layer, certain matching conditions are imposed. The boundary layer velocity at the outer boundary of the inner region must match the edge velocity of the potential flow at the inner boundary of the outer region. Also, at points aft of the trip point, the turbulent viscosity is set to a small but nonzero value at the edge of the boundary layer. The influence of the inner viscous flow on the outer inviscid flow is modelled with a transpiration boundary condition at the wall. A similar transpiration condition is applied on either side of the wake centerline, but also includes the effect of an oscillating wake. To provide closure for the steady and unsteady wake positions, we require that the pressure jump across the wake be zero, and that the wake centerline be attached to the oscillating trailing edge.

Upstream and downstream of the cascade, flow periodicity is enforced. In the case of unsteady flow, the unsteady flow variables satisfy a complex periodicity condition which requires that flow perturbations be equal in magnitude from one blade passage to the next, but shifted in phase by the interblade phase angle σ. This complex periodicity condition allows the computational domain to be reduced to a single blade passage.

Finally far-field boundary conditions must be applied at the upstream and downstream boundaries of the inviscid outer flow computational domain. For steady flows, the total pressure, total density, and circumferential velocity are prescribed upstream while the mass flux is prescribed downstream. For unsteady flows, numerically exact non-reflecting boundary conditions are applied in the conventional analysis (Hall *et al.*, 1993).

2.4. NUMERICAL DISCRETIZATION

We only briefly describe the numerical discretization technique used in this analysis. For more details, the interested reader is referred to Hall (1993), and Florea (1996).

The governing equations are discretized on a computational grid system composed of two grids: an H-grid for the outer potential flow, and a C-grid for the inner viscous flow. Figure 1 shows the computational domains. Note that the inner viscous grid is rectangular in the computational (s, ζ) coordinate system.

For the outer potential flow, a deforming grid variational finite element technique developed by Hall (1993), and similar to a technique developed by Whitehead & Grant (1981), is used to discretize the governing equations. The inner viscous flow, including the modified Spalart-Allmaras turbulence model, is discretized using conventional finite difference techniques. The inner and outer flows are coupled together at common nodes on airfoil surface using a combination of finite element and finite difference techniques. The coupling is *simultaneous* so that flows with fairly large separations can be computed (Cizmas & Hall, 1995).

The steady and unsteady flows are solved sequentially. First, the nonlinear steady flow is computed using Newton iteration. Once the steady flow has been computed, the linearized equations are assembled for each interblade phase angle σ and frequency ω of interest. The resulting linear equations are of the form

$$[\mathbf{A}]\{\mathbf{u}\} = \{\mathbf{b}\} \tag{9}$$

where $\mathbf{A}$ is a sparse $m \times m$ matrix which is a function of the frequency and interblade phase angle, and the vector $\mathbf{b}$ depends on the frequency, interblade phase angle, and the particular type of flutter problem (e.g. bending or torsion) being solved. Equation (9) is solved efficiently using LU decomposition.

2.5. COMPUTATIONAL RESULTS

In this section, we present computational results for a low-speed compressor geometry. The cascade geometry is the same as the so-called Tenth Standard geometry (Fransson & Verdon, 1993), but the flow conditions have been modified. The cascade is composed of airfoils with a slightly modified NACA 0006 thickness distribution, but with a circular arc camber line with 5% camber. The airfoils have a stagger angle Θ of 45°, and a gap-to-chord ratio G of 1.0. For all cases, the inflow Mach number M is 0.3, but the upstream steady flow angle Ω measured from the axial direction is varied.

A composite computational grid is used in the calculation of the steady and unsteady flows. The results presented in this paper were computed on a computational grid with 2112 nodes in the outer inviscid grid and 5700 nodes in the viscous grid, resulting in approximately 26,120 unknowns. Although not presented here, this grids was found to be fine enough so that the computed results are nearly grid independent.

We first consider the steady flow through the compressor cascade. Figure 2 shows the computed pressure coefficient, boundary layer displacement thickness, and skin friction coefficient on the surface of the airfoil and in the wake for four different inflow angles, $\Omega = 55°$, $58°$, $60°$, and $62°$. Note from the skin friction plot that the flow is separated for all of these inflow angles, although the separation region is quite small for the $\Omega = 55°$ case. As the inflow angle is increased, the size of the flow separation, i.e. the axial extent of the separation as well as the thickness of the boundary layer, grows dramatically. Also, at the largest inflow angle shown in Fig. 2, $\Omega = 62°$, a small leading edge separation bubble is clearly evident.

Next, we consider the unsteady aerodynamic moment M acting on the airfoils in the cascade as a result of pitching motion about a point near the the airfoil midchord, i.e. $(x/c, y/c) = (0.5, 0.05)$. Figure 3 shows the computed unsteady moment coefficient ($C_M = M/\rho V^2 c^2$) as a function of interblade phase angle σ for several different inflow angles Ω at two reduced frequencies, $\overline{\omega} = \omega c/V$. Using our notation, a structurally tuned cascade will flutter whenever the imaginary part of the unsteady aerodynamic moment is positive. Note that the (viscous) cascade is generally less stable at the lower reduced frequency ($\overline{\omega} = 0.5$), and there is a dramatic decrease in stability as the incidence increases. This instability may be properly called "stall flutter," since there is a significant loss in stability with only a small change in incidence. Furthermore, this loss in stability is coincident with the rapid increase in flow separation. Although not shown here, an inviscid analysis predicts the rotor will be aeroelastically stable over the range of parameters examined.

3. Eigenmode Analysis and Reduced Order Modelling

3.1. FLUID DYNAMIC EIGENMODES

Except for the nonreflecting boundary conditions in their original form, all of the terms in the discretized time-linearized equations, Eq. (9), are

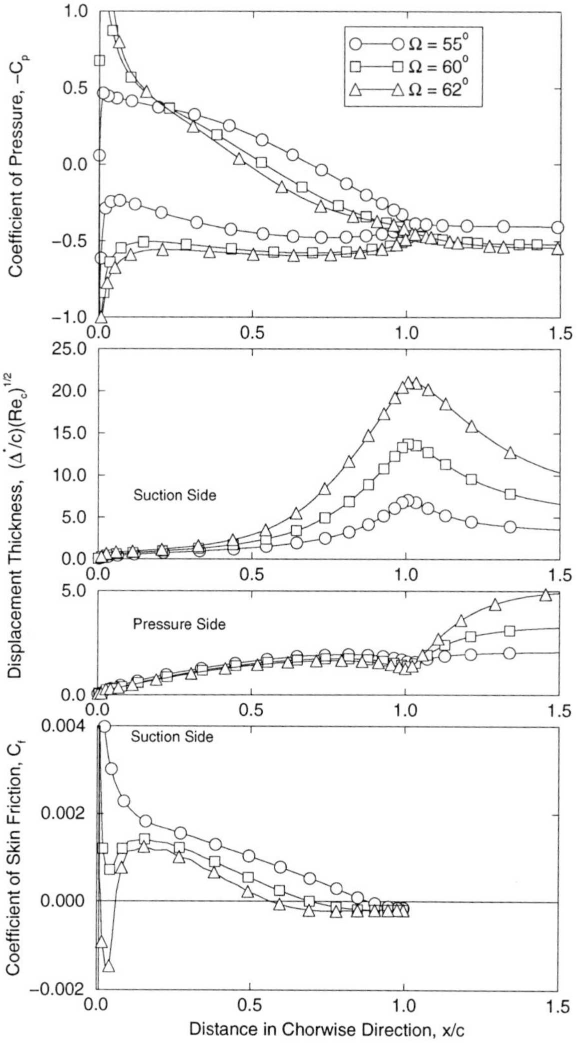

Figure 2. Computed surface quantities for steady flow through modified Tenth Standard Configuration. Top to bottom: coefficient of pressure, boundary layer displacement thickness, coefficient of skin friction.

quadratic in the frequency ω. For convenience, one can approximate the far-field boundary condition terms in terms of a quadratic polynomial in frequency by Taylor expanding the boundary conditions about some nominal frequency (Florea *et al.*, 1996), or alternatively – as in the present analysis – by a quadratic curve fit over a specified frequency range. In any

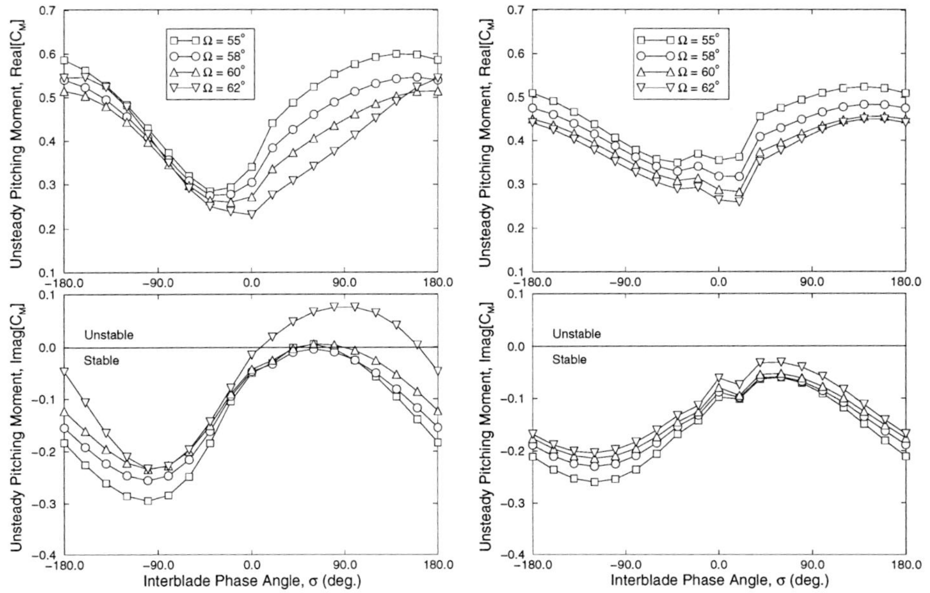

Figure 3. Real and imaginary parts of the aerodynamic pitching moment due to pitching of the airfoils about a point near their midchords. $\mathrm{Re}_c = 100{,}000$. Left: $\overline{\omega} = 0.5$. Right: $\overline{\omega} = 1.0$.

event, the resulting matrix equation takes the form

$$\left[\mathbf{A}_0 + j\omega\mathbf{A}_1 - \omega^2\mathbf{A}_2\right]\mathbf{u} = \mathbf{b} \tag{10}$$

or, in the equivalent first-order form,

$$\mathcal{A}\mathbf{U} - j\omega\mathcal{B}\mathbf{U} = \overline{\mathbf{b}} \tag{11}$$

where

$$\mathbf{U} = \left\{ \begin{array}{c} j\omega\mathbf{u} \\ \mathbf{u} \end{array} \right\}, \quad \mathcal{A} = \left[\begin{array}{cc} \mathbf{A}_0 & \mathbf{0} \\ \mathbf{A}_1 & \mathbf{A}_0 \end{array} \right], \quad \mathcal{B} = \left[\begin{array}{cc} \mathbf{0} & \mathbf{A}_0 \\ \mathbf{A}_2 & \mathbf{0} \end{array} \right], \quad \overline{\mathbf{b}} = \left\{ \begin{array}{c} \mathbf{0} \\ \mathbf{b} \end{array} \right\} \tag{12}$$

The matrix equation, Eq. (11), describes a linear inhomogeneous problem for the vector of unknown perturbations quantities $\mathbf{U}$.

Next, we find the homogeneous solutions, or eigenmodes, of Eq. (11). Thus, we define the eigenvalue problem

$$\mathcal{A}\mathbf{x}_i = \lambda_i \mathcal{B}\mathbf{x}_i, \quad \text{for } i = 1 \text{ to } 2m \tag{13}$$

where λ_i and $\mathbf{x}_i$ are the ith eigenvalue and eigenmode. In the present investigation, a subset of the eigenvalues λ_i and corresponding right and left

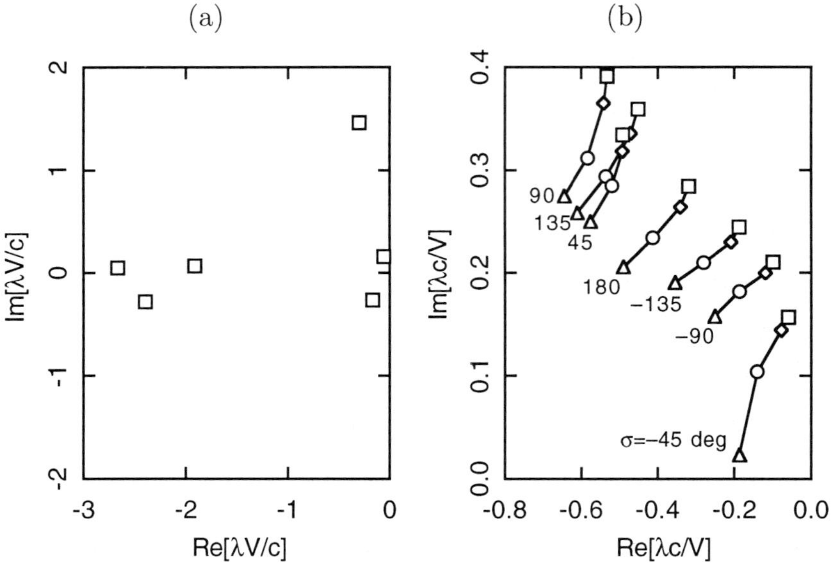

Figure 4. Eigenvalues of low-speed unsteady flow about cascade. (a) Eigenvalues for $\Omega = 62.24°$, $\sigma = -45°$. (b) Eigenvalues for several interblade phase angles and inflow angles. △, ○, ◇, and □: $\Omega = 60°$, $61°$, $62°$, and $62.24°$.

eigenmodes $\mathbf{x}_i$ and $\mathbf{y}_i$ are computed using a nonsymmetric Lanczos algorithm (Florea, 1996).

3.2. COMPUTED EIGENVALUES AND EIGENMODES

The aerodynamic eigenvalues (natural frequencies) and eigenmodes (natural modes) – like those computed for finite element discretizations of structural dynamic systems – provide physical insight into the behavior of the fluid system. For example, consider the compressor cascade previously described in Section 2.5, but now operating with an inflow angle Ω of $62.24°$, the highest inflow angle for which a converged steady solution could be obtained. Figure 4a shows the computed eigenvalues for an interblade phase angle σ of $-45°$. Shown in Fig. 4b is an expanded view of the region near the origin of Fig. 4a, but for several interblade phase angles and inflow angles. Note that as the incidence is increased, the eigenvalues move to the right, that is, toward instability.

Consider the least stable eigenvalue in Fig. 4b, i.e. $\sigma = -45°$ and $\Omega = 62.24°$. The unsteady disturbance associate with the corresponding eigenmode rotates circumferentially around the cascade (up the page as drawn), with a phase speed V_p equal to $-\mathrm{Im}(\lambda)cG/\sigma$ where G is the gap-to-chord ratio. For the eigenmode in question, the phase speed is 0.199 times the upstream velocity V. Although we were unable to obtain a converged

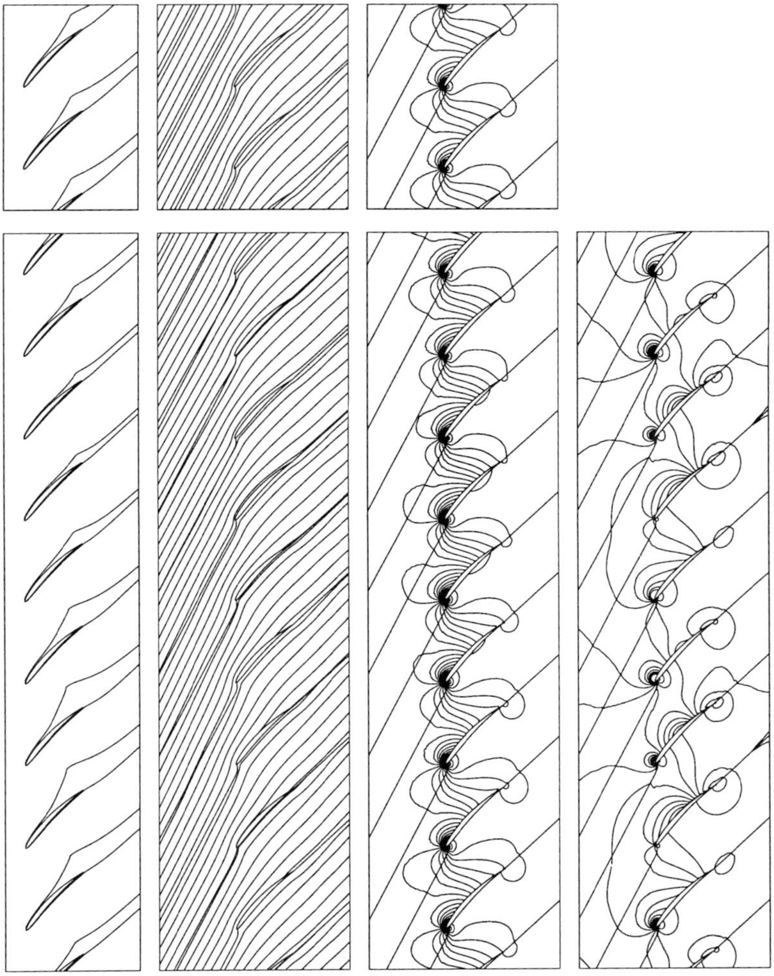

Figure 5. Top, left to right: steady flow quantities. Displacement thickness Δ^* (exaggerated by a factor of three); inviscid stream function Ψ_{inv}; and contours of pressure, P for cascade operating at $\Omega = 62.24°$. Bottom, left to right: unsteady flow quantities. Displacement thickness $\Delta^* + \epsilon{\delta^*}' e^{\lambda t}$ (exaggerated by a factor of three); inviscid stream function $\Psi_{\text{inv}} + 2\epsilon\psi'_{\text{inv}} e^{\lambda t}$; contours of pressure $P + \epsilon p' e^{\lambda t}$; and perturbation pressure, p'.

steady solution at higher inflow angles Ω, if one extrapolates the data to an incidence which causes the $\sigma = -45°$ mode to be on the verge on instability, the phase speed is found to increase to about $0.24 \times V$. The authors believe this eigenmode to be the linear precursor to rotating stall. Figure 4 shows the computed unsteady eigenmode corresponding this eigenvalue. This eigenmode is qualitatively similar to the model of rotating stall proposed by Emmons *et al.* (1955), with a stall cell rotating circumferentially through the cascade in the direction opposite to the direction of rotation (in the rotor frame of reference) at a fraction of the relative inflow velocity. Note in Fig. 5 that the streamlines are slightly diverted around the passages with the largest flow separations. This causes the incidence to increase on the airfoils above the blockage, and decrease on the airfoils below the blockage. This in turn causes the separation to grow on the airfoils above, and diminish on the airfoils below. Thus, the pattern tends to propagate upward through the cascade.

3.3. REDUCED ORDER MODEL

Next, we use the computed eigenmodes as basis vectors to reconstruct the solution of Eq. (13). We assume the the solution $\mathbf{U}$ may be represented as a sum of the eigenmodes, so that

$$\mathbf{U} = \sum_i w_i \mathbf{x}_i \tag{14}$$

where w_i is the so-called participation factor which describes how much of each eigenmode is present in the solution. Note that the series is truncated, retaining just a few of the eigenmodes, generally those with the smallest eigenvalues. Equation (14) is then substituted into Eq. (13) to obtain

$$\sum_k w_k \left(\mathcal{A}\mathbf{x}_k - j\omega\mathcal{B}\mathbf{x}_k\right) = \overline{\mathbf{b}} \tag{15}$$

Pre-multiplying Eq. (15) by $\mathbf{y}_i^T$ and making use of orthogonality (the right and left eigenvectors are orthonormal with respect to $\mathcal{B}$) gives

$$\lambda_i w_i - j\omega w_i = \kappa_i, \qquad i = 1 \text{ to } 2m \tag{16}$$

where $\kappa_i = \mathbf{y}_i^T \cdot \overline{\mathbf{b}}$. Therefore,

$$\mathbf{U} = \sum_i w_i \mathbf{x}_i = \sum_i \frac{\kappa_i}{\lambda_i - j\omega}\mathbf{x}_i \tag{17}$$

Equation (17) is exact if all $2m$ eigenmodes are retained in the summation, since the eigenvectors form a complete basis. However, for computational efficiency, one would like to use only a few of the many modes, say the n smallest eigenvalues. Unfortunately, the series in Eq. (17) is slow to converge. To speed convergence of this series, we have developed the method of multiple static corrections – a technique similar to one used in structural dynamics (Bisplinghoff & Ashley, 1962) – that includes the effects of the neglected eigenmodes in a quasi-static way. Specifically, we let

$$\mathbf{U} \approx \sum_{i=1}^{n} \left(\frac{j\omega}{\lambda_i}\right)^{M_c} \frac{\kappa_i}{\lambda_i - j\omega}\mathbf{x}_i + \sum_{r=1}^{M_c} (j\omega)^{r-1}\mathbf{U}^r_{\text{static}} \tag{18}$$

where

$$\mathcal{A}\mathbf{U}^1_{\text{static}} = \overline{\mathbf{b}} \quad \text{and} \quad \mathcal{A}\mathbf{U}^r_{\text{static}} = \mathcal{B}\mathbf{U}^{r-1}_{\text{static}} \quad \text{for } r > 1$$

For more details regarding the method of multiple corrections, see Florea & Hall (1994).

Finally, it is sometimes convenient to construct the reduced order model using a "shift" in the frequency. For example, let $\tilde{\omega} = \omega - \omega_0$, where ω_0 is the a frequency located approximately in the middle of the frequency range of interest. Substitution into Eq. (10) gives

$$\left[\tilde{\mathbf{A}}_0 + j\tilde{\omega}\tilde{\mathbf{A}}_1 - \tilde{\omega}^2\tilde{\mathbf{A}}_2\right]\mathbf{u} = \mathbf{b} \tag{19}$$

where, for example,

$$\tilde{\mathbf{A}}_0 = \mathbf{A}_0 + j\omega_0\mathbf{A}_1 - \omega_0^2\mathbf{A}_2 \tag{20}$$

The shifted reduced order model is constructed as before, working with $\tilde{\omega}$ in place of ω. The advantage of this approach is that the accuracy of the reduced order model using, for example, three static corrections is improved by about one order of magnitude compared to the unshifted reduced order model.

3.4. REDUCED ORDER MODEL COMPUTATIONS

Using the reduced order modelling technique described in the previous section, the unsteady aerodynamic response of the compressor cascade due to pitching is computed for several interblade phase angles over a range of reduced frequencies. For each interblade phase angle considered, a new reduced order model must be constructed. For the cases presented here, a frequency shift $\omega_0 = 0.5$ is used, and all the eigenmodes with reduced eigenvalues $(\lambda c/V)$ within a radius of 1.5 of $j\omega_0$ are retained in the reduced order model. Also, several static corrections are usually performed to improve the accuracy of the model.

Figure 6 illustrates a typical case. Shown is the unsteady surface pressure and displacement thickness for the case of pitching motion with a reduced frequency $\overline{\omega}$ of 0.3, an interblade phase angle σ of 90°, and a steady inflow angle Ω of 62°. Using just two eigenmodes and one static correction, a fair approximation of the exact solution is obtained. Using three static corrections, the reduced order model and the exact solution are nearly identical.

Next, the unsteady surface pressure can be integrated to obtain the unsteady aerodynamic moment. Figure 7 shows the computed aerodynamic moment over a range of interblade phase angles. Again, just two eigenmodes (three for the $\sigma = 45°$ case) and one static correction, the reduced order

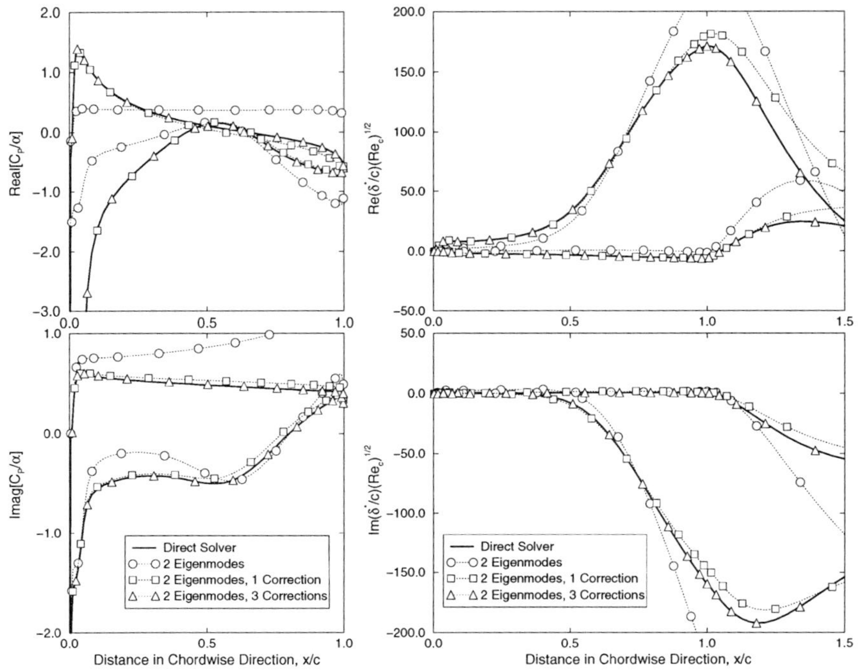

Figure 6. Real and imaginary parts of unsteady coefficient of pressure (left) and unsteady displacement thickness (right) for airfoils pitching about a point near midchord. $\Omega = 62°$, $\sigma = 90°$, $\overline{\omega} = 0.3$.

model results have a similar trend as the exact solution, but are quantitatively different, especially away form the shift frequency ω_0. With three static corrections, however, the reduced order model results and the exact solution are nearly indistinguishable.

A comment regarding computational efficiency is in order. The computational time required to compute the steady flow solution for the previous example was 40 minutes on a Silicon Graphics Origin 2000 computer using a single processor. A single conventional unsteady flow calculation (a single interblade phase angle and frequency) requires about 160 seconds. By way of comparison, approximately 360 seconds of computer time was required to generate a single reduced order model (a single interblade phase angle). However, once computed, the unsteady aerodynamic response can be computed in less than 1 second for each desired reduced frequency.

Finally, we note that the form of the reduced order model presented in this paper is ideally suited for certain specialized applications where low-order models are required – such as the design of active control strategies for controlling rotating stall or stall flutter.

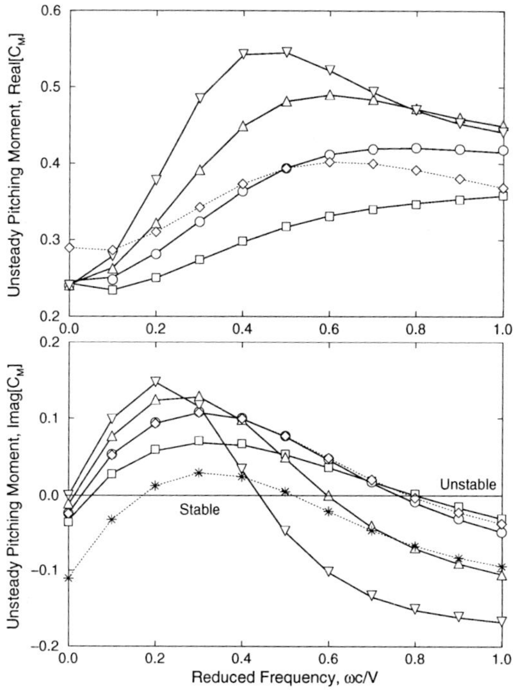

Figure 7. Real and imaginary parts of the aerodynamic pitching moment due to pitching of the airfoils about a point near their midchords. $\Omega = 62°$. □, ○, △, and ▽: reduced order model with three static corrections for $\sigma = 45°$, $90°$, $135°$, and $180°$, respectively; ———: exact solution; - -∗- -, - -◇- -: reduced order model for $\sigma = 90°$ case with zero and one static corrections, respectively.

4. Summary

In this paper, we have developed an efficient reduced order model using the numerically computed eigenmodes of a linearized viscous flow solver. The eigenmodes themselves provide important physical insight into the unsteady fluid flow. For example, an eigenmode thought to correspond to the onset of rotating stall was identified. The reduced order model is constructed using just a handful of eigenmodes, along with a static correction technique similar to that used in structural dynamic problems. The resulting model is computationally efficient, especially when the response at many frequencies is required.

Acknowledgments

This material is based upon work supported by the National Science Foundation under Grant No. CTS-9157908. The United States government has certain rights in this material.

References

Bisplinghoff, R.L., & Ashley, H. (1962) *Principles of Aeroelasticity*, John Wiley and Sons, New York, p. 350.

Cizmas, P.G.A., & Hall, K.C. (1995) "Computation of Steady and Unsteady Viscous Flows Using a Simultaneously Coupled Inviscid-Viscous Interaction Technique," *Journal of Fluids and Structures*, Vol. 9, No. 6, pp. 639–657.

Dowell, E.H., Hall, K.C., & Romanowski, M.C. (1997) "Eigenmode Analysis in Unsteady Aerodynamics: Reduced Order Models," *Applied Mechanics Review*, Vol. 50, No. 6, pp. 371–386.

Emmons, H.W., Pearson, C.E., & Grant, H.P. (1955) "Compressor Surge and Stall Propagation," *Transactions of the ASME*, Vol. 77, May, pp. 450–469.

Florea, R. (1996) "Reduced Order Modelling and Eigenvalue Analysis of Unsteady Flows About Oscillating Airfoils and Turbomachinery Cascades," Ph.D. thesis, Duke University.

Florea, R., & Hall, K.C. (1994) "Reduced Order Modeling of Unsteady Flows About Airfoils," in *AD-Vol. 44, Aeroelasticity and Fluid Structure Interaction Problems*, eds. P.P. Friedman and J.C.I. Chang, ASME.

Florea, R., Hall, K.C., & Cizmas, P.G.A. (1996) "Reduced Order Modelling of Unsteady Viscous Flow in a Compressor Cascade," AIAA Paper 96-2572.

Fransson, T.H., & Verdon, J.M. (1993) "Panel Discussion on Standard Configurations for Unsteady Flow Through Vibrating Axial-Flow Turbomachine Cascades," in *Unsteady Aerodynamics, Aeroacoustics, and Aeroelasticity of Turbomachines and Propellers*, ed. H.M. Atassi, Springer-Verlag, New York, pp. 859–889.

Hall, K.C. (1993) "Deforming Grid Variational Principle for Unsteady Small Disturbance Flows in Cascades," *AIAA Journal*, Vol. 31, No. 5, pp. 891–900.

Hall, K.C. (1994) "Eigenanalysis of Unsteady Flows About Airfoils, Cascades, and Wings," AIAA Paper 94–1427–CP.

Hall, K.C., Clark, W.S., & Lorence, C.B. (1993) "Nonreflecting Boundary Conditions for Linearized Unsteady Aerodynamic Calculations," AIAA Paper 93-0882.

Hall, K.C., Florea, R., & Lanzkron, P.J. (1994) "A Reduced Order Model of Unsteady Flows in Turbomachinery," ASME Paper 94–GT–291.

Mahajan, A.J., Bakhle, M.A., & Dowell, E.H. (1993) "An Efficient Procedure for Cascade Aeroelastic Stability Determination Using Nonlinear, Time-Marching Aerodynamic Solvers," AIAA Paper 93–1631.

Mahajan, A.J., Dowell, E.H., & Bliss, D.B. (1991) "Eigenvalue Calculation Procedure for Euler/Navier-Stokes Solvers with Application to Flows over Airfoils," *Journal of Computational Physics*, Vol. 97, No. 2, pp. 398–413.

Romanowski, M.C. (1996) "Reduced Order Unsteady Aerodynamic and Aeroelastic Models Using Karhunen-Lovee Eigenmodes," presented at the 6th AIAA/USAF/NASA/ISSMO Symposium on Multidisciplinary Analysis and Optimization, Bellevue, WA, Sept. 4–6.

Romanowski, M.C., & Dowell, E.H. (1994) "Using Eigenmodes to Form an Efficient Euler Based Unsteady Aerodynamic Analysis," in *AD-Vol. 44, Aeroelasticity and Fluid Structure Interaction Problems*, eds. P.P. Friedman and J.C.I. Chang, ASME.

Sisto, F. (1987) "Stall Flutter," in *AGARD Manual on Aeroelasticity in Axial Flow Turbomachines, Unsteady Turbomachinery Aerodynamics* (eds. Platzer, M.F., & Carta, F.O.), AGARD–AG–298, Vol. 1, Neuilly sur Seine, France.

Spalart, P.R., & Allmaras, S.R. (1994) "A One-Equation Turbulence Model for Aerodynamics Flows," *La Recherche Aérospatiale*, No. 1, pp. 5–21.

Whitehead, D.S., & Grant, R.J. (1981) "Force and Moment Coefficient of High Deflection Cascades," in *Proceedings of the 2nd International Symposium on Aeroelasticity in Turbomachines*, ed. P. Suter, Juris-Verlag, Zurich, pp.85–126.

TRANSONIC CASCADE FLUTTER IN COMBINED BENDING-CHORDWISE TRANSLATIONAL MODE

Shojiro KAJI

Department of Aeronautics and Astronautics, University of Tokyo,

Bunkyo-ku, Tokyo 113 JAPAN

ABSTRACT

Blade flutter in fan and compressor stages is one of the most serious problems in the development of aeroengines. Especially, when the cascade blades are working supersonically at high loading conditions, the combined bending-chordwise translational mode flutter becomes critical. This paper treats supersonic flutter of cascades with a strong passage normal shock in the blade channel by use of the semi-actuator disk theory. This method enables us to treat correctly the pressure wave radiation in the supersonic upstream field and also the entropy production in the downstream of the normal shock. Results indicate the possibility of flutter at comparatively high frequencies in conjunction with inter-blade phase. The effect of chordwise translation in vibration mode is shown to be significant for the flutter boundary.

INTRODUCTION

In modern fans and compressors of aeroengines, the rotor blades are working under supersonic inlet conditions, although the axial component of the velocity is subsonic. For such a flow, one branch of oblique shocks starting from the leading edge of each blade extends in the upstream field, and usually a strong normal shock whose position is determined by the downstream pressure condition exists inside the blade channel. It is known that bending mode flutter is probable under such conditions, and the problem was once treated by Goldstein et al (1977). But in their analysis it was assumed that no disturbance due to blade vibration existed upstream from the cascade because of the supersonic inlet velocity. This assumption may be valid for one-dimensional supersonic flow, but it is not the case for the present situation where the flow is actually two-dimensional. As we see the oblique shock extends upstream from the blade leading edge, the pressure disturbance due to blade vibration should propagate upstream as long as the axial velocity is subsonic.

In the semi-actuator disk analysis, the flow field upstream from, inside, and downstream from the cascade are combined consistently by solving the linearized governing

T.H. Fransson (ed.), Unsteady Aerodynamics and Aeroelasticity of Turbomachines, 783-795.

equations (Kaji, 1980). Thus, the pressure wave radiated from the cascade in the supersonic upstream field can be legitimately taken into account, and the effect of entropy wave due to shock oscillation is correctly considered (Kaji, 1994).

The other important point is the direction of bending oscillation. We handle the flow of the blade tip region because the unsteady aerodynamic force in the tip region gives predominant effect on the stability of blade vibration. But the first flexure mode of vibration has the node located at the blade hub region, where the stagger angle is generally low. In this case the tip section performs vibration not normal to chord (pure bending) but in the direction nearer to cascade direction (combined bending-chordwise translational mode). One purpose of the present study is to make clear the effect of this chordwise translation on vibrational instability of cascaded blades.

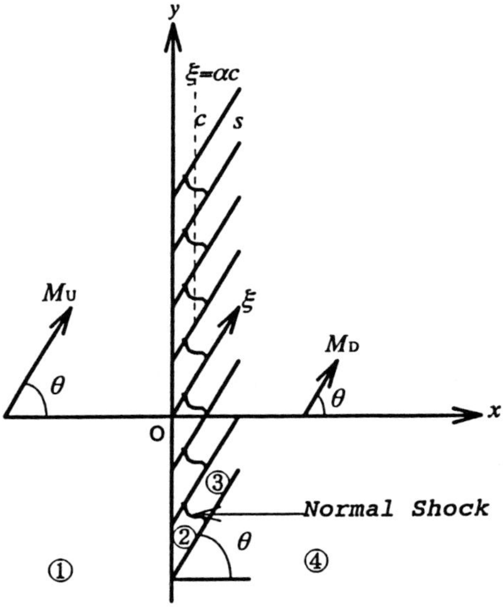

Fig. 1 Cascade geometry and flow field.

FORMULATION

Figure 1 shows the flow field around the vibrating linear cascade which will be modeled by the semi-actuat or disk in the analysis. The cascade is composed of flat plate airfoils of chord length c and placed at stagger angle θ with spacing s. We assume the flow fields upstream and downstream of the cascade are two-dimensional while the flow inside the cascade is one-dimensional. Further, a normal shock is assumed to be located inside the blade channel at $\xi=\alpha c$ ($0<\alpha<1$), where ξ is the chordwise coordinate. Thus in the regions ① and ② of Fig. 1 the flow is supersonic, while it is subsonic in the regions ③ and ④.

In the fields upstream and downstream from the cascade, the perturbation equations of mass, momentum, and energy for small disturbances are (suffixes are omitted):

$$\frac{D\rho}{Dt}+\bar{\rho}\left(\frac{\partial u}{\partial x}+\frac{\partial v}{\partial y}\right)=0 \tag{1}$$

$$\frac{Du}{Dt} = -\frac{1}{\bar{\rho}}\frac{\partial p}{\partial x} \tag{2}$$

$$\frac{Dv}{Dt} = -\frac{1}{\bar{\rho}}\frac{\partial p}{\partial y} \tag{3}$$

$$\frac{1}{\bar{p}}\frac{Dp}{Dt} - \frac{\kappa}{\bar{\rho}}\frac{D\rho}{Dt} = 0 \tag{4}$$

where κ is the ratio of specific heats, and

$$\frac{D}{Dt} = \frac{\partial}{\partial t} + U\frac{\partial}{\partial x} + V\frac{\partial}{\partial y}$$

Here, t is time and the coordinates (x, y) are taken in the axial and cascade direction, respectively. (U, V) are the time-averaged uniform velocity components, $\bar{p}$ and $\bar{\rho}$ are the time-averaged uniform pressure and density. (u, v) are the perturbation velocity components, and p and ρ are the perturbation pressure and density. These perturbation quantities are induced by blade vibration, thus their variations in the cascade direction are governed by the flutter mode. The flutter mode is assumed in the form

$$\exp\left\{i\omega\left(t - \frac{y}{V_s}\right)\right\}$$

where ω is the angular frequency of blade vibration, and V_s is the phase velocity of the flutter mode. Then, assuming the variation in the x-direction in the form of $\exp(\lambda x)$, we obtain four eigen values with respect to λ.

$$\lambda_{\substack{1\\2}} = \frac{a\omega}{a^2 - U^2}\left[\frac{iU(V_s - V)}{aV_s} \pm \sqrt{\frac{a^2 - U^2 - (V_s - V)^2}{V_s^2}}\right] \tag{5}$$

$$\lambda_3 = -\frac{i\omega}{U}\frac{V_s - V}{V_s} \tag{6}$$

Here, a expresses the speed of sound. λ_1 represents the pressure wave propagating upstream, and λ_2 that propagating downstream. λ_3 represents the convective waves such as vorticity waves and entropy waves. In the upstream field we have only the pressure wave radiated by the vibrating blades, whereas we have all three waves in the downstream field, i.e., the pressure wave, the vorticity wave and the entropy wave. The vorticity wave comes from the shed vortexes due to blade vibration, and the entropy wave is attributed to shock oscillation in the cascade.

In order to analyze the flow inside the cascade, we investigate the field confined by two blades adjacent each other as shown in Fig.2.

Since the upper blades is vibrating with the phase difference of $\exp(-i\omega s/V_s)$, the relative displacement s' in the vibrational direction ψ can be expressed by

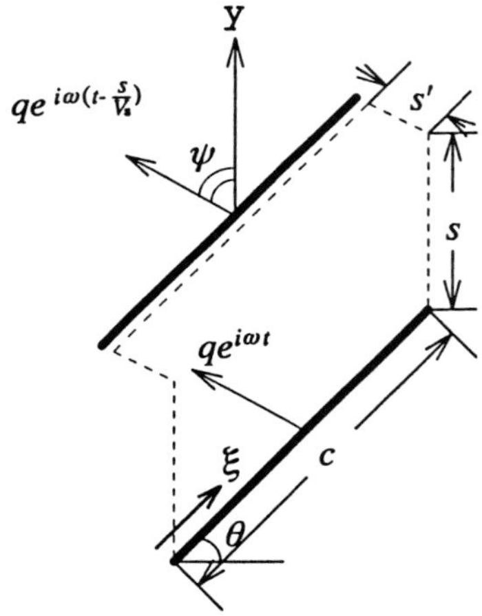

Fig. 2 Cascade channel and control volume.

$$s' = \frac{q}{i\omega} e^{i\omega t}\left[e^{\frac{i\omega s}{V_s}} - 1\right] \cong -\frac{qs}{V_s} e^{i\omega t} \tag{7}$$

where q is the velocity amplitude of blade vibration.

Thus the governing equations inside the blade channel when treated one-dimensionally can be expressed as follows:

$$\frac{D\rho}{Dt} + \bar{\rho}\frac{\partial w}{\partial \xi} = -\bar{\rho}\frac{\cos(\psi - \theta)}{s\cos\theta}\frac{ds'}{dt} \tag{8}$$

$$\frac{Dw}{Dt} = -\frac{1}{\bar{\rho}}\frac{\partial p}{\partial \xi} \tag{9}$$

$$\frac{1}{\bar{p}}\frac{Dp}{Dt} - \frac{\kappa}{\bar{\rho}}\frac{D\rho}{Dt} = 0 \tag{10}$$

where

$$\frac{D}{Dt} = \frac{\partial}{\partial t} + W\frac{\partial}{\partial \xi}$$

ξ is the chordwise coordinate. W is the time- averaged state velocity and w is the perturbation velocity in the ξ direction. The right hand side of Eq. (8) represents the rate of mass flowing out of the control volume which should be considered because the upper blade departs from the control volume although the lower blade is attached to it. This departure is due to the fact that the vibrating velocities are different for two blades.

Here, we introduce the following wave form.

$$\exp\left[i\omega\left(t - \frac{y}{V_s}\right) + \lambda\xi\right]$$

Although we treat the cascade channel flow one-dimensionally, the phase difference from channel to channel should be considered in accordance with the upstream and downstream flow fields. For this purpose the first term in the exponential function is included. Then, we obtain

$$\lambda_{\substack{4\\5}} = -\frac{i\omega}{W \mp a} \tag{11}$$

$$\lambda_6 = -\frac{i\omega}{W} \tag{12}$$

λ_4 and λ_5 represent pressure waves while λ_6 represents an entropy wave.

In the region ② in front of the normal shock, we have two pressure waves. On the other hand, in the region ③ behind the normal shock we have two pressure waves and, in addition, an entropy wave due to shock oscillation. When the normal shock is oscillated, the shock strength changes, and the entropy production changes accordingly. The oscillation velocity of the normal shock is also an unknown quantity to be determined.

Therefore, we have ten unknowns in total. In the region ① there is one pressure wave, in the region ② there are two pressure waves, in the region ③ there are two pressure waves and an entropy wave, and in the region ④ there are one pressure wave, a vorticity wave and an entropy wave. In addition to the oscillation velocity of the normal shock, the amplitudes of these nine waves should be determined in relation to the vibration velocity of blades.

For these ten unknowns we need ten boundary conditions which combine four regions ①, ②, ③ and ④ consistently. In order to specify boundary conditions, we first define the control volume which properly indicates the flow region confined by adjacent two blades in vibration. This is shown in Fig.2 by broken lines. The control volume moves at a constant speed which the lower blade has at the moment of consideration. We must note that the upper blade has a different speed because of the phase difference. For this control volume, the conditions of conservation of mass and conservation of energy should be considered.

Considering the mass flux entering the front part of the control volume and those going out of the rear part and the upper part of it, which should balance with the mass increase in the control volume, we obtain the equation of mass conservation with respect to the control volume as follows.

$$i\omega\cos\theta\left\{\int_0^{\alpha c}\rho_{②}d\xi+\int_{\alpha c}^{c}\rho_{③}d\xi+d\left(\rho_U-\rho_D\right)\right\}+\frac{i\omega s'c}{s}\cos(\psi-\theta)\left\{\alpha\rho_U+(1-\alpha)\rho_D\right\}$$

$$+\left(\rho_D u_{④}-\rho_U u_{①}\right)+\left(\rho_D-\rho_U\right)q\sin\psi+\left(\rho_{④}U_D-\rho_{①}U_U\right)=0 \tag{13}$$

where suffix U and D represent values in front of and behind the normal shock, respectively, the number expresses the value in each region or at the region boundary. d expresses the displacement distance of the normal shock.

Considering the total enthalpy flow into the front part and those out of the rear part and the upper part of the control volume, which balance with the internal energy increase in the control volume, we obtain the equation of energy conservation with respect to the control volume as follows.

$$\rho_D U_D\left\{\frac{\kappa}{\kappa-1}\frac{p_D}{\rho_D}\left(\frac{p_④}{p_D}-\frac{\rho_④}{\rho_D}\right)+U_D\left(u_④+q\sin\psi\right)+V_D\left(v_④-q\cos\psi\right)\right\}$$

$$-\rho_U U_U\left\{\frac{\kappa}{\kappa-1}\frac{p_U}{\rho_U}\left(\frac{p_①}{p_U}-\frac{\rho_①}{\rho_U}\right)+U_U\left(u_①+q\sin\psi\right)+V_U\left(v_①-q\cos\psi\right)\right\}$$

$$-i\omega\cos\theta\left\{\int_0^{\alpha c}\rho_② d\xi+\int_{\alpha c}^{c}\rho_③ d\xi+d\left(\rho_U-\rho_D\right)\right\}\left\{\frac{\kappa}{\kappa-1}\frac{p_U}{\rho_U}+\frac{1}{2}W_U^{\,2}\right\}+i\omega\cos\theta\times$$

$$\left[\int_0^{\alpha c}\left(\frac{W_U^{\,2}}{2}\rho_②+\frac{p_②}{\kappa-1}+\rho_U W_U w_②\right)d\xi+\int_{\alpha c}^{c}\left(\frac{W_D^{\,2}}{2}\rho_③+\frac{p_③}{\kappa-1}+\rho_D W_D w_③\right)d\xi\right]$$

$$+i\omega d\cos\theta\left\{\frac{1}{\kappa-1}\left(p_U-p_D\right)+\frac{1}{2}\rho_U W_U\left(W_U-W_D\right)\right\}=0 \tag{14}$$

where κ is the ratio of specific heats.

At the leading edge plane of the cascade, it is assumed that the unique incidence condition should be satisfied. This condition is usually applied for the steady supersonic cascade flow, but here we apply it also for the unsteady supersonic cascade flow. This condition is described as

$$v_①-q\cos\psi=\left(u_①+q\sin\psi\right)\tan\theta \tag{15}$$

At the position of the oscillating normal shock, usual conservation conditions of mass, momentum, and energy should be satisfied. These are described as follows. The conservation of mass flow is

$$\rho_U w_②+W_U\rho_②-\left(\rho_D w_③+W_D\rho_③\right)-\left(\rho_U-\rho_D\right)i\omega d=0 \tag{16}$$

The conservation of momentum is

$$p_②+2\rho_U W_U w_②+W_U^{\,2}\rho_②-\left(p_③+2\rho_D W_D w_③+W_D^{\,2}\rho_③\right)=0 \tag{17}$$

The conservation of energy is

$$\rho_U W_U\left\{\frac{\kappa}{\kappa-1}\frac{p_U}{\rho_U}\left(\frac{p_②}{p_U}-\frac{\rho_②}{\rho_U}\right)+W_U w_②\right\}$$

$$-\rho_D W_D\left\{\frac{\kappa}{\kappa-1}\frac{p_D}{\rho_D}\left(\frac{p_③}{p_D}-\frac{\rho_③}{\rho_D}\right)+W_D w_③\right\}-\left(p_D-p_U\right)i\omega d=0 \tag{18}$$

Finally, at the trailing edge plane of the cascade we assume that the flow inside the cascade continues smoothly to the flow downstream of the cascade. That means the continuation of two components of velocity, the continuations of pressure and entropy. They are described as follows.

$$p_③=p_④ \tag{19}$$

$$\rho_③=\rho_④ \tag{20}$$

$$w_③\cos\theta=u_④+q\sin\psi \tag{21}$$

$$w_{③}\sin\theta = v_{④} - q\cos\psi \tag{22}$$

Each quantity should be evaluated at the region boundary, i.e., at the trailing edge plane.

Concludingly, we apply the unique incidence condition for the supersonic leading edge, and the Kutta condition for the subsonic trailing edge. These ten boundary conditions determine amplitudes of unknown waves in relation to the vibrational velocity of blades.

Applying the momentum principle to the control volume shown in Fig.2, we can find the x- and y-component of the aerodynamic force acting on vibrating blades. Considering the momentum entering the front part of the control volume, and those getting out of the rear part and the upper part of it in addition to the forces acting on the control surface, which all together should be in balance with the momentum increase in the control volume, we obtain the momentum equation for each component. The x-wise momentum equation becomes

$$p_{①} - p_{④} + (p_U - p_D)\frac{s'}{s}\cos\psi - \frac{1}{s}\left\{f_U + f_L e^{-\frac{i\omega s}{V_s}}\right\}_x - i\omega\cos\theta \times$$

$$\left[\int_0^{\alpha c+d}\{\rho_{②}U_U + \rho_U(w_{②}\cos\theta - q\sin\psi)\}d\xi + \int_{\alpha c+d}^{c}\{\rho_{③}U_D + \rho_D(w_{③}\cos\theta - q\sin\psi)\}d\xi\right]$$

$$\left[+\rho_{①}U_U^{\,2} - \rho_{④}U_D^{\,2} + \rho_U U_U\left\{2(u_{①} - u_{④}) + \frac{s'}{s}\cos(\psi - \theta)(W_U - W_D)\right\}\right]$$

$$-i\omega c\frac{s'}{s}\cos(\psi - \theta)\rho_U U_U = 0 \tag{23}$$

where, f_U and f_L express the unsteady force acting on the upper surface and the lower surface of a blade, respectively. The y-wise momentum equation becomes

$$(p_U - p_D)\frac{s'}{s}\sin\psi - \frac{1}{s}\left\{f_U + f_L e^{-\frac{i\omega s}{V_s}}\right\}_y - i\omega\cos\theta \times$$

$$\left[\int_0^{\alpha c+d}\{\rho_{②}V_U + \rho_U(w_{②}\sin\theta + q\cos\psi)\}d\xi + \int_{\alpha c+d}^{c}\{\rho_{③}V_D + \rho_D(w_{③}\sin\theta + q\cos\psi)\}d\xi\right]$$

$$+\rho_U U_U\tan\theta\left\{u_{①} - u_{④} + (W_U - W_D)\frac{s'}{s}\cos(\psi - \theta)\right\} + \rho_{①}U_U V_U - \rho_{④}U_D V_D$$

$$+\rho_U U_U(v_{①} - v_{④}) - i\omega c\frac{s'}{s}\rho_U U_U\tan\theta\cos(\psi - \theta) = 0 \tag{24}$$

From these two components the aerodynamic force acting on blades in the direction of vibration, f_ψ, can be determined.

$$\frac{f_\psi}{\rho_U a_U s q} = C_{fr} + iC_{fi} \tag{25}$$

If the force component, C_{fr}, in phase with the vibration velocity of blades, q, is positive, then the vibration energy is enhanced and flutter will occur.

RESULTS AND DISCUSSION

The aerodynamic force acting on blades in the direction of blade vibration is the function of the inter-blade phase parameter $\sigma = \omega c / V_s$ and the reduced frequency $k = \omega c / W_U$ for the specified cascade flow and vibration conditions such as M_U, θ and ψ. The normal shock position inside the blade channel may depend on the downstream pressure and the details of blade profile, but it is given by the parameter α here, which expresses the distance from the leading edge to the normal shock relative to blade chord. The cascade with stagger angle $\theta = 60^\circ$ was chosen, and not only pure bending vibration of $\psi = 60^\circ$ but also combined bending-chordwise translational mode, in which $\psi < 60^\circ$ was investigated.

Figure 3 shows the variation of the real part of the aerodynamic force, C_{fr}, against the reduced frequency k for various inter-blade phase parameter σ. The angle of blade vibration is $\psi = 60^\circ$ (pure bending mode). The upstream Mach number $M_U = 1.2$ and the normal shock position parameter $\alpha = 0.3$. We find each curve has a sharp peak in the lower frequency region and a relatively dull hump in the higher frequency region. These peak and hump occur at or near the cascade resonance conditions.

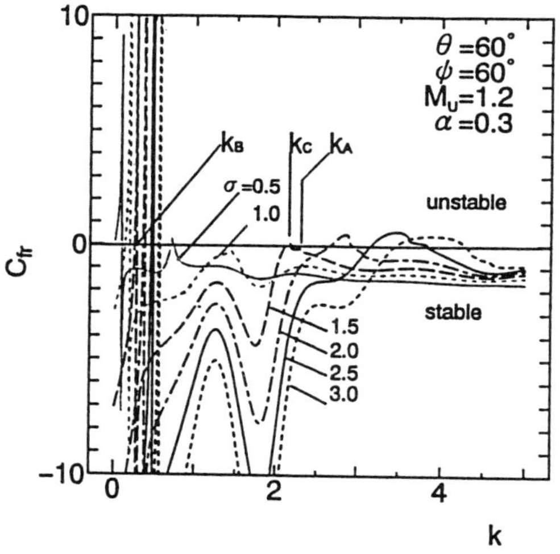

Fig. 3 Aerodynamic force coefficient C_{fr} versus reduced frequency k for various inter-blade phase parameter σ.

Figure 4 explains the situation where the cascade resonance occurs. Physically, the cascade resonance corresponds to the condition at which the wave pattern of the upstream field or the downstream field changes from decaying to propagating. That means mathematically the value inside the square root of λ_1 or λ_2 changes its sign. In Fig.4 the circular wave front indicates the disturbance emitted by the vibrating blade located at the origin. This disturbance spread at the velocity of sound with its center being convected by the free stream velocity. The same disturbance propagates along the cascade at the phase velocity V_s. If V_s coincides with the velocity at which the

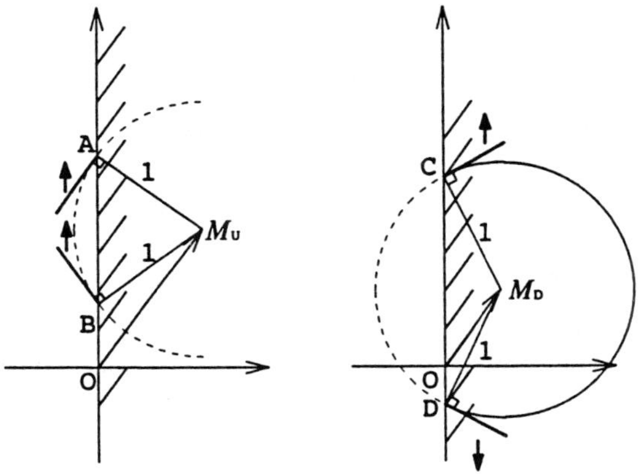

Fig. 4 Cascade resonance conditions.

circular wave front sweeps the cascade, then the cascade resonance will result. In the upstream field we have two cases of cascade resonance because the main stream velocity is supersonic thus the circular wave front crosses the cascade twice.

The sharp peak in Fig. 3 in the lower frequency region corresponds to the case of the lower phase velocity which signifies the cascade resonance occurring at the point B in Fig. 4. The case of the higher phase velocity occurring at the point A corresponds to the cascade resonance of a higher frequency, which is indicated by k_A on the curve for σ=1.5 in Fig. 3. In the downstream field we have one case of cascade resonance for positive V_s occurring at the point C and another cascade resonance for negative V_s occurring at the point D. But as we see Fig. 5, for negative V_s the vibration at k_D is stabilized.

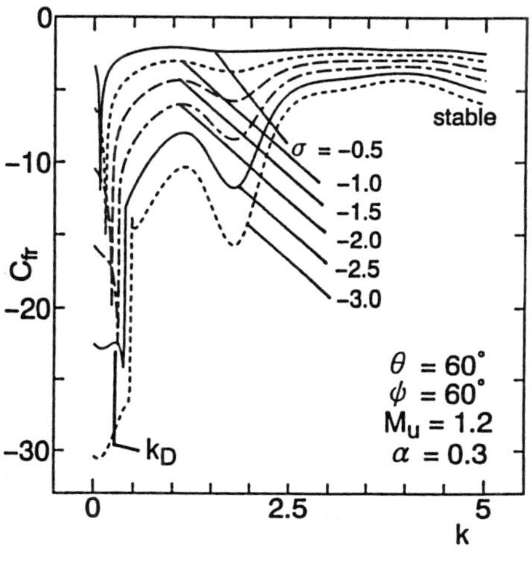

Fig. 5 C_{fr} versus k for negative σ.

Figure 6 and 7 show the results for the combined bending-chordwise translational mode of $\psi=15^{\circ}$, for positve σ and negative σ, respectively. In this case the real part of the aerodynamic force becomes positive at lower frequencies for negative values of σ, i.e., in the forward traveling mode. We can recognize the highest reduced frequency of instability occurs at $\sigma=1.0$ at around k=0.33. This reduced frequency is called flutter boundary. As seen is Fig. 5, there is no flutter boundary for the case of pure bending mode of $\psi=60^{\circ}$. But if we reduce the angle ψ to 45°, the flutter boundary appears at around k=0.06. At $\psi=30^{\circ}$, the flutter boundary is around k=0.13. Therefore, we can conclude that as the degree of chordwise translation increases the flutter boundary extends to higher frequencies. Because of spanwise twist, this situation is the case for the tip region of actual fan blades. The blade tip region contributes greatly to the exciting force of blade vibration.

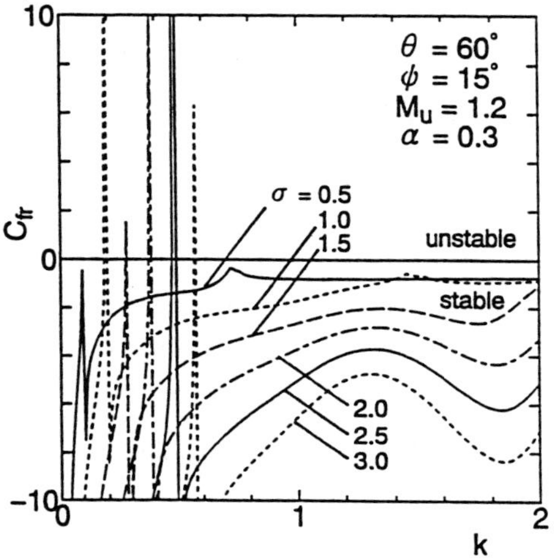

Fig. 6 C_{fr} versus k for combined bending-chordwise translational mode, $\sigma>0$.

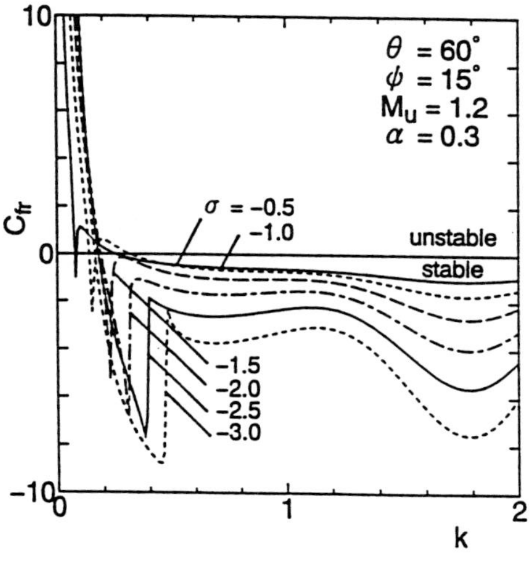

Fig. 7 C_{fr} versus k for combined bending-chordwise translational mode, $\sigma<0$.

Figure 8 shows the results for ψ=15°, in which the inlet flow Mach number is changed. The value of σ is kept constant at 1.0. We can recognize that as the inlet flow Mach number increases, the flutter boundary extends to higher frequencies.

Figure 9 shows the effect of the normal shock position in the cascade channel. Although the general tendency does not change much, the flutter boundary of frequency is actually affected by the shock position.

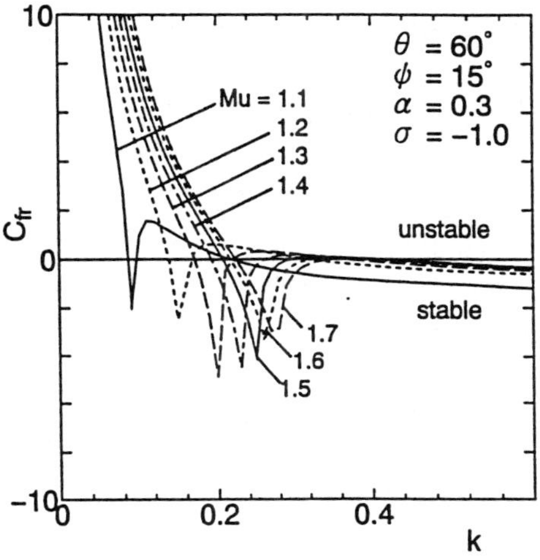

Fig. 8 Flutter boundary change due to inlet flow Mach number.

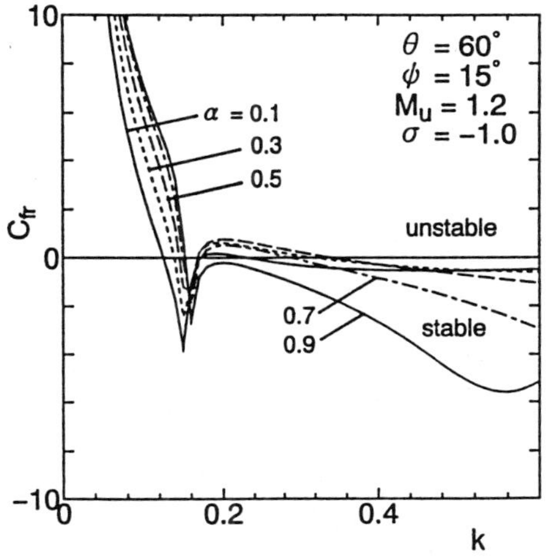

Fig. 9 Effect of normal shock position on C_{fr} versus k relation.

If there is no throat in the cascade channel, the normal shock may not be produced. Then, the time-averaged flow field becomes uniformly supersonic. Figure 10 shows the results for the uniformly supersonic case for positive σ, and Fig. 11 for negative σ. We can recognize that the blade vibration is generally stabilized. The high peak at lower frequencies in Fig. 10 corresponds to the resonance of type B in Fig. 4. At the other resonance of type A, C_{fr} seems to become zero.

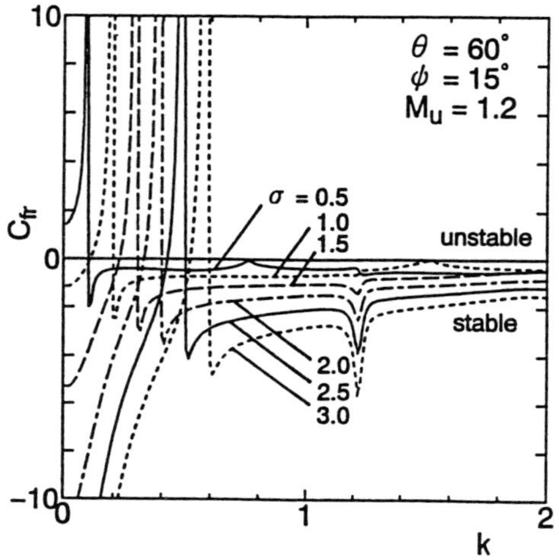

Fig. 10 C_{fr} versus k for uniformly supersonic case (no shock in cascade), $\sigma>0$.

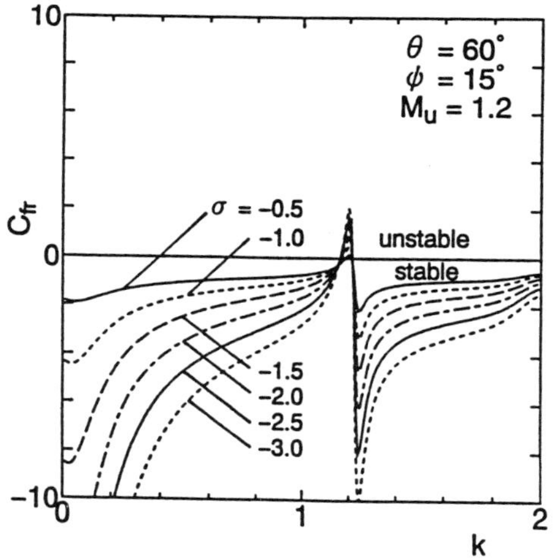

Fig. 11 C_{fr} versus k for uniformly supersonic case (no shock in cascade), $\sigma<0$.

CONCLUSION

Bending mode flutter with a strong normal shock in the cascade channel was investigated by the semi-actuator disk method. Possible disturbances induced by shock and blade vibration such as the pressure wave in the supersonic upstream field and the entropy wave downstream of the normal shock are all taken into consideration.

The aerodynamic force component which determines whether the blade vibration is excited or damped out shows a complex variation against the increase in reduced frequency. The curve for this force component rises at or near cascade resonance conditions, and the possibility of flutter in a fairly wide range in a high frequency region is predicted.

Although in the pure bending mode of vibration the flutter boundary of frequency does not appear, it actually does as the chordwise translation component is included in the vibration mode. The flutter boundary grows as the degree of chordwise translation increases. In the actual turbomachinery, twisted blades have a possibility of flutter because of this chordwise translation effect.

The existence of the normal shock in the cascade channel contributes the occurrence of flutter. In the uniformly supersonic flow field, flutter in usual sense would not take place.

REFERENCES

M. E. Goldstein et al., “Unsteady Flow in a Supersonic Cascade with Strong In-Passage Shocks”, J. Fluid Mech. 83-3(1977), 569.

S. Kaji, “Stall Flutter of Linear Cascade in Compressible Flow”, C282/80, Vibrations in rotating machinery, I Mech E. (1980), 209.

S. Kaji, “Analysis of Cascade Flutter with Strong Shocks by the Semi-Actuator Disk Method”, Proc. Asian-Pacific Conference on Aerospace Technology and Science (1994), 810.

ON THE PREDICTION OF DYNAMIC STALL ONSET ON AIR-FOILS IN LOW SPEED FLOW

K. D. JONES and M. F. PLATZER
Naval Postgraduate School
Code AA/Jo, 699 Dyer Rd., Monterey, CA 93943-5106

1. Abstract

A computational approach is described for the rapid and systematic prediction and evaluation of the onset of dynamic stall due to rapid incidence changes or unsteady pitch or plunge motions. The method combines an unsteady, two-dimensional panel code with a two-dimensional boundary-layer code. The panel code provides incompressible, inviscid flowfields about arbitrary airfoils undergoing prescribed motions. The boundary-layer code computes laminar, transitional and turbulent regimes.

Presented results demonstrate that the delay in dynamic stall onset is directly related to the dynamic pressure lag, in agreement with previous Navier-Stokes simulations. Also, results are presented showing the effect of airfoil shape and Reynolds number on the onset of dynamic stall.

2. Nomenclature

c = chord length
C_p = pressure coefficient, $(p - p_\infty)/q_\infty$
f = frequency in Hz
k = for ramp motion: nondimensional pitch rate, $2\pi/\tau_c$
for oscillating motion: reduced frequency, $2\pi fc/V_\infty$
p = pressure
p_∞ = freestream pressure
q_∞ = freestream dynamic pressure, $1/2\rho_\infty V_\infty^2$
R_L = chord Reynolds number, $V_\infty c/\nu_\infty$
t = time
U = tangent velocity in the boundary layer
U_e = boundary layer edge velocity
V_∞ = freestream velocity
α = angle of attack, positive clockwise

T.H. Fransson (ed.), Unsteady Aerodynamics and Aeroelasticity of Turbomachines, 797-812.

α_0 = angle of attack at $\tau = 0$
$\Delta\alpha$ = pitch amplitude
ν_∞ = freestream kinematic viscosity
ρ_∞ = freestream density
τ = nondimensional time, tV_∞/c
τ_c = nondimensional time-to-rise for ramp motion, $2\pi/k$

3. Introduction

Rapid incidence changes or unsteady airfoil motions are known to cause the phenomenon of dynamic stall. For flow moving from left to right this phenomenon involves the generation of a clockwise vortex that is produced by the large amount of coherent vorticity created near the airfoil's leading edge region. As the incidence angle is increased well past the static stall angle, the vortex begins to grow and convect over the suction surface. A survey of the available literature, for example the recent review paper of Carr and Chandrasekhara[1], reveals that the process of dynamic stall is strongly dependent on the airfoil geometry, especially on the leading-edge shape, Mach number, degree of flow unsteadiness or nondimensional pitch rate, state of the airfoil boundary layer, type of airfoil motion (pitch or plunge), location of pitch axis, etc. As a result of this dependence on many parameters, the computational prediction of dynamic stall using Navier-Stokes codes is a very time-consuming task and, as pointed out by Ekaterinaris and Platzer[2], is hampered by the need to include the proper modeling of the boundary-layer transition process in addition to adequate turbulence modeling. On the other hand, semi-empirical dynamic-stall models require little computational effort but they tend to obscure the major physical mechanisms which influence the dynamic-stall process. Ericsson and Reding[3], in their review of the fluid dynamics of unsteady, separated flow, emphasized the need for the modeling of the proper physical phenomena, especially for the inclusion of time-lag effects due to circulation lag, boundary-layer convection lag, and lag due to moving flow separation points. In particular, they introduced the moving-wall or leading-edge jet effect by pointing out that an upward pitching airfoil creates fuller, even jet-like boundary layer profiles which therefore delay dynamic stall.

In this paper, we attempt to quantify the time-lag effects by using a combined panel and boundary-layer code for the analysis of unsteady, incompressible, attached airfoil flows with laminar, transitional or turbulent boundary layers. The use of such a code has the advantage of computational efficiency, but it limits the analysis to low-speed flows. Nevertheless, it enables the prediction of the onset of dynamic stall because, for many airfoils, the dynamic-stall process is initiated by the formation of a laminar separation bubble near the airfoil leading edge at a relatively low angle of attack. Hence predicting the onset of dynamic stall becomes a problem of predicting laminar flow separation.

In recent experiments, Chandrasekhara *et al.* [4] and Chandrasekhara and Carr [5] have provided conclusive evidence of the importance and magnitude of the time-lag effect caused by transiently or sinusoidally pitching airfoils. The effect of pitch rate or reduced frequency on the initiation of dynamic stall on a NACA 0012 airfoil in flows with freestream Mach numbers of 0.2 and 0.25, as presented in Refs. 4 and 5, are shown in Figs. 1 and 2, respectively.

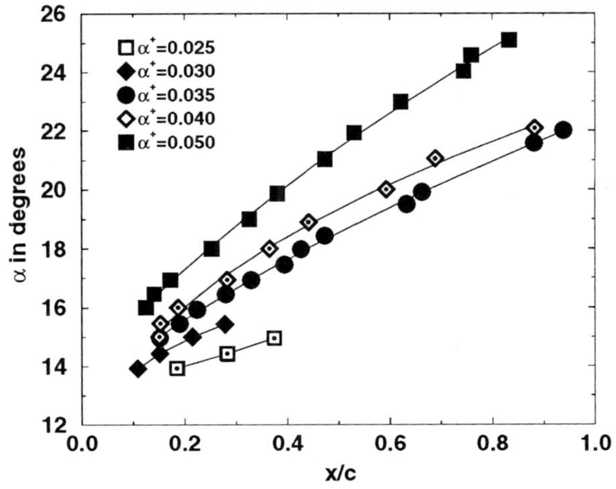

Figure 1. Dynamic-stall onset measurements at Mach 0.20 (Ref. 4.). The nondimensional pitch-rate is $\alpha^{+}=\dot{\alpha}c/V_{\infty}$ where $\dot{\alpha}$ is the pitch-rate in deg/s.

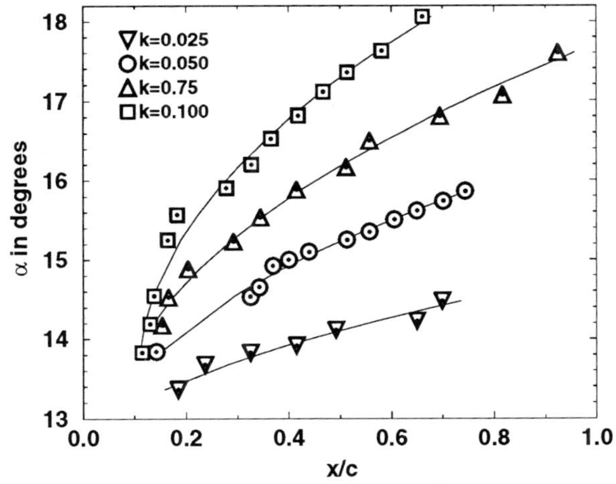

Figure 2. Dynamic-stall onset measurements at Mach 0.25 (Ref. 5). Here the reduced frequency is given by $k=2\pi c/V_{\infty}$.

The Reynolds number was 400,000. It is seen that an increase in pitch rate or reduced frequency caused a delay in the dynamic stall vortex formation and propagation. It must be emphasized that these experiments were based on flow-visualization experiments using a stroboscopic Schlieren system. Therefore, the identification of the dynamic-stall vortex is subject to some error and, most importantly, the details of the dynamic-stall initiation process cannot be ascertained with such a method. In particular, the incidence angle at which a small separation bubble is formed cannot be determined, but it is obvious that it occurs at incidence angles which are much smaller than the angles plotted in Figs. 1 and 2.

The hybrid panel/boundary-layer code used in this paper allows the rapid and precise determination of the pressure distributions, pressure-lag effects and boundary-layer characteristics due to airfoil incidence changes, oscillation mode, frequency and amplitude, as well as the change of these characteristics with airfoil geometry. The major elements of this hybrid code are described in the next section, which is followed by a description of representative results.

4. Approach

The numerical methods utilized in the panel code and the boundary-layer code are briefly summarized in the following subsections, with details and validations of the methods available in the cited references.

4.1. PANEL CODE

Flow solutions are computed using an unsteady, potential-flow code originally developed by Teng [6]. The basic, steady panel code follows the approach of Hess and Smith [7], where the airfoil is approximated by a finite number of panels, each with a local, uniform, distributed source strength and all with a global, uniform, distributed vorticity strength. For n panels there are n unknown source strengths, q_j, and an unknown vorticity strength, γ. Boundary conditions include flow tangency at the midpoint of the n panels and the Kutta condition which postulates that the pressure on the upper and lower surfaces of the airfoil at the trailing edge must be equal.

The unsteady panel code adopts the procedure of Basu and Hancock [8], where a wake panel is attached to the trailing edge through which vorticity is shed into the flow, as shown in Fig. 3.

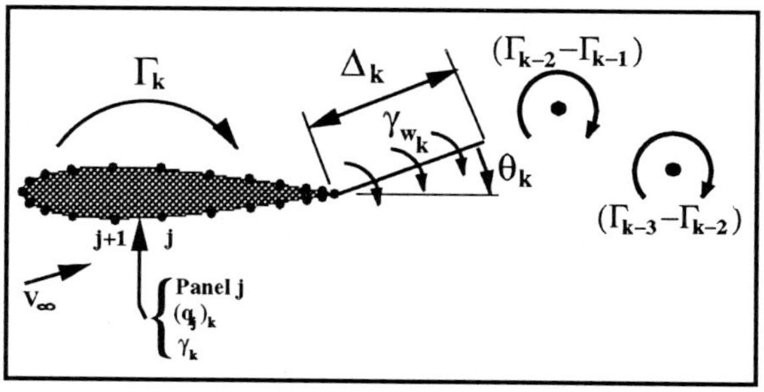

Figure 3. Schematic of the panel code wake model.

The Helmholtz theorem states that the bound vorticity in a flow remains constant, thus a change in circulation about the airfoil must result in the release of vorticity into the wake equal in magnitude and opposite in direction, given numerically by

$$\Delta_k(\gamma_W)_k + \Gamma_k = \Gamma_{k-1} \tag{1}$$

where Δ is the wake panel length, γ_W is the distributed vorticity strength on the wake panel and Γ is the circulation about the airfoil, and where the subscript k indicates the current time step, and $k-1$ indicates the previous time step.

The wake panel introduces two additional unknowns; the wake panel length and its orientation, θ_k. Thus, two additional conditions must be specified for closure;

1. The wake panel is oriented in the direction of the local resultant velocity at the panel midpoint.
2. The length of the wake panel is proportional to the magnitude of the local resultant velocity at the panel midpoint and the time-step size.

At the end of each time step the vorticity contained in the wake panel is concentrated into a point vortex which is shed into the wake and convected downstream with the flow, influencing and being influenced by the other shed vortices and the airfoil. Note, implementation of this approach requires an iterative scheme, since the velocity direction and magnitude used to define the wake panel are not initially known. Note also that this wake model is nonlinear. The unsteady panel code has been extensively documented in Refs. 6 and 9-14.

4.2. BOUNDARY-LAYER CODE

Flow properties in the boundary layer are computed using the Keller-Cebeci box method [15]. The code was generated and combined with a steady panel code by Nowak [16]. The general algorithm and several modifications used in the current implementation are discussed below.

The boundary layer code treats the airfoil surface as a flat plate with a variable pressure gradient, and steady conditions are assumed within the boundary layer. Laminar, transitional and turbulent regions are considered, and the turbulent region is computed using the Cebeci-Smith (CS) eddy-viscosity model. Like all eddy-viscosity methods, the CS model leaves the basic boundary layer equations unchanged but modifies the viscosity term by adding a local eddy viscosity, $\nu = \nu_\infty + \epsilon_m$. The CS model divides the viscous region into an inner and an outer layer with the eddy viscosity in each region empirically formulated. The inner region is modeled by

$$\left(\frac{\epsilon_m}{\nu}\right)_i = 0.16\sqrt{Re_x}\Big[1 - \exp(-y/A)\Big]^2 \eta^2 \nu_\infty \gamma_{tr} \tag{2}$$

and the outer region is modeled by

$$\left(\frac{\epsilon_m}{\nu}\right)_o = 0.0168\sqrt{Re_x}\Big[\eta_e - f_e\Big]\gamma_{tr} \tag{3}$$

where

$$Re_x = \frac{U_e}{V_\infty}\xi R_L \tag{4}$$

$$\frac{y}{A} = \frac{\eta}{26} \sqrt[4]{Re_x} \sqrt{\nu_w} \tag{5}$$

and where ξ, η and f are the Falkner-Skan variables. The term γ_{tr} models the length of the transition or intermittency region, and its formulation is discussed below.

The point of transition onset is of critical importance for the prediction of dynamic stall. For the steady implementation of the code developed by Nowak the transition point is specified as input, presumably determined from experimental data. This is of little use in the present unsteady approach, as transition points would need to be specified for an infinite variety of conditions. Thus Michel's criterion is used to predict transition onset, where transition is initiated when the Reynolds number based on momentum thickness, R_θ, and the Reynolds number based on x, Re_x, satisfy the equation

$$R_{\theta_{tr}} = 1.174\left(1 + \frac{22,400}{Re_{x_{tr}}}\right) Re_{x_{tr}}^{0.46} \tag{6}$$

where $R_\theta = U_e\theta/\nu_\infty$.

The Chen-Thyson intermittency model is used to predict the transition length where

$$\gamma_{tr} = 1 - \exp\left[-G(x - x_{tr}) \int_{x_{tr}}^{x} \frac{1}{U_e} dx\right] \tag{7}$$

and

$$G = \frac{1}{G_{tr}} \left(\frac{U_e}{V_\infty}\right)^3 R_L^2 Re_{x_{tr}}^{-1.34} \tag{8}$$

In the original Chen-Thyson formulation G_{tr} is set to a constant value of 1200, but in the present implementation it is given by Cebeci [17] as a function of the transition Reynolds number

$$G_{tr} = 71\left[\ln(Re_{x_{tr}}) - 4.732\right] \tag{9}$$

The use of Michel's criterion for the prediction of transition onset and the use of the Chen-Thyson model for the transition region has been validated in Refs. 17 and 18.

Note, that while the boundary-layer routine is steady, it has been shown that, for low reduced frequencies, changes in the boundary layer occur much more quickly than changes in the external flow, thus a steady boundary-layer analysis is sufficient [19].

The present combination of panel and boundary-layer codes is not an inviscid/viscous interaction approach (eg. Cebeci *et al.* [20]); that is, information is only passed from the panel code to the boundary-layer code; never the other way. Therefore, it is not possible to predict the viscous flow region beyond separation, as the flow separation would have a significant influence on the effective body shape seen by the panel code. However, the hybrid code can be used to predict the point in the cycle and on the airfoil surface where separation first occurs, and by looking in detail at the velocity profiles in the boundary layer, in particular in the region surrounding the suction peak, much insight may be obtained into the initial stages of dynamic stall. Note, the present code runs simulations in a matter of seconds on a workstation, and does not have convergence problems sometimes encountered with inviscid/viscous interaction methods.

5. Results

It is important to note that the only information the boundary layer algorithm is given is the surface point distribution and the external velocity distribution computed by the panel code at each time step. From the velocity distribution the local pressure and pressure gradient are computed, and it is really the pressure gradient that determines the nature of the boundary layer. Thus, it is essential to compute the surface pressure distribution accurately, and by looking at the pressure distributions, as well as the predicted position of the stagnation point, much can be determined about the flowfield without the aid of the boundary layer code.

As an additional consideration, note that in the unsteady panel code the pressure distribution is computed from the unsteady Bernoulli equation and depends on both the local velocity and the time-rate-of-change of the potential. In the boundary layer code this second term is not considered, but for the frequencies of interest in this study the term is negligible.

It is instructive to consider the pressure and boundary layer response of a NACA 0012 airfoil to a ramp change in angle of attack given by the function

$$\alpha(\tau) = \alpha_0 + \Delta\alpha \left[10\left(\frac{\tau}{\tau_c}\right)^3 - 15\left(\frac{\tau}{\tau_c}\right)^4 + 6\left(\frac{\tau}{\tau_c}\right)^5 \right] \tag{10}$$

where

$$\tau_c = \frac{2\pi}{k} \ . \tag{11}$$

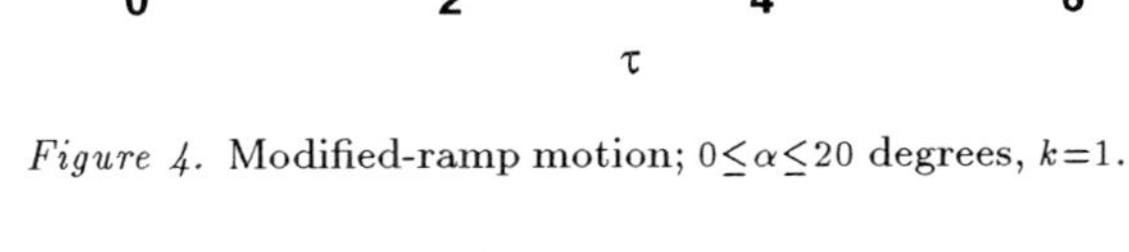

Figure 4. Modified-ramp motion; $0 \leq \alpha \leq 20$ degrees, $k=1$.

This function is depicted in Fig. 4. The constant k controls the pitch rate of the airfoil about the leading edge and hence the rate of incidence change. It is seen that the pressure build-up increasingly lags behind the steady-state pressure at a given angle of attack as the rate of incidence change, k, is increased. The pressure distributions in the vicinity of the leading edge are shown in Fig. 5 as a function of k at an incidence of 4.53 degrees.

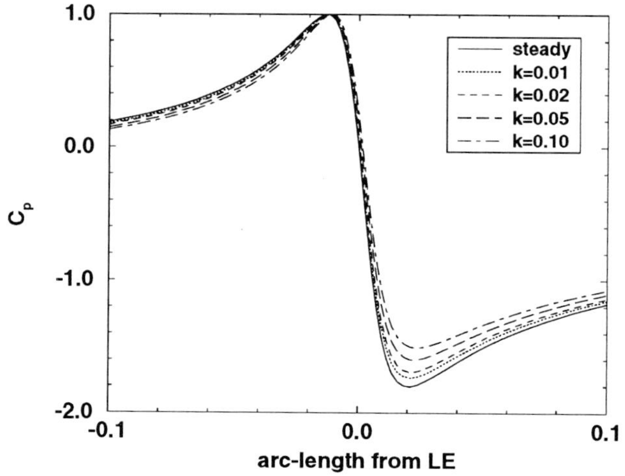

Figure 5. Pressure distribution near the LE.

The boundary-layer profiles at the same incidence angle of 4.53 degrees at a position 10 percent downstream from the leading edge on the suction surface are shown in Fig. 6. It is seen that the profiles are less developed as k is increased.

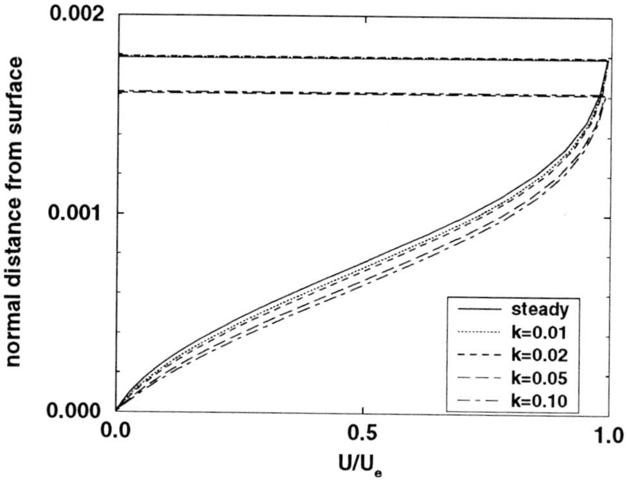

Figure 6. Boundary layer profiles.

This is consistent with the lag in migration of the stagnation point, as shown in Fig. 7, as k is increased because it produces a shorter running length for the boundary-layer build-up from the stagnation point to the 10 percent chord position.

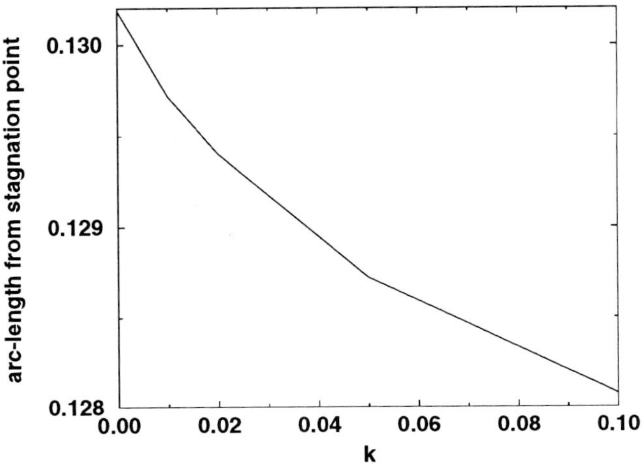

Figure 7. Distance from stagnation point to $0.1c$.

Using the ramp change in angle of attack given by Eq. (10) between an initial value of 0 degree to a final value of 20 degrees it is possible to search for the incidence angle where laminar flow separation is first predicted. The dynamic stall onset angle for a NACA 0012 airfoil for a range of Reynolds numbers and frequencies is shown in Fig. 8.

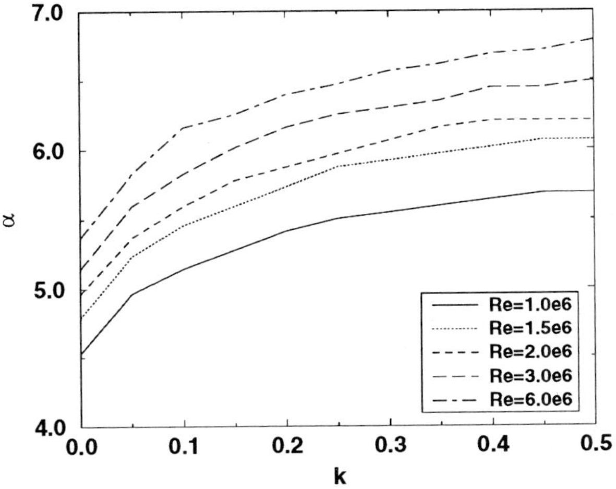

Figure 8. Stall onset AOA variation for a NACA 0012.

As expected, laminar separation is delayed as the rate of incidence change is increased. The pressure distributions and the pressure gradient distributions, shown in Figs. 9 and 10 for a Reynolds number of 1 million, are virtually identical in each case. An increase in Reynolds number from 1 to 6 million delays the dynamic stall onset by approximately 1 degree.

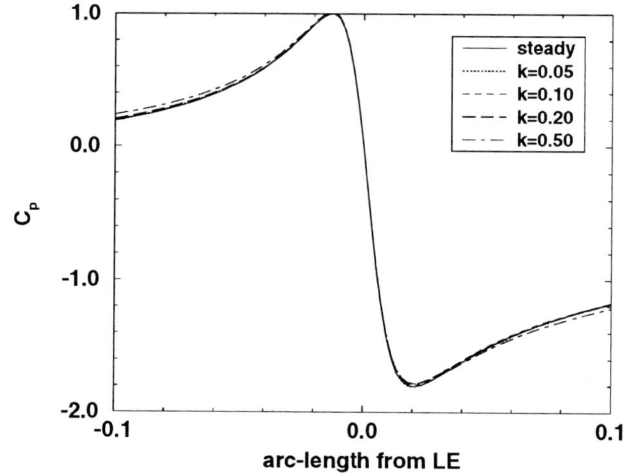

Figure 9. Pressure distribution near the LE at stall onset.

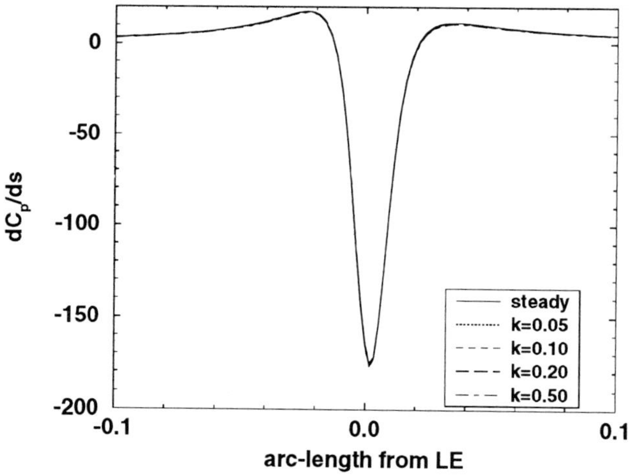

Figure 10. Pressure gradient distribution near the LE at stall onset.

A sharper leading edge reduces the dynamic stall onset angle, as shown in Fig. 11 for the NACA 0009 airfoil, whereas a rounder leading edge delays it, as depicted in Fig. 12 for the NACA 0015 airfoil. Additional results are shown in Figs. 13 and 14 for the NACA 1412 and 2412 airfoils.

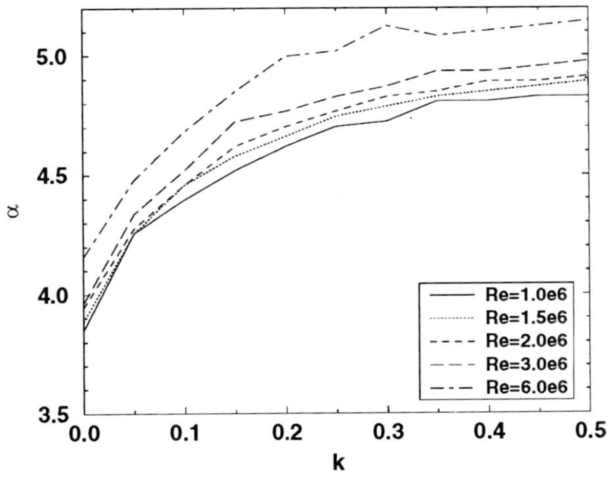

Figure 11. Stall onset AOA variation for a NACA 0009.

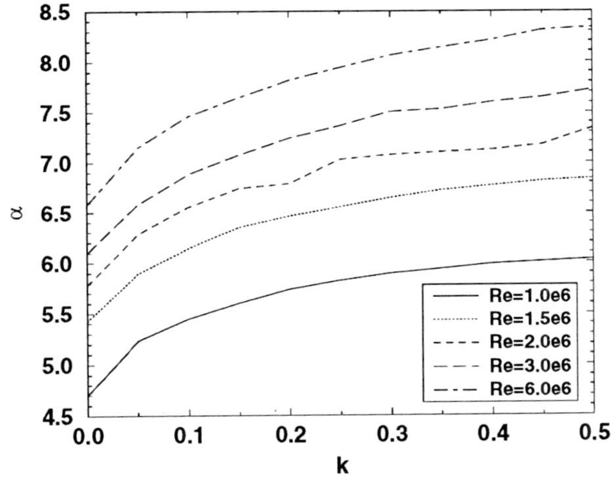

Figure 12. Stall onset AOA variation for a NACA 0015.

Ekaterinaris and Platzer [2] showed in their Navier-Stokes analysis of the dynamic stall behavior of a NACA 0012 airfoil which oscillates with an amplitude of 5 degrees about a mean angle of attack of 9 degrees in a low subsonic flow (M=0.3) of 4 million Reynolds number that it is essential to incorporate the boundary layer transition near the leading edge in order to obtain the experimentally measured lift, drag and moment hysteresis loops. It is instructive to study the boundary-layer development with the hybrid panel/boundary layer code and to compare with the Navier-Stokes calculations. This is shown in Fig. 15 where the onset of dynamic stall is again found to be delayed to a larger angle of attack in comparison to the steady-state stall onset if natural boundary-layer transition is assumed to occur.

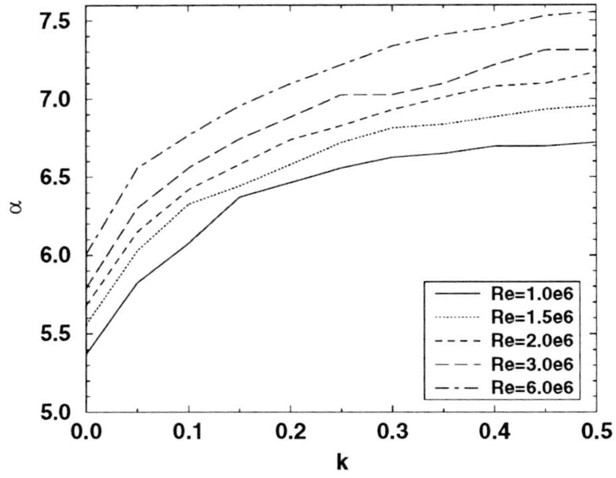

Figure 13. Stall onset AOA variation for a NACA 1412.

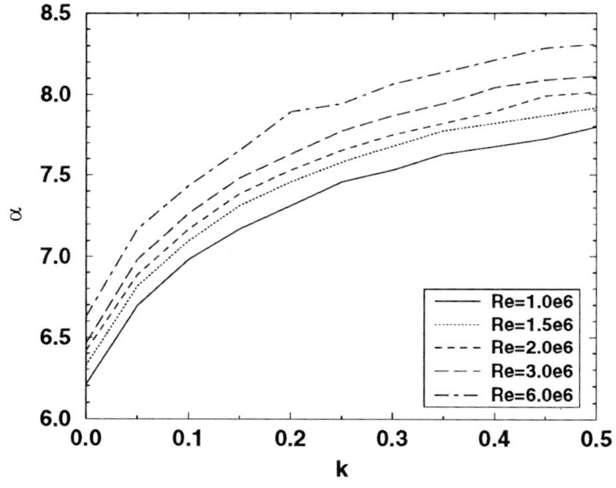

Figure 14. Stall onset AOA variation for a NACA 2412.

On the other hand, if the boundary layer is tripped into turbulent flow at the leading edge, the panel/boundary-layer code predicts turbulent separation from the trailing edge occurring it a much larger angle of attack. This is again in agreement with the fully turbulent Navier-Stokes calculations of Ref. 2.

A similar Navier-Stokes analysis has recently been carried out by Geissler *et al.* [21] for a NACA 0012 airfoil which oscillates in pitch with an amplitude of 10 degrees about a mean angle of attack of 10 degrees. For this case Chandrasekhara *et al.* [22] have provided experimental results. The Navier-Stokes computations again showed that it is crucial to account for the effect of boundary-layer transition. The fully turbulent Navier-Stokes computation predicts dynamic stall to occur at an incidence of about 17 degrees, whereas the transitional flow calculation puts it at about 14 degrees, in agreement with the measurement. The transitional hybrid code predicts dynamic-stall onset to occur at about 5 degrees, as shown in Fig. 16, whereas a fully turbulent calculation delays it to about 18 degrees.

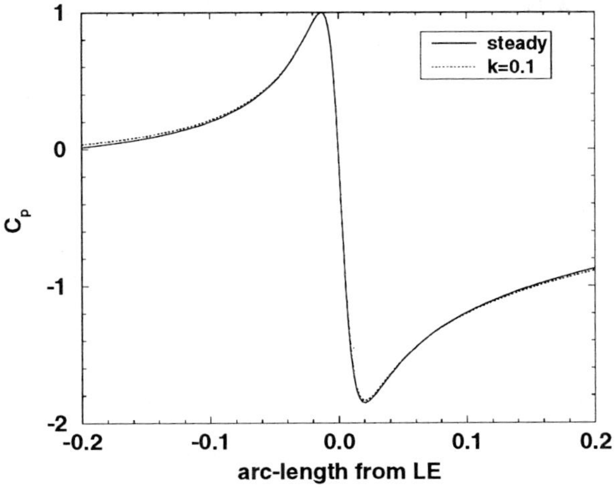

Figure 15. Pressure distribution at dynamic-stall onset for a sinusoidally pitching NACA 0012 airfoil; α=9±5 degrees, Re=4×10^6.

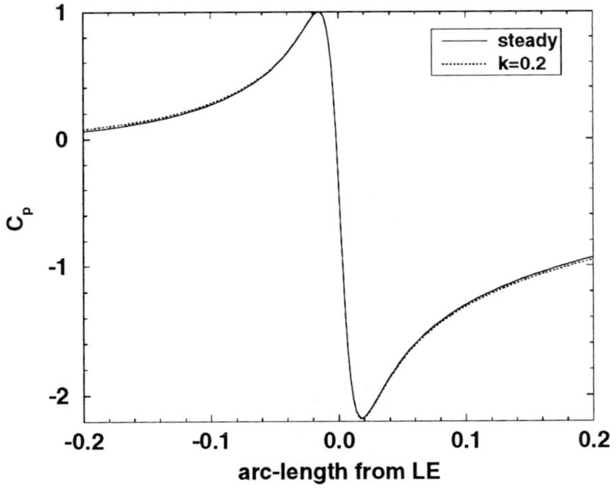

Figure 16. Pressure distribution at dynamic-stall onset for a sinusoidally pitching NACA 0012 airfoil; α=10±10 degrees, Re=1.1×10^6.

The incidence where stall onset is predicted by the hybrid code is plotted as a function of the reduced frequency in Figs. 17 and 18, corresponding to the configurations of Figs. 15 and 16, respectively. For these sinusoidally pitching calculations, typically several periods are required for the transients to fade and a periodic solution to form. The predicted stall angles are plotted for the first three cycles. The results for the first cycle are very much like the ramp motion, as expected, but the results during the later cycles are quite different and require further investigation.

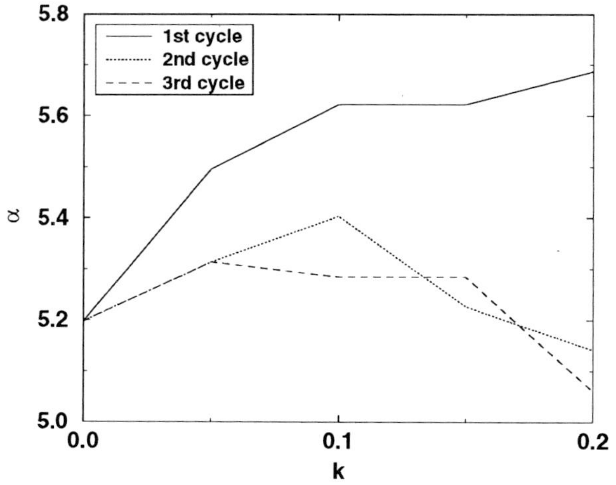

Figure 17. Incidence for dynamic-stall onset for a sinusoidally pitching NACA 0012 airfoil; α=9±5 degrees, Re=4×10^6.

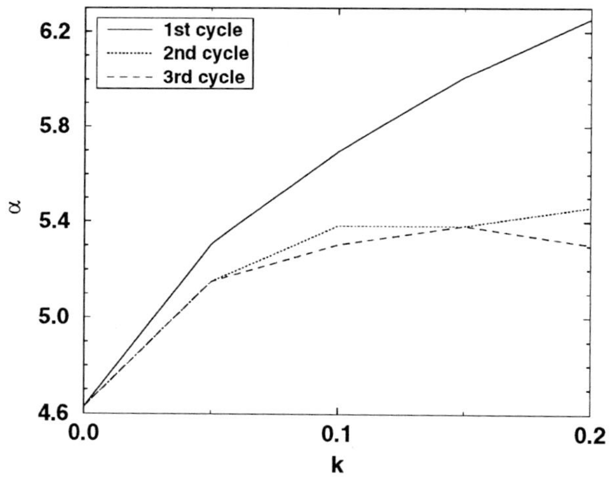

Figure 18. Incidence for dynamic-stall onset for a sinusoidally pitching NACA 0012 airfoil; α=10±10 degrees, Re=1.1×10^6.

Further results, showing the effect of oscillation mode (pitch versus plunge) on the pressure distribution and boundary-layer development, have recently been given by Jones and Platzer [23] using the hybrid panel/boundary-layer code.

6. Conclusions

An efficient computational approach was presented for the fast prediction of the onset of dynamic stall due to rapid incidence changes or due to pitch / plunge motions of blades or airfoils. The method provides a means for the systematic evaluation of flow properties leading to dynamic stall over a broad parameter space.

Presented results suggest that the primary factor in dynamic stall onset is the phase lag in the surface pressure, data that may be obtained to the point of separation using a purely inviscid analysis. Viscous results obtained from the hybrid code illustrate that the pressure lag effect delays flow separation in dynamic cases even at very low frequencies, and that the onset of dynamic stall occurs when a critical pressure gradient distribution is obtained. Furthermore, in agreement with previous Navier-Stokes simulations [24], the critical pressure distribution leading to dynamic stall was found to be independent of mode and frequency.

7. Acknowledgments

This investigation was partially supported by the Naval Research Laboratory.

8. References

1. Carr, L.W. and Chandrasekhara, M.S., "Compressibility Effects on Dynamic Stall", *Progress in Aerospace Sciences* **32** (1996), 523-573.
2. Ekaterinaris, J.A. and Platzer, M.F., "Numerical Investigation of Stall Flutter", *J. Turbomachinery* **118-4** (1996), 197-203.
3. Ericsson, L.E. and Reding, J.P., "Fluid Dynamics of Unsteady Separated Flow, Part II, Lifting Surfaces", *Progress in Aerospace Sciences* **24** (1987), 249-356.
4. Chandrasekhara, M.S., Ahmed, S. and Carr, L.W., "Schlieren Studies of Compressibility Effects on Dynamic Stall of Transiently Pitching Airfoils", *J. Aircraft* **30-2** (1993), 213-220.
5. Chandrasekhara, M.S. and Carr, L.W., "Flow Visualization Studies of the Mach Number Effects on Dynamic Stall of an Oscillating Airfoil", *J. Aircraft* **27-6** (1990), 516-522.
6. Teng, N.H., "The Development of a Computer Code for the Numerical Solution of Unsteady, Inviscid and Incompressible Flow over an Airfoil," Master's Thesis, Naval Postgraduate School, Monterey, CA, June 1987.
7. Hess, J.L. and Smith, A.M.O., "Calculation of Potential Flow about Arbitrary Bodies," *Progress in Aeronautical Sciences* **8** (1966), 1-138.
8. Basu, B.C. and Hancock, G.J., "The Unsteady Motion of a Two-Dimensional Aerofoil in Incompressible Inviscid Flow," *J. Fluid Mechanics* **87** (1978), 159-168.
9. Platzer, M.F., Neace, K.S. and Pang, C.K., "Aerodynamic Analysis of Flapping Wing Propulsion," AIAA Paper No. 93-0484, Jan. 1993.
10. Riester, P.J., "A Computational and Experimental Investigation of Incompressible Oscillatory Airfoil Flow and Flutter Problems," Master's Thesis, Naval Postgraduate School, Monterey, CA, June 1993.

11. Turner, M., "A Computational Investigation of Wake-Induced Airfoil Flutter in Incompressible Flow and Active Flutter Control," Master's Thesis, Naval Postgraduate School, Monterey, CA, March, 1994.
12. Tuncer, I.H., Platzer, M.F. and Ekaterinaris, J.A., "Computational Analysis of Flapping Airfoil Aerodynamics," ASME Fluids Engineering Division, Summer Meeting, June, 1994.
13. Jones, K.D. and Platzer, M.F., "Time-Domain Aeroelastic Analysis of a two Airfoil System with Application to Unsteady Rotary Wing Flowfields," AIAA Paper No. 95-0337, Jan., 1995.
14. Jones, K.D., Dohring, C.M. and Platzer, M.F., "Wake Structures Behind Plunging Airfoils: A Comparison of Numerical and Experimental Results," AIAA Paper No. 96-0078, Jan., 1996.
15. Keller, H.B. and Cebeci, T., "Accurate Numerical Methods for Boundary-Layer Flows, pt. 2, Two-Dimensional Turbulent Flows," *AIAA Journal* **10** (1972), 1193.
16. Nowak, L.M., "Computational Investigations of a NACA 0012 Airfoil in Low Reynolds Number Flows," Master's Thesis, Naval Postgraduate School, Monterey, CA, Sept., 1992.
17. Cebeci, T., "Essential Ingredients of a Method for Low Reynolds-Number Airfoils," *AIAA Journal* **27-12** (1989), 1680-1688.
18. VanDyken, R.D., Ekaterinaris, J.A., Chandrasekhara, M.S. and Platzer, M.F., "Analysis of Compressible Light Dynamic Stall Flow at Transitional Reynolds Numbers," *AIAA Journal* **34-7** (1996), 1420-1427.
19. Hancock, G.J. and Mabey, D.G., "Unsteady Aerodynamics of Controls," AGARD CP-465, April, 1990.
20. Cebeci, T., Platzer, M.F., Jang, H.M. and Chen, H.H., "An Inviscid-Viscous Interaction Approach to the Calculation of Dynamic Stall Initiation on Airfoils," *J. Turbomachinery* **115-4** (1993), 714-723.
21. Geissler, W., Chandrasekhara, M.S., Platzer, M.F. and Carr, L.W., "The Effect of Transition Modeling on the Prediction of Compressible Deep Dynamic Stall", Proc. 7th Asian Congress of Fluid Mechanics, Madras, India, Dec. 8-12, 1997.
22. Chandrasekhara, M.S., Wilder, M.C. and Carr, L.W., "Reynolds Number Influence on 2-D Compressible Dynamic Stall", AIAA Paper No. 96-0073, Jan. 1996.
23. Jones, K.D. and Platzer, M.F., "A Fast Method for the Prediction of Dynamic Stall Onset on Turbomachinery Blades," ASME Paper No. 97-GT-101, June 1997.
24. Grohsmeyer, S.P., Ekaterinaris, J.A. and Platzer, M.F., "Numerical Investigation of the Effect of Leading Edge Geometry on Dynamic Stall of Airfoils," AIAA Paper No. 91-1798, June, 1991.

15: Theoretical Studies of Unsteady Flow Around Oscillating Blades - 6

FLUTTER OF AIRCRAFT ENGINE LOW PRESSURE TURBINE BLADES: OSCILLATING CASCADE EXPERIMENTS AND ANALYSIS

J. Panovsky
GE Aircraft Engines
Cincinnati, Ohio USA

M. Nowinski and A. Bölcs
Ecole Polytechnique Fédérale de Lausanne
Lausanne, Switzerland

Abstract
This paper describes a portion of a research effort to eliminate occurrences of flutter of low pressure turbine blades in aircraft engines. Fundamental unsteady aerodynamic experiments in an annular cascade plus correlating analyses are conducted to improve the understanding of the flutter mechanism in these blades and to identify the key flutter parameters. Mode shape is identified as the most important contributor to determining the stability of a blade design, while reduced frequency has a secondary influence. Low pressure turbine flutter is shown to be a linear, inviscid phenomenon.

1. Introduction

A current design issue for low pressure turbine (LPT) blades is torsional flutter, which has been encountered on several recent engines. The need for increased performance had pushed these blades into lower reduced frequencies, and existing design guidelines were found to be inadequate. It became apparent that improved design approaches are needed to meet the requirements for lighter-weight, higher-performance blading.

A research project was initiated to address these concerns. The initial step, which is the focus of the present paper, is to determine the critical physical contributors to LPT flutter by combining fundamental experimental data generated in an annular cascade with correlating analysis using computational fluid dynamics (CFD) codes. Subsequent papers will describe other aspects of the research effort, such as the experimental verification of the influence coefficient technique and a direct demonstration of the beneficial stabilizing effects of mistuning. In addition, the use of the validated analytical tools in conducting studies to identify the key design parameters will be discussed. Finally, the results will be used as a basis for developing a new design approach for flutter, including both the preliminary and detailed design analysis phases. Additional details of the results summarized here can be found in Nowinski (1995, 1996) and Panovsky (1997).

T.H. Fransson (ed.), Unsteady Aerodynamics and Aeroelasticity of Turbomachines, 815-829.

2. Experimental Program

2.1. TEST FACILITY

The experimental measurements were conducted in the non-rotating annular test facility (Bölcs, 1983) at the Ecole Polytechnique Fédérale de Lausanne (EPFL). This is a continuous flow, open cycle facility with a test section of 160 mm inner radius and 200 mm outer radius. A schematic of the cascade facility is shown in Figure 1. Air enters the facility through two independent inlet valves and is turned through a pair of pre-swirl vanes. These devices allow for the precise control of the inlet flow conditions over a wide range of Mach numbers (0.3 to 1.6) and flow angles (20° to 70°). The flow is turned vertically and spirals upward through the test section and then into a plenum before exiting the facility to the atmosphere through an outlet valve. The facility also includes boundary layer suction capability (locations are denoted by arrows in Figure 1), but this capability was not used during these experiments. The overall intent of the facility is to provide nearly two-dimensional flow conditions. In addition, it provides some important advantages with regards to oscillating blade measurements due to its inherent circumferential flow periodicity and absence of lateral boundaries (see Buffum and Fleeter, 1994, for a discussion of the effects of wall reflections).

The cascade is composed of an annular arrangement of twenty blades. Each blade is imbedded within a base which also forms the inner boundary of the test section. Gaps between the bases and the fixed upstream and downstream walls are required to allow the blades to vibrate freely. The design of the bases is fairly complex and includes several overlapping features to minimize leakage of the airflow. The base is attached to mass and spring elements, and the resulting blade assemblies are mounted into a common core or inner ring. Each blade in the cascade is excited using a periodic forcing function applied to the transformer plates by individual electromagnets.

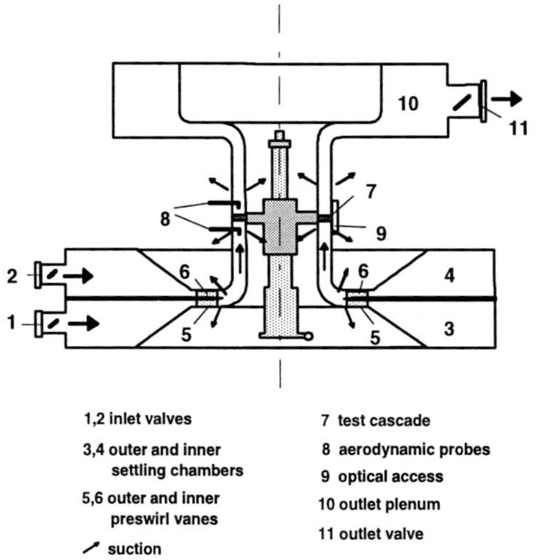

Figure 1. Schematic of annular cascade facility.

To ensure a sufficient vibratory amplitude, the excitation frequency typically coincides with the natural frequency of the blade assembly. Control of the vibration amplitude and interblade phase angle (IBPA) is accomplished electronically based on feedback signals provided by inductive displacement transducers placed over each individual blade assembly. The errors on the controlled values of blade vibration amplitude and interblade phase angle are ±2% and ±3°, respectively.

2.2. BLADE DESCRIPTION

The LPT blade used for this series of tests corresponds to a latter stage of a modern commercial turbofan engine. These blades are characterized by airfoils with high aspect ratios and thin, highly-cambered profiles. The cascade airfoil is a simple radial extrusion of the profile from the 60% span of the blade. While the actual blade is tip-shrouded, a tip gap of 0.5 mm is used at the outer radius for the cascade blades. Additional features of the cascade airfoil are summarized in Table 1.

Geometric Parameters	
Chord	75 mm
Camber	115°
Stagger	35°
Max thickness	5% of chord
Solidity	0.9
Aerodynamic Parameters	
Inlet Mach	0.40
Inlet Flow Angle	45°
Exit Mach	0.80
Exit Flow Angle	65°

TABLE 1. Physical and aerodynamic parameters at design point for cascade LPT blade.

The frequency and mode shape are based on the torsional mode of the actual blade. To permit the investigation of the effects of changes in the mode shape, provisions for two locations of the spring axis were made. These locations are shown in the blade schematic of Figure 2. Both of the spring axis locations are along the meanline of the blade, one at 15% and the other at 35% as measured along the true chord. A photograph of the complete cascade assembly is shown in Figure 3.

2.3. MEASUREMENTS

The upstream and downstream steady-state flow conditions are determined from spanwise-circumferential traverse data obtained using 3-hole aerodynamic probes ("8" in Figure 1). These traverses are typically performed over 1.5 blade passages in steps of 1° to 2° and over the middle 70% of the blade span. In addition, throughflow information is provided by a series of static pressure taps located along the outer wall of the test section. A parallel line of taps are available for unsteady pressure measurements as well. For all of these measurements, a circumferential distribution is obtained by taking a series of readings as the cascade is incrementally rotated.

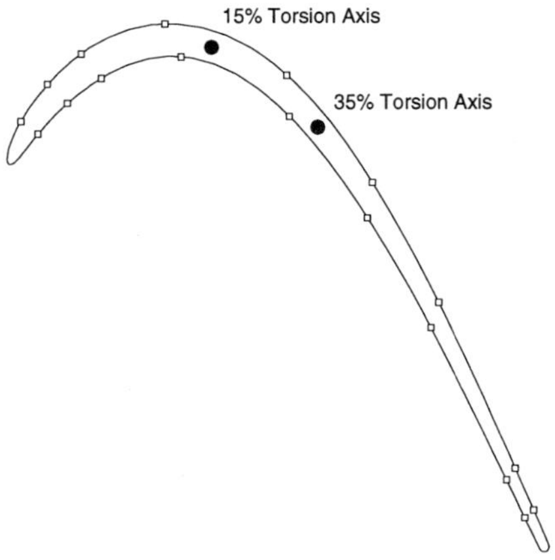

Figure 2. Blade profile with torsion axis and transducer locations.

Steady pressure taps and unsteady pressure transducers are installed at the pitchline radius of the blade. A total of four blades are instrumented, with one passage (two blades) dedicated to steady measurements and another passage to unsteady measurements. Nine taps are used on each blade surface, with locations as indicated in Figure 2. The steady taps and unsteady transducers are located at corresponding positions on their respective blades.

The calibration of the pressure transducers is performed dynamically using a device which generates a periodically varying pressure. The pressure transducer signals are corrected for the presence of false content arising from acceleration of the blade. During an unsteady measurement, approximately 25,000 data points are obtained for each channel and then filtered. This data is then sub-divided into overlapping blocks and each is Fourier decomposed. The spectral components corresponding to the blade vi-

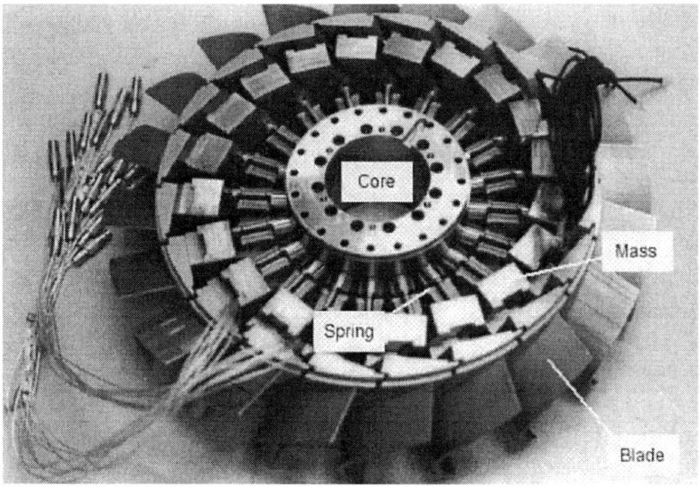

Figure 3. Photograph of cascade assembly.

bration frequency are extracted and averaged over the individual blocks. Statistical quantities are calculated using the average values and the individual data block values. Additional details can be found in Nowinski (1995) and Panovsky (1997).

3. Steady Flow Results

3.1. EXPERIMENTAL DATA

The emphasis of the experimental program is clearly on obtaining unsteady data for stability analysis. However, the codes used for the unsteady correlations are linearized harmonic methods, so that the unsteadiness is considered a small perturbation from the steady flow. As a result, it is critical to understand the steady flow characteristics of the cascade and to contrast those results to the behavior for an actual engine application. Four or five steady flow conditions were investigated for each of the two torsion axis locations shown in Figure 2. The sets of tests corresponding to each of these torsion axis locations will be collectively referred to as "Configurations". The steady conditions covered the subsonic to transonic range, and emphasized the effect of incidence angle.

Radial profiles as well as overall passage averages of the steady flow variables are obtained by mass averaging the values from the upstream and downstream probe traverses. Figure 4 shows the radial profiles of static and total pressure and the calculated mass flow for design point conditions. A significant loss in total pressure near the inner wall at the exit is evident. The mass flow indicates that much of the flow is migrating from the inner wall toward the middle of the passage through the blade row. Even at the inlet, though, there is a fairly significant bias away from the walls.

The source of the losses is not obvious from the experimental results. Possible causes include separation of the flow from the airfoil or the inner wall of the cascade, disruption of the flow by the gaps between the blade bases, large profile losses, etc. Comparison of the exit flow angle measurement versus the metal angle indicates that there is significant under-turning of the flow, especially near the pitch section.

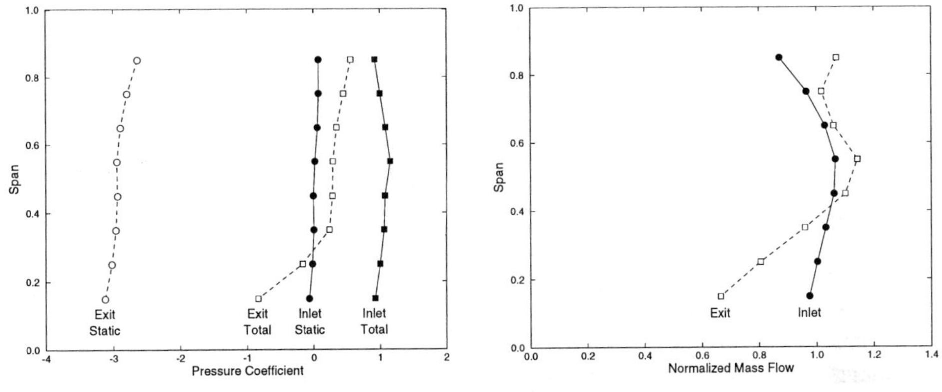

(a) (b)

Figure 4. Radial profiles for design point conditions: (a) Pressures; (b) Mass flow.

3.2. ANALYTICAL CORRELATIONS

A number of 2D and 3D CFD codes, both inviscid and viscous, were used to correlate to the experimental results. The 2D inviscid code NOVAK (Holmes, Lamson, and Connell, 1988) has been used to analyze each of the steady conditions tested. In addition, 3D analyses were conducted at design point conditions to help understand key features of the steady flowfield. Three codes - ViscousEuler (Turner and Jennions, 1992), STAR-CD (1995), and TACOMA (Holmes, Mitchell, and Lorence, 1997) - are used for this work because there are features and/or limitations in each that are needed as different aspects of the steady flow are considered. Each can be used for steady viscous analysis, but TACOMA can also be run in a linearized harmonic Euler mode for unsteady analyses to determine stability.

In each case analyzed, the computational domain extends from the upstream to the downstream probe locations. Aligning the computational boundaries with the measurement planes seems to definitively set the flow conditions but, unfortunately, this is not the case. Because the inviscid analyses cannot model the total pressure losses as measured, there is an inconsistency between the static pressure and the Mach number at the exit for the experimental and analytical results. Matching Mach number was chosen as the general approach, since this leads to consistency in reduced frequency between test and analysis. Note that this results in an airfoil with slightly less loading than the experimental data indicates.

For the 3D analyses, the radial profiles obtained from the test data are prescribed to allow a radial variation in the boundary conditions at the inlet and exit. Because the probes only measured between 15% and 85% span, it is necessary to extrapolate the radial profile values from the closest measurement station to each wall. Several approaches were used for this extrapolation, but none significantly changed the results. A simple tip gap model is also included.

Figure 5 shows a comparison of the mid-span surface Mach number distributions from the 2D and 3D analyses compared to the experimental data. This plot shows rea-

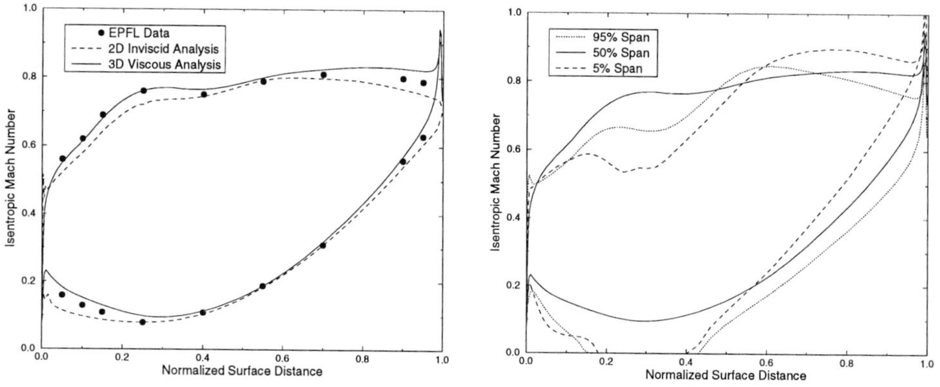

(a) (b)

Figure 5. Blade surface Mach number distributions: (a) Comparison of analytical and test results, (b) Tip, pitch, and hub sections from 3D analysis.

sonable agreement between the 2D inviscid analysis and the data, but significant improvement with the 3D viscous code. The tip, pitch, and hub distributions from the 3D analysis are also compared to indicate the variation with radius.

3.3. NEAR-SEPARATION AND PASSAGE VORTEX

Streamlines resulting from the 3D viscous analysis for the pitch and hub sections are shown in Figure 6. These are more properly termed quasi-streamlines, since they are constructed using only the flow components in the plane shown. The pitch section plot shows that the flow does not follow the surface of the blade on the suction side near the trailing edge. Rather, the streamlines migrate toward the pressure side of the adjacent blade, resulting in a lower overall average exit angle. This is interpreted as the flow approaching separation, an effect that was not predicted by 2D or 3D inviscid codes.

The behavior at the root is clarified by similar streamline plots for the meridional suction side and pressure side surfaces shown in Figure 7. These plots indicate significant migration of the flow as it traverses from the leading edge (vertical line on the left) to the trailing edge (vertical line on the right). Because this is a viscous analysis, which has no-slip conditions at the walls, meridional surfaces a few grid points away from the actual blade surfaces were used to construct the quasi-streamline plots. On the suction side, the flow migrates toward the mid-span, while on the pressure side it moves toward the outer walls.

Combining Figures 6 and 7, the flow actually twists into a corkscrew pattern and there is an overall migration away from the inner and outer cascade walls and toward the mid-span of the passage. This follows the classical behavior expected for secondary flows (Horlock, 1966, Sieverding, 1985), which are generated by the boundary

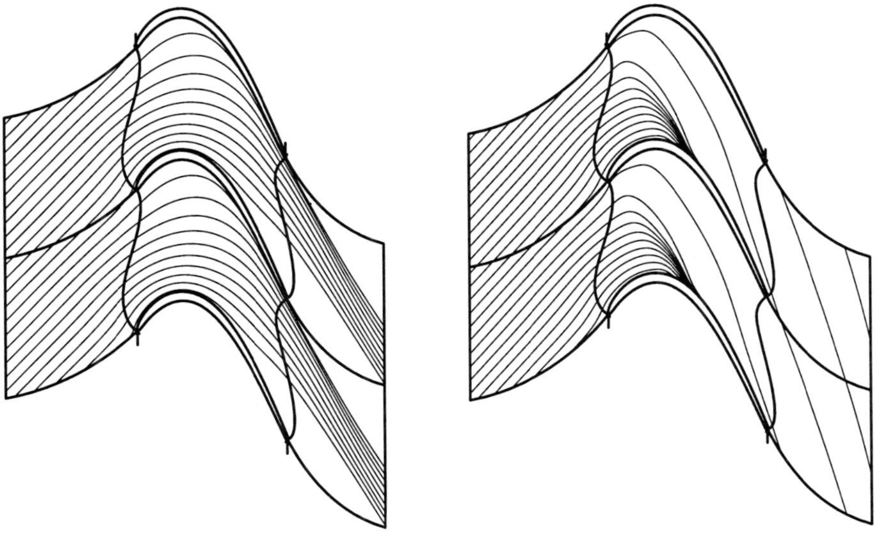

(a) (b)

Figure 6. Streamlines from 3D analysis in blade-to-blade view: (a) Pitch; (b) Hub.

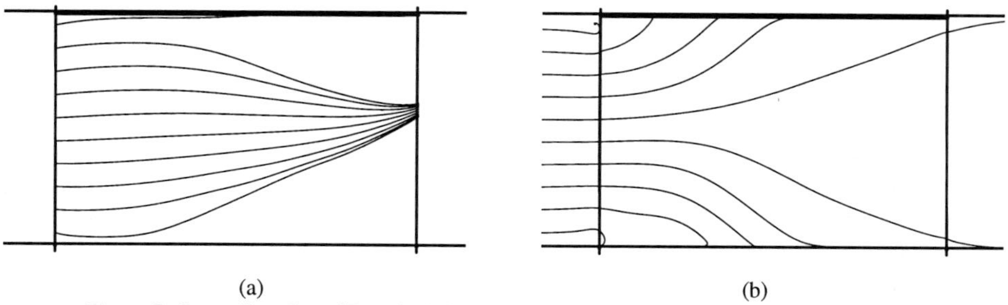

Figure 7. Streamlines from 3D analysis in meridional view: (a) Suction side; (b) Pressure side.

layer entering and being turned by the blade row. While the existence of these vortices is expected, the extent of this behavior well into the main flowfield near the trailing edge is surprising, and raises concern as to its effects on the unsteady measurements.

The cause of this observed secondary flow behavior has been found to be due to small variations in the inlet total pressure which are present in the radial profiles, as in Figure 4. Replacing the actual profile with a constant average value eliminates the vortex formation. The vortex formation has been found to be a primarily inviscid effect, though viscosity does somewhat affect their strength and location. Also playing a significant part is the high turning of this blade profile.

3.4. HUB LOSSES

One of the major concerns from the test data is the pressure loss noted at the inner wall. The 3D viscous analysis compares well to the pressure probe measurements at the exit in the outer portion of the blade but does not predict the large losses at the hub. Many attempts at changing parameters in the boundary layer model were unsuccessful in predicting the experimental data, including variations in the assumptions used for the un-measured portion of the passage. While separation from the inner wall is considered a strong possibility, the CFD computations do not indicate this type of behavior.

Another possible cause of the pressure loss is the gaps which are present between the blade bases. The total pressure profile, though, shows no effect when the downstream gap is included in the 3D CFD calculations. The effect of the gaps on all sides of the base, those upstream and downstream as well as the inter-blade gaps, were also experimentally checked in the cascade by filling these voids with clay. Again, there was little effect on the measured loss.

Flow visualization techniques were utilized in the cascade facility to compare to the 3D computational results and to help determine the source of the hub loss. The overall flowfield inferred from these visualization results is in very good agreement with the analytical predictions, confirming the presence of the strong passage vortices. In addition, deviation of the flow angle from the metal angle is evident at the blade midspan. The source of the hub loss, though, was not identified. In these results, the gaps at the inner wall appear to be more significant than the analysis or other testing indicates, and so it is likely that the gaps have a significant contribution to these losses.

3.5. SUMMARY OF STEADY FLOW RESULTS

Significant effort has been directed toward understanding the steady flowfield. Each of the nine steady conditions has been analyzed using a 2D nonlinear inviscid CFD code, giving results that agree reasonably well with the measurements. In addition, design point conditions were studied in detail using 3D inviscid and viscous codes. Experimental flow visualization techniques were also utilized to better understand the steady flow and to confirm observations from the 3D analyses.

Overall, 3D effects were found to be quite significant, which is surprising because the intent of the cascade is to provide nearly 2D flow. A near-separation on the suction side of the blade was identified, as was a strong passage vortex. The primary reason for this behavior is the high turning of the blade profile coupled with the small radius of the inner cascade wall. Slight variations in the inlet total pressure profile serve as a catalyst for much of the 3D behavior. The identification of these vortices is important for any future work with similar high-turning blades. The presence of these vortices must be recognized and accounted for in the experimental measurements and any correlating analyses.

The 3D inviscid code predicts the existence of the large passage vortices and a reasonable indication of their extent into the primary flowfield. The inclusion of viscosity in the 3D analysis results in an improvement in the prediction of the static pressure on the blade suction side near the trailing edge, where the flow is just approaching separation from the blade surface. This also explains the deviation of the exit flow angle from the trailing edge metal angle. Viscous effects also tended to smear the vortices and alter the location slightly, resulting in better agreement with the experimental data.

The only unresolved issue for the steady flowfield is determination of the source of the losses in total pressure measured near the inner wall of the cascade by the downstream probe. While it is strongly suspected that this behavior is due to a flow separation from the inner wall or the gaps between the bases, this has not yet been confirmed through flow visualization techniques or 3D viscous analysis. This is thought to be an effect that is peculiar to the cascade and would not be present in an actual blade row.

4. Unsteady Results

Several types of unsteady experiments have been conducted but only results for the traveling wave mode, where each blade is driven to a constant amplitude with a constant temporal phase difference between adjacent blades, are included here. This mode, of course, simulates the system mode behavior of the actual rotor assembly (Lane, 1957). A total of twelve interblade phase angles are used for each of the steady conditions.

During all of the testing, the raw signals from the unsteady pressure transducers show significant noise in the time domain. But the predominance of the fundamental vibratory frequency is obvious in the harmonic decomposition, with the content at other frequencies reaching a maximum of 15% of that at the fundamental frequency. The magnitude and phase of this component were also shown to be very consistent between sample blocks, with the corresponding 95% confidence intervals calculated as approximately ±3% for the amplitude and ±4° for the phase lag. The absence of integer multiples of this fundamental frequency suggests that the unsteady pressures are linearly related to the blade motion.

Computational results using the 2D inviscid NOVAK code, now in its linearized unsteady mode (Holmes and Chuang, 1993), have been obtained for all of the experimental cases. In this type of method, the unsteadiness in the flowfield is assumed to be a small harmonic perturbation to the nonlinear steady flow governed by the Euler equations. These predictions are compared to the experimental results in terms of the unsteady pressure, work, and damping coefficients (Bölcs and Fransson, 1986). A completely consistent calculation approach was utilized for these comparisons, to assure that any discrepancies will be solely due to differences between the measured and calculated unsteady pressures.

Comparisons are also made in terms of the influence coefficients (Crawley, 1988), which give the contribution from each blade in the cascade to the total unsteady pressure or work. The influence coefficients were obtained by utilizing a Fourier Transform of the traveling wave results, which are equally spaced in terms of IBPA.

4.1. CONFIGURATION 1

The torsion axis for Configuration 1 is at the 35% chord location shown in Figure 2, and reduced frequencies in the 0.16 to 0.19 range were investigated. The distributions of unsteady pressure magnitude and damping coefficient are summarized in Figure 8 for design point conditions at IBPA=+90°. The predicted pressures are in reasonable agreement with the measured data, especially trendwise, though there are some discrepancies on the suction side of the blade. Comparison of the damping coefficient, which involves only the imaginary part of the unsteady pressure, is considerably better. These results are typical of those at other IBPA though, in general, the largest unsteady pressures occur near +90° and -90°.

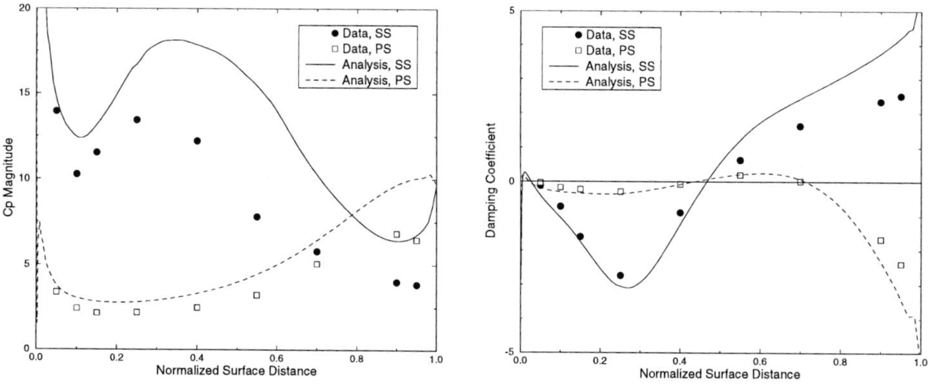

(a) (b)

Figure 8. Sample distributions for Configuration 1 at IBPA=+90°: (a) Magnitude of unsteady pressure coefficient; (b) Local damping coefficient.

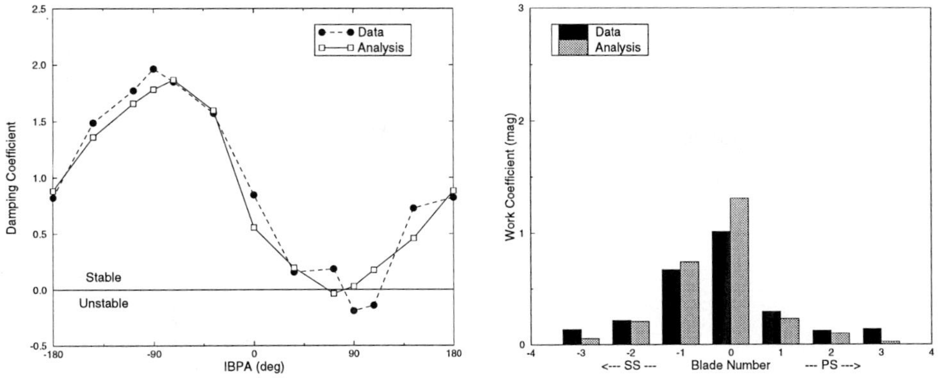

(a) (b)

Figure 9. Summary of work coefficient for Configuration 1: (a) Traveling wave representation; (b) Influence coefficients.

In Figure 9, the comparison of the overall damping coefficient versus IBPA and the influence coefficients for the work coefficient at this steady condition are summarized. Agreement between the experimental and computational results is very good for both the traveling wave and the corresponding influence coefficient representations. From this figure, the contribution from the *-1* blade (on the suction side of the reference blade) is nearly as large as the reference blade itself. Similar comparisons have been conducted for the three other steady conditions. In each case, agreement of the experimental and computational results is excellent.

4.2. CONFIGURATION 2

The torsion axis for Configuration 2 is at 15% chord, also shown in Figure 2. Reduced frequencies for this Configuration ranged from 0.17 to 0.25. Unsteady pressure and damping distributions are again shown in Figure 10 for a single case at

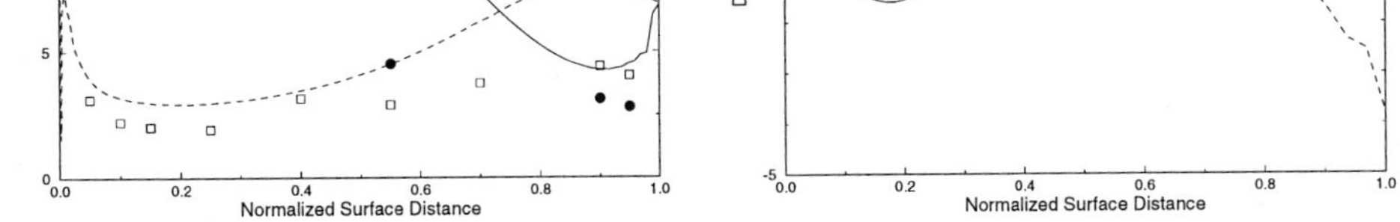

(a) (b)

Figure 10. Sample distributions for Configuration 2 at IBPA=+90°: (a) Magnitude of unsteady pressure coefficient; (b) Local damping coefficient.

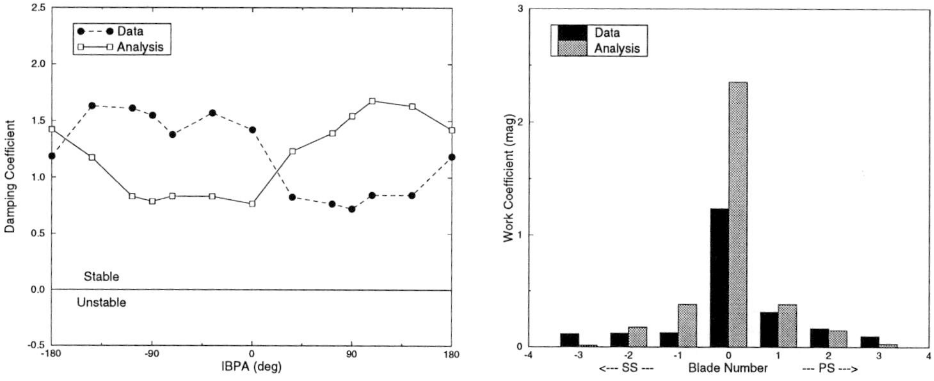

(a) (b)

Figure 11. Summary of work coefficient for Configuration 2: (a) Traveling wave representation; (b) Influence coefficients.

IBPA=+90° near design point. In this case, the unsteady pressures are in good agreement trend-wise, but the analysis again over-predicts the level over much of the suction side. Because of the position of the torsion axis, the motion of the trailing edge region is now even more dominant than before, which is evidenced by the increased discrepancies in the local damping coefficient there.

This difference is also obvious in the summary plots shown in Figure 11. Unlike Configuration 1, there is now little variation of the damping with IBPA. This is evident from the plots of the influence coefficients, where the reference blade clearly dominates the other blades. This plot is quite different than the results of Figure 9. Thus, a change of torsion axis of just 20% of chord dramatically changes the character of the stability behavior. Note that while the analysis does a fairly poor job of predicting the small variation of the damping with IBPA for Configuration 2, it correctly predicts the overall stability of this Configuration. To a great extent, this Configuration is dominated by the blade-on-itself contributions, and cascade effects are small. Similar comparisons were repeated for the four other steady conditions tested, which cover a wide range of steady flow parameters, with similar levels of agreement attained.

4.3. 3D UNSTEADY ANALYSIS

As noted previously, the 3D TACOMA code can also be used for unsteady analysis, again as a linearized Euler algorithm. Results for the Configuration 1 design point case are presented here to compare to the 2D predictions. The unsteady pressures at the pitch section are compared for TACOMA, NOVAK, and the experimental data in Figure 12, and good agreement is observed. In this figure, the imaginary part of the pressure is shown since only this portion contributes to the damping. The unsteady pressures at three different spans from the 3D analysis are also presented. This indicates that unsteady 3D effects are small for this blade. The effect of the passage vortex was also found to be small. It is noteworthy that the 3D code does not improve the

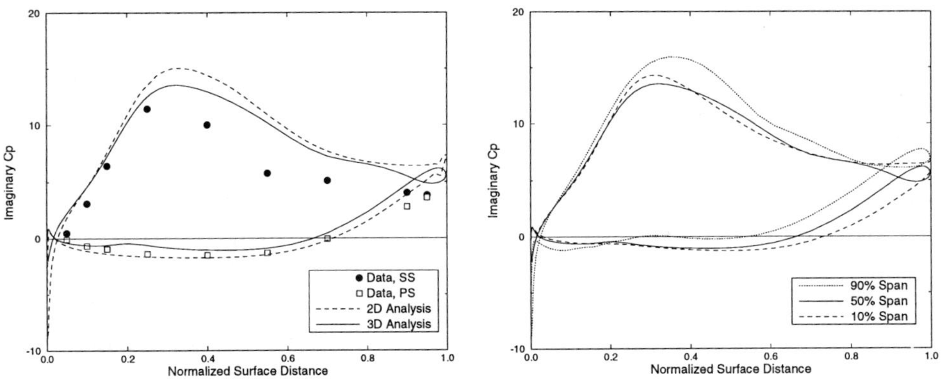

(a) (b)

Figure 12. Comparison of unsteady pressures for Configuration 1: (a) Pitch section from 3D TACOMA compared to 2D NOVAK and experimental data; (b) Tip, pitch, and hub from TACOMA.

agreement substantially for the suction surface. Since this is also the region where the near-separation of the steady flow is occurring, it is likely that viscosity will have to be included to reduce the remaining discrepancies between analysis and the unsteady test data.

4.4. SUMMARY OF UNSTEADY FLOW RESULTS

Several significant conclusions can be reached by assessing the results from this series of tests. The torsion axis location is the most important parameter in determining the stability behavior of the blade. While all of the tests use a torsion mode, small changes to the location of the torsion axis have a dramatic effect on the stability of the blade. This is an extremely important conclusion, because the primary design parameter up to this time has been reduced frequency. While a dependence on reduced frequency was also found, its influence is of secondary importance. Blade loading, in general, also has an effect though a single correlating parameter could not be identified. Incidence angle has a negligible effect on the stability behavior of LPT blades. The results also show that there is a fairly small range of IBPA where flutter can occur for this LPT blade.

Correlating analyses for all of the unsteady data have been completed using a 2D linearized Euler code. Comparison of the experimental data to these predictions ranged from fair to excellent. Interestingly, the agreement is dictated to a large extent by the Configuration being analyzed. Overall trends are always well captured by the analysis, but detailed comparison shows discrepancies, with the largest errors typically occurring on the aft portion of the suction surface. In a practical sense, the 2D linearized Euler analysis correctly predicts whether the cascade is stable or unstable for all of the steady conditions considered, including those not shown in the present paper. The importance of this "bottom line" assessment should not be under-estimated.

Analyses using a 3D linearized Euler code were also conducted. The results are in very good agreement with the 2D results, and show little 3D influence in the unsteady behavior of this blade. This is surprising given the substantial 3D effects found in the steady flow. The unsteadiness due to the presence of the strong passage vortex could be identified. But the vortex does not have a substantial influence at the experimental measurement section, and it is not the cause of the discrepancies between the data and the analytical predictions on the suction side of the blade.

The influence coefficients of the local transducer pressures and work coefficients and the overall work coefficient were very useful in identifying the contributions from each blade. It was shown that the important contributions come from the reference blade and its immediate neighbors. It is perhaps significant that the analytical correlation is much better when the influence of the adjacent blades increases. The remaining discrepancies may be due to viscosity, which is not included in the analytical approach.

5. Summary and Conclusions

Detailed measurements of the unsteady pressures generated by the torsional vibration of a modern LPT blade in an annular cascade have been completed. A large database for LPT flutter has been generated through the experimental program, and several key conclusions can be reached by studying this data. In addition, correlating analyses for the steady flow have been conducted using 2D and 3D Euler and Navier-Stokes methods, and for the unsteady flow using 2D and 3D linearized Euler methods. The key findings can be summarized as follows:

1) Mode shape is the key parameter affecting the stability characteristics of LPT blades. Reduced frequency and steady blade loading also have an influence, but are of secondary importance.
2) Linearized Euler methods are sufficient for the prediction of LPT flutter, which has been shown to be a predominantly linear, inviscid phenomenon.
3) Three-dimensional steady flow effects in the annular cascade are much more significant than anticipated. The source of this three-dimensionality has been identified. However, the corresponding unsteady pressures are primarily two-dimensional in nature.
4) The unsteady pressures measured on the blade surface due to the vibratory motion are dominated by the frequency corresponding to that motion, with negligible content in higher harmonics.

As noted in the introduction, a substantial amount of additional work, both experimental and computational, has been completed or is currently in progress as a part of this research effort. Additional results will be presented in subsequent papers.

Acknowledgements

The authors wish to thank GE Aircraft Engines and Ecole Polytechnique Fédérale de Lausanne for permission to publish this paper. The first author gratefully acknowledges the significant technical contributions of Dr. Graham Holmes, Dr. Christopher Lorence, Dr. Steven Manwaring, and especially Dr. Robert Kielb; thanks are also extended to Prof. Torsten Fransson of the Royal Institute of Technology in Sweden. The second author wishes to thank Dr. Peter Ott and Dr. Heiko Körbächer for their valuable technical comments.

References

Bölcs, A., "A Test Facility for the Investigation of Steady and Unsteady Transonic Flows in Annular Cascades," ASME Paper 83-GT-34, 1983.

Bölcs, A., and Fransson, T.H., "Aeroelasticity in Turbomachines: Comparison of Theoretical and Experimental Cascade Results," Air Force Office of Scientific Research, AFOSR-TR-87-0605, 1986.

Buffum, D.H., and Fleeter, S., "Effect of Wind Tunnel Acoustic Modes on Linear Oscillating Cascade Aerodynamics," ASME Journal of Turbomachinery, Vol. 116, pp. 513-524, 1994.

Crawley, E.F., "Aeroelastic Formulation for Tuned and Mistuned Rotors," Chapter 19 in AGARD Manual on Aeroelasticity in Axial-Flow Turbomachines, Vol. 2, Structural Dynamics and Aeroelasticity, M.F. Platzer and F.O. Carta, eds., AGARD-AG-298, 1988.

Holmes, D.G., and Chuang, H.A., "2D Linearized Harmonic Euler Flow Analysis for Flutter and Forced Response," in Unsteady Aerodynamics, Aeroacoustics, and Aeroelasticity of Turbomachines and Propellers, H.M Atassi, ed., Springer-Verlag, New York, 1993.

Holmes, D.G., Lamson, S.H., and Connell, S.D., "Quasi-3D Solutions for Transonic, Inviscid Flows by Adaptive Triangulation," ASME Paper 88-GT-83, 1988.

Holmes, D.G., Mitchell, B.E., and Lorence, C.B., "Three-Dimensional Linearized Navier-Stokes Calculations for Flutter and Forced Response," to be presented at The 8th International Symposium of Unsteady Aerodynamics and Aeroelasticity of Turbomachines, Stockholm, Sweden, Sept. 14-18, 1997.

Horlock, J.H., *Axial Flow Turbines: Fluid Mechanics and Thermodynamics,* Butterworth & Co., Ltd., 1966.

Lane, F., "System Mode Shapes in the Flutter of Compressor Blade Rows," Journal of the Aeronautical Sciences, pp. 54-66, Jan., 1956.

Nowinski, M., "Construction Instrumentation, and Assembly of an Annular, Low-Pressure Turbine Cascade for Torsional Flutter Studies, Status Report 1A," Ecole Polytechnique Federale de Lausanne, LTT-95-15, 1995.

Nowinski, M., "Investigation of the Steady-State and Unsteady Flow through the LPT Annular Cascade (Cfgs. 1b and 2b), Status Report 2," Ecole Polytechnique Federale de Lausanne, LTT-96-14, 1996.

Panovsky, J., "Flutter of Aircraft Engine Turbine Blades," Ph.D. Thesis, University of Cincinnati, Cincinnati, Ohio, 1997.

Sieverding, C.H., "Recent Progress in the Understanding of Basic Aspects of Secondary Flows in Turbine Blade Passages," ASME Journal of Engineering for Gas Turbines and Power, Vol. 107, pp. 248-257, April, 1985.

STAR-CD Reference Manual, Computational Dynamics, Ltd., 1995.

Turner, M.G., and Jennions, I.K., "An Investigation of Turbulence Modeling in Transonic Fans Including a Novel Implementation of an Implicit k-ε Turbulence Model," ASME Journal of Turbomachinery, Vol. 115, pp. 249-260, 1992.

COMPARISON OF DLT AND CFD PREDICTIONS OF UNSTEADY AERODYNAMIC FORCE ON VIBRATING SUPERSONIC THROUGH-FLOW FAN CASCADE

M. NAMBA, N. YAMASAKI AND T. OTSUKA
Department of Aeronautics and Astronautics
Kyushu University
6-10-1 Hakozaki, Higashi-ku, Fukuoka 812, Japan

Abstract. A two-dimensional linearized Euler solver for computing unsteady small disturbance flows and time-accurate non-linear Euler and Navier-Stokes solvers for computing unsteady flows in supersonic through-flow fan cascades vibrating with finite amplitudes have been developed to validate the predictions based on the double linearization theory (DLT). Some example calculations indicate that DLT can well predict the general trend of dependences of the unsteady aerodynamic forces on the parameters relevant to the steady loading as far as it is not large. Also it is found that the linearized Euler solutions which assume an infinitesimally small vibration amplitude agree well with the time accurate Euler solutions, and that the viscous effects on the moment coefficient for the normal translational oscillation are small.

1. Introduction

The self-excited blade vibration will be a crucial problem for a supersonic through-flow fan (STFF) which is to be composed of very thin blades and to operate under high loading conditions. The first systematic study on STFF cascade flutter was conducted by Kielb and Ramsey (1989) based on the two-dimensional linearized theory which deals with a cascade of flat plates oscillating at zero mean incidence angle. It is, however, widely recognized that the cascade flutter is seriously influenced by the steady loading, and

T.H. Fransson (ed.), Unsteady Aerodynamics and Aeroelasticity of Turbomachines, 831-845.

Toshimitsu *et al.* and Namba and Hanada have revealed that STFF cascades are no exceptions by their double linearization theory developed for linear cascades (Toshimitsu *et al.*, 1995) and rotating annular cascades (Hanada and Namba, 1995) operating at supersonic axial velocities.

The utility of the double linearization theory, however, is inherently limited to lightly loaded cascades with thin and small cambered blades at small angles of attack (Namba and Toshimitsu, 1991). The effects of thickness and camber on STFF cascade flutter were studied by Ramsey (1991) on the basis of the non-linear piston theory. His theory, however, fails to deal with the effect of chordwise motion of shock reflection points, which will play a crucial role on the unsteady blade loading. Therefore validation of the DLT predictions for the STFF cascade flutter problem remains unstudied yet.

DLT is based on the assumption that both the steady and the superposed unsteady components are potential flows, the former is sufficiently small to be linearized, and the latter is far smaller than the former, i.e., $O(\epsilon) << O(\bar{\epsilon}) << 1$. Here ϵ and $\bar{\epsilon}$ denote orders of magnitude of the unsteady and steady disturbances, respectively.

Nowadays Computational Fluid Dynamics (CFD) has become the major tool in unsteady cascade aerodynamics. There are essentially three CFD approaches to the unsteady aerodynamic analysis of oscillating cascades.

The first approach, which is least sophisticated among them, is to use the potential equations for the steady flows and the linearized potential equations for the unsteady flows (Whitehead, 1982), (Yang *et al.*, 1988). This approach is applicable only to cases of irrotational steady flows and $O(\epsilon) << O(\bar{\epsilon}) \simeq 1$, but has not been applied to STFF cascades as far as the authors are aware.

The second approach is to assume the flow field to be composed of a steady flow governed by the non-linear Euler equations and an unsteady small perturbation flow governed by the linear Euler equations of variable coefficients due to the underlying nonuniform steady flow. Although $O(\epsilon) << O(\bar{\epsilon}) \simeq 1$ should remain to be held, it is applicable even to the rotational flows. This method can not deal with non-linear unsteady phenomena, but it is still expected to model the essentially important physics dominating flutter phenomena. Hall *et al.* (1993) successfully applied this method to subsonic flows. Above all, the method is quite time saving in comparison with the most sophisticated model which will be discussed next because the local time stepping scheme can be used for time marching computations of both steady and unsteady equations and because the expressions of unsteady variables in terms of simple harmonic time dependent complex numbers enable the computation domain to be confined to a single blade passage. However it is still time consuming in comparison with

DLT. DLT and the first and second CFD approaches discussed so far are restricted to the vibrations with infinitesimally small amplitude and cannot handle the vibrating blades with a finite amplitude.

The third CFD approach, which is most sophisticated, is to solve the discretized non-linear full Euler or Navier-Stokes equations by a time-accurate time marching method. The method can deal with non-linear unsteady effects generating high harmonic and subharmonic modes when applied to the multi-passage flows. Thus there exists no restriction on the magnitude of the steady and unsteady disturbances. The calculations, however, are extremely expensive with highly long computational times and huge computer resources due to the requirements of time-accurate calculation of periodic flows, numerical stability and computing multiple blade passages to deal with oscillating cascade blades at nonzero interblade phase angles. In particular the Navier-Stokes calculations must be conducted with highly refined grids in order to accurately resolve the viscous regions, which inevitably leads to small time steps and subsequently long computations per the period of the vibration since the maximum CFL number is limited.

The present study aims at quantitative evaluation of utility and limitation of DLT for predictions of unsteady aerodynamic loading on oscillating STFF cascades by comparing the results with more sophisticated CFD predictions. The concepts and details of DLT are given in reference (Namba and Toshimitsu, 1991).

Since DLT is based on the potential flows, one of the main limitations in applying DLT is that the entropy increase due to the generation of the shock waves must be small. The restriction will not be serious for STFF cascade, which assumes the flow field of no normal shock waves. If Mach numbers normal to existing oblique shock waves are not far from one, the entropy increase in the flow field is expected to be small. Another restriction of DLT is the assumption of the small disturbance flow, which enables the linear superposition of effects of the various parameters on the resultant flow field and simplifies the calculation. In the DLT formulation, disturbances are convected at a uniform time- and space-averaged flow velocity and propagated at a uniform time- and space-averaged speed of sound. Yamasaki *et al.* (1996b), (1996a) have developed non-linear full Euler and Navier-Stokes solvers for computing steady flows in STFF cascades, and they confirmed the validity of the linearized theory for steady flow field in STFF cascades as far as the perturbation is small.

The Euler solver based on the second approach of linear unsteady perturbation is mainly used in the present study. This code is essentially a finite volume method based on Ni's Lax-Wendroff scheme (Ni, 1982) modified by Giles (1988) and uses the deforming computational grid technique by Hall and Clark (1993) to overcome the numerical inaccuracy which would arise from computing highly steep gradients of steady variables at the blade surface boundaries on the fixed-grid schemes.

The Euler and Navier-Stokes solvers based on the third approach of time-accurate solution are also used for comparison. Although many papers on the direct time-accurate calculation of the vibrating cascade flow fields, e.g., most of the papers in the last ISUAAT Proceedings (Tanida and Namba, 1995), treat multi-passage flow fields by limiting the interblade phase angles to 0, 90, 180, 270, 120 or 240 degrees, a formulation to solve the single passage is used in the present paper. The advantages of the single passage calculation are a smaller number of arithmetic operations due to the smaller extent of the computational domain and the capability of dealing with the blade vibrations under arbitrary interblade phase angles (e.g., near zero degrees). The disadvantages are a large number of iterations required to obtain the converged solution because matching the time lagging upper and lower boundary conditions is iteratively attained, inability to deal with subharmonic phenomena, and difficulty in adapting the code to paralell processor computers.

2. The Euler and Navier-Stokes Equations in Physical Plane

Consider a two-dimensional cascade of blades which are oscillating at a common frequency ω and a constant interblade phase angle $\hat{\sigma} = 2\pi\sigma$ in (x, y, t) coordinate system. It is assumed that the fluid is air and a calorically perfect gas. Then the Reynolds-averaged Navier-Stokes or Euler equations in conservative form governing the flow field are given by

$$\frac{\partial \boldsymbol{U}}{\partial t} + \frac{\partial \boldsymbol{F}}{\partial x} + \frac{\partial \boldsymbol{G}}{\partial y} = \frac{\partial \boldsymbol{F}_V}{\partial x} + \frac{\partial \boldsymbol{G}_V}{\partial y}, \tag{1}$$

Here terms with subscript V denote the diffusion or viscous terms, which vanish in the Euler equations since the viscosity and the thermal conductivity are neglected. Expressions of $\boldsymbol{U}$, $\boldsymbol{F}$, $\boldsymbol{G}$, $\boldsymbol{F}_V$ and $\boldsymbol{G}_V$ are omitted due to the limited space of the paper.

3. Mathematical Formulation of Linear Unsteady Calculation

3.1. DECOMPOSITION AND COORDINATE TRANSFORMATION

In this formulation, blades are assumed to be oscillating with a common small amplitude. A computational coordinate system (ξ, η, t) in which blades are relatively fixed is defined in the transformation form as follows:

$$x = \xi + f(\xi, \eta)e^{i\omega t}, \quad y = \eta + g(\xi, \eta)e^{i\omega t}, \quad t = t. \tag{2}$$

The definition of the functions $f(\xi, \eta)$ and $g(\xi, \eta)$ are rather arbitrary except that they must coincide with x- and y-wise displacements of a blade motion respectively at the blade surface points.

Let the unsteady flow field be represented by the superposition of the mean or steady flow field $\bar{\boldsymbol{U}}$ and the first order perturbation of a simple harmonic time dependence $\tilde{\boldsymbol{U}} e^{i\omega t}$: Expansion of dependent variables up to the first order gives

$$\boldsymbol{U}(x, y, t) = \bar{\boldsymbol{U}}(\xi, \eta) + \tilde{\boldsymbol{U}}(\xi, \eta, t) e^{i\omega t}. \tag{3}$$

Here properties with overbar denote steady components and those with tilde denote the complex amplitude of time-varying harmonic components. The complex amplitude $\tilde{\boldsymbol{U}}(\xi, \eta, t)$ is expressed as if it were dependent on time in order to apply the pseude-time marching method. It will ultimately become independent of time. Expanding dependent variables up to the first order, one can resolve the full Euler equation into the zero-th order steady part,

$$\frac{\partial \bar{\boldsymbol{U}}}{\partial t} + \frac{\partial \bar{\boldsymbol{F}}}{\partial \xi} + \frac{\partial \bar{\boldsymbol{G}}}{\partial \eta} = 0, \tag{4}$$

and the first order unsteady part,

$$\frac{\partial \tilde{\boldsymbol{U}}}{\partial t} + i\omega \tilde{\boldsymbol{U}} + \frac{\partial}{\partial \xi}\left(\frac{\partial \bar{\boldsymbol{F}}}{\partial \bar{\boldsymbol{U}}} \tilde{\boldsymbol{U}}\right) + \frac{\partial}{\partial \eta}\left(\frac{\partial \bar{\boldsymbol{G}}}{\partial \bar{\boldsymbol{U}}} \tilde{\boldsymbol{U}}\right)$$

$$= i\omega\left(f\frac{\partial}{\partial \xi} + g\frac{\partial}{\partial \eta}\right)\bar{\boldsymbol{U}} + \left(\frac{\partial f}{\partial \xi}\frac{\partial}{\partial \xi} + \frac{\partial g}{\partial \xi}\frac{\partial}{\partial \eta}\right)\bar{\boldsymbol{F}} + \left(\frac{\partial f}{\partial \eta}\frac{\partial}{\partial \xi} + \frac{\partial g}{\partial \eta}\frac{\partial}{\partial \eta}\right)\bar{\boldsymbol{G}}, \tag{5}$$

where $\bar{\boldsymbol{F}} = \boldsymbol{F}(\bar{\boldsymbol{U}})$ and $\bar{\boldsymbol{G}} = \boldsymbol{G}(\bar{\boldsymbol{U}})$. Note that in equation (4) for the steady part the dependent variables are expressed as if they were dependent on time in order to apply the time marching method. They are, however, independent of time at the stage of solving equation (5) for the unsteady part.

3.2. FINITE VOLUME FORMULATION AND BOUNDARY CONDITIONS

Equations (4) and (5) are solved by the finite volume method base on the Ni's Lax-Wendroff scheme (Ni, 1982). Implicit treatment of the time varying amplitude of unsteady components is introduced. This scheme may allow one to use larger time step Δt resulting in higher rate of convergence than otherwise. To stabilize shock calculations the second difference smoothing is added. Also a fourth difference damping is added to suppress unwanted high frequency waves. Details are omitted.

Let $\boldsymbol{V}$ be fluid velocity with respect to the absolute coordinate system and $\boldsymbol{n}$ be the unit normal vector at a blade surface point. Then the flow tangency condition is expressed by $(\boldsymbol{V} - i\omega \boldsymbol{a} e^{i\omega t}) \cdot \boldsymbol{n} = 0$ at an instantaneous blade surface position. Noting $\boldsymbol{V} = \bar{\boldsymbol{V}} + \tilde{\boldsymbol{V}} e^{i\omega t}$ and $\boldsymbol{n} = \bar{\boldsymbol{n}} + \tilde{\boldsymbol{n}} e^{i\omega t}$, one can resolve the flow tangency condition into a steady part: $\bar{\boldsymbol{V}} \cdot \bar{\boldsymbol{n}} = 0$ at a mean blade surface position, and an unsteady part: $\tilde{\boldsymbol{V}} \cdot \bar{\boldsymbol{n}} = i\omega \boldsymbol{a} \cdot \bar{\boldsymbol{n}} - \bar{\boldsymbol{V}} \cdot \tilde{\boldsymbol{n}}$ at an instantaneous blade surface position. The difficulty in applying the flow tangency condition on the moving blade surface can be overcome by describing the condition in terms of the transformed coordinate system (ξ, η, t) in which blades are stationary.

The periodic boundary condition corresponding to the interblade phase angle is imposed on the top and bottom boundaries of the computational domain of a single blade-to-blade channel.

4. Mathematical Formulation of Non-Linear and Time-Accurate Unsteady Calculation

4.1. GOVERNING EQUATIONS IN GENERAL COORDINATE SYSTEM

In the present formulation, the generalized (computational) coordinate system is used. The computational coordinate (ξ, η, t) is related to the physical coordinate (x, y, t) by

$$\xi = \xi(x, y, t), \quad \eta = \eta(x, y, t), \quad t = t. \tag{6}$$

Then Eq.(1) is rewritten as

$$\frac{\partial \hat{\boldsymbol{U}}}{\partial t} + \frac{\partial \hat{\boldsymbol{F}}}{\partial \xi} + \frac{\partial \hat{\boldsymbol{G}}}{\partial \eta} = \frac{\partial \hat{\boldsymbol{F}}_V}{\partial \xi} + \frac{\partial \hat{\boldsymbol{G}}_V}{\partial \eta}, \tag{7}$$

where

$$\hat{\boldsymbol{U}} = \boldsymbol{U}/J,$$
$$\hat{\boldsymbol{F}} = (\xi_t \boldsymbol{U} + \xi_x \boldsymbol{F} + \xi_y \boldsymbol{G})/J, \quad \hat{\boldsymbol{G}} = (\eta_t \boldsymbol{U} + \eta_x \boldsymbol{F} + \eta_y \boldsymbol{G})/J,$$
$$\hat{\boldsymbol{F}}_V = (\xi_x \boldsymbol{F}_V + \xi_y \boldsymbol{G}_V)/J, \quad \hat{\boldsymbol{G}}_V = (\eta_x \boldsymbol{F}_V + \eta_y \boldsymbol{G}_V)/J,$$
$$J = \xi_x \eta_y - \xi_y \eta_x = 1/(x_\xi y_\eta - x_\eta y_\xi).$$

4.2. DIFFUSION MODELS

In the Navier-Stokes calculations, the viscosity is introduced as follows: The laminar viscosity μ_l is given by Sutherland's formula. The algebraic modelling of Baldwin-Lomax (Baldwin and Lomax, 1978) is employed for the calculation of the eddy viscosity.

For flows in the wake, the Baldwin-Lomax turbulence model with the relaxation factor is used (Waskiexicz *et al.*, 1980), i.e., the local turbulent viscosity μ_t in the wake region is given by $(\mu_t - \mu_{\text{TE}})/(\mu_{\text{local eq.}} - \mu_{\text{TE}}) = 1 - \exp[-\Delta x/(\lambda_0 \delta_{\text{TE}})]$, where μ_{TE} is the calculated turbulent viscosity at the trailing edge of the blade, $\mu_{\text{local eq.}}$ is the calculated local equilibrium properties given by the Baldwin-Lomax turbulence model, Δx is the x-wise distance from the blade trailing-edge and λ_0 is the relaxation length scale. We put $\lambda_0 = 20$ in the present calculation. The displacement thickness deduced from 1/7-th power law is given by $\delta_{\text{TE}} = 0.37(\text{chord length})/Re^{0.2}$ (Schlichting, 1960).

In unsteady calculations, the turbulence model is applied in a quasi-steady fashion, i.e., the instantaneous flow field relative to the moving blade is used to calculate the turbulent viscosity.

4.3. TIME DISCRETIZATION AND TVD SCHEME

The solution is marched in time using the difference formula. For the present compressible Navier-Stokes or Euler equations, the three-point backward implicit difference scheme, which is second order in time, is used for the unsteady (oscillatory) flow calculations.

In discretizing the convective part, Yee-Harten's non-MUSCLE type (Yee, 1989) of the fully implicit upwind TVD scheme is applied, while the second order central finite difference scheme is used for the evaluation of viscous terms in the Navier-Stokes calculations. LCI (Linearized Conservative Implicit) form (Yee, 1989) and ADI (Alternative Direction Implicit) method are employed.

4.4. BOUNDARY CONDITIONS AND MISCELLANEOUS

The boundary condition on the blade is specified as the non-slip condition $U = 0$ and $V = 0$ for the viscous flows, or the slip (flow tangency) condition $V = 0$ (U is extrapolated) for the inviscid flows. Here U and V are contravarient velocities defined by $U = \xi_t + \xi_x u + \xi_y v$ and $V = \eta_t + \eta_x u + \eta_y v$, and u and v are velocity components in the physical (x, y) plane. In addition, in both viscous and inviscid calculations, zero normal derivative of pressure $\partial p/\partial \eta = 0$, and adiabatic condition $\partial T/\partial \eta = 0$ are used on the blade, where η=constant denotes the blade surface.

As to the boundary condition in the blade-to-blade direction, the time-lagging periodical condition stipulated by the interblade phase angle is applied by bookkeeping the Fourier transformed properties (Wolff and Fleeter, 1994) but in a little different way. Details are omitted.

Grids in the present calculation are generated using the elliptic solver (Thompson *et al.*, 1985). The generated instantaneous grids with the constant parameter relevant to the concentration of mesh, etc., are also stored in the Fourier transformed form.

Geometrical conservation law (GCL) is fully installed in the code as is described by Fujii (1994). Non-reflecting upstream or downstream boundary condition (Giles, 1988) has not been implemented at present, since it is not essential for STFF calculations, but the computational domain should be adequately extended spatially for the non-linear unsteady calculations.

The present code is pretty robust. CFL number of as high as 15 in Euler calculations and CFL number of 5 in Navier-Stokes calculations are stably used.

5. Numerical Calculations

In the followings, the linear unsteady calculation based on the finite volume method discussed in Section 3 is to be referred to as FVM, and the non-linear and time-accurate calculation based on the finite difference method discussed in Section 4 is to be referred to as FDM (Euler) or FDM (N-S), where the former is based on the Euler equations and the latter is based on the Navier-Stokes equations. In subsections 5.1 and 5.2, the comparison between DLT and FVM is to be discussed, while, in subsection 5.3, the FDM (Euler) and FDM (N-S) calculations applied to some special cases are to be discussed.

In the following numerical examples, the axial Mach number $M_a = M\cos\gamma = 2.3$, the angle of stagger $\gamma = 28°$, the pitch/chord ratio $s/c = 0.311$, and the reduced frequency $\omega c/U_0 = 0.1$ are fixed, where c and U_0 denote the chord length and the undisturbed upstream flow speed, respectively. Here a cascade of blades of double parabolic profile given by

$$\eta = \phi_{\pm}(\xi) = c(\mu_0 \pm \tau_0/2)\{1 - 4[(\xi - \xi_0)/c]^2\},$$

is assumed where μ_0 and τ_0 are the maximum camber and thickness ratios, respectively. The mean angle of attack is denoted by α. Geometric parameters and oscillating modes of four cases to be dealt with in the present paper are shown in TABLE 1. The blade motion is defined by the angular displacement $\tilde{\alpha}e^{i\omega t}$ about the elastic axis $\mathrm{E}(\xi_e, 0)$ and/or the tranlational displacement $\tilde{\boldsymbol{h}}(\tilde{h}_x, \tilde{h}_y)e^{i\omega t}$ of the elastic axis E. The lift and drag forces are defined as components perpendicular to and parallel to the blade chord, respectively, and lift, drag and moment coefficients are non-dimensionalized by $\rho_0 U_0^2 c$, $\rho_0 U_0^2 c$, and $\rho_0 U_0^2 c^2$ and denoted by $\tilde{C}_L$, $\tilde{C}_D$ and $\tilde{C}_M$, respectively, where ρ_0 denotes the undisturbed upstream fluid density.

The computational domain for the FVM calculation is confined to a single blade passage with streamwise extent of twice the chord length, and consists of 152 (streamwise) $\times$ 65 (blade to blade) grid points. The convergence of the steady and unsteady solutions is judged to be numerically

Case No.	μ_0	τ_0	α [degrees]	Oscillating mode
3	0.00	0.00	0.0	Translational oscillation normal to chord
5	0.00	0.00	2.0	Translational oscillation normal to chord
6	0.03	0.03	0.0	Translational oscillation normal to chord
8	0.06	0.03	0.0	Translational oscillation normal to chord

TABLE 1. Geometric Parameters and Oscillating Modes

attained in the FVM calculations when the maximum absolute of the residual of the second component of the conserved variables is less than the threshold value (1.0×10^{-6} in the present paper).

5.1. BLADES WITH NO STEADY LOADING

In the present subsection, comparison of the unsteady aerodynamic characteristics between the linear unsteady predictions (FVM) and the DLT predictions applied to oscillations of a cascade of flat plates at zero mean angle of attack (Case 3) is discussed. Note that, in such no loading cases, the steady flow field is exactly same as a uniform flow with no steady perturbations in the DLT and FVM calculations and DLT reduces to the linear unsteady potential theory. Shock waves and expansion fans are generated when the overlaid unsteadiness due to the oscillation of blades is taken into account.

Figure 1 (left) shows the unsteady pressure difference distribution under normal translational oscillation ($\tilde{h}_x = 0,\ \tilde{h}_y = 1,\ \tilde{\alpha} = 0$) at the interblade phase angle of $\hat{\sigma} = 90°$. A good agreement is obtained between DLT and FVM. Although figures are omitted, a good coincidence between DLT and FVM is also obtained in all interblade phase angles and oscillating modes. Although the second order dissipation is introduced in FVM, the numerical overshooting and oscillation around shock waves are observed. The overall lift or moment coefficient is, however, not affected by such oscillations as will be discussed next because the oscillations are canceled out each other.

Figure 1 (right) shows a polar diagram of the unsteady lift coefficient $C_L/\tilde{h}_y$ for the normal translational oscillation with the interblade phase angle as a variable parameter. Scattered plots constitute a loop with the interblade phase angles of every 30 degrees. These results demonstrate that the agreement between FVM and DLT is quite satisfactory in the case of oscillations under zero mean loading, as it should be. The correspondence is good between DLT and FVM in torsional oscillations as well.

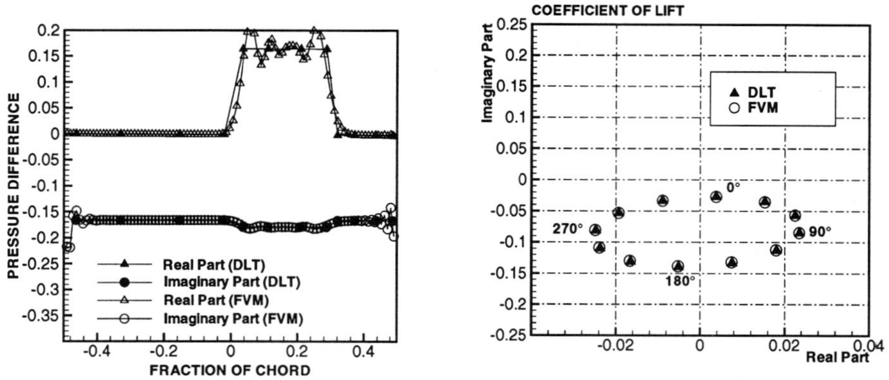

Figure 1. Unsteady pressure difference distribution at $\hat{\sigma} = 90^\circ$ (LEFT) and unsteady lift coefficient $\tilde{C}_L$ (RIGHT) for normal translational oscillation of flat plate cascade at zero mean angle of attack ($\alpha = \mu_0 = \tau_0 = 0$).

5.2. BLADES WITH STEADY LOADING

In the present subsection, comparison of the unsteady aerodynamic characteristics between the FVM and DLT predictions applied to oscillations of a cascade with steady loading are discussed. It has first been confirmed that the steady components of pressure and velocity fields predicted by the present FVM code are in good agreement with those by another code based on the much sophisticated finite difference method given in reference (Yamasaki *et al.*, 1996a), (Yamasaki *et al.*, 1996b). The minor differences between these solutions are due to the different artificial viscosity. The flow field obtained by FVM calculations is more dissipative than that obtained by the steady finite difference calculation, since the finite difference calculation uses the TVD scheme less dissipative than the second and fourth order smoothings employed in FVM.

Before discussing the unsteady characteristics, the comparison of the steady characteristics between DLT and FVM is discussed first. In general, as the camber or thickness of blades becomes large, the differences in shock wave reflection location and shock strength between FVM and DLT (more precisely, the linear potential theory in steady cases) become conspicuous, and the DLT solutions deviate from the FVM solutions. In DLT, a shock wave implies an edge Mach wave, and an expansion fan is represented by a single Mach wave with a finite pressure jump. Agreement of the steady flow field predicted by DLT and FVM is fairly good for the cascade of flat plate with a finite angle of attack (Case 5) but not so good for the cambered cascade blade (Case 6) since the non-linear behavior is prominent in the latter case. As far as the total steady lift coefficient is concerned, however, the difference between the FVM and DLT predictions is small for the conditions in the present subsection.

Figure 2 shows the unsteady lift coefficient for normal translational oscillation ($\tilde{\alpha} = 0$, $\tilde{h}_x = 0$, $\tilde{h}_y = 1$) of the flat plate cascade at a mean angle of attack $\bar{\alpha} = 2°$ (Case 5) and the cambered thick cascade blades ($\mu_0 = 0.03$ and $\tau_0 = 0.03$) at zero mean angle of attack (Case 6). Figure 3 shows the unsteady lift and moment coefficients of the cambered thick cascade blades ($\mu_0 = 0.03$ and $\tau_0 = 0.03$) at a mean angle of attack $\bar{\alpha} = 2°$ (Case 8) under tranlational oscillation ($\tilde{\alpha} = 0$, $\tilde{h}_x = 0$, $\tilde{h}_y = 1$). The discrepancies between DLT and FVM predictions are not large, and DLT can suitably predict the trend of the parametric dependences. Both results demonstrate that the translational oscillation is aerodynamically unstable (the imaginary part of C_L is positive) due to nonzero mean loading in an interblade phase angle region for Case 5 and Case 7. For Case 6, on the other hand, DLT predicts that the translational oscillation is aerodynamically unstable in an interblade phase angle region $\hat{\sigma} = 300 - 330°$, but FVM predicts that the tranlational oscillation is aerodynamically stable in the whole interblade phase angle region.

Figure 4 shows the unsteady pressure difference distribution for cambered thick cascade blades under normal translational oscillation ($\tilde{h}_x = 0$, $\tilde{h}_y = 1$, $\tilde{\alpha} = 0$) with non-zero mean loading ($\bar{\alpha} = 2°$) (Case 8) at the interblade phase angle of $\hat{\sigma} = 90°$. In the DLT formulations, the delta-function-type singularity is introduced in the blade surface pressure to correctly model the effect of shock wave motion due to the oscillation of blades. In Figure 4, the contribution of such delta-function-type singularities to the unsteady pressure difference distribution is represented by the equivalent triangular pulses. In FVM the effect of the shock motion on the pressure distribution appears as a sharp peak or pit at the shock point like triangular pulses. It should be noted that the effect of the shock motion overwhelmingly dominates the unsteady aerodynamic force on the blades. It is observed that the difference between the DLT and FVM calculations becomes conspicuous as the camber or thickness is increased, and, in Case 8, the differences between the shock reflection location and shock strength in the FVM calculations and the Mach wave reflection location and its strength in DLT are fairly large. In particular a delta-function-type singularity also appears at the reflection point of the expansion Mach wave from the leading edge in DLT. Such a peculiarity is absent in FVM, where the expansion waves from the edges appear as expansion fans. Nevertheless the correspondence in the total lift or moment coefficient between DLT and FVM is satisfactory as shown in Figure 3.

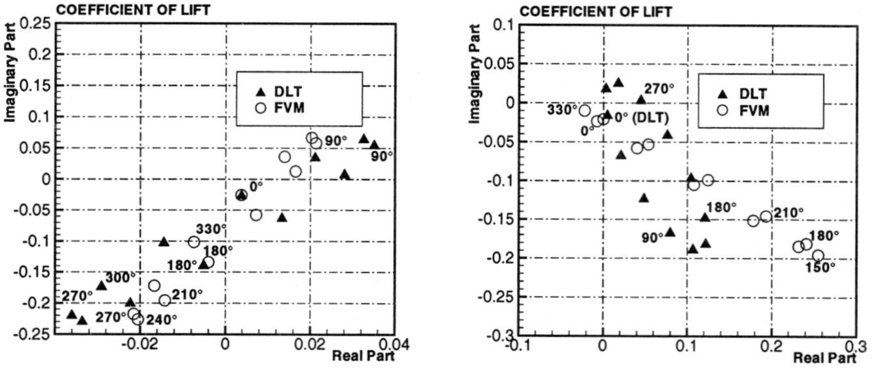

Figure 2. Unsteady lift coefficient $\tilde{C}_L$ for normal translational oscillation. LEFT: Flat plate cascade at a finite mean angle of attack ($\alpha = 2°$, $\mu_0 = \tau_0 = 0$). RIGHT: Cascade of blades with a finite camber and thickness at zero mean angle of attack ($\alpha = 0$, $\mu_0 = 0.03$, $\tau_0 = 0.03$).

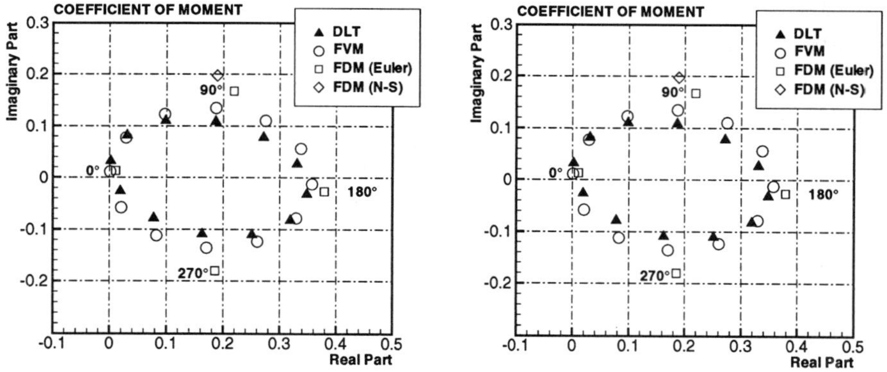

Figure 3. Unsteady lift coefficient $\tilde{C}_L$ (LEFT) and moment coefficient $\tilde{C}_M$ (RIGHT) for normal translational oscillation of cascade of blades with a finite camber and thickness at a finite mean angle of attack ($\alpha = 2°$, $\mu_0 = 0.06$, $\tau_0 = 0.03$).

5.3. NON-LINEAR AND TIME-ACCURATE UNSTEADY CALCULATION

In this subsection, the non-linear time-accurate Euler and Navier-Stokes solutions for Case 8 (the angle of attack of $\bar{\alpha} = 2°$, camber ratio of μ_0 =0.06 and thickness ratio of τ_0 =0.03, translational oscillation normal to the chord line) are to be discussed. The dimensional parameters required to be specified additionally for non-linear calculations are the chord length of

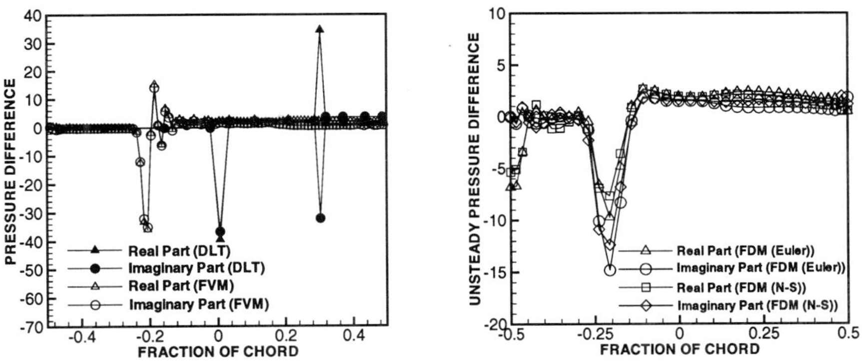

Figure 4. Unsteady pressure difference distribution at $\hat{\sigma} = 90°$ for normal translational oscillation. LEFT: DLT and FVM. RIGHT: FDM (Euler) and FDM (N-S). Cascade of blades with a finite camber and thickness at a finite mean angle of attack ($\alpha = 2°$, $\mu_0 = 0.06$, $\tau_0 = 0.03$).

c =0.0326m, the inflow total pressure of p_0 =0.1MPa, and the inflow total temperature of T_0 =510K. The interblade phase angles of $\hat{\sigma} = 0$, 90, 180 and 270° are calculated in the Euler calculations (FDM (Euler)) and $\hat{\sigma} = 90°$ in the Navier-Stokes calculation (FDM(N-S)). The vibration amplitude is fixed to 1% of chord length ($\tilde{\alpha} = 0$, $\tilde{h}_x = 0$ and $\tilde{h}_y/c = 0.01$). Calculations are executed on the grid of 97×33 (upstream boundary at about 0.1 chord upstream of the leading edge locus and downstream boundary at about 1.7 chord downstream of trailing edge locus). Iteration was repeated for 120 oscillation periods.

It is useful to compute the first order Fourier coefficients from the results of the non-linear time-accurate unsteady calculations because they correspond to the complex magnitude in the linear unsteady calculations (FVM or DLT). Note that the zero-th order Fourier coefficient corresponds to the steady part. Higher order mode coefficients are usually smaller by several orders of magnitude compared with the first order coefficients.

Figure 4 (Right) shows the unsteady pressure difference distribution by the time-accurate Euler (FDM (Euler)) and Navier-Stokes (FDM (N-S)) finite difference calculations at the interblade phase angle of $\hat{\sigma} = 90°$. Both unsteady pressure distributions by the FDM (Euler) and FDM (N-S) calculations along the chord show the same tendencies as the time-linear FVM calculations except near the leading edge, where the FDM (Euler) and FDM (N-S) results show the large variation in their real part components. In particular the effect of the shock motion is well represented by triangular pulses without any oscillating overshoot. Minor difference in the unsteady pressure distribution between FDM (Euler) and FDM (N-S) gives rise to a pretty large difference in the lift coefficient between FDM (Euler) and FDM(N-S) as is discussed below.

In Figure 3 are also plotted the unsteady lift (Left) and moment coefficients (Right) predicted by FDM (Euler) and FDM (N-S). The lift coefficients predicted by the FDM (Euler) calculations show good agreement with those by the DLT and FVM calculations at the interblade phase angles of $\sigma \simeq 0°$ and $90°$, but a poor correspondence is observed at the interblade phase angles of $\sigma \simeq 180°$ and $270°$. As far as the moment coefficient is concerned, however, the coincidence between the FVM and FDM (N-S) calculations is good. The time-accurate Navier-Stokes predictions (FDM (N-S)) at the interblade phase angle of $\sigma = 90°$ is in poor agreement with other calculations in the lift coefficient and in good agreement in the moment coefficient. At the present stage the convergence of the time-accurate FDM calculations is still open to question. The detail should be addressed in the future work.

It is also found that the required CPU time per case for the non-linear unsteady calculation is an order of 100 times as large as the linear unsteady calculation, which is an order of 10000 times as large as that for DLT.

6. Conclusions

A two-dimensional linearized Euler solver for computing unsteady small disturbance flows and time-accurate non-linear Euler and Navier-Stokes solvers for computing unsteady flows in supersonic through-flow fan cascades vibrating with finite amplitudes have been developed to validate the predictions based on the double linearization theory (DLT). The results obtained from numerical calculations are summarized as follows:

- DLT can well predict the general trend of dependences of the unsteady aerodynamic forces on the parameters relevant to the steady loading as far as it is not large.
- The linearized Euler solutions which assume an infinitesimally small vibration amplitude well correspond with the time accurate Euler solutions, and that the viscous effects on the moment coefficient for the normal translational oscillation are small.
- For the oscillation of cascade of blades with steady loading, the total unsteady aerodynamic force on the blade is overwhelmingly dominated by the effect of the shock motion which appears as triangular peaks in the blade surface pressure distribution.

References

Baldwin, B. S., and Lomax H. (1978) *Thin Layer Approximation and Algebraic Model for Separated Turbulent Flows*, AIAA Paper 78-257.

Fujii, K. (1994) *Numerical Methods for Computational Fluid Dynamics*, University of Tokyo Press (in Japanese).

Giles, M. (1988), UNSFLO: A Numerical Method for Unsteady Inviscid Flow in Turbomachinery, *Gas Turbine Laboratory Report*, **No. 195**, MIT.

Hall, K.C. and Clark, W.S. (1993) Linearized Euler Predictions of Unsteady Aerodynamic Loads in Cascades, *AIAA Journal*, **Vol. 31, No. 3**, pp. 540-550.

Hanada, T. and Namba, M. (1995) Mean Loading Effect on the Flutter of Supersonic Through-Flow Fan Blade, *Unsteady Aerodynamics and Aeroelasticity of Turbomachines, Proceedings of the Seventh Symposium*, Tanida, Y. and Namba, M., eds., Elsevier Science B.V., pp. 547-564.

Kielb, R.E. and Ramsey, J.K. (1989) Flutter of a Fan Blade in Supersonic Axial Flow, *ASME Journal of Turbomachinery*, **Vol. 111**, pp. 462-467.

Namba, M. and Toshimitsu, K. (1991) Double Linearization Theory for Prediction of Mean Loading Effect on Cascade Flutter, I: Two-Dimensional Subsonic Cascade, II: Two-Dimensional Supersonic Cascade, *Journal of Sound and Vibration*, **Vol. 147**, pp. 203-231, **Vol. 147**, pp. 233-254.

Ni, Ron-Ho (1982) A Multiple-Grid Scheme for Solving the Euler Equations, *AIAA Journal*, **Vol. 20, No. 10**, pp. 1565-1571.

Ramsey, J. K. (1991) Influence of Thickness and Camber on the Aeroelastic Stability of Supersonic Throughflow Fans, *Journal of Propulsion and Power*, **Vol. 7**, pp. 462-467.

Schlichting, H. (1960) *Boundary Layer Theory (Fourth Edition)*, Translated by Kestin, J., McGraw Hill.

Tanida, Y. and Namba, M., eds. (1994) *The Proceedings of the Seventh International Symposium on the Unsteady Aerodynamics and Aeroelasticity of Turbomachines*, Elsevier Publishers BV.

Thompson, J.F., Warsi, Z.U.A., and Mastin, C. W. (1985) *Numerical Grid Generation*, North-Holland.

Toshimitsu, K., Iwai, S., and Namba, M. (1995) Double Linearization Theory Applied to Three-Dimensional Cascades Oscillating under Supersonic Through-Flow Fan, *JSME International Journal, Series B*, **Vol. 38, No. 4**, pp. 582 - 592.

Waskiexicz J. D., Shang J. S., and Hankey, W. L. (1980) Numerical Simulation of Near Wakes Utilizing a Relaxation Turbulent Model, *AIAA Journal*, **Vol. 18, No. 12**, pp. 1440-1445.

Whitehead, D. S. (1982) The Calculation of Steady and Unsteady Transonic Flow in Cascades, Cambridge University Engineering Department Report *CUED/A-Turbo/TR*, **98**, Cambridge University.

Wolff, J.M., and Fleeter, S. (1994) Single Passage Euler Analysis of Oscillating Cascade Aerodynamics for Arbitrary Interblade Phase, *Journal of Propulsion and Power*, **Vol. 10, No. 5**, pp. 690-697.

Yamasaki, N., Namba M., and Tashiro, K. (1996a) Flow Field in the Supersonic Through-Flow Fan (Comparison of the Solution of the Linear Potential Theory and the Numerical Solution of the Euler Equations), *Technology Reports of Kyushu University* **Vol. 69, No. 2**, pp. 189-193 (in Japanese).

Yamasaki, N., Namba, M., and Seet, K-K., (1996b) Flow Field Analysis for the Supersonic Through-Flow Fan, *Transactions of the Japan Society of Mechanical Engineers, Series B*, **Vol. 62, No. 595**, pp. 1045-1052 (in Japanese).

Yang, K-C., Yamasaki, N. and Namba, M. (1988) Numerical Calculation of Aerodynamic Forces for Two-Dimensional Subsonic Oscillating Cascade by a Finite Element Method, *Memoirs of Faculty of Engineering, Kyushu University*, **Vol. 48, No. 4**, pp. 253-280.

Yee, H.C. (1989), *A Class of High-Resolution Explicit and Implicit Shock Capturing Methods*, NASA TM-101088.

CHAIRPERSON INDEX

Chairpersons & Vice Chairpersons	Session	Page
Atassi, H.M.	14	765
Bölcs, A.	05	241
Ferrand, P.	04	181
Ffowcs Williams, J.	01	1
Fleeter, S.	02	51
Fransson, T.	06	301
Gallus, H.E.	07	361
Hall, K.C.	08	425
He, L.	13	705
Höhn, W.	12	647
Imregun, M.	15	813
Isomura, K.	01	1
Kahl, G.	10	539
Kaji, S.	09	493
Kielb, R.	06	301
Körbächer, H.	12	647
Manwaring, S.R.	10	539
Marshall, J.G.	13	705
Montgomery, M.D.	14	765
Moyroud, F.	11	583
Mårtensson, H.	11	583
Namba, M.	04	181
Platzer, M.F.	15	813
Reiss, W.	08	425
Sajben, M.	02	51
Tanida, Y.	07	361
Tinoco, H.	09	493
Verdon, J.M.	03	117
Whitehead, D.S.	03	117
Zhou, S.	05	241

AUTHOR INDEX

A

Abo El lail, A.S. 495
Acton, P. 393
Adachi, T. 363
Aotsuka, M. 679
Atassi, H.M. 131, 147
Aubert, S. 35, 749

B

Bell, D.L. 53
Bölcs, A. 103, 815

C

Cizmas, P. 633, 767
Chernysheva, O.L. 119
Chew, J.W. 707
Chiba, K. 511

D

Debard, M. 585
Dorney, D.J. 407

F

Ferrand, P. 35, 585, 749
Fleeter, S. 67, 465
Florea, R. 767
Fottner, L. 243, 393
Fransson, T.H. 195, 649
Freund, D.D. 287
Frey, K.K. 67
Fujimoto, I. 83

G

Gallus, H.E. 477
Gnesis, V. 317, 613
Giles, M.B. 225
Golubev, V.V. 131, 147
Green, J.S. 347
Groth, P. 565

H

Hall, K.C. 767
Hassan, A.S.A. 495
He, L. 53, 183
Hirano, T. 83
Holmes, D.G. 211
Höhn, W. 195

I

Imregun, M. 347, 541, 707
Isomura, K. 725
Iwamoto, Y. 525

J

Jahnen, W. 243
Johansson, U. 649
Jones, K.D. 797

K

Kahl, G. 335
Kaji, S. 733, 783
Kamijo, K. 3
Kato, D. 511
Kemppainen, A. 649
Kirkeng, K.L. 379
Korobeinikov, S. 303
Kovalyov, A. 317
Kurzin, V.B. 303

L

Leboeuf, F. 585
Lorence, C.B. 211

M

Machida, Y. 679
Manwaring, S.R. 379
Marshall, J.G. 225, 707
Marx, Y. 599
Massão, F. 749

Matsushita, M. 693
Mitchell, B.E. 211
Montgomery, M.D. 427, 445
Mårtensson, H. 565

N

Nagai, K. 165
Nagashima, T. 259
Namba, M. 165, 831
Ning, W. 183
Noguchi, Y. 693
Norryd, M. 103
Nowinski, M. 665, 815

O

Oliveira, G. 35
Otsuka, T. 831
Ott, P. 665
Outa, E. 511

P

Panovsky, J. 815
Peitsch, D. 477
Peters, Th. 243
Platzer, M.F. 797

R

Riess, W. 273
Ryabchenko, V.P. 303
Rządkowski, R. 317, 613

S

Sajben, M. 287
Saren, V.E. 407
Savin, N.M. 407
Sayma, A.I. 347
Sbardella, L. 541
Sentker, A. 273
Shiratori, T. 693
Shibata, T. 733
Smati, L. 35, 749
Subramanya, R. 633
Svensdotter, S. 649

T

Tanaka, H. 83
Tkacheva, L.A. 303
Tsujimoto, Y. 3, 525

V

Vahdati, M. 347, 707
Verdon, J.M. 427, 445

W

Watanabe, S. 3, 679
Weber, S. 477
Wiercinski, Z. 19
Wolff, J.M. 465

Y

Yamane, T. 259
Yamasaki, N. 831
Yamashita, Y. 363
Yokota, K. 3, 525
Yudin, V.A. 119

Z

Zacharias, R.M. 407
Zhou, S. 555

Abhari, Reza
Professor
The Ohio State University
Gas Türbine Lab
2300 West Case Road
COLUMBUS, OHIO 43235
UNITED STATES

Abo El-Lail, Aly Sabry
Mr
Assiut University
Faculty of Engineering
71516 ASSIUT
EGYPT

Acton, Paola
Ms
Univ der Bundeswehr München
Werner-Heisenberg-Weg 39
85577 NEUBIBERG
GERMANY

Adachi, Tsutomu
Professor
Osaka Sangyo University
Faculty of Engineering
3-1-1 Nakagaito, Daito
OSAKA 574
JAPAN

Ahlinder, Stefan
Mr
Forsmarks Kraftgrupp AB
742 03 ÖSTHAMMAR
SWEDEN

Arvidsson, Ulf
Mr
Elforsk AB
101 53 STOCKHOLM
SWEDEN

Atassi, Hafiz
Professor
University of Notre Dame
Aerospace & Mechanical Engineering
NOTRE DAME, IN 46556
UNITED STATES

Aubert, Stéphane
Mr
Ecole Centrale de Lyon
Lab de Mecanique des Fluides et d'Ac
36 avenue Guy de Collongue, BP 163
69131 ECULLY Cedex
FRANCE

Bauer, Holger
Mr
University of Stuttgart
ITSM
Pfaffenwaldring 6
70550 STUTTGART
GERMANY

Bell, David
Mr
University of Durham
School of Engineering
South Road
DURHAM DH1 3LE
UNITED KINGDOM

Burgos Olmos, Manuel Antonio
Mr
Industria de Turbopropulsores S A
Parque Tecnologico 300
48170 ZAMUDIO
SPAIN

Bölcs, Albin
Professor
Swiss Federal Inst of Technology-LTT
1015 LAUSANNE
SWITZERLAND

Börjesson, Tomas
Royal Inst of Technology
Heat and Power Technology
Brinellvägen 60
100 44 STOCKHOLM
SWEDEN

Casanovas, Virginie
Royal Inst of Technology
Heat and Power Technology
Brinellvägen 60
100 44 STOCKHOLM
SWEDEN

Chernysheva, Olga
Ms
Royal Inst of Technology
Brinellvägen 60
100 44 STOCKHOLM
SWEDEN

Cizmas, Paul
Dr
Westinghouse Electric Corp
Science & Technology Center
1310 Beulah Road, MS 401-4814
PITTSBURGH, PA
UNITED STATES

Dewhurst, Stephen
Mr
GEC Alsthom
Aerodynamics Group
Newbold Road
RUGBY Warwickshire CV21 2NH
UNITED KINGDOM

Dorney, Daniel
Dr
GMI Engineering & Management Inst
1700 West Third Ave
FLINT, MI 48504-4898
UNITED STATES

Einarsson, Tommy
Mr
Forsmarks Kraftgrupp AB
742 03 ÖSTHAMMAR
SWEDEN

Ferrand, Pascal
Dr
Ecole Centrale de Lyon
Lab de Mecanique des Fluides et d'Ac
36 avenue Guy de Collongue, BP 163
69131 ECULLY Cedex
FRANCE

Ffowcs Williams, John
Professor
University of Cambridge
Dept of Engineering
Trumpington Street
CAMBRIDGE CB2 1PZ
UNITED KINGDOM

Fleeter, Sanford
Professor
Purdue University
1288 Mechanical Engineering Bldg
WEST LAFAYETTE, IN 47907-1288
UNITED STATES

Fransson, Torsten
Professor
Royal Inst of Technology
Heat and Power Technology
Brinellvägen 60
100 44 STOCKHOLM
SWEDEN

Freudenreich, Kai
Royal Inst of Technology
Heat and Power Technology
Brinellvägen 60
100 44 STOCKHOLM
SWEDEN

Frey, Kuk
Dr
Allison Engine Co
Speed Code T10B
P O Box 420
INDIANAPOLIS, IN 46206-0420
UNITED STATES

Fujimoto, Ichiro
Professor
Takushoku University
815-1 Tatemachi, Hachioji-shi
TOKYO 193
JAPAN

Gallus, Heinz E
Professor
Strahlantriebe & Turboarbeitsmaschin
RWTH Aachen
Templergraben 55
52062 AACHEN
GERMANY

Golubev, Vladimir
Dr
University of Notre Dame
705 Riverside Drive, Apt 2
SOUTH BEND, IN 46601
UNITED STATES

Grisval, Jean-Pierre
Dr
ONERA
Structures Department
BP 72
92322 CHATILLON Cedex
FRANCE

Gysling, Daniel
Dr
United Techn Research Center
411 Silver Lane
EAST HARTFORD, CT 06108
UNITED STATES

Hall, Kenneth
Professor
Duke University
Dept of Mechanical Eng
Box 90300
DURHAM, NC 27708-0300
UNITED STATES

Hassan, Ahmed Sayed
Dr
Assiut University
Faculty of Engineering
71516 ASSIUT
EGYPT

He, Li
Dr
University of Durham
School of Engineering
South Road
DURHAM DH1 3LE
UNITED KINGDOM

Henry, Veronique
Ms
Royal Inst of Technology
Flygteknik
100 44 STOCKHOLM
SWEDEN

Hirano, Takanori
Mr
Takushoku University
815-1 Tatemachi, Hachioji-shi
TOKYO 193
JAPAN

Höhn, Wolfgang
Mr
Royal Inst of Technology
Chair of Heat & Power
Brinellvägen 60
100 44 STOCKHOLM
SWEDEN

Imregun, Mehmet
Dr
Imperial College
Centre of Vibration Engineering
Dept of Mech Eng
LONDON SW7 2BX
UNITED KINGDOM

Isomura, Kousuke
Dr
IHI
IHI R&D Dept
229 Tonogaya Mizuho-machi
TOKYO 190-12
JAPAN

Jahnen, Werner
Mr
Univ der Bundeswehr München
Werner-Heisenberg-Weg 39
85577 NEUBIBERG
GERMANY

Jeanpierre, Jerome
Royal Inst of Technology
Heat and Power Technology
Brinellvägen 60
100 44 STOCKHOLM
SWEDEN

Johansson, Ulf
Mr
Volvo Aero Corporation
461 81 TROLLHÄTTAN
SWEDEN

Jöcker, Markus
Mr
Royal Inst of Technology
Heat and Power Technology
Brinellvägen 60
100 44 STOCKHOLM
SWEDEN

Kahl, Gerhard
Mr
MTU München GmbH
P O Box 500640
80976 MÜNCHEN
GERMANY

Kaji, Shojiro
Professor
University of Tokyo
Dept of Aeronautics & Astronautics
7-3-1 Hongo, Bunkyo-ku
TOKYO 113
JAPAN

Kato, Dai
Mr
Waseda University
Dept of Mechanical Engineering
3-4-1 Okubo, Shinjuku-ku
TOKYO 169
JAPAN

Kielb, Robert
Dr
GE Aircraft Engines
One Neumann Way, MD A412
CINCINNATI, OH 45215-1988
UNITED STATES

Kim, Young-Nam
Dr
Pratt & Whitney
MS 163-17
400 Main St
EAST HARTFORD, CT 06108
UNITED STATES

Kurzin, Vladimir
Professor
Lavrentyev Inst of Hydrodynamics
Lavrentyev prospekt 15
NOVOSIBIRSK 630 090
RUSSIAN FEDERATION

Körbächer, Heiko
Dr
Swiss Federal Inst of Technology-LTT
1015 LAUSANNE
SWITZERLAND

Lorence, Christopher
Dr
GE Aircraft Engines
One Neumann Way, MD A412
CINCINNATI, OH 45215-1988
UNITED STATES

Manwaring, Steven
Dr
GE Aircraft Engines
One Neumann Way, MD A411
CINCINNATI, OH 45215-1988
UNITED STATES

Marshall, John
Dr
Rolls Royce Plc
Moor Lane 83
P O Box 31
DERBY DE24 8BJ
UNITED KINGDOM

Marx, Yves
Dr
Swiss Federal Inst of Technology-LTT
1015 LAUSANNE
SWITZERLAND

Maurice, Jean Marc
Dr
Thermodyn / Framatome
BP 119
71200 LE CREUSOT Cedex
FRANCE

Mayer, Juergen F
Dr
University of Stuttgart
ITSM
Pfaffenwaldring 6
70550 STUTTGART
GERMANY

Montgomery, Matthew D
Dr
Pratt & Whitney
400 Main St MS 163-17
EAST HARTFORD, CT 06108
UNITED STATES

Moreno, Monica
Ms
Forsmarks Kraftgrupp AB
742 03 ÖSTHAMMAR
SWEDEN

Moyroud, Francois
Royal Inst of Technology
Heat and Power Technology
Brinellvägen 60
100 44 STOCKHOLM
SWEDEN

Mundinger, Gerd
Dr
ABB Turbo Systems Ltd
ZX
5401 BADEN
SWITZERLAND

Mårtensson, Hans
Mr
Volvo Aero Corporation
461 81 TROLLHÄTTAN
SWEDEN

Nagashima, Toshio
Professor
University of Tokyo
Dept of Aeronautics & Astronautics
7-3-1 Hongo, Bunkyo-ku
TOKYO 113
JAPAN

Namba, Masanobu
Professor
Kyushu University
Dept of Aeronautics & Astronautics
6-10-1 Hakozaki, Higashi-ku
FUKUOKA 812
JAPAN

Ning, Wei
Mr
University of Durham
School of Engineering
South Road
DURHAM DH1 3LE
UNITED KINGDOM

Norén, Bjarne
Dr
OKG AB
572 83 OSKARSHAMN
SWEDEN

Norryd, Magnus
Dr
Swiss Federal Inst of Technology-LTT
1015 LAUSANNE
SWITZERLAND

Nowinski, Matthew
Mr
Swiss Federal Inst of Technology-LTT
1015 LAUSANNE
SWITZERLAND

Ott, Peter
Dr
Swiss Federal Inst of Technology-LTT
1015 LAUSANNE
SWITZERLAND

Outa, Eisuke
Professor
Waseda University
Dept of Mechanical Engineering
Okubo 3-4-1, Shinjuku-ku
TOKYO
JAPAN

Panovsky, Josef
Dr
GE Aircraft Engines
One Neumann Way, MD A406
CINCINNATI, OH 45215-1988
UNITED STATES

Peters, Thomas
Mr
Univ der Bundeswehr München
Werner-Heisenberg-Weg 39
85577 NEUBIBERG
GERMANY

Pitkanen, Harri
Lappeenranta Univ of Technology
Dept of Energy Technology
P O Box 20
53851 LAPPEENRANTA
FINLAND

Platzer, Max
Professor
Naval Postgraduate School
Dept of Aeronautics & Astronautics
MONTEREY, CA
UNITED STATES

Rangwalla, Akil
Dr
Pratt & Whitney
MS 163-17
400 Main St
EAST HARTFORD, CT 06108
UNITED STATES

Riess, Walter
Professor
Universität Hannover
Inst for Turbomachinery
Appelstr 9
30167 HANNOVER
GERMANY

Rzadkowski, Romvald
Dr
Polish Academy of Sciences
Institute of Fluid-Flow Machinery
Fiszera 14
80-952 GDANSK
POLAND

Sajben, Miklos
Professor
University of Cincinnati
Dept of Aerospace Engineering
ML 70
CINCINNATI, OH 45221-0070
UNITED STATES

Saren, Victor E
Professor
Central Inst of Aviation Motors
CIAM 2 Aviamotornaya St
MOSCOW 111250
RUSSIAN FEDERATION

Sayma, Abdulnaser I
Dr
Imperial College
Centre of Vibration Engineering
Dept of Mech Eng
LONDON SW7 2BX
UNITED KINGDOM

Sbardella, Luca
Mr
Imperial College
Centre of Vibration Engineering
Dept of Mech Eng
LONDON SW7 2BX
UNITED KINGDOM

Sentker, Alexandra
Ms
Universität Hannover
Inst for Turbomachinery
Appelstr 9
30167 HANNOVER
GERMANY

Servaty, Stephan
Mr
MTU München GmbH
P O Box 500640
80976 MÜNCHEN
GERMANY

Shibata, Takanori
Dr
Hitachi Ltd
Computational Engineering
7-2-1 Omika-cho, Hitachi-shi
IBARAKI 319-12
JAPAN

Shiratori, Toshimasa
Ass Professor
Tokyo Metropolitan Inst of Technolog
6-6 Asahigaoka, Hino
TOKYO 191
JAPAN

Smati, Laurent
Mr
Ecole Centrale de Lyon
Lab de Mecanique des Fluides et d'Ac
36 avenue Guy de Collongue, BP 163
69131 ECULLY Cedex
FRANCE

Sommer, Thomas
Dr
ABB Power Generation Ltd
5401 BADEN
SWITZERLAND

Svensdotter, Susanne
Ms
Royal Inst of Technology
Heat and Power Technology
Brinellvägen 60
100 44 STOCKHOLM
SWEDEN

Tanaka, Hideo
Professor
University of Tokyo
6-15-11 Minami Oizumi, Nerima-ku
TOKYO 178
JAPAN

Tanida, Yoshimichi
Professor
Tokai University
7-17-33 Konan-dai, Konan-ku
YOKOHAMA 234
JAPAN

Teusch, Reinhold
Mr
Univ der Bundeswehr München
Werner-Heisenberg-Weg 39
85577 NEUBIBERG
GERMANY

Tinoco, Hernan
Dr
Forsmarks Kraftgrupp AB
742 03 ÖSTHAMMAR
SWEDEN

Tsujimoto, Yoshinobu
Professor
Osaka University
Div of Mechanical Science
1-3 Machikaneyama, Toyonaka
OSAKA 560
JAPAN

Watanabe, Satoshi
Dr
Osaka University
Div of Mechanical Sci
1-3 Machikaneyama, Toyonaka
OSAKA 560
JAPAN

Watanabe, Toshinori
Professor
University of Tokyo
Dept of Aeronautics & Astronautics
7-3-1 Hongo, Bunkyo-ku
TOKYO 113
JAPAN

Weber, Stefan
Mr
Strahlantriebe & Turboarbeitsmaschin
RWTH Aachen
Templergraben 55
52062 AACHEN
GERMANY

Verdon, Joseph
Dr
United Technologies Research Center
411 Silver Lane, MS 129-20
EAST HARTFORD, CT 06108
UNITED STATES

Whitehead, Denis S
Dr
Imperial College Consultant
Inwoods
Farleigh Wick
BRADFORD ON AVON Wilts BA15 2PU
UNITED KINGDOM

Wiercinski, Zygmunt
Dr
Inst of Fluid-Flow Machinery
14 Fiszera Str
80-952 GDANSK
POLAND

Wiers, Sven-Hendrik
Mr
Royal Inst of Technology
Heat and Power Technology
Brinellvägen 60
100 44 STOCKHOLM
SWEDEN

Vincent de Paul, Michel
Mr
GEC Alsthom
141 Rue Rateau
93126 LA COURNEUVE Cedex
FRANCE

Wolff, J Mitch
Professor
Wright State University
Dept of Mechanical Engineering
DAYTON, OH 45435
UNITED STATES

Yamamoto, Kazuomi
Dr
DLR, German Aerospace Res Establishm
Institute for Propulsion Technology
Linder Höhe
51147 KÖLN
GERMANY

Yamane, Takashi
Dr
National Aerospace Laboratory
Thermofluid Dynamics Div
7-44-1 Jindaiji-higashi, Chofu
TOKYO 182
JAPAN

Yamasaki, Nobuhiko
Dr
Kyushu University
Dept of Aeronautics & Astronautics
6-10-1 Hakozaki, Higashi-ku
FUKUOKA 812
JAPAN

Yokota, Kazuhiko
Dr
Osaka University
Div of Mechanical Sci
1-3 Machikaneyama, Toyonaka
OSAKA 560
JAPAN

Zhou, Sheng
Professor
Apt. High Building No. 1 room 1203
Tsing_Hua univ.
Beijing 100084
China